Lecture Notes in Artificial Intelligence 9284

Subseries of Lecture Notes in Computer Science

More information about this series at http://www.springer.com/series/1244

Annalisa Appice · Pedro Pereira Rodrigues
Vitor Santos Costa · Carlos Soares
João Gama · Alípio Jorge (Eds.)

Machine Learning and Knowledge Discovery in Databases

European Conference, ECML PKDD 2015
Porto, Portugal, September 7–11, 2015
Proceedings, Part I

 Springer

Editors

Annalisa Appice
University of Bari Aldo Moro
Bari
Italy

Carlos Soares
University of Porto - INESC TEC
Porto
Portugal

Pedro Pereira Rodrigues
University of Porto
Porto
Portugal

João Gama
University of Porto - INESC TEC
Porto
Portugal

Vitor Santos Costa
University of Porto - CRACS/INESC TEC
Porto
Portugal

Alípio Jorge
University of Porto - INESC TEC
Porto
Portugal

ISSN 0302-9743   ISSN 1611-3349   (electronic)
Lecture Notes in Artificial Intelligence
ISBN 978-3-319-23527-1   ISBN 978-3-319-23528-8   (eBook)
DOI 10.1007/978-3-319-23528-8

Library of Congress Control Number: 2015947118

LNCS Sublibrary: SL7 – Artificial Intelligence

Springer Cham Heidelberg New York Dordrecht London

Printed on acid-free paper

Springer International Publishing AG Switzerland is part of Springer Science+Business Media
(www.springer.com)

Preface

We are delighted to introduce the proceedings of the 2015 edition of the European Conference on Machine Learning and Principles and Practice of Knowledge Discovery in Databases, or ECML PKDD for short. This conference stems from the former ECML and PKDD conferences, the two premier European conferences on, respectively, Machine Learning and Knowledge Discovery in Databases. Originally independent events, the two conferences were organized jointly for the first time in 2001. The sinergy between the two led to increasing integration, and eventually the two merged in 2008. Today, ECML PKDD is a world-wide leading scientific event that aims at exploiting the synergies between Machine Learning and Data Mining, focusing on the development and application of methods and tools capable of solving real-life problems.

ECML PKDD 2015 was held in Porto, Portugal, during September 7–11. This was the third time Porto hosted the major European Machine Learning event. In 1991, Porto was host to the fifth EWSL, the precursor of ECML. More recently, in 2005, Porto was host to a very successful ECML PKDD. We were honored that the community chose to again have ECML PKDD 2015 in Porto, just ten years later. The 2015 ECML PKDD was co-located with "Intelligent System Applications to Power Systems", ISAP 2015, a well-established forum for scientific and technical discussion, aiming at fostering the widespread application of intelligent tools and techniques to the power system network and business. Moreover, it was collocated, for the first time, with the Summer School on "Data Sciences for Big Data."

ECML PKDD traditionally combines the research-oriented extensive program of the scientific and journal tracks, which aim at being a forum for high quality, novel research in Machine Learning and Data Mining, with the more focused programs of the demo track, dedicated to presenting real systems to the community, the PhD track, which supports young researchers, and the nectar track, dedicated to bringing relevant work to the community. The program further includes an industrial track, which brings together participants from academia, industry, government, and non-governmental organizations in a venue that highlights practical and real-world studies of machine learning, knowledge discovery, and data mining. The industrial track of ECML PKDD 2015 has a separate Program Committee and separate proceedings volume. Moreover, the conference program included a doctoral consortium, three discovery challenges, and various workshops and tutorials.

The research program included five plenary talks by invited speakers, namely, Hendrik Blockeel (University of Leuven and Leiden University), Pedro Domingos (University of Washington), Jure Leskovec (Stanford University), Nataša Milić-Frayling (Microsoft Research), and Dino Pedreschi (Università di Pisa), as well as one ISAP +ECML PKDD joint plenary talk by Chen-Ching Liu (Washington State University). Three invited speakers contributed to the industrial track: Andreas Antrup (Zalando and

University of Edinburgh), Wei Fan (Baidu Big Data Lab), and Hang Li (Noah's Ark Lab, Huawei Technologies).

Three discovery challenges were announced this year. They focused on "MoRe-BikeS: Model Reuse with Bike rental Station data," "On Learning from Taxi GPS Traces," and "Activity Detection Based on Non-GPS Mobility Data," respectively.

Twelve workshops were held, providing an opportunity to discuss current topics in a small and interactive atmosphere: "MetaSel - Meta-learning and Algorithm Selection," "Parallel and Distributed Computing for Knowledge Discovery in Databases," "Interactions between Data Mining and Natural Language Processing," "New Frontiers in Mining Complex Patterns," "Mining Ubiquitous and Social Environments," "Advanced Analytics and Learning on Temporal Data," "Learning Models over Multiple Contexts," "Linked Data for Knowledge Discovery," "Sports Analytics," "BigTargets: Big Multi-target Prediction," "DARE: Data Analytics for Renewable Energy Integration," and "Machine Learning in Life Sciences."

Ten tutorials were included in the conference program, providing a comprehensive introduction to core techniques and areas of interest for the scientific community: "Similarity and Distance Metric Learning with Applications to Computer Vision," "Scalable Learning of Graphical Models," "Meta-learning and Algorithm Selection," "Machine Reading the Web - Beyond Named Entity Recognition and Relation Extraction," "VC-Dimension and Rademacher Averages: From Statistical Learning Theory to Sampling Algorithms," "Making Sense of (Multi-)Relational Data," "Collaborative Filtering with Binary, Positive-Only Data," "Predictive Maintenance," "Eureka! - How to Build Accurate Predictors for Real-Valued Outputs from Simple Methods," and "The Space of Online Learning Problems."

The main track received 380 paper submissions, of which 89 were accepted. Such a high volume of scientific work required a tremendous effort by the Area Chairs, Program Committee members, and many additional reviewers. We managed to collect three highly qualified independent reviews per paper and one additional overall input from one of the Area Chairs. Papers were evaluated on the basis of significance of contribution, novelty, technical quality, scientific, and technological impact, clarity, repeatability, and scholarship. The industrial, demo, and nectar tracks were equally successful, attracting 42, 32, and 29 paper submissions, respectively.

For the third time, the conference used a double submission model: next to the regular conference tracks, papers submitted to the Springer journals Machine Learning (MACH) and Data Mining and Knowledge Discovery (DAMI) were considered for presentation at the conference. These papers were submitted to the ECML PKDD 2015 special issue of the respective journals, and underwent the normal editorial process of these journals. Those papers accepted for one of these journals were assigned a presentation slot at the ECML PKDD 2015 conference. A total of 191 original manuscripts were submitted to the journal track during this year. Some of these papers are still being refereed. Of the fully refereed papers, 10 were accepted in DAMI and 15 in MACH, together with 4+4 papers from last year's call, which were also scheduled for presentation at this conference. Overall, this resulted in a number of 613 submissions (to the scientific track, industrial track and journal track), of which 126 were selected for presentation at the conference, making an overall acceptance rate of about 21%.

Part I and Part II of the proceedings of the ECML PKDD 2015 conference contain the full papers of the contributions presented in the scientific track, the abstracts of the scientific plenary talks, and the abstract of the ISAP+ECML PKDD joint plenary talk. Part III of the proceedings of the ECML PKDD 2015 conference contains the full papers of the contributions presented in the industrial track, short papers describing the demonstrations, the nectar papers, and the abstracts of the industrial plenary talks.

The scientific track program results from continuous collaboration between the scientific tracks and the general chairs. Throughout we had the unfaltering support of the Local Chairs, Carlos Ferreira, Rita Ribeiro, and João Moreira, who managed this event in a thoroughly competent and professional way. We thank the Social Media Chairs, Dunja Mladenić and Márcia Oliveira, for tweeting the new face of ECML PKDD, and the Publicity Chairs, Ricardo Campos and Carlos Ferreira, for their excellent work in spreading the news. The beautiful design and quick response time of the web site is due to the work of our Web Chairs, Sylwia Bugla, Rita Ribeiro, and João Rodrigues. The beautiful image on all the conference materials is based on the logo designed by Joana Amaral e João Cravo, inspired by Porto landmarks. It has been a pleasure to collaborate with the Journal, Industrial, Demo, Nectar, and PhD Track Chairs. ECML PKDD would not be complete if not for the efforts of the Tutorial Chairs, Fazel Famili, Mykola Pechenizkiy, and Nikolaj Tatti, the Workshop Chairs, Stan Matwin, Bernhard Pfahringer, and Luís Torgo, and the Discovery Challenge Chairs, Michel Ferreira, Hillol Kargupta, Luís Moreira-Matias, and João Moreira. We thank the Awards Committee Chairs, Pavel Brazdil, Sašo Džerosky, Hiroshi Motoda, and Michèle Sebag, for their hard work in selecting papers for awards. A special meta thanks to Pavel: ECML PKDD at Porto is only possible thanks to you. We gratefully acknowledge the work of the Sponsorship Chairs, Albert Bifet and André Carvalho, for their key work. Special thanks go to the Proceedings Chairs, Michelangelo Ceci and Paulo Cortez, for the difficult task of putting these proceedings together. We appreciate the support of Artur Aiguzhinov, Catarina Félix Oliveira, and Mohammad Nozari (U. Porto) for helping to check this front matter. We thank the ECML PKDD Steering Committee for kindly sharing their experience, and particularly the General Steering Committe Chair, Fosca Giannotti. The quality of ECML PKDD is only possible due to the tremendous efforts of the Program Committee; our sincere thanks for all the great work in improving the quality of these proceedings. Throughout, we relied on the exceptional quality of the Area Chairs. Our most sincere thanks for their support, with a special thanks to the members who contributed in difficult personal situations, and to Paulo Azevedo for stepping in when the need was there. Last but not least, we would like to sincerely thank all the authors who submitted their work to the conference.

July 2015

Annalisa Appice
Pedro Pereira Rodrigues
Vítor Santos Costa
Carlos Soares
João Gama
Alípio Jorge

Organization

ECML/PKDD 2015 Organization

Conference Co-chairs

João Gama	University of Porto, INESC TEC, Portugal
Alipío Jorge	University of Porto, INESC TEC, Portugal

Program Co-chairs

Annalisa Appice	University of Bari Aldo Moro, Italy
Pedro Pereira Rodrigues	University of Porto, CINTESIS, INESC TEC, Portugal
Vtor SantosCosta	University of Porto, INESC TEC, Portugal
Carlos Soares	University of Porto, INESC TEC, Portugal

Journal Track Chairs

Concha Bielza	Technical University of Madrid, Spain
João Gama	University of Porto, INESC TEC, Portugal
Alipío Jorge	University of Porto, INESC TEC, Portugal
Indré Žliobaité	Aalto University and University of Helsinki, Finland

Industrial Track Chairs

Albert Bifet	Huawei Noah's Ark Lab, China
Michael May	Siemens, Germany
Bianca Zadrozny	IBM Research, Brazil

Local Organization Chairs

Carlos Ferreira	Oporto Polytechnic Institute, INESC TEC, Portugal
João Moreira	University of Porto, INESC TEC, Portugal
Rita Ribeiro	University of Porto, INESC TEC, Portugal

Tutorial Chairs

Fazel Famili	CNRC, France
Mykola Pechenizkiy	TU Eindhoven, The Netherland
Nikolaj Tatti	Aalto University, Finland

Workshop Chairs

Stan Matwin   Dalhousie University, NS, Canada
Bernhard Pfahringer   University of Waikato, New Zealand
Luís Torgo   University of Porto, INESC TEC, Portugal

Awards Committee Chairs

Pavel Brazdil   INESC TEC, Portugal
Sašo Džeroski   Jožef Stefan Institute, Slovenia
Hiroshi Motoda   Osaka University, Japan
Michèle Sebag   Université Paris Sud, France

Nectar Track Chairs

Ricard Gavaldà   UPC, Spain
Dino Pedreschi   Università di Pisa, Italy

Demo Track Chairs

Francesco Bonchi   Yahoo! Labs, Spain
Jaime Cardoso   University of Porto, INESC TEC, Portugal
Myra Spiliopoulou   Otto-von-Guericke University Magdeburg, Germany

PhD Chairs

Jaakko Hollmén   Aalto University, Finland
Panagiotis Papapetrou Stockholm University, Sweden

Proceedings Chairs

Michelangelo Ceci   University of Bari, Italy
Paulo Cortez   University of Minho, Portugal

Discovery Challenge Chairs

Michel Ferreira   University of Porto, INESC TEC, Geolink, Portugal
Hillol Kargupta   Agnik, MD, USA
Luís Moreira-Matias   NEC Research Labs, Germany
João Moreira   University of Porto, INESC TEC, Portugal

Sponsorship Chairs

Albert Bifet   Huawei Noah's Ark Lab, China
André Carvlho   University of São Paulo, Brazil
Pedro Pereira Rodrigues University of Porto, Portugal

Publicity Chairs

Ricardo Campos   Polytechnic Institute of Tomar, INESC TEC, Portugal
Carlos Ferreira   Oporto Polytechnic Institute, INESC TEC, Portugal

Social Media Chairs

Dunja Mladenić   JSI, Slovenia
Márcia Oliveira   University of Porto, INESC TEC, Portugal

Web Chairs

Sylwia Bugla   INESC TEC, Portugal
Rita Ribeiro   University of Porto, INESC TEC, Portugal
João Rodrigues   INESC TEC, Portugal

ECML PKDD Steering Committee

Fosca Giannotti   ISTI-CNR Pisa, Italy
Michèle Sebag   Université Paris Sud, France
Francesco Bonchi   Yahoo! Research, Spain
Hendrik Blockeel   KU Leuven, Belgium and Leiden University,
                         The Netherlands
Katharina Morik   University of Dortmund, Germany
Tobias Scheffer   University of Potsdam, Germany
Arno Siebes   Utrecht University, The Netherlands
Peter Flach   University of Bristol, UK
Tijl De Bie   University of Bristol, UK
Nello Cristianini   University of Bristol, Uk
Filip Železný   Czech Technical University in Prague, Czech Republic
Siegfried Nijssen   LIACS, Leiden University, The Netherlands
Kristian Kersting   Technical University of Dortmund, Germany
Rosa Meo   Università di Torino, Italy
Toon Calders   Eindhoven University of Technology, The Netherlands
Chedy Raïssi   INRIA Nancy Grand-Est, France

Area Chairs

Paulo Azevedo   University of Minho
Michael Berthold   Universität Konstanz
Francesco Bonchi   Yahoo Labs Barcelona
Henrik Boström   University of Stockholm
Jean-Françis Boulicaut   Institut National des Sciences Appliquées de Lyon, LIRIS
Pavel Brazdil   University of Porto
André Carvalho   University of São Paulo
Michelangelo Ceci   Università degli Studi di Bari Aldo Moro

Jesse Davis	Katholieke Universiteit Leuven
Luc De Raedt	Katholieke Universiteit Leuven
Peter Flach	University of Bristol
Johannes Fürnkranz	TU Darmstadt
Thomas Gaertner	Fraunhofer IAIS
Bart Goethals	University of Antwerp
Andreas Hotho	University of Kassel
Eyke Hüllermeier	University of Paderborn
George Karypis	University of Minnesota
Kristian Kersting	Technical University of Dortmund
Arno Knobbe	Universiteit Leiden
Pedro Larrañaga	Technical University of Madrid
Peter Lucas	Radboud University Nijmegen
Donato Malerba	Università degli Studi di Bari Aldo Moro
Stan Matwin	Dalhousie University
Katharina Morik	TU Dortmund
Sriraam Natarajan	Indiana University
Eugénio Oliveira	University of Porto
Mykola Pechenizkiy	Eindhoven University of Technology
Bernhard Pfahringer	University of Waikato
Michèle Sebag	CNRS
Myra Spiliopoulou	Otto-von-Guericke University Magdeburg
Jerzy Stefanowski	Poznań University of Technology
Luís Torgo	University of Porto
Stefan Wrobel	Fraunhofer IAIS, Germany
Philip Yu	University of Illinois at Chicago

Program Committee

Leman Akoglu	Narayanaswamy	Jerzy Blaszczynski
Mehmet Sabih Aksoy	Balakrishnan	Konstantinos Blekas
Mohammad Al Hasan	Elena Baralis	Mario Boley
Omar Alonso	Daniel Barbará	Gianluca Bontempi
Aijun An	Gustavo Batista	Christian Borgelt
Aris Anagnostopoulos	Christian Bauckhage	José Luís Borges
Marta Arias	Roberto Bayardo	Marc Boullé
Rubén Armañanzas	Vaishak Belle	Ulf Brefeld
Ira Assent	András Benczúr	Róbert Busa-Fekete
Martin Atzmueller	Bettina Berendt	Toon Calders
Chloé-Agathe Azencott	Michele Berlingerio	Rui Camacho
Paulo Azevedo	Indrajit Bhattacharya	Longbing Cao
Antonio Bahamonde	Marenglen Biba	Henrique Lopes Cardoso
James Bailey	Enrico Blanzieri	Francisco Casacuberta

Gladys Castillo
Loic Cerf
Tania Cerquitelli
Edward Chang
Duen Horng Chau
Sanjay Chawla
Keke Chen
Ling Chen
Weiwei Cheng
Silvia Chiusano
Frans Coenen
Fabrizio Costa
Germán Creamer
Bruno Crémilleux
Marco Cristo
Tom Croonenborghs
Boris Cule
Tomaž Curk
James Cussens
Alfredo Cuzzocrea
Claudia d'Amato
Sašo Džeroski
Maria Damiani
Jeroen De Knijf
Gerard de Melo
Marcílio de Souto
Kurt DeGrave
Juan del Coz
Krzysztof Dembczyński
François Denis
Anne Denton
Mohamed Dermouche
Christian Desrosiers
Luigi Di Caro
Nicola Di Mauro
Jana Diesner
Ivica Dimitrovski
Ying Ding
Stephan Doerfel
Anne Driemel
Chris Drummond
Brett Drury
Devdatt Dubhashi
Wouter Duivesteijn
Bob Durrant
Inês Dutra

Tapio Elomaa
Floriana Esposito
Roberto Esposito
Hadi Fanaee-T
Nicola Fanizzi
Elaine Faria
Fabio Fassetti
Hakan Ferhatosmanoglou
Stefano Ferilli
Carlos Ferreira
Hugo Ferreira
Cèsar Ferri
George Fletcher
Eibe Frank
Élisa Fromont
Fabio Fumarola
Mohamed Medhat Gaber
Fábio Gagliardi Cozman
Patrick Gallinari
José A. Gámez
Jing Gao
Byron Gao
Paolo Garza
Éric Gaussier
Pierre Geurts
Fosca Giannotti
Christophe Giraud-Carrier
Aris Gkoulalas-Divanis
Marco Gori
Pablo Granitto
Michael Granitzer
Maria Halkidi
Jiawei Han
Daniel Hernández Lobato
José Hernández-Orallo
Thanh Lam Hoang
Frank Hoeppner
Geoff Holmes
Arjen Hommersom
Estevam Hruschka
Xiaohua Hu
Minlie Huang
Dino Ienco
Iñaki Inza
Frederik Janssen
Nathalie Japkowicz

Szymon Jaroszewicz
Ulf Johansson
Tobias Jung
Hachem Kadri
Theodore Kalamboukis
Alexandros Kalousis
U. Kang
Andreas Karwath
Hisashi Kashima
Ioannis Katakis
Mehdi Kaytoue
John Keane
Latifur Khan
Dragi Kocev
Levente Kocsis
Alek Kolcz
Irena Koprinska
Jacek Koronacki
Nitish Korula
Petr Kosina
Walter Kosters
Lars Kottof
Georg Krempl
Artus Krohn-Grimberghe
Marzena Kryszkiewicz
Matjaž Kukar
Meelis Kull
Sergei Kuznetsov
Nicolas Lachiche
Helge Langseth
Mark Last
Silvio Lattanzi
Niklas Lavesson
Nada Lavrač
Gianluca Lax
Gregor Leban
Sangkyun Lee
Wang Lee
Florian Lemmerich
Philippe Lenca
Philippe Leray
Carson Leung
Lei Li
Jiuyong Li
Juanzi Li
Edo Liberty

Hsuan-Tien Lin
Shou-de Lin
Yan Liu
Lei Liu
Corrado Loglisci
Eneldo Loza Mencía
Jose A. Lozano
Chang-Tien Lu
Panagis Magdalinos
Giuseppe Manco
Yannis Manolopoulos
Enrique Martinez
Elio Masciari
Florent Masseglia
Luís Matias
Oleksiy Mazhelis
Wannes Meert
Wagner Meira
Ernestina Menasalvas
Corrado Mencar
Rosa Meo
Pauli Miettinen
Dunja Mladenić
Anna Monreale
João Moreira
Emmanuel Müller
Mohamed Nadif
Mirco Nanni
Amedeo Napoli
Houssam Nassif
Benjamin Nguyen
Thomas Niebler
Thomas Nielsen
Siegfried Nijssen
Xia Ning
Niklas Norén
Kjetil Nørvåg
Eirini Ntoutsi
Andreas Nürnberger
Irene Ong
Salvatore Orlando
Gerhard Paaß
David Page
George Paliouras
Pance Panov
Spiros Papadimitriou

Apostolos Papadopoulos
Panagiotis Papapetrou
Ioannis Partalas
Andrea Passerini
Dino Pedreschi
Nikos Pelekis
Jing Peng
Yonghong Peng
Ruggero Pensa
Andrea Pietracaprina
Fabio Pinelli
Marc Plantevit
Pascal Poncelet
Lubos Popelinksky
George Potamias
Ronaldo Prati
Doina Precup
Ricardo Prudêncio
Kai Puolamäki
Buyue Qian
Chedy Raïssi
Liva Ralaivola
Karthik Raman
Jan Ramon
Huzefa Rangwala
Zbigniew Ras
Chotirat Ann
    Ratanamahatana
Jan Rauch
Soumya Ray
Jesse Read
Steffen Rendle
Achim Rettinger
Rita Ribeiro
Fabrizio Riguzzi
Céline Robardet
Marko Robnik-Šikonja
Juan Rodriguez
Irene Rodríguez Luján
André Rossi
Fabrice Rossi
Juho Rousu
Céline Rouveirol
Salvatore Ruggieri
Stefan Rüping
Y. van Saeys

Alan Said
Lorenza Saitta
Ansaf Salleb-Aouissi
Jose S. Sanchez
Raul Santos-Rodriguez
Sam Sarjant
Claudio Sartori
Yücel Saygin
Erik Schmidt
Lars Schmidt-Thieme
Christoph Schommer
Matthias Schubert
Marco Scutari
Thomas Seidl
Nazha Selmaoui
Giovanni Semeraro
Junming Shao
Yun Sing Koh
Andrzej Skowron
Kevin Small
Tomislav Šmuc
Yangqiu Song
Cheng Soon Ong
Arnaud Soulet
Mauro Sozio
Alessandro Sperduti
Eirini Spyropoulou
Steffen Staab
Gregor Stiglic
Markus Strohmaier
Enrique Sucar
Mahito Sugiyama
Johan Suykens
Einoshin Suzuki
Panagiotis Symeonidis
Sándor Szedmák
Andrea Tagarelli
Domenico Talia
Letizia Tanca
Dacheng Tao
Nikolaj Tatti
Maguelonne Teisseire
Alexandre Termier
Evimaria Terzi
Ljupco Todorovski
Vicenç Torra

Roberto Trasarti
Brigitte Trousse
Panayiotis Tsaparas
Vincent Tseng
Grigorios Tsoumakas
Theodoros Tzouramanis
Antti Ukkonen
Takeaki Uno
Athina Vakali
Wil van der Aalst
Guy van der Broeck
Maarten van der Heijden
Peter van der Putten
Matthijs van Leeuwen
    Putten
Martijn van Otterlo
Maarten van Someren
Joaquin Vanschoren
Iraklis Varlamis
Raju Vatsavai
Michalis Vazirgiannis

Julien Velcin
Shankar Vembu
Sicco Verwer
Vassilios Verykios
Herna Viktor
Ricardo Vilalta
Pavlovic Vladimir
Christel Vrain
Jilles Vreeken
Willem Waegeman
Byron Wallace
Fei Wang
Jianyong Wang
Yang Wang
Takashi Washio
Jörg Simon Wicker
Chun-Nam Yu
Jeffrey Yu
Jure Zabkar
Gerson Zaverucha
Demetris Zeinalipour

Filip Železný
Bernard Ženko
Junping Zhang
Kun Zhang
Lei Zhang
Min-Ling Zhang
Nan Zhang
Shichao Zhang
Zhongfei Zhang
Liang Zhao
Ying Zhao
Elena Zheleva
Bin Zhou
Kenny Zhu
Xiaofeng Zhu
Djamel Zighed
Arthur Zimek
Albrecht Zimmermann
Blaž Zupan

Additional Reviewers

Greet Baldewijns
Jessa Bekker
Nuno Castro
Shiyu Chang
Yu Cheng
Paolo Cintia
Heidar Davoudi
Thomas Delacroix
Martin Dimkovski
Michael Färber
Ricky Fok
Emanuele Frandi
Tatiana Gossen
Valerio Grossi
Riccardo Guidotti
Ming Jiang
Nikos Katzouris

Sebastian Kauschke
Jinseok Kim
Jan Kralj
Thomas Low
Stijn Luca
Rafael Mantovani
Pasquale Minervini
Shubhanshu Mishra
Christos Perentis
Fábio Pinto
Dimitrios Rafailidis
Giulio Rossetti
Alexandros Sarafianos
Antonio Vergari
Dimtrios Vogiatzis
Andreas Zioupos

Sponsors

Platinum Sponsors

BNP PARIBAS	http://www.bnpparibas.com/
ONR Global	www.onr.navy.mil/science-technology/onr-global.aspx

Gold Sponsors

Zalando	https://www.zalando.co.uk/
HUAWEI	http://www.huawei.com/en/

Silver Sponsors

Deloitte	http://www2.deloitte.com/
Amazon	http://www.amazon.com/

Bronze Sponsors

Xarevision	http://xarevision.pt/
Farfetch	http://www.farfetch.com/pt/
NOS	http://www.nos.pt/particulares/Pages/home.aspx

Award Sponsor

Machine Learning	http://link.springer.com/journal/10994
Data Mining and Knowledge	http://link.springer.com/journal/10618
Discovery Deloitte	http://www2.deloitte.com/

Lanyard Sponsor

KNIME	http://www.knime.org/

Invited Talk Sponsors

ECCAI	http://www.eccai.org/
Cliqz	https://cliqz.com/
Technicolor	http://www.technicolor.com/
University of Bari Aldo Moro	http://www.uniba.it/english-version

Additional Supporters

INESCTEC	https://www.inesctec.pt/
University of Porto, Faculdade de Economia	http://sigarra.up.pt/fep/pt/web_page.inicial
Springer	http://www.springer.com/
University of Porto	http://www.up.pt/

Official Carrier

TAP	http://www.flytap.com/

Abstracts of Invited Talks

Towards Declarative, Domain-Oriented Data Analysis

Hendrik Blockeel

University of Leuven, Leiden University

Abstract. The need for advanced data analysis now pervades all areas of science, industry and services. A wide variety of theory and techniques from statistics, data mining, and machine learning is available. Addressing a concrete question or problem in a particular application domain requires multiple non-trivial steps: translating the question to a data analysis problem, selecting a suitable approach to solve this problem, correctly applying that approach, and correctly interpreting the results. In this process, specialist knowledge on data analysis needs to be combined with domain expertise. As data analysis becomes ever more advanced, this becomes increasingly difficult. In an ideal world, data analysis would be declarative and domain-oriented: the user should be able to state the question, rather than describing a solution procedure, and the software should decide how to provide an answer. The user then no longer needs to be, or hire, a specialist in data analysis for every step of the knowledge discovery process. This would make data analysis easier, more efficient, and less error-prone. In this talk, I will discuss contemporary research that is bringing the state of the art in data analysis closer to that long-term goal. This includes research on inductive databases, constraint-based data mining, probabilistic-logical modeling, and declarative experimentation.

Bio. Hendrik Blockeel is a professor at the Computer Science department of KU Leuven, Belgium, and part-time associate professor at Leiden University, The Netherlands. His research interests lie mostly in machine learning and data mining. He has made a variety of research contributions in these fields, including work on decision tree learning, inductive logic programming, predictive clustering, probabilistic-logical models, inductive databases, constraint-based data mining, and declarative data analysis. He is an action editor for Machine Learning and serves on the editorial board of several other journals. He has chaired or organized multiple conferences, workshops, and summer schools, including ILP, ECMLPKDD, IDA and ACAI, and he has been vice-chair, area chair, or senior PC member for ECAI, IJCAI, ICML, KDD, ICDM. He was a member of the board of the European Coordinating Committee for Artificial Intelligence from 2004 to 2010, and currently serves as publications chair for the ECMLPKDD steering committee.

Sum-Product Networks: Deep Models with Tractable Inference

Pedro Domingos

University of Washington

Abstract. Big data makes it possible in principle to learn very rich probabilistic models, but inference in them is prohibitively expensive. Since inference is typically a subroutine of learning, in practice learning such models is very hard. Sum-product networks (SPNs) are a new model class that squares this circle by providing maximum flexibility while guaranteeing tractability. In contrast to Bayesian networks and Markov random fields, SPNs can remain tractable even in the absence of conditional independence. SPNs are defined recursively: an SPN is either a univariate distribution, a product of SPNs over disjoint variables, or a weighted sum of SPNs over the same variables. It's easy to show that the partition function, all marginals and all conditional MAP states of an SPN can be computed in time linear in its size. SPNs have most tractable distributions as special cases, including hierarchical mixture models, thin junction trees, and nonrecursive probabilistic context-free grammars. I will present generative and discriminative algorithms for learning SPN weights, and an algorithm for learning SPN structure. SPNs have achieved impressive results in a wide variety of domains, including object recognition, image completion, collaborative filtering, and click prediction. Our algorithms can easily learn SPNs with many layers of latent variables, making them arguably the most powerful type of deep learning to date. (Joint work with Rob Gens and Hoifung Poon.)

Bio. Pedro Domingos is Professor of Computer Science and Engineering at the University of Washington. His research interests are in machine learning, artificial intelligence and data science. He received a PhD in Information and Computer Science from the University of California at Irvine, and is the author or co-author of over 200 technical publications. He is a winner of the SIGKDD Innovation Award, the highest honor in data science. He is a AAAI Fellow, and has received a Sloan Fellowship, an NSF CAREER Award, a Fulbright Scholarship, an IBM Faculty Award, and best paper awards at several leading conferences. He is a member of the editorial board of the Machine Learning journal, co-founder of the International Machine Learning Society, and past associate editor of JAIR. He was program co-chair of KDD-2003 and SRL-2009, and has served on numerous program committees.

Mining Online Networks and Communities

Jure Leskovec

Stanford University

Abstract. The Internet and the Web fundamentally changed how we live our daily lives as well as broadened the scope of computer science. Today the Web is a 'sensor' that captures the pulse of humanity and allows us to observe phenomena that were once essentially invisible to us. These phenomena include the social interactions and collective behavior of hundreds of millions of people, recorded at unprecedented levels of scale and resolution. Analyzing this data offers novel algorithmic as well as computational challenges. Moreover, it offers new insights into the design of information systems in the presence of complex social feedback effects, as well as a new perspective on fundamental questions in the social sciences.

Bio. Jure Leskovec is assistant professor of Computer Science at Stanford University. His research focuses on mining large social and information networks. Problems he investigates are motivated by large scale data, the Web and on-line media. This research has won several awards including a Microsoft Research Faculty Fellowship, the Alfred P. Sloan Fellowship and numerous best paper awards. Leskovec received his bachelor's degree in computer science from University of Ljubljana, Slovenia, and his PhD in in machine learning from the Carnegie Mellon University and postdoctoral training at Cornell University. You can follow him on Twitter @jure.

Learning to Acquire Knowledge in a Smart Grid Environment

Chen-Ching Liu

Washington State University

Abstract. In a smart grid, a massive amount of data is collected by millions of sensors and meters on the transmission, distribution, and customers' facilities. There is a strong dependence of the smart grid on the information and communications technology for its monitoring and control. As a result, the cyber systems are also an important source of information. This presentation will be focused on the opportunities and challenges for machine learning and knowledge discovery in a smart grid environment. The application areas of (1) anomaly detection for cyber and physical security, and (2) intelligent remedial control of power grids will be used as examples.

Bio. Chen-Ching Liu is Boeing Distinguished Professor at Washington State University, Pullman, USA. During 1983-2005, he was a Professor of EE at University of Washington, Seattle. Dr. Liu was Palmer Chair Professor at Iowa State University from 2006 to 2008. From 2008-2011, he served as Acting/Deputy Principal of the College of Engineering, Mathematical and Physical Sciences at University College Dublin, Ireland. Professor Liu received an IEEE Third Millennium Medal in 2000 and the Power and Energy Society Outstanding Power Engineering Educator Award in 2004. In 2013, Dr. Liu received a Doctor Honoris Causa from Polytechnic University of Bucharest, Romania. He is a co-founder of the International Council on Intelligent Systems Application to Power Systems (ISAPs) and served as the founding president. He chaired the IEEE Power and Energy Society Technical Committee on Power System Analysis, Computing and Economics. Dr. Liu served on the U.S. National Academies Board on Global Science and Technology. Professor Liu is a Fellow of the IEEE. He was elected a Member of the Washington State Academy of Sciences in 2014.

Untangling the Web's Invisible Net

Nataša Milić-Frayling

Microsoft Research

Abstract. This presentation will shed light on user tracking and behavioural targeting on the Web. Through empirical studies of cookie tracking practices, we will take an alternative view of the display ad business by observing the network of third party trackers that envelopes the Web. The practice begs a question of how to resolve a dissonance between the current consumer tracking practices and the vendors' desire for consumers' loyalty and trustful long-term engagements. It also makes us aware of how computing designs and techniques, inaccessible to individuals, cause imbalance in the knowledge acquisition and enablement, disempowering the end-users.

Bio. As a Principal Researcher at Microsoft Research (MSR) in Cambridge, Nataša is working on the design, prototyping and evaluation of information and communication systems. She is passionate about innovation in personal and social computing and promotes a dialogue between IT industry, consumers, and policy makers on the issues that arise from the adoption and use of technologies. Her current focus is on digital obsolescence, activity based computing, and privacy respecting systems and applications. Natasa is actively involved with a wider community of academics and practitioners through public speaking, collaborative projects, and serving on advisory boards of academic programs and commercial enterprises. She is a Visiting Professor at the UCL and Queen Mary University of London and a member of the ACM Europe Council. She serves on the Advisory Boards for the Course in Entrepreneurship at the University of Cambridge and the Turing Gateway in Mathematics at the Isaac Newton Institute for Mathematical Sciences (INI).

Towards a Digital Time Machine Fueled by Big Data and Social Mining

Dino Pedreschi

University of Pisa

Abstract. My seminar discusses the novel questions that big data and social mining allow to raise and answer, how a new paradigm for scientific exploration, statistics and policy making is emerging, and the major scientific, technological and societal barriers to be overcome to realize this vision. I will focus on concrete projects with telecom providers and official statistics bureau in Italy and France aimed at measuring, quantifying and possibly predicting key demographic and socio-economic indicators based on nation-wide mobile phone data: the population of different categories of city users (residents, commuters, visitors) in urban spaces, the inter-city mobility, the level of well-being and economic development of geographical units at various scales.

Bio. Dino Pedreschi is a Professor of Computer Science at the University of Pisa, and a pioneering scientist in mobility data mining, social network mining and privacy-preserving data mining. He co-leads with Fosca Giannotti the Pisa KDD Lab - Knowledge Discovery and Data Mining Laboratory, a joint research initiative of the University of Pisa and the Information Science and Technology Institute of the Italian National Research Council, one of the earliest research lab centered on data mining. His research focus is on big data analytics and mining and their impact on society. He is a founder of the Business Informatics MSc program at Univ. Pisa, a course targeted at the education of interdisciplinary data scientists. Dino has been a visiting scientist at Barabasi Lab (Center for Complex Network Research) of Northeastern University, Boston (2009-2010), and earlier at the University of Texas at Austin (1989-90), at CWI Amsterdam (1993) and at UCLA (1995). In 2009, Dino received a Google Research Award for his research on privacy-preserving data mining.

Abstracts of Journal Track Articles

A Bayesian Approach for Comparing Cross-Validated Algorithms on Multiple Data Sets

Giorgio Corani and Alessio Benavoli
Machine Learning
DOI: 10.1007/s10994-015-5486-z

We present a Bayesian approach for making statistical inference about the accuracy (or any other score) of two competing algorithms which have been assessed via cross-validation on multiple data sets. The approach is constituted by two pieces. The first is a novel correlated Bayesian t-test for the analysis of the cross-validation results on a single data set which accounts for the correlation due to the overlapping training sets. The second piece merges the posterior probabilities computed by the Bayesian correlated *t*-test on the different data sets to make inference on multiple data sets. It does so by adopting a Poisson-binomial model. The inferences on multiple data sets account for the different uncertainty of the cross-validation results on the different data sets. It is the first test able to achieve this goal. It is generally more powerful than the signed-rank test if ten runs of cross-validation are performed, as it is anyway generally recommended.

A Decomposition of the Outlier Detection Problem into a Set of Supervised Learning Problems

Heiko Paulheim and Robert Meusel
Machine Learning
DOI: 10.1007/s10994-015-5507-y

Outlier detection methods automatically identify instances that deviate from the majority of the data. In this paper, we propose a novel approach for unsupervised outlier detection, which re-formulates the outlier detection problem in numerical data as a set of supervised regression learning problems. For each attribute, we learn a predictive model which predicts the values of that attribute from the values of all other attributes, and compute the deviations between the predictions and the actual values. From those deviations, we derive both a weight for each attribute, and a final outlier score using those weights. The weights help separating the relevant attributes from the irrelevant ones, and thus make the approach well suitable for discovering outliers otherwise masked in high-dimensional data. An empirical evaluation shows that our approach outperforms existing algorithms, and is particularly robust in datasets with many irrelevant attributes. Furthermore, we show that if a symbolic machine learning method is used to solve the individual learning problems, the approach is also capable of generating concise explanations for the detected outliers.

Assessing the Impact of a Health Intervention via User-Generated Internet Content

Vasileios Lampos, Elad Yom-Tov, Richard Pebody, and Ingemar J. Cox

Data Mining and Knowledge Discovery
DOI: 10.1007/s10618-015-0427-9

Assessing the effect of a health-oriented intervention by traditional epidemiological methods is commonly based only on population segments that use healthcare services. Here we introduce a complementary framework for evaluating the impact of a targeted intervention, such as a vaccination campaign against an infectious disease, through a statistical analysis of user-generated content submitted on web platforms. Using supervised learning, we derive a nonlinear regression model for estimating the prevalence of a health event in a population from Internet data. This model is applied to identify control location groups that correlate historically with the areas, where a specific intervention campaign has taken place. We then determine the impact of the intervention by inferring a projection of the disease rates that could have emerged in the absence of a campaign. Our case study focuses on the influenza vaccination program that was launched in England during the 2013/14 season, and our observations consist of millions of geo-located search queries to the Bing search engine and posts on Twitter. The impact estimates derived from the application of the proposed statistical framework support conventional assessments of the campaign.

Beyond Rankings: Comparing Directed Acyclic Graphs

Eric Malmi, Nikolaj Tatti, Aristides Gionis

Data Mining and Knowledge Discovery
DOI: 10.1007/s10618-015-0406-1

Defining appropriate distance measures among rankings is a classic area of study which has led to many useful applications. In this paper, we propose a more general abstraction of preference data, namely directed acyclic graphs (DAGs), and introduce a measure for comparing DAGs, given that a vertex correspondence between the DAGs is known. We study the properties of this measure and use it to aggregate and cluster a set of DAGs. We show that these problems are NP-hard and present efficient methods to obtain solutions with approximation guarantees. In addition to preference data, these methods turn out to have other interesting applications, such as the analysis of a collection of information cascades in a network. We test the methods on synthetic and real-world datasets, showing that the methods can be used to, e.g., find a set of influential individuals related to a set of topics in a network or to discover meaningful and occasionally surprising clustering structure.

Clustering Boolean Tensors

Saskia Metzler and Pauli Miettinen
Data Mining and Knowledge Discovery
DOI: 10.1007/s10618-015-0420-3

Graphs - such as friendship networks - that evolve over time are an example of data that are naturally represented as binary tensors. Similarly to analysing the adjacency matrix of a graph using a matrix factorization, we can analyse the tensor by factorizing it. Unfortunately, tensor factorizations are computationally hard problems, and in particular, are often significantly harder than their matrix counterparts. In case of Boolean tensor factorizations - where the input tensor and all the factors are required to be binary and we use Boolean algebra - much of that hardness comes from the possibility of overlapping components. Yet, in many applications we are perfectly happy to partition at least one of the modes. For instance, in the aforementioned timeevolving friendship networks, groups of friends might be overlapping, but the time points at which the network was captured are always distinct. In this paper we investigate what consequences this partitioning has on the computational complexity of the Boolean tensor factorizations and present a new algorithm for the resulting clustering problem. This algorithm can alternatively be seen as a particularly regularized clustering algorithm that can handle extremely high-dimensional observations. We analyse our algorithm with the goal of maximizing the similarity and argue that this is more meaningful than minimizing the dissimilarity. As a by-product we obtain a PTAS and an efficient 0.828-approximation algorithm for rank-1 binary factorizations. Our algorithm for Boolean tensor clustering achieves high scalability, high similarity, and good generalization to unseen data with both synthetic and realworld data sets.

Consensus Hashing

Cong Leng and Jian Cheng
Machine Learning
DOI: 10.1007/s10994-015-5496-x

Hashing techniques have been widely used in many machine learning applications because of their efficiency in both computation and storage. Although a variety of hashing methods have been proposed, most of them make some implicit assumptions about the statistical or geometrical structure of data. In fact, few hashing algorithms can adequately handle all kinds of data with different structures. When considering hybrid structure datasets, different hashing algorithms might produce different and possibly inconsistent binary codes. Inspired by the successes of classifier combination and clustering ensembles, in this paper, we present a novel combination strategy for multiple hashing results, named Consensus Hashing (CH). By defining the measure of consensus of two hashing results, we put forward a simple yet effective model to learn consensus hash functions which generate binary codes consistent with the existing ones. Extensive experiments on several large scale benchmarks demonstrate the overall superiority of the proposed method compared with state-of-the art hashing algorithms.

Convex Relaxations of Penalties for Sparse Correlated Variables With Bounded Total Variation

Eugene Belilovsky, Andreas Argyriou, Gael Varoquaux, Matthew B. Blaschko
Machine Learning

DOI: 10.1007/s10994-015-5511-2

We study the problem of statistical estimation with a signal known to be sparse, spatially contiguous, and containing many highly correlated variables. We take inspiration from the recently introduced k-support norm, which has been successfully applied to sparse prediction problems with correlated features, but lacks any explicit structural constraints commonly found in machine learning and image processing. We address this problem by incorporating a total variation penalty in the k-support framework. We introduce the (k,s) support total variation norm as the tightest convex relaxation of the intersection of a set of sparsity and total variation constraints. We show that this norm leads to an intractable combinatorial graph optimization problem, which we prove to be NP-hard. We then introduce a tractable relaxation with approximation guarantees that scale well for grid structured graphs. We devise several first-order optimization strategies for statistical parameterestimation with the described penalty. We demonstrate the effectiveness of this penalty on classification in the low sample regime, classification with M/EEG neuroimaging data, and image recovery with synthetic and real data background subtracted image recovery tasks. We extensively analyse the application of our penalty on the complex task of identifying predictive regions from low-sample high-dimensional fMRI brain data, we show that our method is particularly useful compared to existing methods in terms of accuracy, interpretability, and stability.

Direct Conditional Probability Density Estimation with Sparse Feature Selection

Motoki Shiga, Voot Tangkaratt, and Masashi Sugiyama
Machine Learning

DOI: 10.1007/s10994-014-5472-x

Regression is a fundamental problem in statistical data analysis, which aims at estimating the conditional mean of output given input. However, regression is not informative enough if the conditional probability density is multi-modal, asymmetric, and heteroscedastic. To overcome this limitation, various estimators of conditional densities themselves have been developed, and a kernel-based approach called leastsquares conditional density estimation (LS-CDE) was demonstrated to be promising. However, LS-CDE still suffers from large estimation error if input contains many irrelevant features. In this paper, we therefore propose an extension of LS-CDE called sparse additive CDE (SA-CDE), which allows automatic feature selection in CDE. SACDE applies kernel LS-CDE to each input feature in an additive manner and penalizes the whole solution by a group-sparse regularizer. We also give a subgradient-based optimization method for SA-CDE training that scales well to high-dimensional large data sets. Through experiments with benchmark and humanoid robot transition datasets, we demonstrate the usefulness of SA-CDE in noisy CDE problems.

DRESS: Dimensionality Reduction for Efficient Sequence Search

Alexios Kotsifakos, Alexandra Stefan, Vassilis Athitsos, Gautam Das,
and Panagiotis Papapetrou
Data Mining and Knowledge Discovery
DOI: 10.1007/s10618-015-0413-2

Similarity search in large sequence databases is a problem ubiquitous in a wide range of application domains, including searching biological sequences. In this paper we focus on protein and DNA data, and we propose a novel approximate method method for speeding up range queries under the edit distance. Our method works in a filter-and-refine manner, and its key novelty is a query-sensitive mapping that transforms the original string space to a new string space of reduced dimensionality. Specifically, it first identifies the most frequent codewords in the query, and then uses these codewords to convert both the query and the database to a more compact representation. This is achieved by replacing every occurrence of each codeword with a new letter and by removing the remaining parts of the strings. Using this new representation, our method identifies a set of candidate matches that are likely to satisfy the range query, and finally refines these candidates in the original space. The main advantage of our method, compared to alternative methods for whole sequence matching under the edit distance, is that it does not require any training to create the mapping, and it can handle large query lengths with negligible losses in accuracy. Our experimental evaluation demonstrates that, for higher range values and large query sizes, our method produces significantly lower costs and runtimes compared to two state-of-the-art competitor methods.

Dynamic Inference of Social Roles in Information Cascade

Sarvenaz Choobdar, Pedro Ribeiro, Srinivasan Parthasarathy,
and Fernando Silva
Data Mining and Knowledge Discovery
DOI: 10.1007/s10618-015-0402-5

Nodes in complex networks inherently represent different kinds of functional or organizational roles. In the dynamic process of an information cascade, users play different roles in spreading the information: some act as seeds to initiate the process, some limit the propagation and others are in-between. Understanding the roles of users is crucial in modeling the cascades. Previous research mainly focuses on modeling users behavior based upon the dynamic exchange of information with neighbors. We argue however that the structural patterns in the neighborhood of nodes may already contain enough information to infer users' roles, independently from the information flow in itself. To approach this possibility, we examine how network characteristics of users affect their actions in the cascade. We also advocate that temporal information is very important. With this in mind, we propose an unsupervised methodology based on ensemble clustering to classify users into their social roles in a network, using not only their current topological positions, but also considering their history over time. Our experiments on two social networks, Flickr and Digg, show that topological metrics indeed possess discriminatory power and that different structural patterns correspond to different parts in the process. We observe that user commitment in the neighborhood affects considerably the influence score of users. In addition, we discover that the cohesion of neighborhood is important in the blocking behavior of users. With this we can construct topological fingerprints that can help us in identifying social roles, based solely on structural social ties, and independently from nodes activity and how information flows.

Efficient and Effective Community Search

Nicola Barbieri, Francesco Bonchi, Edoardo Galimberti,
and Francesco Gullo
Data Mining and Knowledge Discovery
DOI: 10.1007/s10618-015-0422-1

Community search is the problem of finding a good community for a given set of query vertices. One of the most studied formulations of community search asks for a connected subgraph that contains all query vertices and maximizes the minimum degree. All existing approaches to min-degree-based community search suffer from limitations concerning efficiency, as they need to visit (large part of) the whole input graph, as well as accuracy, as they output communities quite large and not really cohesive. Moreover, some existing methods lack generality: they handle only single-vertex queries, find communities that are not optimal in terms of minimum degree, and/or require input parameters. In this work we advance the state of the art on community search by proposing a novel method that overcomes all these limitations: it is in general more efficient and effective—one/two orders of magnitude on average, it can handle multiple query vertices, it yields optimal communities, and it is parameter-free. These properties are confirmed by an extensive experimental analysis performed on various real-world graphs.

Finding the Longest Common Sub-Pattern in Sequences of Temporal Intervals

Orestis Kostakis and Panagiotis Papapetrou
Data Mining and Knowledge Discovery
DOI: 10.1007/s10618-015-0404-3

We study the problem of finding the Longest Common Sub-Pattern (LCSP) shared by two sequences of temporal intervals. In particular we are interested in finding the LCSP of the corresponding arrangements. Arrangements of temporal intervals are a powerful way to encode multiple concurrent labeled events that have a time duration. Discovering commonalities among such arrangements is useful for a wide range of scientific fields and applications, as it can be seen by the number and diversity of the datasets we use in our experiments. In this paper, we define the problem of LCSP and prove that it is NP-complete by demonstrating a connection between graphs and arrangements of temporal intervals, which leads to a series of interesting open problems. In addition, we provide an exact algorithm to solve the LCSP problem, and also propose and experiment with three polynomial time and space underapproximation techniques. Finally, we introduce two upper bounds for LCSP and study their suitability for speeding up 1-NN search. Experiments are performed on seven datasets taken from a wide range of real application domains, plus two synthetic datasets.

Generalization Bounds for Learning with Linear, Polygonal, Quadratic and Conic Side Knowledge

Theja Tulabandhula and Cynthia Rudin
Machine Learning
DOI: 10.1007/s10994-014-5478-4

In this paper, we consider a supervised learning setting where side knowledge is provided about the labels of unlabeled examples. The side knowledge has the effect of reducing the hypothesis space, leading to tighter generalization bounds, and thus possibly better generalization. We consider several types of side knowledge, the first leading to linear and polygonal constraints on the hypothesis space, the second leading to quadratic constraints, and the last leading to conic constraints. We show how different types of domain knowledge can lead directly to these kinds of side knowledge.

We prove bounds on complexity measures of the hypothesis space for quadratic and conic side knowledge, and show that these bounds are tight in a specific sense for the quadratic case.

Generalization of Clustering Agreements and Distances for Overlapping Clusters and Network Communities

Reihaneh Rabbany and Osmar R. Zaiane
Data Mining and Knowledge Discovery
DOI: 10.1007/s10618-015-0426-x

A measure of distance between two clusterings has important applications, including clustering validation and ensemble clustering. Generally, such distance measure provides navigation through the space of possible clusterings. Mostly used in cluster validation, a normalized clustering distance, a.k.a. agreement measure, compares a given clustering result against the ground-truth clustering. The two widely-used clustering agreement measures are Adjusted Rand Index (ARI) and Normalized Mutual Information (NMI). In this paper, we present a generalized clustering distance from which these two measures can be derived. We then use this generalization to construct new measures specific for comparing (dis)agreement of clusterings in networks, a.k.a. communities. Further, we discuss the difficulty of extending the current, contingency based, formulations to overlapping cases, and present an alternative algebraic formulation for these (dis)agreement measures. Unlike the original measures, the new co-membership based formulation is easily extendable for different cases, including overlapping clusters and clusters of inter-related data. These two extensions are, in particular, important in the context of finding communities in complex networks.

Generalized Twin Gaussian Processes Using Sharma-Mittal Divergence

Mohamed Elhoseiny and Ahmed Elgammal
Machine Learning
DOI: 10.1007/s10994-015-5497-9

There has been a growing interest in mutual information measures due to its wide range of applications in Machine Learning and Computer Vision. In this manuscript, we present a generalized structured regression framework based on Shama-Mittal divergence, a relative entropy measure, firstly addressed in the Machine Learning community, in this work. Sharma-Mittal (SM) divergence is a generalized mutual information measure for the widely used Rényi, Tsallis, Bhattacharyya, and Kullback-Leibler (KL) relative entropies. Specifically, we study Sharma-Mittal divergence as a cost function in the context of the Twin Gaussian Processes, which generalizes over the KL-divergence without computational penalty. We show interesting properties of Sharma-Mittal TGP (SMTGP) through a theoretical analysis, which covers missing insights in the traditional TGP formulation. However, we generalize this theory based on SM-divergence instead of KL-divergence which is a special case. Experimentally, we evaluated the proposed SMTGP framework on several datasets. The results show that SMTGP reaches better predictions than KL-based TGP (KLTGP), since it offers a bigger class of models through its parameters that we learn from the data.

Half-Space Mass: A Maximally Robust and Efficient Data Depth Method

Bo Chen, Kai Ming Ting, Takashi Washio, and Gholamreza Haffari
Machine Learning
DOI: 10.1007/s10994-015-5524-x

Data depth is a statistical method which models data distribution in terms of centeroutward ranking rather than density or linear ranking. While there are a lot of academic interests, its applications are hampered by the lack of a method which is both robust and efficient. This paper introduces Half-Space Mass which is a significantly improved version of half-space data depth. Half-Space Mass is the only data depth method which is both robust and efficient, as far as we know. We also reveal four theoretical properties of Half-Space Mass: (i) its resultant mass distribution is concave regardless of the underlying density distribution, (ii) its maximum point is unique which can be considered as median, (iii) the median is maximally robust, and (iv) its estimation extends to a higher dimensional space in which the convex hull of the dataset occupies zero volume. We demonstrate the power of Half-Space Mass through its applications in two tasks. In anomaly detection, being a maximally robust location estimator leads directly to a robust anomaly detector that yields a better detection accuracy than halfspace depth; and it runs orders of magnitude faster than L2 depth, an existing maximally robust location estimator. In clustering, the Half-Space Mass version of Kmeans overcomes three weaknesses of K-means.

Improving Classification Performance Through Selective Instance Completion

Amit Dhurandhar and Karthik Sankarnarayanan
Machine Learning
DOI: 10.1007/s10994-015-5500-5

In multiple domains, actively acquiring missing input information at a reasonable cost in order to improve our understanding of the input-output relationships is of increasing importance. This problem has gained prominence in healthcare, public policy making, education, and in the targeted advertising industry which tries to best match people to products. In this paper we tackle an important variant of this problem: Instance Completion, where we want to choose the best k incomplete instances to query from a much larger universe of N(>>k) incomplete instances so as to learn the most accurate classifier. We propose a principled framework which motivates a generally applicable yet efficient meta-technique for choosing k such instances. Since we cannot know *a priori* the classifier that will result from the completed dataset, i.e. the final classifier, our method chooses the k instances based on a derived upper bound on the expectation of the distance between the next classifier and the final classifier. We additionally derive a sufficient condition for these two solutions to match. We then empirically evaluate the performance of our method relative to the state-of-the-art methods on 4 UCI datasets as well as 3 proprietary e-commerce datasets used in previous studies. In these experiments, we also demonstrate how close we are likely to be to the optimal solution, by quantifying the extent to which our sufficient condition is satisfied. Lastly, we show that our method is easily extensible to the setting where we have a non uniform cost associated with acquiring the missing information.

Incremental Learning of Event Definitions with Inductive Logic Programming

Nikos Katzouris, Alexander Artikis, and Georgios Paliouras
Machine Learning
DOI: 10.1007/s10994-015-5512-1

Event recognition systems rely on knowledge bases of event definitions to infer occurrences of events in time. Using a logical framework for representing and reasoning about events offers direct connections to machine learning, via Inductive Logic Programming (ILP), thus allowing to avoid the tedious and error-prone task of manual knowledge construction. However, learning temporal logical formalisms, which are typically utilized by logic-based event recognition systems is a challenging task, which most ILP systems cannot fully undertake. In addition, event-based data is usually massive and collected at different times and under various circumstances. Ideally, systems that learn from temporal data should be able to operate in an incremental mode, that is, revise prior constructed knowledge in the face of new evidence. In this work we present an incremental method for learning and revising event-based knowledge, in the form of Event Calculus programs. The proposed algorithmrelies on abductive-inductive learning and comprises a scalable clause refinement methodology, based on a compressive summarization of clause coverage in a stream of examples. We present an empirical evaluation of our approach on real and synthetic data from activity recognition and city transport applications.

Knowledge Base Completion by Learning Pairwise-Interaction Differentiated Embeddings

Yu Zhao, Sheng Gao, Patrick Gallinari, and Jun Guo
Data Mining and Knowledge Discovery
DOI: 10.1007/s10618-015-0430-1

Knowledge base consisting of triple like (subject entity, predicate relation, object entity) is a very important database for knowledge management. It is very useful for humanlike reasoning, query expansion, question answering (Siri) and other related AI tasks. However, knowledge base often suffers from incompleteness due to a large volume of increasing knowledge in the real world and a lack of reasoning capability. In this paper, we propose a Pairwise-interaction Differentiated Embeddings (PIDE) model to embed entities and relations in the knowledge base to low dimensional vector representations and then predict the possible truth of additional facts to extend the knowledge base. In addition, we present a probability-based objective function to improve the model optimization. Finally, we evaluate the model by considering the problem of computing how likely the additional triple is true for the task of knowledge base completion.Experiments on WordNet and Freebase dataset show the excellent performance of our model and algorithm.

Learning from Evolving Video Streams in a Multi-camera Scenario

Samaneh Khoshrou, Jaime dos Santos Cardoso, and Luís Filipe Teixeira
Machine Learning
DOI: 10.1007/s10994-015-5515-y

Nowadays, video surveillance systems are taking the first steps toward automation, in order to ease the burden on human resources as well as to avoid human error. As the underlying data distribution and the number of concepts change over time, the conventional learning algorithms fail to provide reliable solutions for this setting. Herein, we formalize a learning concept suitable for multi-camera video surveillance and propose a learning methodology adapted to that new paradigm. The proposed framework resorts to the universal background model to robustly learn individual object models from small samples and to more effectively detect novel classes. The individual models are incrementally updated in an ensemble based approach, with older models being progressively forgotten. The framework is designed to detect and label new concepts automatically. The system is also designed to exploit active learning strategies, in order to interact wisely with operator, requesting assistance in the most ambiguous to classify observations. The experimental results obtained both on real and synthetic data sets verify the usefulness of the proposed approach.

Learning Relational Dependency Networks in Hybrid Domains

Irma Ravkic, Jan Ramon, and Jesse Davis
Machine Learning
DOI: 10.1007/s10994-015-5483-2

Statistical Relational Learning (SRL) is concerned with developing formalisms for representing and learning from data that exhibit both uncertainty and complex, relational structure. Most of the work in SRL has focused on modeling and learning from data that only contain discrete variables. As many important problems are characterized by the presence of both continuous and discrete variables, there has been a growing interest in developing hybrid SRL formalisms. Most of these formalisms focus on reasoning and representational issues and, in some cases, parameter learning. What has received little attention is learning the structure of a hybrid SRL model from data. In this paper, we fill that gap and make the following contributions. First, we propose Hybrid Relational Dependency Networks (HRDNs), an extension to Relational Dependency Networks that are able to model continuous variables. Second, we propose an algorithm for learning both the structure and parameters of an HRDN from data. Third, we provide an empirical evaluation that demonstrates that explicitly modeling continuous variables results in more accurate learned models than discretizing them prior to learning.

MassExodus: Modeling Evolving Networks in Harsh Environments

Saket Navlakha, Christos Faloutsos, and Ziv Bar-Joseph
Data Mining and Knowledge Discovery
DOI: 10.1007/s10618-014-0399-1

Defining appropriate distance measures among rankings is a classic area of study which has led to many useful applications. In this paper, we propose a more general abstraction of preference data, namely directed acyclic graphs (DAGs), and introduce a measure for comparing DAGs, given that a vertex correspondence between the DAGs is known. We study the properties of this measure and use it to aggregate and cluster a set of DAGs. We show that these problems are NP-hard and present efficient methods to obtain solutions with approximation guarantees. In addition to preference data, these methods turn out to have other interesting applications, such as the analysis of a collection of information cascades in a network. We test the methods on synthetic and real-world datasets, showing that the methods can be used to, e.g., find a set of influential individuals related to a set of topics in a network or to discover meaningful and occasionally surprising clustering structure.

Minimum Message Length Estimation of Mixtures of Multivariate Gaussian and von Mises-Fisher Distribution

Parthan Kasarapu and Lloyd Allison
Machine Learning
DOI: 10.1007/s10994-015-5493-0

Mixture modelling involves explaining some observed evidence using a combination of probability distributions. The crux of the problem is the inference of an optimal number of mixture components and their corresponding parameters. This paper discusses unsupervised learning of mixture models using the Bayesian Minimum Message Length (MML) criterion. To demonstrate the effectiveness of search and inference of mixture parameters using the proposed approach, we select two key probability distributions, each handling fundamentally different types of data: the multivariate Gaussian distribution to address mixture modelling of data distributed in Euclidean space, and the multivariate von Mises-Fisher (vMF) distribution to address mixture modelling of directional data distributed on a unit hypersphere. The key contributions of this paper, in addition to the general search and inference methodology, include the derivation of MML expressions for encoding the data using multivariate Gaussian and von Mises-Fisher distributions, and the analytical derivation of the MML estimates of the parameters of the two distributions. Our approach is tested on simulated and real world data sets. For instance, we infer vMF mixtures that concisely explain experimentally determined three dimensional protein conformations, providing an effective null model description of protein structures that is central to many inference problems in structural bioinformatics. The experimental results demonstrate that the performance of our proposed search and inference method along with the encoding schemes improve on the state of the art mixture modelling techniques.

Mining Outlying Aspects on Numeric Data

Lei Duan, Guanting Tang, Jian Pei, James Bailey,
Akiko Campbell, and Changjie Tang
Data Mining and Knowledge Discovery
DOI: 10.1007/s10618-014-0398-2

When we are investigating an object in a data set, which itself may or may not be an outlier, can we identify unusual (i.e., outlying) aspects of the object? In this paper, we identify the novel problem of mining outlying aspects on numeric data. Given a query object o in a multidimensional numeric data set O, in which subspace is o most outlying? Technically, we use the rank of the probability density of an object in a subspace to measure the outlyingness of the object in the subspace. A minimal subspace where the query object is ranked the best is an outlying aspect. Computing the outlying aspects of a query object is far from trivial. A naïve method has to calculate the probability densities of all objects and rank them in every subspace, which is very costly when the dimensionality is high. We systematically develop a heuristic method that is capable of searching data sets with tens of dimensions efficiently. Our empirical study using both real data and synthetic data demonstrates that our method is effective and efficient.

Multiscale Event Detection in Social Media

Xiaowen Dong, Dimitrios Mavroeidis, Francesco Calabrese,
Pascal Frossard
Data Mining and Knowledge Discovery
DOI: 10.1007/s10618-015-0421-2

Event detection has been one of the most important research topics in social media analysis. Most of the traditional approaches detect events based on fixed temporal and spatial resolutions, while in reality events of different scales usually occur simultaneously, namely, they span different intervals in time and space. In this paper, we propose a novel approach towards multiscale event detection using social media data, which takes into account different temporal and spatial scales of events in the data. Specifically, we explore the properties of the wavelet transform, which is a welldeveloped multiscale transform in signal processing, to enable automatic handling of the interaction between temporal and spatial scales. We then propose a novel algorithm to compute a data similarity graph at appropriate scales and detect events of different scales simultaneously by a single graph-based clustering process. Furthermore, we present spatiotemporal statistical analysis of the noisy information present in the data stream, which allows us to define a novel term-filtering procedure for the proposed event detection algorithm and helps us study its behavior using simulated noisy data. Experimental results on both synthetically generated data and real world data collected from Twitter demonstrate the meaningfulness and effectiveness of the proposed approach. Our framework further extends to numerous application domains that involve multiscale and multiresolution data analysis.

Optimised Probabilistic Active Learning (OPAL) for Fast, Non-Myopic, Cost-Sensitive Active Classification

Georg Krempl, Daniel Kottke, and Vincent Lemaire
Machine Learning
DOI: 10.1007/s10994-015-5504-1

In contrast to ever increasing volumes of automatically generated data, human annotation capacities remain limited. Thus, fast active learning approaches that allow the efficient allocation of annotation efforts gain in importance. Furthermore, cost-sensitive applications such as fraud detection pose the additional challenge of differing misclassification costs between classes. Unfortunately, the few existing cost-sensitive active learning approaches rely on time-consuming steps, such as performing self labelling or tedious evaluations over samples. We propose a fast, non-myopic, and cost-sensitive probabilistic active learning approach for binary classification. Our approach computes the expected reduction in misclassification loss in a labelling candidate's neighbourhood. We derive and use a closed-form solution for this expectation, which considers the possible values of the true posterior of the positive class at the candidate's position, its possible label realisations, and the given labelling budget. The resulting myopic algorithm runs in the same linear asymptotic time as uncertainty sampling, while its non-myopic counterpart requires an additional factor of $O(m \log m)$ in the budget size. The experimental evaluation on several synthetic and real-world data sets shows competitive or better classification performance and runtime, compared to several uncertainty sampling- and error-reduction-based active learning strategies, both in cost-sensitive and cost-insensitive settings.

Poisson Dependency Networks - Gradient Boosted Models for Multivariate Count Data

Fabian Hadiji, Alejandro Molina, Sriraam Natarajan, and Kristian Kersting
Machine Learning
DOI: 10.1007/s10994-015-5506-z

Although count data are increasingly ubiquitous, surprisingly little work has employed probabilistic graphical models for modeling count data. Indeed the univariate case has been well studied, however, in many situations counts influence each other and should not be considered independently. Standard graphical models such as multinomial or Gaussian ones are also often ill-suited, too, since they disregard either the infinite range over the natural numbers or the potentially asymmetric shape of the distribution of count variables. Existing classes of Poisson graphical models can only model negative conditional dependencies or neglect the prediction of counts or do not scale well. To ease the modeling of multivariate count data, we therefore introduce a novel family of Poisson graphical models, called Poisson Dependency Networks (PDNs). A PDN consists of a set of local conditional Poisson distributions, each representing the probability of a single count variable given the others, that naturally facilities a simple Gibbs sampling inference. In contrast to existing Poisson graphical models, PDNs are non-parametric and trained using functional gradient ascent, i.e., boosting. The particularly simple form of the Poisson distribution allows us to develop the first multiplicative boosting approach: starting from an initial constant value, alternatively a log-linear Poisson model, or a Poisson regression tree, a PDN is represented as products of regression models grown in a stage-wise optimization. We demonstrate on several real world datasets that PDNs can model positive and negative dependencies and scale well while often outperforming state-of-the-art, in particular when using multiplicative updates.

Policy Gradient in Lipschitz Markov Decision Processes

Matteo Pirotta, Marcello Restelli, and Luca Bascetta
Machine Learning
DOI: 10.1007/s10994-015-5484-1

This paper is about the exploitation of Lipschitz continuity properties for Markov Decision Processes (MDPs) to safely speed up policy-gradient algorithms.Starting from assumptions about the Lipschitz continuity of the state-transition model, the reward function, and the policies considered in the learning process, we show that both the expected return of a policy and its gradient are Lipschitz continuous w.r.t. policy parameters.By leveraging such properties, we define policy-parameter updates that guarantee a performance improvement at each iteration. The proposed methods are empirically evaluated and compared to other related approaches using different configurations of three popular control scenarios: the linear quadratic regulator, the mass-spring-damper system and the ship-steering control.

Probabilistic Clustering of Time-Evolving Distance Data

*Julia Vogt, Marius Kloft, Stefan Stark, Sudhir S. Raman,
Sandhya Prabhakaran, Volker Roth, and Gunnar Rätsch*
Machine Learning
DOI: 10.1007/s10994-015-5516-x

We present a novel probabilistic clustering model for objects that are represented via pairwise distances and observed at different time points. The proposed method utilizes the information given by adjacent time points to find the underlying cluster structure and obtain a smooth cluster evolution. This approach allows the number of objects and clusters to differ at every time point, and no identification on the identities of the objects is needed. Further, the model does not require the number of clusters being specified in advance – they are instead determined automatically using a Dirichlet process prior. We validate our model on synthetic data showing that the proposed method is more accurate than state-of-the-art clustering methods. Finally, we use our dynamic clustering model to analyze and illustrate the evolution of brain cancer patients over time.

Ranking Episodes Using a Partition Model

Nikolaj Tatti
Data Mining and Knowledge Discovery
DOI: 10.1007/s10618-015-0419-9

One of the biggest setbacks in traditional frequent pattern mining is that overwhelmingly many of the discovered patterns are redundant. A prototypical example of such redundancy is a freerider pattern where the pattern contains a true pattern and some additional noise events. A technique for filtering freerider patterns that has proved to be efficient in ranking itemsets is to use a partition model where a pattern is divided into two subpatterns and the observed support is compared to the expected support under the assumption that these two subpatterns occur independently. In this paper we develop a partition model for episodes, patterns discovered from sequential data. An episode is essentially a set of events, with possible restrictions on the order of events. Unlike with itemset mining, computing the expected support of an episode requires surprisingly sophisticated methods. In order to construct the model, we partition the episode into two subepisodes. We then model how likely the events in each subepisode occur close to each other. If this probability is high—which is often the case if the subepisode has a high support—then we can expect that when one event from a subepisode occurs, then the remaining events occur also close by. This approach increases the expected support of the episode, and if this increase explains the observed support, then we can deem the episode uninteresting. We demonstrate in our experiments that using the partition model can effectively and efficiently reduce the redundancy in episodes.

Regularized Feature Selection in Reinforcement Learning

Dean Stephen Wookey and George Dimitri Konidaris
Machine Learning
DOI: 10.1007/s10994-015-5518-8

We introduce feature regularization during feature selection for value function approximation. Feature regularization introduces a prior into the selection process, improving function approximation accuracy and reducing overfitting. We show that the smoothness prior is effective in the incremental feature selection setting and present closed-form smoothness regularizers for the Fourier and RBF bases. We present two methods for feature regularization which extend the temporal difference orthogonal matching pursuit (OMP-TD) algorithm and demonstrate the effectiveness of the smoothness prior; smooth Tikhonov OMP-TD and smoothness scaled OMP-TD. We compare these methods against OMP-TD, regularized OMP-TD and least squares TD with random projections, across six benchmark domains using two different types of basis functions.

Soft-max Boosting

Matthieu Geist
Machine Learning
DOI: 10.1007/s10994-015-5491-2

The standard multi-class classification risk, based on the binary loss, is rarely directly minimized. This is due to (i) the lack of convexity and (ii) the lack of smoothness (and even continuity). The classic approach consists in minimizing instead a convex surrogate. In this paper, we propose to replace the usually considered deterministic decision rule by a stochastic one, which allows obtaining a smooth risk (generalizing the expected binary loss, and more generally the cost-sensitive loss). Practically, this (empirical) risk is minimized by performing a gradient descent in the function space linearly spanned by a base learner (a.k.a. boosting). We provide a convergence analysis of the resulting algorithm and experiment it on a bunch of synthetic and real world data sets (with noiseless and noisy domains, compared to convex and non convex boosters).

Tractome: A Visual Data Mining Tool for Brain Connectivity Analysis

Diana Porro-Munoz, Emanuele Olivetti, Nusrat Sharmin,
Thien Bao Nguyen, Eleftherios Garyfallidis, and Paolo Avesani
Data Mining and Knowledge Discovery
DOI: 10.1007/s10618-015-0408-z

Diffusion magnetic resonance imaging data allows reconstructing the neural pathways of the white matter of the brain as a set of 3D polylines. This kind of data sets provides a means of study of the anatomical structures within the white matter, in order to detect neurologic diseases and understand the anatomical connectivity of the brain. To the best of our knowledge, there is still not an effective or satisfactory method for automatic processing of these data. Therefore, a manually guided visual exploration of experts is crucial for the purpose. However, because of the large size of these data sets, visual exploration and analysis has also become intractable. In order to make use of the advantages of both manual and automatic analysis, we have developed a new visual data mining tool for the analysis of human brain anatomical connectivity. With such tool, humans and automatic algorithms capabilities are integrated in an interactive data exploration and analysis process. A very important aspect to take into account when designing this tool, was to provide the user with comfortable interaction. For this purpose, we tackle the scalability issue in the different stages of the system, including the automatic algorithm and the visualization and interaction techniques that are used.

Contents – Part I

Clustering and Unsupervised Learning

Data Preprocessing

Data Streams and Online Learning

Deep Learning

Contents – Part II

Preference Learning and Label Ranking

Probabilistic, Statistical, and Graphical Approaches

Rich Data

Social and Graphs

Contents – Part III

Demo Track

Research Track

Classification, Regression and Supervised Learning

Data Split Strategies
for Evolving Predictive Models

Vikas C. Raykar[✉] and Amrita Saha

IBM Research, Bangalore, India
{viraykar,amrsaha4}@in.ibm.com

Abstract. A conventional textbook prescription for building good predictive models is to split the data into three parts: training set (for
model fitting), validation set (for model selection), and test set (for final
model assessment). Predictive models can potentially evolve over time
as developers improve their performance either by acquiring new data
or improving the existing model. The main contribution of this paper
is to discuss problems encountered and propose workflows to manage
the allocation of newly acquired data into different sets in such dynamic
model building and updating scenarios. Specifically we propose three
different workflows (parallel dump, serial waterfall, and hybrid) for allocating new data into the existing training, validation, and test splits.
Particular emphasis is laid on avoiding the bias due to the repeated use
of the existing validation or the test set.

Keywords: Data splits · Model assessment · Predictive models

1 Introduction

A common data mining task is to build a good predictive model which generalizes
well on future unseen data. Based on the annotated data collected so far the goal
for a machine learning practitioner is to search for the best predictive model
(known as *supervised learning*) and at the same time have a reasonably good
estimate of the performance (or risk) of the model on future unseen data. It is well
known that the performance of the model on the data used to learn the model
(training set) is an overly optimistic estimate of the performance on unseen data.
For this reason it is a common practice to sequester a portion of the data to assess
the model performance and never use it during the actual model building process.
When we are in a data rich situation a conventional textbook prescription (for
example refer to Chapter 7 in [6]) is to split the data into three parts: *training
set, validation set,* and *test set* (See Figure 1). The training set is used for
model fitting, that is, estimate the parameters of the model. The validation set
is used for *model selection*, that is, we use the performance of the model on
the validation set to select among various competing models (e.g. should we
use a linear classifier like logistic regression or a non-linear neural network) or
to choose the hyperparameters of the model (e.g. choosing the regularization

© Springer International Publishing Switzerland 2015
A. Appice et al. (Eds.): ECML PKDD 2015, Part I, LNAI 9284, pp. 3–19, 2015.
DOI: 10.1007/978-3-319-23528-8_1

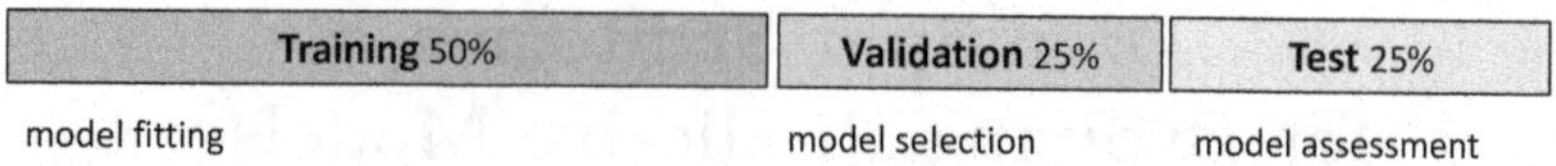

Training 50%	Validation 25%	Test 25%
model fitting	model selection	model assessment

Fig. 1. *Data splits for model fitting, selection, and assessment.* The *training split* is used to estimate the model parameters. The *validation split* is used to estimate prediction error for model selection. The *test split* is used to estimate the performance of the final chosen model.

parameter for logistic regression or the number of nodes in the hidden layer for a neural network). The test set is then used for final *model assessment*, that is, to estimate the performance of the estimated model.

However in practice searching for the best predictive model is often an iterative and continuous process. A major bottleneck typically encountered in many learning tasks is to collect the data and annotate them. Due to various constraints (either time or financial) very often the best model based on the data available so far is deployed in practice. At the same time the data collection and annotation process will continue so that the model can be improved at a later stage. Once we have reasonably enough data we refit the model to the new data to make it more accurate and then release this new model. Sometimes after the model has been deployed in practice we find that the model does not perform well on a new kind of data which we do not have in our current training set. So we redirect our efforts into collecting more data on which our model fails. The main contribution of this paper is to discuss problems encountered and propose various workflows to manage the allocation of newly acquired data into different sets in such dynamic model building/updating scenarios.

With the advent of increased computing power it is very easy to come up with a model that performs best on the validation set by searching over an extremely large range of diverse models. This procedure can lead to non-trivial bias (or over-fitting to the validation set) in the estimated model parameters. It is very likely that we found the best model on the validation set by chance. The same applies to the testing set. One way to think of this is that every time we use the test set to estimate the performance the dataset becomes less fresh and can increase the risk of over-fitting. The proposed data allocation workflows are designed with a particular emphasis on avoiding this bias.

2 Data Splits for Model Fitting, Selection, and Assessment

A typical supervised learning scenario consists of an annotated data set $\mathcal{T} = \{(\boldsymbol{x}_i, y_i)\}_{i=1}^{n}$ containing n instances, where $\boldsymbol{x}_i \in \mathcal{X}$ is an instance (typically a d-dimensional feature vector) and $y_i \in \mathcal{Y}$ is the corresponding known label. The task is to learn a function $f : \mathcal{X} \rightarrow \mathcal{Y}$ which performs well on an independent test data and also have a reasonably good estimate of the performance (also known as the test error of the model). Let $\widehat{f}(\boldsymbol{x})$ be the prediction model/function that has been learnt/estimated using the training data $\mathcal{T}$. Let $L(y, \widehat{f}(\boldsymbol{x}))$ be the *loss function* [1] for measuring errors between the target response y and the prediction from the learnt model $\widehat{f}(\boldsymbol{x})$. The (conditional) *test error*, also referred to as *generalization error*, is the prediction error over an independent test sample, that is, $\mathrm{Err}_\mathcal{T} = E_{(\boldsymbol{x}, y)}[L(y, \widehat{f}(\boldsymbol{x}))|\mathcal{T}]$, where $(\boldsymbol{x}, y)$ are drawn randomly from their joint distribution. Since the training set $\mathcal{T}$ is fixed, and test error refers to the error obtained with this specific training set. Assessment of this test error is very important in practice since it gives us a measure of the quality of the ultimately chosen model (referred to as *model assessment*) and also guides the choice of learning method or model (also known as *model selection*). Typically our model will also have tuning parameters (for example the regularization parameter in lasso or the number of trees in random forest) and we write our predictions as $\widehat{f}_\theta(\boldsymbol{x})$. The tuning parameter θ varies the complexity of our model, and we wish to find the value of θ that minimizes the test error. The *training error* is the average loss over the entire training sample, that is, $\overline{\mathrm{err}} = \frac{1}{n} \sum_{i=1}^{n} L(y_i, \widehat{f}_\theta(\boldsymbol{x}_i))$. Unfortunately *training error is not a good estimate of the test error*. A learning method typically adapts to the training data, and hence the training error will be an overly optimistic estimate of the test error. Training error consistently decreases with model complexity, typically dropping to zero if we increase the model complexity large enough. However, a model with zero training error is overfit to the training data and will typically generalize poorly.

If we are in a data-rich situation, the best approach to estimate the test error is to randomly divide the dataset into three parts [2,4,6]: a *training split* $\mathcal{T}$, a *validation split* $\mathcal{V}$, and a *test split* $\mathcal{U}$. While it is difficult to give a general rule on the split proportions a typical split suggested in [6] is to use 50% for training, and 25% each for validation and testing (see Figure 1). The *training split* $\mathcal{T}$ is used to fit the model (*i.e.* estimate the parameters of the model for a fixed set of tuning parameters). The *validation split* $\mathcal{V}$ is used to estimate prediction error for model selection. We use the performance on the validation split to select among various competing models or to choose the tuning parameters of the model. The *test split* $\mathcal{U}$ is used to estimate the performance of the final chosen model. Ideally, the test set should be sequestered and be brought out only at the end of the data analysis.

In this paper we specifically assume that we are in a *data-rich situation*, that is we have a reasonably large amount of data. In data poor situations where we do not have the luxury of reserving a separate test set, it does not seem possible to estimate conditional error effectively, given only the information in the same training set. A related quantity sometimes used in data poor situations is the expected test error $\mathrm{Err} = E[\mathrm{Err}_\mathcal{T}]$. While the estimation of the conditional test error $\mathrm{Err}_\mathcal{T}$ will be our goal the expected test Err is more amenable to statistical analysis, and most methods like cross-validation [15] and bootstrap [3] effectively estimate the expected error [6].

---

[1] Typical loss functions include the 0-1 loss ($L(y, \widehat{f}(\boldsymbol{x})) = I(y \neq \widehat{f}(\boldsymbol{x}))$,where I is the indicator function) or the log-likelihood loss for classification and the squared error ($L(y, \widehat{f}(\boldsymbol{x})) = (y - \widehat{f}(\boldsymbol{x}))^2$) or the absolute error ($L(y, \widehat{f}(\boldsymbol{x})) = |y - \widehat{f}(\boldsymbol{x})|$) for regression problems.

3   Issues with Evolving Models

Arrival of New Data After Model Deployment. Having done model selection using the validation split the parameters of the model are estimated using the training split and the performance on unseen data is assessed using the test split. If this performance is reasonable enough the model is finally deployed in practice. This scenario has implicitly assumed that we start the data analysis after all the data is collected. However in practice data collection and annotation is a continuous ongoing process, often in a non-iid fashion. After the final model is frozen and deployed in practice after a few months let us say we have more annotated data. Now the question that arises is how do we use this data. Should we dump all this data into the training split to improve the model performance? Or should we dump this into the test split so that we have a better estimate of the model performance?

Model Driven Data Collection. Very often it so happens that once the model is deployed in practice we discover that the model performs poorly on a certain class of data. The most likely cause of this is that we did not have enough data of that particular kind in our dataset. This drives the collection of specific kind of data on which our model performs poorly. Having collected the data similar questions arise. What options should we pursue for allocating the new data to different splits ?

Test Set Reuse. When developing predictive models in many domains (and especially in medical domains) it is a common practice to completely sequester the test set from the data mining practitioners. Once the model has been finalized a review board (such as the Food and drug administration) evaluates the model and then gives a final decision as to whether the model passed the test or not. The feedback could be binary(pass/fail) or more detailed like the error or the kind of mistakes made. If the model failed the test then they have to go back and build a better model and test it again. With the advent of increased computing power it is very easy to come up with a model that performs best on the test set by searching over an extremely large range of diverse models. In such a scenario multiple reuse of the test set can often lead to overfitting on the test set. The ideal solution is to completely replace the test set, but having a new test set every time can be expensive given that the data mining practitioners keep churning out new models extremely fast. Another approach to avoid this is that the test set has to be kept fresh by supplementing it with new data every time a developer requests it for testing. At the same time the practitioners would like to learn from the mistakes in the test set. This can be achieved by releasing some part of the test set to the practitioners to improve the model.

These issues arise across different domains. In natural language processing tasks textual resources are acquired one at a time and annotated. The resource acquisition is often guided by various non-technical constraints and often arrives in a non-iid fashion. In the medical domain a common task is to build a predictive model to predict whether a suspicious region on a medical image is malignant or benign. In order to train such a classifier, a set of medical images is collected from hospitals. These scans are then read by expert radiologists who mark the suspicious locations. Ideally we would like our model to handle all possible variations—different scanners, acquisition protocols, hospitals, and patients. Collecting such data is a time consuming process. Typically we can have contracts with different hospitals to collect and process data. Each hospital has a specific kind of scanners, acquisition protocols, and patient demographics. While most learning methods assume that data is randomly sampled from the population in reality due to various constraints data does not arrive in a random fashion. Based on the contracts at the end of a year we have data from say around five hospitals and the data from the another hospital may arrive a year later. Based on the data from five hospitals we can deploy a model and later update the model when we acquire the data from the other hospital.

These kind of issues also arise in data mining competitions which traditionally operate in a similar setup. Kaggle [1], for example, is a platform for data prediction competitions that allows organizations to post their data and have it scrutinized by thousands of data scientists. The training set along with the labels is released to the public to develop the predictive model. Another set for which the labels have been withheld is used to track the performance of the competitors on a public leader board. Very often is happens that the competitors try to overfit the model on this leader board set. For this reason only a part of this set is used for the leader board and the remaining data is used to decide the final rankings. An important feature of our proposed workflows is that there is a movement of data across different sets at regular intervals and this can help avoid the competitors trying to overfit their models to the leader board.

4   Data Splits for Evolving Models

Based on the data collected so far let us assume we start with a 50% training-25% validation-25% test split as described earlier. In this paper we propose a workflow to allocate newly acquired data into the existing splits. Any workflow to split new data should balance the following desired objectives:

1. *Exploit large portion for training quickly.* We want to exploit as much of the new data as quickly as possible for training our final predictive model. For most models, the larger the dataset the more accurate are the estimated parameters.

2. *Reserve sufficient amount for testing.* However, at the same time we want to reserve a sufficient amount of the data (in the validation and testing set) for getting an unbiased estimate of the performance of the learnt model. It is very likely that the new data is a different kind of data not existing in the current splits and hence we want to have a sufficient representation of the new data in all the three splits.

3. *Keep the test set fresh.* We want to keep the testing/validation sets fresh to avoid the bias due to the reuse of the test set.

4. *Learn from your mistakes.* At the same time we do not want to completely sequester the test set and make sure that data mining practitioners

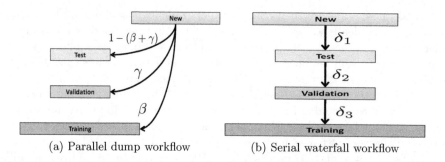

(a) Parallel dump workflow   (b) Serial waterfall workflow

Fig. 2. (a) *Parallel dump workflow* The new data is split into three parts (according to the ratio $\beta : \gamma : 1 - (\beta + \gamma)$) and directly dumped into the existing training, validation, and test splits. (b) *Serial waterfall workflow* A δ_3 fraction of the validation set moves to the training set, a δ_2 part of the test set moves to the validation set, and a δ_1 fraction of the new data is allocated to the test set.

learn from our mistakes in the test set. This is especially useful in scenarios where then the data mining practitioners have to go back to their drawing boards and re-design their model because it failed on a sequestered test set.

In the next section we describe two workflows: the *parallel dump* (§ 4.1) and the *serial waterfall* (§ 4.2) each of which can address some of these objectives. In § 4.3 we describe the proposed *hybrid workflow* which can balance all the four objectives described above.

4.1   Parallel Dump Workflow

The most obvious method is to split the new data into three parts and directly dump each part into the existing test, validation, and training splits as shown in Figure 2(a). One could use a 50% training-25% validation-25% test split as earlier or any desired ratio $\beta : \gamma : 1 - (\beta + \gamma)$. The main advantage of this method is that we have immediate access to the new data in the training set and the model can be improved quickly. Also a sufficient portion of the new data goes into the validation and the test set immediately. However once the new data is allocated we end up using the validation and test splits again one more time (the split is now no longer as fresh as the first time) thus leading to the model selection bias. In this workflow the splits are static and there is no movement across the splits. As a result we do not have a chance to learn from mistakes in the validation and testing set. Generally it may not be to our advantage to let the test split to keep growing without learning from the errors the model makes on the test set. So it makes sense to move some part of the data from the test set to either the training or validation set.

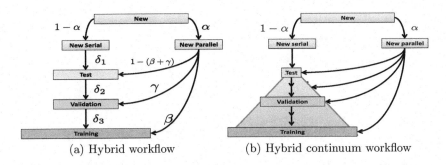

(a) Hybrid workflow   (b) Hybrid continuum workflow

Fig. 3. (a) *The hybrid workflow* There is a movement of data among different sets in the serial mode and at the same time there is an immediate fresh influx of data from the parallel mode. (b) *The continuum workflow* In this setting instead of having 3 sets (training, testing, unseen) we will in theory have a continuum of sets (with samples trickling to the lower levels), with the lowest one for training and the top most one to give the most unbiased estimate of performance.

4.2 Serial Waterfall Workflow

In this workflow data keeps trickling from one level to the other as illustrated in Figure 2(b). Once new data arrives, a δ_3 fraction of the validation set moves to the training set, a δ_2 part of the test set moves to the validation set, and a δ_1 fraction of the new data is allocated to the test set. The training set always keeps getting bigger and once a data moves to the training set it stays there forever. This mode has the following advantages: (1) the test and validation sets are always kept fresh. This avoids over fitting due to extensive model search since the validation and test sets are always refreshed. (2) Since part of the data from the validation and the test set eventually moves to the training we have a chance to learn from the mistakes. The disadvantage of this serial workflow is that the new data takes some time to move to the training set, depending on how often we refresh the existing sets. This restricts us from exploiting the new data as quickly as possible as it takes some time for the data to trickle to the training set.

4.3 Hybrid Workflow

Our proposed workflow as illustrated in Figure 3(a) is a combination of the above two modes of dataflow. It combines the advantages of both these modes. The key feature of the proposed workflow is that there is a movement of data among different sets in the serial mode and at the same time there is an immediate fresh influx of data from the parallel mode. We split the new data randomly into two sets according to the ratio $1 - \alpha : \alpha$. One split is used for the Serial Waterfall mode and the other split is used for the Parallel Dump mode. A value

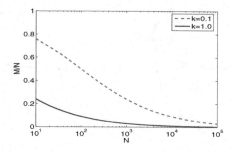

Fig. 4. The ratio $M/N = 1/(k\sqrt{N} + 1)$ as a function of N for two different k.

of $\alpha = 0.5$ seems to be a reasonable choice [2]. The value of α can be increased if a more dominant parallel workflow is desired. Note that $\alpha = 0$ corresponds to the serial workflow and $\alpha = 1$ corresponding to the parallel workflow. The value of parameter α is more of design choice and can be chosen based on the domain and various constraints.

In principle the proposed workflow can be extended to more than 3 levels (see Figure 3(b)). In this setting instead of having 3 sets (training, validation, test) we will in theory have a continuum of sets (with samples trickling to the lower levels), with the lowest one to be used for training and the top most one to give the most unbiased estimate of performance of the model. Presumably levels could be: training/tweaking data, cross validation data, hold out testing data, up to highly unbiased very fresh test data. Lower levels would obviously be used more often than higher ones, and testing on higher levels would presumably only happen when tests on lower levels were successful.

5   Bias Due to Test Set Reuse

A key idea in the serial mode is to move data from one level to another to avoid bias due to multiple reuse. We derive a simple rule to decide how much of the new data has to be moved to the test set to avoid the bias due to reuse of the test set multiple times. The same can be used to move data from the test set to the validation set to keep the validation set fresh. This analysis is based on ideas in [14]. Our goal is not to get a exact expression but to get the nature of dependence on the set size. We will consider a scenario where we have a test set of N examples. At each reuse we supplement the test set with M new examples

---

[2] The parameter α decides the proportion of the incoming data that will go into the train and the test splits. Without any prior knowledge or assumption, the reasonable value of α is a constant fixed at 0.5. But α can be further made to vary every time a new batch of data arrives. For example consider a batch scenario where for every incoming batch of data we can compute the similarity of the current batch with the previously seen batches (for example using the Kullback-Leilbler (KL) divergence). The parameter α can then be made to vary based on the estimated KL divergence.

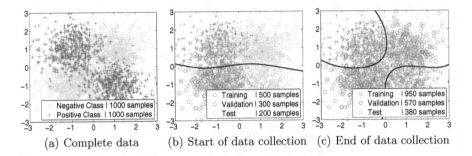

(a) Complete data   (b) Start of data collection   (c) End of data collection

Fig. 5. *Workflow simulation setup* (a) A two-dimensional binary classification simulation setup with data sampled from four Gaussians. (b) The decision boundary obtained by a neural network at the start of the workflow simulation. At each time step we add 50 new examples into the data pool, allocate the new data according to different workflows (either the parallel dump, serial waterfall, or the hybrid workflow) and repeat the model building process. (c) The final decision boundary obtained when all the data has been acquired.

and remove M of the oldest examples. Specifically we will prove the following result (see the appendix for the proof):

After each reuse if we supplement the test set (consisting of N samples) with $M > N/(k\sqrt{N} + 1)$ new samples for some small constant k then the bias due to test set reuse can be safely ignored.

Figure 4 plots the ratio M/N as a function of N for two values of k. We need to replace a smaller fraction of the data when N is large. Small datasets lead to a larger bias after each use and hence need to be substantially supplemented.

6   Illustration on Synthetic Data

We first illustrate the advantages and the disadvantages of the three different workflows on a two-dimensional binary classification problem shown in Figure 5(a) with data sampled from a mixture of four Gaussians. The positive class consists of 1000 examples from two Gaussians centered at [-1,1] and [1,-1]. The negative class consists of 1000 examples from a mixture of two Gaussians centered at [1,1] and [-1,-1] respectively. We use a multi-layer neural network with one hidden layer and trained via back propagation. The number of units in the hidden layer is considered as the tuning parameter and selected using the validation split. In order to see the effect of having a non-representative set of data during typical model building scenarios, we consider a scenario where at the beginning of the model building process we have *collected data only from two Gaussians* (positive class centered at [-1,1] and negative class centered at [-1,-1]) as shown in Figure 5(b). Based on the data collected so far we will use the validation split to tune the number of hidden units in the neural network, the training split to train the neural network via back propagation, and the test split to compute the misclassification error of the final trained model. Figure 5(b) shows the decision boundary obtained for the trained neural network using such splits at the start of the model building process. At each time step we add 50 new examples into the data pool, allocate the new data according to different workflows (either the parallel dump, the serial waterfall, or the hybrid workflow) and repeat the model building process. *The new data does not arrive in a random fashion.* We first sample data from the Gaussian centered at [1,1] and then the data from the remaining Gaussian centered at [1,-1]. While this may not be a fully realistic scenario it helps us to illustrate the different workflows. One way to think of this is to visualize that each Gaussian represents data from different hospital/scanner and the data collection may not be designed such that data arrives in a random fashion. Figure 5(c) shows the final decision boundary obtained when all the data has been acquired. Once we have all the data all different workflows reach the same performance. Here we are interested in the model performance and our estimate of it at different stages as new data arrives.

Actual Performance of The Model. Figure 6 plots the misclassification error at each time point (until all the data has been used) for the parallel dump (with parameters $\beta = 0.5$ and $\gamma = 0.3$), the serial waterfall (with δ parameters automatically chosen), and the hybrid workflow (with $\alpha = 0.1$). The error *is computed on the entire dataset* [3]. If we had the entire dataset the final model should have an error of around 0.15. It can be seen that the parallel dump workflow exploits the new data quickly and reaches this performance in around 20 time steps. The serial waterfall moves the data to the training set slowly and achieves the same performance in around 40 time steps. The hybrid workflow can be considered as a compromise between these two workflows and exploits all the new data in around 25 time steps. The sharp drops in the curve occur when the decision boundary changes abruptly because we have now started collecting data from a new cluster.

Estimate of The Performance of The Model. We want to exploit as much of the new data as quickly as possible for training our final predictive model. However at the same time we want to keep the test set fresh in order to get the most unbiased estimate of the performance of the final model. Note that Figure 6 plotted the test error assuming that some oracle gave us the actual data distribution. However in practice we do not have access to this distribution and should use the existing data collected so far (more precisely the test split) to also get a reasonably good estimate of the test error. Figure 7(a), (b), and (c) compares the training error, test error, and the actual error for three different workflows. The results are averaged over 100 repetitions and the standard deviation is also shown. While the test error for all the three workflows approaches the true error when all the data has been collected we want to track how the test error changes during any stage of the model building process. While the

---

[3] We have shown how to select the delta parameter automatically in the serial waterfall model. The other parameters are more of design choice and have to be chosen based on the various constraints (time, financial, etc.).

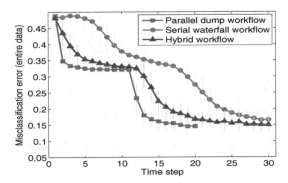

Fig. 6. *Comparison of workflows* The error (computed on the entire dataset) at each time point (until all the data has been used) for the parallel dump (with parameters $\beta = 0.5$ and $\gamma = 0.3$), the serial waterfall (with δ automatically chosen), and the hybrid workflow (with $\alpha = 0.1$).

parallel dump workflow (see Figure 7(a)) gave us a predictive model quickly the test error is highly optimistic (and close to the training error) and is close to the true error only at the end when we have access to all the data. The test error for serial waterfall workflow (see Figure 7(b)) does not track the training error and is better reflection of the risk of the model than the parallel dump workflow. These variations in the test error can be explained because the composition of the test set is continuously changed at each time step and it takes some time for the new data to finally reach the training set. However the serial waterfall workflow can be overly pessimistic and the proposed hybrid workflow (see Figure 7(c)) can be a good compromise between the two—it can give a good model reasonably fast and at the same time produce a reasonably good estimate of the performance of the model. By varying the parameter α one can obtain a desired compromise between the serial and the parallel workflow. Figure 7(d) compares the hybrid workflow for two different values of the split parameters $\alpha = 0.1$ and $\alpha = 0.5$ with $\alpha = 0$ corresponding to the serial workflow and $\alpha = 1$ corresponding to the parallel workflow. The value of parameter α is more of design choice and can be chosen based on the domain and various constraints.

7   Case Study: Paraphrase Detection

We demonstrate the tradeoffs of the different workflows on a natural language processing task of *paraphrase detection* [8]. Given a pair of sentences, for example, *Video game violence is not related to serious aggressive behavior in real life.* and *Violence in video games is not causally linked with aggressive tendencies.*, we would like to learn a model which predicts whether the two sentences are semantically equivalent (paraphrases) or not. We take a supervised learning approach to this problem by first manually collecting a labeled data, extracting features from the sentences, and then training a binary classifier. Given a pair

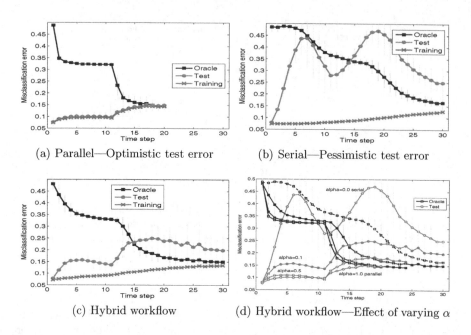

(a) Parallel—Optimistic test error   (b) Serial—Pessimistic test error

(c) Hybrid workflow   (d) Hybrid workflow—Effect of varying α

Fig. 7. The estimated training error, test error, and the actual error (oracle) for (a) the parallel dump, (b) the serial waterfall and (c) the hybrid workflow. (d) The effect of varying the split parameter α.

of sentences one can construct various features which quantify the dissimilarity between two sentences. One of the most import set of features are the based on machine translation (MT) metrics. For example the BLEU score [11] (which measures n-gram overlap) which is an widely used evaluation metric for MT systems is an important feature for paraphrase detection. In our system we used a total of 14 such features and then trained a binary decision tree using the labeled data to train the classifier.

To collect the labeled data we show a pair of sentences to three in-house annotators and ask them to label them as either semantically equivalent or not. The sentences were taken from wikipedia articles corresponding to a specified topic. Due to the overall project design the labeling proceeded one topic at a time. We currently have a labeled data of 715 sentence pairs from a total of 6 topics (56, 76, 88, 108, 140, 247 sentence for each of the six topics) of which 112 were annotated as semantically equivalent by a majority of the annotators. We analyse a situation where the data arrives one topic at a time. The new data is allocated into train, validation, and test splits according the parallel dump, serial waterfall, and the hybrid workflow. At each round a binary decision tree is trained using the train split, the decision tree parameters are chosen using the validation split, and the model performance is assessed using the test split.

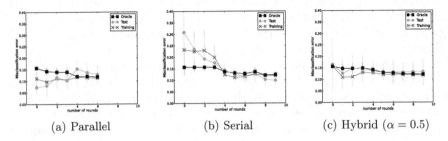

(a) Parallel   (b) Serial   (c) Hybrid ($\alpha = 0.5$)

Fig. 8. *Paraphrase detection* (see § 7) (a) The misclassification error at each round (until all data has been used) on the oracle, train and test split for (a) the parallel dump (with parameters $\beta = 0.6$ and $\gamma = 0.2$), (b) the serial waterfall (with $\delta_1 = 0.5$, $\delta_2 = 0.3$, and $\delta_3 = 0.2$), and (b) the hybrid workflow (with $\alpha = 0.5$). The oracle is a surrogate for the true model performance which is evaluated on 30 % of entire original data.

Figure 8 shows the misclassification error on the train and the test splits as a function of the number of rounds, here each round refers to a point in time when we acquire a new labeled data and data reallocation/movement happens. The results are averaged over 50 replications, where for each replication the order of the topics is randomly permuted. We would like to see how close is the model performance assessed using the test split to the true model performance on the entire data (which we call the oracle). Since we do not have access to the true data distribution we sequester 30 % of original data (which includes data from all topics) and use this as a surrogate for the true performance. The following observations can be made (see Figure 8): (1) The test error for all the three workflows approaches the true oracle error at steady state when a large amount of data has been collected. (2) However in the early stages the performance of the model on the test split as assessed by the parallel workflow (Figure 8(a)) is relatively optimistic while that of the serial workflow(Figure 8(b)) is highly pessimistic. (3) The proposed hybrid workflow (see Figure 8(d)) estimates the test error much closer to the oracle error.

8 Related Work

There is not much related work in this area in the machine learning literature. Most earlier research has focussed on settings with either unlimited data or finite fixed data, while this paper proposes data flow strategies for finite but growing datasets. The bias due to repeated use of the test set has been pointed out in a few papers in the cross-validation setting [10,13]. However main focus of this paper is on data rich situations where we estimate the prediction capability of a classifier on a independent test data set (called the *conditional test error*). In data poor situations techniques like cross-validation [7] and bootstrapping [3] are widely used as a surrogate to this, but they can only estimate the *expected test error* (averaged over multiple training sets).

There is a rich literature in the area of learning under concept drift [5] and dataset shifts [12]. Concept drift primarily refers to an online supervised learning scenario when the relation between the input data and the target variable changes over time. Dataset shift is a common problem in predictive modeling that occurs when the joint distribution of inputs and outputs differs between training and test stages. Covariate shift, a particular case of dataset shift, occurs when only the input distribution changes. The other kinds of concept-drifts are *prior probability shift* where the distribution over true label y changes, *sample selection bias* where the data distributions varies over time because of an unknown sample rejection bias and *source component shift* where the datastream can be thought to be originating from different unknown sources at different time points.

Various strategies have been proposed to correct for these shifts in test distribution [5,12]. In general the field of concept drift seeks to develop shift-aware models that can capture these specific types of variations or a combination of the different modes of variations or do model selection to assess whether dataset shift is an issue in particular circumstances [9]. In our current work the focus is to investigate into different kinds of dynamic workflows for assigning data into train, test and validation splits to reduce the effect of bias in the scenario of a time-shifting dataset. In our setting the drift arises as a consequence of that data arriving in an non-iid fashion. Hence this body of work mainly deals with the source component shift of the datasets where for example in the medical domain where a classifier is built from the training data obtained from various hospitals, each hospital may have a different machine with different bias, each producing different ranges of values for the covariate x and possibly even the true label y.

We are mainly concerned with allocating the new data to the existing splits and not with modifying any particular learning method to account for the shifts. One of our motivations was not to make the allocation strategies model or data distribution specific. We wanted to come up with strategies that can be used with any model or data distribution. The main contribution of our work is to propose these strategies and empirically analyze them. These kind of strategies have not been discussed in the concept drift literature.

9   Conclusions

We analysed three workflows for allocating new data into the existing training, validation, and test splits. The parallel dump workflow splits the data into three parts and directly dumps them into the existing splits. While it can exploit the new data quickly to build a good model the estimate of the model performance is optimistic especially when the new data does not arrive in a random fashion. The serial waterfall workflow which trickles the data from one level to another avoids this problem by keeping the test set fresh and prevents the bias due to multiple reuse of the test set. However it takes a long time for the new data to reach the training split. The proposed hybrid workflow which balances both the workflows seems to be a good compromise—it can give a good model reasonably fast and at the same time produce a reasonably good estimate of the model performance.

A   Appendix: Bias Due to Test Set Reuse

We will consider a scenario where we have a test set of N examples. At each reuse we supplement the test set with M new examples and remove M of the oldest examples. In this appendix we prove the following result:

After each reuse if we supplement the test set (consisting of N samples) with $M > N/(k\sqrt{N} + 1)$ new samples for some small constant k then the bias due to test set reuse can be safely ignored.

Maximum Bias Due to Test Set Reuse: Let E_j be the estimated error of the model $\widehat{f}_j$ on a test set consisting of N samples after the test set having been reused j times, *i.e.*,

$$\mathrm{E}_j = \frac{1}{N} \sum_{i=1}^{N} L(y_i, \widehat{y}_{ij}) = \frac{1}{N} \sum_{i=1}^{N} L(y_i, \widehat{f}_j(\boldsymbol{x}_i)), \tag{1}$$

where y_i is the true response, $\widehat{y}_{ij} = \widehat{f}_j(\boldsymbol{x}_i)$ is the response predicted by the learnt model after the test set has been reused j times, and L is the loss function used to measure the error. If the test set is used multiple times the final model will overfit to the testing set. In other words the performance of the model on the test set will be biased and will not reflect the true performance of the model.

Let $\mathrm{Bias}_{max}(\mathrm{E}_j)$ be the maximum possible bias in the estimate of the error E_j caused due to multiple reuse of the test set. The first time the test set is used the bias is zero, *i.e.*, $\mathrm{Bias}_{max}(\mathrm{E}_1) = 0$. Every subsequent use increases the bias. The worst case scenario (the perhaps the easiest for the developer) is when the developer directly observes the predictions $\widehat{y}_{ij}$ and learns a model on these predictions to match the desired response y_j for all examples in the test set. Since we have N examples in the test set by reusing the test set $N + 1$ times one can actually drive the error $\mathrm{E}_{N+1} = 0$ since we will have N unknowns to estimate $(y_1, \ldots, y_N)$ and N new tests. Hence $\mathrm{Bias}_{max}(\mathrm{E}_{N+1}) = \mathrm{E}_1$. We will further assume that at each test we lose one degree of freedom and hence approximate

$$\mathrm{Bias}_{max}(\mathrm{E}_j) = \frac{j-1}{N} E_1, \quad j = 1, \ldots, N. \tag{2}$$

Supplement the Test Set to Avoid the Bias: In order to avoid this bias our strategy is to supplement the test set with M new examples and move the oldest M examples to the lower validation set. We do this every time we reuse the test set. If we keep doing this then in the long run we will have a set of N examples which has been reused N/M times. Hence if we supplement the test set with M examples at each reuse the bias in the test set at steady state will be

$$\mathrm{Bias}_{max}(\mathrm{E}_\infty) = \frac{N-M}{NM} E_1, \tag{3}$$

where E_∞ is the error in this steady state scenario.

How Much to Supplement?: If we require that for some small constant k, $|\text{Bias}_{max}(\text{E}_\infty)| < \text{sd}(\text{E}_\infty)k$ then the bias can be safely ignored, where $\text{sd}(\text{E}_\infty)$ is the standard deviation of the error estimate. For analytical tractability we will assume a squared loss error function, *i.e.*, $L(y, \widehat{y}) = (y - \widehat{y})^2$. Hence at steady state

$$\text{E}_\infty = \frac{1}{N} \sum_{i=1}^{N} (y_i - \widehat{y}_{i\infty})^2. \tag{4}$$

Since we have a sum of squares we assume $\text{E}_\infty \sim \Gamma(N, \sigma^2/N)$, a gamma distribution with N degrees of freedom and $\sigma^2 = \text{Var}(y_i - \widehat{y}_{i\infty})$. Hence

$$\text{sd}(\text{E}_\infty) = \sigma^2/\sqrt{N}. \tag{5}$$

We also approximate E_1 by its expected value $E_1 \approx E[E_1] = \sigma^2$.

Hence $|\text{Bias}_{max}(\text{E}_\infty)| < \text{sd}(\text{E}_\infty)k$ implies $\frac{N-M}{NM}\sigma^2 < \frac{\sigma^2}{\sqrt{N}}k$ that is $M > \frac{N}{k\sqrt{N}+1}$. Hence after each reuse if we supplement the test set (consisting of N samples) with $M > N/(k\sqrt{N}+1)$ new samples for some small constant k then the bias due to test set reuse can be safely ignored. Note that M has approximately $\sqrt{N}$ dependency.

References

1. Kaggle. www.kaggle.com
2. Chatfield, C.: Model uncertainty, data mining and statistical inference. Journal of the Royal Statistical Society. Series A (Statistics in Society) **158**(3), 419–466 (1995)
3. Efron, B., Tibshirani, R.: An introduction to the bootstrap. Chapman and Hall (1993)
4. Faraway, J.: Data splitting strategies for reducing the effect of model selection on inference. Computing Science and Statistics **30**, 332–341 (1998)
5. Gama, J., Zliobaite, I., Bifet, A., Pechenizkiy, M., Bouchachia, A.: A survey on concept drift adaptation. ACM Computing Surveys **46**(4) (2014)
6. Hastie, T., Tibshirani, R., Friedman, J.: The Elements of Statistical Learning: Data Mining, Inference, and Prediction. Springer Series in Statistics (2009)
7. Kohavi, R.: A study of cross-validation and bootstrap for accuracy estimation and model selection. IJCAI **14**, 1137–1145 (1995)
8. Madnani, N., Tetreault, J., Chodorow, M.: Re-examining machine translation metrics for paraphrase identification. In: Proceedings of 2012 Conference of the North American Chapter of the Association for Computational Linguistics (NAACL 2012), pp. 182–190 (2012)
9. Mohri, M., Rostamizadeh, A.: Stability bounds for non-i.i.d. processes. In: Advances in Neural Information Processing Systems, vol. 20, pp. 1025–1032 (2008)
10. Ng, A.Y.: Preventing overfitting of crossvalidation data. In: Proceedings of the 14th International Conference on Machine Learning, pp. 245–253 (1997)
11. Papineni, K., Roukos, S., Ward, T., Zhu, W.J.: BLEU: a method for automatic evaluation of machine translation. In: Proceedings of the 40th Annual meeting of the Association for Computational Linguistics (ACL-2102), pp. 311–318 (2002)

12. Quinonero-Candela, J., Sugiyama, M., Schwaighofer, A., Lawrence, N.D. (eds.): Dataset Shift in Machine Learning. Neural Information Processing series. MIT Press (2008)
13. Rao, R.B., Fung, G.: On the dangers of cross-validation. An experimental evaluation. In: Proceedings of the SIAM Conference on Data Mining, pp. 588–596 (2008)
14. Samuelson, F.: Supplementing a validation test sample. In: 2009 Joint Statistical Meetings (2009)
15. Stone, M.: Asymptotics for and against crossvalidation. Biometrika **64**, 29–35 (1977)

Discriminative Interpolation for Classification of Functional Data

Rana Haber[1]([✉]), Anand Rangarajan[2], and Adrian M. Peter[3]

[1] Mathematical Sciences Department,
Florida Institute of Technology, Melbourne, FL, USA
rhaber2010@my.fit.edu
[2] Department of Computer and Information Science and Engineering,
University of Florida, Gainesville, FL, USA
anand@cise.ufl.edu
[3] Systems Engineering Department,
Florida Institute of Technology, Melbourne, FL, USA
apeter@fit.edu

Abstract. The *modus operandi* for machine learning is to represent data as feature vectors and then proceed with training algorithms that seek to optimally partition the feature space $S \subset \mathbb{R}^n$ into labeled regions. This holds true even when the original data are functional in nature, i.e. curves or surfaces that are inherently varying over a continuum such as time or space. Functional data are often reduced to summary statistics, locally-sensitive characteristics, and global signatures with the objective of building comprehensive feature vectors that uniquely characterize each function. The present work directly addresses representational issues of functional data for supervised learning. We propose a novel *classification by discriminative interpolation (CDI)* framework wherein functional data in the same class are adaptively reconstructed to be more similar to each other, while simultaneously repelling nearest neighbor functional data in other classes. Akin to other recent nearest-neighbor metric learning paradigms like stochastic k-neighborhood selection and large margin nearest neighbors, our technique uses class-specific representations which gerrymander similar functional data in an appropriate parameter space. Experimental validation on several time series datasets establish the proposed discriminative interpolation framework as competitive or better in comparison to recent state-of-the-art techniques which continue to rely on the standard feature vector representation.

Keywords: Functional data classification · Wavelets · Discriminative Interpolation

1 Introduction

The choice of data representation is foundational to all supervised and unsupervised analysis. The *de facto* standard in machine learning is to use feature representations that treat data as n-dimensional vectors in Euclidean space and then

---

This work is partially supported by NSF IIS 1065081 and 1263011.

A. Appice et al. (Eds.): ECML PKDD 2015, Part I, LNAI 9284, pp. 20–36, 2015.
DOI: 10.1007/978-3-319-23528-8_2

proceed with multivariate analysis in $\mathbb{R}^n$. This approach is uniformly adopted even for sensor-acquired data which naturally come in functional form—most sensor measurements are identifiable with real-valued functions collected over a discretized continuum like time or space. Familiar examples include time series data, digital images, video, LiDAR, weather, and multi-/hyperspectral volumes. The present work proposes a framework where we directly leverage functional representations to develop a k-nearest neighbor (kNN) supervised learner capable of discriminatively interpolating functions in the same class to appear more similar to each other than functions from other classes—we refer to this throughout as *Classification by Discriminative Interpolation* (CDI).

Over the last 25 years, the attention to statistical techniques with functional representations has resulted in the subfield commonly referred to as Functional Data Analysis (FDA) [20]. In FDA, the above mentioned data modalities are represented by their functional form: $f : \mathbb{R}^p \to \mathbb{R}^q$. To perform data analysis, one relies on the machinery of a Hilbert space structure endowed with a collection of functions. On a Hilbert space $\mathcal{H}$, many useful tools like bases, inner products, addition, and scalar multiplication allow us to mimic multivariate analysis on $\mathbb{R}^n$. However, since function spaces are in general infinite dimensional, special care must be taken to ensure theoretical concerns like convergence and set measures are defined properly. To bypass these additional concerns, some have resorted to treating real-valued functions of a real variable $f : \mathbb{R} \to \mathbb{R}$, sampled at m discrete points, $\mathbf{f} = \{f(t_i)\}_{i=1}^m$, as a vector $\mathbf{f} \in \mathbb{R}^m$ and subsequently apply standard multivariate techniques. Clearly, this is not a principled approach and unnecessarily strips the function properties from the data. The predominant method, when working with functional data, is to move to a feature representation where filters, summary statistics, and signatures are all derived from the input functions and then analysis proceeds per the norm. To avoid this route, several existing FDA works have revisited popular machine learning algorithms and reformulated them rigorously to handle functional data [9,12,21,22]—showing both theoretical and practical value in retaining the function properties. Here we also demonstrate how utilizing the well-known and simple interpolation property of functions can lead to a novel classification framework. Intuitively, given A classes and labeled exemplar functions $\{\mathbf{f}^j, y^j\}$, where labels $y^j = \{1, \ldots, A\}$ are available during training, we expand each $\mathbf{f}^j$ in an appropriate basis representation to yield a faithful reconstruction of the original curve. (Note: we use *function* and *curve* synonymously throughout the exposition.) In a supervised framework, we then allow the interpolants to morph the functions such that they resemble others from the same class, and incorporate a counter-force to repel similarities to other classes. The use of basis interpolants critically depends on the functional nature of the data and cannot be replicated efficiently in a feature vector representation where no underlying continuum is present. Though admittedly, this would become more cumbersome if for a function with domain dimension p and co-domain dimension q both were high dimensional, but for most practical datasets we have $p \leq 4$. In the current work, we detail the theory and

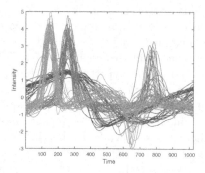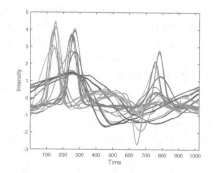

Fig. 1. Three-class *classification by discriminative interpolation* (CDI). (a) Original training functions from three different classes, showing 30 curves each. (b) Post training via the proposed CDI method, functions in each class morphed to resemble k-nearest neighbors [same 30 curves per class as in (a)].

algorithms for real-valued functions over the real line expanded in wavelet bases with extensions to higher dimensions following in a straightforward manner.

The motivation for a competing push-pull framework is built on several recent successful efforts in metric learning [23, 24, 26, 29] where Mahalanobis-style metrics are learned by incorporating optimization terms that promote purity of local neighborhoods. The metric is learned using a kNN approach that is locally sensitive to the the k neighbors around each exemplar and optimized such that the metric moves data with the same class labels closer in proximity (pulling in good neighbors) and neighbors with differing labels from the exemplar are moved out of the neighborhood (pushing away bad neighbors). As these are kNN approaches, they can inherently handle nonlinear classification tasks and have been proven to work well in many situations [16] due to their ability to contextualize learning through the use of neighborhoods. In these previous efforts, the metric learning framework is well-suited for feature vectors in $\mathbb{R}^n$. In our current situation of working with functional data, we propose an analogue that allows the warping of the function data to visually resemble others in the same class (within a k-neighborhood) and penalize similarity to bad neighbors from other classes. We call this gerrymandered morphing of functions based on local neighborhood label characteristics *discriminative interpolation*. Figure 1 illustrates this neighborhood-based, supervised deforming on a three-class problem. In Fig. 1(a), we see the original training curves from three classes, colored magenta, green, and blue; each class has been sub-sampled to 30 curves for display purposes. Notice the high variability among the classes which can lead to misclassifications. Fig. 1(b) shows the effects of CDI post training. Now the curves in each class more closely resemble each other, and ultimately, this leads to better classification of test curves.

Learning and generalization properties have yet to be worked out for the CDI framework. Here, we make a few qualitative comments to aid better understanding of the formulation. Clearly, we can overfit during the training stage by forcing the basis representation of all functions belonging to the same class to be identical. During the testing stage, in this overfitting regime, training is irrelevant and the method devolves into a nearest neighbor strategy (using function distances). Likewise, we can underfit during the training stage by forcing the basis representation of each function to be without error (or residual). During testing, in this underfitting regime, the basis representation coefficients are likely to be far (in terms of a suitable distance measure on the coefficients) from members in each class since no effort was made during training to conform to any class. We think a happy medium exists where the basis coefficients for each training stage function strike a reasonable compromise between overfitting and underfitting— or in other words, try to reconstruct the original function to some extent while simultaneously attempting to draw closer to nearest neighbors in each class. Similarly, during testing, the classifier fits testing stage function coefficients while attempting to place the function pattern close to nearest neighbors in each class with the eventual class label assigned to that class with the smallest compromise value. From an overall perspective, CDI marries function reconstruction with neighborhood gerrymandering (with the latter concept explained above).

To the best of our knowledge, this is the first effort to develop a FDA approach that leverages function properties to achieve kNN-margin-based learning in a *fully multi-class framework*. Below, we begin by briefly covering the requisite background on function spaces and wavelets (Section 2). Related works are detailed in Section 3. This is followed by the derivation of the proposed CDI framework in Section 4. Section 5 demonstrates extensive experimental validations on several functional datasets and shows our method to be competitive with other functional and feature-vector based algorithms—in many cases, demonstrating the highest performance measures to date. The article concludes with Section 6 where recommendations and future extensions are discussed.

2 Function Representations and Wavelets

Most FDA techniques are developed under the assumption that the given set of labeled functional data can be suitably approximated by and represented in an infinite dimensional Hilbert space $\mathcal{H}$. Ubiquitous examples of $\mathcal{H}$ include the space of square-integrable functions $\mathcal{L}^2([a,b] \subset \mathbb{R})$ and square-summable series $l^2(\mathbb{Z})$. This premise allows us to transition the analysis from the functions themselves to the coefficients of their basis expansion. Moving to a basis expansion also allows us to seamlessly handle irregularly sampled functions, missing data, and interpolate functions (the most important property for our approach). We now provide a brief exposition of working in the setting of $\mathcal{H}$. The reader is referred to many suitable functional analysis references [17] for further details.

The infinite dimensional representation of $f \in \mathcal{H}$ comes from the fact that there are a countably infinite number of basis vectors required to produce an exact representation of f, i.e.

$$f(t) = \sum_{l=1}^{\infty} \alpha_l \phi_l(t) \tag{1}$$

where ϕ is one of infinitely possible bases for $\mathcal{H}$. Familiar examples of ϕ include the Fourier basis, appropriate polynomials, and in the more contemporary setting—wavelets. In computational applications, we cannot consider infinite expansions and must settle for a projection into a finite d-dimensional subspace $\mathcal{P}$. Given a discretely sampled function $\mathbf{f} = \{f(t_i)\}_{1 \leq i \leq m}$, the coefficients for this projection are given by minimizing the quadratic objective function

$$\min_{\{\alpha_l\}} \sum_{i=1}^{m} \left(f(t_i) - \sum_{l=1}^{d} \alpha_l \phi_l(t_i) \right)^2$$

or in matrix form

$$\min_{\alpha} \|\mathbf{f} - \boldsymbol{\phi}\boldsymbol{\alpha}\|^2, \tag{2}$$

where ϕ is an $m \times d$ matrix with entries $\phi_{i,l} = \phi_l(t_i)$ and $\boldsymbol{\alpha}$ is an $d \times 1$ column vector of the coefficients. Estimation is computationally efficient with complexity $O(md^2)$ [20]. For an orthonormal basis set, this is readily obtainable via the inner product of the space $\alpha_l = \langle f, \phi_i \rangle$. Once the function is represented in the subspace, we can shift our analysis from working directly with the function to instead working with the coefficients. Relevant results utilized later in our optimization framework include

$$\langle \mathbf{f}^h, \mathbf{f}^j \rangle = \left(\boldsymbol{\alpha}^h \right)^T \boldsymbol{\Phi} \boldsymbol{\alpha}^j, \tag{3}$$

where the $d \times d$ matrix $\boldsymbol{\Phi}$ is defined by $\boldsymbol{\Phi}_{r,s} = \langle \phi_r, \phi_s \rangle$ (not dependent on $\mathbf{f}^h$ and $\mathbf{f}^j$), and

$$\|\mathbf{f}^h - \mathbf{f}^j\|_2^2 = \left(\boldsymbol{\alpha}^h - \boldsymbol{\alpha}^j \right)^T \boldsymbol{\Phi} \left(\boldsymbol{\alpha}^h - \boldsymbol{\alpha}^j \right). \tag{4}$$

For orthonormal basis expansions, such as many useful wavelet families, eq. (4) reduces to $\|\boldsymbol{\alpha}^h - \boldsymbol{\alpha}^j\|_2^2$.

Though many different bases exist for $\mathcal{H} = \mathcal{L}^2([a, b])$, wavelet expansions are now widely accepted as the most flexible in providing compact representations and faithful reconstructions. Functions $f \in \mathcal{L}^2([a, b])$ can be represented as a linear combination of wavelet bases [7]

$$f(t) = \sum_{k} \alpha_{j_0,k} \phi_{j_0,k}(t) + \sum_{j \geq j_0, k}^{\infty} \beta_{j,k} \psi_{j,k}(t) \tag{5}$$

where $t \in \mathbb{R}$, $\phi(x)$ and $\psi(x)$ are the *scaling* (a.k.a. father) and *wavelet* (a.k.a. mother) basis functions respectively, and $\alpha_{j_0,k}$ and $\beta_{j,k}$ are scaling and wavelet basis function coefficients; the j-index represents the current resolution level and the k-index the integer translation value. (The translation range of k can be computed from the span of the data and basis function support size at the different resolution levels. Also, when there is no need to distinguish between scaling or wavelet coefficients, we simply let $\mathbf{c} = \left[\alpha, \beta\right]^{T}$.) The linear combination in eq. (5) is known as a multiresolution expansion. The key idea behind multiresolution theory is the existence of a sequence of nested subspaces V_j $j \in \mathbb{Z}$ such that

$$\cdots V_{-2} \subset V_{-1} \subset V_0 \subset V_1 \subset V_2 \cdots \tag{6}$$

and which satisfy the properties $\bigcap V_j = \{0\}$ and $\overline{\bigcup V_j} = \mathcal{L}^2$ (completeness). The resolution increases as $j \to \infty$ and decreases as $j \to -\infty$ (some references show this order reversed due to the fact they invert the scale [7]). At any particular level $j + 1$, we have the following relationship

$$V_j \bigoplus W_j = V_{j+1} \tag{7}$$

where W_j is a space orthogonal to V_j, i.e. $V_j \bigcap W_j = \{0\}$. The father wavelet $\phi(x)$ and its integer translations form a basis for V_0. The mother wavelet $\psi(x)$ and its integer translates span W_0. These spaces decompose the function into its smooth and detail parts; this is akin to viewing the function at different scales and at each scale having a low pass and high pass version of the function. The primary usefulness of a full multiresolution expansion given by eq. (5) is the ability to threshold the wavelet coefficients and obtain a sparse representation of the signal. If sparse representations are not required, functions can be expanded using strictly scaling functions ϕ. Our CDI technique adopts this approach and selects an appropriate resolution level j_0 based on empirical cross-validation on the training data. Given a set of functional data, coefficients for each of the functions can be computed using eq. (2). Once the coefficients are estimated, as previously discussed, all subsequent analysis can be transitioned to working directly with the coefficients. There are also a number wavelet families from which we can select ϕ. With our desire to work with orthonormal bases, we limit ourselves to the compactly supported families of Daubechies, Symlets, and Coiflets [7].

3   Related Work

Several authors have previously realized the importance of taking the functional aspect of the data into account. These include representing the data in a different basis such as B-Splines [1], Fourier [6] and wavelets [3,4] or utilizing the continuous aspects [2] and differentiability of the functional data [18]. Abraham *et al.* [1] fit the data using B-Splines and then use the K-means algorithm to cluster the data. Biau *et al.* [6] reduce the infinite dimension of the space of functions to the first d dimensional coefficients of a Fourier series expansion of each function and then they apply a k-nearest neighborhood for classification. Antoniadis *et al.* [3] use the wavelet transform to detect clusters in high dimensional functional data. They present two different measures, a similarity and a dissimilarity, that are then used to apply the k-centroid clustering algorithm. The similarity measure is based on the distribution of energy across scales generating a number of features and the dissimilarity measure are based on wavelet-coherence tools.

In [4], Berlinet *et al.* expand the observations on a wavelet basis and then apply a classification rule on the non-zero coefficients that is not restricted to the *k*-nearest neighbors. Alonso *et al.* [2] introduce a classification technique that uses a weighted distance that utilizes the first, second and third derivative of the functional data. López-Pintado and Romo [18] propose two *depth*-based classification techniques that take into account the continuous aspect of the functional data. Though these previous attempts have demonstrated the utility of basis expansions and other machine learning techniques on functional data, none of them formulate a neighborhood, margin-based learning technique as proposed by our current CDI framework.

In the literature, many attempts have been made to find the *best* neighborhood and/or define a good metric to get better classification results [10,19,24,26,29]. Large Margin Nearest Neighbor (LMNN) [8,26] is a locally adaptive metric classification method that uses margin maximization to estimate a local flexible metric. The main intuition of LMNN is to learn a metric such that at least *k* of its closest neighbors are from the same class. It pre-defines *k* neighbors and identifies them as *target* neighbors or *impostors*—the same class or different class neighbors respectively. It aims at readjusting the margin around the data such that the *impostors* are outside that margin and the *k* data points inside the margin are of the same class. Prekopcsk and Lemire [19] classify times series data by learning a Mahalanobis distance by taking the pseudoinverse of the covariance matrix, limiting the Mahalanobis matrix to a diagonal matrix or by applying covariance shrinkage. They claim that these metric learning techniques are comparable or even better than LMNN and Dynamic Time Warping (DTW) [5] when one nearest neighbor is used to classify functional data. We show in the experiments in Section 5 that our CDI method performs better. In [24], Trivedi *et al.* introduce a metric learning algorithm by selecting the neighborhood based on a gerrymandering concept; redefining the margins such that the majority of the nearest neighbors are of the same class. Unlike many other algorithms, in [24], the choice of neighbors is a latent variable which is learned at the same time as it is learning the optimal metric—the metric that gives the best classification accuracy. These neighborhood-based approaches are pioneering approaches that inherently incorporate context (via the neighbor relationships) while remaining competitive with more established techniques like SVMs [25] or deep learning [13]. Building on their successes, we adapt a similar concept of redefining the class margins through pushing away *impostors* and pulling *target* neighbors closer.

4   Classification by Discriminative Interpolation

In this section, we introduce the CDI method for classifying functional data. Before embarking on the detailed development of the formulation, we first provide an overall sketch.

In the training stage, the principal task of discriminative interpolation is to learn a basis representation of all training set functions while simultaneously pulling each function representation toward a set of nearest neighbors in the same class and pushing each function representation away from nearest neighbors in the other classes. This can be abstractly characterized as

$$E_{\text{CDI}}(\mathbf{c}^i) = \sum_i \left[D_{\text{rep}}(\mathbf{f}^i, \phi \mathbf{c}^i) + \lambda \sum_{j \in \mathcal{N}(i)} D_{\text{pull}}(\mathbf{c}^i, \mathbf{c}^j) - \mu \sum_{k \in \mathcal{M}(i)} D_{\text{push}}(\mathbf{c}^i, \mathbf{c}^k) \right] \quad (8)$$

where D_{rep} is the representation error between the actual data (training set function sample) and its basis representation, D_{pull} is the distance between the coefficients of the basis representation in the same class and D_{push} the distance between coefficients in different classes (with the latter two distances often chosen to be the same). The parameters λ and μ weigh the pull and push terms respectively. $\mathcal{N}(i)$ and $\mathcal{M}(i)$ are the sets of nearest neighbors in the same class and different classes respectively.

Upon completion of training, the functions belonging to each class have been discriminatively interpolated such that they are more similar to their neighbors in the same class. This contextualized representation is reflected by the coefficients of the wavelet basis for each of the curves. We now turn our attention to classifying incoming test curves. Our labeling strategy focuses on selecting the class that is able to best represent the test curve under the pulling influence of its nearest neighbors in the same class. In the testing stage, the principal task of discriminative interpolation is to learn a basis representation for just the test set function while simultaneously pulling the function representation toward a set of nearest neighbors in the chosen class. This procedure is repeated for all classes with class assignment performed by picking that class which has the lowest compromise between the basis representation and pull distances. This can be abstractly characterized as

$$\hat{a} = \arg\min_a \min_{\mathbf{c}} D_{\text{rep}}(\mathbf{f}, \phi \mathbf{c}) + \lambda \sum_{k \in \mathcal{K}(a)} D_{\text{pull}}(\mathbf{c}, \mathbf{c}^k_{(a)}) \quad (9)$$

where $\mathcal{K}(a)$ is the set of nearest neighbors in class a of the incoming test pattern's coefficient vector $\mathbf{c}$. Note the absence of the push mechanism during testing. Further, note that we have to solve an optimization problem during the testing stage since the basis representation of each function is *a priori* unknown (for both training and testing).

4.1 Training Formulation

Given labeled functional data $\{(\mathbf{f}^i, y^i)\}_{i=1}^N$ where $\mathbf{f}^i \in \mathcal{H}$ and $y^i = \{1, \ldots, A\}$ are the labels, A is the number of classes. We can express $\mathbf{f}^i$ as a series expansion

$$\mathbf{f}^i = \sum_{l=1}^{\infty} c_l^i \phi_l \quad (10)$$

where $\{\phi_d\}_{d=1}^{\infty}$ form a complete, orthonormal system of $\mathcal{H}$. As mentioned in Section 2, we approximate the discretized data $\mathbf{f}^i = \{f^i(t_j)\}_{1 \leq j \leq m}$ in a d-dimensional space. Let $\mathbf{c}^i = [c_1^i, c_2^i, \cdots, c_d^i]^T$ be the $d \times 1$ vector of coefficients associated with the approximation $\hat{\mathbf{f}}^i$ of $\mathbf{f}^i$ and let $\boldsymbol{\phi} = [\phi_1, \phi_2, \cdots, \phi_d]$ be the $m \times d$ matrix of the orthonormal basis. Then the approximation to eq. 10 can be written in matrix form as

$$\hat{\mathbf{f}}^i = \boldsymbol{\phi}\mathbf{c}^i. \tag{11}$$

Getting the best approximation to $\mathbf{f}^i$ requires minimizing eq. 2, but in CDI we want to find a weighted approximation $\hat{\mathbf{f}}^i$ such that the function resembles more the functions in its class; i.e. we seek to minimize

$$\sum_{j \text{ s.t. } y_i = y_j} M_{ij} \|\hat{\mathbf{f}}^i - \hat{\mathbf{f}}^j\|^2 \tag{12}$$

while reducing the resemblance to functions in other classes; i.e. we seek to maximize

$$\sum_{j \text{ s.t. } y_i \neq y_j} M'_{ij} \|\hat{\mathbf{f}}^i - \hat{\mathbf{f}}^j\|^2 \tag{13}$$

where M_{ij} and M'_{ij} are some weight functions—our implementation of the adapted push-pull concept. Combining the three objective function terms, we attempt to get the best approximation by *pulling* similarly labeled data together while *pushing* different labeled data away. This yields the following objective function and optimization problem:

$$\min_{\hat{\mathbf{f}}, M, M'} E = \min_{\hat{\mathbf{f}}, M, M'} \sum_{i=1}^{N} \|\mathbf{f}^i - \hat{\mathbf{f}}^i\|^2 + \lambda \sum_{i,j \text{ s.t. } y_i = y_j} M_{ij} \|\hat{\mathbf{f}}^i - \hat{\mathbf{f}}^j\|^2 - \mu \sum_{i,j \text{ s.t. } y_i \neq y_j} M'_{ij} \|\hat{\mathbf{f}}^i - \hat{\mathbf{f}}^j\|^2 \tag{14}$$

where $M_{ij} \in (0,1)$ and $\sum_j M_{ij} = 1$ is the nearest neighbor constraint for $y_i = y_j$ where $j \neq i$, similarly, $M'_{ij} \in (0,1)$ and $\sum_j M'_{ij} = 1$ is the nearest neighbor constraint for $y_i \neq y_j$ where $j \neq i$. From Section 2, we showed that given an orthonormal basis, $\|\hat{\mathbf{f}}^i - \hat{\mathbf{f}}^j\|^2 = \|\mathbf{c}^i - \mathbf{c}^j\|^2$, eq. 14 can be expressed in terms of the coefficients as

$$\min_{\mathbf{c}, M, M'} E = \min_{\mathbf{c}, M, M'} \sum_{i=1}^{N} \|\mathbf{f}^i - \boldsymbol{\phi}\mathbf{c}^i\|^2 + \lambda \sum_{i,j \text{ s.t. } y_i = y_j} M_{ij} \|\mathbf{c}^i - \mathbf{c}^j\|^2 - \mu \sum_{i,j \text{ s.t. } y_i \neq y_j} M'_{ij} \|\mathbf{c}^i - \mathbf{c}^j\|^2. \tag{15}$$

As we saw in Section 3, there are many different ways to set the nearest neighbor. The simplest case is to find only one nearest neighbor from the same class and one from the different classes. Here, we set $M_{ij} \in \{0,1\}$ and likewise for M'_{ij}, which helps us obtain an analytic update for $\mathbf{c}^i$ (while keeping M, M' fixed). An extension to this approach would be to find the k-nearest neighbors allowing $k > 1$. A third alternative would be to have graded membership of neighbors with a free parameter deciding the degree of membership. We adopt this strategy—widely prevalent in the literature [27]—as a softmax winner-take-all. In this approach, the M_{ij} and M'_{ij} are "integrated out" to yield

Algorithm 1. Functional Classification by Discriminative Interpolation (CDI)

Training

Input: $\mathbf{f}_{\text{train}} \in \mathbb{R}^{N \times m}$, $y_{\text{train}} \in \mathbb{R}^{N}$, λ (CV*), μ (CV), k, k_p, η (stepsize)
Output: $\mathbf{c}^i$ (optimal interpolation)
1. **For** $i \leftarrow 1$ to N
2.   **Repeat**
3.   Find $\mathcal{N}(i)$ and $\mathcal{M}(i)$ using kNN
4.   Compute M_{ij} $\forall i, j$ pairs (eq. 18)
5.   Compute $\nabla E^i_{\text{interp}} = \frac{\partial E}{\partial \mathbf{c}^i}$ (eq. 17)
6. $\mathbf{c}^i \leftarrow \mathbf{c}^i - \eta \nabla E^i_{\text{interp}}$
8.   **Until** convergence
7. **End For**
*Obtained via cross validation (CV).

Testing

Input: $\mathbf{f}_{\text{test}} \in \mathbb{R}^{L \times m}$, $\mathbf{c}^i$ (from training), λ (CV), μ (CV), k, η' (stepsize)
Output: $\hat{y}$ (labels for all testing data)
1. **For** $l \leftarrow 1$ to L
2.   **For** $a \leftarrow 1$ to A
3.   **Repeat**
4.   Find $\mathcal{N}(a)$ for $\tilde{\mathbf{c}}^l$ using kNN
5.   Compute M^a_i $\forall i$ neighborhood (22)
6.   Compute $\nabla E^l_a = \frac{\partial E^a}{\partial \tilde{\mathbf{c}}^l}$ (eq. 19)
7. $\tilde{\mathbf{c}}^l \leftarrow \tilde{\mathbf{c}}^l - \eta' \nabla E^l_a$
8.   **Until** convergence
9.   compute E^l_a (eq. 21)
10.   **End For**
11. $\tilde{y}^l \leftarrow \{a | \min E^l_a \, \forall a\}$
12. **End For**

$$\min_{\mathbf{c}} \sum_{i=1}^{N} \|\mathbf{f}^i - \phi \mathbf{c}^i\|^2 - \frac{\lambda}{\beta} \sum_i \log \sum_{j \, \text{s.t.} \, y_i = y_j} e^{-\beta \|\mathbf{c}^i - \mathbf{c}^j\|^2} + \frac{\mu}{\beta} \sum_i \log \sum_{r \, \text{s.t.} \, y_i \neq y_r} e^{-\beta \|\mathbf{c}^i - \mathbf{c}^r\|^2} \quad (16)$$

where β is a free parameter deciding the degree of membership. This will allow curves to have a weighted vote in the CDI of $\mathbf{f}^i$ (for example).

An update equation for the objective in eq. 16 can be found by taking the gradient with respect to $\mathbf{c}^i$, which yields

$$\frac{\partial E}{\partial \mathbf{c}^i} = -2 \left(\phi^T \mathbf{f}^i - \phi^T \phi \mathbf{c}^i - \lambda \sum_{j \, \text{s.t.} \, y_i = y_j} M_{ij}(\mathbf{c}^i - \mathbf{c}^j) + \lambda \sum_{k \, \text{s.t.} \, y_k = y_i} M_{ki}(\mathbf{c}^k - \mathbf{c}^i) \cdots \right.$$

$$\left. + \mu \sum_{r \, \text{s.t.} \, y_i \neq y_r} M_{ir}(\mathbf{c}^i - \mathbf{c}^r) - \mu \sum_{s \, \text{s.t.} \, y_s \neq y_i} M_{si}(\mathbf{c}^s - \mathbf{c}^i) \right) \quad (17)$$

where

$$M_{ij} = \frac{\exp\left(-\beta\|\mathbf{c}^i - \mathbf{c}^j\|^2\right)}{\sum\limits_{t,\,\text{s.t. } y_i = y_t} \exp\left(-\beta\|\mathbf{c}^i - \mathbf{c}^t\|^2\right)}. \tag{18}$$

The computational complexity of this approach can be prohibitive. Since the gradient w.r.t. the coefficients involves a graded membership to *all* the datasets, the worst case complexity is $O(Nd)$. In order to reduce this complexity, we use an approximation to eq. 17 via an expectation-maximization (EM) style heuristic. We first find the nearest neighbor sets, $\mathcal{N}(i)$ (same class) and $\mathcal{M}(i)$ (different classes), compute the softmax in eq. 18 using these nearest neighbor sets and then use gradient descent to find the coefficients. Details of the procedure are found in Algorithm 1. Future work will focus on developing efficient *and* convergent optimization strategies that leverage these nearest neighbor sets.

4.2   Testing Formulation

We have estimated a set of coefficients which in turn give us the best approximation to the training curves in a discriminative setting. We now turn to the testing stage. In contrast to the feature vector-based classifiers, this stage is not straightforward. When a test function appears, we don't know its wavelet coefficients. In order to determine the best set of coefficients for each test function, we minimize an objective function which is very similar to the training stage objective function. To test membership in each class, we minimize the sum of the wavelet reconstruction error and a suitable distance between the unknown coefficients and its nearest neighbors in the chosen class. The test function is assigned to the class that yields the minimum value of the objective function. This overall testing procedure is formalized as

$$\arg\min_a \left(\min_{\tilde{\mathbf{c}}} E^a(\tilde{\mathbf{c}}) \right) = \arg\min_a \left(\min_{\tilde{\mathbf{c}}} \|\tilde{\mathbf{f}} - \phi\tilde{\mathbf{c}}\|^2 + \lambda \sum_{i\,\text{s.t. } y_i = a} M_i^a \|\tilde{\mathbf{c}} - \mathbf{c}^i\|^2 \right) \tag{19}$$

where $\tilde{\mathbf{f}}$ is the test set function and $\tilde{\mathbf{c}}$ is its vector of reconstruction coefficients. M_i^a is the nearest neighbor in the set of class a patterns. As before, the membership can be "integrated out" to get

$$E^a(\tilde{\mathbf{c}}) = \|\tilde{\mathbf{f}} - \phi\tilde{\mathbf{c}}\|^2 - \frac{\lambda}{\beta} \log \sum_{i\,\text{s.t. } y_i = a} \exp\left\{-\beta\|\tilde{\mathbf{c}} - \mathbf{c}^i\|^2\right\}. \tag{20}$$

This objective function can be minimized using methods similar to those used during training. The testing stage algorithm comprises the following steps.

1. Solve $\Gamma(a) = \min_{\tilde{\mathbf{c}}} E^a(\tilde{\mathbf{c}})$ for every class a using the objective function gradient

$$\frac{\partial E^a}{\partial \tilde{\mathbf{c}}} = -2\phi^T\tilde{\mathbf{f}} + 2\phi^T\phi\tilde{\mathbf{c}} + \lambda \sum_{i\,\text{s.t. } y_i = a} M_i^a \left(2\tilde{\mathbf{c}} - 2\mathbf{c}^i\right) \tag{21}$$

where

$$M_i^a = \frac{\exp\left\{-\beta\|\tilde{\mathbf{c}} - \mathbf{c}^i\|^2\right\}}{\sum_{j\in\mathcal{N}(a)} \exp\left\{-\beta\|\tilde{\mathbf{c}} - \mathbf{c}^j\|^2\right\}}. \tag{22}$$

2. Assign the label $\tilde{y}$ to $\tilde{\mathbf{f}}$ by finding the class with the smallest value of $\Gamma(a)$, namely $\arg\min_a(\Gamma(a))$.

5   Experiments

In this section, we discuss the performance of the CDI algorithm using publicly available functional datasets, also known as time series datasets from the "UCR Time Series Data Mining Archive" [15]. The multi-class datasets are divided into training and testing sets with detailed information such as the number of classes, number of curves in each of the testing sets and training sets and the length of the curves shown in Table 1. The datasets that we have chosen to run the experiments on range from 2 class datasets—the Gun Point dataset, and up to 37 classes—the ADIAC dataset. Learning is also exercised under a considerable mix of balanced and unbalanced classes, and minimal training versus testing exemplars, all designed to rigorously validate the generalization capabilities of our approach.

For comparison against competing techniques, we selected four other leading methods based on reported results on the selected datasets. Three out of the four algorithms are classification techniques based on support vector machines (SVM) with extensions to Dynamic Time Warping (DTW). DTW has shown to be a very promising similarity measurement for functional data, supporting warping of functions to determine closeness. Gudmundsson *et al.* [11] demonstrate the feasibility of the DTW approach to get a positive semi-definite kernel for classification with SVM. The approach in Zhang *et al.* [28] is one of many that use a vectorized method to classify functional data instead of using functional properties of the dataset. They develop several kernels for SVM known as elastic kernels—Gaussian elastic metric kernel (GEMK) to be exact and introduce several extensions to GEMK with different measurements. In [14], another SVM

Table 1. Functional Datasets. Datasets contain a good mix of multiple classes, class imbalances, and varying number of training versus testing curves.

Dataset	Number of Classes	Size of Training Set	Size of Testing Set	Length
Synthetic Control	6	300	300	60
Gun Point	2	50	150	150
ADIAC	37	390	391	176
Swedish Leaf	15	500	625	128
ECG200	2	100	100	96
Yoga	2	300	3000	426
Coffee	2	28	28	286
Olive Oil	4	30	30	570

Table 2. Experimental parameters. Free parameter settings via cross-validation for each of the datasets. $|\cdot|$ represents set cardinality.

Dataset	λ	μ	$|\mathcal{N}|$	$|\mathcal{M}|$
Synthetic Control	2	0.01	10	5
Gun Point	0.1	0.001	7	2
ADIAC	0.8	0.012	5	5
Swedish Leaf	0.1	0.009	7	2
ECG200	0.9	0.01	3	3
Yoga	0.08	0.002	2	2
Coffee	0.1	0.005	1	1
Olive Oil	0.1	0.005	1	1

classification technique with a DTW kernel is employed but this time a weight is added to the kernel to provide more flexibility and robustness to the kernel function. Prekopcsák *et al.* [19] do not utilize the functional properties of the data. Instead they learn a Mahalanobis metric followed by standard nearest neighbors. For brevity, we have assigned the following abbreviations to these techniques: SD [11], SG [28], SW [14], and MD [19]. In addition to these published works, we also evaluated a standard kNN approach directly on the wavelet coefficients obtained from eq. 2, i.e. direct representation of functions in a wavelet basis without neighborhood gerrymandering. This was done so that we can comprehensively evaluate if the neighborhood adaptation aspect of CDI truly impacted generalization (with this approach abbreviated as kNN).

A k-fold cross-validation is performed on the training datasets to find the optimal values for each of our free parameters (λ and μ being the most prominent). Since the datasets were first standardized (mean subtraction, followed by standard deviation normalization), the free parameters λ and μ became more uniform across all the datasets. λ ranges from $(0.05, 2.0)$ while μ ranges from $(10^{-3}, 0.01)$. Table 2 has detailed information on the optimal parameters found for each of the datasets. In all our experiments, β is set to 1 and Daubechies 4 (DB4) at $j_0 = 0$ was used as the wavelets basis (i.e. only scaling functions used). We presently do not investigate the effects of β on the classification accuracy as we perform well with it set at unity. The comprehensive results are given in Table 3, with the error percentage being calculated per the usual:

$$\text{Error} = 100 \frac{\text{\# of misclassified curves}}{\text{Total Number of Curves}}. \tag{23}$$

The experiments show very promising results for the proposed CDI method in comparison with the other algorithms, with our error rates as good or better in most datasets. CDI performs best on the ADIAC dataset compared to the other techniques, with an order of magnitude improvement over the current state-of-the-art. This is a particularly difficult dataset having 37 classes where the class sizes are very small, only ranging from 4 to 13 curves. Figure 2(a) illustrates all original curves from the 37 classes which are very similar to each other. Having many classes with only a few training samples in each presents a significant

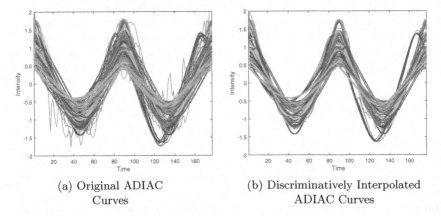

(a) Original ADIAC
Curves

(b) Discriminatively Interpolated
ADIAC Curves

Fig. 2. ADIAC dataset 37 classes - 300 Curves. The proposed CDI method is an order of magnitude better than the best reported competitor. Original curves in (a) are uniquely colored by class. The curves in (b) are more similar to their respective classes and are smoothed by the neighborhood regularization—unique properties of the CDI.

Table 3. Classification Errors. The proposed CDI method achieves state-of-the-art performance on half of the datasets, and is competitive in almost all others. The ECG200 exception, with kNN outperforming everyone, is discussed in text.

Dataset	SD [11]	SG [28]	SW [14]	MD [19]	kNN	CDI
Synthetic Control	0.67 (2)	0.7 (4)	0.67 (2)	1 (5)	9.67 (6)	0.33 (1)
Gun Point	4.67 (3)	0 (1)	2 (2)	5 (4)	9 (6)	6 (5)
ADIAC	32.48 (5)	24(3)	24.8 (4)	23 (2)	38.87 (6)	3.84 (1)
Swedish Leaf	14.72 (3)	5.3 (1)	22.5 (6)	15 (4)	16.97 (5)	8.8 (2)
ECG200	16 (6)	7 (2)	14 (5)	8 (3)	0 (1)	8 (3)
Yoga	16.37 (5)	11 (2)	18 (6)	16 (4)	10 (1)	15 (3)
Coffee	10.71 (4)	0 (1)	23 (5)	-	8.67 (3)	0 (1)
Olive Oil	13.33 (4)	10 (1)	17.3 (5)	13 (3)	20.96 (6)	10 (1)

classification challenge and correlates with why the competing techniques have a high classification error. In Figure 2(b), we show how CDI brings curves within the same class together making them more "pure" (such as the orange curves) while also managing to separate classes from each other. Regularized, discriminative learning also has the added benefit of smoothing the functions. The competing SVM-based approaches suffer in accuracy due to the heavily unbalanced classes. In some datasets (e.g. Swedish Leaf or Yoga) where we are not the leader, we are competitive with the others. Comparison with the standard kNN resulted in valuable insights. CDI fared better in 6 out of the 8 datasets, solidifying the utility of our push-pull neighborhood adaptation (encoded by the learned μ and λ), clearly showing CDI is going beyond vanilla kNN on the coefficient vectors. However, it is interesting that in two of the datasets that kNN beat not only CDI but all other competitors. For example, kNN obtained a perfect score on

ECG200. The previously published works never reported a simple kNN score on this dataset, but as it can occur, a simple method can often beat more advanced methods on particular datasets. Further investigation into this dataset showed that the test curves contained variability versus the training, which may have contributed to the errors. Our error is on par with all other competing methods.

6   Conclusion

The large margin k-nearest neighbor functional data classification framework proposed in this work leverages class-specific neighborhood relationships to *discriminatively interpolate* functions in a manner that morphs curves from the same class to become more similar in their appearance, while simultaneously pushing away neighbors from competing classes. Even when the data naturally occur as functions, the norm in machine learning is to move to a feature vector representation, where the burden of achieving better performance is transferred from the representation to the selection of discriminating features. Here, we have demonstrated that such a move can be replaced by a more principled approach that takes advantage of the functional nature of data.

Our CDI objective uses a wavelet expansion to produce faithful approximations of the original dataset and concurrently incorporates localized push-pull terms that promote neighborhood class purity. The detailed training optimization strategy uses a familiar iterative, alternating descent algorithm whereby first the coefficients of the basis expansions are adapted in the context of their labeled, softmax-weighted neighbors, and then, the curves' k-neighborhoods are updated. Test functional data are classified by the cost to represent them in each of the morphed training classes, with a minimal cost correct classification reflecting both wavelet basis reconstruction accuracy and nearest-neighbor influence. We have extensively validated this simple, yet effective, technique on several datasets, achieving competitive or state-of-the-art performance on most of them. In the present work, we have taken advantage of the interpolation characteristic of functions. In the future, we intend to investigate other functional properties such as derivatives, hybrid functional-feature representations, and extensions to higher dimensional functions such as images. We also anticipate improvements to our optimization strategy which was not the focus of the present work.

References

1. Abraham, C., Cornillon, P.A., Matzner-Lober, E., Molinari, N.: Unsupervised curve clustering using B-splines. Scandinavian J. of Stat. **30**(3), 581–595 (2003)
2. Alonso, A.M., Casado, D., Romo, J.: Supervised classification for functional data: A weighted distance approach. Computational Statistics & Data Analysis **56**(7), 2334–2346 (2012)
3. Antoniadis, A., Brossat, X., Cugliari, J., Poggi, J.M.: Clustering functional data using wavelets. International Journal of Wavelets, Multiresolution and Information Processing **11**(1), 1350003 (30 pages) (2013)

4. Berlinet, A., Biau, G., Rouviere, L.: Functional supervised classification with wavelets. Annale de l'ISUP 52 (2008)
5. Berndt, D., Clifford, J.: Using dynamic time warping to find patterns in time series. In: AAAI Workshop on Knowledge Discovery in Databases, pp. 229–248 (1994)
6. Biau, G., Bunea, F., Wegkamp, M.: Functional classification in Hilbert spaces. IEEE Transactions on Information Theory 51(6), 2163–2172 (2005)
7. Daubechies, I.: Ten Lectures on Wavelets. CBMS-NSF Reg. Conf. Series in Applied Math., SIAM (1992)
8. Domeniconi, C., Gunopulos, D., Peng, J.: Large margin nearest neighbor classifiers. IEEE Transactions on Neural Networks 16(4), 899–909 (2005)
9. Garcia, M.L.L., Garcia-Rodenas, R., Gomez, A.G.: K-means algorithms for functional data. Neurocomputing 151(Part 1), 231–245 (2015)
10. Globerson, A., Roweis, S.T.: Metric learning by collapsing classes. In: Advances in Neural Information Processing Systems, vol. 18, pp. 451–458. MIT Press (2006)
11. Gudmundsson, S., Runarsson, T., Sigurdsson, S.: Support vector machines and dynamic time warping for time series. In: IEEE International Joint Conference on Neural Networks (IJCNN), pp. 2772–2776, June 2008
12. Hall, P., Hosseini-Nasab, M.: On properties of functional principal components analysis. Journal of the Royal Statistical Society: Series B (Statistical Methodology) 68(1), 109–126 (2006)
13. Hinton, G., Osindero, S., Teh, Y.W.: A fast learning algorithm for deep belief nets. Neural Computation 18(7), 1527–1554 (2006)
14. Jeong, Y.S., Jayaramam, R.: Support vector-based algorithms with weighted dynamic time warping kernel function for time series classification. Knowledge-Based Systems 75, 184–191 (2014)
15. Keogh, E., Xi, X., Wei, L., Ratanamahatana, C.: The UCR time series classification/clustering homepage (2011). http://www.cs.ucr.edu/eamonn/time_series_data/
16. Keysers, D., Deselaers, T., Gollan, C., Ney, H.: Deformation models for image recognition. IEEE Transactions on PAMI 29(8), 1422–1435 (2007)
17. Kreyszig, E.: Introductory Functional Analysis with Applications. John Wiley and Sons (1978)
18. López-Pintado, S., Romo, J.: Depth-based classification for functional data. DIMACS Series in Discrete Mathematics and Theoretical Comp. Sci. 72, 103 (2006)
19. Prekopcsák, Z., Lemire, D.: Time series classification by class-specific Mahalanobis distance measures. Adv. in Data Analysis and Classification 6(3), 185–200 (2012)
20. Ramsay, J., Silverman, B.: Functional Data Analysis, 2nd edn. Springer (2005)
21. Rossi, F., Delannay, N., Conan-Guez, B., Verleysen, M.: Representation of functional data in neural networks. Neurocomputing 64, 183–210 (2005)
22. Rossi, F., Villa, N.: Support vector machine for functional data classification. Neurocomputing 69(7–9), 730–742 (2006)
23. Tarlow, D., Swersky, K., Charlin, L., Sutskever, I., Zemel, R.S.: Stochastic k-neighborhood selection for supervised and unsupervised learning. In: Proc. of the 30th International Conference on Machine Learning (ICML), vol. 28, no. 3, pp. 199–207 (2013)
24. Trivedi, S., McAllester, D., Shakhnarovich, G.: Discriminative metric learning by neighborhood gerrymandering. In: Advances in Neural Information Processing Systems (NIPS), vol. 27, pp. 3392–3400. Curran Associates, Inc. (2014)
25. Vapnik, V.N.: The nature of statistical learning theory. Springer, New York (1995)

26. Weinberger, K.Q., Blitzer, J., Saul, L.K.: Distance metric learning for large margin nearest neighbor classification. In: Advances in Neural Information Processing Systems (NIPS), vol. 18, pp. 1473–1480. MIT Press (2006)
27. Yuille, A., Kosowsky, J.: Statistical physics algorithms that converge. Neural Computation **6**(3), 341–356 (1994)
28. Zhang, D., Zuo, W., Zhang, D., Zhang, H.: Time series classification using support vector machine with Gaussian elastic metric kernel. In: 20th International Conference on Pattern Recognition (ICPR), pp. 29–32 (2010)
29. Zhu, P., Hu, Q., Zuo, W., Yang, M.: Multi-granularity distance metric learning via neighborhood granule margin maximization. Inf. Sci. **282**, 321–331 (2014)

Fast Label Embeddings
via Randomized Linear Algebra

Paul Mineiro[✉] and Nikos Karampatziakis

Microsoft Cloud Information Services Lab, Redmond, USA
{pmineiro,nikosk}@microsoft.com

Abstract. Many modern multiclass and multilabel problems are characterized by increasingly large output spaces. For these problems, label embeddings have been shown to be a useful primitive that can improve computational and statistical efficiency. In this work we utilize a correspondence between rank constrained estimation and low dimensional label embeddings that uncovers a fast label embedding algorithm which works in both the multiclass and multilabel settings. The result is a randomized algorithm whose running time is exponentially faster than naive algorithms. We demonstrate our techniques on two large-scale public datasets, from the Large Scale Hierarchical Text Challenge and the Open Directory Project, where we obtain state of the art results.

1 Introduction

Recent years have witnessed the emergence of many multiclass and multilabel datasets with increasing number of possible labels, such as ImageNet [12] and the Large Scale Hierarchical Text Classification (LSHTC) datasets [25]. One could argue that all problems of vision and language in the wild have extremely large output spaces.

When the number of possible outputs is modest, multiclass and multilabel problems can be dealt with directly (via a max or softmax layer) or with a reduction to binary classification. However, when the output space is large, these strategies are too generic and do not fully exploit some of the common properties that these problems exhibit. For example, often the alternatives in the output space have varying degrees of similarity between them so that typical examples from similar classes tend to be closer[1] to each other than from dissimilar classes. More concretely, classifying an image of a Labrador retriever as a golden retriever is a more benign mistake than classifying it as a rowboat.

Shouldn't these problems then be studied as structured prediction problems, where an algorithm can take advantage of the structure? That would be the case if for every problem there was an unequivocal structure (e.g. a hierarchy) that everyone agreed on and that structure was designed with the goal of being beneficial to a classifier. When this is not the case, we can instead let the algorithm uncover a structure that matches its own capabilities.

---

[1] Or more confusable, by machines and humans alike.

© Springer International Publishing Switzerland 2015
A. Appice et al. (Eds.): ECML PKDD 2015, Part I, LNAI 9284, pp. 37–51, 2015.
DOI: 10.1007/978-3-319-23528-8_3

In this paper we use label embeddings as the underlying structure that can help us tackle problems with large output spaces, also known as extreme classification problems. Label embeddings can offer improved computational efficiency because the embedding dimension is much smaller than the dimension of the output space. If designed carefully and applied judiciously, embeddings can also offer statistical efficiency because the number of parameters can be greatly reduced without increasing, or even reducing, generalization error.

1.1   Contributions

We motivate a particular label embedding defined by the low-rank approximation of a particular matrix, based upon a correspondence between label embedding and the optimal rank-constrained least squares estimator. Assuming realizability and infinite data, the matrix being decomposed is the expected outer product of the conditional label probabilities. In particular, this indicates two labels are similar when their conditional probabilities are linearly dependent across the dataset. This unifies prior work utilizing the confusion matrix for multiclass [5] and the empirical label covariance for multilabel [41].

We apply techniques from randomized linear algebra [19] to develop an efficient and scalable algorithm for constructing the embeddings, essentially via a novel randomized algorithm for rank-constrained squared loss regression. Intuitively, this technique implicitly decomposes the prediction matrix of a model which would be prohibitively expensive to form explicitly. The first step of our algorithm resembles compressed sensing approaches to extreme classification that use random matrices [21]. However our subsequent steps tune the embeddings to the data at hand, providing the opportunity for empirical superiority.

2   Algorithm Derivation

2.1   Notation

We denote vectors by lowercase letters x, y etc. and matrices by uppercase letters W, Z etc. The input dimension is denoted by d, the output dimension by c and the embedding dimension by k. For multiclass problems y is a one hot (row) vector (i.e. a vertex of the $c-1$ unit simplex) while for multilabel problems y is a binary vector (i.e. a vertex of the unit c-cube). For an $m \times n$ matrix $X \in \mathbb{R}^{m \times n}$ we use $||X||_F$ for its Frobenius norm, $X^\dagger$ for the pseudoinverse, $\Pi_{X,L}$ for the projection onto the left singular subspace of X, and $X_{1:k}$ for the matrix resulting by taking the first k columns of X. We use X^* to denote a matrix obtained by solving an optimization problem over matrix parameter X. The expectation of a random variable v is denoted by $\mathbb{E}[v]$.

2.2   Background

In this section we offer an informal discussion of randomized algorithms for approximating the principal components analysis of a data matrix $X \in \mathbb{R}^{n \times d}$

Algorithm 1. Randomized PCA

```
 1: function RPCA(k, X ∈ ℝ^{n×d})
 2:     (p, q) ← (20, 1)              ▷ These hyperparameters rarely need adjustment.
 3:     Q ← randn(d, k + p)
 4:     for i ∈ {1, . . . , q} do              ▷ Randomized range finder for X⊤X
 5:         Ψ ← X⊤XQ              ▷ Ψ can be computed in one pass over the data
 6:         Q ← orthogonalize(Ψ)      ▷ orth complexity O(dk²) is independent of n
 7:     end for              ▷ NB: total of (q + 1) data passes, including next line
 8:     F ← (XQ)⊤(XQ)              ▷ F ∈ ℝ^{(k+p)×(k+p)} is "small'
 9:     (V̂, Σ²) ← eig(F, k)              ▷ Exact optimization on small matrix
10:     V ← QV̂              ▷ Back out the solution
11:     return (V, Σ)
12: end function
```

---

with n examples and d features. For a very thorough and more formal discussion see [19].

Algorithm 1 shows a recipe for performing randomized PCA. In both theory and practice, the algorithm is insensitive to the parameters p and q as long as they are large enough (in our experiments we use $p = 20$ and $q = 1$). We start with a set of $k+p$ random vectors and use them to probe the range of $X^\top X$. Since principal eigenvectors can be thought as "frequent directions" [28], the range of Ψ will tend to be more aligned with the space spanned by the top eigenvectors of $X^\top X$. We compute an orthogonal basis for the range of Ψ and repeat the process q times. This can also be thought as orthogonal (aka subspace) iteration for finding eigenvectors with the caveat that we early stop (i.e., q is small). Once we are done and we have a good approximation for the principal subspace of $X^\top X$, we optimize fully over that subspace and back out the solution. The last few steps are cheap because we are only working with a $(k + p) \times (k + p)$ matrix and the largest bottleneck is either the computation of Ψ in a single machine setting or the orthogonalization step if parallelization is employed. An important observation we use below is that X or $X^\top X$ need not be available explicitly; to run the algorithm we only need to be able to compute the result of multiplying with $X^\top X$.

2.3   Rank-Constrained Estimation and Embedding

We begin with a setting superficially unrelated to label embedding. Suppose we seek an optimal squared loss predictor of a high-cardinality target vector $y \in \mathbb{R}^c$ which is linear in a high dimensional feature vector $x \in \mathbb{R}^d$. Due to sample complexity concerns, we impose a low-rank constraint on the weight matrix. In matrix form,

$$W^* = \underset{W \in \mathbb{R}^{d \times c} \mid \operatorname{rank}(W) \leq k}{\arg\min} \|Y - XW\|_F^2, \tag{1}$$

where $Y \in \mathbb{R}^{n \times c}$ and $X \in \mathbb{R}^{n \times d}$ are the target and design matrices respectively. This is a special case of a more general problem studied by [14]; specializing

---

Algorithm 2. Rembrandt: Response EMBedding via RANDomized Techniques

---

1: **function** REMBRANDT($k, X \in \mathbb{R}^{n \times d}, Y \in \mathbb{R}^{n \times c}$)
2:     $(p, q) \leftarrow (20, 1)$ ▷ These hyperparameters rarely need adjustment.
3:     $Q \leftarrow \text{randn}(c, k + p)$
4:     **for** $i \in \{1, \ldots, q\}$ **do**   ▷ Randomized range finder for $Y^\top \Pi_{X,L} Y$
5:         $Z \leftarrow \arg\min \|YQ - XZ\|_F^2$
6:         $Q \leftarrow \text{orthogonalize}(Y^\top X Z)$
7:     **end for**   ▷ NB: total of $(q + 1)$ data passes, including next line
8:     $F \leftarrow (Y^\top X Q)^\top (Y^\top X Q)$ ▷ $F \in \mathbb{R}^{(k+p) \times (k+p)}$ is "small"
9:     $(V, \Sigma^2) \leftarrow \text{eig}(F, k)$
10:    $V \leftarrow QV$ ▷ $V \in \mathbb{R}^{c \times k}$ is the embedding
11:    **return** (V, Σ)
12: **end function**

---

their result yields the solution $W^* = X^\dagger (\Pi_{X,L} Y)_k$, where $\Pi_{X,L}$ projects onto the left singular subspace of X, and $(\cdot)_k$ denotes optimal Frobenius norm rank-k approximation, which can be computed[2] via SVD. The expression for W^* can be written in terms of the SVD $\Pi_{X,L} Y = U \Sigma V^\top$, which, after simple algebra, yields $W^* = \left(X^\dagger (Y V_{1:k})\right) V_{1:k}^\top$. This is equivalent to the following procedure:

1. $Y V_{1:k}$: Project Y down to k dimensions using the top right singular vectors of $\Pi_{X,L} Y$.
2. $X^\dagger (Y V_{1:k})$ Least squares fit the projected labels using X and predict them.
3. $\left(X^\dagger (Y V_{1:k})\right) V_{1:k}^\top$: Map predictions to the original output space, using the transpose of the top right singular vectors of $\Pi_{X,L} Y$.

This motivates the use of the right singular vectors of $\Pi_{X,L} Y$ as a label embedding. The $\Pi_{X,L} Y$ term can be demystified: it corresponds to the predictions of the optimal unconstrained model,

$$Z^* = \underset{Z \in \mathbb{R}^{d \times c}}{\arg\min} \|Y - XZ\|_F^2,$$

$$\Pi_{X,L} Y = X Z^* \overset{\text{def}}{=} \hat{Y}.$$

The right singular vectors V of $\Pi_{X,L} Y$ are therefore the eigenvectors of $\hat{Y}^\top \hat{Y}$, i.e., the matrix formed by the sum of outer products of the optimal unconstrained model's predictions on each example. Note that actually computing and materializing $Z^* \in \mathbb{R}^{d \times c}$ would be expensive; a key aspect of the randomized algorithm is that we get the same result while avoiding this intermediate. In particular we can find the product of $\Pi_{X,L} Y$ with another matrix $Q \in \mathbb{R}^{c \times k}$ via

$$Z^* Q = \underset{Z \in \mathbb{R}^{d \times k}}{\arg\min} \|YQ - XZ\|_F^2,$$

$$\Pi_{X,L} Y Q = X Z^* Q. \tag{2}$$

---

[2] if $X = U_X \Sigma_X V_X^\top$ is the SVD of X, then $\Pi_{X,L} = U_X U_X^\top$.

Because squared loss is a proper scoring rule it is minimized at the conditional mean. In the limit of infinite training data ($n \rightarrow \infty$) and sufficient model flexibility (so that $\hat{y} = \mathbb{E}[y|x]$) we have that

$$\frac{1}{n}\hat{Y}^{\top}\hat{Y} \xrightarrow{\text{a.s.}} \mathbb{E}[\mathbb{E}[y|x]^{\top}\mathbb{E}[y|x]] \tag{3}$$

by the strong law of large numbers. An embedding based upon the eigendecomposition of $\mathbb{E}[\mathbb{E}[y|x]^{\top}\mathbb{E}[y|x]]$ is not practically actionable, but does provide valuable insights. For example, the principal label space transformation of [41] is an eigendecomposition of the empirical label covariance $Y^{\top}Y$. This is a plausible approximation to $\mathbb{E}[\mathbb{E}[y|x]^{\top}\mathbb{E}[y|x]]$ in the multilabel case. However, for multiclass (or multilabel where most examples have at most one nonzero component), the low-rank constraint alone cannot produce good generalization if the input representation is sufficiently flexible; the eigendecomposition of the prediction covariance will merely select a basis for the k most frequent labels due to the absence of empirical cooccurence statistics. Under these conditions we must further regularize (i.e., tradeoff variance for bias) beyond the low-rank constraint, so that $\hat{Y}$ better approximates $\mathbb{E}[Y|X]$ rather than the observed Y. Our procedure admits tuning the bias-variance tradeoff via choice of model (features) used in line 5 of Algorithm 2.

2.4 Rembrandt

Our proposal is Rembrandt, described in Algorithm 2. In the previous section, we motivated the use of the top right singular space of $\Pi_{X,L}Y$ as a label embedding, or equivalently, the top principal components of $Y^{\top}\Pi_{X,L}Y$ (leveraging the fact that the projection is idempotent). Using randomized techniques, we can decompose this matrix without explicitly forming it, because we can compute the product of $\Pi_{X,L}Y$ with another matrix Q via equation 2. Algorithm 2 is a specialization of randomized PCA to this particular form of the matrix multiplication operator. Starting from a random label embedding which satisfies the conditions for randomized PCA (e.g., a Gaussian random matrix), the algorithm first fits the embedding, outer products the embedding with the labels, orthogonalizes and repeats for some number of iterations. Then a final exact eigendecomposition is used to remove the additional dimensions of the embedding that were added to improve convergence. Note that the optimization of 2 is over $\mathbb{R}^{d\times(k+p)}$, not $\mathbb{R}^{d\times c}$, although the result is equivalent; this is the main computational advantage of our technique.

The connection to compressed sensing approaches to extreme classification is now clear, as the random sensing matrix corresponds to the starting point of the iterations in Algorithm 2. In other words, compressed sensing corresponds to Algorithm 2 with $q = 0$ and $p = 0$, which results in a whitened random projection of the labels as the embedding. Additional iterations ($q > 0$) and oversampling ($p > 0$) improve the approximation of the top eigenspace, hence the potential for improved performance. However when the model is sufficiently flexible, an embedding matrix which ignores the training data might be superior to one which overfits the training data.

Equation (2) is inexpensive to compute. The matrix vector product YQ is a sparse matrix-vector product so complexity $O(nsk)$ depends only on the average (label) sparsity per example s and the embedding dimension k, and is independent of the number of classes c. The fit is done in the embedding space and therefore is independent of the number of classes c, and the outer product with the predicted embedding is again a sparse product with complexity $O(nsk)$. The orthogonalization step is $O(ck^2)$, but this is amortized over the data set and essentially irrelevant as long as $n > c$. While random projection theory suggests k should grow logarithmically with c, this is only a mild dependence on the number of classes.

3   Related Work

Low-dimensional dense embeddings of sparse high-cardinality output spaces have been leveraged extensively in the literature, due to their beneficial impact on multiple algorithmic desiderata. As this work emphasizes, there are potential statistical (i.e., regularization) benefits to label embeddings, corresponding to the rich literature of low-rank regression regularization [22]. Another common motivation is to mitigate space or time complexity at training or inference time. Finally, embeddings can be part of a strategy for zero-shot learning [34], i.e., designing a classifier which is extensible in the output space.

[21], motivated by advances in compressed sensing, utilized a random embedding of the labels along with greedy sparse decoding strategy. For the multilabel case, [41] construct a low-dimensional embedding using principal components on the empirical label covariance, which they utilize along with a greedy sparse decoding strategy. For multivariate regression, [7] use the principal components of the empirical label covariance to define a shrinkage estimator which exploits correlations between the labels to improve accuracy. In these works, the motivation for embeddings was primarily statistical benefit. Conversely, [44] motivate their ranking-loss optimized embeddings solely by computational considerations of inference time and space complexity.

Multiple authors leverage side information about the classes, such as a taxonomy or graph, in order to learn a label representation which is felicitous for classification, e.g. when composed with online learning [11]; Bayesian learning [10]; support vector machines [6]; and decision tree ensembles [38]. Our embedding approach neither requires nor exploits such side information, and is therefore applicable to different scenarios, but is potentially suboptimal when side information is present. However, our embeddings can be complementary to such techniques when side information is not present, as some approaches condense side information into a similarity matrix between classes, e.g., the sub-linear inference approach of [9] and the large margin approach of [43]. Our embeddings provide a low-rank similarity matrix between classes in factored form, i.e., represented in $O(kc)$ rather than $O(c^2)$ space, which can be composed with these techniques. Analogously, [5] utilize a surrogate classifier rather than side information to define a similarity matrix between classes; our procedure can efficiently produce a similarity matrix which can ease the computational burden of this portion of their procedure.

Another intriguing use of side information about the classes is to enable zero-shot learning. To this end, several authors have exploited the textual nature of classes in image annotation to learn an embedding over the classes which generalizes to novel classes, e.g., [15] and [39]. Our embedding technique does not address this problem.

[18] focus nearly exclusively on the statistical benefit of incorporating label structure by overcoming the space and time complexity of large-scale one-against-all classification via distributed training and inference. Specifically, they utilize side information about the classes to regularize a set of one-against-all classifiers towards each other. This leads to state-of-the-art predictive performance, but the resulting model has high space complexity, e.g., terabytes of parameters for the LSHTC [24] dataset we utilize in section 4.3. This necessitates distributed learning and distributed inference, the latter being a more serious objection in practice. In contrast, our embedding technique mitigates space complexity and avoids model parallelism.

Our objective in equation (1) is highly related to that of partial least squares [16], as Algorithm 2 corresponds to a randomized algorithm for PLS if the features have been whitened.[3] Unsurprisingly, supervised dimensionality reduction techniques such as PLS can be much better than unsupervised dimensionality reduction techniques such as PCA regression in the discriminative setting if the features vary in ways irrelevant to the classification task [2].

Two other classical procedures for supervised dimensionality reduction are Fisher Linear Discriminant [37] and Canonical Correlation Analysis [20]. For multiclass problems these two techniques yield the same result [2,3], although for multilabel problems they are distinct. Indeed, extension of FLD to the multilabel case is a relatively recent development [42] whose straightforward implementation does not appear to be computationally viable for large number of classes. CCA and PLS are highly related, as CCA maximizes latent correlation and PLS maximizes latent covariance [2]. Furthermore, CCA produces equivalent results to PLS if the features are whitened [40]. Therefore, there is no obvious statistical reason to prefer CCA to our proposal in this context.

Regarding computational considerations, scalable CCA algorithms are available [30,32], but it remains open how to specialize them to this context to leverage the equivalent of equation (2); whereas, if CCA is desired, Algorithm 2 can be utilized in conjunction with whitening pre-processing.

Text is one the common input domains over which large-scale multiclass and multilabel problems are defined. There has been substantial recent work on text embeddings, e.g., word2vec [31], which (empirically) provide analogous statistical and computational benefits despite being unsupervised. The text embedding technique of [27] is a particularly interesting comparison because it is a variant of Hellinger PCA which leverages sequential information. This suggests that unsupervised dimensionality reduction approaches can work well when additional

---

[3] More precisely, if the feature covariance is a rotation.

Algorithm 3. Stagewise classification algorithm utilized for experiments. Loss is either log loss (multiclass) or independent log loss per class (multilabel).

1: **function** DECODERTRAIN($k, X \in \mathbb{R}^{n \times d}, Y \in \mathbb{R}^{n \times c}, \phi$)
2:    $(R, \sim) \leftarrow$ Rembrandt(k, X, Y)   ▷ or other comparison embedding
3:    $W^* \leftarrow \arg\min_W \|YR - XW\|_F^2$
4:    $\hat{Y} \leftarrow \phi(XW^*)$ ▷ ϕ is an optional random feature map
5:    $Q^* \leftarrow \arg\min_Q \text{loss}(Y, \hat{Y}Q)$ ▷ early-stopped, see text
6:    **return** (W^*, Q^*)
7: **end function**

---

structure of the input domain is incorporated, in this case by modeling word burstiness with the square root nonlinearity [23] and word order via decomposing neighborhood statistics. Nonetheless [27] note that when maximum statistical performance is desired, the embeddings must be fine-tuned to the particular task, i.e., supervised dimensionality reduction is required.

Another plausible regularization technique which mitigates inference space and time complexity is L1 regularization [29]. One reason to prefer low-rank regularization to L1 regularization is if the prediction covariance of equation (3) is well-modeled by a low-rank matrix.

4   Experiments

The goal of these experiments is to demonstrate the computational viability and statistical benefits of the embedding algorithm, not to advocate for a particular classification algorithm per se. We utilize classification tasks for demonstration, and utilize our embedding strategy as part of algorithm 3, but focus our attention on the impact of the embedding on the result.

Table 1. Data sets used for experimentation and times to compute an embedding.

Dataset	Type	Modality	Examples	Features	Classes	Rembrandt	
						k	Time (sec)
ALOI	Multiclass	Vision	108K	128	1000	50	4
ODP	Multiclass	Text	$\sim$ 1.5M	$\sim$ 0.5M	$\sim$ 100K	300	6,530
LSHTC	Multilabel	Text	$\sim$ 2.4M	$\sim$ 1.6M	$\sim$ 325K	500	8,006

In table 1 we present some statistics about the datasets we use in this section as well as times required to compute an embedding for the dataset. Unless otherwise indicated, all timings presented in the experiments section are for a Matlab implementation running on a standard desktop, which has dual 3.2Ghz Xeon E5-1650 CPU and 48Gb of RAM.

4.1   ALOI

ALOI is a color image collection of one-thousand small objects recorded for scientific purposes [17]. The number of classes in this data set does not qualify as extreme by current standards, but we begin with it as it will facilitate comparison with techniques which in our other experiments are intractable on a single machine. For these experiments we will consider test classification accuracy utilizing the same train-test split and features from [8]. Specifically there is a fixed train-test split of 90:10 for all experiments and the representation is linear in 128 raw pixel values.

Algorithm 2 produces an embedding matrix whose transpose is a squared-loss optimal decoder. In practice, optimizing the decode matrix for logistic loss as described in Algorithm 3 gives much better results. This is by far the most computationally demanding step in this experiment, e.g., it takes 4 seconds to compute the embedding but 300 seconds to perform the logistic regression. Fortunately the number of features (i.e., embedding dimensionality) for this logistic regression is modest so the second order techniques of [1] are applicable (in particular, their Algorithm 1 with a simple modification to include acceleration [4, 33]). We determine the number of fit iterations for the logistic regression by extracting a hold-out set from the training set and monitoring held-out loss. We do not use a random feature map, i.e., ϕ in line 4 of Algorithm 3 is the identity function.

Table 2. ALOI results. $k = 50$ for all embedding strategies.

Method	RE + LR	PCA + LR	CS + LR	LR	OAA	LT
Test Error	9.7%	9.7%	10.8%	10.8%	11.5%	16.5%

We compare to several different strategies in table 2. OAA is the one-against-all reduction of multiclass to binary. LR is a standard logistic regression, i.e., learning directly from the original features. Both of these options are intractable on a single machine for our other data sets. We also compare against Lomtree (LT), which has training and test time complexity logarithmic in the number of classes [8]. Both OAA and LT are provided by the Vowpal Wabbit [26] machine learning tool.

The remaining techniques are variants of Algorithm 3 using different embedding strategies. PCA + LR refers to logistic regression after first projecting the features onto their top principal components. CS + LR refers to logistic regression on a label embedding which is a random Gaussian matrix suitable for compressed sensing. Finally RE + LR is Rembrandt composed with logistic regression. These techniques were all implemented in Matlab.

Interestingly, OAA underperforms the full logistic regression. Rembrandt combined with logistic regression outperforms logistic regression, suggesting a beneficial effect from low-rank regularization. Compressed sensing is able to match the performance of the full logistic regression while being computationally

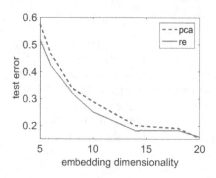
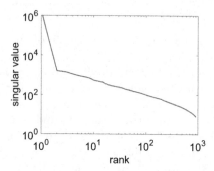

(a) Performance of logistic regression on ALOI when combined with either a feature embedding (PCA) or label embedding (RE).

(b) The empirical label covariance spectrum for LSHTC.

more tractable, but underperforms Rembrandt. Lomtree has the worst prediction performance but the lowest computational overhead when the number of classes is large.

At $k = 50$, there is no difference in quality between using the Rembrandt (label) embedding and the PCA (feature) embedding. This is not surprising considering the effective rank of the covariance matrix of ALOI is 70. For small embedding dimensionalities, however, PCA underperforms Rembrandt as indicated in Figure 1a. For larger numbers of output classes, where the embedding dimension will be a small fraction of the number of classes by computational necessity, we anticipate PCA regression will not be competitive.

Note that, in addition to better statistical performance, all of the "embedding + LR" approaches have lower space complexity $O(k(c + d))$ than direct logistic regression $O(cd)$. For ALOI the savings are modest (255600 bytes vs. 516000 bytes) because the input dimensionality is only $d = 128$, but for larger problems the space savings are necessary for feasible implementation on a single commodity computer. Inference time on ALOI is identical for embedding and direct approaches in practice (both achieving $\approx$ 170k examples/sec).

4.2　ODP

The Open Directory Project [13] is a public human-edited directory of the web which was processed by [6] into a multiclass data set. For these experiments we will consider test classification error utilizing the same train-test split, features, and labels from [8]. Specifically there is a fixed train-test split of 2:1 for all experiments, the representation of document is a bag of words, and the unique class assignment for each document is the most specific category associated with the document.

Table 3. ODP results. $k = 300$ for all embedding strategies.

Method	RE + LR	CS + LR	PCA + LR	LT
Test Error	83.15%	85.14%	90.37%	93.46%

The procedures are the same as in the previous experiment, except that we do not compare to OAA or full logistic regression due to intractability on a single machine.

The combination of Rembrandt and logistic regression result is, to the best of our knowledge, the best published result on this dataset. PCA logistic regression has a performance gap compared to Rembrandt and logistic regression. The poor performance of PCA logistic regression is doubly unsurprising, both for general reasons previously discussed, and due to the fact that covariance matrices of text data typically have a long plateau of weak spectral decay. In other words, for text problems projection dimensions quickly become nearly equivalent in terms of input reconstruction error, and common words and word combinations are not discriminative. In contrast, Rembrandt leverages the spectral properties of the prediction covariance of equation (3), rather than the spectral properties of the input features.

Finally, we remark the following: although inference (i.e., finding the maximum output) is linear in the number of classes, the constant factors are favorable due to modern vectorized processors, and therefore proceeds at ≈ 1700 examples/sec for the embedding based approaches.

4.3   LSHTC

The Large Scale Hierarchical Text Classification Challenge (version 4) was a public competition involving multilabel classification of documents into approximately 300,000 categories [24]. The training and test files are available from the Kaggle platform. The features are bag of words representations of each document.

Table 4. Embedding quality for LSHTC. $k = 100$ for all embedding strategies.

Method	Most fraternal	CS	PLST	Rembrandt
Sibling Fraction	0.32%	3.08%	19.65%	23.61%

Embedding Quality Assessment. A representation of a DAG hierarchy associated with the classes is also available. We used this to assess the quality of various embedding strategies independent of classification performance. In particular, we computed the fraction of class embeddings whose nearest neighbor was also a sibling in the DAG, as shown in Table 4. "Most fraternal" refers to an embedding which arranges for every category's nearest neighbor in the embedding to be the node in the DAG with the most siblings, i.e., the constant

Table 5. LSHTC results.

Method	RE ($k = 800$) + ILR	RE ($k = 500$) + ILR	FastXML	LPSR-NB
Precision-at-1	53.39%	52.84%	49.78%	27.91%

predictor baseline for this task. PLST [41] has performance close to Rembrandt according to this metric, so the 3.2 average nonzero classes per example is apparently enough for the approximation underlying PLST to be reasonable.

Empirical Label Covariance Spectrum. Our embedding approach is based upon a low-rank assumption for the (unobservable) prediction covariance of equation (3). Because LSHTC is a multi-label dataset, we can use the empirical label covariance as a proxy to investigate the spectral properties of the prediction covariance and test our assumption. We used Algorithm 1 (i.e., two pass randomized PCA) to estimate the spectrum of the empirical label covariance, shown in Figure 1b. The spectrum decays modestly and suggests that an embedding dimension of $k \approx 1000$ or more might be necessary for good classification performance.

Classification Performance. We built an end-to-end classifier using an approximate kernelized variant of Algorithm 3, where we processed the embeddings with Random Fourier Features [36], i.e., in line 4 of Algorithm 3 we use a random cosine feature map for ϕ. We found Cauchy distributed random vectors, corresponding to the Laplacian kernel, gave good results. We used 4,000 random features and tuned the kernel bandwidth via cross-validation on the training set.

The LSHTC competition metric is macro-averaged F1, which emphasizes performance on rare classes. However, we are using a multilabel classification algorithm which maximizes accuracy of predictions without regard to the importance of rare classes. Therefore we compare with published results of [35], who report example-averaged precision-at-k on the label ordering induced for each example. To facilitate comparison we do a 75:25 train-test split of the public training set, which is the same proportions as in their experiments (albeit a different split).

Based upon the previous spectral analysis, we anticipate a large embedding dimension is required for best results. With our current implementation, up to the limit of available memory in our desktop machine ($k = 800$) we found increasing embedding dimensionality improved performance.

"RE ($k = \dots$) + ILR" corresponds for *Rembrandt* coupled with independent (kernel) logistic regression, i.e., Algorithm 3. LPSR-NB is the Label Partitioning by Sub-linear Ranking algorithm of [45] composed with a Naive Bayes base learner, as reported in [35], where they also introduce and report precision for the multilabel tree learning algorithm FastXML. Inference for our best model proceeds at ≈ 60 examples/sec, substantially slower than for ODP, due to the larger output space, larger embedding dimensionality, and the use of random Fourier features.

5 Discussion

In this paper we identify a correspondence between rank constrained regression and label embedding, and we exploit that correspondence along with randomized matrix decomposition techniques to develop a fast label embedding algorithm.

To facilitate analysis and implementation, we focused on linear prediction, which is equivalent to a simple neural network architecture with a single linear hidden layer bottleneck. Because linear predictors perform well for text classification, we obtained excellent experimental results, but more sophistication is required for tasks where deep architectures are state-of-the-art. Although the analysis presented herein would not strictly be applicable, it is plausible that replacing line 5 in Algorithm 2 with an optimization over a deep architecture could yield good embeddings. This would be computationally beneficial as reducing the number of outputs (i.e., predicting embeddings rather than labels) would mitigate space constraints for GPU training.

Our technique leverages the (putative) low-rank structure of the prediction covariance of equation (3). For some problems a low-rank plus sparse assumption might be more appropriate. In such cases combining our technique with L1 regularization, e.g., on a classification residual or on separately regularized direct connections from the original inputs, might yield superior results.

Acknowledgments. We thank John Langford for providing the ALOI and ODP data sets.

References

1. Agarwal, A., Kakade, S.M., Karampatziakis, N., Song, L., Valiant, G.: Least squares revisited: Scalable approaches for multi-class prediction. In: Proceedings of the 31st International Conference on Machine Learning, pp. 541–549 (2014)
2. Barker, M., Rayens, W.: Partial least squares for discrimination. Journal of chemometrics **17**(3), 166–173 (2003)
3. Bartlett, M.S.: Further aspects of the theory of multiple regression. In: Mathematical Proceedings of the Cambridge Philosophical Society, vol. 34, pp. 33–40. Cambridge Univ. Press (1938)
4. Beck, A., Teboulle, M.: A fast iterative shrinkage-thresholding algorithm for linear inverse problems. SIAM Journal on Imaging Sciences **2**(1), 183–202 (2009)
5. Bengio, S., Weston, J., Grangier, D.: Label embedding trees for large multi-class tasks. In: Advances in Neural Information Processing Systems, pp. 163–171 (2010)
6. Bennett, P.N., Nguyen, N.: Refined experts: improving classification in large taxonomies. In: Proceedings of the 32nd International ACM SIGIR Conference on Research and Development in Information Retrieval, pp. 11–18. ACM (2009)
7. Breiman, L., Friedman, J.H.: Predicting multivariate responses in multiple linear regression. Journal of the Royal Statistical Society: Series B (Statistical Methodology) **59**(1), 3–54 (1997)

8. Choromanska, A., Langford, J.: Logarithmic time online multiclass prediction. arXiv preprint arXiv:1406.1822 (2014)
9. Cissé, M., Artières, T., Gallinari, P.: Learning compact class codes for fast inference in large multi class classification. In: Flach, P.A., De Bie, T., Cristianini, N. (eds.) ECML PKDD 2012, Part I. LNCS, vol. 7523, pp. 506–520. Springer, Heidelberg (2012)
10. DeCoro, C., Barutcuoglu, Z., Fiebrink, R.: Bayesian aggregation for hierarchical genre classification. In: ISMIR, pp. 77–80 (2007)
11. Dekel, O., Keshet, J., Singer, Y.: Large margin hierarchical classification. In: Proceedings of the Twenty-First International Conference on Machine Learning, p. 27. ACM (2004)
12. Deng, J., Dong, W., Socher, R., Li, L.J., Li, K., Fei-Fei, L.: Imagenet: a large-scale hierarchical image database. In: IEEE Conference on Computer Vision and Pattern Recognition, CVPR 2009, pp. 248–255. IEEE (2009)
13. DMOZ: The open directory project (2014). http://dmoz.org/
14. Friedland, S., Torokhti, A.: Generalized rank-constrained matrix approximations. SIAM Journal on Matrix Analysis and Applications $29(2)$, 656–659 (2007)
15. Frome, A., Corrado, G.S., Shlens, J., Bengio, S., Dean, J., Mikolov, T., et al.: Devise: a deep visual-semantic embedding model. In: Advances in Neural Information Processing Systems, pp. 2121–2129 (2013)
16. Geladi, P., Kowalski, B.R.: Partial least-squares regression: a tutorial. Analytica Chimica Acta 185, 1–17 (1986)
17. Geusebroek, J.M., Burghouts, G.J., Smeulders, A.W.: The Amsterdam library of object images. International Journal of Computer Vision $61(1)$, 103–112 (2005)
18. Gopal, S., Yang, Y.: Recursive regularization for large-scale classification with hierarchical and graphical dependencies. In: Proceedings of the 19th ACM SIGKDD International Conference on Knowledge Discovery and Data Mining, pp. 257–265. ACM (2013)
19. Halko, N., Martinsson, P.G., Tropp, J.A.: Finding structure with randomness: Probabilistic algorithms for constructing approximate matrix decompositions. SIAM Review $53(2)$, 217–288 (2011)
20. Hotelling, H.: Relations between two sets of variates. Biometrika, 321–377 (1936)
21. Hsu, D., Kakade, S., Langford, J., Zhang, T.: Multi-label prediction via compressed sensing. In: NIPS, vol. 22, pp. 772–780 (2009)
22. Izenman, A.J.: Reduced-rank regression for the multivariate linear model. Journal of Multivariate Analysis $5(2)$, 248–264 (1975)
23. Jégou, H., Douze, M., Schmid, C.: On the burstiness of visual elements. In: IEEE Conference on Computer Vision and Pattern Recognition, CVPR 2009, pp. 1169–1176. IEEE (2009)
24. Kaggle: Large scale hierarchical text classification (2014). http://www.kaggle.com/c/lshtc
25. Kosmopoulos, A., Gaussier, E., Paliouras, G., Aseervatham, S.: The ECIR 2010 large scale hierarchical classification workshop. In: ACM SIGIR Forum, vol. 44, pp. 23–32. ACM (2010)
26. Langford, J.: Vowpal Wabbit (2007). https://github.com/JohnLangford/vowpal_wabbit/wiki
27. Lebret, R., Collobert, R.: Word emdeddings through hellinger pca. arXiv preprint arXiv:1312.5542 (2013)
28. Liberty, E.: Simple and deterministic matrix sketching. In: Proceedings of the 19th ACM SIGKDD International Conference on Knowledge Discovery and Data Mining, pp. 581–588. ACM (2013)

29. Lokhorst, J.: The lasso and generalised linear models. Tech. rep., University of Adelaide, Adelaide (1999)
30. Lu, Y., Foster, D.P.: Large scale canonical correlation analysis with iterative least squares. arXiv preprint arXiv:1407.4508 (2014)
31. Mikolov, T., Chen, K., Corrado, G., Dean, J.: Efficient estimation of word representations in vector space. arXiv preprint arXiv:1301.3781 (2013)
32. Mineiro, P., Karampatziakis, N.: A randomized algorithm for CCA. arXiv preprint arXiv:1411.3409 (2014)
33. Nesterov, Y.: A method of solving a convex programming problem with convergence rate $O(1/k^2)$. Dokl. Akad. Nauk SSSR **269**, 543–547 (1983)
34. Palatucci, M., Pomerleau, D., Hinton, G.E., Mitchell, T.M.: Zero-shot learning with semantic output codes. In: Advances in Neural Information Processing Systems, pp. 1410–1418 (2009)
35. Prabhu, Y., Varma, M.: Fastxml: a fast, accurate and stable tree-classifier for extreme multi-label learning. In: Proceedings of the 20th ACM SIGKDD International Conference on Knowledge Discovery and Data Mining, pp. 263–272. ACM (2014)
36. Rahimi, A., Recht, B.: Random features for large-scale kernel machines. In: Advances in Neural Information Processing Systems, pp. 1177–1184 (2007)
37. Rao, C.R.: The utilization of multiple measurements in problems of biological classification. Journal of the Royal Statistical Society. Series B (Methodological) **10**(2), 159–203 (1948). http://www.jstor.org/stable/2983775
38. Schietgat, L., Vens, C., Struyf, J., Blockeel, H., Kocev, D., Džeroski, S.: Predicting gene function using hierarchical multi-label decision tree ensembles. BMC Bioinformatics **11**, 2 (2010)
39. Socher, R., Ganjoo, M., Manning, C.D., Ng, A.: Zero-shot learning through cross-modal transfer. In: Advances in Neural Information Processing Systems, pp. 935–943 (2013)
40. Sun, L., Ji, S., Yu, S., Ye, J.: On the equivalence between canonical correlation analysis and orthonormalized partial least squares. In: IJCAI, vol. 9, pp. 1230–1235 (2009)
41. Tai, F., Lin, H.T.: Multilabel classification with principal label space transformation. Neural Computation **24**(9), 2508–2542 (2012)
42. Wang, H., Ding, C., Huang, H.: Multi-label linear discriminant analysis. In: Daniilidis, K., Maragos, P., Paragios, N. (eds.) ECCV 2010, Part VI. LNCS, vol. 6316, pp. 126–139. Springer, Heidelberg (2010)
43. Weinberger, K.Q., Chapelle, O.: Large margin taxonomy embedding for document categorization. In: Advances in Neural Information Processing Systems, pp. 1737–1744 (2009)
44. Weston, J., Bengio, S., Usunier, N.: Wsabie: scaling up to large vocabulary image annotation. In: IJCAI, vol. 11, pp. 2764–2770 (2011)
45. Weston, J., Makadia, A., Yee, H.: Label partitioning for sublinear ranking. In: Proceedings of the 30th International Conference on Machine Learning (ICML 2013), pp. 181–189 (2013)

Maximum Entropy Linear Manifold for Learning Discriminative Low-Dimensional Representation

Wojciech Marian Czarnecki[1]([✉]), Rafal Jozefowicz[2], and Jacek Tabor[1]

[1] Faculty of Mathematics and Computer Science,
Jagiellonian University, Krakow, Poland
{wojciech.czarnecki,jacek.tabor}@uj.edu.pl
[2] Google, New York, USA
rafjoz@gmail.com

Abstract. Representation learning is currently a very hot topic in modern machine learning, mostly due to the great success of the deep learning methods. In particular low-dimensional representation which discriminates classes can not only enhance the classification procedure, but also make it faster, while contrary to the high-dimensional embeddings can be efficiently used for visual based exploratory data analysis.

In this paper we propose Maximum Entropy Linear Manifold (MELM), a multidimensional generalization of Multithreshold Entropy Linear Classifier model which is able to find a low-dimensional linear data projection maximizing discriminativeness of projected classes. As a result we obtain a linear embedding which can be used for classification, class aware dimensionality reduction and data visualization. MELM provides highly discriminative 2D projections of the data which can be used as a method for constructing robust classifiers.

We provide both empirical evaluation as well as some interesting theoretical properties of our objective function such us scale and affine transformation invariance, connections with PCA and bounding of the expected balanced accuracy error.

Keywords: Dense representation learning · Data visualization · Entropy · Supervised dimensionality reduction

1 Introduction

Correct representation of the data, consistent with the problem and used classification method, is crucial for the efficiency of the machine learning models. In practice it is a very hard task to find suitable embedding of many real-life objects in $\mathbb{R}^d$ space used by most of the algorithms. In particular for natural language processing [12], cheminformatics or even image recognition tasks it is still an open problem. As a result there is a growing interest in methods of representation learning [8], suited for finding better embedding of our data, which may be further used for classification, clustering or other analysis purposes. Recent

© Springer International Publishing Switzerland 2015
A. Appice et al. (Eds.): ECML PKDD 2015, Part I, LNAI 9284, pp. 52–67, 2015.
DOI: 10.1007/978-3-319-23528-8_4

years brought many success stories, such as dictionary learning [13] or deep learning [9]. Many of them look for a sparse [7], highly dimensional embedding which simplify linear separation at a cost of making visual analysis nontrivial. A dual approach is to look for low-dimensional linear embedding, which has advantage of easy visualiation, interpretation and manipulation at a cost of much weaker (in terms of models complexity) space of transformations.

In this work we focus on the scenario where we are given labeled dataset in $\mathbb{R}^d$ and we are looking for such low-dimensional linear embedding which allows to easily distinguish each of the classes. In other words we are looking for a highly discriminative, low-dimensional representation of the given data.

Our basic idea follows from the observation [15] that the density estimation is credible only in the low-dimensional spaces. Consequently, we first project the data onto an arbitrary k-dimensional affine submanifold $\mathcal{V}$ (where k is fixed), and search for the $\mathcal{V}$ for which the estimated densities of the projected classes are orthogonal to each other, where the Cauchy-Schwarz Divergence is applied as a measure of discriminativeness of the projection, see Fig. 1 for an example of such projection preserving classes' separation. The work presented in this paper is a natural extension of our earlier results [6], where we considered the one-dimensional case. However, we would like to emphasize that the used approach needed a nontrivial modification. In the one-dimensional case we could identify subspaces with elements of the unit sphere in a natural way. For higher dimensional subspaces such an identification is no longer possible.

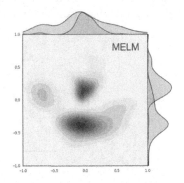

Fig. 1. Visualizatoin of *sonar* dataset using Maximum Entropy Linear Manifold with $k = 2$.

To the authors best knowledge the presented idea is novel, and has not been earlier considered as a method of classification and data visualization. As one of its benefits is the fact that it does not depend on affine rescaling of the data, which is a rare feature of the common classification tools. What is also interesting, we show that as its simple limiting one-class case we obtain the classical PCA projection. Moreover, from the theoretical standpoint the Cauchy-Schwarz Divergence factor can be decomposed into the fitting term, bounding the expected balanced misclassification error, and regularizing term, simplifying the resulting model. We compute its value and derivative so one can use first-order optimization to find a solution even though the true optimization should be performed on a Steifel manifold. Empirical tests show that such a method not only in some cases improves the classification score over learning from raw data but, more importantly, consistently finds highly discriminative representation which can be easily visualized. In particular, we show that resulting projections' discriminativeness is much higher than many popular linear methods, even recently proposed GEM model [11]. For the sake of completness we also include the full source code of proposed method in the supplementary material.

2   General Idea

In order to visualize dataset in $\mathbb{R}^d$ we need to project it onto $\mathbb{R}^k$ for very small k (typically 2 or 3). One can use either linear transformation or some complex embedding, however choosing the second option in general leads to hard inter-pretability of the results. Linear projections have a tempting characteristics of being both easy to understand (from both theoretical perspective and practical implications of the obtained results) as well as they are highly robust in further application of this transformation.

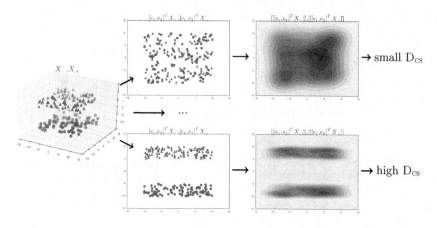

Fig. 2. Visualization of the MELM idea. For given dataset X_-, X_+ we search through various linear projections V and analyze how divergent are their density estimations in order to select the most discriminative.

In this work we focus on such class of projections so in practise we are looking for some matrix $V \in \mathbb{R}^{d \times k}$, such that for a given dataset $X \in \mathbb{R}^{d \times N}$ projection $V^T X$ preserves as much of the important information about X as possible (sometimes additionally under additional constraints). The choice of the definition of *information measure* IM together with the set of constraints φ_i defines a particular reduction method.

$$\underset{V \in \mathbb{R}^{d \times k}}{\text{maximize}} \quad \text{IM}(V^T X; X, Y)$$

$$\text{subject to} \quad \varphi_i(V), \ i = 1, \ldots, m.$$

There are many transformations which can achieve such results. For example, the well known Principal Component Analysis defines important information as data scattering so it looks for V which preserves as much of the X variance as possible and requires V to be orthogonal. In information bottleneck method one defines this measure as amount of mutual information between X and some additional Y (such as set of labels) which has to be preserved. Similar approaches are adapted in recently proposed Generalized Eigenvectors for Discriminative Features (GEM) where one tries to preserve the signal to noise ratio between samples from different classes. In case of Maximum Entropy Linear Manifold (MELM), introduced in this paper, important information is defined as the discriminativness of the samples from different classes with orthonormal V. In other words we work with labeled samples (in general, binary labeled) and wish to preserve the ability to distinguish one class (X_-) from another (X_+). In more formal terms, our optimization problem is to

$$\underset{V \in \mathbb{R}^{d \times k}}{\text{maximize}} \quad D_{CS}(\llbracket V^T X_- \rrbracket, \llbracket V^T X_+ \rrbracket)$$
$$\text{subject to} \quad V^T V = I,$$

where $D_{CS}(\cdot, \cdot)$ denotes the Cauchy-Schwarz Divergence, the measure of how divergent are given probability distributions; $\llbracket \cdot \rrbracket$ denotes some density estimator which, given samples, returns a probability distribution. The general idea is also visualized on Fig. 2.

3 Theory

We first discuss the one class case which has mainly introductory character as it shows the simplified version of our main idea.

Suppose that we have unlabeled data $X \subset \mathbb{R}^d$ and that we want to reduce the dimension of the data (for example to visualize it, reduce outliers, etc.) to $k < d$. One of the possible approaches is to use information theory and search for such k-dimensional subspace $\mathcal{V} \subset \mathbb{R}^d$ for which the orthogonal projection of X onto $\mathcal{V}$ preserves as much information about X as possible.

One can clearly choose various measures of information. In our case, due to computational simplicity, we have decided to use Renyi's quadratic entropy, which for the density f on $\mathbb{R}^k$ is given by

$$H_2(f) = -\log \int_{\mathbb{R}^k} f^2(x)dx.$$

One can equivalently use information potential [14], which is given as the L^2 norm of the density $ip(f) = \int_{\mathbb{R}^k} f^2(x)dx$. We need an easy observation that one can compute the Renyi's quadratic entropy for the normal density $\mathcal{N}(m, \Sigma)$ in $\mathbb{R}^k$ [4]:

$$H_2(\mathcal{N}(m, \Sigma)) = \tfrac{k}{2}\log(4\pi) + \tfrac{1}{2}\log(\det \Sigma). \tag{1}$$

However, in order to compute the Renyi's quadratic entropy of the discrete data we first need to apply some density estimation technique. By joining all the above mentioned steps together we are able to pose the basic optimization problem we are interested in.

Optimization problem 1. *Suppose that we are given data* X, *and* k *which denotes the dimension reduction. Find the orthonormal base* V *of the* k-*dimensional subspace*[1] V *for which the value of* $H_2([\![V^T X]\!])$ *is maximal, where* $[\![\cdot]\!]$ *denotes a given fixed method of density estimation.*

If we have data X with mean m and covariance Σ in $\mathbb{R}^d$ and k orthonormal vectors $V = [V_1, \ldots, V_k]$ then the we can ask what will be the mean and covariance of the orthogonal projection of X onto the space spanned by V. It is easy to show that it is given by $V^T m$ and $V^T \Sigma V$. In other words, if we consider data in the base given by orthonormal extension of V to the whole $\mathbb{R}^d$, the covariance of the projected data corresponds to the left upper $k \times k$ block submatrix of the original covariance.

We are going to show that if we apply the simplest density estimation of the underlying density for projected data given by the maximal likelihood estimator over the family of normal densities[2] then our optimization problem is equivalent to taking first k elements of the base given by PCA.

Theorem 1. *Let* $X \subset \mathbb{R}^d$ *be a given dataset with mean* m *and covariance* Σ *and let* $[\![\cdot]\!]_{\mathcal{N}}$ *denote the density estimation which returns the maximum likelihood estimator over Gaussian densities. Then*

$$\max\{H_2([\![V^T X]\!]_{\mathcal{N}}) : V \in \mathbb{R}^{d \times k}, V^T V = I\}$$

is realized for the first k *orthonormal vectors given by the PCA and*

$$\min\{H_2([\![V^T X]\!]_{\mathcal{N}}) : V \in \mathbb{R}^{d \times k}, V^T V = I\}$$

is realized for the last k *orthonormal vectors defined by the PCA.*

Proof. By the comments before and (1) we have

$$H_2([\![V^T X]\!]_{\mathcal{N}}) = \tfrac{k}{2} \log(4\pi) + \tfrac{1}{2} \log(\det(V^T \Sigma V)).$$

In other words we search for these V for which the value of $\det(V^T \Sigma V)$ is maximized. Now by Cauchy interlacing theory [2] eigenvalues of $V^T \Sigma V$ (ordered decreasingly) are bounded above by the eigenvalues of Σ. Consequently, the maximum is obtained in the case when V denotes the orthonormal eigenvectors of Σ corresponding to the biggest eigenvalues of Σ. However, this is exactly the first k elements of the orthonormal base constructed by the PCA. Proof of the second part is fully analogous.

As a result we obtain some general intuition that maximization of the Renyi's quadratic entropy leads to the selection of highly spread data, while its minimization selects projection where image is very condensed.

---

[1] We identify those vectors with a linear space spanned over them.
[2] That is for $A \subset V$ we put $[\![A]\!]_{\mathcal{N}} = \mathcal{N}(m_A, \text{cov}_A) : V \to \mathbb{R}_+$.

Let us now proceed to the binary labeled data. Recall that D_{CS} can be equivalently expressed in terms of Renyi's quadratic entropy (H_2) and Renyi's quadratic cross entropy ($H_2^\times$):

$$D_{CS}(V) = \log \int [\![V^T X_+]\!]^2 + \log \int [\![V^T X_-]\!]^2 - 2 \log \int [\![V^T X_+]\!][\![V^T X_-]\!]$$

$$= -H_2([\![V^T X_-]\!]) - H_2([\![V^T X_+]\!]) + 2H_2^\times([\![V^T X_+]\!], [\![V^T X_-]\!]).$$

Let us recall that our optimization aim is to find a sequence V consisting of k orthonormal vectors for which $D_{CS}(V)$ is maximized.

Observation 1. *Assume that the density estimator $[\![\cdot]\!]$ does not change under the affine change of the coordinate system[3]. One can show, by an easy modification of the theorem by Czarnecki and Tabor [6, Theorem 4.1], that the maximum of $D_{CS}(\cdot)$ is independent of the affine change of data. Namely, for an arbitrary affine invertible map M we have:*

$$\max\{D_{CS}(V; X_+, X_-) : V \text{ orthonormal}\}$$

$$= \max\{D_{CS}(V; X_+, X_-) : V \text{ linearly independent}\}$$

$$= \max\{D_{CS}(V; MX_+, MX_-) : V \text{ orthonormal}\}.$$

The above feature, although typical in the density estimation, is rather uncommon in modern classification tools.

Similarly to the one-dimensional case, when $V \in \mathbb{R}^d$, we can decompose the objective function into fitting and regularizing terms:

$$D_{CS}(V) = \underbrace{2H_2^\times([\![V^T X_+]\!], [\![V^T X_-]\!])}_{\text{fitting term}} - \underbrace{(H_2([\![V^T X_-]\!]) + H_2([\![V^T X_+]\!]))}_{\text{regularizing term}}.$$

Regularizing term has a slightly different meaning than in most of the machine learning models. Here it controls number of disjoint regions which appear after performing density based classification in the projected space. For one dimensional case it is a number of thresholds in the multithreshold linear classifier, for $k = 2$ it is the number of disjoint curves defining decision boundary, and so on. Renyi's quadratic entropy is minimized when each class is as condensed as possible (as we show in Theorem 1), intuitively resulting in a small number of disjoint regions.

It is worth noting that, despite similarities, it is not the common classification objective which can be written as an additive loss function and a regularization term

$$L(V) = \sum_{i=1}^{N} \ell(V^T x_i, y_i, x_i) + \Omega(V),$$

as the error depends on the relations between each pair of points instead of each point independently. One can easily prove that there are no ℓ, Ω for which

---

[3] This happens in particular for the kernel density estimation we apply in the paper.

$D_{cs}(v) = L(V; \ell, \Omega)$. Such choice of the objective function might lead to the lack of connections with optimization of any reasonable accuracy related metric, as those are based on the point-wise loss functions. However it appears that D_{cs} bounds the expected balanced accuracy[4] similarly to how hinge loss bounds 0/1 loss. This can be formalized in the following way.

Theorem 2. *Negative log-likelihood of balanced misclassification in k-dimensional linear projection of any non-separable densities $f_\pm$ onto V is bounded by half of the Renyi's quadratic cross entropy of these projections.*

Proof. Likelihood of balanced misclassification over a k-dimensional hypercube after projection through V equals

$$\int_{[0,1]^k} \min\{(V^T f_+)(x), (V^T f_-)(x)\}dx.$$

Using analogous reasoning to the one presented by Czarnecki [5], using Cauchy and other basic inequalities, one can show that

$$-\log \int_{[0,1]^k} \min\{(V^T f_+)(x), (V^T f_-)(x)\}dx \geq \tfrac{1}{2}H_2^\times(V^T f_+, V^T f_-).$$

□

As a result we might expect that maximizing of the D_{CS} leads to the selection of the projection which on one hand maximizes the balanced accuracy over the training set (minimizes empirical error) and on the other fights with overfitting by minimizing the number of disjoint classification regions (minimizes model complexity).

4   Closed form Solution for Objective and its Gradient

Let us now investigate more practical aspects of proposed approach. We show the exact formulas of both D_{CS} and its gradient as functions of finite, labeled samples (binary datasets) so one can easily plug it in to any first-order optimization software.

Let X_+, X_- be fixed subsets of $\mathbb{R}^d$. Let $\mathcal{V}$ denote the k-dimensional subspace generated by $V = [V_1, \ldots, V_k] \in \mathbb{R}^{d \times k}$ (we consider only the case when the sequence V is linearly independent). We project sets $X_\pm$ orthogonally on V, and compute the Cauchy-Schwarz Divergence of the kernel density estimations (using Silverman's rule) of the resulting projections:

$$G^{-1}(V)[\![V^T X_+]\!] \text{ and } G^{-1}(V)[\![V^T X_-]\!],$$

where $G(V) = V^T V$ denotes the grassmanian. We search for such V for which the Cauchy-Schwarz Divergence is maximal. Recall that the scalar product in the space of matrices is given by $\langle V_1, V_2 \rangle = \text{tr}(V_1^T V_2)$.

---

[4] Accuracy without class priors $BAC(TP, FP, TN, FN) = \frac{1}{2}\left(\frac{TP}{TP+FN} + \frac{TN}{TN+FP}\right)$.

There are basically two possible approaches one can apply: either search for the solution in the set of orthonormal V which generate $\mathcal{V}$, or allow all V with a penalty function. The first method is possible[5], but does not allow use of most of the existing numerical libraries as the space we work in is highly nonlinear. This is the reason why we use the second approach which we describe below.

Since, as we have observed in the previous section, the result does not depend on the affine transformation of data, we can restrict to the analogous formula for the sets

$$V^T X_+ \text{ and } V^T X_-,$$

where V consists of linearly independent vectors. Consequently, we need to compute the gradient of the function

$$D_{CS}(V) = D_{CS}([\![V^T X_+]\!], [\![V^T X_-]\!])$$

$$= \log \int [\![V^T X_+]\!]^2 + \log \int [\![V^T X_-]\!]^2 - 2\log \int [\![V^T X_+]\!][\![V^T X_-]\!],$$

where we consider the space consisting only of linearly independent vectors. Since as the base of the space V we can always take orthonormal vectors, the maximum is realized for orthonormal sequence, and therefore we can add a penalty term for being non-orthonormal sequence, which helps avoiding numerical instabilities:

$$D_{CS}(V) - \|V^T V - I\|^2,$$

where as we recall the sequence V is orthonormal iff $V^T V = I$. We denote above augmented D_{CS} by the *maximum entropy linear manifold* objective function

$$\text{MELM}(V) = D_{CS}(V) - \|V^T V - I\|^2. \tag{2}$$

Besides MELM($\cdot$) value we need the formula for its gradient ∇MELM($\cdot$). For the second term we obviously have

$$\nabla \|V^T V - I\|^2 = 4VV^T V - 4V.$$

We consider the first term. Let us first provide the formula for the computation of the product of kernel density estimations of two sets.

Assume that we are given set $A \subset \mathcal{V}$ (in our case A will be the projection of $X_\pm$ onto $\mathcal{V}$), where $\mathcal{V}$ is k-dimensional. Then the formula for the kernel density estimation with Gaussian kernel, is given by [15]:

$$[\![A]\!] = \frac{1}{|A|} \sum_{a \in A} \mathcal{N}(a, \Sigma_A),$$

where $\Sigma_A = (h_A^\gamma)^2 \text{cov}_A$ and (for γ being a scaling hyperparameter [6]) $h_A^\gamma = \gamma(\frac{4}{k+2})^{1/(k+4)}|A|^{-1/(k+4)}$.

---

[5] And has advantage of having smaller number of parameters.

Now we need the formula for $\int [\![A]\!] [\![B]\!]$, which is calculated [6] with the use of

$$\int \mathcal{N}(a, \Sigma_A) \mathcal{N}(b, \Sigma_B) = \mathcal{N}(a - b, \Sigma_A + \Sigma_B)(0).$$

Then we get

$$\int [\![A]\!] [\![B]\!] = \frac{1}{|A||B|} \sum_{w \in A - B} \mathcal{N}(w, \Sigma_A + \Sigma_B)(0)$$

$$= \frac{1}{(2\pi)^{k/2} \det^{1/2}(\Sigma_{AB}) |A||B|} \sum_{w \in A - B} \exp(-\tfrac{1}{2} \|w\|^2_{\Sigma_{AB}}),$$

where $A - B = \{a - b : a \in A, b \in B\}$ and Σ_{AB} is defined by

$$\Sigma_{AB} = (h_A^\gamma)^2 \mathrm{cov}_A + (h_B^\gamma)^2 \mathrm{cov}_B$$
$$= \gamma^2 (\tfrac{4}{k+2})^{2/(k+4)} (|A|^{-2/(k+4)} \mathrm{cov}_A + |B|^{-2/(k+4)} \mathrm{cov}_B).$$

For a sequence $V = [V_1, \ldots, V_k] \in \mathbb{R}^{d \times k}$ of linearly independent vectors we put

$$\Sigma_{AB}(V) = V^T \Sigma_{AB} V \text{ and } S_{AB}(V) = \Sigma_{AB}(V)^{-1}.$$

Observe that $\Sigma_{AB}(V)$ and $S_{AB}(V)$ are square symmetric matrices which represent the properties of the projection of the data onto the space spanned over V. We put

$$\phi_{AB}(V) = \frac{1}{(2\pi)^{k/2} \det^{1/2}(\Sigma_{AB}(V)) |A||B|},$$

thus

$$\nabla \phi_{AB}(V) = -\phi_{AB}(V) \cdot \Sigma_{AB} \cdot V \cdot S_{AB}(V).$$

Consequently to compute the final formula, we need the gradient of the function $V \to \det(\Sigma_{AB}(V))$, which as one can easily verify, is given by the formula

$$\nabla \det(\Sigma_{AB}(V)) = 2 \det(V^T \Sigma_{AB} V) \cdot \Sigma_{AB} V (V^T \Sigma_{AB} V)^{-1}. \tag{3}$$

One can also easily check that for

$$\psi_{AB}^w(V) = \exp(-\tfrac{1}{2} \|V^T w\|^2_{\Sigma_{AB}(V)}),$$

where w arbitrarily fixed, we get

$$\nabla \psi_{AB}^w(V) = -\psi_{AB}^w(V) \cdot (w w^T V S_{AB}(V) - \Sigma_{AB} V S_{AB}(V) V^T w w^T V S_{AB}(V)).$$

To present the final form for the gradient of $D_{cs}(V)$ we need the gradient of the cross information potential

$$\mathrm{ip}_{AB}^\times(V) = \phi_{AB}(V) \sum_{w \in A - B} \psi_{AB}^w(V),$$

$$\nabla \mathrm{ip}_{AB}^\times(V) = \phi_{AB}(V) \sum_{w \in A - B} \nabla \psi_{AB}^w(V) + \left(\sum_{w \in A - B} \psi_{AB}^w(V) \right) \cdot \nabla \phi_{AB}(V).$$

Since

$$D_{cs}(V) = \log(ip^{\times}_{X_+X_+}(V)) + \log(ip^{\times}_{X_-X_-}(V)) - 2\log(ip^{\times}_{X_+X_-}(V)),$$

we finally get

$$\nabla D_{cs}(V) = \frac{1}{ip^{\times}_{X_+X_+}(V)}\nabla ip^{\times}_{X_+X_+}(V) + \frac{1}{ip^{\times}_{X_-X_-}(V)}\nabla ip^{\times}_{X_-X_-}(V)$$
$$- 2\frac{1}{ip^{\times}_{X_+X_-}(V)}\nabla ip^{\times}_{X_+X_-}(V).$$

Given

$$MELM(V) = D_{cs}(V) - \|V^TV - I\|^2,$$
$$\nabla MELM(V) = \nabla D_{cs}(V) - (4VV^TV - 4V),$$

one can run any first-order optimization method to find vectors V spanning k-dimensional subspace $\mathcal{V}$ representing low-dimensional, discriminative manifold of the input space.

As one can notice from the above equations, the computational complexity of both function evaluation and its gradient are quadratic in terms of training set size. For big datasets this can a serious bottleneck. One of the possible solutions is to use approximation schemes for the computation of the Cauchy-Schwarz diviergence, which are known to significantly reduce the computational time without sacrificing the accuracy [10]. One other option is to use analogous of stochastic gradient descent where we define function value on a random sample of $\mathcal{O}(\sqrt{N})$ points (resampled in each iteration) from each class, leading to linear complexity and given that training set is big enough, one can get theoretical guarantees on the quality of approximation [15]. Finally, it is possible to first build a Gaussian Mixture Model (GMM) of each class distribution [17] and perform optimization on such density estimator. Computational complexity would be reduced to constant time per-iteration (due to fixing number of components during clustering) trading speed for accuracy.

5 Experiments

We use ten binary classification datasets from UCI repository [1] and libSVM repository [3], which are briefly summarized in Table 1. These are moderate size problems.

Code was written in Python with the use of scikit-learn, numpy and scipy. Besides MELM we use 8 other linear dimensionality reduction techniques, namely: Principal Component Analysis (PCA), class PCA (cPCA[6]), two ellipsoid PCA (2ePCA[7]), per class PCA (pPCA[8]), Independent Component Analysis (ICA), Factor Analysis (FA), Nonnegative Matrix Factorization (NMF[9]),

---

[6] cPCA uses sum of each classes covariances, weighted by classes sizes, instead of whole data covariance.

[7] 2ePCA is cPCA without weights, so it is a balanced counterpart.

[8] pPCA uses as V_i the first principal component of ith class.

[9] In order to use NMF we first transform dataset so it does not contain negative values.

Disriminative Learning using Generalized Eigenvectors (GEM [11]). PCA, ICA, NMF and FA are implemented in scikit-learn, cPCA, pPCA and 2ePCA were coded by authors and for GEM we use publically available code[10]. Implementation of MELM as a model compatible with scikit-learn classifiers and transformers is available both in supplementary materials and online[11].

Table 1. Summary of used datasets. N denote number of points, d dimensionality, $|X_l|$ number of samples with l label, $\bar{m}$ mean density (number of nonzero elements) and d_l^t denotes number of dimensions which we have to include during PCA to keep t of label l variance.

dataset	N	d	$|X_-|$	$|X_+|$	$\hat{m}$	$d^{.95}$	$d_-^{.95}$	$d_+^{.95}$
australian	690	14	383	307	0.80	1	2	1
breast-cancer	683	10	444	239	1.00	1	1	1
diabetes	768	8	268	500	0.88	2	2	3
fourclass	862	2	555	307	1.00	2	2	2
german.numer	1000	24	700	300	0.75	3	3	3
heart	270	13	150	120	0.75	3	3	3
ionosphere	351	34	126	225	0.88	24	26	7
liver-disorders	345	6	145	200	1.00	3	3	3
sonar	208	60	111	97	1.00	28	24	24
splice	1000	60	483	517	1.00	55	52	54

In order to estimate how hard to optimize is the MELM objective function we plot in Fig. 3 histograms of D_{CS} values obtained during 500 random starts for each of the dataset. First, one can easily notice that D_{CS} have multiple local extrema (see for example *heart* or *liver-disorders* histograms). It also appears that in some of the considered datasets it is not easy to obtain maximum by the use of completely random starting point (see *ionosphere* and *australian* datasets), which suggests that one should probably use some more advanced initialization techniques.

To further investigate how hard it is to find a good solution when selecting maximum of D_{CS} we estimate the expected value of D_{CS} after s random starts from matrices $V^{(1)}, \ldots, V^{(s)}$

$$\mathbb{E}\left[\max_{V=V^{(1)},\ldots,V^{(s)}} D_{CS}(\text{L-BFGS}(\text{MELM}|V))\right].$$

As one can see on Fig. 4 for 8 out of 10 considered datasets one can expect to find the maximum (with 5% error) after just 16 random starts. Obviously this cannot be used as a general heuristics as it is heavily dependent on the dataset size, dimensionality as well as its discriminativness. However, this experiment

---

[10] forked at http://gist.github.com/lejlot/3ab46c7a249d4f375536

[11] http://github.com/gmum/melm

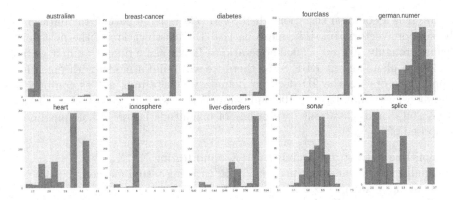

Fig. 3. Histograms of D_{CS} values obtained for each dataset during 500 random starts using L-BFGS.

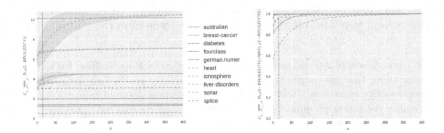

Fig. 4. Expected value of Cauchy-Schwarz Divergence after MELM optimization for s random starts using L-BFGS algorithm (on the left) and its ratio to the maximum obtainable Cauchy-Schwarz Divergence (on the right). Dotted black line shows 16 starts threshold.

shows that for moderate size problems (hundreds to thousands samples with dozens of dimensions) MELM can be relatively easily optimized even though it is a rather complex function with possibly many local maxima.

It is worth noting that truly complex optimization problem is only given by *ionosphere* dataset. One can refer to Table 1 to see that this is a very specific problem where positive class is located in a very low-dimensional linear manifold (approximately 7 dimensional) while the negative class is scattered over nearly 4 times more dimensions.

We check how well MELM behaves when used in a classification pipeline. There are two main reasons for such approach, first if the discriminative manifold is low-dimensional, searching for it may boost the classification accuracy. Second, even if it decreases classification score as compared to non-linear methods applied directly in the input space, the resulting model will be much simpler and more robust. For comparison consider training a RBF SVM in $\mathbb{R}^{60}$ using 1000 data points. It is a common situation when SVM selects large part of the dataset as the support vectors [16], [18], meaning that the classification of the new point requires roughly $500 \cdot 60 = 30000$ operations. In the same time if we first embed space in a plane and fit RBF SVM there we will build a model with much less support vectors (as the 2D decision boundary generally is not as complex as 60-dimensional one), lets say 100 and consequently we will need $60 \cdot 2 + 2 \cdot 100 = 120 + 200 = 320$ operations, two orders of magnitude faster. Whole pipeline is composed of:

1. Splitting dataset into training X_-, X_+ and testing $\hat{X}_-, \hat{X}_+$.
2. Finding plane embedding matrix $V \in \mathbb{R}^{d \times 2}$ using tested method.
3. Training a classifier cl on $V^T X_-, V^T X_+$.
4. Testing cl on $V^T \hat{X}_-, V^T \hat{X}_+$.

Table 2 summarizes BAC scores obtained by each method on each of the considered datasets in 5-fold cross validation. For the classifier module we used SVM RBF, KNN and KDE-based density classification. Each of them was fitted using internal cross-validation to find the best parameters. GEM and MELM γ hyperperameters were also fitted. Reported results come from the best classifier.

Table 2. Comparison of 2-dimensional reduction followed by the classifier. I stands for Identity, meaning that we simply trained classifiers directly on the raw data, without any dimensionality reduction. Bold values indicate the best score obtained across all dimensionality reduction pipelines. If the classifier trained on raw data is better than any of the reduced models than its score is also bolded.

	MELM	FA	ICA	GEM	NMF	2ePCA	cPCA	PCA	pPCA	I
australian	**0.866**	0.847	0.829	0.791	0.817	0.764	0.756	0.825	0.769	0.860
breast-cancer	**0.976**	0.973	**0.976**	0.930	**0.976**	0.966	0.967	**0.976**	0.961	0.966
diabetes	**0.744**	0.682	0.705	0.637	0.704	0.689	0.695	0.705	0.646	0.728
fourclass	**1.0**	0.720	**1.0**	**1.0**	0.999	**1.0**	**1.0**	**1.0**	**1.0**	**1.0**
german.numer	**0.705**	0.588	0.648	0.653	0.63	0.588	0.602	0.650	0.619	**0.728**
heart	**0.831**	0.792	0.818	0.675	0.811	0.793	0.782	0.817	0.787	**0.837**
ionosphere	**0.892**	0.794	0.757	0.763	0.799	0.783	0.780	0.757	0.826	**0.944**
liver-disorders	**0.710**	0.546	0.545	0.681	0.553	0.531	0.548	0.531	0.557	0.705
sonar	0.766	0.558	0.600	**0.889**	0.657	0.593	0.575	0.600	0.676	0.862
splice	**0.862**	0.718	0.697	0.799	0.691	0.686	0.686	0.697	0.694	**0.887**

In four datasets, MELM based embeding led to the construction of better classifier than both other dimensionality reduction techniques as well as training models on raw data. This suggests that for these datasets the discriminative manifold is truly at most 2-dimensional. At the same time in nearly all (besides *sonar*) datasets the pipeline consisting of MELM yielded significantly better classification results than any other embeding considered.

One of the main applications of MELM is to visualize the dataset through linear projection in such a way that classes do not overlap. One can see comparisons of *heart* dataset projections using all considered approaches in Fig. 5. As one can notice our method finds plane projection where classes are nearly perfectly discriminated. Interestingly, this separation is only obtainable in two dimensions, as neither marginal distributions nor any other one-dimensional projection can construct such separation.

While visual inspection is crucial for such tasks, to truly compare competetive methods we need some metric to measure quality of the visualization. In order to do so, we propose to assign a *visual separability* score as the mean BAC score over three considered classifiers (SVM RBF, KNN, KDE) trained and tested in 5-fold cross validation of the projected data. The only difference between this test and the previous one is that we use whole data to find a projection (so each projection technique uses all datapoints) and only further visualization testing is performed using train-test splits. This way we can capture "how easy to discriminate are points in this projection" rather than "how useful for data discrimination is using the projection". Experiments are repeated using various random subsets of samples and mean results are reported.

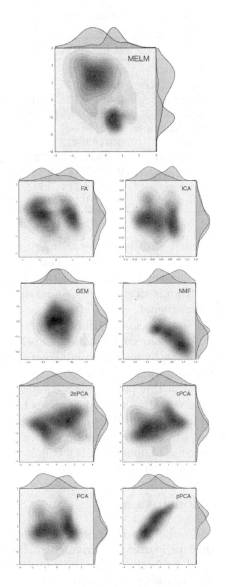

Fig. 5. Comparison of heart dataset 2D projections by analyzed methods. Visualization uses kernel density estimation.

During these experiments MELM achieved essentially better scores than any other tested method (see Table 3). Solutions were about 10% better under our metric and this difference is consistent over all considered datasets. In other words MELM finds two-dimensional representations of our data using just linear projection where classes overlap to a

Table 3. Comparison of 2-dimensional projections discriminativeness.

	MELM	FA	ICA	GEM	NMF	2ePCA	cPCA	PCA	pPCA
australian	**0.888**	0.856	0.845	0.792	0.838	0.782	0.781	0.845	0.792
breast-cancer	**0.985**	0.979	0.979	0.942	0.979	0.967	0.969	0.978	0.965
diabetes	**0.806**	0.732	0.737	0.691	0.737	0.734	0.733	0.734	0.697
fourclass	**0.988**	0.665	**0.988**	**0.988**	**0.988**	**0.988**	**0.988**	**0.988**	**0.988**
german.numer	**0.819**	0.640	0.687	0.686	0.672	0.665	0.657	0.686	0.692
heart	**0.918**	0.822	0.834	0.751	0.839	0.787	0.783	0.833	0.799
ionosphere	**0.990**	0.810	0.798	0.763	0.849	0.804	0.814	0.798	0.863
liver-disorders	**0.763**	0.682	0.659	0.707	0.698	0.691	0.676	0.688	0.715
sonar	**0.996**	0.714	0.717	0.892	0.729	0.702	0.709	0.717	0.735
splice	**0.927**	0.738	0.724	0.829	0.716	0.717	0.718	0.723	0.742

significantly smaller degree than using PCA, cPCA, 2ePCA, pPCA, ICA, NMF, FA or GEM. It is also worth noting that Factor Analysis, as the only method which does not require orthogonality of resulting projection vectors did a really bad job while working with fourclass data even though these samples are just two-dimensional.

As stated before, MELM is best suited for low-dimensional embedings and one of its main applications is supervised data visualization. However in general one can be interested in higher dimensional subspaces. During preliminary studies we tested model behavior up to $k = 5$ and results were similar to the one reported in this paper (when compared to the same methods, with analogous k). It is worth noting that methods like PCA also use a density estimator - one big Gaussian fitted through maximum likelihood estimation. Consequently even though from theoretical point of view MELM should not be used for $k > 5$ (due to the curse of dimensionality [15], it works fine as long as one uses good density estimator (such as a well fitted GMM [17]).

6   Conclusions

In this paper we construct Maximum Entropy Linear Manifold (MELM), a method of learning discriminative low-dimensional representation which can be used for both classification purposes as well as a visualization preserving classes separation. Proposed model has important theoretical properties including affine transformations invariance, connections with PCA as well as bounding the expected balanced misclassification error. During evaluation we show that for moderate size problems MELM can be efficiently optimized using simple first-order optimization techniques. Obtained results confirm that such an approach leads to highly discriminative transformation, better than obtained by 8 compared solutions.

Acknowledgments. The work has been partially financed by National Science Centre Poland grant no. 2014/13/B/ST6/01792.

References

1. Bache, K., Lichman, M.: UCI machine learning repository (2013). http://archive. ics.uci.edu/ml
2. Bhatia, R.: Matrix analysis, vol. 169. Springer Science & Business Media (1997)
3. Chang, C.C., Lin, C.J.: Libsvm: a library for support vector machines. ACM Transactions on Intelligent Systems and Technology (TIST) **2**(3), 27 (2011)
4. Cover, T.M., Thomas, J.A.: Elements of information theory, 2nd edn. Willey-Interscience, NJ (2006)
5. Czarnecki, W.M.: On the consistency of multithreshold entropy linear classifier. Schedae Informaticae (2015)
6. Czarnecki, W.M., Tabor, J.: Multithreshold entropy linear classifier: Theory and applications. Expert Systems with Applications (2015)
7. Geng, Q., Wright, J.: On the local correctness of 1-minimization for dictionary learning. In: 2014 IEEE International Symposium on Information Theory (ISIT), pp. 3180–3184. IEEE (2014)
8. Goodfellow, I.J., et al.: Challenges in representation learning: a report on three machine learning contests. In: Lee, M., Hirose, A., Hou, Z.-G., Kil, R.M. (eds.) ICONIP 2013, Part III. LNCS, vol. 8228, pp. 117–124. Springer, Heidelberg (2013)
9. Hinton, G., Osindero, S., Teh, Y.W.: A fast learning algorithm for deep belief nets. Neural Computation **18**(7), 1527–1554 (2006)
10. Jozefowicz, R., Czarnecki, W.M.: Fast optimization of multithreshold entropy linear classifier (2015). arXiv preprint arXiv:1504.04739
11. Karampatziakis, N., Mineiro, P.: Discriminative features via generalized eigenvectors. In: Proceedings of the 31st International Conference on Machine Learning (ICML 2014), pp. 494–502 (2014)
12. Levy, O., Goldberg, Y.: Neural word embedding as implicit matrix factorization. In: Advances in Neural Information Processing Systems (NIPS 2014), pp. 2177–2185 (2014)
13. Mairal, J., Bach, F., Ponce, J., Sapiro, G.: Online dictionary learning for sparse coding. In: Proceedings of the 26th Annual International Conference on Machine Learning, pp. 689–696. ACM (2009)
14. Principe, J.C., Xu, D., Fisher, J.: Information theoretic learning. Unsupervised Adaptive Filtering **1**, 265–319 (2000)
15. Silverman, B.W.: Density estimation for statistics and data analysis, vol. 26. CRC Press (1986)
16. Suykens, J.A., Van Gestel, T., De Brabanter, J., De Moor, B., Vandewalle, J.: Least squares support vector machines, vol. 4. World Scientific (2002)
17. Tabor, J., Spurek, P.: Cross-entropy clustering. Pattern Recognition **47**(9), 3046–3059 (2014)
18. Wang, L.: Support Vector Machines: theory and applications, vol. 177. Springer Science & Business Media (2005)

Novel Decompositions of Proper Scoring Rules for Classification: Score Adjustment as Precursor to Calibration

Meelis Kull[(✉)] and Peter Flach

Intelligent Systems Laboratory, University of Bristol, Bristol, UK
{meelis.kull,peter.flach}@bristol.ac.uk

Abstract. There are several reasons to evaluate a multi-class classifier on other measures than just error rate. Perhaps most importantly, there can be uncertainty about the exact context of classifier deployment, requiring the classifier to perform well with respect to a variety of contexts. This is commonly achieved by creating a scoring classifier which outputs posterior class probability estimates. Proper scoring rules are loss evaluation measures of scoring classifiers which are minimised at the true posterior probabilities. The well-known decomposition of the proper scoring rules into calibration loss and refinement loss has facilitated the development of methods to reduce these losses, thus leading to better classifiers. We propose multiple novel decompositions including one with four terms: adjustment loss, post-adjustment calibration loss, grouping loss and irreducible loss. The separation of adjustment loss from calibration loss requires extra assumptions which we prove to be satisfied for the most frequently used proper scoring rules: Brier score and log-loss. We propose algorithms to perform adjustment as a simpler alternative to calibration.

1 Introduction

Classifier evaluation is crucial for building better classifiers. Selecting the best from a pool of models requires evaluation of models on either hold-out data or through cross-validation with respect to some evaluation measure. An obvious choice is the same evaluation measure which is later going to be relevant in the model deployment context.

However, there are situations where the deployment measure is not necessarily the best choice, as in model construction by optimisation. Optimisation searches through the model space to find ways to improve an existing model according to some evaluation measure. If this evaluation measure is simply the error rate, then the model fitness space becomes discrete in the sense that there are improvements only if some previously wrongly classified instance crosses the decision boundary. In this case, surrogate losses such as quadratic loss, hinge loss or log-loss enable SVMs, logistic regression or boosting to converge towards better models.

© Springer International Publishing Switzerland 2015
A. Appice et al. (Eds.): ECML PKDD 2015, Part I, LNAI 9284, pp. 68–85, 2015.
DOI: 10.1007/978-3-319-23528-8_5

The second situation where the choice of evaluation measure is non-trivial is when the exact context of model deployment is unknown during model training. For instance, the misclassification costs or deployment class distribution might be unknown. In such cases a scoring classifier is more versatile than a crisp classifier, because once the deployment context becomes known, the best decision can be made using ROC analysis by finding the optimal score threshold. Particularly useful are scoring classifiers which estimate class probabilities, because these are easiest to adapt to different contexts.

Proper scoring rules are loss measures which give the lowest losses to the ideal model outputting the true class posterior probabilities. Therefore, using a proper scoring rule as model evaluation measure helps to develop models which are good class probability estimators, and hence easy to adapt to different contexts. The best known proper scoring rules are log-loss and Brier score, both of which we are concentrating on in this paper. These two are also frequently used as surrogate losses for optimisation.

In practice it can be hard to decide which proper scoring rule to use. According to one view this choice could be based on the assumptions about the probability distribution over possible deployment contexts. For example, [6] shows that the Brier score can be derived from a particular additive cost model.

Once the loss measure is fixed, the best model has to be found with respect to that measure. The decomposition of expected loss corresponding to any proper scoring rule into calibration loss and refinement loss has facilitated the development of calibration methods (*i.e.* calibration loss reduction methods) which have been shown to be beneficial for classification performance [2]. Another decomposition[1] splits refinement loss into *uncertainty* minus *resolution* [5,9]. Interestingly, none of the decompositions relates to the loss of the optimal model. This inspires our first novel decomposition of any proper scoring rule loss into epistemic loss and irreducible (aleatoric[2]) loss. Irreducible loss is the loss of the optimal model which outputs the true posterior class probability given the instance.

For our second decomposition we introduce a novel *adjustment loss*, which is extra loss due to the difference between the average of estimated scores and the class distribution. For both Brier score and log-loss we propose a corresponding adjustment procedure, which reduces this loss to zero, and hence decreases the overall loss. This procedure uses only the output scores and class distribution and not the feature values. Therefore, it can easily be used in any context, whereas a calibration procedure needs to make extra assumptions about the shape of the calibration map.[3]

Finally, we propose a four-way decomposition by combining the decompositions relating to the notions of optimality, calibration and adjustment. The separation of adjustment loss from calibration loss is specific to the proper scoring

---

[1] Note that the commonly used bias-variance decompositions apply to the loss of a learning algorithm, whereas we are studying the loss of a particular model.

[2] Our terminology here relates to epistemic and aleatoric uncertainty [10].

[3] In some literature a classifier has been called *calibrated* when it is actually only *adjusted*, a confusion that we hope to help remove by giving a name for the latter.

rule (*i.e.* it relies on the existence of an adjustment procedure) whereas the remainder of the decomposition applies to any proper scoring rule. The decomposition has the following terms: adjustment loss (AL), post-adjustment calibration loss (PCL), grouping loss (GL) and irreducible loss (IL). Grouping loss is the divergence of calibrated probabilities from the true posterior probabilities and intuitively measures the loss due to the model assigning the same score to (*i.e.* grouping together) instances which have different posterior class probabilities (cf. refinement loss is the loss due to the same scores being assigned to instances from different classes). Grouping loss has earlier been introduced in [7] where it facilitated the improvement of probability estimation and classification using reliability maps, which quantify conditional grouping loss given the model output score. Our proposed decompositions aim to provide deeper insight into the causes behind losses and facilitate development of better classification methods.

The structure of the paper is as follows: Section 2 defines proper scoring rules and introduces notation. Section 3 provides two decompositions using ideal scores and calibrated scores, respectively. Section 4 introduces the notion of adjustment and a decomposition using adjusted scores. Section 5 provides two theorems from which all decompositions follow, and provides terminology for the obtained decomposed losses. Section 6 describes two proposed algorithms and the results of convergence experiments. Section 7 discusses related work and Section 8 concludes.

2   Proper Scoring Rules

2.1   Scoring Rules

Consider the task of multi-class classification with k classes. We represent the true class of an instance as a vector $y = (y_1, \ldots, y_k)$ where $y_j = 1$ if the true class is j, and $y_j = 0$ otherwise. Let $p = (p_1, \ldots, p_k)$ be an estimated class probability vector for an instance, *i.e.* $p_j \geq 0$, $j = 1, \ldots, k$ and $\sum_{j=1}^{k} p_j = 1$. A scoring rule $\phi(p, y)$ is a non-negative measure measuring the goodness of match between the estimated probability vector p and the true class y.

Two well known scoring rules are log-loss ϕ^{LL} (also known as ignorance score) and Brier score ϕ^{BS} (also known as squared loss or quadratic score), defined as follows:

$$\phi^{\mathsf{LL}}(p, y) := -\log p_y \qquad\qquad \text{log-loss,}$$

$$\phi^{\mathsf{BS}}(p, y) := \sum_{i=1}^{k} (p_i - y_i)^2 \qquad\qquad \text{Brier score}^4,$$

where by a slight abuse of notation p_y denotes p_j for j such that $y_j = 1$. Both these rules are *proper* in the sense defined in the following subsection. Note that the scoring rules apply for a single instance, application to a dataset is by averaging across all instances.

---

[4] This Brier score definition agrees with the original definition by Brier [3]. Since it ranges between 0 and 2, sometimes half of this quantity is also referred to as Brier score.

2.2 Divergence, Entropy and Properness

Suppose now that the true class y is being sampled from a distribution q over classes (*i.e.* q is a probability vector). We denote by $\mathsf{s}(p,q)$ the expected score with rule ϕ on probability vector p with respect to the class label drawn according to q:

$$\mathsf{s}(p,q) := \mathbb{E}_{Y \sim q}\phi(p, Y) = \sum_{j=1}^{k} \phi(p, e_j)q_j \, ,$$

where e_j denotes a vector of length k with 1 at position j and 0 everywhere else. We define divergence $\mathsf{d}(p,q)$ of p from q and entropy $\mathsf{e}(q)$ of q as follows:

$$\mathsf{d}(p,q) := \mathsf{s}(p,q) - \mathsf{s}(q,q) \, , \qquad\qquad \mathsf{e}(q) := \mathsf{s}(q,q) \, .$$

A scoring rule ϕ is called proper if the respective divergence is always non-negative, and strictly proper if additionally $\mathsf{d}(p,q) = 0$ implies $p = q$. It is easy to show that both log-loss and Brier score are strictly proper scoring rules.

For the scoring rules ϕ^{LL} and ϕ^{BS} the respective divergence and entropy measures can easily be shown to be the following:

$$\mathsf{d}^{\mathsf{LL}}(p,q) = \sum_{j=1}^{k} q_j \log \frac{q_j}{p_j} \quad \text{KL-divergence};$$

$$\mathsf{e}^{\mathsf{LL}}(q) = -\sum_{j=1}^{k} q_j \log q_j \quad \text{information entropy};$$

$$\mathsf{d}^{\mathsf{BS}}(p,q) = \sum_{j=1}^{k} (p_j - q_j)^2 \quad \text{mean squared difference};$$

$$\mathsf{e}^{\mathsf{BS}}(q) = \sum_{j=1}^{k} q_j(1 - q_j) \quad \text{Gini index}.$$

In the particular case where q is equal to the true class label y, divergence is equal to the proper scoring rule itself, *i.e.* $\mathsf{d}(p,y) = \phi(p,y)$. In the following we refer to proper scoring rules as $\mathsf{d}(p,y)$ because this makes the decompositions more intuitive.

2.3 Expected Loss and Empirical Loss

Proper scoring rules define the loss of a class probability estimator on a single instance. In practice, we are interested in the performance of the model on test data. Once the test data are fixed and known, the proper scoring rules provide the performance measure as the average of instance-wise losses across the test data. We refer to this as *empirical loss*. If the test data are drawn randomly from a (potentially infinite) labelled instance space, then the performance measure can be defined as the expected loss on a randomly drawn labelled instance. We refer to this as *expected loss*.

Empirical loss can be thought of as a special case of expected loss with uniform distribution over the test instances and zero probability elsewhere. Indeed, suppose that the generative model is uniformly randomly picking and outputting one of the test instances. The empirical loss on the (original) test data and the expected loss with this generative model are then equal. Therefore, all decompositions that we derive for the expected loss naturally apply to the empirical loss as well, assuming that test data represent the whole population.

Next we introduce our notation in terms of random variables. Let X be a random variable (a vector) representing the attributes of a randomly picked instance, and $Y = (Y_1, \ldots, Y_k)$ be a random vector specifying the class of that instance, where $Y_j = 1$ if X is of class j, and $Y_j = 0$ otherwise, for $j = 1, 2, \ldots, k$. Let now f be a fixed scoring classifier (or class probability estimator), then we denote by $S = (S_1, S_2, \ldots, S_k) = f(X)$ the score vector output by the classifier on instance X. Note that S is now a random vector, as it depends on the random variable X. The expected loss of S with respect to Y under the proper scoring rule d is $\mathbb{E}[d(S, Y)]$.

Example 1. Consider a binary ($k = 2$) classification test set of 8 instances with 2 features, as shown in column $X^{(i)}$ of Table 1. Suppose the instances with indices 1,2,3,5,6 are positives (class 1) and the rest are negatives (class 2). This information is represented in column $Y_1^{(i)}$, where 1 means 'class 1' and 0 means 'not class 1'.

Suppose we have two models predicting both 0.9 as the probability of class 1 for the first 4 instances, but differ in probability estimates for the remaining 4 instances with 0.3 predicted by the first and 0.4 by the second model. This information is represented in the columns labelled $S_1^{(i)}$ for both models.

Table 1. Example dataset with 2 classes, with information shown for class 1 only. The score for class 1 is $S_1 = 0.3X_1$ by Model 1 and $S_1 = 0.25X_1 + 0.15$ by Model 2, whereas the optimal model is $Q_1 = 0.5X_2$ (or any other model which outputs 1 for first two instances and 0.5 for the rest). Columns $A_{+,1}$, $A_{*,1}$ and C_1 represent additively adjusted, multiplicatively adjusted, and calibrated scores, respectively. The average of each column is presented (mean), as well as log-loss (LL) and Brier score (BS) with respect to the true labels ($Y_1 = 1$ stands for class 1).

	Task			Model 1				Model 2			
i	$X^{(i)}$ $Y_1^{(i)}$		$Q_1^{(i)}$	$S_1^{(i)}$	$A_{+,1}^{(i)}$ $A_{*,1}^{(i)}$ $C_1^{(i)}$			$S_1^{(i)}$	$A_{+,1}^{(i)}$ $A_{*,1}^{(i)}$ $C_1^{(i)}$		
1	(3,2) 1		1.0	0.9	0.925 0.914 0.75			0.9	0.875 0.886 0.75		
2	(3,2) 1		1.0	0.9	0.925 0.914 0.75			0.9	0.875 0.886 0.75		
3	(3,1) 1		0.5	0.9	0.925 0.914 0.75			0.9	0.875 0.886 0.75		
4	(3,1) 0		0.5	0.9	0.925 0.914 0.75			0.9	0.875 0.886 0.75		
5	(1,1) 1		0.5	0.3	0.325 0.336 0.50			0.4	0.375 0.364 0.50		
6	(1,1) 1		0.5	0.3	0.325 0.336 0.50			0.4	0.375 0.364 0.50		
7	(1,1) 0		0.5	0.3	0.325 0.336 0.50			0.4	0.375 0.364 0.50		
8	(1,1) 0		0.5	0.3	0.325 0.336 0.50			0.4	0.375 0.364 0.50		
mean:	0.625 0.625			0.6	0.625 0.625 0.625			0.65	0.625 0.625 0.625		
LL:	0		0.520	0.717 0.732 0.715 0.628				0.684 0.673 0.683 0.628			
BS:	0		0.375	0.5	0.499 0.491 0.438			0.47	0.469 0.474 0.438		

The second model is better according to both log-loss ($0.684 < 0.717$) and Brier score ($0.47 < 0.5$). These can equivalently be considered either as empirical losses (as they are averages over 8 instances) or as expected losses (if the generative model picks one of the 8 instances uniformly randomly). The meaning of the remaining columns in Table 1 will become clear in the following sections.

3 Decompositions with Ideal Scores and Calibrated Scores

In this paper, all decompositions of proper scoring rules are built on procedures to map the estimated scores to new scores such that the loss is guaranteed to decrease. We start from an idealistic procedure requiring an optimal model and move towards realistic procedures.

3.1 Ideal Scores Q and the Decomposition $L = EL + IL$

Our first novel decomposition is determined by a procedure which changes the estimated scores into true posterior class probabilities (which is clearly impossible to do in practice). We denote the true posterior probability vector by $Q = (Q_1, Q_2, \ldots, Q_k)$ where $Q_j := \mathbb{E}[Y_j|X]$. Variable Q_j can be interpreted as the true proportion of class j among the instances with feature values X, and hence it is independent of the model. For our running example in Table 1 the true posterior probabilities for class 1 are given in column $Q_1^{(i)}$.

Our decomposition states that the expected loss corresponding to any proper scoring rule is the sum of expected divergence of S from Q and the expected divergence of Q from Y:

$$\mathbb{E}[\mathsf{d}(S, Y)] = \mathbb{E}[\mathsf{d}(S, Q)] + \mathbb{E}[\mathsf{d}(Q, Y)] \ .$$

This can be proved as a direct corollary of Theorem 2 in Section 5. As all these expected divergences are non-negative (due to properness of the scoring rule) and Q is the same regardless of the scoring model S, it immediately follows that $S := Q$ is the optimal model with respect to any proper scoring rule (it is a model because it is a function of X). This justifies the following terminology:

- **Epistemic Loss** $EL = \mathbb{E}[\mathsf{d}(S, Q)]$ is the extra loss due to the model not being optimal, and equals zero if and only if the model is optimal. The term relates to epistemic uncertainty (as opposed to aleatoric uncertainty) [10] and is due to our mistreatment of the evidence X with respect to the ideal model.
- **Irreducible Loss** $IL = \mathbb{E}[\mathsf{d}(Q, Y)]$ is the loss due to inherent uncertainty in the classification task, the loss which is the same for all models. This type of uncertainty is called aleatoric [10] so the loss could also be called *aleatoric loss*. It is the loss of the optimal model and equals zero only if the attributes of the instance X provide enough information to uniquely determine the right label Y (with probability 1).

For our running example the epistemic log-loss EL^{LL} for the two models is 0.198 and 0.164 (not shown in Table 1) and the (model-independent) irreducible log-loss is $IL^{LL} = 0.520$, which (as expected) sum up to the total expected log-loss of 0.717 and 0.684, respectively (with the rounding effect in the last digit of 0.717). For Brier score the decomposition for the two models is $0.5 = 0.125 + 0.375$ and $0.47 = 0.095 + 0.375$, respectively.

3.2   Calibrated Scores C and the Decomposition $L = CL + RL$

The second, well-known decomposition [5] is determined by a procedure which changes the estimated scores into calibrated probabilities. We denote the calibrated probability vector by $C = (C_1, C_2, \ldots, C_k)$ where $C_j := \mathbb{E}[Y_j|S]$. Variable C_j can be interpreted as the true proportion of class j among the instances for which the model has output the same estimate S, and hence calibration is model-dependent. For our running example in Table 1 the calibrated probabilities of class 1 for the two models are given in columns $C_1^{(i)}$. Note that the columns for the two models are identical. This is only because for any two instances in our example, the first model gives them the same estimate if and only if the second model does so.

The standard calibration-refinement decomposition [4] states[5] that the expected loss according to any proper scoring rule is the sum of expected divergence of S from C and the expected divergence of C from Y:

$$\mathbb{E}[\mathsf{d}(S, Y)] = \mathbb{E}[\mathsf{d}(S, C)] + \mathbb{E}[\mathsf{d}(C, Y)] \ .$$

This is another direct corollary of Theorem 2 in Section 5. The standard terminology is as follows:

- **Calibration Loss** $CL = \mathbb{E}[\mathsf{d}(S, C)]$ is the loss due to the difference between the model output score S and the proportion of positives among instances with the same output (calibrated score).
- **Refinement Loss** $RL = \mathbb{E}[\mathsf{d}(C, Y)]$ is the loss due to the presence of instances from multiple classes among the instances with the same estimate S.

For our running example the calibration loss for Brier score CL^{BS} for the two models is 0.062 and 0.033 (not shown in Table 1) and the refinement loss is for both equal to $RL^{BS} = 0.438$, which sum up to the total expected Brier scores of 0.5 and 0.47, respectively (with the rounding effect in the last digit, we omit this comment in the following cases). For log-loss the decomposition for the two models is $0.717 = 0.090 + 0.628$ and $0.684 = 0.056 + 0.628$, respectively.

In practice, calibration has proved to be an efficient way of decreasing proper scoring rule loss [2]. Calibrating a model means learning a calibration mapping from the model output scores to the respective calibrated probability scores. Calibration is simple to perform if the model has only a few possible output

---

[5] Actually, in [4] the calibration-refinement decomposition is stated as $\mathbb{E}[\mathsf{s}(S, Y)] = \mathbb{E}[\mathsf{d}(S, C)] + \mathbb{E}[\mathsf{e}(C)]$ but this can easily be shown to be equivalent to our statement.

scores, each covered by many training examples. Then the empirical class distribution among training instances with the same output scores can be used as calibrated score vector. However, in general, there might be a single or even no training instances with the same score vector as the model outputs on a test instance. Then the calibration procedure needs to make additional assumptions (inductive bias) about the shape of the calibration map, such as monotonicity and smoothness.

Regardless of the method, calibration is almost never perfect. Even if perfectly calibrated on the training data, the model can suffer some calibration loss on test data. In the next section we propose an adjustment procedure as a precursor of calibration. Adjustment does not make any additional assumptions and is guaranteed to decrease loss if the test class distribution is known exactly.

4   Adjusted Scores A and the Decomposition $L = AL + PL$

Ideal scores cannot be obtained in practice, and calibrated scores are hard to obtain, requiring extra assumptions about the shape of the calibration map. Here we propose two procedures which take as input the estimated scores and output *adjusted scores* such that the mean matches with the given target class distribution. As opposed to calibration, no labels are required for learning how to adjust, only the scores and target class distribution are needed. We prove that *additive adjustment* is guaranteed to decrease Brier score, and *multiplicative adjustment* is guaranteed to decrease log-loss. In both cases we can decompose the expected loss in a novel way.

4.1   Adjustment

Suppose we are given the class distribution of the test data, represented as a vector π of length k, with non-negative entries and adding up to 1. It turns out that if the average of the model output scores on the test data does not match with the given distribution then for both log-loss and Brier score it is possible to adjust the scores with guaranteed reduction of loss. First we define what we mean by adjusted scores.

Definition 1. *Let π be a class distribution with k classes and A be a random real-valued vector of length k. If $\mathbb{E}[A_j] = \pi_j$ for $j = 1, \ldots, k$, then we say that A is adjusted to the class distribution π.*

If the scores are not adjusted, then they can be adjusted using one of the two following procedures.

Additive (score) adjustment is a procedure applying the following function α_+:

$$\alpha_+(s) = (s_1 + b_1, \ldots, s_k + b_k) \qquad \forall s \in \mathbb{R}^k ,$$

where $b_j = \pi_j - \mathbb{E}[S_j]$, for $j = 1, \ldots, k$. Hence, the function is different depending on what the model output scores and class distribution are. It is easy to prove that the scores $\alpha_+(S)$ are adjusted: $\mathbb{E}[S_j + b_j] = \mathbb{E}[S_j] + b_j = \mathbb{E}[S_j] + \pi_j - \mathbb{E}[S_j] = \pi_j$, for $j = 1, \ldots, k$.

Multiplicative (score) adjustment is a procedure applying the function α_*:

$$\alpha_*(s) = \left(\frac{w_1 s_1}{\sum_{j=1}^k w_j s_j}, \ldots, \frac{w_k s_k}{\sum_{j=1}^k w_j s_j} \right) \qquad \forall s \in \mathbb{R}^k ,$$

where w_j are suitably chosen non-negative weights such that $\alpha_*(S)$ is adjusted to π. It is not obvious that such weights exist because of the required renormalisation, but the following theorem gives this guarantee.

Theorem 1 (Existence of weights for multiplicative adjustment). *Let π be a class distribution with $k \geq 2$ classes and S be a random positive real vector of length k. Then there exist non-negative weights $w_1, \ldots, w_k$ such that*

$$\mathbb{E}\left[\frac{w_i S_i}{\sum_{j=1}^k w_j S_j} \right] = \pi_i \text{ for } i = 1, \ldots, k.$$

Proof. All the proofs are in the Appendix and the extended proofs are available at http://www.cs.bris.ac.uk/~flach/Kull_Flach_ECMLPKDD2015_Supplementary.pdf.

For our running example in Table 1 the additively adjusted and multiplicatively adjusted scores for class 1 are shown in columns $A_{+,1}^{(i)}$ and $A_{*,1}^{(i)}$, respectively. The shift b for additive adjustment was $(+0.025, -0.025)$ for Model 1 and $(-0.025, +0.025)$ for Model 2. The weights w for multiplicative adjustment were $(1.18, 1)$ for Model 1 and $(1, 1.16)$ for Model 2. For example, for Model 1 the scores $(0.9, 0.1)$ (of first four instances) become $(1.062, 0.1)$ after weighting and $(0.914, 0.086)$ after renormalising (dividing by $1.062 + 0.1 = 1.162$). The average score for class 1 becomes 0.625 for both additive and multiplicative adjustment and both models, confirming the correctness of these procedures.

4.2   The Right Adjustment Procedure Guarantees Decreased Loss

The existence of multiple adjustment procedures raises a question of which one to use. As seen from the losses after adjustment in Table 1, multiplicative adjustment achieves a lower loss for Model 1 and additive adjustment achieves a lower loss for Model 2, for both log-loss and Brier score. This shows that neither procedure is better than the other across all models.

Further inspection of Table 1 shows that for Model 1 the log-loss increased after additive adjustment and for Model 2 the Brier score increased after multiplicative adjustment. Interestingly, we can guarantee decreased loss if the right adjustment procedure is used: multiplicative adjustment always decreases logloss, and additive adjustment always decreases Brier score. Of course, the exception is when the scores are already adjusted, in which case there is no change in the loss. The guarantee is due to the following novel loss-specific decompositions (and non-negativity of divergence):

$$\mathbb{E}[\mathsf{d}^{\mathsf{BS}}(S, Y)] = \mathbb{E}[\mathsf{d}^{\mathsf{BS}}(S, A_+)] + \mathbb{E}[\mathsf{d}^{\mathsf{BS}}(A_+, Y)] \,,$$
$$\mathbb{E}[\mathsf{d}^{\mathsf{LL}}(S, Y)] = \mathbb{E}[\mathsf{d}^{\mathsf{LL}}(S, A_*)] + \mathbb{E}[\mathsf{d}^{\mathsf{LL}}(A_*, Y)] \,,$$

where $A_+ = \alpha_+(S)$ and $A_* = \alpha_*(S)$ are obtained from the scores S using additive and multiplicative adjustment, respectively. Note that the additive adjustment procedure can produce values out of the range $[0, 1]$ but Brier score is defined for these as well. The decompositions follow from Theorem 4 in Section 5, which provides a unified decomposition:

$$\mathbb{E}[\mathsf{d}(S, Y)] = \mathbb{E}[\mathsf{d}(S, A)] + \mathbb{E}[\mathsf{d}(A, Y)]$$

under an extra assumption which links the adjustment method and the loss measure. Due to this unification we propose the following terminology for the losses:

- **Adjustment Loss** $AL = \mathbb{E}[\mathsf{d}(S, A)]$ is the loss due to the difference between the mean model output $\mathbb{E}[S]$ and the overall class distribution $\pi := \mathbb{E}[Y]$. This loss is zero if the scores are adjusted.
- **Post-adjustment Loss** $PL = \mathbb{E}[\mathsf{d}(A, Y)]$ is the loss after adjusting the model output with the method corresponding to the loss measure.

For our running example the adjustment log-loss AL^{LL} for the two models is 0.0021 and 0.0019 (not shown in Table 1) and the respective post-adjustment losses PL^{LL} are 0.7154 and 0.6822, which sum up to the total expected log-loss of 0.7175 and 0.6841, respectively. For Brier score the decomposition for the two models is $0.5 = 0.00125 + 0.49875$ and $0.47 = 0.00125 + 0.46875$, respectively.

In practice, the class distribution is usually not given, and has to be estimated from training data. Therefore, if the difference between the average output scores and class distribution is small (*i.e.* adjustment loss is small), then the benefit of adjustment might be subsumed by class distribution estimation errors. Experiments about this remain as future work.

So far we have given three different two-term decompositions of expected loss: epistemic loss plus irreducible loss, calibration loss plus refinement loss, and adjustment loss plus post-adjustment loss. In the following section we show that these can all be obtained from a single four-term decomposition, and provide more terminology and intuition.

5 Decomposition Theorems and Terminology

In the previous sections we had the following decompositions of expected loss using a proper scoring rule (with extra assumptions for the last decomposition):

$$\mathbb{E}[\mathsf{d}(S, Y)] = \mathbb{E}[\mathsf{d}(S, Q)] + \mathbb{E}[\mathsf{d}(Q, Y)] = \mathbb{E}[\mathsf{d}(S, C)] + \mathbb{E}[\mathsf{d}(C, Y)] = \mathbb{E}[\mathsf{d}(S, A)] + \mathbb{E}[\mathsf{d}(A, Y)]$$

All these decompositions follow a pattern $\mathbb{E}[\mathsf{d}(S,Y)] = \mathbb{E}[\mathsf{d}(S,V)] + \mathbb{E}[\mathsf{d}(V,Y)]$ for some random variable V. In this section we generalise further, and introduce decompositions $\mathbb{E}[\mathsf{d}(V_1,V_3)] = \mathbb{E}[\mathsf{d}(V_1,V_2)] + \mathbb{E}[\mathsf{d}(V_2,V_3)]$ for some random variables V_1, V_2, V_3. The random variables will always be from the list S, A, C, Q, Y, and always in the same order. Actually, we will prove that the decomposition holds for any subset of 3 variables out of these 5, as long as the ordering is preserved. For decompositions involving adjusted scores A there is an extra assumption required, this is introduced in Section 5.2. First we provide decompositions without A.

5.1 Decompositions with S, C, Q, Y

Theorem 2. *Let (X,Y) be random variables representing features and labels for a k-class classification task, f be a scoring classifier, and d be the divergence function of a strictly proper scoring rule. Denote $S = f(X)$, $C_j = \mathbb{E}[Y_j|S]$, and $Q_j = \mathbb{E}[Y_j|X]$ for $j = 1, \ldots, k$. Then for any subsequence V_1, V_2, V_3 of the random variables S, C, Q, Y the following holds:*

$$\mathbb{E}[\mathsf{d}(V_1,V_3)] = \mathbb{E}[\mathsf{d}(V_1,V_2)] + \mathbb{E}[\mathsf{d}(V_2,V_3)] \ .$$

This theorem proves the decompositions of Section 3 but adds two more:

$$\mathbb{E}[\mathsf{d}(S,Q)] = \mathbb{E}[\mathsf{d}(S,C)] + \mathbb{E}[\mathsf{d}(C,Q)] \ , \qquad EL = CL + GL \ ;$$
$$\mathbb{E}[\mathsf{d}(C,Y)] = \mathbb{E}[\mathsf{d}(C,Q)] + \mathbb{E}[\mathsf{d}(Q,Y)] \ , \qquad RL = GL + IL \ .$$

These decompositions introduce the following new quantity:

- **Grouping Loss** $GL = \mathbb{E}[\mathsf{d}(C,Q)]$ is the loss due to many instances being grouped under the same estimate S while having different true posterior probabilities Q.

The above decompositions together imply the following three-term decomposition:

$$\mathbb{E}[\mathsf{d}(S,Y)] = \mathbb{E}[\mathsf{d}(S,C)] + \mathbb{E}[\mathsf{d}(C,Q)] + \mathbb{E}[\mathsf{d}(Q,Y)] \ , \quad L = CL + GL + IL \ .$$

5.2 Decompositions with S, A, C, Q, Y and Terminology

As discussed in Section 4, the decomposition of expected loss into adjustment loss and post-adjustment loss requires a link between the adjustment procedure and loss measure. The following definition presents the required link formally.

Definition 2. *Let (X,Y) be random variables representing features and labels for a k-class classification task, f be a scoring classifier, and ϕ be a strictly proper scoring rule. Denote $S = f(X)$. Let $\alpha = (\alpha_1, \ldots, \alpha_k)$ be a vector function with $\alpha_j : \mathbb{R} \to \mathbb{R}$ and let us denote $A = (A_1, \ldots, A_k)$ with $A_j = \alpha_j(S)$. We say that α provides coherent adjustment of S for proper scoring rule d if A is* *adjusted to the class distribution* $\mathbb{E}[Y]$ *and the following quantity is a constant (not a random variable), depending on* i, j *only:*

$$\phi(A, e_i) - \phi(A, e_j) - \phi(S, e_i) + \phi(S, e_j) = const_{i,j} \qquad (1)$$

where e_m *is a vector of length* k *with 1 at position* m *and 0 everywhere else.*

Intuitively, (1) requires α to apply in some sense the same adjustment to different scores, with respect to the scoring rule. In Appendix we prove the following theorem:

Theorem 3. *Additive adjustment is coherent with Brier score and multiplicative adjustment is coherent with log-loss.*

Now we are ready to present our most general decomposition theorem:

Theorem 4. *Let* (X, Y) *be random variables representing features and labels for a k-class classification task, f be a scoring classifier, and* d *be the divergence function of a strictly proper scoring rule. Denote* $S = f(X)$, $C_j = \mathbb{E}[Y_j|S]$, *and* $Q_j = \mathbb{E}[Y_j|X]$ *for* $j = 1, \ldots, k$. *Let* $A = \alpha(S)$ *where* α *provides coherent adjustment of* S *for proper scoring rule* d. *Then for any subsequence* V_1, V_2, V_3 *of the random variables* S, A, C, Q, Y *the following holds:*

$$\mathbb{E}[d(V_1, V_3)] = \mathbb{E}[d(V_1, V_2)] + \mathbb{E}[d(V_2, V_3)] .$$

Note that coherent adjustment might not exist for all proper scoring rules: then the decompositions involving A do not work, falling back to Theorem 2. Theorem 4 proves the decompositions in Section 4 and also provides the following decompositions:

$$\begin{aligned}
\mathbb{E}[d(S, C)] &= \mathbb{E}[d(S, A)] + \mathbb{E}[d(A, C)] , & CL &= AL + PCL ; \\
\mathbb{E}[d(S, Q)] &= \mathbb{E}[d(S, A)] + \mathbb{E}[d(A, Q)] , & EL &= AL + PEL ; \\
\mathbb{E}[d(A, Q)] &= \mathbb{E}[d(A, C)] + \mathbb{E}[d(C, Q)] , & PEL &= PCL + GL ; \\
\mathbb{E}[d(A, Y)] &= \mathbb{E}[d(A, Q)] + \mathbb{E}[d(Q, Y)] , & PL &= PEL + IL ,
\end{aligned}$$

which introduce new quantities PCL and PEL.

- **Post-adjustment Calibration Loss** $PCL = \mathbb{E}[d(A, C)]$ is the loss due to the remaining calibration loss after perfect adjustment.
- **Post-adjustment Epistemic Loss** $PEL = \mathbb{E}[d(A, Q)]$ is the loss due to the remaining epistemic loss after perfect adjustment.

Now we have introduced all pairwise divergences between two variables from the ordered list S, A, C, Q, Y. Table 2 summarises our proposed terminology.

A direct corollary from Theorem 4 is that if we choose 4 or 5 out of 5 variables from S, A, C, Q, Y, then we get a 3- or 4-term decomposition, respectively. In particular, the full 4-term decomposition involving all 5 variables is as follows:

$$\mathbb{E}[d(S, Y)] = \mathbb{E}[d(S, A)] + \mathbb{E}[d(A, C)] + \mathbb{E}[d(C, Q)] + \mathbb{E}[d(Q, Y)] , \quad L = AL + PCL + GL + IL .$$

Table 2. Proposed terminology

	Definition	Visual	Name	Description
L	$\mathbb{E}[d(S,Y)]$	`S...Y`	Loss	total expected loss
AL	$\mathbb{E}[d(S,A)]$	`SA...`	Adjustment Loss	loss due to lack of adjustment
PCL	$\mathbb{E}[d(A,C)]$	`.AC..`	Post-adjustment Calibration Loss	calibration loss after adjustment
GL	$\mathbb{E}[d(C,Q)]$	`..CQ.`	Grouping Loss	loss due to grouping
IL	$\mathbb{E}[d(Q,Y)]$	`...QY`	Irreducible Loss	loss of the optimal model
CL	$\mathbb{E}[d(S,C)]$	`S.C..`	Calibration Loss	loss due to lack of calibration
PEL	$\mathbb{E}[d(A,Q)]$	`.A.Q.`	Post-adjustment Epistemic Loss	epistemic loss after adjustment
RL	$\mathbb{E}[d(C,Y)]$	`..C.Y`	Refinement Loss	loss after calibration
EL	$\mathbb{E}[d(S,Q)]$	`S..Q.`	Epistemic Loss	loss due to non-optimal model
PL	$\mathbb{E}[d(A,Y)]$	`.A..Y`	Post-adjustment Loss	loss after adjustment

Table 3. The decomposed losses (left) and their values for model 1 of the running example using log-loss (middle) and Brier score (right).

	S	A	C	Q	Y
S	0	AL	CL	EL	L
A		0	PCL	PEL	PL
C			0	GL	RL
Q				0	IL
Y					0

LL	S	A_*	C	Q	Y
S	0	0.002	0.090	0.198	0.717
A_*		0	0.088	0.196	0.715
C			0	0.108	0.628
Q				0	0.520
Y					0

BS	S	A_+	C	Q	Y
S	0	0.001	0.062	0.125	0.5
A_+		0	0.061	0.124	0.499
C			0	0.062	0.438
Q				0	0.375
Y					0

Table 3 provides numerical values for all 10 losses of Table 2 for Model 1 in our running example data (Table 1). The 4-term decomposition proves that the numbers right above the main diagonal (AL, PCL, GL, IL) add up to the total loss at the top right corner (L). All other decompositions can be checked numerically from the table (taking into account the accumulating rounding errors).

6   Algorithms and Experiments

We have proposed two new procedures in the paper: additive and multiplicative adjustment. Here we provide algorithms to perform these procedures. Both procedures first require estimation of the parameter vectors: b for additive and w for multiplicative adjustment. If the test instances are all given together in batch, then the scores of the model on test data can be used to estimate these parameter vectors. Otherwise, these need to be estimated on training (or validation) data.

Additive adjustment is algorithmically very easy. Parameter b_j is the difference of proportion π_j of class j and the mean $\mathbb{E}[S_j]$, calculated as the average output score for class j over all instances. This is exact if test data are given in batch and π_j is the true proportion, and it is approximate if π_j is estimated from training data. Finally, adjusted scores can be calculated by adding b to the model output scores, for each test instance.

Table 4. Average number of rounds to convergence of multiplicative adjustment across 10000 synthetic tasks with k classes and n instances. The number in parentheses shows the count of failures to converge out of 10000.

	$k = 2$	$k = 3$	$k = 4$	$k = 5$	$k = 10$	$k = 20$	$k = 30$	$k = 50$
$n = 10$	1.00 (21)	3.66 (9)	3.95 (3)	3.97 (3)	3.88 (3)	3.66 (4)	3.49 (23)	3.25 (99)
$n = 100$	1.00 (4)	3.64 (2)	3.95 (2)	3.97 (0)	3.86 (1)	3.63 (0)	3.44 (2)	3.22 (48)
$n = 1000$	1.00 (6)	3.64 (0)	3.95 (1)	3.96 (0)	3.85 (0)	3.62 (0)	3.44 (4)	3.22 (43)

For multiplicative adjustment the hard part is to obtain the parameter (weight) vector w, whereas applying adjustment using the weights is straightforward. The weight vector w can be obtained by the coordinate descent optimisation algorithm where for coordinate j the task is to minimise the difference between $\mathbb{E}[w_j s_j / \sum_{i=1}^{k} w_i s_i]$ and π_j, by changing only w_j. The minimisation in one coordinate can be done by binary search, since the expected value is monotonically increasing with respect to w_j. It is clear that if coordinate descent algorithm converges, then the obtained w is the right one. However, the algorithm can fail to converge.

We have performed experiments with synthetic tasks with $k = 2, 3, 4, 5, 10, 20, 50$ classes and $n = 10, 100, 1000$ instances to check convergence. Each task is a pair of a $n \times k$ model score matrix and class distribution vector of length k, all filled with uniformly random entries between 0 and 1, and each row is normalised to add up to 1. Table 4 reports the number of cycles through the coordinates to convergence, averaged over 10000 tasks for each k, n pair. As expected, the results have almost no dependence on the number of instances. The maximal number of rounds to convergence was 6. However, on average in 10 out of 10000 times there was no convergence. Further improvement of this result remains as future work.

7   Related Work

Proper scoring rules have a long history of research, with Brier score introduced in 1950 in the context of weather forecasting [3], and the general presentation of proper scoring rules soon after, see *e.g.* [11]. The decomposition of Brier score into calibration and refinement loss (which were back then called reliability and resolution) was introduced by Murphy [8] and was generalised for proper scoring rules by DeGroot and Fienberg [5]. The decompositions with three terms were introduced by Murphy [9] with uncertainty, reliability and resolution (Murphy reused the same name for a different quantity), later generalised to all proper scoring rules as well [4]. In our notation these can be stated as $\mathbb{E}[d(S, Y)] = REL + UNC - RES = \mathbb{E}[d(S, C)] + \mathbb{E}[d(\pi, Y)] - \mathbb{E}[d(\pi, C)]$. This can easily be proved by taking into account that the last term can be viewed as calibration loss for constant estimator π but segmented in the same way as S.

In machine learning proper scoring rules are often treated as surrogate loss functions, which are used instead of the 0-1 loss to facilitate optimisation [1]. An important question in practice is which proper scoring rule to use. One possible viewpoint is to assume a particular distribution over anticipated deployment contexts and derive the expected loss from that assumption. Hernández-Orallo *et al.* have shown that the Brier score can be derived from a particular additive cost model [6].

8   Conclusions

This paper proposes novel decompositions of proper scoring rules. All presented decompositions are sums of expected divergences between original scores S, adjusted scores A, calibrated scores C, true posterior probabilities Q and true labels Y. Each such divergence stands for one part of the total expected loss. Calibration and refinement loss are known losses of this form, the paper proposes names for the other 7 losses and provides underlying intuition. In particular, we have introduced adjustment loss, which arises from the difference between mean estimated scores and true class distribution. While it is a part of calibration loss, it is easier to eliminate or decrease than calibration loss. We have proposed first algorithms for additive and multiplicative adjustment, which we prove to be coherent with (decomposing) Brier score and log-loss, respectively. More algorithm development is needed for multiplicative adjustment, as the current algorithm can sometimes fail to converge. An open question is whether there are other, potentially better coherent adjustment procedures for these losses. We hope that the proposed decompositions provide deeper insight into the causes behind losses and facilitate development of better classification methods, as knowledge about calibration loss has already delivered several calibration methods, see *e.g.* [2].

Acknowledgments. This work was supported by the REFRAME project granted by the European Coordinated Research on Long-term Challenges in Information and Communication Sciences & Technologies ERA-Net (CHIST-ERA), and funded by the Engineering and Physical Sciences Research Council in the UK under grant EP/K018728/1.

References

1. Bartlett, P.L., Jordan, M.I., McAuliffe, J.D.: Convexity, Classification, and Risk Bounds. Journal of the American Statistical Association **101**(473), 138–156 (2006)
2. Bella, A., Ferri, C., Hernández-Orallo, J., Ramírez-Quintana, M.J.: On the effect of calibration in classifier combination. Applied Intelligence **38**(4), 566–585 (2012)
3. Brier, G.W.: Verification of forecasts expressed in terms of probability. Monthly weather review **78**(1), 1–3 (1950)
4. Bröcker, J.: Reliability, sufficiency, and the decomposition of proper scores. Quarterly Journal of the Royal Meteorological Society **135**(643), 1512–1519 (2009)
5. De Groot, M.H., Fienberg, S.E.: The Comparison and Evaluation of Forecasters. Journal of the Royal Statistical Society. Series D (The Statistician) **32**(1/2), 12–22 (1983)

6. Hernández-Orallo, J., Flach, P., Ferri, C.: A unified view of performance metrics: translating threshold choice into expected classification loss. The Journal of Machine Learning Research **13**(1), 2813–2869 (2012)
7. Kull, M., Flach, P.A.: Reliability maps: a tool to enhance probability estimates and improve classification accuracy. In: Calders, T., Esposito, F., Hüllermeier, E., Meo, R. (eds.) ECML PKDD 2014, Part II. LNCS, vol. 8725, pp. 18–33. Springer, Heidelberg (2014)
8. Murphy, A.H.: Scalar and vector partitions of the probability score: Part I. Two-state situation. Journal of Applied Meteorology **11**(2), 273–282 (1972)
9. Murphy, A.H.: A new vector partition of the probability score. Journal of Applied Meteorology **12**(4), 595–600 (1973)
10. Senge, R., Bösner, S., Dembczynski, K., Haasenritter, J., Hirsch, O., Donner-Banzhoff, N., Hüllermeier, E.: Reliable classification: Learning classifiers that distinguish aleatoric and epistemic uncertainty. Information Sciences **255**, 16–29 (2014)
11. Winkler, R.L.: Scoring Rules and the Evaluation of Probability Assessors. Journal of the American Statistical Association **64**(327), 1073–1078 (1969)

Appendix: Proofs of the Theorems

Here we prove the theorems presented in the paper, extended proofs are available at http://www.cs.bris.ac.uk/~flach/Kull_Flach_ECMLPKDD2015_Supplementary.pdf.

Proof of Theorem 1: If there are any zeros in the vector π, then we can set the respective positions in the weight vector also to zero and solve the problem with the remaining classes. Therefore, from now on we assume that all entries in π are positive.

Let $\mathbb{W}$ denote the set of all non-negative (weight) vectors of length k with at least one non-zero component. We introduce functions $t_i : \mathbb{W} \to \mathbb{R}$ with $t_i(w) = \mathbb{E}[w_i S_i / \sum_{j=1}^{k} w_j S_j]$. Then we need to find w^* such that $t_i(w^*) = \pi_i$ for $i = 1, \ldots, k$. For this we prove the existence of increasingly better functions $h_0, h_1, \ldots, h_{k-1} : \mathbb{W} \to \mathbb{W}$ such that for $m = 0, \ldots, k-1$ the function h_m satisfies $t_i(h_m(w)) = \pi_i$ for $i = 1, \ldots, m$ for any w. Then $w^* = h_{k-1}(w)$ is the desired solution, where $w \in \mathbb{W}$ is any weight vector, such as the vector of all ones. Indeed, it satisfies $t_i(w^*) = \pi_i$ for $i = 1, \ldots, k-1$ and hence for $i = k$.

We choose h_0 to be the identity function and prove the existence of other functions h_m by induction. Let h_m for $m < k-1$ be such that for any w the vector $h_m(w)$ does not differ from w in positions $m+1, \ldots, k$ and $t_i(h_m(w)) = \pi_i$ for $i = 1, \ldots, m$. For a fixed w it is now sufficient to prove the existence of w' such that it does not differ from w in positions $m + 2, \ldots, k$ and $t_i(w') = \pi_i$ for $i = 1, \ldots, m+1$. We search for such w' among the vectors $h_m(w[m+1 : x])$ with $x \in [0, \infty)$ where $w[m+1 : x]$ denotes the vector w with the element at position $m + 1$ changed into x. The chosen form of w' guarantees that it does not differ from w in positions $m+2, \ldots, k$ and $t_i(w') = \pi_i$ for $i = 1, \ldots, m$. It only remains to choose x such that $t_{m+1}(w') = \pi_{m+1}$. For this we note that for $x = 0$ we have $t_{m+1}(h_m(w[m+1 : 0])) = 0$ because the weight at position $m + 1$ is zero. In the limit process $x \to \infty$ we have $t_{m+1}(h_m(w[m+1:x])) \to 1 - \sum_{i=1}^{m} \pi_i$ because the weight x at position $m+1$ will dominate over weights at $m+2, \ldots, k$, whereas the weights at $1, \ldots, m$ ensure that $t_i(h_m(w[m+1:x])) = \pi_i$ for $i = 1, \ldots, m$. Since $0 < \pi_{m+1} < 1 - \sum_{i=1}^{m} \pi_i$ then according to the intermediate value theorem there exists x such that $t_{m+1}(h_m(w[m+1:x])) = \pi_{m+1}$. By this we have proved the existence of a suitable function h_{m+1}, proving the step of induction, which concludes the proof. □

Lemma 1. *Let* V_1, V_2, V_3, W *be real-valued random vectors with length* k *where* $V_{2,j} = \mathbb{E}[V_{3,j}|W]$ *for* $j = 1, \ldots, k$, *and* V_1 *is functionally dependent on* W. *If* d *is divergence of a proper scoring rule, then* $\mathbb{E}[d(V_1, V_3)] = \mathbb{E}[d(V_1, V_2)] + \mathbb{E}[d(V_2, V_3)]$.

Proof. Due to the law of total expectation it is enough to prove that $\mathbb{E}[d(V_1, V_3)|W] = \mathbb{E}[d(V_1, V_2)|W] + \mathbb{E}[d(V_2, V_3)|W]$. After expressing each d as a difference of two s terms, all obtained terms are sums over $j = 1, \ldots, k$ and it is enough to prove that for each j the equality holds. Also, as we are conditioning on W, all terms that do not involve V_3 are constants with respect to conditional expectation. Therefore, we need to prove that $\phi(V_1, e_j)\mathbb{E}[V_{3,j}|W] - \mathbb{E}[s(V_3, V_3)|W]$ equals $\phi(V_1, e_j)V_{2,j} - \phi(V_2, e_j)V_{2,j} + \phi(V_2, e_j)\mathbb{E}[V_{3,j}|W] - \mathbb{E}[s(V_3, V_3)|W]$. This holds due to $\mathbb{E}[V_{3,j}|W] = V_{2,j}$. □

Proof of Theorem 2: We consider the following two possibilities:

1. $V_2 = C$. Let us take $W = S$ in Lemma 1. Then $V_1 = S$ and it is functionally dependent on itself, W. Also, $V_{2,j} = \mathbb{E}[V_{3,j}|W]$ regardless of whether $V_3 = Y$ or $V_3 = Q$ because $C_j = \mathbb{E}[Y_j|S] = \mathbb{E}[\mathbb{E}[Y_j|X, S]|S] = \mathbb{E}[Q_j|S]$, where the second equality is due to the law of iterated expectations. The result now follows from Lemma 1.

2. $V_2 = Q$. Then $V_3 = Y$ and the result follows from Lemma 1 with $W = X$ because $V_{2,j} = Q_j = \mathbb{E}[Y_j|X] = \mathbb{E}[V_{3,j}|W]$ and both candidates S and C for V_1 are functionally dependent on $W = X$. □

Proof of Theorem 3: In Section 4 we proved that both methods provide adjusted scores, so we only need to prove Eq.(1). For log-loss we need to prove that $-\log A_i + \log A_j + \log S_i - \log S_j$ is a constant. For this it is enough to show that $(A_j/A_i)/(S_j/S_i)$ is constant. According to the definition of multiplicative adjustment this quantity equals $((w_j S_j)/(w_i S_i))/(S_j/S_i) = w_j/w_i$ which is a constant, proving that multiplicative adjustment is coherent with log-loss. For Brier score we need to prove that

$$\sum_{m=1}^{k}(A_m - \delta_{mi})^2 - \sum_{m=1}^{k}(A_m - \delta_{mj})^2 - \sum_{m=1}^{k}(S_m - \delta_{mi})^2 + \sum_{m=1}^{k}(S_m - \delta_{mj})^2 = const_{ij} \, ,$$

where δ_{mi} is 1 if $m = i$ and 0 otherwise. For $m \notin \{i, j\}$ the respective terms in the first and second sums and in the third and fourth sums are equal and therefore cancel each other. For $m = i$ the respective terms together give

$(A_i - 1)^2 - A_i^2 - (S_i - 1)^2 + S_i^2$, for additive adjustment this equals to the constant $-2b_i$ due to $A_i = S_i + b_i$. A similar argument holds for $m = j$ and as a result we have proved that the requirement (1) holds and additive adjustment is coherent with Brier score. $\qquad\square$

Proof of Theorem 4: If none of V_1, V_2, V_3 is A, then the result follows from Theorem 2. If $V_1 = A$, then the result follows from Theorem 2 with $f^{NEW} = \alpha \circ f$ because then $S^{NEW} = A$, $C^{NEW} = C$, $Q^{NEW} = Q$. It remains to prove the result for the case where $V_1 = S$ and $V_2 = A$. Denote $\beta_j = \phi(A, e_1) - \phi(A, e_j) - \phi(S, e_1) + \phi(S, e_j)$ for $j = 1, \ldots, k$, then β_j are all constants. Now it is enough to prove that the following quantity is zero:

$$\mathbb{E}[\mathsf{d}(S, V_3)] - \mathbb{E}[\mathsf{d}(S, A)] - \mathbb{E}[\mathsf{d}(A, V_3)] =$$

$$= \mathbb{E}\Big[\sum_{j=1}^{k}\big(\phi(S, e_j)V_{3,j} - \phi(S, e_j)A_j + \phi(A, e_j)A_j - \phi(A, e_j)V_{3,j}\big) - \mathsf{s}(V_3, V_3) + \mathsf{s}(V_3, V_3)\Big]$$

$$= \mathbb{E}\Big[\sum_{j=1}^{k}\big(\phi(S, e_j) - \phi(A, e_j)\big)\big(V_{3,j} - A_j\big)\Big] = \mathbb{E}\Big[\sum_{j=1}^{k}\big(\beta_j + \phi(S, e_1) - \phi(A, e_1)\big)\big(V_{3,j} - A_j\big)\Big]$$

$$= \sum_{j=1}^{k}\beta_j\Big(\mathbb{E}[V_{3,j}] - \mathbb{E}[A_j]\Big) + \mathbb{E}\Big[\big(\phi(S, e_1) - \phi(A, e_1)\big)\Big(\sum_{j=1}^{k}V_{3,j} - \sum_{j=1}^{k}A_j\Big)\Big].$$

The first term is equal to zero regardless of whether V_3 is Y or Q or C since $\mathbb{E}[A_j] = \mathbb{E}[Y_j] = \mathbb{E}[Q_j] = \mathbb{E}[C_j]$. The second term is equal to zero because both $V_{3,j}$ and A_j for $j = 1, \ldots, k$ add up to 1. $\qquad\square$

Parameter Learning of Bayesian Network Classifiers Under Computational Constraints

Sebastian Tschiatschek[1]($\boxtimes$) and Franz Pernkopf[2]

[1] Learning and Adaptive Systems Group, ETH Zurich, Zürich, Switzerland
sebastian.tschiatschek@inf.ethz.ch
[2] Signal Processing and Speech Communication Laboratory,
Graz University of Technology, Graz, Austria
pernkopf@tugraz.at

Abstract. We consider online learning of Bayesian network classifiers (BNCs) with reduced-precision parameters, i.e. the conditional-probability tables parameterizing the BNCs are represented by low bit-width fixed-point numbers. In contrast to previous work, we analyze the learning of these parameters using reduced-precision arithmetic only which is important for computationally constrained platforms, e.g. embedded- and ambient-systems, as well as power-aware systems. This requires specialized algorithms since naive implementations of the projection for ensuring the sum-to-one constraint of the parameters in gradient-based learning are not sufficiently accurate. In particular, we present generative and discriminative learning algorithms for BNCs relying only on reduced-precision arithmetic. For several standard benchmark datasets, these algorithms achieve classification-rate performance close to that of BNCs with parameters learned by conventional algorithms using double-precision floating-point arithmetic. Our results facilitate the utilization of BNCs in the foresaid systems.

Keywords: Bayesian network classifiers · Reduced-precision · Resource-constrained computation · Generative/discriminative learning

1 Introduction

Most commonly Bayesian network classifiers (BNCs) are implemented on nowadays desktop computers, where double-precision floating-point numbers are used for parameter representation and arithmetic operations. In these BNCs, inference and classification is typically performed using the same precision for parameters and operations, and the executed computations are considered as exact. However, there is a need for BNCs working with limited computational resources. Such resource-constrained BNCs are important in domains such as ambient computing, on-satellite computations[1] or acoustic environment classification in hearing

---

F. Pernkopf—This work was supported by the Austrian Science Fund (FWF) under the project number P25244-N15.

[1] Computational capabilities on satellites are still severely limited due to power constraints and restricted availability of hardware satisfying the demanding requirements with respect to radiation tolerance.

© Springer International Publishing Switzerland 2015
A. Appice et al. (Eds.): ECML PKDD 2015, Part I, LNAI 9284, pp. 86–101, 2015.
DOI: 10.1007/978-3-319-23528-8_6

aids, machine learning for prosthetic control, e.g. a brain implant to control hand movements, amongst others. In all these applications, a trade-off between accuracy and required computational resources is essential.

In this paper, we investigate BNCs with limited computational demands by considering BNCs with reduced-precision parameters, i.e. fixed-point parameters with limited precision.[2] Using reduced-precision parameters is advantageous in many ways, e.g. power consumption compared to full-precision implementations can be reduced [20] and reduced-precision parameters enable one to implement many BNCs in parallel on field programmable gate arrays (FPGAs), i.e. the circuit area requirements on the FPGA correlate with the parameter precision [9]. Our investigations are similar to those performed in digital signal-processing, where reduced-precision implementations for digital signal processors are of great importance [10]. Note that there is also increased interest in implementing other machine learning models, e.g. neural networks, using reduced-precision parameters/computations to achieve faster training and to facilitate the implementation of larger models [2,18].

We are especially interested in learning the reduced-precision parameters using as little computational resources as possible. To decide on how to perform this learning, several questions should be answered. Should reduced-precision parameters be learned in a pre-computation step in which we can exploit the full computational power of nowadays computers? Or is it necessary to learn/adopt parameters using reduced-precision arithmetic only? The answers to these questions depend on the application of interest and identify several learning scenarios that are summarized in Figure 1. In the following, we discuss these scenarios briefly:

(a) **Training and testing using full-precision arithmetic.** This corresponds to what machine learners typically do, i.e. all computations are performed using full-precision arithmetic.

(b) **Training using reduced-precision and testing using full-precision arithmetic.** A rash thought rejects this option. But it might be interesting in the vicinity of big-data where the amount of data is so huge that it can only be processed in a compressed form, i.e. in reduced-precision.

(c) **Training using full-precision and testing using reduced-precision arithmetic.** This describes an application scenario where BNCs with pre-computed parameters can be used, e.g. hearing-aids for auditory scene classification. This scenario enables one to exploit large computational resources for parameter learning, while limiting computational demands at test time. Recent work considered this for BNCs [22].

(d) **Training and testing using reduced-precision arithmetic.** This is the scenario considered within this paper. It opens the door to many interesting applications, e.g. continuous parameter adaptation in hearing-aids using reduced-precision computations only. Another example could be a

---

[2] We are interested in fixed-point arithmetic and not in floating-point arithmetic, because typically the implementation of fixed-point processing units requires less resources than the implementation of floating-point processing units.

TRAINING

	full-precision	reduced-precision
full-precision	classical scenario, e.g. training and testing on PCs	potentially relevant for big-data applications
reduced-precision	full-precision pre-computation of parameters	e.g. parameter adaptation during testing

TESTING

Fig. 1. Combinations of training/testing using full-precision/reduced-precision arithmetic.

satellite-based system for remote sensing that tunes its parameter according to changing atmospheric conditions.

We start our investigation of parameter learning using reduced-precision computations by analyzing the effect of approximate computations on *online* parameter learning. This leads to the observation that the approximate projections needed in the used projected gradient ascent/descent algorithms to ensure the sum-to-one normalization constraints of the parameters can severely affect the learning process. We circumvent the need for these projections by proposing special purpose learning algorithms for generative maximum likelihood (ML) and discriminative maximum margin (MM) parameters.

This paper is structured as follows: In Section 2 we consider related work, followed by an introduction of the used notation and some background on parameter learning in Bayesian networks (BNs) in Section 3. We derive our proposed algorithms in Section 4 and test them in experiments in Section 5. In Section 6 we conclude the paper.

2   Related Work

For undirected graphical models, approximate inference *and* learning using integer parameters has been proposed [16]. While undirected graphical models are more amenable to integer approximations mainly due to the absence of sum-to-one constraints, there are domains where probability distributions represented by directed graphical models are desirable, e.g. in expert systems in the medical domain.

Directly related work can be summarized as follows:

- The *feasibility of BNCs with reduced-precision floating-point parameters* has been empirically investigated in [14,24]. These papers analyzed (i) the effect of precision-reduction of the parameters on the classification performance of BNCs, and (ii) how BNCs with reduced-precision parameters can be implemented using integer computations only.
- The above mentioned experimental studies where extended by a thorough theoretical analysis of using fixed-point parameters in BNCs [23]. The authors used fixed-point numbers for the following two reasons: First, because fixed-point parameters can even be used on computing platforms without floating-point processing capabilities. Second, because summation of fixed-point numbers is exact (neglecting the possibility of overflows), while summation of floating-point numbers is in general not exact.

  In particular, theoretical bounds on the classification performance when using reduced-precision fixed-point parameters have been analyzed in [21, 23]. The authors derived worst-case and probabilistic bounds on the classification rate (CR) for different bit-widths. Furthermore, they compared the classification performance and the robustness of BNCs with generatively and discriminatively optimized parameters, i.e. parameters optimized for high data likelihood and parameters optimized for classification, with respect to parameter quantization.
- In [22], *learning of reduced-precision parameters using full-precision computations* was addressed while the work mentioned above considers only rounding of double-precision parameters. An algorithm for the computation of MM reduced-precision parameters was presented and its efficiency was demonstrated. The resulting parameters had superior classification performance compared to parameters obtained by simple rounding of double-precision parameters, particularly for very low numbers of bits.

3   Background and Notation

Probabilistic Classification. Probabilistic classifiers are embedded in the framework of probability theory. One assumes a random variable (RV) C denoting the class and RVs $X_1, \ldots, X_L$ representing the attributes/features of the classifier. Each X_i can take one value in the set $\mathbf{val}(X_i)$. Similarly, C can assume values in $\mathbf{val}(C)$, i.e. $\mathbf{val}(C)$ is the set of classes. We denote the random vector consisting of $X_1, \ldots, X_L$ as $\mathbf{X} = (X_1, \ldots, X_L)$. Instantiations of RVs are denoted using lower case letters, i.e. $\mathbf{x}$ is an instantiation of $\mathbf{X}$ and c an instantiation of C, respectively. The RVs $C, X_1, \ldots, X_L$ are assumed to be jointly distributed according to the distribution $\mathrm{P}^*(C, \mathbf{X})$. In typical settings, $\mathrm{P}^*(C, \mathbf{X})$ is unknown, but a number of samples drawn iid from $\mathrm{P}^*(C, \mathbf{X})$ is at hand, i.e. a training set $\mathcal{D} = ((c^{(n)}, \mathbf{x}^{(n)}) \mid 1 \leq n \leq N)$, where $c^{(n)}$ denotes the instantiation of the RV C and $\mathbf{x}^{(n)}$ the instantiation of $\mathbf{X}$ in the nth training sample. The aim is to induce *good* classifiers provided the training set, i.e. classifiers with low generalization error. Any probability distribution $\mathrm{P}(C, \mathbf{X})$

naturally induces a classifier $h_{P(C,\mathbf{X})}$ according to $h_{P(C,\mathbf{X})}$: $\mathbf{val}(\mathbf{X}) \to \mathbf{val}(C)$, $\mathbf{x} \mapsto \arg\max_{c' \in \mathbf{val}(C)} P(c'|\mathbf{x})$. In this way, each instantiation $\mathbf{x}$ of $\mathbf{X}$ is classified by the maximum a-posteriori (MAP) estimate of C given $\mathbf{x}$ under $P(C, \mathbf{X})$. Note that $\arg\max_{c' \in \mathbf{val}(C)} P(c'|\mathbf{x}) = \arg\max_{c' \in \mathbf{val}(C)} P(c', \mathbf{x})$.

Bayesian Networks and Bayesian Network Classifiers. We consider probability distributions represented by BNs [7,11]. A BN $\mathcal{B} = (\mathcal{G}, \mathcal{P}_\mathcal{G})$ consists of a directed acyclic graph (DAG) $\mathcal{G} = (\mathbf{Z}, \mathbf{E})$ and a collection of conditional probability distributions $\mathcal{P}_\mathcal{G} = (P(X_0|\mathbf{Pa}(X_0)), \ldots, P(X_L|\mathbf{Pa}(X_L)))$, where the terms $\mathbf{Pa}(X_i)$ denote the set of parents of X_i in $\mathcal{G}$. The nodes $\mathbf{Z} = (X_0, \ldots, X_L)$ correspond to RVs and the edges $\mathbf{E}$ encode conditional independencies among these RVs. Throughout this paper, we often denote X_0 as C, i.e. X_0 represents the class. Then, a BN defines the joint distribution

$$P^\mathcal{B}(C, X_1, \ldots, X_L) = P(C|\mathbf{Pa}(C)) \prod_{i=1}^{L} P(X_i|\mathbf{Pa}(X_i)). \tag{1}$$

According to the joint distribution, a BN $\mathcal{B}$ induces the classifier $h_\mathcal{B} = h_{P^\mathcal{B}(C,\mathbf{X})}$.

In this paper, we assume discrete valued RVs only. Then, a general representation of $\mathcal{P}_\mathcal{G}$ is a collection of conditional probability tables (CPTs), i.e. $\mathcal{P}_\mathcal{G} = (\boldsymbol{\theta}^0, \ldots, \boldsymbol{\theta}^L)$, with $\boldsymbol{\theta}^i = (\theta^i_{j|\mathbf{h}}|j \in \mathbf{val}(X_i), \mathbf{h} \in \mathbf{val}(\mathbf{Pa}(X_i)))$, where $\theta^i_{j|\mathbf{h}} = P(X_i = j|\mathbf{Pa}(X_i) = \mathbf{h})$. The BN distribution can then be written as

$$P^\mathcal{B}(C = c, \mathbf{X} = \mathbf{x}) = \prod_{i=0}^{L} \prod_{j \in \mathbf{val}(X_i)} \prod_{\mathbf{h} \in \mathbf{val}(\mathbf{Pa}(X_i))} {\theta^i_{j|\mathbf{h}}}^{\nu^i_{j|\mathbf{h}}}, \tag{2}$$

where $\nu^i_{j|\mathbf{h}} = \mathbf{1}_{([c,\mathbf{x}](X_i)=j \text{ and } [c,\mathbf{x}](\mathbf{Pa}(X_i))=\mathbf{h})}$.[3] We typically represent the BN parameters in the logarithmic domain, i.e. $w^i_{j|\mathbf{h}} = \log\theta^i_{j|\mathbf{h}}$, $\mathbf{w}^i = (w^i_{j|\mathbf{h}}|j \in \mathbf{val}(X_i), \mathbf{h} \in \mathbf{val}(\mathbf{Pa}(X_i)))$, and $\mathbf{w} = (\mathbf{w}^0, \ldots, \mathbf{w}^L)$. In general, we will interpret $\mathbf{w}$ as a vector, whose elements are addressed as $w^i_{j|\mathbf{h}}$. We define a vector-valued function $\boldsymbol{\phi}(c, \mathbf{x})$ of the same length as $\mathbf{w}$, collecting $\nu^i_{j|\mathbf{h}}$, analog to the entries $w^i_{j|\mathbf{h}}$ in $\mathbf{w}$. In that way, we can express the logarithm of (2) as

$$\log P^\mathcal{B}(C = c, \mathbf{X} = \mathbf{x}) = \boldsymbol{\phi}(c, \mathbf{x})^T \mathbf{w}. \tag{3}$$

Consequently, classification, can be performed by simply adding the log-probabilities corresponding to an instantiation $[c, \mathbf{x}]$ for all $c \in \mathbf{val}(C)$.[4]

Fixed-Point Numbers. Fixed-point numbers are essentially integers scaled by a constant factor, i.e. the fractional part has a fixed number of digits. We characterize fixed-point numbers by the number of integer bits b_i and the number

---

[3] Note that $[c, \mathbf{x}]$ denotes the joint instantiation of C and $\mathbf{X}$ and $[c, \mathbf{x}](\mathbf{A})$ corresponds to the subset of values of $[c, \mathbf{x}]$ indexed by $\mathbf{A} \subseteq \{X_0, \ldots, X_L\}$.

[4] In general graphs, potentially with latent variables, the needed inference can be performed using *max-sum* message passing [7,13].

of fractional bits b_f. The addition of two fixed-point numbers can be easily and accurately performed, while the multiplication of two fixed-point numbers often leads to overflows and requires truncation to achieve results in the same format.

Learning Bayesian Network Classifiers

BNs for classification can be optimized in two ways: firstly, one can select the graph structure $\mathcal{G}$ (*structure learning*), and secondly, one can learn the conditional probability distributions $\mathcal{P}_\mathcal{G}$ (*parameter learning*). In this paper, we consider fixed structures of the BNCs, namely naive Bayes (NB) and tree augmented network (TAN) structures [4], i.e. 1-tree among the attributes. The NB structure implies conditional independence of the features, given the class. Obviously, this conditional independence assumption is often violated in practice. TAN structures relax these strong independence assumptions, enabling better classification performance.[5]

Parameter Learning. The conditional probability densities (CPDs) $\mathcal{P}_\mathcal{G}$ of BNs can be optimized either generatively or discriminatively. Two standard approaches for optimizing $\mathcal{P}_\mathcal{G}$ are:

- **Generative Maximum Likelihood Parameters.** In generative parameter learning one aims at identifying parameters modeling the generative process that results in the data of the training set, i.e. generative parameters are based on the idea of *approximating* $\mathrm{P}^*(C, \mathbf{X})$ by a distribution $\mathrm{P}^\mathcal{B}(C, \mathbf{X})$. An example of this paradigm is *maximum likelihood (ML)* learning. Its objective is maximization of the likelihood of the training data given the parameters, i.e.

$$\mathcal{P}_\mathcal{G}^{\mathrm{ML}} = \arg\max_{\mathcal{P}_\mathcal{G}} \prod_{n=1}^{N} \mathrm{P}^\mathcal{B}(c^{(n)}, \mathbf{x}^{(n)}). \tag{4}$$

Note that the above optimization problem implicitly includes sum-to-one constraints because the learned parameters in $\mathcal{P}_\mathcal{G}^{\mathrm{ML}}$ must represent normalized probabilities. Maximum likelihood parameters minimize the Kullback-Leibler (KL)-divergence between $\mathrm{P}^\mathcal{B}(C, \mathbf{X})$ and $\mathrm{P}^*(C, \mathbf{X})$ [7].
- **Discriminative Maximum Margin Parameters [5, 12, 15].** In discriminative learning one aims at identifying parameters leading to good classification performance on new samples from $\mathrm{P}^*(C, \mathbf{X})$. This type of learning is for example advantageous in cases where the assumed model distribution $\mathrm{P}^\mathcal{B}(C, \mathbf{X})$ cannot approximate $\mathrm{P}^*(C, \mathbf{X})$ well, for example because of a too limited BN structure [17].
  Discriminative MM parameters $\mathcal{P}_\mathcal{G}^{\mathrm{MM}}$ are found as

$$\mathcal{P}_\mathcal{G}^{\mathrm{MM}} = \arg\max_{\mathcal{P}_\mathcal{G}} \prod_{n=1}^{N} \min\left(\gamma, d^\mathcal{B}(c^{(n)}, \mathbf{x}^{(n)})\right), \tag{5}$$

---

[5] Note that the parameter learning approach can be applied to more complex structures, e.g. k-trees among the attributes.

where $d^{\mathcal{B}}(c^{(n)}, \mathbf{x}^{(n)})$ is the margin of the nth sample given as

$$d^{\mathcal{B}}(c^{(n)}, \mathbf{x}^{(n)}) = \frac{P^{\mathcal{B}}(c^{(n)}, \mathbf{x}^{(n)})}{\max_{c \neq c^{(n)}} P^{\mathcal{B}}(c, \mathbf{x}^{(n)})}, \tag{6}$$

and where the hinge loss function is denoted as $\min(\gamma, d^{\mathcal{B}}(c^{(n)}, \mathbf{x}^{(n)}))$. The parameter $\gamma > 1$ controls the margin. In this way, the margin *measures* the ratio of the likelihood of the nth sample belonging to the correct class $c^{(n)}$ to the likelihood of belonging to the most likely competing class. The nth sample is correctly classified iff $d^{\mathcal{B}}(c^{(n)}, \mathbf{x}^{(n)}) > 1$ and vice versa.

4   Algorithms for Online Learning of Reduced-Precision Parameters

We start by considering learning ML parameters in Section 4.1 and then move on to learning MM parameters in Section 4.2. We claim that learning using reduced-precision arithmetic is most useful in online settings, i.e. parameters are updated on a per-sample basis. This online learning scenario captures the important case in which initially pre-computed parameters are used and these parameters are updated online as new samples become available, e.g. adaptation of a hearing-aid to a new acoustic environment. In this setting, learning using reduced-precision computations requires specialized algorithms, i.e. gradient-descent (or gradient-ascent) procedures using reduced-precision arithmetic do not perform well. The reason is that the necessary exact projections of the parameters onto the sum-to-one constraints cannot be accurately performed. Another issue is the limited resolution of the learning rate. However, we find this issue less important as the inexact projections.

4.1   Learning Maximum Likelihood Parameters

We consider an online algorithm for learning ML parameters. The ML objective (4) for the offline scenario can be equivalently written as

$$\mathbf{w}^{\text{ML}} = \arg\max_{\mathbf{w}} \sum_{n=1}^{N} \phi(c^{(n)}, \mathbf{x}^{(n)})^T \mathbf{w} \quad \text{s.t.} \sum_{j} \exp(w_{j|\mathbf{h}}^i) = 1, \forall i, j, \mathbf{h}, \tag{7}$$

where optimisation is performed over the log-parameters $\mathbf{w}$. In an online scenario, not all samples are available for learning at once but are available one at a time; the parameters $\mathbf{w}^{\text{ML},t}$ at time-step t are updated according to the gradient of a single sample $(c, \mathbf{x})$ (or, alternatively, a batch of samples) and projected such that they satisfy the sum-to-one constraints, i.e.

$$\mathbf{w}^{\text{ML},t+1} = \Pi \left[\mathbf{w}^{\text{ML},t} + \eta \left(\nabla_{\mathbf{w}} \phi(c, \mathbf{x})^T \mathbf{w} \right) (\mathbf{w}^{\text{ML},t}) \right] \tag{8}$$

$$= \Pi \left[\mathbf{w}^{\text{ML},t} + \eta \phi(c, \mathbf{x}) \right], \tag{9}$$

where η is the learning rate, $\nabla_{\mathbf{w}}(f)(\mathbf{a})$ denotes the gradient of f with respect to $\mathbf{w}$ at $\mathbf{a}$, and $\Pi[\mathbf{w}]$ denotes the ℓ_2-norm projection of the parameter vector $\mathbf{w}$ onto the set of normalized parameter vectors. Note that the gradient has a simple form: it consists only of zeros and ones, where the ones are *indicators* of *active* entries in the CPTs of sample $(c, \mathbf{x})$. Furthermore, assuming normalized parameters at time-step t, the direction of the gradient is always such that the parameters $\mathbf{w}^{\mathrm{ML},t+1}$ are super-normalized. Consequently, after (exact) projection the parameters satisfy the sum-to-one constraints.

We continue by analyzing the effect of using reduced-precision arithmetic on the online learning algorithm. Therefore, we performed the following experiment: Assume that the projection can only be approximately performed. We *simulate* the approximate projection by performing an exact projection and subsequently adding quantization noise (this is similar to reduced-precision analysis in signal processing [10]). We sample the noise from a Gaussian distribution with zero mean and with variance $\sigma^2 = q^2/12$, where $q = 2^{-b_f}$. For the satimage dataset from the UCI repository [1] we construct BNCs with TAN structure. As initial parameters we use rounded ML parameters computed from one tenth of the training data. Then, we present the classifier further samples in an online manner and update the parameters according to (9). During learning, we set the learning rate η to $\eta = \eta_0/\sqrt{1+t}$, where η_0 is some constant (η_0 is tuned by hand such that the test set performance is maximized). The resulting classification performance is shown in Figures 2a and 2b for the exact and the approximate projection, respectively. One can observe, that the algorithm does not properly learn using the approximate projection. Thus, it seems crucial to perform the projections rather accurately. To circumvent the need for accurate projections, we propose a method that avoids computing a projection at all in the following.

Consider again the offline parameter learning case. ML parameters can be computed in closed-form by computing relative frequencies, i.e.

$$\theta^i_{j|\mathbf{h}} = \frac{m^i_{j|\mathbf{h}}}{m^i_{\mathbf{h}}}, \tag{10}$$

where

$$m^i_{j|\mathbf{h}} = \sum_{n=1}^{N} \phi(c^{(n)}, \mathbf{x}^{(n)})^i_{j|\mathbf{h}}, \text{ and } m^i_{\mathbf{h}} = \sum_{j} m^i_{j|\mathbf{h}}. \tag{11}$$

This can be easily extended to online learning. Assume that the counts $m^{i,t}_{j|\mathbf{h}}$ at time t are given and that a sample $(c^t, \mathbf{x}^t)$ is presented to the learning algorithm. Then, the counts are updated according to

$$m^{i,t+1}_{j|\mathbf{h}} = m^{i,t}_{j|\mathbf{h}} + \phi(c^t, \mathbf{x}^t)^i_{j|\mathbf{h}}. \tag{12}$$

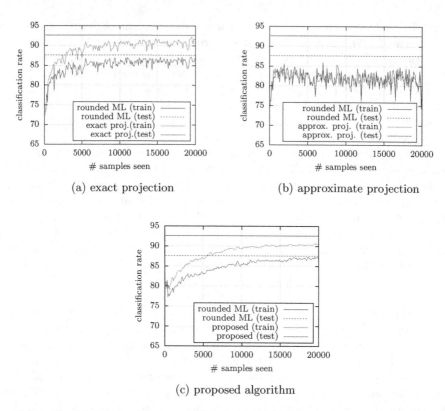

(a) exact projection   (b) approximate projection

(c) proposed algorithm

Fig. 2. Classification performance of BNCs with TAN structure for satimage data in an online learning scenario; (a) Online ML parameter learning with exact projection after each parameter update, (b) online ML parameter learning with approximate projection after each parameter update (see text for details), (c) proposed algorithm for online ML parameter learning.

Exploiting these counts, the logarithm of the ML parameters $\theta_{j|\mathbf{h}}^{i,t}$ at time t can be computed as

$$w_{j|\mathbf{h}}^{i,t} = \log\left(\frac{m_{j|\mathbf{h}}^{i,t}}{m_{\mathbf{h}}^{i,t}}\right), \qquad (13)$$

where similarly to before $m_{\mathbf{h}}^{i,t} = \sum_j m_{j|\mathbf{h}}^{i,t}$. A straightforward approximation of (13) is to (approximately) compute the counts $m_{j|\mathbf{h}}^{i,t}$ and $m_{\mathbf{h}}^{i,t}$, respectively, and to use a lookup table to determine $w_{j|\mathbf{h}}^{i,t}$. The lookup table can be indexed in terms of $m_{j|\mathbf{h}}^{i,t}$ and $m_{\mathbf{h}}^{i,t}$ and stores values for $w_{j|\mathbf{h}}^{i,t}$ in the desired reduced-precision format. To limit the maximum size of the lookup table and the bit-width required for the counters for $m_{j|\mathbf{h}}^{i,t}$ and $m_{\mathbf{h}}^{i,t}$, we assume some maximum integer number M. We pre-compute the lookup table L such that

$$L(i,j) = \left\lceil \frac{\log_2(i/j)}{q} \right\rfloor_R \cdot q, \tag{14}$$

where $[\cdot]_R$ denotes rounding to the closest the integer, q is the quantization interval of the desired fixed-point representation, $\log_2(\cdot)$ denotes the base-2 logarithm, and where i and j are in the range $0, \ldots, M-1$. Given sample $(c^t, \mathbf{x}^t)$, the counts $m_{j|\mathbf{h}}^{i,t+1}$ and $m_{\mathbf{h}}^{i,t+1}$ are computed according to Algorithm 1 from the counts $m_{j|\mathbf{h}}^{i,t}$ and $m_{\mathbf{h}}^{i,t}$. To guarantee that the counts stay in range, the algorithm identifies counters that reach their maximum value, and halfs these counters as well as all other counters corresponding to the same CPTs. This division by 2 can be implemented as a bitwise shift operation.

Algorithm 1. Reduced-precision ML online learning

Require: Old counts $m_{j|\mathbf{h}}^{i,t}$; sample $(c^t, \mathbf{x}^t)$

$\quad m_{j|\mathbf{h}}^{i,t+1} \leftarrow m_{j|\mathbf{h}}^{i,t} + \phi(c^t, \mathbf{x}^t)_{j|\mathbf{h}}^i \quad \forall i, j, \mathbf{h}$ ▷ update counts

$\quad$ **for** $i, j, \mathbf{h}$ **do**

$\quad\quad$ **if** $m_{j|\mathbf{h}}^{i,t+1} = M$ **then**   ▷ maximum value of counter reached?

$\quad\quad\quad m_{j|\mathbf{h}}^{i,t+1} \leftarrow \lfloor m_{j|\mathbf{h}}^{i,t+1}/2 \rfloor \quad \forall j$ ▷ half counters of considered CPT (round down)

$\quad\quad$ **end if**

$\quad$ **end for**

$\quad$ **return** $m_{j|\mathbf{h}}^{i,t+1}$

Initially, we set all counts to zero, i.e. $m_{j|\mathbf{h}}^{i,0} = 0$, respectively. For the cumulative counts, i.e. $m_{\mathbf{h}}^{i,t}$ in (13), we did not limit the number of bits (for real implementations the necessary number of bits for this counter can be computed from the bit-width of the individual counters that are summed up and the graph structure of the considered BNC). Logarithmic parameters $w_{j|\mathbf{h}}^{i,t}$ are computed using the lookup table described above and using Algorithm 2. The classification performance during online learning is shown in Figure 2c. We can observe, that the algorithm behaves pleasant and the limited range of the used counters does not seem to affect classification performance (compared to the classification performance using rounded ML parameters computed using full-precision computations and all training samples). Further experimental results can be found in Section 5.

4.2 Learning Maximum Margin Parameters

In this section, we consider a variant of the MM objective proposed in [12] that balances the MM objective (5) against the ML objective (4), i.e. the objective is to maximize

---

Algorithm 2. Computation of logarithmic probabilities from lookup table

---

Require: Counts $m_{j|h}^{i,t}$ and $m_h^{i,t}$; lookup table L of size $M \times M$
    $\text{div} \leftarrow 0$
    while $m_h^{i,t} \geq M$ **do**   ▷ ensure that index into lookup table is in range
        $m_h^{i,t} \leftarrow \lfloor m_h^{i,t}/2 \rfloor$ ▷ half and round down
        $\text{div} \leftarrow \text{div} + 1$
    end while
    $w_{j|h}^{i,t} \leftarrow L(m_{j,h}^{i,t}, m_h^{i,t}) \quad \forall j$ ▷ get log-probability from lookup table
    while $\text{div} > 0$ and $\forall j: w_{j|h}^{i,t} > (-2^{b_i} + 2^{b_f}) + 1$ **do**   ▷ revise index correction
        $w_{j|h}^{i,t} \leftarrow w_{j|h}^{i,t} - 1 \quad \forall j$
        $\text{div} \leftarrow \text{div} - 1$
    end while
    return $w_{j|h}^{i,t}$

$$\underbrace{\log\left[\prod_{n=1}^{N} P^{\mathcal{B}}(c^{(n)}, \mathbf{x}^{(n)})\right]}_{\text{ML}} + \underbrace{\lambda \log\left[\prod_{n=1}^{N} \min\left(\gamma, \frac{P^{\mathcal{B}}(c^{(n)}, \mathbf{x}^{(n)})}{\max_{c \neq c^{(n)}} P^{\mathcal{B}}(c, \mathbf{x}^{(n)})}\right)\right]}_{\text{MM}}. \quad (15)$$

In this way, generative properties, e.g. the ability to marginalize over missing features, are combined with good discriminative performance. This variant of the MM objective can be easily written in the form

$$\mathbf{w}^{\text{MM}} = \arg\max_{\mathbf{w}} \left[\sum_{n=1}^{N} \boldsymbol{\phi}(c^{(n)}, \mathbf{x}^{(n)})^T \mathbf{w} + \right. \quad (16)$$

$$\left. \lambda \sum_{n=1}^{N} \min\left(\gamma, \min_{c \neq c^{(n)}} \left[(\boldsymbol{\phi}(c^{(n)}, \mathbf{x}^{(n)}) - \boldsymbol{\phi}(c, \mathbf{x}^{(n)}))^T \mathbf{w}\right]\right)\right)$$

and, for simplicity, we will refer to this modified objective as the MM objective. Note that there are implicit sum-to-one constraints in problem (16), i.e. any feasible solution $\mathbf{w}$ must satisfy $\sum_j \exp(w_{j|h}^i) = 1$ for all i, j, h. In the online learning case, given sample $(c, \mathbf{x})$, the parameters $\mathbf{w}^{\text{MM},t+1}$ at time $t + 1$ are computed from the parameters $\mathbf{w}^{\text{MM},t}$ at time t as

$$\mathbf{w}^{\text{MM},t+1} = \Pi\left[\mathbf{w}^{\text{MM},t} + \eta\boldsymbol{\phi}(c, \mathbf{x}) + \eta\lambda\mathbf{g}(c, \mathbf{x})\right], \quad (17)$$

where

$$\mathbf{g}(c, \mathbf{x}) = \begin{cases} 0 & \min_{c' \neq c}\left[(\boldsymbol{\phi}(c, \mathbf{x}) - \boldsymbol{\phi}(c', \mathbf{x}))^T \mathbf{w}\right] \geq \gamma, \\ \boldsymbol{\phi}(c, \mathbf{x}) - \boldsymbol{\phi}(c', \mathbf{x}) & \text{o.w., } c' = \arg\min_{c'}\left[(\boldsymbol{\phi}(c, \mathbf{x}) - \boldsymbol{\phi}(c', \mathbf{x}))^T \mathbf{w}\right] \end{cases} \quad (18)$$

and where similar as before $\Pi[\mathbf{w}]$ denotes the projection.

For learning MM parameters, a similar observation with respect to the accuracy of the projection can be made as for ML parameters. But we cannot proceed exactly as in the case of learning ML parameters because we cannot compute MM parameters in closed-form. As in the ML parameter learning case, the gradient for the parameter update has a rather simple form, but the projection to satisfy the sum-to-one constraints is difficult to compute. Therefore, for online MM parameter learning, we propose Algorithm 3 that is similar to Algorithm 1 in Section 4.1, i.e. we avoid to compute the projection explicitly. From the counts computed by the algorithm, log-probabilities can be computed using Algorithm 2. Note that the proposed algorithm does not exactly optimize (16) but a, not explicitly defined, surrogate. The idea behind the algorithm is (1) to optimize the likelihood term in (16) as in the algorithm for ML parameter learning, and (2) to optimize the margin term by increasing the likelihood for the correct class and simultaneously decreasing the likelihood for the strongest competitor class. Note that the idea of optimizing the margin term as explained above is similar in spirit to that of discriminative frequency estimates [19]. However, discriminative frequency estimates do not optimize a margin term but a term more closely related to the class-conditional likelihood.

5   Experiments

5.1   Datasets

In our experiments, we considered the following datasets.

1. **UCI data [1].** This is in fact a large collection of datasets, with small to medium number of samples. Features are discretized as needed using the algorithm proposed in [3]. If not stated otherwise, in case of the datasets *chess, letter, mofn-3-7-10, segment, shuttle-small, waveform-21, abalone, adult, car, mushroom, nursery, and spambase*, a test set was used to estimate the accuracy of the classifiers. For all other datasets, classification accuracy was estimated by 5-fold cross-validation. Information on the number of samples, classes and features for each dataset can be found in [1].

2. **USPS data [6].** This data set contains 11000 handwritten digit images from zip codes of mail envelopes. The data set is split into 8000 images for training and 3000 for testing. Each digit is represented as a 16×16 greyscale image. These greyscale values are discriminatively quantized [3] and each pixel is considered as feature.

3. **MNIST Data [8].** This dataset contains 70000 samples of handwritten digits. In the standard setting, 60000 samples are used for training and 10000 for testing. The digits represented by grey-level images were down-sampled by a factor of two resulting in a resolution of 16×16 pixels, i.e. 196 features.

5.2   Results

We performed experiments using $M = 1024$, i.e. we used counters with 10 bits $(b_i + b_f = 10)$. The splitting of the available bits into integer bits and fractional bits was set using 10-fold cross-validation. Experimental results for BNCs

---

Algorithm 3. Reduced-precision MM online learning

Require: Old counts $m_{j|\mathbf{h}}^{i,t}$; sample $(c^t, \mathbf{x}^t)$; hyper-parameters $\gamma, \lambda \in \mathbb{N}_+$ for MM formulation

$m_{j|\mathbf{h}}^{i,t+1} \leftarrow m_{j|\mathbf{h}}^{i,t} + \phi(c^t, \mathbf{x}^t)_{j|\mathbf{h}}^i \quad \forall i, j, \mathbf{h}$ $\triangleright$ update counts (likelihood term)

for $i, j, \mathbf{h}$ **do** $\triangleright$ ensure that parameters stay in range

     if $m_{j|\mathbf{h}}^{i,t+1} = M$ **then**

         $m_{j|\mathbf{h}}^{i,t+1} \leftarrow \lfloor m_{j|\mathbf{h}}^{i,t+1}/2 \rfloor \quad \forall j$

     end if

end for

$c' \leftarrow$ strongest competitor of class c for features $\mathbf{x}$

if $\left[(\phi(c^t, \mathbf{x}^{(n)}) - \phi(c', \mathbf{x}^{(n)}))^T \mathbf{w} < \gamma \right]$ **then**

     $m_{j|\mathbf{h}}^{i,t+1} \leftarrow m_{j|\mathbf{h}}^{i,t} \quad \forall i, j, \mathbf{h}$

     for $k = 1, \dots, \lambda$ **do** $\triangleright$ Add-up gradient in λ steps

         $m_{j|\mathbf{h}}^{i,t+1} \leftarrow m_{j|\mathbf{h}}^{i,t+1} + \phi(c^t, \mathbf{x}^t)_{j|\mathbf{h}}^i \quad \forall i, j, \mathbf{h}$ $\triangleright$ update counts (margin term)

         $m_{j|\mathbf{h}}^{i,t+1} \leftarrow m_{j|\mathbf{h}}^{i,t+1} - \phi(c', \mathbf{x}^t)_{j|\mathbf{h}}^i \quad \forall i, j, \mathbf{h}$ $\triangleright$ update counts (margin term)

         for $i, j, \mathbf{h}$ **do** $\triangleright$ ensure that parameters stay in range

            if $m_{j|\mathbf{h}}^{i,t+1} = 0$ **then**

                $m_{j|\mathbf{h}}^{i,t+1} \leftarrow m_{j|\mathbf{h}}^{i,t+1} + 1 \quad \forall j$

            end if

            if $m_{j|\mathbf{h}}^{i,t+1} = M$ **then**

                $m_{j|\mathbf{h}}^{i,t+1} \leftarrow \lfloor m_{j|\mathbf{h}}^{i,t+1}/2 \rfloor \quad \forall j$

            end if

         end for

     end for

end if

return $m_{j|\mathbf{h}}^{i,t+1}$

---

with NB and TAN structures are shown in Table 1 for the datasets described above. All samples from the training set were presented to the proposed algorithm twenty times in random order. The absolute reduction in classification rate (CR) compared to the *exact* CR, i.e. using BNCs with the optimal double-precision parameters, for the considered datasets is, with few exceptions, relatively small. Thus the proposed reduced-precision computation scheme seems to be sufficiently accurate to yield good classification performance while employing only range-limited counters and a lookup table of size $M \times M$. Clearly, the performance of the proposed method can be improved by using larger and more accurate lookup tables and counters with larger bit-width.

For discriminative parameter learning, we set the hyper-parameters $\lambda \in \{0, 1, 2, 4, 8, 16\}$ and $\gamma \in \{0.25, 0.5, 1, 2, 4, 8.\}$ using 10-fold cross-validation. For this setup, we observed the classification performance summarized in Table 1. While the results are not as good as those of the exact MM solution, in terms of the absolute reduction in CR, we can clearly observe an improvement in classification performance using the proposed MM parameter learning method over the proposed ML parameter learning method for many datasets. The performance of BNCs using

Table 1. Classification performance. CRs using ML/MM parameters according to (10)/(16) in double-precision are denoted as *ML exact/MM exact*. CRs using reduced-precision ML/MM parameters computed according to Algorithm 1/Algorithm 3 using only reduced-precision arithmetic are denoted as *ML prop./MM prop.*; *ML abs./MM abs.* denote the absolute reduction in CR for double-precision ML/MM parameters to reduced-precision ML/MM parameters.

Dataset	Structure	ML – CR [%]			MM – CR [%]		
		exact	prop.	abs.	exact	prop.	abs.
USPS	NB	86.89	86.34	0.55	93.91	93.17	0.74
	TAN	91.39	90.05	1.34	93.01	93.50	−0.49
MNIST	NB	82.88	80.61	2.26	93.11	93.00	0.11
	TAN	90.49	87.92	2.57	93.49	93.83	−0.34
australian	NB	85.92	85.48	0.44	87.24	85.63	1.61
	TAN	81.97	84.46	−2.49	84.76	83.58	1.18
breast	NB	97.63	97.48	0.15	97.04	97.63	−0.59
	TAN	95.85	96.15	−0.30	96.00	94.52	1.48
chess	NB	87.45	86.20	1.25	97.68	94.32	3.36
	TAN	92.19	92.13	0.06	97.99	96.27	1.73
cleve	NB	82.87	83.55	−0.68	82.53	80.84	1.69
	TAN	79.09	80.47	−1.37	80.79	75.69	5.10
corral	NB	89.16	89.22	−0.07	93.36	93.36	0.00
	TAN	97.53	94.96	2.57	100.00	99.20	0.80
crx	NB	86.84	86.22	0.62	86.06	86.68	−0.62
	TAN	83.73	84.04	−0.31	84.20	83.58	0.62
diabetes	NB	73.96	72.65	1.31	74.87	75.01	−0.14
	TAN	73.83	73.44	0.39	74.35	71.73	2.62
flare	NB	77.16	75.81	1.34	83.11	83.97	−0.86
	TAN	83.59	79.46	4.13	83.30	83.20	0.10
german	NB	74.50	72.90	1.60	75.30	73.80	1.50
	TAN	72.60	71.80	0.80	72.60	72.10	0.50
glass	NB	71.16	71.66	−0.50	70.61	71.08	−0.47
	TAN	71.11	69.58	1.53	72.61	69.55	3.05
heart	NB	81.85	82.96	−1.11	83.33	84.44	−1.11
	TAN	81.48	81.11	0.37	81.48	81.48	0.00
hepatitis	NB	89.83	89.83	0.00	92.33	88.67	3.67
	TAN	84.83	87.33	−2.50	86.17	88.58	−2.42
letter	NB	74.95	74.41	0.54	85.79	81.50	4.30
	TAN	86.26	85.93	0.33	88.57	88.43	0.14
lymphography	NB	84.23	85.71	−1.48	82.80	87.31	−4.51
	TAN	82.20	82.86	−0.66	76.92	80.66	−3.74
nursery	NB	89.97	89.63	0.35	93.03	93.05	−0.02
	TAN	92.87	92.87	0.00	98.68	98.12	0.56
satimage	NB	81.56	82.02	−0.45	88.41	86.96	1.45
	TAN	85.85	86.40	−0.55	86.98	87.44	−0.47
segment	NB	92.68	91.90	0.78	95.37	93.85	1.52
	TAN	94.85	94.89	−0.04	95.76	95.63	0.13
shuttle	NB	99.66	99.10	0.56	99.95	99.86	0.09
	TAN	99.88	99.71	0.17	99.93	99.87	0.06
soybean-large	NB	93.35	92.80	0.56	91.50	92.05	−0.55
	TAN	91.14	89.12	2.02	91.87	92.61	−0.74
spambase	NB	90.03	89.88	0.15	94.08	93.19	0.89
	TAN	92.97	92.79	0.17	94.03	93.73	0.31
vehicle	NB	61.57	61.93	−0.36	67.95	69.16	−1.21
	TAN	71.09	68.91	2.18	69.88	69.16	0.72
vote	NB	90.16	90.63	−0.47	94.61	94.61	0.00
	TAN	94.61	94.60	0.01	95.31	94.60	0.71
waveform-21	NB	81.14	81.18	−0.04	85.14	84.16	0.98
	TAN	82.52	82.20	0.32	83.48	83.94	−0.46

the optimal double-precision parameters is in many cases not significantly better. Note that the hyper-parameters used for determining double-precision parameters are different than those used for determining reduced-precision parameters, i.e. a larger range of values is used (details are provided in [12]). The larger range of values cannot be used in case of reduced-precision parameters because of the limited parameter resolution.

6   Discussions

We proposed online algorithms for learning BNCs with reduced-precision fixed-point parameters using reduced-precision computations only. This facilitates the utilization of BNCs in computationally constrained platforms, e.g. embedded- and ambient-systems, as well as power-aware systems. The algorithms differ from naive implementations of conventional algorithms by avoiding error-prone parameter projections commonly used in gradient ascent/descent algorithms. In experiments, we demonstrated that our algorithms yield parameters that achieve classification performances close to that of optimal double-precision parameters for many of the investigated datasets.

Our algorithms have similarities with a very simple method for learning discriminative parameters of BNCs known as *discriminative frequency estimates* [19]. According to this method, parameters are estimated using a perceptron-like algorithm, where parameters are updated by the prediction loss, i.e. the difference of the class posterior of the correct class (which is assumed to be 1 for the data in the training set) and the class posterior according to the model using the current parameters.

References

1. Bache, K., Lichman, M.: UCI Machine Learning Repository (2013). http://archive. ics.uci.edu/ml
2. Courbariaux, M., Bengio, Y., David, J.: Low precision arithmetic for deep learning. CoRR abs/1412.7024 (2014). http://arxiv.org/abs/1412.7024
3. Fayyad, U.M., Irani, K.B.: Multi-Interval discretization of continuous-valued attributes for classification learning. In: International Conference on Artificial Intelligence (IJCAI), pp. 1022–1029 (2003)
4. Friedman, N., Geiger, D., Goldszmidt, M.: Bayesian Network Classifiers. Machine Learning **29**, 131–163 (1997)
5. Guo, Y., Wilkinson, D., Schuurmans, D.: Maximum Margin Bayesian Networks. In: Uncertainty in Artificial Intelligence (UAI), pp. 233–242 (2005)
6. Hull, J.J.: A Database for Handwritten Text Recognition Research. IEEE Transactions on Pattern Analysis and Machine Intelligence (TPAMI) **16**(5), 550–554 (1994)
7. Koller, D., Friedman, N.: Probabilistic Graphical Models: Principles and Techniques. MIT Press (2009)
8. LeCun, Y., Bottou, L., Bengio, Y., Haffner, P.: Gradient-based Learning Applied to Document Recognition. Proceedings of the IEEE **86**(11), 2278–2324 (1998)

9. Lee, D.U., Gaffar, A.A., Cheung, R.C.C., Mencer, O., Luk, W., Constantinides, G.A.: Accuracy-Guaranteed Bit-Width Optimization. IEEE Transactions on Computer-Aided Design of Integrated Circuits and Systems **25**(10), 1990–2000 (2006)
10. Oppenheim, A.V., Schafer, R.W., Buck, J.R.: Discrete-time Signal Processing, 2nd edn., Prentice-Hall Inc. (1999)
11. Pearl, J.: Probabilistic Reasoning in Intelligent Systems: Networks of Plausible Inference. Morgan Kaufmann Publishers Inc. (1988)
12. Peharz, R., Tschiatschek, S., Pernkopf, F.: The most generative maximum margin bayesian networks. In: International Conference on Machine Learning (ICML), vol. 28, pp. 235–243 (2013)
13. Pernkopf, F., Peharz, R., Tschiatschek, S.: Introduction to Probabilistic Graphical Models, vol. 1, chap. 18, pp. 989–1064. Elsevier (2014)
14. Pernkopf, F., Wohlmayr, M., Mücke, M.: Maximum margin structure learning of bayesian network classifiers. In: International Conference on Acoustics, Speech and Signal Processing (ICASSP), pp. 2076–2079 (2011)
15. Pernkopf, F., Wohlmayr, M., Tschiatschek, S.: Maximum Margin Bayesian Network Classifiers. IEEE Transactions on Pattern Analysis and Machine Intelligence (TPAMI) **34**(3), 521–531 (2012)
16. Piatkowski, N., Sangkyun, L., Morik, K.: The integer approximation of undirected graphical models. In: International Conference on Pattern Recognition Applications and Methods (ICPRAM) (2014)
17. Roos, T., Wettig, H., Grünwald, P., Myllymäki, P., Tirri, H.: On Discriminative Bayesian Network Classifiers and Logistic Regression. Journal of Machine Learning Research **59**(3), 267–296 (2005)
18. Soudry, D., Hubara, I., Meir, R.: Expectation backpropagation: parameter-free training of multilayer neural networks with continuous or discrete weights. In: Advances in Neural Information Processing Systems, pp. 963–971 (2014)
19. Su, J., Zhang, H., Ling, C.X., Matwin, S.: Discriminative parameter learning for bayesian networks. In: International Conference on Machine Learning (ICML), pp. 1016–1023. ACM (2008)
20. Tong, J.Y.F., Nagle, D., Rutenbar, R.A.: Reducing Power by Optimizing the Necessary Precision/Range of Floating-point Arithmetic. IEEE Transactions on Very Large Scale Integration Systems **8**(3), 273–285 (2000)
21. Tschiatschek, S., Cancino Chacón, C.E., Pernkopf, F.: Bounds for bayesian network classifiers with reduced precision parameters. In: International Conference on Acoustics, Speech and Signal Processing (ICASSP), pp. 3357–3361 (2013)
22. Tschiatschek, S., Paul, K., Pernkopf, F.: Integer bayesian network classifiers. In: Calders, T., Esposito, F., Hüllermeier, E., Meo, R. (eds.) ECML PKDD 2014, Part III. LNCS, vol. 8726, pp. 209–224. Springer, Heidelberg (2014)
23. Tschiatschek, S., Pernkopf, F.: On Reduced Precision Bayesian Network Classifiers. IEEE Transactions on Pattern Analysis and Machine Intelligence (TPAMI) (to be published)
24. Tschiatschek, S., Reinprecht, P., Mücke, M., Pernkopf, F.: Bayesian network classifiers with reduced precision parameters. In: Flach, P.A., De Bie, T., Cristianini, N. (eds.) ECML PKDD 2012, Part I. LNCS, vol. 7523, pp. 74–89. Springer, Heidelberg (2012)

Predicting Unseen Labels Using Label Hierarchies in Large-Scale Multi-label Learning

Jinseok Nam[1,2,3]([✉]), Eneldo Loza Mencía[2,3],
Hyunwoo J. Kim[4], and Johannes Fürnkranz[2,3]

[1] Knowledge Discovery in Scientific Literature, TU Darmstadt, Darmstadt, Germany
nam@cs.tu-darmstadt.de
[2] Research Training Group AIPHES, TU Darmstadt, Darmstadt, Germany
[3] Knowledge Engineering Group, TU Darmstadt, Darmstadt, Germany
[4] Department of Computer Sciences,
University of Wisconsin-Madison, Madison, USA

Abstract. An important problem in multi-label classification is to capture label patterns or underlying structures that have an impact on such patterns. One way of learning underlying structures over labels is to project both instances and labels into the same space where an instance and its relevant labels tend to have similar representations. In this paper, we present a novel method to learn a joint space of instances and labels by leveraging a hierarchy of labels. We also present an efficient method for pretraining vector representations of labels, namely label embeddings, from large amounts of label co-occurrence patterns and hierarchical structures of labels. This approach also allows us to make predictions on labels that have not been seen during training. We empirically show that the use of pretrained label embeddings allows us to obtain higher accuracies on unseen labels even when the number of labels are quite large. Our experimental results also demonstrate qualitatively that the proposed method is able to learn regularities among labels by exploiting a label hierarchy as well as label co-occurrences.

1 Introduction

Multi-label classification is an area of machine learning which aims to learn a function that maps instances to a label space. In contrast to multiclass classification, each instance is assumed to be associated with more than one label. One of the goals in multi-label classification is to model the underlying structure of the label space because in many such problems, the occurrences of labels are not independent of each other.

Recent developments in multi-label classification can be roughly divided into two bodies of research. One is to build a classifier in favor of statistical dependencies between labels, and the other is devoted to making use of prior information over the label space. In the former area, many attempts have been made to exploit label patterns [6,9,24]. As the number of possible configurations of labels grows exponentially with respect to the number of labels, it is required for multi-label classifiers to handle many labels efficiently [4] or to reduce the dimensionality of

© Springer International Publishing Switzerland 2015
A. Appice et al. (Eds.): ECML PKDD 2015, Part I, LNAI 9284, pp. 102–118, 2015.
DOI: 10.1007/978-3-319-23528-8_7

a label space by exploiting properties of label structures such as sparsity [17] and co-occurrence patterns [7]. Label space dimensionality reduction (LSDR) methods allow to make use of latent information on a label space as well as to reduce computational cost. Another way of exploiting information on a label space is to use its underlying structures as a prior. Many methods have been developed to use hierarchical output structures in machine learning [27]. In particular, several researchers have looked into utilizing the hierarchical structure of the label space for improved predictions in multi-label classification [26, 30, 32].

Although extensive research has been devoted to techniques for utilizing implicitly or explicitly given label structures, there remain the scalability issues of previous approaches in terms of both the number of labels and documents in large feature spaces. Consider a very large collection of scientific documents covering a wide range of research interests. In an emerging research area, it can be expected that the number of publications per year grows rapidly. Moreover, new topics will emerge, so that the set of indexing terms, which has initially been provided by domain experts or authors to describe publications with few words for potential readers, will grow as well.

Interestingly, similar problems have been faced recently in a different domain, namely *representation learning* [2]. In language modeling, for instance, a word is traditionally represented by a K-dimensional vector where K is the number of unique words, typically hundreds of thousands or several millions. Clearly, it is desirable to reduce this dimensionality to much smaller values $d \ll K$. This can, e.g., be achieved with a simple log-linear model [21], which can efficiently compute a so-called *word embedding*, i.e., a lower-dimensional vector representations for words. Another example for representation learning is a technique for learning a joint embedding space of instances and labels [31]. This approach maximizes the similarity between vector representations of instances and relevant labels while projecting them into the same space.

Inspired by the log-linear model and the joint space embedding, we address large-scale multi-label classification problems, in which both hierarchical label structures are given *a priori* as well as label patterns occur in the training data. The mapping functions in the joint space embedding method can be used to rank labels for a given instance, so that relevant labels are placed at the top of the ranking. In other words, the quality of such a ranking depends on the mapping functions. As mentioned, two types of information on label spaces are expected to help us to train better joint embedding spaces, so that the performance on unseen data can be improved. We focus on exploiting such information so as to learn a mapping function projecting labels into the joint space. The vector representations of labels by using this function will be referred to as *label embeddings*. While *label embeddings* are usually initialized randomly, it will be beneficial to learn the joint space embedding method taking label hierarchies into consideration when label structures are known. To this end, we adopt the above-mentioned log-linear model which has been successfully used to learn *word embeddings*.

Learning *word embeddings* relies fundamentally on the use of the context information, that is, a fixed number of words surrounding that word in a sentence or a document. In order to adapt this idea to learning *label embeddings*, we need to define context information in a label space, where, unlike in textual documents, there is no sequence information which can be used to define the context of words. We use, instead, *pairwise* relationships in label hierarchies and in label co-occurrence patterns.

There are two major contributions of this work: 1) We build efficient multi-label classifiers which employ label hierarchies so as to predict unseen labels. 2) We provide a novel method to efficiently learn label representations from hierarchical structures over labels as well as their co-occurrence patterns.

2   Multi-label Classification

In *multi-label classification*, assuming that we are given a set of training examples $\mathcal{D} = \{(\mathbf{x}_n, \mathcal{Y}_n)\}_{n=1}^N$, our goal is to learn a classification function $f : \mathbf{x} \to \mathcal{Y}$ which maps an instance $\mathbf{x}$ to a set of *relevant* labels $\mathcal{Y} \subseteq \{1, 2, \cdots, L\}$. All other labels $\bar{\mathcal{Y}} = \{1, 2, \cdots, L\} \setminus \mathcal{Y}$ are called *irrelevant*. Often, it is sufficient, or even required, to obtain a list of labels ordered according to some relevance scoring functions.

In *hierarchical multi-label classification* (HMLC) labels are explicitly organized in a tree usually denoting a *is-a* or *composed-of* relation. Several approaches to HMLC have been proposed which replicate this structure with a hierarchy of classifiers which predict the paths to the correct labels [5,30,32]. Although there is evidence that exploiting the hierarchical structure in this way has advantages over the flat approach [3,5,30], some authors unexpectedly found that ignoring the hierarchical structure gives better results. For example, in [32] it is claimed that if a strong flat classification algorithm is used the lead vanishes. Similarly, in [30] it was found that learning a single decision tree which predicts probability distributions at the leaves outperforms a hierarchy of decision trees. One of the reasons may be that hierarchical relations in the output space are often not in accordance with the input space, as claimed by [15] and [32]. Our proposed approach aims at overcoming this problem as it learns an embedding space where similarities in the input, output and label hierarchies are jointly respected.

3   Model Description

3.1   Joint Space Embeddings

Weston et al. [31] proposed an efficient online method to learn ranking functions in a joint space of instances and labels, namely *Wsabie*. Under the assumption that instances which have similar representation in a feature space tend to be associated with similar label sets, we find joint spaces of both instances and labels where the relevant labels for an instance can be separated from the irrelevant ones with high probability.

Formally, consider an instance $\mathbf{x}$ of dimension D and a set of labels $\mathcal{Y}$ associated with $\mathbf{x}$. Let $\phi(\mathbf{x}) = \mathbf{W}\mathbf{x}$ denote a linear function which projects the original feature representations of an instance $\mathbf{x}$ to a d-dimensional joint space, where $\mathbf{W} \in \mathbb{R}^{d \times D}$ is a transformation matrix. Similarly, let $\mathbf{U}$ be a $d \times L$ matrix that maps labels into the same joint d-dimensional space. A label $i \in \mathcal{Y}$ can then be represented as a d-dimensional vector $\mathbf{u}_i$, which is the i-th column vector of $\mathbf{U}$. We will refer to the matrix $\mathbf{U} = [\mathbf{u}_1, \mathbf{u}_2, \cdots, \mathbf{u}_L]$ as *label embeddings*. The objective function is given by

$$\mathcal{L}\left(\mathbf{\Theta}_F; \mathcal{D}\right) = \sum_{n=1}^{N} \sum_{i \in \mathcal{Y}_n} \sum_{j \in \bar{\mathcal{Y}}_n} h(r_i(\mathbf{x}_n)) \, \ell\left(\mathbf{x}_n, y_i, y_j\right) \tag{1}$$

with the *pairwise hinge loss* function $\ell\left(\mathbf{x}_n, y_i, y_j\right) = \left[m_a - \mathbf{u}_i^T \phi(\mathbf{x}_n) + \mathbf{u}_j^T \phi(\mathbf{x}_n)\right]_+$ where $r_i(\cdot)$ denotes the rank of label i for a given instance $\mathbf{x}_n$, $h(\cdot)$ is a function that maps this rank to a real number (to be introduced shortly in more detail), $\bar{\mathcal{Y}}_n$ is the complement of $\mathcal{Y}_n$, $[x]_+$ is defined as x if $x > 0$ and 0 otherwise, $\mathbf{\Theta}_F = \{\mathbf{W}, \mathbf{U}\}$ are model parameters, and m_a is a real-valued parameter, namely the *margin*. The relevance scores $\mathbf{s}(\mathbf{x}) = [s_1(\mathbf{x}), s_2(\mathbf{x}), \cdots, s_L(\mathbf{x})]$ of labels for a given instance $\mathbf{x}$ can be computed as $s_i(\mathbf{x}) = \mathbf{u}_i^T \phi(\mathbf{x}) \in \mathbb{R}$. Then, the rank of label i with respect to an instance $\mathbf{x}$ can be determined based on the relevance scores $r_i(\mathbf{x}) = \sum_{j \in \bar{\mathcal{y}}, j \neq i} \left[m_a - s_i(\mathbf{x}) + s_j(\mathbf{x})\right]_+$. It is prohibitively expensive to compute such rankings exactly when L is large. We use instead its approximation to update parameters $\mathbf{\Theta}_F$ given by $r_i(\mathbf{x}) \approx \lfloor \frac{L - |\mathcal{Y}|}{T_i} \rfloor$ where $\lfloor \cdot \rfloor$ denotes the floor function and T_i is the number of trials to sample an index j yielding incorrect ranking against label i such that $m_a - s_i(\mathbf{x}) + s_j(\mathbf{x}) > 0$ during stochastic parameter update steps. Having an approximate rank $r_i(\mathbf{x})$, we can obtain a weighted ranking function $h(r_i(\mathbf{x})) = \sum_{k=1}^{r_i(\mathbf{x})} \frac{1}{k}$, which is shown to be an effective way of optimizing precision at the top of rankings.

3.2   Learning with Hierarchical Structures Over Labels

Wsabie is trained in a way that the margin of similarity scores between positive associations $\mathbf{u}_p^T \phi(\mathbf{x})$ and negative associations $\mathbf{u}_n^T \phi(\mathbf{x})$ is maximized, where $\mathbf{u}_p$ and $\mathbf{u}_n$ denote the embeddings of relevant and irrelevant labels, respectively, for an instance $\mathbf{x}$. In practice, this approach works well if label patterns of test instances appear in training label patterns. If there are few or no training instances for some labels, the model may fail to make predictions accurately on test instances associated with those labels. In such cases, a joint space learning method could benefit from label hierarchies. In this section, we introduce a simple and efficient joint space learning method by adding a regularization term which employs label hierarchies, hereafter referred to as *Wsabie$_H$*.

Notations. Consider multi-label problems where label hierarchies exist. Label graphs are a natural way to represent such hierarchical structures. Because it is possible for a label to have more than one parent node, we represent a hierarchy of labels in a directed acyclic graph (DAG). Consider a graph $\mathcal{G} = \{V, E\}$ where V denotes a set of nodes and E represent a set of connections between nodes. A

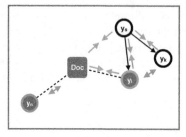

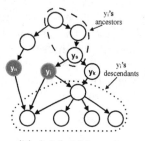

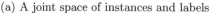

(a) A joint space of instances and labels   (b) A label hierarchy

Fig. 1. An illustrative example of our proposed method. A label y_i (green circle) indicates a relevant label for a document (rectangle) while y_n (red circle) is one of the irrelevant labels. In the joint space, we learn representations for the relevant label, its ancestor y_s, and the document to be similar whereas the distance between the document and the irrelevant label is maximized. Also, the parent label, y_s, and its children are forced to be similar while sibling labels of y_i, i.e. y_k, are kept away from each other.

node $u \in V$ corresponds to a label. A directed edge from a node u to a node v is denoted as $e_{u,v}$, in which case we say that u is a parent of v and v is a child of u. The set of all parents / children of v is denoted with $\mathcal{S}_P(v)$ / $\mathcal{S}_C(v)$. If there exists a directed path from u to v, u is an ancestor of v and v is a descendant of u, the set of all ancestors / descendants is denoted as $\mathcal{S}_A$ / $\mathcal{S}_D(u)$.

Label structures as regularizers. As an example, let us consider three labels, "computer science" (CS), "artificial intelligence" (AI), and "software engineering" (SE). The label CS can be viewed as a parent label of AI and SE. Given a paper dealing with problems in artificial intelligence and having AI as a label, we wish to learn a joint embedding model in a way that it is also highly probable to predict CS as a relevant label. Following our hypothesis in label spaces, even though we have no paper of software engineering, a label hierarchy allows us to make *reasonable* predictions on such a label by representing label SE close to label CS in a vector space. In order to prevent the model from converging to trivial solutions that representations of all three labels are identical, it is desired that sibling labels such as AI and SE in the hierarchy are well separated from each other in a joint embedding space. For an illustration of our method, see Fig. 1.

Formally, we can achieve this by defining a regularization term Ω, which takes into account the hierarchical label structure

$$\Omega(\mathbf{\Theta}_H) = \sum_{n=1}^{N} \frac{1}{\mathcal{Z}_A} \sum_{i \in \mathcal{Y}_n} \sum_{s \in \mathcal{S}_A(i)} -\log p(y_s | y_i, \mathbf{x}_n)$$
$$+ \sum_{l=1}^{L} \sum_{q \in \mathcal{S}_P(l)} \sum_{\substack{k \in \mathcal{S}_C(q) \\ k \neq l}} h(r_q(\mathbf{u}_l)) \left[m_b - \mathbf{u}_q^T \mathbf{u}_l + \mathbf{u}_k^T \mathbf{u}_l \right]_+ \qquad (2)$$

where m_b is the margin, $\mathcal{Z}_A = |\mathcal{Y}_n||\mathcal{S}_A(i)|$, and $p(y_s|y_i, \mathbf{x}_n)$ denotes the probability of predicting an ancestor label s of a label i given i and an instance $\mathbf{x}_n$ for which i is relevant. More specifically, the probability $p(y_s|y_i, \mathbf{x}_n)$ can be defined as

$$p(y_s|y_i, \mathbf{x}_n) = \frac{\exp(\mathbf{u}_s^T \hat{\mathbf{u}}_i^{(n)})}{\sum_{v \in L} \exp(\mathbf{u}_v^T \hat{\mathbf{u}}_i^{(n)})}, \tag{3}$$

where $\hat{\mathbf{u}}_i^{(n)} = \frac{1}{2} (\mathbf{u}_i + \phi(\mathbf{x}_n))$ is the averaged-representation of a label i and the n-th instance in a joint space. Intuitively, this regularizer forces labels, which share the same parent label, to have similar vector representations as much as possible while keeping them separated from each other. Moreover, an instance $\mathbf{x}$ has the potential to make good predictions on some labels even though they do not appear in the training set only if their descendants are associated with training instances.

Adding Ω to Eq. 1 results in the objective function of $Wsabie_H$

$$\mathcal{L}(\mathbf{\Theta}_H; \mathcal{D}) = \sum_{n=1}^{N} \sum_{i \in \mathcal{Y}_n} \sum_{j \in \bar{\mathcal{Y}}_n} h(r_i(\mathbf{x}_n)) \, \ell(\mathbf{x}_n, y_i, y_j) + \lambda \Omega(\mathbf{\Theta}_H) \tag{4}$$

where λ is a control parameter of the regularization term. If we set $\lambda = 0$, then the above objective function is equivalent to the objective function of $Wsabie$ in Eq 1.

3.3 Efficient Gradients Computation

Due to the high computational cost for computing gradients for the softmax function in Eq. 3, we use *hierarchical softmax* [22,23] which reduces the gradient computing cost from $\mathcal{O}(L)$ to $\mathcal{O}(\log L)$. Similar to [21], in order to make use of the *hierarchical softmax*, a binary tree is constructed by Huffman coding, which yields binary codes with variable length to each label according to $|\mathcal{S}_D(\cdot)|$. Note that by definition of the Huffman coding all L labels correspond to leaf nodes in a binary tree, called the Huffman tree. Instead of computing L outputs, the *hierarchical softmax* computes a probability of $\lceil \log L \rceil$ binary decisions over a path from the root node of the tree to the leaf node corresponding to a target label, say, y_j in Eq. 3.

More specifically, let $C(y)$ be a codeword of a label y by the Huffman coding, where each bit can be either 0 or 1, and $I(C(y))$ be the number of bits in the codeword for that label. $C_l(y)$ is the l-th bit in y's codeword. Unlike for *softmax*, for computing the *hierarchical softmax* we use the output label representations $\mathbf{U}'$ as vector representations for inner nodes in the Huffman tree. The *hierarchical softmax* is then given by

$$p(y_j|y_i) = \prod_{l=1}^{I(C(y_j))} \sigma([\![C_l(y_j) = 0]\!] \, \mathbf{u}'^T_{n(l, y_j)} \mathbf{u}_i) \tag{5}$$

where $\sigma(\cdot)$ is the logistic function, $[\![\cdot]\!]$ is 1 if its argument is true and -1 otherwise, and $\mathbf{u}'_{n(l,y_j)}$ is a vector representation for the l-th node in the path from the root node to the node corresponding to the label y_j in the Huffman tree. While L inner products are required to compute the normalization term in Eq. 3, the *hierarchical softmax* needs $I(C(\cdot))$ computations. Hence, the *hierarchical softmax* allows substantial improvements in computing gradients if $\mathbb{E}\left[I(C(\cdot))\right] \ll L$.

3.4 Label Ranking to Binary Predictions

It is often sufficient in practice to just predict a ranking of labels instead of a bipartition of labels, especially in settings where the learning system is comprehended as supportive [8]. On the other hand, there are several ways to convert ranking results into a bipartition. Basically all of them split the ranking at a certain position depending on a predetermined or predicted threshold or amount of relevant labels.

Instead of experimenting with different threshold techniques, we took a pragmatic stance, and simply assume that there is an oracle which tells us the actual number of relevant labels for an unseen instance. This allows us to evaluate and compare the ranking quality of our approaches independently of the performance of an underlying thresholding technique. The bipartition measures obtained by this method could be interpreted as a (soft) upper bound for any thresholding approach.

4   Experimental Setup

Datasets. We benchmark our proposed method on two textual corpora consisting of a large number of documents and with label hierarchies provided.

The RCV1-v2 dataset [19] is a collection of newswire articles. There are 103 labels and they are organized in a tree. Each label belongs to one of four major categories. The original train/test split in the RCV1-v2 dataset consists of 23,149 training documents and 781,265 test documents. In our experiments, we switched the training and the test data, and selected the top 20,000 words according to the document frequency. We chose randomly 10,000 training documents as the validation set.

The second corpus is the OHSUMED dataset [16] consisting of 348,565 scientific articles from MEDLINE. Each article has multiple index terms known as Medical Subject Headings (MeSH). In this dataset, the training set contains articles from year 1987 while articles from 1988 to 1991 belong to the test set. We map all MeSH terms in the OHSUMED dataset to 2015 MeSH vocabulary[1] in which 27,483 MeSH terms are organized in a DAG hierarchy. Originally, the OHSUMED collection consists of 54,710 training documents and 293,856 test documents. Having removed all MeSH terms that do not appear in the 2015 MeSH vocabulary, we excluded all documents that have no label from the corpus. To represent documents in a vector space, we selected unigram words that

---

[1] http://www.nlm.nih.gov/pubs/techbull/so14/so14_2015_mesh_avail.html

Table 1. Number of instances (M), Size of vocabulary (D), Number of labels (L), Average number of labels per instance (C), and the type of label hierarchy (HS). L subscripted k and u denote the number of *known* and *unseen* labels, respectively.

	Original datasets			Modified datasets				D	HS
	M	L	C	M	L_k	L_u	C		
RCV1-v2	804 414	103	3.24	700 628	82	21	1.43	20 000	Tree
OHSUMED	233 369	27 483	9.07	100 735	9570	17 913	3.87	25 892	DAG

occur more than 5 times in the training set. These pre-processing steps left us with 36,883 train documents and 196,486 test documents. Then, 10% of the training documents were randomly set aside for the validation set. Finally, for both datasets, we applied *log tf-idf* term-weighting and then normalized document vectors to unit length.

Preparation of the datasets in zero-shot settings. We hypothesize that label hierarchies provide possibilities of learning representations of unseen labels, thereby improving predictive performance for unseen data. To test our hypothesis, we modified the datasets. For the RCV1-v2 dataset, we removed all labels corresponding to non-terminals in the label hierarchy from training data and validation data while these non-terminal labels remain intact in the test set. In other words, we train models with labels corresponding to the leaves in the label hierarchy, then test them on the modified test set which only contains *unseen* labels.

Since the train and test examples of the OHSUMED dataset was split by year, the training data does not cover all labels in the test set. More specifically, there are 27,483 labels in the label hierarchy (cf. Table 1), of which only 9,570 occur in both training and test sets, which will be referred to as the set of *known* labels. Of the 12,568 labels that occur in the test set, 2,998 cannot be found in the *known* labels set, and thus form a set of *unseen* labels together with the 14,915 labels which are only available in the label hierarchy, but not present in the label patterns. In order to test predictive performance on these unseen labels, we omitted all labels in the *known* label set from the test examples. This resulted in some test examples having an empty set of labels, which were ignored for the evaluation. Finally, the above preprocessing steps left us 67,391 test examples.

The statistics of the datasets and the modified ones are summarized in Table 1.

Representing parent-child pairs of MeSH terms in a DAG. As mentioned earlier, we use parent-child pairs of MeSH terms in the 2015 MeSH vocabulary as the label hierarchy for the OHSUMED dataset. If we represent parent-child pairs of labels as a graph, it may contain cycles. Hence, we removed edges resulting in cycles as follows: 1) Pick a node that has no parent as a starting node. 2) Run Depth-First Search (DFS) from the starting node in order to detect edges pointing to nodes visited already, then remove such edges. 3) Repeat the 1 & 2 steps until all nodes having no parents are visited. There are 16 major categories in the MeSH vocabulary. In contrast to RCV1-v2, the MeSH terms are formed in complex structures so that a label can have more than one parent.

Baselines. We compare our algorithm, Wsabie$_H$, which uses hierarchical information for label embeddings, to Wsabie ignoring label hierarchies and several other benchmark algorithms. For *binary relevance* (BR), which decomposes a multi-label problem into L binary problems, we use LIBLINEAR [12] as a base learner which is a good compromise between efficiency and effectiveness in multi-label text document classification.

To address the limitations of BR, specifically, when L is large, dimensionality reduction method on label spaces, namely *Principal Label Space Transformation* (PLST) and *Conditional Principal Label Space Transformation* (CPLST), have been proposed [7,29] which try to capture label correlations before learning per-label classifiers. Instead of directly predicting labels for given instances, the *LSDR approach* learns d-output linear predictors in a reduced label space. Then, the original label space is reconstructed from the outputs of the linear predictors using the transformation matrix for reducing the label dimensionality. We use ridge regression as a linear predictor.

Pairwise decomposition has been already successfully applied for multi-label text classification [14,20]. Here, one classifier is trained for each pair of classes, i.e., a problem with L different classes is decomposed into $\frac{L(L-1)}{2}$ subproblems. At test time, all of the $\frac{L(L-1)}{2}$ base classifiers make a prediction for one of its two corresponding classes, which is interpreted as a full vote (0 or 1) for this label. Adding these up results in a ranking over the labels. To convert the ranking into a multi-label prediction, we use the *calibrated label ranking* (CLR) approach. Though CLR is able to predict cutting points of ranked lists, in this work, in order to allow a fair comparison, it also relies on an oracle to predict the number of relevant labels for a given instance (cf Section 3.4). We denote by CLR$_{svm}$ the use of CLR in combination with SVMs.

Evaluation measures. There are several measures to evaluate multi-label algorithms and they can be split into two groups; *ranking* and *bipartition* measures. If an algorithm generates a list of labels, which is sorted by relevance scores, for a given instance, ranking measures need to be considered. The most widely used ranking measures are *rank loss* (RL) and *average precision* (AvgP). The *rank loss* accounts for the ratio of the number of mis-ordered pairs between relevant and irrelevant labels in a ranked list of labels to all possible pairs, which defined as $RL = \frac{1}{|\mathcal{Y}||\bar{\mathcal{Y}}|} \sum_{i,j \in \mathcal{Y} \times \bar{\mathcal{Y}}} [r(i) > r(j)]_+ + \frac{1}{2}[r(i) = r(j)]_+$ where $r(\cdot)$ denotes the position of a label in the ranked list. The *average precision* quantifies the average precision at each point in the ranking where a relevant label is placed and is computed as $AvgP = \frac{1}{|\mathcal{Y}|} \sum_{i \in \mathcal{Y}} (|\{j \in \mathcal{Y} | r(j) \leq r(i)\}|/r(i))$. Label-based measures for binary predictions can be also considered. In this work, we report the micro- and macro-averaged F-score defined as: MiF $= (\sum_{l=1}^{L} 2tp_l)/(\sum_{l=1}^{L} 2tp_l + fp_l + fn_l)$, MaF $= \frac{1}{L} \sum_{l=1}^{L} 2tp_l/(2tp_l + fp_l + fn_l)$ where tp_l, fp_l and fn_l are the number of true positives, false positives and

Table 2. Comparison of Wsabie$_H$ to baselines on the benchmarks. (Best in bold)

	RCV1-v2						OHSUMED			
	BR	PLST	CPLST	CLR$_{svm}$	Wsabie	Wsabie$_H$	BR	PLST	Wsabie	Wsabie$_H$
AvgP	94.20	92.75	92.76	**94.76**	94.34	94.39	45.00	26.50	45.72	**45.76**
RL	0.46	0.78	0.76	**0.40**	0.44	0.44	4.48	15.06	4.09	**3.72**

false negatives for label l, respectively. Throughout the paper, we present the evaluation scores of these measures multiplied by 100.

Training Details. All hyperparameters were empirically chosen based on AvgP on validation sets. The dimensionality of the joint space d was selected in a range of $\{16, 32, 64, 128\}$ for the RCV1-v2 dataset and $\{128, 256, 512\}$ for the OHSUMED dataset. The margins m_a and m_b were chosen ranging from $\{10^{-3}, 10^{-2}, 10^{-1}, 10^0, 10^1, 10^2\}$. We used Adagrad [10] to optimize parameters Θ in Eq. 1 and 4. Let $\Delta_{i,\tau}$ be the gradient of the objective function in Eq. 4 with respect to a parameter $\theta_i \in \Theta$ at time τ. Then, the update rule for parameters indexed i at time τ is given by $\theta_i^{(\tau+1)} = \theta_i^{(\tau)} - \eta_i^{(\tau)} \Delta_{i,t}$ with an adaptive learning rate per parameter $\eta_i^{(\tau)} = \eta_0 / \sqrt{\sum_{t=1}^{\tau} \Delta_{i,t}^2}$ where $\eta_0 \in \{10^{-4}, 10^{-3}, 10^{-2}, 10^{-1}\}$ denotes a base learning rate which decrease by a factor of 0.99 per epoch. We implemented our proposed methods using a lock-free parallel gradient update scheme [25], namely *Hogwild!*, in a shared memory system since the number of parameters involved during updates is sparse even though the whole parameter space is large. For BR and CLR$_{svm}$, LIBLINEAR[12] was used as a base learner and the regularization parameter $C = \{10^{-2}, 10^0, 10^2, 10^4, 10^6\}$ was chosen by validation sets.

5 Experimental Results

5.1 Learning All Labels Together

Table 2 compares our proposed algorithm, *Wsabie$_H$*, with the baselines on the benchmark datasets in terms of two ranking measures. It can be seen that CLR$_{svm}$ outperforms the others including *Wsabie$_H$* on the RCV1-v2 dataset, but the performance gap across all algorithms in our experiments is not large. Even BR ignoring label relationship works competitively on this dataset. Also, no difference between *Wsabie* and *Wsabie$_H$* was observed. This is attributed to characteristics of the RCV1-v2 dataset that if a label corresponding to one of the leaf nodes in the label hierarchy is associated with an instance, (almost) all nodes in a path from the root node to that node are also present, so that the hierarchical information is implicitly present in the training data.

Let us now turn to the experimental results on the OHSUMED dataset which are shown on the right-hand side of Table 2. Since the dataset consists of many labels, as an LSDR approach, we include PLST only in this experiment because

Table 3. The performance of $Wsabie_H$ compared to its baseline on the benchmarks in *zero-shot* learning settings.

	RCV1-v2				OHSUMED			
	AvgP	RL	MiF	MaF	AvgP	RL	MiF	MaF
Wsabie	2.31	62.29	0.00	0.00	0.01	56.37	0.00	0.00
$Wsabie_H$	**9.47**	**30.39**	**0.50**	**1.64**	**0.06**	**39.91**	0.00	0.00

CPLST is computationally more expensive than PLST, but no significant difference was observed. Similarly, due to the computational cost of CLR_{svm} with respect to the number of labels, we excluded it from the experiment. It can be seen that regardless of the choice of the regularization term, the Wsabie approaches perform better than the other methods. PLST performed poorly under the settings where the density of labels, i.e., C/L in Table 1, is very low. Moreover, we projected the original label space $L = 27{,}483$ into a much smaller dimension $d = 512$ using a small amount of training examples. Although the difference to BR is rather small, the margin is more pronounced that on RCV1.

5.2 Learning to Predict Unseen Labels

Over the last few years, there has been an increasing interest in *zero-shot learning*, which aims to learn a function that maps instances to classes or labels that have not been seen during training. Visual attributes of an image [18] or textual description of labels [13, 28] may serve as additional information for zero-shot learning algorithms. In contrast, in this work, we focus on how to exploit label hierarchies and co-occurrence patterns of labels to make predictions on such *unseen* labels. The reason is that in many cases it is difficult to get additional information for some specific labels from external sources. In particular, while using a semantic space of labels' textual description is a promising way to learn vector representations of labels, sometimes it is not straightforward to find suitable mappings of specialized labels.

Table 3 shows the results of *Wsabie* against $Wsabie_H$ on the modified datasets which do not contain any known label in the test set (cf. Sec. 4). As can be seen, $Wsabie_H$ clearly outperforms *Wsabie* on both datasets across all measures except for MiF and MaF on the OHSUMED dataset. Note that the key difference between $Wsabie_H$ and *Wsabie* is the use of hierarchical structures over labels during the training phase. Since the labels in the test set do not appear during training, *Wsabie* can basically only make random predictions for the unknown labels. Hence, the comparison shows that taking only the hierarchical relations into account already enables a considerable improvement over the baseline. Unfortunately, the effect is not substantial enough in order to be reflected w.r.t. MiF and MaF on OHSUMED. Note, however, that a relevant, completely unknown label must be ranked approximately as one of the top 4 labels out of 17,913 in order to count for bipartition measures in this particular setting.

In summary, these results show that the regularization of joint embedding methods is an effective way of learning representations for *unseen* labels in a tree-structured hierarchy of a small number of labels . However, if a label hierarchy is defined on more complex structures and while a fewer number of training examples exists per label, it might be difficult for *Wsabie_H* to work well on unseen data.

6   Pretrained Label Embeddings as Good Initial Guess

From the previous experiments, we see that the regularization of *Wsabie_H* using the hierarchical structure of labels allows us to obtain better performance for *unseen* labels. The objective function (Eq. 4) penalizes parameters of observable labels in the training data by the negative log probability of predicting their ancestors in a hierarchy. If we initialize label spaces parameterized by $\mathbf{U}$ at random, presumably, the regularizer may rather act as noise at a beginning stage of the training. Especially for OHSUMED, the label hierarchy is complex and positive documents are very few for some labels.

We address this by exploiting both label hierarchies and co-occurrence patterns between labels in the training data. Apart from feature representations of a training instance, it is possible to capture underlying structures of the label space based on the co-occurrence patterns. Hence, we propose a method to learn label embeddings from hierarchical information and *pairwise* label relationships.

The basic idea of pretraining label embeddings is to maximize the probability of predicting an ancestor given a particular label in a hierarchy as well as predicting co-occurring labels with it. Given the labels of N training instances $\mathcal{D}_\mathcal{Y} = \{\mathcal{Y}_1, \mathcal{Y}_2, \cdots, \mathcal{Y}_N\}$, the objective function is to maximize the average log probability given by

$$\sum_{n=1}^{N} \left[\frac{(1-\alpha)}{\mathcal{Z}_\mathcal{A}} \sum_{i \in \mathcal{Y}_n} \sum_{j \in \mathcal{S}_\mathcal{A}(i)} \log p\,(y_j|y_i) + \frac{\alpha}{\mathcal{Z}_\mathcal{N}} \sum_{i \in \mathcal{Y}_n} \sum_{\substack{k \in \mathcal{Y}_n \\ k \neq i}} \log p\,(y_k|y_i) \right] \qquad (6)$$

where α determines the importance of each term ranging from 0 to 1, $\mathcal{Z}_\mathcal{A} = |\mathcal{Y}_n||\mathcal{S}_\mathcal{A}(\cdot)|$ and $\mathcal{Z}_\mathcal{N} = |\mathcal{Y}_n|(|\mathcal{Y}_n| - 1)$. The probability of predicting an ancestor label j of a label i, i.e., $p(y_j|y_i)$, can be computed similarly to Eq. 3 by using *softmax* and slight modifications. Thus, the log-probability can be defined by

$$p(y_j|y_i) = \frac{\exp(\mathbf{u'}_j^T \mathbf{u}_i)}{\sum_{v \in L} \exp(\mathbf{u'}_v^T \mathbf{u}_i)} \qquad (7)$$

where $\mathbf{u}_i$ is the i-th column vector of $\mathbf{U} \in \mathbb{R}^{d \times L}$ and $\mathbf{u'}_j$ is a vector representation for label j and the j-th column vector of $\mathbf{U'} \in \mathbb{R}^{d \times L}$. The softmax function in Eq. 7 can be viewed as an objective function of a neural network consisting of a linear activation function in the hidden layer and two weights $\{\mathbf{U}, \mathbf{U'}\}$, where $\mathbf{U}$ connects the input layer to the hidden layer while $\mathbf{U'}$ is used to convey

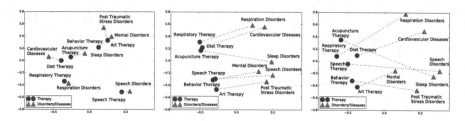

Fig. 2. Visualization of learned label embeddings by the log-linear model (Eq 6). (*left*) using only label co-occurrence patterns $\alpha = 1$ (*middle*) using a hierachy as well as co-occurrences $\alpha = 0.5$ (*right*) using only a hierarchy $\alpha = 0$.

the hidden activations to the output layer. Here, $\mathbf{U}$ and $\mathbf{U}'$ correspond to vector representations for input labels and output labels, respectively. Like Eq. 3, we use *hierarchical softmax* instead of Eq. 7 to speed up pre-training label embeddings.

6.1   Understanding Label Embeddings

We begin by qualitatively demonstrating label embeddings trained on label co-occurrence patterns from the BioASQ dataset [1], which is one of the largest datasets for multi-label text classification, and label hierarchies in the 2015 MeSH vocabulary. The BioASQ dataset consists of more than 10 millions of documents. Note that we only use its label co-occurrence patterns. Its labels are also defined over the same MeSH vocabulary, so that we can use it for obtaining knowledge about the OHSUMED labels (cf Section 6). We trained the label embeddings using Eq. 6 by setting the dimensionality of the label embeddings to $d = \{128, 256, 512\}$ with different weighting values $\alpha = \{0, 0.5, 1\}$ for 100 epoch using SGD with a fixed learning rate of 0.1. If we set $d = 128$, training took about 6 hours on a machine with dual Xeon E5-2620 CPUs.

Analysis on learned label representations. Fig. 2 shows vector representations of labels related to Disorders/Diseases and their therapy in the 2015 MeSH vocabulary in 2D space.[2] It is likely that label pairs that co-occur frequently are close to each other. Particularly, on the *left* in Fig. 2, each therapy is close to a disorder for which the therapy is an effective treatment. If we make use of hierarchical information as well as co-occurrence label patterns during training, i.e., $\alpha = 0.5$ in Eq. 6, more interesting relationships are revealed which are not observed from the model trained only on co-occurrences ($\alpha = 1$). We can say that the learned vector representations has identified *Therapy-Disorders/Diseases* relationships (on the *middle* in Fig. 2). We also present label embeddings trained using only label hierarchies ($\alpha = 0$) on the *right* in Fig. 2.

Analogical reasoning in label spaces. One way to evaluate representation quality is analogical reasoning as shown in [21]. Upon the above observations (on

---

[2] Projection of 128-dim label embeddings into 2D was done by Principal Component Analysis.

Table 4. Analogical reasoning on learned vector representations of MeSH vocabulary

On learned representations *using* the hierarchy	
Analogy questions	Most probable answers
Cardiovascular Diseases : Diet Therapy ≈ Respiration Disorders : ?	Diet Therapy Enteral Nutrition Gastrointestinal Intubation Total Parenteral Nutrition Parenteral Nutrition Respiratory Therapy
Mental Disorders : Behavior Therapy ≈ PTSD : ?	Behavior Therapy Cognitive Therapy Rational-Emotive Psychotherapy Brief Psychotherapy Psychologic Desensitization Implosive Therapy
On learned representations *without using* the hierarchy	
Analogy questions	Most probable answers
Cardiovascular Diseases : Diet Therapy ≈ Respiration Disorders : ?	Respiration Disorders Respiratory Tract Diseases Respiratory Sounds Airway Obstruction Hypoventilation Croup
Mental Disorders : Behavior Therapy ≈ PTSD : ?	Behavior Therapy Psychologic Desensitization Internal-External Control PTSD Phobic Disorders Anger

the *middle* in Fig. 2), we performed analogical reasoning on both the representations trained with the hierarchy and ones without the hierarchy, specifically, regarding *Therapy-Disorders/Diseases* relationships (Table 4). As expected, it seems like the label representations trained with the hierarchy are clearly advantageous to the ones trained without the hierarchy on analogical reasoning. To be more specific, consider the first example, where we want to know what kinds of therapies are effective on "Respiration Disorders" as the relationship between "Diet Therapy" and "Cardiovascular Diseases". When we perform such analogical reasoning using learned embeddings with the hierarchy, the most probable answers to this analogy question are therapies that can be used to treat "Respiration Disorders" including nutritional therapies. Unlike the learned embeddings with the hierarchy, the label embeddings without the hierarchy perform poorly. In the bottom-right of Table 4, "Phobic Disorders" can be considered as a type of anxiety disorders that occur commonly together with "Post-traumatic Stress Disorders (PTSD)" rather than a treatment of it.

6.2 Results

The results on the modified zero-shot learning datasets in Table 5 show that we can obtain substantial improvements by the pretrained label embeddings. Please note that the scores obtained by using *random* label embeddings on the left in Table 5 are the same as those of *Wsabie* and *Wsabie$_H$* in Table 3. In this experiment, we used very small base learning rates (i.e., $\eta_0 = 10^{-4}$ chosen by validation) for updating label embeddings in Eq. 4 after being initialized

Table 5. Initialization of label embeddings on OHSUMED under *zero-shot* settings.

	random label embeddings				*pretrained* label embeddings			
	AvgP	RL	MiF	MaF	AvgP	RL	MiF	MaF
Wsabie	0.01	56.37	0.00	0.00	**1.64**	**2.82**	0.03	0.06
Wsabie$_H$	0.06	39.91	0.00	0.00	1.36	5.33	**0.08**	**0.14**

Table 6. Evaluation on the *full* test data of the OHSUMED dataset. Numbers in parentheses are standard deviation over 5 runs. Subscript P denotes the use of pretrained label embeddings.

	Wsabie	Wsaibe$_H$	Wsabie$_P$	Wsabie$_{HP}$
AvgP	45.72 (0.04)	45.76 (0.06)	45.88 (0.09)	**45.92** (0.02)
RL	4.09 (0.18)	3.72 (0.11)	3.44 (0.13)	**3.11** (0.10)
MiF	46.32 (0.04)	46.34 (0.04)	46.45 (0.07)	**46.50** (0.01)
MaF	13.93 (0.03)	13.96 (0.07)	14.19 (0.05)	**14.25** (0.02)

by the pretrained ones. This means that our proposed method is trained in a way that maps a document into the some point of label embeddings while the label embeddings hardly change. In fact, the pretrained label embeddings have interesting properties shown in Section 6.1, so that *Wsabie* starts learning at good initial parameter spaces. Interestingly, it was observed that some of the unseen labels are placed at the top of rankings for test instances, so that relatively higher scores of bipartition measures are obtained even for Wsabie. We also performed an experiment on the *full* OHSUMED dataset. The experimental results are given in Table 6. *Wsabie$_{HP}$* combining pretrained label embeddings with hierarchical label structures is able to further improve, outperforming both extensions by its own across all measures.

7   Conclusions

We have presented a method that learns a joint space of instances and labels taking hierarchical structures of labels into account. This method is able to learn representations of labels, which are not presented during the training phase, by leveraging label hierarchies. We have also proposed a way of pretraining label embeddings from huge amounts of label patterns and hierarchical structures of labels.

We demonstrated the joint space learning method on two multi-label text corpora that have different types of label hierarchies. The empirical results showed that our approach can be used to place relevant unseen labels on the top of the ranked list of labels. In addition to the quantitative evaluation, we also analyzed label representations qualitatively via a 2D-visualization of label representations. This analysis showed that using hierarchical structures of labels allows us to assess vector representations of labels by analogical reasoning. Further studies should be carried out to make use of such regularities in label embeddings at testing time.

References

1. Balikas, G., Partalas, I., Ngomo, A.N., Krithara, A., Paliouras, G.: Results of the BioASQ track of the question answering lab at CLEF 2014. In: Working Notes for CLEF 2014 Conference, pp. 1181–1193 (2014)
2. Bengio, Y., Courville, A., Vincent, P.: Representation learning: A review and new perspectives. IEEE Transactions on Pattern Analysis and Machine Intelligence **35**(8), 1798–1828 (2013)
3. Bi, W., Kwok, J.T.: Multilabel classification on tree- and DAG-structured hierarchies. In: Proceedings of the 28th International Conference on Machine Learning, pp. 17–24 (2011)
4. Bi, W., Kwok, J.T.: Efficient multi-label classification with many labels. In: Proc. of the 30th International Conference on Machine Learning, pp. 405–413 (2013)
5. Cesa-Bianchi, N., Gentile, C., Zaniboni, L.: Incremental algorithms for hierarchical classification. Journal of Machine Learning Research **7**, 31–54 (2006)
6. Chekina, L., Gutfreund, D., Kontorovich, A., Rokach, L., Shapira, B.: Exploiting label dependencies for improved sample complexity. Machine Learning **91**(1), 1–42 (2013)
7. Chen, Y.N., Lin, H.T.: Feature-aware label space dimension reduction for multilabel classification. In: Advances in Neural Information Processing Systems, pp. 1529–1537 (2012)
8. Crammer, K., Singer, Y.: A family of additive online algorithms for category ranking. The Journal of Machine Learning Research **3**, 1025–1058 (2003)
9. Dembczyński, K., Waegeman, W., Cheng, W., Hüllermeier, E.: On label dependence and loss minimization in multi-label classification. Machine Learning **88**(1-2), 5–45 (2012)
10. Duchi, J., Hazan, E., Singer, Y.: Adaptive subgradient methods for online learning and stochastic optimization. The Journal of Machine Learning Research **12**, 2121–2159 (2011)
11. Elisseeff, A., Weston, J.: A kernel method for multi-labelled classification. Advances in Neural Information Processing Systems **14**, 681–687 (2001)
12. Fan, R.E., Chang, K.W., Hsieh, C.J., Wang, X.R., Lin, C.J.: LIBLINEAR: A library for large linear classification. The Journal of Machine Learning Research **9**, 1871–1874 (2008)
13. Frome, A., Corrado, G.S., Shlens, J., Bengio, S., Dean, J., Ranzato, M., Mikolov, T.: Devise: A deep visual-semantic embedding model. Advances in Neural Information Processing Systems **26**, 2121–2129 (2013)
14. Fürnkranz, J., Hüllermeier, E., Loza Mencía, E., Brinker, K.: Multilabel classification via calibrated label ranking. Machine Learning **73**(2), 133–153 (2008)
15. Fürnkranz, J., Sima, J.F.: On exploiting hierarchical label structure with pairwise classifiers. SIGKDD Explorations **12**(2), 21–25 (2010)
16. Hersh, W., Buckley, C., Leone, T.J., Hickam, D.: Ohsumed: an interactive retrieval evaluation and new large test collection for research. In: Proceedings of the 17th Annual International ACM SIGIR Conference, pp. 192–201 (1994)
17. Hsu, D., Kakade, S., Langford, J., Zhang, T.: Multi-label prediction via compressed sensing. In: Advances in Neural Information Processing Systems 22, vol. 22, pp. 772–780 (2009)
18. Lampert, C.H., Nickisch, H., Harmeling, S.: Attribute-based classification for zero-shot visual object categorization. IEEE Transactions on Pattern Analysis and Machine Intelligence **36**(3), 453–465 (2014)

19. Lewis, D.D., Yang, Y., Rose, T.G., Li, F.: RCV1: A new benchmark collection for text categorization research. The Journal of Machine Learning Research **5**, 361–397 (2004)
20. Loza Mencía, E., Fürnkranz, J.: Pairwise learning of multilabel classifications with perceptrons. In: Proceedings of the International Joint Conference on Neural Networks, pp. 2899–2906 (2008)
21. Mikolov, T., Sutskever, I., Chen, K., Corrado, G.S., Dean, J.: Distributed representations of words and phrases and their compositionality. Advances in Neural Information Processing Systems **26**, 3111–3119 (2013)
22. Mnih, A., Hinton, G.E.: A scalable hierarchical distributed language model. Advances in Neural Information Processing Systems **22**, 1081–1088 (2009)
23. Morin, F., Bengio, Y.: Hierarchical probabilistic neural network language model. In: Proceedings of the 10th International Workshop on Artificial Intelligence and Statistics, pp. 246–252 (2005)
24. Read, J., Pfahringer, B., Holmes, G., Frank, E.: Classifier chains for multi-label classification. Machine Learning **85**(3), 333–359 (2011)
25. Recht, B., Re, C., Wright, S., Niu, F.: Hogwild: A lock-free approach to parallelizing stochastic gradient descent. In: Advances in Neural Information Processing Systems, pp. 693–701 (2011)
26. Rousu, J., Saunders, C., Szedmák, S., Shawe-Taylor, J.: Kernel-based learning of hierarchical multilabel classification models. Journal of Machine Learning Research **7**, 1601–1626 (2006)
27. Silla Jr, C.N., Freitas, A.A.: A survey of hierarchical classification across different application domains. Data Mining and Knowledge Discovery **22**(1–2), 31–72 (2011)
28. Socher, R., Ganjoo, M., Manning, C.D., Ng, A.: Zero-shot learning through cross-modal transfer. In: Advances in Neural Information Processing Systems, pp. 935–943 (2013)
29. Tai, F., Lin, H.T.: Multilabel classification with principal label space transformation. Neural Computation **24**(9), 2508–2542 (2012)
30. Vens, C., Struyf, J., Schietgat, L., Džeroski, S., Blockeel, H.: Decision trees for hierarchical multi-label classification. Machine Learning **73**(2), 185–214 (2008)
31. Weston, J., Bengio, S., Usunier, N.: Wsabie: scaling up to large vocabulary image annotation. In: Proceedings of the 22nd International Joint Conference on Artificial Intelligence, pp. 2764–2770 (2011)
32. Zimek, A., Buchwald, F., Frank, E., Kramer, S.: A study of hierarchical and flat classification of proteins. IEEE/ACM Transactions on Computational Biology and Bioinformatics **7**, 563–571 (2010)

Regression with Linear Factored Functions

Wendelin Böhmer$^{(\boxtimes)}$ and Klaus Obermayer

Neural Information Processing Group, Technische Universität Berlin,
Sekr. MAR5-6, Marchstr. 23, 10587 Berlin, Germany
{wendelin,oby}@ni.tu-berlin.de
http://www.ni.tu-berlin.de

Abstract. Many applications that use empirically estimated functions face a *curse of dimensionality*, because integrals over most function classes must be approximated by sampling. This paper introduces a novel *regression*-algorithm that learns *linear factored functions* (LFF). This class of functions has structural properties that allow to analytically solve certain integrals and to calculate point-wise products. Applications like *belief propagation* and *reinforcement learning* can exploit these properties to break the curse and speed up computation. We derive a regularized greedy optimization scheme, that learns factored basis functions during training. The novel regression algorithm performs competitively to *Gaussian processes* on benchmark tasks, and the learned LFF functions are with 4-9 factored basis functions on average very compact.

Keywords: Regression · Factored functions · Curse of dimensionality

1 Introduction

This paper introduces a novel regression-algorithm, which performs competitive to *Gaussian processes*, but yields *linear factored functions* (LFF). These have outstanding properties like analytical *point-wise products* and *marginalization*.

Regression is a well known problem, which can be solved by many non-linear architectures like *kernel methods* (Shawe-Taylor and Cristianini 2004) or *neural networks* (Haykin 1998). While these perform well, the estimated functions often suffer a *curse of dimensionality* in later applications. For example, computing an integral over a neural network or kernel function requires to sample the entire input space. Applications like *belief propagation* (Pearl 1988) and *reinforcement learning* (Kaelbling et al. 1996), on the other hand, face large input spaces and require therefore efficient computations. We propose LFF for this purpose and showcase its properties in comparison to kernel functions.

1.1 Kernel Regression

In the last 20 years, kernel methods like *support vector machines* (SVM, Boser et al. 1992; Vapnik 1995) have become a de facto standard in various practical applications. This is mainly due to a sparse representation of the learned classifiers with so called *support vectors* (SV). The most popular kernel method for

© Springer International Publishing Switzerland 2015
A. Appice et al. (Eds.): ECML PKDD 2015, Part I, LNAI 9284, pp. 119–134, 2015.
DOI: 10.1007/978-3-319-23528-8_8

regression, *Gaussian processes* (GP, see Bishop 2006; Rasmussen and Williams 2006), on the other hand, requires as many SV as training samples. Sparse versions of GP aim thus for a small subset of SV. Some select this set based on constraints similar to SVM (Tipping 2001; Vapnik 1995), while others try to conserve the spanned linear function space (*sparse GP*, Csató and Opper 2002; Rasmussen and Williams 2006). There exist also attempts to construct new SV by averaging similar training samples (e.g. Wang et al. 2012).

Well chosen SV for regression are usually not sparsely concentrated on a decision boundary as they are for SVM. In fact, many practical applications report that they are distributed uniformly in the input space (e.g. in Böhmer et al. 2013). Regression tasks restricted to a small region of the input space may tolerate this, but some applications require predictions everywhere. For example, the *value function* in reinforcement learning must be generalized to each state. The number of SV required to *represent* this function equally well in each state grows exponentially in the number of input-space dimensions, leading to Bellman's famous curse of dimensionality (Bellman 1957).

Kernel methods derive their effectiveness from linear optimization in a nonlinear *Hilbert space* of functions. Kernel-functions parameterized by SV are the non-linear *basis functions* in this space. Due to the functional form of the kernel, this can be a very ineffective way to select basis functions. Even in relatively small input spaces, it often takes hundreds or thousands SV to approximate a function sufficiently. To alleviate the problem, one can construct complex kernels out of simple prototypes (see a recent review in Gönen and Alpaydın 2011).

1.2 Factored Basis Functions

Diverging from all above arguments, this article proposes a more radical approach: to construct the non-linear basis functions directly during training, without the detour over kernel functions and support vectors. This poses two main challenges: to select a *suitable functions space* and to *regularize the optimization* properly. The former is critical, as a small set of basis functions must be able to approximate any target function, but should also be easy to compute in practice.

We propose *factored functions* $\psi_i = \prod_k \psi_i^k \in \mathcal{F}$ as basis functions for regression, and call the linear combination of m of those bases a *linear factored function* $f \in \mathcal{F}^m$ (LFF, Section 3). For example, *generalized linear models* (Nelder and Wedderburn 1972) and *multivariate adaptive regression splines* (MARS, Friedman 1991) are both LFF. Computation remains feasible by using *hinge functions* $\psi_i^k(x_k) = \max(0, x_k - c)$ and restricting the scope of each factored function ψ_i. In contrast, we assume the general case without restrictions to functions or scope.

Due to their structure, LFF can solve certain integrals analytically and allow very efficient computation of point-wise products and marginalization. We show that our LFF are universal function approximators and derive an appropriate *regularization* term. This regularization promotes smoothness, but also retains a high degree of variability in densely sampled regions by linking smoothness to uncertainty about the sampling distribution. Finally, we derive a novel regression algorithm for LFF based on a greedy optimization scheme.

Functions learned by this algorithm (Algorithm 1, see pages 125 and 133) are very compact (between 3 and 12 bases on standard benchmarks) and perform competitive with Gaussian processes (Section 4). The paper finishes with a discussion of the computational possibilities of LFF in potential areas of application and possible extensions to *sparse regression* with LFF (Section 5).

2   Regression

Let $\{x_t \in \mathcal{X}\}_{t=1}^n$ be a set of n *input samples,* i.i.d. drawn from an input set $\mathcal{X} \subset \mathbb{R}^d$. Each so called "training sample" is *labeled* with a real number $\{y_t \in \mathbb{R}\}_{t=1}^n$. *Regression* aims to find a function $f : \mathcal{X} \to \mathbb{R}$, that predicts the labels to all (previously unseen) test samples as well as possible. Labels may be afflicted by *noise* and f must thus approximate the mean label of each sample, i.e., the function $\mu : \mathcal{X} \to \mathbb{R}$. It is important to notice that *conceptually* the noise is introduced by two (non observable) sources: noisy labels y_t and noisy samples x_t. The latter will play an important role for regularization. We define the conditional distribution χ of observable samples $x \in \mathcal{X}$ given the non-observable "true" samples $z \in \mathcal{X}$, which are drawn by a distribution ξ. In the limit of infinite samples, the *least squares* cost-function $\mathcal{C}[f|\chi, \mu]$ can thus be written as

$$\lim_{n \to \infty} \inf_f \frac{1}{n} \sum_{t=1}^n \Big(f(x_t) - y_t \Big)^2 \;=\; \inf_f \iint \xi(dz)\,\chi(dx|z)\Big(f(x) - \mu(z) \Big)^2. \quad (1)$$

The cost function $\mathcal{C}$ can never be computed *exactly,* but *approximated* using the training samples[1] and assumptions about the unknown noise distribution χ.

3   Linear Factored Functions

Any non-linear function can be expressed as a linear function $f(x) = a^\top \psi(x)$, $\forall x \in \mathcal{X}$, with m non-linear basis functions $\psi_i : \mathcal{X} \to \mathbb{R}$, $\forall i \in \{1 \ldots, m\}$. In this section we will define *linear factored functions* (LFF), that have *factored basis functions* $\psi_i(x) := \psi_i^1(x_1) \cdot \ldots \cdot \psi_i^d(x_d) \in \mathcal{F}$, a regularization method for this function class and an algorithm for regression with LFF.

3.1   Function Class

We define the class of linear factored functions $f \in \mathcal{F}^m$ as a linear combination (with linear parameters $a \in \mathbb{R}^m$) of m factored basis functions $\psi_i : \mathcal{X} \to \mathbb{R}$

---

[1] The distribution ξ of "true" samples z can *not* be observed. We approximate in the following ξ with the training-sample distribution. This may be justified if the sample-noise χ is comparatively small. Although not strictly rigorous, the presented formalism helps to put the regularization derived in Proposition 2 into perspective.

(with parameters $\{\mathbf{B}^k \in \mathbb{R}^{m_k \times m}\}_{k=1}^d$):

$$f(\boldsymbol{x}) \; := \; \boldsymbol{a}^\top \boldsymbol{\psi}(\boldsymbol{x}) \; := \; \boldsymbol{a}^\top \left[\prod_{k=1}^d \boldsymbol{\psi}^k(x_k) \right] \; := \; \sum_{i=1}^m a_i \prod_{k=1}^d \sum_{j=1}^{m_k} B_{ji}^k \, \phi_j^k(x_k). \qquad (2)$$

LFF are formally defined in Appendix A. In short, a basis function ψ_i is the *point-wise product* of one-dimensional functions ψ_i^k in each input dimension k. These are themselves constructed as linear functions of a corresponding one-dimensional base $\{\phi_j^k\}_{j=1}^{m_k}$ over that dimension and ideally can approximate arbitrary functions[2]. Although each factored function ψ_i is very restricted, a linear combination of them can be very powerful:

Corollary 1. *Let $\mathcal{X}_k$ be a bounded continuous set and ϕ_j^k the j'th Fourier base over $\mathcal{X}_k$. In the limit of $m_k \to \infty, \forall k \in \{1, \ldots, d\}$, holds $\mathcal{F}^\infty = L^2(\mathcal{X}, \vartheta)$.*

Strictly this holds in the limit of infinitely many basis functions ψ_i, but we will show empirically that there exist close approximations with a small number m of factored functions. One can make similar statements for other bases $\{\phi_j^k\}_{j=1}^\infty$. For example, for Gaussian kernels one can show that the space $\mathcal{F}^\infty$ is in the limit equivalent to the corresponding *reproducing kernel Hilbert space* $\mathcal{H}$.

LFF offer some structural advantages over other universal function approximation classes like neural networks or reproducing kernel Hilbert spaces. Firstly, the *inner product* of two LFF in $L^2(\mathcal{X}, \vartheta)$ can be computed as products of one-dimensional integrals. For some bases[3], these integrals can be calculated analytically without any sampling. This could in principle break the curse of dimensionality for algorithms that have to approximate these inner products numerically. For example, input variables can be *marginalized* (integrated) out analytically (Equation 9 on Page 130). Secondly, the *point-wise product* of two LFF is a LFF as well[4] (Equation 10 on Page 131). See Appendix A for details. These properties are very useful, for example in *belief propagation* (Pearl 1988) and *factored reinforcement learning* (Böhmer and Obermayer 2013).

3.2 Constraints

LFF have some degrees of freedom that can impede optimization. For example, the norm of $\psi_i \in \mathcal{F}$ does not influence function $f \in \mathcal{F}^m$, as the corresponding linear coefficients a_i can be scaled accordingly. We can therefore introduce the *constraints* $\|\psi_i\|_\vartheta = 1, \forall i$, without restriction to the function class. The factorization of inner products (see Appendix A on Page 130) allows us furthermore to rewrite the constraints as $\|\psi_i\|_\vartheta = \prod_k \|\psi_i^k\|_{\vartheta^k} = 1$. This holds as long as the product is one, which exposes another unnecessary degree of freedom. To

---

[2] Examples are Fourier bases, Gaussian kernels or hinge-functions as in MARS.

[3] E.g. Fourier bases for continuous, and Kronecker-delta bases for discrete variables.

[4] One can use the trigonometric product-to-sum identities for Fourier bases or the Kronecker delta for discrete bases to construct LFF from a point-wise product without changing the underlying basis $\{\{\phi_i^k\}_{i=1}^{m_k}\}_{k=1}^d$.

finally make the solution unique (up to permutation), we define the constraints as $\|\psi_i^k\|_{\vartheta^k} = 1, \forall k, \forall i$. Minimizing some $\mathcal{C}[f]$ w.r.t. $f \in \mathcal{F}^m$ is thus equivalent to

$$\inf_{f \in \mathcal{F}^m} \mathcal{C}[f] \qquad \text{s.t.} \quad \|\psi_i^k\|_{\vartheta^k} = 1, \quad \forall k \in \{1, \ldots, d\}, \quad \forall i \in \{1, \ldots, m\}. \quad (3)$$

The *cost function* $\mathcal{C}[f|\chi, \mu]$ of Equation 1 with the constraints in Equation 3 is equivalent to *ordinary least squares* (OLS) w.r.t. linear parameters $\boldsymbol{a} \in \mathbb{R}^m$. However, the optimization problem is *not* convex w.r.t. the parameter space $\{\mathbf{B}^k \in \mathbb{R}^{m_k \times m}\}_{k=1}^d$, due to the nonlinearity of products.

Instead of tackling the global optimization problem induced by Equation 3, we propose a *greedy* approximation algorithm. Here we optimize at iteration $\hat{\imath}$ one linear basis function $\psi_{\hat{\imath}} =: g =: \prod_k g^k \in \mathcal{F}$, with $g^k(x_k) =: \boldsymbol{b}^{k\top} \boldsymbol{\phi}^k(x_k)$, at a time, to fit the residual $\mu - f$ between the true *mean label* function $\mu \in L^2(\mathcal{X}, \vartheta)$ and the current regression estimate $f \in \mathcal{F}^{\hat{\imath}-1}$, based on all $\hat{\imath} - 1$ previously constructed factored basis functions $\{\psi_i\}_{i=1}^{\hat{\imath}-1}$:

$$\inf_{g \in \mathcal{F}} \mathcal{C}[f + g|\chi, \mu] \qquad \text{s.t.} \quad \|g^k\|_{\vartheta^k} = 1, \quad \forall k \in \{1, \ldots, d\}. \quad (4)$$

3.3 Regularization

Regression with any powerful function class requires regularization to avoid overfitting. Examples are *weight decay* for neural networks (Haykin 1998) or parameterized *priors* for Gaussian processes. It is, however, not immediately obvious how to regularize the parameters of a LFF and we will derive a regularization term from a Taylor approximation of the cost function in Equation 1.

We aim to enforce smooth functions, especially in those regions our knowledge is limited due to a lack of training samples. This *uncertainty* can be expressed as the *Radon-Nikodym derivative*[5] $\frac{\vartheta}{\xi} : \mathcal{X} \to [0, \infty)$ of our factored measure ϑ (see Appendix A) w.r.t. the sampling distribution ξ. Figure 1 demonstrates at the example of a uniform distribution ϑ how $\frac{\vartheta}{\xi}$ reflects our empirical knowledge of the input space $\mathcal{X}$.

We use this uncertainty to modulate the *sample noise distribution* χ in Equation 1. This means that frequently sampled regions of $\mathcal{X}$ shall yield low, while scarcely sampled regions shall yield high variance. Formally, we assume $\chi(d\boldsymbol{x}|\boldsymbol{z})$ to be a Gaussian probability measure over $\mathcal{X}$ with mean $\boldsymbol{z}$ and a *covariance matrix* $\boldsymbol{\Sigma} \in \mathbb{R}^{d \times d}$, scaled by the local uncertainty in $\boldsymbol{z}$ (modeled as $\frac{\vartheta}{\xi}(\boldsymbol{z})$):

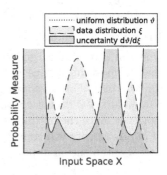

Fig. 1. We interpret the Radon-Nikodym derivative $\frac{d\vartheta}{d\xi}$ as *uncertainty measure* for our knowledge of $\mathcal{X}$. Regularization enforces smoothness in uncertain regions.

---

[5] Technically we have to assume that ϑ is *absolutely continuous* in respect to ξ. For "well-behaving" distributions ϑ, like the uniform or Gaussian distributions we discuss in Appendix A, this is equivalent to the assumption that in the limit of infinite samples, each sample $\boldsymbol{z} \in \mathcal{X}$ will *eventually* be drawn by ξ.

$$\int \chi(d\boldsymbol{x}|\boldsymbol{z})(\boldsymbol{x}-\boldsymbol{z}) = 0, \quad \int \chi(d\boldsymbol{x}|\boldsymbol{z})(\boldsymbol{x}-\boldsymbol{z})(\boldsymbol{x}-\boldsymbol{z})^\top = \tfrac{\vartheta}{\xi}(\boldsymbol{z}) \cdot \boldsymbol{\Sigma}, \quad \forall \boldsymbol{z} \in \mathcal{X}. \quad (5)$$

In the following we assume without loss of generality[6] the matrix $\boldsymbol{\Sigma}$ to be diagonal, with the diagonal elements called $\sigma_k^2 := \Sigma_{kk}$.

Proposition 2. *Under the assumptions of Equation 5 and a diagonal covariance matrix $\boldsymbol{\Sigma}$, the first order Taylor approximation of the cost $\mathcal{C}$ in Equation 4 is*

$$\tilde{\mathcal{C}}[g] \quad := \quad \underbrace{\|g - (\mu - f)\|_\xi^2}_{\text{sample-noise free cost}} + \underbrace{\sum_{k=1}^d \sigma_k^2 \, \|\tfrac{\partial}{\partial x_k}g + \tfrac{\partial}{\partial x_k}f\|_\vartheta^2}_{\text{smoothness in dimension } k}. \quad (6)$$

Proof: see Appendix C on Page 132.   □

Note that the approximated cost $\tilde{\mathcal{C}}[g]$ consists of the sample-noise free cost (measured w.r.t. training distribution ξ) and d regularization terms. Each term prefers functions that are smooth[7] in one input dimension. This enforces smoothness everywhere, but allows exceptions where enough data is available. To avoid a cluttered notation, in the following we will use the symbol $\nabla_k f := \tfrac{\partial}{\partial x_k} f$.

3.4   Optimization

Another advantage of cost function $\tilde{\mathcal{C}}[g]$ is that one can optimize one factor function g^k of $g(\boldsymbol{x}) = g^1(x_1) \cdot \ldots \cdot g^d(x_d) \in \mathcal{F}$ at a time, instead of time consuming *gradient descend* over the entire parameter space of g. To be more precise:

Proposition 3. *If all but one factor function g^k are considered constant, Equation 6 has an analytical solution. If $\{\phi_j^k\}_{j=1}^{m_k}$ is a Fourier base, $\sigma_k^2 > 0$ and $\vartheta \ll \xi$, then the solution is also unique.*

Proof: see Appendix C on Page 133.   □

One can give similar guarantees for other bases, e.g. Gaussian kernels. Note that Proposition 3 does *not* state that the optimization problem has a unique solution in $\mathcal{F}$. Formal convergence statements are not trivial and empirically the parameters of g do not converge, but evolve around orbits of equal cost instead. However, since the optimization of *any* g^k cannot increase the cost, any sequence of improvements will converge to (and stay in) a *local minimum*. This implies a *nested* optimization approach, that is formulated in Algorithm 1 on Page 133:

- An *inner loop* that optimizes one factored basis function $g(\boldsymbol{x}) = g^1(x_1) \cdot \ldots \cdot g^d(x_d)$ by selecting an input dimension k in each iteration and solve Equation 6 for the corresponding g^k. A detailed derivation of the optimization steps of

---

[6] Non-diagonal covariance matrices $\boldsymbol{\Sigma}$ can be cast in this framework by projecting the input samples into the eigenspace of $\boldsymbol{\Sigma}$ (thus diagonalizing the input) and use the corresponding eigenvalues λ_k instead of the regularization parameters σ_k^2's.

[7] Each regularization term is measured w.r.t. the factored distribution ϑ. We also tested the algorithm without consideration of "uncertainty" $\tfrac{\vartheta}{\xi}$, i.e., by measuring each term w.r.t. ξ. As a result, regions outside the hypercube containing the training set were no longer regularized and predicted arbitrary (often extreme) values.

Algorithm 1. (abstract) – a detailed version can be found on Page 133

---

while new factored basis function can improve solution **do**
    initialize new basis function g as constant function
    while optimization improves cost in Equation 6 **do**
        for random input dimension k **do**
            calculate optimal solution for g^k without changing $g^l, \forall l \neq k$
        end for
    end while   // new basis function g has converged
    add g to set of factored basis functions and solve OLS
end while   // regression has converged

---

the inner loop is given in Appendix B on Page 131. The choice of k influences the solution in a non-trivial way and further research is needed to build up a rationale for any meaningful decision. For the purpose of this paper, we assume k to be chosen randomly by permuting the order of updates.
The *computational complexity* of the inner loop is $\mathcal{O}(m_k^2 n + d^2 m_k m)$. Memory complexity is $\mathcal{O}(d\, m_k m)$, or $\mathcal{O}(d\, m_k n)$ with the optional cache speedup of Algorithm 1. The loop is repeated for random k until the cost-improvements of all dimensions k fall below some small ϵ.

- After convergence of the inner loop in (outer) iteration $\hat{\imath}$, the new basis function is $\psi_{\hat{\imath}} := g$. As the basis has changed, the linear parameters $\boldsymbol{a} \in \mathbb{R}^{\hat{\imath}}$ have to be readjusted by solving the ordinary least squares problem

$$\boldsymbol{a} = (\boldsymbol{\Psi}\boldsymbol{\Psi}^\top)^{-1}\boldsymbol{\Psi}\boldsymbol{y}, \quad \text{with } \Psi_{it} := \psi_i(\boldsymbol{x}_t), \ \forall i \in \{1, \dots, \hat{\imath}\}, \ \forall t \in \{1, \dots, n\}.$$

We propose to stop the approximation when the newly found basis function $\psi_{\hat{\imath}}$ is no longer *linearly independent* of the current basis $\{\psi_i\}_{i=1}^{\hat{\imath}-1}$. This can for example be tested by comparing the *determinant* $\det(\frac{1}{n}\boldsymbol{\Psi}\boldsymbol{\Psi}^\top) < \varepsilon$, for some very small ε.

4 Empirical Evaluation

In this section we will evaluate the novel LFF regression Algorithm 1, printed in detail on Page 133. We will analyze its properties on low dimensional toy-data, and compare its performance with sparse and traditional Gaussian processes (GP, see Bishop 2006; Rasmussen and Williams 2006).

4.1 Demonstration

To showcase the novel Algorithm 1, we tested it on an artificial two-dimensional regression toy-data set. The $n = 1000$ training samples were drawn from a noisy spiral and labeled with a sinus. The variance of the Gaussian sample-noise grew with the spiral as well:

$$\boldsymbol{x}_t = 6\frac{t}{n}\begin{bmatrix} \cos\left(6\frac{t}{n}\pi\right) \\ \sin\left(6\frac{t}{n}\pi\right) \end{bmatrix} + \mathcal{N}\left(\boldsymbol{0}, \frac{t^2}{4n^2}\mathbf{I}\right), \ y_t = \sin\left(4\frac{t}{n}\pi\right), \ \forall t \in \{1, \dots, n\}. \tag{7}$$

Fig. 2. Two LFF functions learned from the same 1000 training samples (white circles). The color inside a circle represents the training label. Outside the circles, the color represents the prediction of the LFF function. The differences between both functions are rooted in the randomized order in which the factor functions g^k are updated. However, the similarity of the sampled region indicates that poor initial choices can be compensated by subsequently constructed basis functions.

Figure 2 shows one training set plotted over two learned[8] functions $f \in \mathcal{F}^m$ with $m = 21$ and $m = 24$ factored basis functions, respectively. Regularization constants were in both cases $\sigma_k^2 = 0.0005, \forall k$. The differences between the functions stem from the randomized order in which the factor functions g^k are updated. Note that the sampled regions have similar predictions. Regions with strong differences, for example the upper right corner, are never seen during training.

In all our experiments, Algorithm 1 always converged. Runtime was mainly influenced by the input dimensionality ($\mathcal{O}(d^2)$), the number of training samples ($\mathcal{O}(n)$) and the eventual number of basis functions ($\mathcal{O}(m)$). The latter was strongly correlated with approximation quality, i.e., bad approximations converged fast. Cross-validation was therefore able to find good parameters efficiently and the resulting LFF were always very similar near the training data.

4.2 Evaluation

We compared the regression performance of LFF and GP with cross-validation on five regression benchmarks from the *UCI Manchine Learning Repository*[9]:

- The *concrete compressive strength* data set (*concrete*, Yeh 1998) consists of $n = 1030$ samples with $d = 8$ dimensions describing various concrete

---

[8] Here (and in the rest of the paper), each variable was encoded with 50 Fourier cosine bases. We tested other sizes as well. Few cosine bases result effectively in a low-pass filtered function, whereas every experiment with more than 20 or 30 bases behaved very similar. We tested up to $m_k = 1000$ bases and did not experience over-fitting.

[9] https://archive.ics.uci.edu/ml/index.html

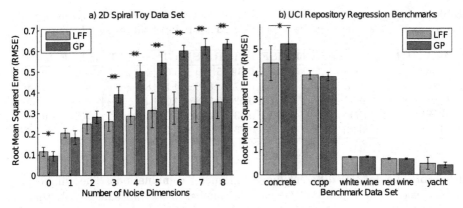

Fig. 3. Mean and standard deviation within a 10-fold cross-validation of a) the toy data set with additional independent noise input dimensions and b) all tested UCI benchmark data sets. The stars mark *significantly* different distribution of RMSE over all folds in both a *paired-sample t-test* and a *Wilcoxon signed rank test*. Significance levels are: one star $p < 0.05$, two stars $p < 0.005$.

mixture-components. The target variable is the real-valued compression strength of the mixture after it hardened.

- The *combined cycle power plant* data set (*ccpp*, Tüfekci 2014) consists of $n = 9568$ samples with $d = 4$ dimensions describing 6 years worth of measurements from a combined gas and steam turbine. The real-valued target variable is the energy output of the system.
- The *wine quality* data set (Cortez et al. 2009) consists of two subsets with $d = 11$ dimensions each, which describe physical attributes of various white and red wines: the set contains $n = 4898$ samples of *white wine* and $n = 1599$ samples of *red wine*. The target variable is the estimated wine quality on a discrete scale from 0 to 10.
- The *yacht hydrodynamics* data set (*yacht*, Gerritsma et al. 1981) consists of $n = 308$ samples with $d = 6$ dimensions describing parameters of the *Delft yacht hull* ship-series. The real-valued target variable is the residuary resistance measured in full-scale experiments.

To demonstrate the advantage of factored basis functions, we also used the 2d-spiral toy-data set of the previous section with a varying number of additional input dimensions. Additional values were drawn i.i.d. from a Gaussian distribution and are thus independent of the target labels. As the input space $\mathcal{X}$ grows, kernel methods will increasingly face the curse of dimensionality during training.

Every data-dimension (except the labels) have been translated and scaled to zero mean and unit-variance before training. Hyper-parameters were chosen w.r.t. the mean of a 10-fold cross-validation. LFF-regression was tested for the uniform noise-parameters $\sigma_k^2 \in \{10^{-10}, 10^{-9.75}, 10^{-9.5}, \ldots, 10^{10}\}, \forall k$, i.e. for 81 different hyper-parameters. GP were tested with Gaussian kernels $\kappa(\boldsymbol{x}, \boldsymbol{y}) = \exp(-\frac{1}{2\bar{\sigma}^2}\|\boldsymbol{x} - \boldsymbol{y}\|_2^2)$ using kernel parameters $\bar{\sigma} \in \{10^{-1}, 10^{-3/4}, 10^{-1/2}, \ldots, 3\}$ and prior-parameters $\beta \in \{10^{-2}, 10^{-1}, \ldots, 10^{10}\}$ (Bishop 2006, see for the

Table 1. 10-fold cross-validation RMSE for benchmark data sets with d dimensions and n samples, resulting in m *basis functions*. The cross-validation took h hours.

DATA SET	d	n	#SV	RMSE LFF	RMSE GP	m LFF	h LFF	h GP
Concrete	8	1030	927	**4.429 ± 0.69**	5.196 ± 0.64	4.2 ± 0.8	3.00	0.05
CCPP	4	9568	2000	3.957 ± 0.17	3.888 ± 0.17	8.8 ± 2.0	1.96	1.14
White Wine	11	4898	2000	0.707 ± 0.02	0.708 ± 0.03	4.2 ± 0.4	4.21	0.69
Red Wine	11	1599	1440	0.632 ± 0.03	0.625 ± 0.03	4.7 ± 0.7	3.25	0.13
Yacht	6	308	278	0.446 ± 0.23	0.383 ± 0.11	4.2 ± 0.6	0.43	0.005

definition), i.e. for 221 different hyper-parameter combinations. The number of support vectors in standard GP equals the number of training samples. As this is not feasible for larger data sets, we used the MP-MAH algorithm (Böhmer et al. 2012) to select a uniformly distributed subset of 2000 training samples for sparse GP (Rasmussen and Williams 2006).

Figure 3a demonstrates the advantage of factored basis functions over kernel methods during training. The plot shows the *root mean squared errors*[10] (RMSE) of the two dimensional spiral toy-data set with an increasing number of independent noise dimensions. GP solves the initial task better, but clearly succumbs to the curse of dimensionality, as the size of the input space $\mathcal{X}$ grows. LFF, on the other hand, significantly overtake GP from 3 noise dimensions on, as the factored basis functions appear to be less affected by the curse. Another difference to GP is that decreasing performance automatically yields less factored basis functions (from 19.9 ± 2.18 with 0, to 6.3 ± 0.48 bases with 8 noise dimensions).

Figure 3b and Table 1 show that our LFF algorithm performs on all evaluated real-world benchmark data sets comparable to (sparse) GP. RMSE distributions over all folds were statistically indistinguishable, except for an advantage of LFF regression in the concrete compressive strength data set ($p < 0.01$ in a *t-test* and $p < 0.02$ in a *signed rank test*). As each basis function requries many iterations to converge, LFF regression runs considerably longer than standard approaches. However, LFF require between 3 and 12 factored basis functions to achieve the *same* performance as GP with 278-2000 kernel basis functions.

5   Discussion

We presented a novel algorithm for regression, which constructs factored basis functions during training. As *linear factored functions* (LFF) can in principle approximate any function in $L^2(\mathcal{X}, \vartheta)$, a regularization is necessary to avoid over-fitting. Here we rely on a regularization scheme that has been motivated by a Taylor approximation of the least-squares cost function with (an approximation of) virtual sample-noise. RMSE performance appears comparable to Gaussian

---

[10] RMSE are not a common performance metric for GP, which represent a *distribution* of solutions. However, RMSE reflect the objective of regression and are well suited to compare our algorithm with the *mean* of a GP.

processes on real-world benchmark data sets, but the factored representation is considerably more compact and seems to be less affected by distractors.

At the moment, LFF optimization faces two challenges. (i) The optimized cost function is not convex, but the local minimum of the solution may be controlled by selecting the next factor function to optimize. For example, MARS successively adds factor functions. Generalizing this will require further research, but may also allow some performance guarantees. (ii) The large number of inner-loop iterations make the algorithm slow. This problem should be mostly solved by addressing (i), but finding a trade-off between approximation quality and runtime may also provide a less compact shortcut with similar performance.

Preliminary experiments also demonstrated the viability of LFF in a *sparse regression* approach. Sparsity refers here to a limited number of input-dimensions that affect the prediction, which can be implemented by adjusting the sample-noise parameters σ_k^2 during training for each variable $\mathcal{X}_k$ individually. This is of particular interest, as factored functions are ideally suited to represent sparse functions and are in principle *unaffected* by the curse of dimensionality in function representation. Our approach modified the cost function to enforce LFF functions that were constant in all noise-dimensions. We did not include our results in this paper, as choosing the first updated factor functions g^k poorly resulted in basis functions that rather fitted noise than predicted labels. When we enforce sparseness, this initial mistake can afterwards no longer be rectified by other basis functions, in difference to the presented Algorithm 1. However, if this can be controlled by a sensible order in the updates, the resulting algorithm should be much faster and more robust than the presented version.

There are many application areas that may exploit the structural advantages of LLF. In *reinforcement learning* (Kaelbling et al. 1996), one can exploit the factorizing inner products to break the curse of dimensionality of the state space (Böhmer and Obermayer 2013). Factored transition models also need to be learned from experience, which is essentially a sparse regression task. Another possible field of application are *junction trees* (for Bayesian inference, see e.g. Bishop 2006) over continuous variables, where sparse regression may estimate the conditional probabilities. In each node one must also marginalize out variables, or calculate the point-wise product over multiple functions. Both operations can be performed analytically with LFF, the latter at the expense of more basis functions in the resulting LFF. However, one can use our framework to *compress* these functions after multiplication. This would allow junction-tree inference over mixed continuous and discrete variables.

In summary, we believe our approach to approximate functions by constructing non-linear factored basis functions (LFF) to be very promising. The presented algorithm performs comparable with Gaussian processes, but appears less sensitive to large input spaces than kernel methods. We also discussed some potential extensions for sparse regression that should improve upon that, in particular on runtime, and gave some fields of application that would benefit greatly from the algebraic structure of LFF.

Acknowledgments. The authors thank Yun Shen and the anonymous reviewers for their helpful comments. This work was funded by the *German science foundation* (DFG) within SPP 1527 *autonomous learning*.

Appendix A   LFF Definition and Properties

Let $\mathcal{X}_k$ denote the subset of $\mathbb{R}$ associated with the k'th variable of input space $\mathcal{X} \subset \mathbb{R}^d$, such that $\mathcal{X} := \mathcal{X}_1 \times \ldots \times \mathcal{X}_d$. To avoid the curse of dimensionality in this space, one can integrate w.r.t. a *factored probability measure* ϑ, i.e. $\vartheta(d\boldsymbol{x}) = \prod_{k=1}^d \vartheta^k(dx_k), \int \vartheta^k(dx_k) = 1, \forall k$. For example, ϑ^k could be uniform or Gaussian distributions over $\mathcal{X}_k$ and the resulting ϑ would be a uniform or Gaussian distribution over the input space $\mathcal{X}$.

A function $g : \mathcal{X} \to \mathbb{R}$ is called a *factored function* if it can be written as a product of one-dimensional *factor functions* $g^k : \mathcal{X}_k \to \mathbb{R}$, i.e. $g(\boldsymbol{x}) = \prod_{k=1}^d g^k(x_k)$. We only consider factored functions g that are twice integrable w.r.t. measure ϑ, i.e. $g \in L^2(\mathcal{X}, \vartheta)$. Note that not all functions $f \in L^2(\mathcal{X}, \vartheta)$ are factored, though. Due to *Fubini's theorem* the d-dimensional inner product between two factored functions $g, g' \in L^2(\mathcal{X}, \vartheta)$ can be written as the product of d one-dimensional inner products:

$$\langle g, g' \rangle_\vartheta = \int \vartheta(d\boldsymbol{x}) \, g(\boldsymbol{x}) \, g'(\boldsymbol{x}) = \int \prod_{k=1}^d \vartheta^k(dx_k) \, g^k(dx_k) \, g'^k(dx_k) = \prod_{k=1}^d \langle g^k, g'^k \rangle_{\vartheta^k} \,.$$

This trick can be used to solve the integrals at the heart of many least-squares algorithms. Our aim is to *learn* factored basis functions ψ_i. To this end, let $\{\phi_j^k : \mathcal{X}_k \to \mathbb{R}\}_{j=1}^{m_k}$ be a well-chosen[11] (i.e. universal) basis on $\mathcal{X}_k$, with the space of linear combinations denoted by $\mathcal{L}_\phi^k := \{\boldsymbol{b}^\top \boldsymbol{\phi}^k | \boldsymbol{b} \in \mathbb{R}^{m_k}\}$. One can thus approximate factor functions of ψ_i in $\mathcal{L}_\phi^k$, i.e., as linear functions

$$\psi_i^k(x_k) := \sum_{j=1}^{m_k} B_{ji}^k \phi_j^k(x_k) \quad \in \quad \mathcal{L}_\phi^k, \qquad\qquad \mathbf{B}^k \quad \in \quad \mathbb{R}^{m_k \times m}. \qquad (8)$$

Let $\mathcal{F}$ be the space of all factored basis functions ψ_i defined by the factor functions ψ_i^k above, and $\mathcal{F}^m$ be the space of all linear combinations of those m factored basis functions (Equation 2).

Marginalization of LFF can be performed analytically with Fourier bases ϕ_j^k and uniform distribution ϑ (many other bases can be analytically solved as well):

$$\int \vartheta^l(dx_l) \, f(\boldsymbol{x}) = \sum_{i=1}^m \Big(a_i \underbrace{\sum_{j=1}^{m_l} B_{ji}^l \langle \phi_j^l, 1 \rangle_{\vartheta^l}}_{\text{mean of } \phi_j^l} \Big) \Big[\prod_{k \neq l}^d \psi_i^k \Big] \overset{\text{Fourier}}{=} \sum_{i=1}^m \underbrace{a_i B_{1i}^l}_{\text{new } a_i} \Big[\prod_{k \neq l}^d \psi_i^k \Big] \tag{9}$$

---

[11] Examples for continuous variables $\mathcal{X}_k$ are Fourier cosine bases $\phi_j^k(x_k) \sim \cos\big((j-1)\pi x_k\big)$, and Gaussian bases $\phi_j^k(x_k) = \exp\big(\frac{1}{2\sigma^2}(x_k - s_{kj})^2\big)$. Discrete variables may be represented with Kronecker-delta bases $\phi_j^k(x_k = i) = \delta_{ij}$.

Using the trigonometric *product-to-sum* identity $\cos(x) \cdot \cos(y) = \frac{1}{2}\big(\cos(x-y) + \cos(x+y)\big)$, one can also compute the point-wise product between two LFF f and $\bar{f}$ with cosine-Fourier base (solutions to other Fourier bases are less elegant):

$$
\tilde{f}(\boldsymbol{x}) \;\; := \;\; f(\boldsymbol{x}) \cdot \bar{f}(\boldsymbol{x})
$$

$$
\overset{\text{Fourier}}{=} \underbrace{\sum_{i,j=1}^{m\bar{m}} a_i \bar{a}_j}_{\text{new } \tilde{a}_t} \prod_{k=1}^{d} \sum_{l=1}^{2m_k} \Big(\overbrace{\tfrac{1}{2}\sum_{q=1}^{l-1} B_{qi}^k \bar{B}_{(l-q)j}^k + \tfrac{1}{2}\sum_{q=l+1}^{m_k} B_{qi}^k \bar{B}_{(q-l)j}^k}^{\text{new } \tilde{B}_{lt}^k} \Big) \phi_l^k(x_k), \quad (10)
$$

where $t := (i-1)\,\bar{m} + j$, and $B_{ji}^k := 0, \forall j > m_k$, for both f and $\bar{f}$. Note that this increases the number of basis functions $\tilde{m} = m\bar{m}$, and the number of bases $\tilde{m}_k = 2m_k$ for each respective input dimension. The latter can be counteracted by *low-pass filtering*, i.e., by setting $\tilde{B}_{ji}^k := 0, \forall j > m_k$.

Appendix B   Inner Loop Derivation

Here we will optimize the problem in Equation 6 for one variable $\mathcal{X}_k$ at a time, by describing the update step $g^k \leftarrow g'^k$. This is repeated with randomly chosen variables k, until convergence the cost $\tilde{\mathcal{C}}[g]$, that is, until all possible updates decrease the cost less than some small ϵ.

Let in the following $\mathbf{C}^k := \langle \boldsymbol{\phi}^k, \boldsymbol{\phi}^{k\top}\rangle_{\vartheta^k}$ and $\dot{\mathbf{C}}^k := \langle \nabla_k \boldsymbol{\phi}^k, \nabla_k \boldsymbol{\phi}^{k\top}\rangle_{\vartheta^k}$ denote covariance matrices, and $\boldsymbol{R}_l^k := \frac{\partial}{\partial \boldsymbol{b}^k}\langle \nabla_l g, \nabla_l f\rangle_\vartheta$ denote the derivative of one regularization term. Note that for some choices of bases $\{\phi_j^k\}_{j=1}^{m_k}$, one can compute the covariance matrices analytically before the main algorithm starts, e.g. Fourier cosine bases have $C_{ij}^k = \delta_{ij}$ and $\dot{C}_{ij}^k = (i-1)^2\,\pi^2\,\delta_{ij}$.

The approximated cost function in Equation 6 is

$$
\tilde{\mathcal{C}}[g] \;\;=\;\; \|g\|_\xi^2 - 2\langle g, \mu - f\rangle_\xi + \|\mu - f\|_\xi^2 + \sum_{k=1}^{d}\sigma_k^2\Big(\|\nabla_k g\|_\vartheta^2 + 2\langle \nabla_k g, \nabla_k f\rangle_\vartheta + \|\nabla_k f\|_\vartheta^2\Big).
$$

The non-zero gradients of all inner products of this equation w.r.t. parameter vector $\boldsymbol{b}^k \in \mathbb{R}^{m_k}$ are

$$
\frac{\partial}{\partial \boldsymbol{b}^k}\langle g, g\rangle_\xi = 2\,\langle \boldsymbol{\phi}^k \cdot \prod_{l\neq k}g^l, \prod_{l\neq k}g^l \cdot \boldsymbol{\phi}^{k\top}\rangle_\xi \boldsymbol{b}^k,
$$

$$
\frac{\partial}{\partial \boldsymbol{b}^k}\langle g, \mu - f\rangle_\xi = \langle \boldsymbol{\phi}^k \cdot \prod_{l\neq k}g^l, \mu - f\rangle_\xi,
$$

$$
\frac{\partial}{\partial \boldsymbol{b}^k}\langle \nabla_l g, \nabla_l g\rangle_\vartheta = \frac{\partial}{\partial \boldsymbol{b}^k}\langle \nabla_l g^l, \nabla_l g^l\rangle_{\vartheta^l}\prod_{s\neq l}\overbrace{\langle g^s, g^s\rangle_{\vartheta^s}}^{1} \;\;=\;\; 2\,\delta_{kl}\,\dot{\mathbf{C}}^k \boldsymbol{b}^k,
$$

$$
\boldsymbol{R}_l^k \;\;:=\;\; \frac{\partial}{\partial \boldsymbol{b}^k}\langle \nabla_l g, \nabla_l f\rangle_\vartheta = \begin{cases} \dot{\mathbf{C}}^k \mathbf{B}^k\Big[a \cdot \prod_{s\neq k}\mathbf{B}^{s\top}\mathbf{C}^s \boldsymbol{b}^s\Big] & \text{, if } k = l \\[2mm] \mathbf{C}^k \mathbf{B}^k\Big[a \cdot \mathbf{B}^{l\top}\dot{\mathbf{C}}^l \boldsymbol{b}^l \prod_{s\neq k\neq l}\mathbf{B}^{s\top}\mathbf{C}^s \boldsymbol{b}^s\Big] & \text{, if } k \neq l \end{cases}.
$$

Setting this to zero yields the unconstrained solution g_{uc}^k,

$$
\boldsymbol{b}_{uc}^k = \Big(\overbrace{\langle \boldsymbol{\phi}^k \cdot \prod_{l\neq k}g^l, \prod_{l\neq k}g^l \cdot \boldsymbol{\phi}^{k\top}\rangle_\xi + \sigma_k^2\dot{\mathbf{C}}^k}^{\text{regularized covariance matrix } \bar{\mathbf{C}}^k}\Big)^{-1}\Big(\langle \boldsymbol{\phi}^k \cdot \prod_{l\neq k}g^l, \mu - f\rangle_\xi - \sum_{l=1}^{d}\boldsymbol{R}_l^k \sigma_l^2\Big). \quad (11)
$$

However, these parameters do not satisfy to the constraint $\|g'^k\|_{\vartheta^k} \overset{!}{=} 1$, and have to be normalized:

$$b'^k \quad := \quad \frac{b_{uc}^k}{\|g_{uc}^k\|_{\vartheta^k}} \quad = \quad \frac{b_{uc}^k}{\sqrt{b_{uc}^{k\top} C^k b_{uc}^k}}. \tag{12}$$

The inner loop finishes when for all k the improvement[12] from g^k to g'^k drops below some very small threshold ϵ, i.e. $\tilde{C}[g] - \tilde{C}[g'] < \epsilon$. Using $g'^l = g^l, \forall l \neq k$, one can calculate the left hand side:

$$\tilde{C}[g] - \tilde{C}[g'] = \|g\|_\xi^2 - \|g'\|_\xi^2 - 2\langle g - g', \mu - f\rangle_\xi$$
$$+ \sum_{l=1}^d \sigma_l^2 \Big[\underbrace{\|\nabla_l g\|_\vartheta^2}_{b^{l\top} \dot{C}^l b^l} - \underbrace{\|\nabla_l g'\|_\vartheta^2}_{b'^{l\top} \dot{C}^l b'^l} - 2 \underbrace{\langle \nabla_l g - \nabla_l g', \nabla_l f\rangle_\vartheta}_{(b^k - b'^k)^\top R_l^k} \Big] \tag{13}$$
$$= 2\langle g - g', \mu - f\rangle_\xi + b^{k\top} \bar{C}^k b^{k\top} - b'^{k\top} \bar{C}^k b'^{k\top} - 2(b^k - b'^k)^\top \Big(\sum_{l=1}^d R_l^k \sigma_l^2\Big).$$

Appendix C Proofs of the Propositions

Proof of Proposition 2: The 1st order Taylor approximation of any $g, f \in L^2(\mathcal{X}, \xi\chi)$ around $z \in \mathcal{X}$ is $f(x) = f(z + x - z) \approx f(z) + (x - z)^\top \nabla f(z)$. For the Hilbert space $L^2(\mathcal{X}, \xi\chi)$ we can thus approximate:

$$\langle g, f\rangle_{\xi\chi} = \int \xi(dz) \int \chi(dx|z) g(x) f(x)$$
$$\approx \int \xi(dz) \Big(g(z) f(z) \overbrace{\int \xi(dx|z)}^{1} + g(z) \overbrace{\int \chi(dx|z) (x - z)^\top}^{0 \text{ due to (eq.5)}} \nabla f(z)$$
$$+ \underbrace{\int \chi(dx|z) (x - z)^\top}_{0 \text{ due to (eq.5)}} \nabla g(z) f(z) + \nabla g(z)^\top \underbrace{\int \chi(dx|z) (x - z)(x - z)^\top}_{\frac{\vartheta}{\xi}(z) \cdot \Sigma \text{ due to (eq.5)}} \nabla f(z)\Big)$$
$$= \langle g, f\rangle_\xi + \sum_{k=1}^d \sigma_k^2 \langle \nabla_k g, \nabla_k f\rangle_\vartheta.$$

Using this twice and the zero mean assumption (Eq. 5), we can derive:

$$\inf_{g \in \mathcal{F}} \mathcal{C}[f + g|\chi, \mu] \equiv \inf_{g \in \mathcal{F}} \iint \xi(dz) \chi(dx|z) \Big(g^2(x) - 2g(x)(\mu(z) - f(x))\Big)$$
$$= \inf_{g \in \mathcal{F}} \langle g, g\rangle_{\xi\chi} + 2\langle g, f\rangle_{\xi\chi} - 2\int \xi(dz) \mu(z) \int \chi(dx|z) g(x)$$
$$\approx \inf_{g \in \mathcal{F}} \langle g, g\rangle_\xi - 2\langle g, \mu - f\rangle_\xi + \sum_{k=1}^d \sigma_k^2 \Big(\langle \nabla_k g, \nabla_k g\rangle_\vartheta + 2\langle \nabla_k g, \nabla_k f\rangle_\vartheta\Big)$$
$$\equiv \inf_{g \in \mathcal{F}} \|g - (\mu - f)\|_\xi^2 + \sum_{k=1}^d \sigma_k^2 \|\nabla_k g + \nabla_k f\|_\vartheta^2 \quad = \quad \tilde{C}[g]. \qquad \square$$

---

[12] Anything simpler does not converge, as the parameter vectors often evolve along chaotic orbits in $\mathbb{R}^{m_k}$.

Proof of Proposition 3: The analytical solution to the optimization problem in Equation 6 is derived in Appendix B and has a unique solution if the matrix $\bar{\mathbf{C}}^k$, defined in Equation 11, is of full rank:

$$\bar{\mathbf{C}}^k \quad := \quad \langle \phi^k \cdot \prod_{l \neq k} g^l, \prod_{l \neq k} g^l \cdot \phi^{k\top} \rangle_\xi \quad + \quad \sigma_k^2 \dot{\mathbf{C}}^k.$$

For Fourier bases the matrix $\dot{\mathbf{C}}^k$ is diagonal, with $\dot{C}_{11}^k$ being the only zero entry. $\bar{\mathbf{C}}^k$ is therfore full rank if $\sigma_k^2 > 0$ and $\bar{C}_{11}^k > 0$. Because ϑ is *absolutely continuous* w.r.t. ξ, the constraint $\|g^l\|_\vartheta = 1, \forall l$, implies that there exist no g^l that is zero on *all* training samples.   As the first Fourier base is a constant, $\langle \phi_1^k \cdot \prod_{l \neq k} g^l, \prod_{l \neq k} g^l \cdot \phi_1^{k\top} \rangle_\xi > 0$ and the matrix $\bar{\mathbf{C}}^k$ is therefore of full rank. $\square$

Algorithm 1. (detailed) – LFF-Regression

Input: $\mathbf{X} \in \mathbb{R}^{d \times n}$, $\boldsymbol{y} \in \mathbb{R}^n$, $\boldsymbol{\sigma}^2 \in \mathbb{R}^d$ $\epsilon, \varepsilon \in \mathbb{R}$	
$\mathbf{C}^k := \langle \phi^k, \phi^k \rangle_{\vartheta k}$, $\dot{\mathbf{C}}^k := \langle \nabla\phi^k, \nabla\phi^k \rangle_{\vartheta k}$, $\forall k$	// analytical covariance matrices
$\Phi_{jt}^k := \phi_j^k(X_{kt})$, $\forall k, \forall j, \forall t$	// optional cache of sample-expansion
$\boldsymbol{f} := 0 \in \mathbb{R}^n;$ $\boldsymbol{a} := \emptyset;$ $\mathbf{B}^k := \emptyset,$ $\forall k;$ $\boldsymbol{\Psi} := \infty$	// initialize empty $f \in \mathcal{F}^0$
while $\det\left(\frac{1}{n}\boldsymbol{\Psi}\boldsymbol{\Psi}^\top\right) > \varepsilon$ **do**	
$\quad \boldsymbol{b}^k := \mathbf{1}^k \in \mathbb{R}^{m_k},$ $\forall k;$ $\boldsymbol{g}^k := \mathbf{1} \in \mathbb{R}^n,$ $\forall k$	// initialize all g^k as constant
$\quad \boldsymbol{h} := \infty \in \mathbb{R}^d$	// initialize estimated improvement
$\quad$ **while** $\max(\boldsymbol{h}) > \epsilon$ **do**	
$\quad\quad$ **for** k in randperm$(1, \ldots, d)$ **do**	
$\quad\quad\quad \boldsymbol{R}_k := \dot{\mathbf{C}}^k \mathbf{B}^k \left[\boldsymbol{a} \cdot \prod_{s \neq k} \mathbf{B}^{s\top} \mathbf{C}^s \boldsymbol{b}^s\right]$	// $\boldsymbol{R}_k = \frac{\partial}{\partial \boldsymbol{b}^k} \langle \nabla_k g, \nabla_k f \rangle_\vartheta$
$\quad\quad\quad \boldsymbol{R}_l := \mathbf{C}^k \mathbf{B}^k \left[\boldsymbol{a} \cdot \mathbf{B}^{l\top} \dot{\mathbf{C}}^l \boldsymbol{b}^l \cdot \prod_{s \neq k \neq l} \mathbf{B}^{s\top} \mathbf{C}^s \boldsymbol{b}^s\right],$ $\forall l \neq k$	// $\boldsymbol{R}_l = \frac{\partial}{\partial \boldsymbol{b}^k} \langle \nabla_l g, \nabla_l f \rangle_\vartheta$
$\quad\quad\quad \bar{\mathbf{C}} := \Phi^k \left[\Phi^{k\top} \cdot \prod_{l \neq k} (g^l)^2 \mathbf{1}^\top\right] + \sigma_k^2 \dot{\mathbf{C}}^k$	// regularized cov. matrix (eq. 11)
$\quad\quad\quad \boldsymbol{b}' := \bar{\mathbf{C}}^{-1}\left(\Phi^k\left[(\boldsymbol{y} - \boldsymbol{f}) \cdot \prod_{l \neq k} g^l\right] - \boldsymbol{R}\sigma^2\right)$	// unconstrained g_{uc}^k (eq. 11)
$\quad\quad\quad \boldsymbol{b}' := \boldsymbol{b}' / \sqrt{\boldsymbol{b}'^\top \mathbf{C}^k \boldsymbol{b}'}$	// enforce constraints (eq. 12)
$\quad\quad\quad h_k := \frac{2}{n} (\boldsymbol{b}^k - \boldsymbol{b}')^\top \left(\Phi^k\left[(\boldsymbol{y} - \boldsymbol{f}) \cdot \prod_{l \neq k} g^l\right]\right)$	// approximate $2\langle g - g', \mu - f \rangle_\xi$
$\quad\quad\quad h_k := h_k + \boldsymbol{b}^k \bar{\mathbf{C}} \boldsymbol{b}^k - \boldsymbol{b}' \bar{\mathbf{C}} \boldsymbol{b}' - 2(\boldsymbol{b}^k - \boldsymbol{b}')^\top \boldsymbol{R}\sigma^2$	// cost improvement (eq. 13)
$\quad\quad\quad \boldsymbol{b}^k := \boldsymbol{b}';$ $\boldsymbol{g}^k := \Phi^{k\top} \boldsymbol{b}^k$	// update factor function g^k
$\quad\quad$ **end for** // end function g^k update	
$\quad$ **end while** // end inner loop: cost function converged and thus g optimized	
$\quad \mathbf{B}^k := [\mathbf{B}^k, \boldsymbol{b}^k],$ $\forall k;$ $\boldsymbol{\Psi} := \left[\prod_{k=1}^d \mathbf{B}^{k\top}\Phi^k\right]$	// adding g to the bases functions of f
$\quad \boldsymbol{a} := (\boldsymbol{\Psi}\boldsymbol{\Psi}^\top)^{-1}\boldsymbol{\Psi}\boldsymbol{y};$ $\boldsymbol{f} := \boldsymbol{\Psi}^\top \boldsymbol{a}$	// project μ onto new bases
end while // end outer loop: new g no longer linear independent, thus $f \approx \mu$	
Output: $\boldsymbol{a} \in \mathbb{R}^m$, $\{\mathbf{B}^k \in \mathbb{R}^{m_k \times m}\}_{k=1}^d$	// return parameters of $f \in \mathcal{F}^m$

References

Bellman, R.E.: Dynamic programming. Princeton University Press (1957)

Bishop, C.M.: Pattern Recognition and Machine Learning. Springer-Verlag New York Inc, Secaucus (2006). ISBN 0387310738

Böhmer, W., Obermayer, K.: Towards structural generalization: Factored approximate planning. ICRA Workshop on Autonomous Learning (2013). http://autonomous-learning.org/wp-content/uploads/13-ALW/paper_1.pdf

Böhmer, W., Grünewälder, S., Nickisch, H., Obermayer, K.: Generating feature spaces for linear algorithms with regularized sparse kernel slow feature analysis. Machine Learning 89(1–2), 67–86 (2012)

Böhmer, W., Grünewälder, S., Shen, Y., Musial, M., Obermayer, K.: Construction of approximation spaces for reinforcement learning. Journal of Machine Learning Research 14, 2067–2118 (2013)

Boser, B.E., Guyon, I.M., Vapnik, V.N.: A training algorithm for optimal margin classifiers. In: Proceedings of the Fifth Annual Workshop on Computational Learning Theory, pp. 144–152 (1992)

Cortez, P., Cerdeira, A., Almeida, F., Matos, T., Reis, J.: Modeling wine preferences by data mining from physicochemical properties. Decision Support Systems 47(4), 547–553 (2009)

Csató, L., Opper, M.: Sparse on-line Gaussian processes. Neural Computation 14(3), 641–668 (2002)

Friedman, J.H.: Multivariate adaptive regression splines. The Annals of Statistics 19(1), 1–67 (1991)

Gerritsma, J., Onnink, R., Versluis, A.: Geometry, resistance and stability of the delft systematic yacht hull series. Int. Shipbuilding Progress 28, 276–297 (1981)

Gönen, M., Alpaydın, E.: Multiple kernel learning algorithms. Journal of Machine Learning Research 12, 2211–2268 (2011). ISSN 1532–4435

Haykin, S.: Neural Networks: A Comprehensive Foundation, 2nd edn. Prentice Hall (1998). ISBN 978-0132733502

Kaelbling, L.P., Littman, M.L., Moore, A.W.: Reinforcement learning: a survey. Journal of Artificial Intelligence Research 4, 237–285 (1996)

Nelder, J.A., Wedderburn, R.W.M.: Generalized linear models. Journal of the Royal Statistical Society, Series A, General 135, 370–384 (1972)

Pearl, J.: Probabilistic reasoning in intelligent systems. Morgan Kaufmann (1988)

Rasmussen, C.E., Williams, C.K.I.: Gaussian Processes for Machine Learning. MIT Press (2006)

Shawe-Taylor, J., Cristianini, N.: Kernel Methods for Pattern Analysis. Cambridge University Press (2004)

Tipping, M.E.: Sparse bayesian learning and the relevance vector machine. Journal of Machine Learning Research 1, 211–244 (2001). ISSN 1532–4435

Tüfekci, P.: Prediction of full load electrical power output of a base load operated combined cycle power plant using machine learning methods. International Journal of Electrical Power & Energy Systems 60, 126–140 (2014)

Vapnik, V.N.: The Nature of Statistical Learning Theory. Springer (1995)

Wang, Z., Crammer, K., Vucetic, S.: Breaking the curse of kernelization: Budgeted stochastic gradient descent for large-scale svm training. Journal of Machine Learning Research 13(1), 3103–3131 (2012). ISSN 1532–4435

Yeh, I.-C.: Modeling of strength of high performance concrete using artificial neural networks. Cement and Concrete Research 28(12), 1797–1808 (1998)

Ridge Regression, Hubness, and Zero-Shot Learning

Yutaro Shigeto[1], Ikumi Suzuki[2], Kazuo Hara[3],
Masashi Shimbo[1(✉)], and Yuji Matsumoto[1]

[1] Nara Institute of Science and Technology, Ikoma, Nara, Japan
{yutaro-s,shimbo,matsu}@is.naist.jp
[2] The Institute of Statistical Mathematics, Tachikawa, Tokyo, Japan
suzuki.ikumi@gmail.com
[3] National Institute of Genetics, Mishima, Shizuoka, Japan
kazuo.hara@gmail.com

Abstract. This paper discusses the effect of hubness in zero-shot learning, when ridge regression is used to find a mapping between the example space to the label space. Contrary to the existing approach, which attempts to find a mapping from the example space to the label space, we show that mapping labels into the example space is desirable to suppress the emergence of hubs in the subsequent nearest neighbor search step. Assuming a simple data model, we prove that the proposed approach indeed reduces hubness. This was verified empirically on the tasks of bilingual lexicon extraction and image labeling: hubness was reduced with both of these tasks and the accuracy was improved accordingly.

1 Introduction

1.1 Background

In recent years, *zero-shot learning* (ZSL) [10,14,15,22] has been an active research topic in machine learning, computer vision, and natural language processing. Many practical applications can be formulated as a ZSL task: drug discovery [15], bilingual lexicon extraction [7,8,20], and image labeling [2,11,21,22,25], to name a few. Cross-lingual information retrieval [28] can also be viewed as a ZSL task.

ZSL can be regarded as a type of (multi-class) classification problem, in the sense that the classifier is given a set of known example-class label pairs (training set), with the goal to predict the unknown labels of new examples (test set). However, ZSL differs from the standard classification in that the labels for the test examples are not present in the training set. In standard settings, the classifier chooses, for each test example, a label among those observed in the training set, but this is not the case in ZSL. Moreover, the number of class labels can be huge in ZSL; indeed, in bilingual lexicon extraction, labels correspond to possible translation words, which can range over entire vocabulary of the target language.

Obviously, such a task would be intractable without further assumptions. Labels are thus assumed to be embedded in a metric space (*label space*), and their

© Springer International Publishing Switzerland 2015
A. Appice et al. (Eds.): ECML PKDD 2015, Part I, LNAI 9284, pp. 135–151, 2015.
DOI: 10.1007/978-3-319-23528-8_9

distance (or similarity) can be measured in this space[1]. Such a label space can be built with the help of background knowledge or external resources; in image labeling tasks, for example, labels correspond to annotation keywords, which can be readily represented as vectors in a Euclidean space, either by using corpus statistics in a standard way, or by using the more recent techniques for learning word representations, such as the continuous bag-of-words or skip-gram models [19].

After a label space is established, one natural approach would be to use a regression technique on the training set to obtain a mapping function from the example space to the label space. This function could then be used for mapping unlabeled examples into the label space, where nearest neighbor search is carried out to find the label closest to the mapped example. Finally, this label would be output as the prediction for the example.

To find the mapping function, some researchers use the standard linear ridge regression [7,8,20,22], whereas others use neural networks [11,21,25].

In the machine learning community, meanwhile, the *hubness phenomenon* [23] is attracting attention as a new type of the "curse of dimensionality." This phenomenon is concerned with nearest neighbor methods in high-dimensional space, and states that a small number of objects in the dataset, or *hubs*, may occur as the nearest neighbor of many objects. The emergence of these hubs will diminish the utility of nearest neighbor search, because the list of nearest neighbors often contain the same hub objects regardless of the query object for which the list is computed.

1.2   Research Objective and Contributions

In this paper, we show the interaction between the regression step in ZSL and the subsequent nearest neighbor step has a non-negligible effect on the prediction accuracy.

In ZSL, examples and labels are represented as vectors in high-dimensional space, of which the dimensionality is typically a few hundred. As demonstrated by Dinu and Baroni [8] (see also Sect. 6), when ZSL is formulated as a problem of ridge regression from examples to labels, "hub" labels emerge, which are simultaneously the nearest neighbors of many mapped examples. This has the consequence of incurring bias in the prediction, as these labels are output as the predicted labels for these examples. The presence of hubs are not necessarily disadvantageous in standard classification settings; there may be "good" hubs as well as "bad" hubs [23]. However, in typical ZSL tasks in which the label set is fine-grained and huge, hubs are nearly always harmful to the prediction accuracy.

Therefore, the objective of this study is to investigate ways to suppress hubs, and to improve the ZSL accuracy. Our contributions are as follows.

1. We analyze the mechanism behind the emergence of hubs in ZSL, both with ridge regression and ordinary least squares. It is established that hubness occurs in ZSL not only because of high-dimensional space, but also because

---

[1] Throughout the paper, we assume both the example and label spaces are Euclidean.

ridge regression has conventionally been used in ZSL in a way that *promotes* hubness. To be precise, the distributions of the mapped examples and the labels are different such that hubs are likely to emerge.

2. Drawing on the above analysis, we propose using ridge regression to map labels into the space of examples. This approach is contrary to that followed in existing work on ZSL, in which examples are mapped into label space. Our proposal is therefore to reverse the mapping direction.

   As shown in Sect. 6, our proposed approach outperformed the existing approach in an empirical evaluation using both synthetic and real data.

3. In terms of contributions to the research on hubness, this paper is the first to provide in-depth analysis of the situation in which the query and data follow different distributions, and to show that the variance of data matters to hubness. In particular, in Sect. 3, we provide a proposition in which the degree of bias present in the data, which causes hub formation, is expressed as a function of the data variance. In Sect 4, this proposition serves as the main tool for analyzing hubness in ZSL.

2   Zero-Shot Learning as a Regression Problem

Let X be a set of examples, and Y be a set of class labels. In ZSL, not only examples but also labels are assumed to be vectors. For this reason, examples are sometimes referred to as *source objects*, and labels as *target objects*. In the subsequent sections of this paper, we mostly follow this terminology when referring to the members of X and Y.

Let $X \subset \mathbb{R}^c$ and $Y \subset \mathbb{R}^d$. These spaces, $\mathbb{R}^c$ and $\mathbb{R}^d$, are called *source space* and *target space*, respectively. Although X can be the entire space $\mathbb{R}^c$, Y is usually a finite set of points in $\mathbb{R}^d$, even though its size may be enormous in some problems.

Let $X_{\text{train}} = \{\mathbf{x}_i \mid i = 1, \ldots, n\}$ be the training examples (training source objects), and $Y_{\text{train}} = \{\mathbf{y}_i \mid i = 1, \ldots, n\}$ be their labels (training target objects); i.e., the class label of example $\mathbf{x}_i$ is $\mathbf{y}_i$, for each $i = 1, \ldots, n$. In a standard classification setting, the labels in the training set are equal to the entire set of labels; i.e., $Y_{\text{train}} = Y$. In contrast, this assumption is not made in ZSL, and Y_{train} is a strict subset of Y. Moreover, it is assumed that the true class labels of test examples do not belong to Y_{train}; i.e., they belong to $Y \backslash Y_{\text{train}}$.

In such a situation, it is difficult to find a function f that maps $\mathbf{x} \in X$ directly to a label in Y. Therefore, a popular (and also natural) approach is to learn a projection $m : \mathbb{R}^c \to \mathbb{R}^d$, which can be done with a regression technique. With a projection function m at hand, the label of a new source object $\mathbf{x} \in \mathbb{R}^c$ is predicted to be the one closest to the mapped point $m(\mathbf{x})$ in the target space. The prediction function f is thus given by

$$f(\mathbf{x}) = \arg\min_{\mathbf{y} \in Y} \|m(\mathbf{x}) - \mathbf{y}\|.$$

After a source object $\mathbf{x}$ is projected to $m(\mathbf{x})$, the task is reduced to that of nearest neighbor search in the target space.

3   Hubness Phenomenon and the Variance of Data

The utility of nearest neighbor search would be significantly reduced if the same objects were to appear consistently as the search result, irrespective of the query. Radovanović et al. [23] showed that such objects, termed *hubs*, indeed occur in high-dimensional space. Although this phenomenon may seem counter-intuitive, hubness is observed in a variety of real datasets and distance/similarity measures used in combination [23,24,26].

The aim of this study is to analyze the hubness phenomenon in ZSL, which involves nearest neighbor search in high-dimensional space as the last step. However, as a tool for analyzing ZSL, the existing theory on hubness [23] is inadequate, as it was mainly developed for comparing the emergence of hubness in spaces of different dimensionalities.

In the analysis of ZSL in Sect. 4.2, we aim to compare two distributions in the same space, but which differ in terms of *variance*. To this end, we first present a proposition below, which is similar in spirit to the main theorem of Radovanović et al. [23, Theorem 1], but which distinguishes the query and data distributions, and also expresses the expected difference between the squared distances from queries to database objects in terms of their variance.

The proposition is concerned with nearest neighbor search, in which $\mathbf{x}$ is a query, and $\mathbf{y}_1$ and $\mathbf{y}_2$ are two objects in a dataset. In the context of ZSL as formulated in Sect. 2, $\mathbf{x}$ represents the image of a source object in the target space (through the learned regression function m), and $\mathbf{y}_1$ and $\mathbf{y}_2$ are target objects (labels) lying at different distances from the origin. We are interested in which of $\mathbf{y}_1$ and $\mathbf{y}_2$ are more likely to be closer to $\mathbf{x}$, when $\mathbf{x}$ is sampled from a distribution $\mathcal{X}$ with zero mean.

Let $\mathrm{E}[\cdot]$ and $\mathrm{Var}[\cdot]$ denote the expectation and variance, respectively, and let $\mathcal{N}(\boldsymbol{\mu}, \boldsymbol{\Sigma})$ be a multivariate normal distribution with mean $\boldsymbol{\mu}$ and covariance matrix $\boldsymbol{\Sigma}$.

Proposition 1. *Let* $\mathbf{y} = [y_1, \ldots, y_d]^{\mathrm{T}}$ *be a d-dimensional random vector, with components y_i ($i = 1, \ldots, d$) sampled i.i.d. from a normal distribution with zero mean and variance s^2; i.e., $\mathbf{y} \sim \mathcal{Y}$, where $\mathcal{Y} = \mathcal{N}(\mathbf{0}, s^2 \mathbf{I})$. Further let $\sigma = \sqrt{\mathrm{Var}_{\mathcal{Y}}[\|\mathbf{y}\|^2]}$ be the standard deviation of the squared norm $\|\mathbf{y}\|^2$.*

Consider two fixed samples $\mathbf{y}_1$ and $\mathbf{y}_2$ of random vector $\mathbf{y}$, such that the squared norms of $\mathbf{y}_1$ and $\mathbf{y}_2$ are $\gamma\sigma$ apart. In other words,

$$\|\mathbf{y}_2\|^2 - \|\mathbf{y}_1\|^2 = \gamma\sigma.$$

Let $\mathbf{x}$ be a point sampled from a distribution $\mathcal{X}$ with zero mean. Then, the expected difference Δ between the squared distances from $\mathbf{y}_1$ and $\mathbf{y}_2$ to $\mathbf{x}$, i.e.,

$$\Delta = \mathrm{E}_{\mathcal{X}}\left[\|\mathbf{x} - \mathbf{y}_2\|^2\right] - \mathrm{E}_{\mathcal{X}}\left[\|\mathbf{x} - \mathbf{y}_1\|^2\right] \tag{1}$$

is given by

$$\Delta = \sqrt{2}\gamma d^{1/2} s^2. \tag{2}$$

Proof. For $i = 1, 2$, the distance between a point $\mathbf{x}$ and $\mathbf{y}_i$ is given by

$$\|\mathbf{x} - \mathbf{y}_i\|^2 = \|\mathbf{x}\|^2 + \|\mathbf{y}_i\|^2 - 2\mathbf{x}^{\mathrm{T}}\mathbf{y}_i,$$

and its expected value is

$$\mathrm{E}_{\mathcal{X}}\left[\|\mathbf{x} - \mathbf{y}_i\|^2\right] = \mathrm{E}_{\mathcal{X}}\left[\|\mathbf{x}\|^2\right] + \|\mathbf{y}_i\|^2 - 2\mathrm{E}_{\mathcal{X}}\left[\mathbf{x}\right]^{\mathrm{T}}\mathbf{y}_i = \mathrm{E}_{\mathcal{X}}\left[\|\mathbf{x}\|^2\right] + \|\mathbf{y}_i\|^2,$$

since $\mathrm{E}_{\mathcal{X}}\left[\mathbf{x}\right] = 0$ by assumption. Substituting this equality in (1) yields

$$\Delta = \overbrace{\left(\mathrm{E}_{\mathcal{X}}\left[\|\mathbf{x}\|^2\right] + \|\mathbf{y}_2\|^2\right)}^{\mathrm{E}_{\mathcal{X}}[\|\mathbf{x}-\mathbf{y}_2\|^2]} - \overbrace{\left(\mathrm{E}_{\mathcal{X}}[\|\mathbf{x}\|^2] + \|\mathbf{y}_1\|^2\right)}^{\mathrm{E}_{\mathcal{X}}[\|\mathbf{x}-\mathbf{y}_1\|^2]} = \|\mathbf{y}_2\|^2 - \|\mathbf{y}_1\|^2 = \gamma\sigma. \quad (3)$$

Now, it is well known that if a d-dimensional random vector $\mathbf{z}$ follows the multivariate standard normal distribution $\mathcal{N}(\mathbf{0}, \mathbf{I})$, then its squared norm $\|\mathbf{z}\|^2$ follows the chi-squared distribution with d degrees of freedom, and its variance is $2d$. Since $\mathbf{y} = s\mathbf{z}$, the variance σ^2 of the squared norm $\|\mathbf{y}\|^2$ is

$$\sigma^2 = \mathrm{Var}_{\mathcal{Y}}\left[\|\mathbf{y}\|^2\right] = \mathrm{Var}_{\mathcal{Z}}\left[s^2\|\mathbf{z}\|^2\right] = s^4\,\mathrm{Var}_{\mathcal{Z}}\left[\|\mathbf{z}\|^2\right] = 2ds^4. \quad (4)$$

From (3) and (4), we obtain $\Delta = \gamma s^2\sqrt{2d}$. $\qquad\square$

Note that in Proposition 1, the standard deviation σ is used as a yardstick of measurement to allow for comparison of "similarly" located object pairs across different distributions; two object pairs in different distributions are regarded as similar if objects in each pair are $\gamma\sigma$ apart as measured by the σ for the respective distributions, but has an equal factor γ. This technique is due to Radovanović et al. [23].

Now, Δ represents the expected difference between the squared distances from $\mathbf{x}$ to $\mathbf{y_1}$ and $\mathbf{y_2}$. Equation (2) shows that Δ increases with γ, the factor quantifying the amount of difference $\|\mathbf{y}_2\|^2 - \|\mathbf{y}_1\|^2$. This suggests that a query object sampled from $\mathcal{X}$ is more likely to be closer to object $\mathbf{y}_1$ than to $\mathbf{y}_2$, if $\|\mathbf{y}_1\|^2 < \|\mathbf{y}_2\|^2$; i.e., $\mathbf{y}_1$ is closer to the origin than $\mathbf{y}_2$ is. Because this holds for any pair of objects $\mathbf{y}_1$ and $\mathbf{y}_2$ in the dataset, we can conclude that the objects closest to the origin in the dataset tend to be hubs.

Equation (2) also states the relationship between Δ and the component variance s^2 of distribution $\mathcal{Y}$, by which the following is implied: For a fixed query distribution $\mathcal{X}$, if we have two distributions for $\mathbf{y}$, $\mathcal{Y}_1 = \mathcal{N}(\mathbf{0}, s_1^2\mathbf{I})$ and $\mathcal{Y}_2 = \mathcal{N}(\mathbf{0}, s_2^2\mathbf{I})$ with $s_1^2 < s_2^2$, it is preferable to choose $\mathcal{Y}_1$, i.e., the distribution with a smaller s^2, when attempting to reduce hubness. Indeed, assuming the independence of $\mathcal{X}$ and $\mathcal{Y}$, we can show that the influence of Δ relative to the expected squared distance from $\mathbf{x}$ to $\mathbf{y}$ (which is also subject to whether $\mathbf{y} \sim \mathcal{Y}_1$ or $\mathcal{Y}_2$), is weaker for $\mathcal{Y}_1$ than for $\mathcal{Y}_2$, i.e.,

$$\frac{\Delta(\gamma, d, s_1)}{\mathrm{E}_{\mathcal{X}\mathcal{Y}_1}[\|\mathbf{x} - \mathbf{y}\|^2]} < \frac{\Delta(\gamma, d, s_2)}{\mathrm{E}_{\mathcal{X}\mathcal{Y}_2}[\|\mathbf{x} - \mathbf{y}\|^2]},$$

where we wrote Δ explicitly as a function of γ, d, and s.

4   Hubness in Regression-Based Zero-Shot Learning

In this section, we analyze the emergence of hubs in the nearest neighbor step of ZSL. Through the analysis, it is shown that hubs are promoted by the use of ridge regression in the existing formulation of ZSL, i.e., mapping source objects (examples) into the target (label) space.

As a solution, we propose using ridge regression in a direction opposite to that in existing work. That is, we project target objects in the space of source objects, and carry out nearest neighbor search in the source space. Our argument for this approach consists of three steps.

1. We first show in Sect. 4.1 that, with ridge regression (and ordinary least squares as well), mapped observation data tend to lie closer to the origin than the target responses do. Because the existing work formulates ZSL as a regression problem that projects source objects into the target space, this means that the norm of the projected source objects tends to be smaller than that of target objects.
2. By combining the above result with the discussion of Sect. 3, we then argue that placing source objects closer to the origin is not ideal from the perspective of reducing hubness. On the contrary, placing target objects closer to the origin, as attained with the proposed approach, is more desirable (Sect. 4.2).
3. In Sect. 4.3, we present a simple additional argument against placing source objects closer to the origin; if the data is unimodal, such a configuration increases the possibility of another target object falling closer to the source object. This argument diverges from the discussion on hubness, but again justifies the proposed approach.

4.1   Shrinkage of Projected Objects

We first prove that ridge regression tends to map observation data closer to the origin of the space. This tendency may be easily observed in ridge regression, for which the penalty term shrinks the estimated coefficients towards zero. However, the above tendency is also inherent in ordinary least squares.

Let $\| \cdot \|_F$ and $\| \cdot \|_2$ respectively denote the Frobenius norm and the 2-norm of matrices.

Proposition 2. *Let* $\mathbf{M} \in \mathbb{R}^{d \times c}$ *be the solution for ridge regression with an observation matrix* $\mathbf{A} \in \mathbb{R}^{c \times n}$ *and a response matrix* $\mathbf{B} \in \mathbb{R}^{d \times n}$; *i.e.,*

$$\mathbf{M} = \arg \min_{\mathbf{M}} \left(\|\mathbf{MA} - \mathbf{B}\|_F^2 + \lambda \|\mathbf{M}\|_F \right). \tag{5}$$

where $\lambda \geq 0$ *is a hyperparameter. Then, we have* $\|\mathbf{MA}\|_2 \leq \|\mathbf{B}\|_2$.

Proof (Sketch). It is well known that $\mathbf{M} = \mathbf{BA}^{\mathrm{T}} \left(\mathbf{AA}^{\mathrm{T}} + \lambda \mathbf{I} \right)^{-1}$. Thus we have

$$\|\mathbf{MA}\|_2 = \|\mathbf{BA}^{\mathrm{T}} \left(\mathbf{AA}^{\mathrm{T}} + \lambda \mathbf{I} \right)^{-1} \mathbf{A}\|_2 \leq \|\mathbf{B}\|_2 \|\mathbf{A}^{\mathrm{T}} \left(\mathbf{AA}^{\mathrm{T}} + \lambda \mathbf{I} \right)^{-1} \mathbf{A}\|_2. \tag{6}$$

Let σ be the largest singular value of $\mathbf{A}$. It can be shown that

$$\|\mathbf{A}^{\mathrm{T}}\left(\mathbf{A}\mathbf{A}^{\mathrm{T}} + \lambda\mathbf{I}\right)^{-1}\mathbf{A}\|_2 = \frac{\sigma^2}{\sigma^2 + \lambda} \leq 1.$$

Substituting this inequality in (6) establishes the proposition. $\square$

Recall that if the data is centered, the matrix 2-norm can be interpreted as an indicator of the variance of data along its principal axis. Proposition 2 thus indicates that the variance along the principal axis of the mapped observations $\mathbf{MA}$ tends to be smaller than that of responses $\mathbf{B}$.

Furthermore, this tendency even persists in the ordinary least squares with no penalty term (i.e., $\lambda = 0$), since $\|\mathbf{MA}\|_2 \leq \|\mathbf{B}\|_2$ still holds in this case; note that $\mathbf{A}^{\mathrm{T}}\left(\mathbf{A}\mathbf{A}^{\mathrm{T}}\right)^{-1}\mathbf{A}$ is an orthogonal projection and its 2-norm is 1, but the inequality in (6) holds regardless. This tendency therefore cannot be completely eliminated by simply decreasing the ridge parameter λ towards zero.

In existing work on ZSL, $\mathbf{A}$ represents the (training) source objects $\mathbf{X} = [\mathbf{x}_1 \cdots \mathbf{x}_n] \in \mathbb{R}^{c \times n}$, to be mapped into the space of target objects (by projection matrix $\mathbf{M}$); and $\mathbf{B}$ is the matrix of labels for the training objects, i.e., $\mathbf{B} = \mathbf{Y} = [\mathbf{y}_1 \cdots \mathbf{y}_n] \in \mathbb{R}^{d \times n}$. Although Proposition 2 is thus only concerned with the training set, it suggests that the source objects at the time of testing, which are not in $\mathbf{X}$, are also likely to be mapped closer to the origin of the target space than many of the target objects in $\mathbf{Y}$.

4.2 Influence of Shrinkage on Nearest Neighbor Search

We learned in Sect. 4.1 that ridge regression (and ordinary least squares) shrink the mapped observation data towards the origin of the space, relative to the response. Thus, in existing work on ZSL in which source objects X are projected to the space of target objects Y, the norm of the mapped source objects is likely to be smaller than that of the target objects.

The proposed approach, which was described in the beginning of Sect. 4, follows the opposite direction: target objects Y are projected to the space of source objects X. Thus, in this case, the norm of the mapped target objects is expected to be smaller than that of the source objects.

The question now is which of these configurations is preferable for the subsequent nearest neighbor step, and we provide an answer under the following assumptions: (i) The source space and the target space are of equal dimensions; (ii) the source and target objects are isotropically normally distributed and independent; and (iii) the projected data is also isotropically normally distributed, except that the variance has shrunk.

Let $\mathcal{D}_1 = \mathcal{N}(0, s_1^2\mathbf{I})$ and $\mathcal{D}_2 = \mathcal{N}(0, s_2^2\mathbf{I})$ be two multivariate normal distributions, with $s_1^2 < s_2^2$. We compare two configurations of source object $\mathbf{x}$ and target objects $\mathbf{y}$: (a) the one in which $\mathbf{x} \sim \mathcal{D}_1$ and $\mathbf{y} \sim \mathcal{D}_2$, and (b) the one in which $\mathbf{x}' \sim \mathcal{D}_2$ and $\mathbf{y}' \sim \mathcal{D}_1$ on the other hand; here, the primes in (b) were added to distinguish variables in two configurations.

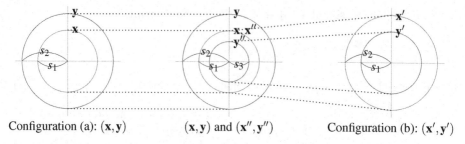

Configuration (a): $(\mathbf{x}, \mathbf{y})$ $(\mathbf{x}, \mathbf{y})$ and $(\mathbf{x}'', \mathbf{y}'')$ Configuration (b): $(\mathbf{x}', \mathbf{y}')$

Fig. 1. Schematic illustration for Sect. 4.2 in two-dimensional space. The left and the right panels depict configurations (a) and (b), respectively, with the center panel showing both configuration (a) and the scaled version of configuration (b) in the same space. A circle represents a distribution, with its radius indicating the standard deviation. The radius of the circles for $\mathbf{y}$ (on the left panel) and $\mathbf{x}'$ (right panel) is s_1, whereas that of the circles for $\mathbf{x}$ (left panel) and $\mathbf{y}'$ (right panel) is s_2, with $s_1 < s_2$. Circles $\mathbf{x}''$ and $\mathbf{y}''$ are the scaled versions of $\mathbf{x}'$ and $\mathbf{y}'$ such that the standard deviation (radius) of $\mathbf{x}''$ is equal to $\mathbf{x}$, which makes the standard deviation of $\mathbf{y}''$ equal to $s_3 = s_1^2/s_2$.

These two configurations are intended to model situations in (a) existing work and (b) our proposal. In configuration (a), $\mathbf{x}$ is shorter in expectation than $\mathbf{y}$, and therefore this approximates the situation that arises from existing work. Configuration (b) represents the opposite situation, and corresponds to our proposal in which $\mathbf{y}$ is the projected vector and thus is shorter in expectation than $\mathbf{x}$.

Now, we aim to verify whether the two configurations differ in terms of the likeliness of hubs emerging, using Proposition 1. First, we scale the entire space of configuration (b) by (s_1/s_2), or equivalently, we consider transformation of the variables by $\mathbf{x}'' = (s_1/s_2)\mathbf{x}'$ and $\mathbf{y}'' = (s_1/s_2)\mathbf{y}'$. Note that because the two variables are scaled equally, this change of variables preserves the nearest neighbor relations among the samples. See Fig. 1 for an illustration of the relationship among $\mathbf{x}$, $\mathbf{y}$, $\mathbf{x}'$, $\mathbf{y}'$, $\mathbf{x}''$, and $\mathbf{y}''$.

Let $\{x_i'\}$ and $\{y_i'\}$ be the components of $\mathbf{x}'$ and $\mathbf{y}'$, respectively, and let $\{x_i''\}$ and $\{y_i''\}$ be those for $\mathbf{x}''$ and $\mathbf{y}''$. Then we have

$$\mathrm{Var}[x_i''] = \mathrm{Var}\left[\frac{s_1}{s_2}x_i'\right] = \left(\frac{s_1}{s_2}\right)^2 \mathrm{Var}[x_i'] = s_1^2,$$

$$\mathrm{Var}[y_i''] = \mathrm{Var}\left[\frac{s_1}{s_2}y_i'\right] = \left(\frac{s_1}{s_2}\right)^2 \mathrm{Var}[y_i'] = \frac{s_1^4}{s_2^2}.$$

Thus, $\mathbf{x}''$ follows $\mathcal{N}(0, s_1^2\mathbf{I})$, and $\mathbf{y}''$ follows $\mathcal{N}(0, (s_1^4/s_2^2)\mathbf{I})$. Since both $\mathbf{x}$ in configuration (a) and $\mathbf{x}''$ above follow the same distribution, it now becomes possible to compare the properties of $\mathbf{y}$ and $\mathbf{y}''$ in light of the discussion at the end of Sect. 3: In order to reduce hubness, the distribution with a smaller variance is preferred to the one with a larger variance, for a fixed distribution of source $\mathbf{x}$ (or equivalently, $\mathbf{x}''$).

It follows that $\mathbf{y}''$ is preferable to $\mathbf{y}$, because the former has a smaller variance. As mentioned above, the nearest neighbor relation between the scaled variables, $\mathbf{y}''$ against $\mathbf{x}''$ (or equivalently $\mathbf{x}$), is identical to $\mathbf{y}'$ against $\mathbf{x}'$ in configuration (b). Therefore, we conclude that configuration (b) is preferable to configuration (a), in the sense that the former is more likely to suppress hubs.

Finally, recall that the preferred configuration (b) models the situation of our proposed approach, which is to map target objects in the space of source objects.

4.3   Additional Argument for Placing Target Objects Closer to the Origin

By assuming a unimodal data distribution of which the probability density function (pdf) $p(\mathbf{z})$ is decreasing in $\|\mathbf{z}\|$, we are able to present the following proposition which also advocates placing the source objects outside the target objects, and not the other way around.

Proposition 3 is concerned with the placement of a source object $\mathbf{x}$ at a fixed distance r from its target object $\mathbf{y}$, for which we have two alternatives $\mathbf{x}_1$ and $\mathbf{x}_2$, located at different distances from the origin of the space.

Proposition 3. *Consider a finite set Y of objects (i.e., points) in a Euclidean space, sampled i.i.d. from a distribution whose pdf $p(\mathbf{z})$ is a decreasing function of $\|\mathbf{z}\|$. Let $\mathbf{y} \in Y$ be an object in the set, and let $r > 0$. Further let $\mathbf{x}_1$ and $\mathbf{x}_2$ be two objects at a distance r apart from $\mathbf{y}$. If $\|\mathbf{x}_1\| < \|\mathbf{x}_2\|$, then the probability that $\mathbf{y}$ is the closest object in Y to $\mathbf{x}_2$ is greater than that of $\mathbf{x}_1$.*

Proof (Sketch). For $i = 1, 2$, if another object in Y appears within distance r of $\mathbf{x}_i$, then $\mathbf{y}$ is not the nearest neighbor of $\mathbf{x}_i$. Thus, we aim to prove that this probability for $\mathbf{x}_2$ is smaller than that for $\mathbf{x}_1$. Since objects in Y are sampled i.i.d, it suffices to prove

$$\int_{\mathbf{z} \in V_2} dp(\mathbf{z}) \leq \int_{\mathbf{z} \in V_1} dp(\mathbf{z}), \tag{7}$$

where V_i $(i = 1, 2)$ denote the balls centered at $\mathbf{x}_i$ with radius r. However, (7) obviously holds because the balls V_1 and V_2 have the same radii, $p(\mathbf{z})$ is a decreasing function of $\|\mathbf{z}\|$, and $\|\mathbf{x}_1\| \leq \|\mathbf{x}_2\|$. See Figure 2 for an illustration with a bivariate standard normal distribution in two-dimensional space.   □

In the context of existing work on ZSL, which uses ridge regression to map source objects in the space of target objects, $\mathbf{x}$ can be regarded as a mapped source object, and $\mathbf{y}$ as its target object. Proposition 3 implies that if we want to make a source object $\mathbf{x}$ the nearest neighbor of a target object $\mathbf{y}$, it should rather be placed farther than $\mathbf{y}$ from the origin, but this idea is not present in the objective function (5) for ridge regression; the first term of the objective allocates the same amount of penalty for $\mathbf{x}_1$ and $\mathbf{x}_2$, as they are equally distant from the target $\mathbf{y}$. On the contrary, the ridge regression actually *promotes* placement of the mapped source object $\mathbf{x}$ closer to the origin, as stated in Proposition 2.

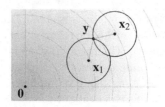

Fig. 2. Illustration of the situation considered in Proposition 3. Here, it is assumed that $\|\mathbf{x}_1\| < \|\mathbf{x}_2\|$ and $\|\mathbf{y} - \mathbf{x}_1\| = \|\mathbf{y} - \mathbf{x}_2\|$. The intensity of the background shading represents the values of the pdf of a bivariate standard normal distribution, from which $\mathbf{y}$ and other objects (not depicted in the figure) in set Y are sampled. The probability mass inside the circle centered at $\mathbf{x}_1$ is greater than that centered at $\mathbf{x}_2$, as the intensity of the shading inside the two circles shows.

4.4   Summary of the Proposed Approach

Drawing on the analysis presented in Sections 4.1–4.3, we propose performing regression that maps *target* objects in the space of *source* objects, and carry out nearest neighbor search in the source space. This opposes the approach followed in existing work on regression-based ZSL [7,8,16,20,22], which maps source objects into the space of target objects.

In the proposed approach, matrix $\mathbf{B}$ in Proposition 2 represents the source objects $\mathbf{X}$, and $\mathbf{A}$ represents the target objects $\mathbf{Y}$. Therefore, $\|\mathbf{MA}\|_2 \leq \|\mathbf{B}\|_2$ means $\|\mathbf{MY}\|_2 \leq \|\mathbf{X}\|_2$, i.e., the mapped target objects tend to be placed closer than the corresponding source objects to the origin.

Admittedly, the above argument for our proposal relies on strong assumptions on data distributions (such as normality), which do not apply to real data. However, the effectiveness of our proposal is verified empirically in Sect. 6 by using real data.

5   Related Work

The first use of ridge regression in ZSL can be found in the work of Palatucci et al. [22]. Ridge regression has since been one of the standard approaches to ZSL, especially for natural language processing tasks: phrase generation [7] and bilingual lexicon extraction [7,8,20]. More recently, neural networks have been used for learning non-linear mapping [11,25]. All of the regression-based methods listed above, including those based on neural networks, map source objects into the target space.

ZSL can also be formulated as a problem of *canonical correlation analysis* (CCA). Hardoon et. al. [12] used CCA and kernelized CCA for image labeling. Lazaridou et. al. [16] compared ridge regression, CCA, singular value decomposition, and neural networks in image labeling. In our experiments (Sect. 6), we use CCA as one of the baseline methods for comparison.

Dinu and Baroni [8] reported the hubness phenomenon in ZSL. They proposed two reweighting techniques to reduce hubness in ZSL, which are applicable to cosine similarity. Tomašev et al. [27] proposed hubness-based instance weighting schemes for CCA. These schemes were applied to classification problems in which multiple instances (vectors) in the target space have the same class label. This setting is different from the one assumed in this paper (see Sect. 2), i.e., we assume that a class label is represented by a single target vector.[2]

Structured output learning [4] addresses a problem setting similar to ZSL, except that the target objects typically have complex structure, and thus the cost of embedding objects in a vector space is prohibitive. *Kernel dependency estimation* [29] is an approach that uses kernel PCA and regression to avoid this issue. In this context, nearest neighbor search in the target space reduces to the *pre-image* problem [18] in the implicit space induced by kernels.

6 Experiments

We evaluated the proposed approach with both synthetic and real datasets. In particular, it was applied to two real ZSL tasks: bilingual lexicon extraction and image labeling.

The main objective of the following experiments is to verify whether our proposed approach is capable of suppressing hub formation and outperforming the existing approach, as claimed in Sect. 4.

6.1 Experimental Setups

Compared Methods. The following methods were compared.

- Ridge$_{X \to Y}$: Linear ridge regression mapping source objects X into the space of target objects Y. This is how ridge regression was used in the existing work on ZSL [7,8,16,20,22].
- Ridge$_{Y \to X}$: Linear ridge regression mapping target objects Y into the source space. This is the proposed approach (Sect. 4.4).
- CCA: Canonical correlation analysis (CCA) for ZSL [12]. We used the code available from http://www.davidroihardoon.com/Professional/Code.html.

We calibrated the hyperparameters, i.e., the regularization parameter in ridge regression and the dimensionality of common feature space in CCA, by cross validation on the training set.

After ridge regression or CCA is applied, both X and Y (or their images) are located in the same space, wherein we find the closest target object for a given source object as measured by the Euclidean distance. In addition to the Euclidean distance, we also tested the *non-iterative contextual dissimilarity measure* (NICDM) [13] in combination with Ridge$_{X \to Y}$ and CCA. NICDM adjusts the Euclidean distance to make the neighborhood relations more symmetrical, and is known to effectively reduce hubness in non-ZSL context [24].

All data were centered before application of regression and CCA, as usual with these methods.

---

[2] Perhaps because of this difference, the method in [27] did not perform well in our experiment, and we do not report its result in Sect. 6.

Evaluation Criteria. The compared methods were evaluated in two respects: (i) the correctness of their prediction, and (ii) the degree of hubness in nearest neighbor search.

Measures of Prediction Correctness. In all our experiments, ZSL was formulated as a ranking task; given a source object, all the target objects were ranked by their likelihood for the source object. As the main evaluation criterion, we used the mean average precision (MAP) [17], which is one of the standard performance metrics for ranking methods. Note that the synthetic and the image labeling experiments are the single-label problems for which MAP is equal to the mean reciprocal rank [17]. We also report the top-k accuracy[3] (Acc_k) for $k = 1$ and 10, which is the percentage of source objects for which the correct target objects are present in their k nearest neighbors.

Measure of Hubness. To measure the degree of hubness, we used the *skewness* of the (empirical) N_k distribution, following the approach in the literature [23, 24, 26, 27]. The N_k distribution is the distribution of the number $N_k(i)$ of times each target object i is found in the top k of the ranking for source objects, and its skewness is defined as follows:

$$(N_k \text{ skewness}) = \frac{\sum_{i=1}^{\ell} \left(N_k(i) - \mathrm{E}\left[N_k\right] \right)^3 / \ell}{\mathrm{Var}\left[N_k\right]^{\frac{3}{2}}}$$

where ℓ is the total number of test objects in Y, $N_k(i)$ is the number of times the ith target object is in the top-k closest target objects of the source objects. A large N_k skewness value indicates the existence of target objects that frequently appear in the k-nearest neighbor lists of source objects; i.e., the emergence of hubs.

6.2   Task Descriptions and Datasets

We tested our method on the following ZSL tasks.

Synthetic Task. To simulate a ZSL task, we need to generate object pairs across two spaces in a way that the configuration of objects is to some extent preserved across the spaces, but is not exactly identical. To this end, we first generated 3000-dimensional (column) vectors $\mathbf{z}_i \in \mathbb{R}^{3000}$ for $i = 1, \ldots, 10000$, whose coordinates were generated from an i.i.d. univariate standard normal distribution. Vectors $\mathbf{z}_i$ were treated as *latent* variables, in the sense that they were not directly observable, but only their images $\mathbf{x}_i$ and $\mathbf{y}_i$ in two different features spaces were. These images were obtained via different random projections, i.e.,

---

[3] In image labeling (only), we report the top-1 accuracy (Acc_1) *macro-averaged* over classes, to allow direct comparison with published results. Note also that Acc_k with a larger k would not be an informative metric for the image labeling task, which only has 10 test labels.

$\mathbf{x}_i = \mathbf{R}_X \mathbf{z}_i$ and $\mathbf{y}_i = \mathbf{R}_Y \mathbf{z}_i$, where $\mathbf{R}_X, \mathbf{R}_Y \in \mathbb{R}^{300 \times 3000}$ are random matrices whose elements were sampled from the uniform distribution over $[-1, 1]$. Because random projections preserve the length and the angle of vectors in the original space with high probability [5,6], the configuration of the projected objects is expected to be similar (but different) across the two spaces.

Finally, we randomly divided object pairs $\{(\mathbf{x}_i, \mathbf{y}_i)\}_{i=1}^{10000}$ into the training set (8000 pairs) and the test set (remaining 2000 pairs).

Bilingual Lexicon Extraction. Our first real ZSL task is bilingual lexicon extraction [7,8,20], formulated as a ranking task: Given a word in the source language, the goal is to rank its gold translation (the one listed in an existing bilingual lexicon as the translation of the source word) higher than other non-translation candidate words.

In this experiment, we evaluated the performance in the tasks of finding the English translations of words in the following source languages: Czech (cs), German (de), French (fr), Russian (ru), Japanese (ja), and Hindi (hi). Thus, in our setting, each of these six languages was used as X alternately, whereas English was the target language Y throughout.[4]

Following related work [7,8,20], we trained a CBOW model [19] on the pre-processed Wikipedia corpus distributed by the Polyglot project[5] (see [3] for corpus statistics), using the word2vec[6] tool. The window size parameter of word2vec was set to 10, with the dimensionality of feature vectors set to 500.

To learn the projection function and measure the accuracy in the test set, we used the bilingual dictionaries[7] of Ács et al. [1] as the gold translation pairs. These gold pairs were randomly split into the training set (80% of the whole pairs) and the test set (20%). We repeated experiments on four different random splits, for which we report the average performance.

Image Labeling. The second real task is image labeling, i.e., the task of finding a suitable word label for a given image. Thus, source objects X are the images and target objects Y are the word labels.

We used the Animal with Attributes (AwA) dataset[8], which consists of 30,475 images of 50 animal classes. For image representation, we used the DeCAF features [9], which are the 4096-dimensional vectors constructed with convolutional neural networks (CNNs). DeCAF is also available from the AwA website. To save computational cost, we used random projection to reduce the dimensionality of DeCAF features to 500.

As with the bilingual lexicon extraction experiment, label features (word representations) were constructed with word2vec, but this time they were trained

---

[4] We also conducted experiments with English as X and other languages as Y. The results are not presented here due to lack of space, but the same trend was observed.

[5] https://sites.google.com/site/rmyeid/projects/polyglot

[6] https://code.google.com/p/word2vec/

[7] http://hlt.sztaki.hu/resources/dict/bylangpair/wiktionary_2013july/

[8] http://attributes.kyb.tuebingen.mpg.de/

Table 1. Experimental results: MAP is the mean average precision. Acc_k is the accuracy of the k-nearest neighbor list. N_k is the skewness of the N_k distribution. A high N_k skewness indicates the emergence of hubs (smaller is better). The bold figure indicates the best performer in each evaluation criteria.

method	MAP	Acc_1	Acc_{10}	N_1	N_{10}
$Ridge_{X \to Y}$	21.5	13.8	36.3	24.19	12.75
$Ridge_{X \to Y}$ + NICDM	58.2	47.6	78.4	13.71	7.94
$Ridge_{Y \to X}$ (proposed)	**91.7**	**87.6**	**98.3**	**0.46**	**1.18**
CCA	78.9	71.6	91.7	12.0	7.56
CCA + NICDM	87.6	82.3	96.5	0.96	2.58

(a) Synthetic data.

method	cs	de	fr	ru	ja	hi
$Ridge_{X \to Y}$	1.7	1.0	0.7	0.5	0.9	5.3
$Ridge_{X \to Y}$ + NICDM	11.3	7.1	5.9	3.8	10.2	21.4
$Ridge_{Y \to X}$ (proposed)	**40.8**	**30.3**	**46.5**	**31.1**	**42.0**	**40.6**
CCA	24.0	18.1	33.7	21.2	27.3	11.8
CCA + NICDM	30.1	23.4	39.7	26.7	35.3	19.3

(b) MAP on bilingual lexicon extraction.

method	cs		de		fr		ru		ja		hi	
	Acc_1	Acc_{10}	Acc_1	Acc_{10}	Acc_1	Acc_{10}	Acc_1	Acc_{10}	Acc_1	Acc_{10}	Acc_1	Acc_{10}
$Ridge_{X \to Y}$	0.7	2.8	0.4	1.6	0.3	1.2	0.2	0.8	0.2	1.3	2.9	8.2
$Ridge_{X \to Y}$ + NICDM	7.2	17.9	4.3	11.4	3.5	9.8	2.1	6.3	6.1	16.8	14.4	32.6
$Ridge_{Y \to X}$ (proposed)	**31.5**	**54.5**	21.6	**43.0**	**36.6**	**58.6**	**21.9**	**43.6**	**31.9**	**56.3**	**31.1**	**55.4**
CCA	17.9	32.7	12.9	25.2	27.0	41.7	15.2	28.8	20.2	37.3	7.4	18.9
CCA + NICDM	21.9	42.3	16.1	33.9	31.1	50.1	18.7	37.0	25.9	48.8	12.4	30.7

(c) Acc_k on bilingual lexicon extraction.

method	cs		de		fr		ru		ja		hi	
	N_1	N_{10}	N_1	N_{10}	N_1	N_{10}	N_1	N_{10}	N_1	N_{10}	N_1	N_{10}
$Ridge_{X \to Y}$	50.29	23.84	43.00	24.37	67.79	35.83	95.05	35.36	62.12	22.78	23.75	10.84
$Ridge_{X \to Y}$ + NICDM	41.56	20.38	39.32	20.82	57.18	25.97	89.08	30.70	57.57	21.62	20.33	9.21
$Ridge_{Y \to X}$ (proposed)	**11.91**	**10.74**	**12.49**	**11.94**	**2.56**	**2.77**	**4.28**	**4.18**	**5.15**	**6.76**	**10.45**	**6.14**
CCA	28.00	18.67	36.66	18.98	30.18	15.95	51.92	21.60	37.73	18.27	22.31	8.95
CCA + NICDM	25.00	17.13	32.94	17.65	25.20	14.65	42.61	20.72	34.66	13.16	22.00	8.46

(d) N_k skewness on bilingual lexicon extraction.

method	MAP	Acc_1	N_1
$Ridge_{X \to Y}$	46.0	22.6	2.61
$Ridge_{X \to Y}$ + NICDM	54.2	34.5	2.17
$Ridge_{Y \to X}$ (proposed)	**62.5**	**41.3**	**0.08**
CCA	26.1	9.2	2.00
CCA + NICDM	26.9	9.3	2.42

(e) Image labeling.

on the English version of Wikipedia (as of March 4, 2015) to cover all AwA labels. Except for the corpus, we used the same word2vec parameters as with bilingual lexicon extraction.

We respected the standard zero-shot setup on AwA provided with the dataset; i.e., the training set contained 40 labels, and test set contained the other 10 labels.

6.3   Experimental Results

Table 1 shows the experimental results. The trends are fairly clear: The proposed approach, Ridge$_{Y \to X}$, outperformed other methods in both MAP and Acc$_k$, over all tasks. Ridge$_{X \to Y}$ and CCA combined with NICDM performed better than those with Euclidean distances, although they still lagged behind the proposed method Ridge$_{Y \to X}$ even with NICDM.

The N_k skewness achieved by Ridge$_{Y \to X}$ was lower (i.e., better) than that of compared methods, meaning that it effectively suppressed the emergence of hub labels. In contrast, Ridge$_{X \to Y}$ produced a high skewness which was in line with its poor prediction accuracy. These results support the expectation we expressed in the discussion in Sect. 4.

The results presented in the tables show that the degree of hubness (N_k) for all tested methods inversely correlates with the correctness of the output rankings, which strongly suggests that hubness is one major factor affecting the prediction accuracy.

For the AwA image dataset, Akata et. al. [2, the fourth row (CNN) and second column (φ^w) of Table 2] reported a 39.7% Acc$_1$ score, using image representations trained with CNNs, and 100-dimensional word representations trained with word2vec. For comparison, our proposed approach, Ridge$_{Y \to X}$, was evaluated in a similar setting: We used the DeCAF features (which were also trained with CNNs) without random projection as the image representation, and 100-dimensional word2vec word vectors. In this setup, Ridge$_{Y \to X}$ achieved a 40.0% Acc$_1$ score. Although the experimental setups are not exactly identical and thus the results are not directly comparable, this suggests that even linear ridge regression can potentially perform as well as more recent methods, such as Akata et al.'s, simply by exchanging the observation and response variables.

7   Conclusion

This paper has presented our formulation of ZSL as a regression problem of finding a mapping from the target space to the source space, which opposes the way in which regression has been applied to ZSL to date. Assuming a simple model in which data follows a multivariate normal distribution, we provided an explanation as to why the proposed direction is preferable, in terms of the emergence of hubs in the subsequent nearest neighbor search step. The experimental results showed that the proposed approach outperforms the existing regression-based and CCA-based approaches to ZSL.

Future research topics include: (i) extending the analysis of Sect. 4 to cover multi-modal data distributions, or other similarity/distance measures such as cosine; (ii) investigating the influence of mapping directions in other regression-based ZSL methods, including neural networks; and (iii) investigating the emergence of hubs in CCA.

Acknowledgments. We thank anonymous reviewers for their valuable comments and suggestions. MS was supported by JSPS Kakenhi Grant no. 15H02749.

References

1. Ács, J., Pajkossy, K., Kornai, A.: Building basic vocabulary across 40 languages. In: Proceedings of the 6th Workshop on Building and Using Comparable Corpora, pp. 52–58 (2013)
2. Akata, Z., Lee, H., Schiele, B.: Zero-shot learning with structured embeddings (2014). arXiv preprint arXiv:1409.8403v1
3. Al-Rfou, R., Perozzi, B., Skiena, S.: Polyglot: Distributed word representations for multilingual NLP. In: CoNLL 2013, pp. 183–192 (2013)
4. Bakir, G., Hofmann, T., Schölkopf, B., Smola, A.J., Taskar, B., Vishwanathan, S.V.N. (eds.): Predicting Structured Data. MIT press (2007)
5. Bingham, E., Mannila, H.: Random projection in dimensionality reduction: applications to image and text data. In: KDD 2001, pp. 245–250 (2001)
6. Dasgupta, S.: Experiments with random projection. In: UAI 2000, pp. 143–151 (2000)
7. Dinu, G., Baroni, M.: How to make words with vectors: phrase generation in distributional semantics. In: ACL 2014, pp. 624–633 (2014)
8. Dinu, G., Baroni, M.: Improving zero-shot learning by mitigating the hubness problem. In: Workshop at ICLR 2015 (2015)
9. Donahue, J., Jia, Y., Vinyals, O., Hoffman, J., Zhang, N., Tzeng, E., Darrell, T.: DeCAF: a deep convolutional activation feature for generic visual recognition (2013). arXiv preprint arXiv:1310.1531
10. Farhadi, A., Endres, I., Hoiem, D., Forsyth, D.: Describing objects by their attributes. In: CVPR 2009, pp. 1778–1785 (2009)
11. Frome, A., Corrado, G.S., Shlens, J., Bengio, S., Dean, J., Ronzato, M., Mikolov, T.: Devise: a deep visual-semantic embedding model. In: NIPS 2013, pp. 2121–2129 (2013)
12. Hardoon, D.R., Szedmak, S., Shawe-Taylor, J.: Canonical correlation analysis: An overview with application to learning methods. Neural Computation **16**, 2639–2664 (2004)
13. Jegou, H., Harzallah, H., Schmid, C.: A contextual dissimilarity measure for accurate and efficient image search. In: CVPR 2007, pp. 1–8 (2007)
14. Lampert, C.H., Nickisch, H., Harmeling, S.: Learning to detect unseen object classes by between-class attribute transfer. In: CVPR 2009. pp. 951–958 (2009)
15. Larochelle, H., Erhan, D., Bengio, Y.: Zero-data learning of new tasks. In: AAAI 2008, pp. 646–651 (2008)
16. Lazaridou, A., Bruni, E., Baroni, M.: Is this a wampimuk? Cross-modal mapping between distributional semantics and the visual world. In: ACL 2014, pp. 1403–1414 (2014)
17. Manning, C.D., Raghavan, P., Schütze, H.: Introduction to Information Retrieval. Cambridge University Press (2008)

18. Mika, S., Schölkopf, B., Smola, A., Müller, K.R., Scholz, M., Rätsch, G.: Kernel PCA and de-noising in feature space. In: NIPS 1998, pp. 536–542 (1998)
19. Mikolov, T., Chen, K., Corrado, G., Dean, J.: Efficient estimation of word representations in vector space. In: Workshop at ICLR 2013 (2013)
20. Mikolov, T., Le, Q.V., Sutskever, I.: Exploiting similarities among languages for machine translation (2013). arXiv preprint arXiv:1309.4168
21. Norouzi, M., Mikolov, T., Bengio, S., Singer, Y., Shlens, J., Frome, A., Corrado, G.S., Dean, J.: Zero-shot learning by convex combination of semantic embeddings. In: ICLR 2014 (2014)
22. Palatucci, M., Pomerleau, D., Hinton, G., Mitchell, T.M.: Zero-shot learning with semantic output codes. In: NIPS 2009, pp. 1410–1418 (2009)
23. Radovanović, M., Nanopoulos, A., Ivanović, M.: Hubs in space: Popular nearest neighbors in high-dimensional data. Journal of Machine Learning Research **11**, 2487–2531 (2010)
24. Schnitzer, D., Flexer, A., Schedl, M., Widmer, G.: Local and global scaling reduce hubs in space. Journal of Machine Learning Research **13**, 2871–2902 (2012)
25. Socher, R., Ganjoo, M., Manning, C.D., Ng, A.Y.: Zero-shot learning through cross-modal transfer. In: NIPS 2013, pp. 935–943 (2013)
26. Suzuki, I., Hara, K., Shimbo, M., Saerens, M., Fukumizu, K.: Centering similarity measures to reduce hubs. In: EMNLP 2013, pp. 613–623 (2013)
27. Tomašev, N., Rupnik, J., Mladenić, D.: The role of hubs in cross-lingual supervised document retrieval. In: Pei, J., Tseng, V.S., Cao, L., Motoda, H., Xu, G. (eds.) PAKDD 2013, Part II. LNCS, vol. 7819, pp. 185–196. Springer, Heidelberg (2013)
28. Vinokourov, A., Shawe-Taylor, J., Cristianini, N.: Inferring a semantic representation of text via cross-language correlation analysis. In: NIPS 2002, pp. 1473–1480 (2002)
29. Weston, J., Chapelle, O., Vapnik, V., Elisseeff, A., Schölkopf, B.: Kernel dependency estimation. In: NIPS 2002, pp. 873–880 (2002)

Solving Prediction Games with Parallel Batch Gradient Descent

Michael Großhans[1] and Tobias Scheffer[2]([⊠])

[1] freiheit.com technologies gmbh, Hamburg, Germany
michael.grosshans@freiheit.com
[2] Department of Computer Science, University of Potsdam, Potsdam, Germany
scheffer@cs.uni-potsdam.de

Abstract. Learning problems in which an adversary can perturb instances at application time can be modeled as games with data-dependent cost functions. In an equilibrium point, the learner's model parameters are the optimal reaction to the data generator's perturbation, and vice versa. Finding an equilibrium point requires the solution of a difficult optimization problem for which both, the learner's model parameters and the possible perturbations are free parameters. We study a perturbation model and derive optimization procedures that use a single iteration of batch-parallel gradient descent and a subsequent aggregation step, thus allowing for parallelization with minimal synchronization overhead.

1 Introduction

In many security-related applications, the assumption that training data and data at application time are identically distributed is routinely violated. For instance, new malware is designed to bypass detection methods which their designers believe virus and malware scanners to employ, and email spamming tools allow their users to develop templates of randomized messages that produce a low spam score with current filters. In these examples, the party that creates the predictive model and the data-generating party factor the possible actions of their opponent into their decisions. This interaction can be modeled as a *prediction game* in which one player controls the predictive model whereas another exercises some control over the process of data generation.

Robust learning methods have been derived under the *zero-sum assumption* that the loss of one player is the gain of the other, for several types of adversarial feature transformations [7,8,11–13,17,23,24]. Settings in which both players have individual cost functions—a fraudster's profit is not the negative of an email service provider's goal of achieving a high spam recognition rate at close-to-zero false positives—cannot adequately be modeled as zero-sum games.

When the learner has to act first and model parameters are disclosed to the data generator, this non-zero-sum interaction can be modeled as a Stackelberg

M. Großhans—This work was done while the author was at the University of Potsdam, Germany.

A. Appice et al. (Eds.): ECML PKDD 2015, Part I, LNAI 9284, pp. 152–167, 2015.
DOI: 10.1007/978-3-319-23528-8_10

competition [4,15]. A Stackelberg competition always has an optimal solution, but generally a difficult bi-level optimization problem has to be solved to find it [4]. For simultaneously acting players, one may resort to the concept of a Nash equilibrium. An equilibrium is a pair of a predictive model and a transformation of the input distribution. In an equilibrium point, the learner's model parameters are the optimal reaction to the data generator's perturbation function, and vice versa. If a game has a unique Nash equilibrium and is played by rational players that aim at minimizing their costs, it may be reasonable for each player to assume that the opponent will play according to the Nash equilibrium strategy as well. If, however, multiple equilibria exist and the players choose their strategy according to distinct ones, then the resulting combination may be arbitrarily disadvantageous for either player. For certain cost functions, the prediction game has been shown to have a unique Nash equilibrium [3].

Finding the equilibrium point of a prediction game requires the solution of a difficult optimization problem: in each iteration of an outer gradient-descent, nested optimization problems have to be solved. This process is two orders of magnitude more expensive than *iid* learning [3]—even more so, if the learner is uncertain about the adversary's cost function [14].

Gradient descent algorithms can be parallelized by distributing the data in batches across multiple worker nodes. Casting gradient descent into the MapReduce programming model [6] offers an almost unlimited potential speed-up, because synchronization is limited to a final *reduce* step, and, unlike multicore or GPU parallelism, MapReduce is not constrained by the limited number of cores that can be fitted into a single unit of computing hardware. In order to conduct gradient descent within the MapReduce model, parallel nodes have to perform gradient descent on subsets of the data. Only in the final step, the local parameter vectors are aggregated [18,26]. This procedure has known convergence bounds [26].

Work on HaLoop [5] and ScalOps [25] aims at allowing for more flexible algorithm design that may include aggregation steps during the parallel optimization process [19]; this, however comes at the cost at higher communication costs which limit the gain of increased parallelization. This paper therefore focuses on rephrasing the search for an equilibrium point of a prediction game within the MapReduce model.

The known analysis and algorithm for finding the equilibrial prediction model [3] are based on a model of the adversarial data transformation that allows the perturbation of each instance to potentially depend on other instances. It is therefore unsuitable for parallelization: When the perturbation of an instance may depend on different instances, a node that does not have access to all interdependent instances cannot anticipate the outcome of the adversary's action. Therefore, we derive a model of adversarial manipulations of the input distribution that is based on the manipulation of individual feature vectors.

The rest of this paper is organized as follows. Section 2 lays out the problem setting and introduces an adversarial data generation model. In Section 3 we study conditions under which a unique equilibrium point exists under this data generation model. We derive a method for finding the unique equilibrium point in a way that can be parallelized in Section 4. Section 5 presents empirical result and Section 6 concludes.

2   Problem Setting and Data Transformation Model

We study static prediction games between two players: The *learner* and its adversary, the *data generator*. For example, in email spam filtering, the learner may be an email service provider whereas the data generator is an amalgamated model of all legitimate and abusive email senders.

At *training time*, the data generator produces a matrix $\mathbf{X}$ of training instances $\mathbf{x}_1, \ldots, \mathbf{x}_n$ and a corresponding vector $\mathbf{y}$ of class labels $y_i \in \mathcal{Y}$. These object-class pairs are drawn according to a training distribution with density function $p(\mathbf{x}, y)$.

The task of the learner is to select the parameters $\mathbf{w} \in \mathcal{W} \subset \mathbb{R}^m$ of a linear model $f_{\mathbf{w}}(\mathbf{x}) = \mathbf{w}^\mathsf{T}\mathbf{x}$. Simultaneously, the data generator can choose a parameterized transformation $g_{\mathbf{A}} : \mathbb{R}^m \to \mathbb{R}^m$, with $\mathbf{A} \in \mathcal{A}$ that perturbs instances; regularizer $\rho_g \Omega_g(\mathbf{A})$ quantifies *transformation costs* which the data generator incurs. For instance, a spammer may obfuscate text messages and remove conspicuous URLs at the cost of reducing the response rate. At test time, instances are drawn according to $p(\mathbf{x}, y)$ and perturbed by $g_{\mathbf{A}}$; this defines the test distribution.

The learner's theoretical *costs* at application time are given by Equation 1; the data generator's theoretical costs by Equation 2.

$$\theta_f(\mathbf{w}, \mathbf{A}) = \sum_{y \in \mathcal{Y}} \int \ell_f(f_{\mathbf{w}}(g_{\mathbf{A}}(\mathbf{x})), y) p(\mathbf{x}, y) \mathrm{d}\mathbf{x} \tag{1}$$

$$\theta_g(\mathbf{w}, \mathbf{A}) = \sum_{y \in \mathcal{Y}} \int \ell_g(f_{\mathbf{w}}(g_{\mathbf{A}}(\mathbf{x})), y) p(\mathbf{x}, y) \mathrm{d}\mathbf{x} + \rho_g \Omega_g(\mathbf{A}) \tag{2}$$

The theoretical costs of both players depend on the unknown test distribution; we will therefore resort to regularized, empirical costs based on the training sample. The empirical costs incurred by the predictive model $f_{\mathbf{w}}$ and transformation $g_{\mathbf{A}}$ are

$$\hat{\theta}_f(\mathbf{w}, \mathbf{A}) = \frac{1}{n} \sum_{i=1}^{n} \ell_f(\mathbf{w}^\mathsf{T} g_{\mathbf{A}}(\mathbf{x}), y_i) + \rho_f \Omega_f(\mathbf{w})$$

$$\hat{\theta}_g(\mathbf{w}, \mathbf{A}) = \frac{1}{n} \sum_{i=1}^{n} \ell_g(\mathbf{w}^\mathsf{T} g_{\mathbf{A}}(\mathbf{x}), y_i) + \rho_g \Omega_g(\mathbf{A}).$$

We employ a linear, parameterized data transformation model of the form $g_{\mathbf{A}}(\mathbf{x}) = \mathbf{x} + \mathbf{A}\mathbf{x}$, where $\mathbf{A} \in \mathbb{R}^{m \times m}$ is the transformation matrix chosen by the adversary. Under this model, the perturbation vector that is added to each instance $\mathbf{x}$ is a linear function of $\mathbf{x}$. For $\mathbf{A} = 0$, instances remain unperturbed.

This model subsumes many relevant data-manipulation operations. For instance, features are scaled by nonzero values at the diagonal elements of $\mathbf{A}$; features i are deleted by $a_{ii} = -1$. Feature i is replaced by feature j (e.g., Viagra $\rightarrow$ Viagra) by a matrix that has entries $a_{ii} = -1$ and $a_{ji} = 1$, and is 0 everywhere else.

We will write the transformation matrix as vector of m-dimensional row vectors $\mathbf{A} = [\mathbf{a}_1, \ldots, \mathbf{a}_m]^\mathsf{T}$ or as an m^2-dimensional vector $\mathbf{a} = [\mathbf{a}_1^\mathsf{T}, \ldots, \mathbf{a}_m^\mathsf{T}]^\mathsf{T}$ whenever this will simplify the notation.

Under this data transformation model, standard ℓ_2 regularization for the data generator [3] would amount to

$$\|(\mathbf{x}_i + \mathbf{A}\mathbf{x}_i) - \mathbf{x}_i\|^2 = \frac{1}{n} \sum_{i=1}^n \|\mathbf{A}\mathbf{x}_i\|^2$$

which is not strongly convex in $\mathbf{A}$ for every data matrix $\mathbf{X}$. Hence, this regularizer can have multiple optima, which should be avoided. Therefore, we use the Frobenius norm of $\mathbf{A}$ as regularizer for the data generator; we use standard ℓ_2 regularization for the learner:

$$\Omega_f(\mathbf{w}) = \|\mathbf{w}\|^2, \tag{3}$$

$$\Omega_g(\mathbf{A}) = \frac{1}{2}\|\mathbf{A}\|_F^2 = \frac{1}{2}\|\mathbf{a}\|^2. \tag{4}$$

3   Analysis of Equilibrium Points

Note that both $\hat{\theta}_f$ and $\hat{\theta}_g$ depend on both players' actions. Neither player can minimize their costs without considering their adversary's options. This motivates the concept of an equilibrium point. Assume that the learner considers using model parameters $\mathbf{w}_1$. The learner can now determine a possible reaction $\mathbf{A}_1$ of the data generator that would minimize $\hat{\theta}_g$ for the given $\mathbf{w}_1$. In turn, the learner can determine model parameters $\mathbf{w}_2$ that would minimize $\hat{\theta}_f$ for this transformation $\mathbf{A}_1$, continue to determine reaction $\mathbf{A}_2$, and so on. It his sequence of reactions reaches a fixed point—a point $(\mathbf{w}^*, \mathbf{A}^*)$ that is the best possible reaction to itself—then this point is a Nash equilibrium and satisfies

$$\mathbf{w}^* = \arg\min_{\mathbf{w}} \hat{\theta}_f(\mathbf{w}, \mathbf{A}^*), \tag{5}$$

$$\mathbf{A}^* = \arg\min_{\mathbf{A}} \hat{\theta}_f(\mathbf{w}^*, \mathbf{A}). \tag{6}$$

In this section, we analyze the prediction game between learner and data generator that we have introduced in the previous section. We will derive conditions under which equilibrium points exist, and conditions under which an equilibrium point is unique.

Known results on the existence of equilibrium points for prediction games [3] do not apply to the data transformation model derived in Section 2: Equation 4 regularizes $\mathbf{A}$ because regularization of $\|\mathbf{X} - g_{\mathbf{A}}(\mathbf{X})\|$ would not be convex in $\mathbf{A}$, and therefore Assumption 3 of [3] is not met.

3.1   Existence of Equilibrium Points

We will now study under which conditions the prediction game between learner and data generator with the data transformation introduced above has at least one equilibrium point. We start by formulating conditions on action spaces and loss functions in the following assumption.

Assumption 1. *The players' action sets $\mathcal{W}$ and $\mathcal{A}$ and loss functions ℓ_f and ℓ_g satisfy the following properties.*

1. *Action spaces $\mathcal{W} \subseteq \mathbb{R}^m$ and $\mathcal{A} \subseteq \mathbb{R}^m \times \ldots \times \mathbb{R}^m$ are non-empty, compact and convex,*
2. *The loss functions $\ell_f(z,y)$ and $\ell_g(z,y)$ are convex and continuous in z for every $y \in \mathcal{Y}$*

Theorem 1. *Under Assumption 1 the game has at least one equilibrium point.*

Proof. By Assumption 1 the loss functions $\ell_f(z_i, y_i)$ and $\ell_g(z_i, y_i)$ are continuous and convex in z_i for any $y_j \in \mathcal{Y}$. Note that $z_i = \mathbf{w}^\mathsf{T}\mathbf{x}_i + \mathbf{w}^\mathsf{T}\mathbf{A}\mathbf{x}_i$ is linear in $\mathbf{w} \in \mathbb{R}^m$ and linear in $\mathbf{A} \in \mathbb{R}^{m^2}$ for any $(\mathbf{x}_i, \mathbf{y}_i) \in \mathcal{X} \times \mathcal{Y}$. Hence, for both $\nu \in \{f, g\}$, the sum of loss terms $\sum_{i=1}^{n} \ell_\nu(z_i, y_i)$ is jointly continuous in $(\mathbf{w}, \mathbf{A}) \in \mathbb{R}^{m \times (m+1)}$ and convex in both $\mathbf{w} \in \mathbb{R}^m$ and $\mathbf{A} \in \mathbb{R}^{m \times m}$. Both regularizers Ω_f and Ω_g are jointly continuous in $(\mathbf{w}, \mathbf{A}) \in \mathbb{R}^{m \times (m+1)}$. Additionally Ω_f is strictly convex in $\mathbf{w} \in \mathbb{R}^m$ and Ω_g is strictly convex in $\mathbf{A} \in \mathbb{R}^{m \times m}$.

Hence, both empirical cost functions $\hat{\theta}_f$ and $\hat{\theta}_g$ are jointly continuous in $(\mathbf{w}, \mathbf{A}) \in \mathbb{R}^{m \times (m+1)}$. Additionally $\hat{\theta}_f$ is strictly convex in $\mathbf{w} \in \mathbb{R}^m$ and $\hat{\theta}_g$ is strictly convex in $\mathbf{A} \in \mathbb{R}^{m \times m}$. Therefore by Theorem 4.3. in [2]—together with the fact that both action spaces are non-empty, compact and convex—at least one equilibrium point exists.

3.2   Uniqueness of Equilibrium Points

In this section, we will derive conditions for uniqueness of equilibrium points. The significance of this result is that an action that is part of an equilibrium point minimizes the costs for either player only if the opponent chooses the same equilibrium point. Otherwise, either player's costs may be arbitrarily high. If multiple equilibria exist, the players cannot determine which action even a perfectly rational opponent will take. We will make use of a theorem of Rosen [22] which states that a unique equilibrium point exists if the Jacobian of the combined loss

$$r_\mathbf{w}\theta_f(\mathbf{w}, \mathbf{A}) + r_\mathbf{A}\theta_g(\mathbf{w}, \mathbf{A})$$

is positive definite for any fixed $r_\mathbf{w} > 0, r_\mathbf{A} > 0$. To prove this condition, we formulate two lemmas. Lemma 1 and Lemma 2 derive two different forms of matrices that are always positive (semi-)definite. In the following, the symbol $\otimes$ denotes the Kronecker-product.

Lemma 1. *For any* $\mathbf{A} \in \mathbb{R}^{m \times m}$ *and* $\mathbf{w} \in \mathbb{R}^m$ *and any positive semi-definite matrix* $\mathbf{X} \in \mathbb{R}^{m \times m}$, *the matrix*

$$\mathbf{M}_1 := \begin{bmatrix} \mathbf{A}\mathbf{X}\mathbf{A}^\mathsf{T} & \mathbf{w}^\mathsf{T} \otimes (\mathbf{A}\mathbf{X}) \\ \mathbf{w} \otimes (\mathbf{X}\mathbf{A}^\mathsf{T}) & (\mathbf{w}\mathbf{w}^\mathsf{T}) \otimes \mathbf{X} \end{bmatrix} \in \mathbb{R}^{(m^2+m) \times (m^2+m)}$$

is positive semi-definite.

Proof. Note that we can rewrite this matrix as a product of three matrices:

$$\begin{bmatrix} \mathbf{A} \\ \mathbf{w} \otimes \mathbf{I}_m \end{bmatrix} \mathbf{X} \begin{bmatrix} \mathbf{A}^\mathsf{T} & \mathbf{w}^\mathsf{T} \otimes \mathbf{I}_m \end{bmatrix}^\mathsf{T}$$

where $\mathbf{I}_m$ denotes the $m \times m$ unit matrix. By Assumption 1 the matrix $\mathbf{X}$ is positive semi-definite and therefore the product $\mathbf{v}_1^\mathsf{T} \mathbf{X} \mathbf{v}_1 \geq 0$ is non-negative for all vectors $\mathbf{v}_1 \in \mathbb{R}^m$. By using the substitution $\mathbf{v}_1 = \begin{bmatrix} \mathbf{A}^\mathsf{T} & \mathbf{w}^\mathsf{T} \otimes \mathbf{I}_m \end{bmatrix} \mathbf{v}_2$, the product

$$\mathbf{v}_2^\mathsf{T} \begin{bmatrix} \mathbf{A} \\ \mathbf{w} \otimes \mathbf{I}_m \end{bmatrix} \mathbf{X} \begin{bmatrix} \mathbf{A}^\mathsf{T} & \mathbf{w}^\mathsf{T} \otimes \mathbf{I}_m \end{bmatrix}^\mathsf{T} \mathbf{v}_2 = \mathbf{v}_1^\mathsf{T} \mathbf{X} \mathbf{v}_1 \geq 0$$

is non-negative. Hence, the matrix $\mathbf{M}_1$ is positive semi-definite, which completes the proof.

Lemma 2. *For any* $\mathbf{x} \in \mathbb{R}^m$ *and any* $a, b \in \mathbb{R}^+$ *the matrix*

$$\mathbf{M}_2 := \begin{bmatrix} a\mathbf{I}_m & \mathbf{I}_m \otimes \mathbf{x}^\mathsf{T} \\ \mathbf{I}_m \otimes \mathbf{x} & b\mathbf{I}_{m^2} \end{bmatrix} \in \mathbb{R}^{(m^2+m) \times (m^2+m)}$$

is positive definite, if and only if $a \cdot b > \mathbf{x}^\mathsf{T} \mathbf{x}$.

Proof. The matrix is a symmetric square matrix. Hence it is positive definite if and only if all eigenvalues $\lambda_i > 0$ for all $i \in \{0, \ldots, m^2 + m\}$. Let $(\mathbf{w}^\mathsf{T}, \mathbf{v}_1^\mathsf{T}, \ldots, \mathbf{v}_m^\mathsf{T})^\mathsf{T}$ be an arbitrary eigenvector with eigenvalue λ and let us define

$$\mathbf{V} = \begin{bmatrix} \mathbf{v}_1^\mathsf{T} \\ \vdots \\ \mathbf{v}_m^\mathsf{T} \end{bmatrix}.$$

Then—by using the definition of eigenvectors—the following equations hold:

$$(\lambda - a)\mathbf{w} = \mathbf{V}\mathbf{x} \tag{7}$$
$$(\lambda - b)\mathbf{V} = \mathbf{w}\mathbf{x}^\mathsf{T}. \tag{8}$$

By combining Equation 7 and Equation 8 the following equation

$$(\lambda - a)(\lambda - b)\mathbf{w} = \mathbf{x}^\mathsf{T}\mathbf{x}\mathbf{w} \tag{9}$$

holds for every eigenvector $\left(\mathbf{w}^\mathsf{T}, \mathbf{v}_1^\mathsf{T}, \ldots, \mathbf{v}_m^\mathsf{T}\right)^\mathsf{T}$ with corresponding eigenvalue λ.

Firstly, assume that $\mathbf{w} = \mathbf{0}$ holds. Due to the definition of an eigenvector, the matrix $\mathbf{V} \neq \mathbf{0}$ is non-zero. By Equation 8, the corresponding eigenvalue would be $\lambda = b$, and hence the corresponding eigenvalue would be positive.

Now assume that $\mathbf{w} \neq \mathbf{0}$ holds. Then, by using Equation 9 it turns out that $(\lambda - a)(\lambda - b) = \mathbf{x}^\mathsf{T}\mathbf{x}$. Solving this Equation for λ results in the following two solutions:

$$\lambda_{1,2} = \frac{a+b}{2} \pm \sqrt{\frac{(a-b)^2}{4} + \mathbf{x}^\mathsf{T}\mathbf{x}}.$$

Therefore, matrix $\mathbf{M}_2$ is positive semi-definite if and only if

$$\frac{a+b}{2} \geq \sqrt{\frac{(a-b)^2}{4} + \mathbf{x}^\mathsf{T}\mathbf{x}} \tag{10}$$

holds, which is equivalent to the inequality

$$\frac{(a+b)^2}{4} \geq \frac{(a-b)^2}{4} + \mathbf{x}^\mathsf{T}\mathbf{x}.$$

Hence, the smallest eigenvalue is non-negative if and only if $a \cdot b > \mathbf{x}^\mathsf{T}\mathbf{x}$ which completes the proof.

We can now formulate Assumptions under which a unique equilibrium point exists.

Assumption 2. *For a given data matrix* $\mathbf{X} \in \mathbb{R}^{m \times n}$ *and labels* $\mathbf{y} \in \mathcal{Y}^n$, *the players' action sets* $\mathcal{W}$ *and* $\mathcal{A}$, *loss functions* ℓ_f *and* ℓ_g, *and regularization parameters* ρ_f, ρ_g *satisfy the following properties.*

1. *the second derivatives of the loss functions are equal for all* $y \in \mathcal{Y}$ *and* $z \in \mathbb{R}$

$$\ell_f''(z, y) = \ell_g''(z, y).$$

2a. The regularization parameters satisfy

$$\rho_f \rho_g > \sup_{(\mathbf{w}, \mathbf{A}) \in \mathcal{W} \times \mathcal{A}} \bar{\mathbf{x}}_{(\mathbf{w}, \mathbf{A}, \mathbf{X}, \mathbf{y})}^\mathsf{T} \bar{\mathbf{x}}_{(\mathbf{w}, \mathbf{A}, \mathbf{X}, \mathbf{y})},$$

where $\bar{\mathbf{x}}$ *is the (derivate-) weighted average over all instances*

$$\bar{\mathbf{x}}_{(\mathbf{w}, \mathbf{A}, \mathbf{X}, \mathbf{y})} = \frac{1}{n} \sum_{i=1}^n \left[\frac{1}{2} \left(\ell_f'(\mathbf{w}^\mathsf{T}\phi_\mathbf{A}(\mathbf{x}_i), y_i) + \ell_g'(\mathbf{w}^\mathsf{T}\phi_\mathbf{A}(\mathbf{x}_i), y_i) \right) \mathbf{x}_i \right].$$

2b. (Sufficient condition for 2a) the regularization parameters satisfy

$$\rho_f \rho_g > \sup_{(\mathbf{w}, \mathbf{A}) \in \mathcal{W} \times \mathcal{A}} \max_{i \in \{1, \ldots, n\}} \tau_i^2(\mathbf{w}, \mathbf{A}) \cdot \mathbf{x}_i^\mathsf{T}\mathbf{x}_i,$$

where $\tau_i(\mathbf{w}, \mathbf{A})$ *is specified by*

$$\tau_i(\mathbf{w}, \mathbf{A}) = \frac{1}{2} \left(\ell_{pL}'(\mathbf{w}^\mathsf{T}\phi_\mathbf{A}(\mathbf{x}_i), y_i) + \ell_g'(\mathbf{w}^\mathsf{T}\phi_\mathbf{A}(\mathbf{x}_i), y_i) \right).$$

Theorem 2. *Under Assumptions 1 and 2, the prediction game between learner and data generator has exactly one equilibrium point.*

The conditions of Assumption 1 impose technical, rather common requirements on the cost functions that can be met in practice. The first condition of Assumption 2 requires the loss function of learner and data generator to have identical curvatures. This can be met, for instance, if both player use a logistic loss function [3]. The second condition imposes a joint bound on the regularization coefficients. Intuitively, if the data generator is allowed to perturb instances strongly, then a unique equilibrium exists only if the learner's cost function has a sufficiently large regularization term.

Proof. By Assumption 1 the game has at least one equilibrium point. We now turn towards the uniqueness of the equilibrium point. Therefore—by following the theorems in [10, 22]—we show that the pseudo-Jacobian

$$
\mathbf{J}_{r_{\mathbf{w}}, r_{\mathbf{A}}}(\mathbf{w}, \mathbf{A}) = \begin{bmatrix} r_{\mathbf{w}} \mathbf{I}_m & \mathbf{0} \\ \mathbf{0} & r_{\mathbf{A}} \mathbf{I}_{m^2} \end{bmatrix} \begin{bmatrix} \nabla^2_{\mathbf{w},\mathbf{w}} \hat{\theta}_f & \nabla^2_{\mathbf{w},\mathbf{a}_1} \hat{\theta}_g & \cdots & \nabla^2_{\mathbf{w},\mathbf{a}_m} \hat{\theta}_f \\ \nabla^2_{\mathbf{a}_1,\mathbf{w}} \hat{\theta}_g & \nabla^2_{\mathbf{a}_1,\mathbf{a}_1} \hat{\theta}_g & \cdots & \nabla^2_{\mathbf{a}_1,\mathbf{a}_m} \hat{\theta}_g \\ \vdots & \vdots & \ddots & \vdots \\ \nabla^2_{\mathbf{a}_m,\mathbf{w}} \hat{\theta}_g & \nabla^2_{\mathbf{a}_m,\mathbf{a}_1} \hat{\theta}_g & \cdots & \nabla^2_{\mathbf{a}_m,\mathbf{a}_m} \hat{\theta}_g \end{bmatrix}
$$

is positive definite at every point $(\mathbf{w}, \mathbf{A}) \in \mathcal{W} \times \mathcal{A}$ for some fixed $r_{\mathbf{w}}, r_{\mathbf{A}} > 0$. We set $r_{\mathbf{w}} = r_{\mathbf{A}} = 1$. Therefore the pseudo-Jacobian (first and second derivations can be found in the Appendix) is given as

$$
\mathbf{J}_r(\mathbf{w}, \mathbf{A}) = \begin{bmatrix} (\mathbf{I}_m + \mathbf{A}) \mathbf{X} \boldsymbol{\Gamma}_f \mathbf{X}^\mathsf{T} (\mathbf{I}_m + \mathbf{A})^\mathsf{T} & \mathbf{w}^\mathsf{T} \otimes \left[(\mathbf{I}_m + \mathbf{A}) \mathbf{X} \boldsymbol{\Gamma}_f \mathbf{X}^\mathsf{T} \right] \\ \mathbf{w} \otimes \left[\mathbf{X} \boldsymbol{\Gamma}_g \mathbf{X}^\mathsf{T} (\mathbf{I}_m + \mathbf{A})^\mathsf{T} \right] & \left[\mathbf{w} \mathbf{w}^\mathsf{T} \right] \otimes \left[\mathbf{X} \boldsymbol{\Gamma}_g \mathbf{X}^\mathsf{T} \right] \end{bmatrix}
$$

$$
+ \begin{bmatrix} \rho_f \mathbf{I}_m & \mathbf{I}_m \otimes \left[\mathbf{X} \boldsymbol{\gamma}_f \right]^\mathsf{T} \\ \mathbf{I}_m \otimes \left[\mathbf{X} \boldsymbol{\gamma}_g \right] & \rho_g \mathbf{I}_{m^2} \end{bmatrix}. \tag{11}
$$

Following Assumption 2(1) the matrices $\boldsymbol{\Gamma}_f = \boldsymbol{\Gamma}_g$ are equal. Additionally, according to Assumption 1(2) the loss functions are convex and, therefore, the matrices $\boldsymbol{\Gamma}_f$ and $\boldsymbol{\Gamma}_g$ are positive semi-definite. Hence, the matrices $\mathbf{X} \boldsymbol{\Gamma}_f \mathbf{X}^\mathsf{T}$ and $\mathbf{X} \boldsymbol{\Gamma}_g \mathbf{X}^\mathsf{T}$ are equal and positive semi-definite. Following Lemma 1 the first summand in Equation 11 is positive semi-definite.

The second summand is positive definite if and only if the square matrix

$$
\begin{bmatrix} \rho_f \mathbf{I}_m & \mathbf{I}_m \otimes \left[\frac{1}{2} \mathbf{X} \boldsymbol{\gamma}_f + \frac{1}{2} \mathbf{X} \boldsymbol{\gamma}_g \right]^\mathsf{T} \\ \mathbf{I}_m \otimes \left[\frac{1}{2} \mathbf{X} \boldsymbol{\gamma}_f + \frac{1}{2} \mathbf{X} \boldsymbol{\gamma}_g \right] & \rho_g \mathbf{I}_{m^2} \end{bmatrix}
$$

is positive definite. According to Lemma 2 this square matrix is positive definite if and only if

$$
\rho_f \rho_g > \left[\frac{1}{2} \mathbf{X} \boldsymbol{\gamma}_f + \frac{1}{2} \mathbf{X} \boldsymbol{\gamma}_g \right]^\mathsf{T} \left[\frac{1}{2} \mathbf{X} \boldsymbol{\gamma}_f + \frac{1}{2} \mathbf{X} \boldsymbol{\gamma}_g \right].
$$

Note that the relation

$$\frac{1}{2}\mathbf{X}\gamma_f + \frac{1}{2}\mathbf{X}\gamma_g$$
$$= \frac{1}{n}\sum_{i=1}^{n}\left[\frac{1}{2}\left(\ell'_f(\mathbf{w}^\mathsf{T}\mathbf{x}_i + \mathbf{w}^\mathsf{T}\mathbf{A}\mathbf{x}_i, y_i) + \ell'_g(\mathbf{w}^\mathsf{T}\mathbf{x}_i + \mathbf{w}^\mathsf{T}\mathbf{A}\mathbf{x}_i, y_i)\right)\mathbf{x}_i\right] \quad (12)$$

holds. Hence, according to Assumption 2(2a) the second summand in Equation 11 and therefore the pseudo-Jacobian $\mathbf{J}_{r_\mathbf{w}, r_\mathbf{A}}(\mathbf{w}, \mathbf{A})$ is positive definite at every point $(\mathbf{w}, \mathbf{A}) \in \mathcal{W} \times \mathcal{A}$, which completes the proof.

4   Finding the Unique Equilibrium Point Efficiently

In the previous section, we derived conditions for the existence of unique equilibrium points. In this section, we will discuss algorithms that find this unique solution and can be phrased as a single iteration of a parallel map step and a reduce step that aggregates the results.

4.1   Inexact Line Search

Equilibrium points can be located by inexact line search [3,16]. In each iteration, the procedure computes the response $\bar{\mathbf{w}}$ of the learner that minimizes $\hat{\theta}_f$ given the previous transformation $\mathbf{A}$, and response $\bar{\mathbf{A}}$ of the data generator that minimizes $\hat{\theta}_g$ given the previous prediction model $\mathbf{w}$ in nested optimization problems using L-BFGS [3]. The descent directions are then given by:

$$\mathbf{d}_f = \bar{\mathbf{w}} - \mathbf{w},$$
$$\mathbf{d}_g = \bar{\mathbf{A}} - \mathbf{A}.$$

Inexact line search tries increasingly large values of the step size t and perform an update by adding $t\mathbf{d}_f$ to the learner's prediction model $\mathbf{w}$ and by adding $t\mathbf{d}_g$ to the data generator's transformation matrix $\mathbf{A}$. This procedure converges to the unique Nash equilibrium—von Heusinger and Kanzow discuss its convergence properties [16].

4.2   Arrow-Hurwicz-Uzawa Method

Inexact line search is computationally expensive because it solves nested optimization problems in each iteration. In this section, we derive an alternative approach without nested optimization problems; it is based on the Arrow-Hurwicz-Uzawa saddle-point method. Equations 5 and 6 define equilibrium points. We start our derivation introducing the Nikaido-Isoda function [21]:

$$\hat{\theta}([\mathbf{w}_1, \mathbf{A}_1], [\mathbf{w}_2, \mathbf{A}_2])$$
$$= \left[\hat{\theta}_f(\mathbf{w}_1, \mathbf{A}_1) - \hat{\theta}_f(\mathbf{w}_2, \mathbf{A}_1)\right] + \left[\hat{\theta}_g(\mathbf{w}_1, \mathbf{A}_1) - \hat{\theta}_g(\mathbf{w}_1, \mathbf{A}_2)\right]. \quad (13)$$

This function quantifies the cost savings that the learner could achieve by unilaterally changing the model from $\mathbf{w}_1$ to $\mathbf{w}_2$ plus the cost savings that the data generator could achieve by unilaterally changing the transformation from $\mathbf{A}_1$ to $\mathbf{A}_2$. Nikaido-Isoda function $\hat{\theta}$ is concave in $(\mathbf{w}_2, \mathbf{A}_2)$ because $\hat{\theta}_f$ and $\hat{\theta}_g$ are convex, and the cost functions for $(\mathbf{w}_2, \mathbf{A}_2)$ enter the function as negatives.

For convex-concave Nikaido-Isoda functions, parameters $[\mathbf{w}^*, \mathbf{A}^*]$ are an equilibrium point if and only if the Nikaido-Isoda function has a saddle point at $([\mathbf{w}^*, \mathbf{A}^*], [\mathbf{w}^*, \mathbf{A}^*])$ [9]. The intuition behind this result is the following. An equilibrium point $(\mathbf{w}^*, \mathbf{A}^*)$ satisfies Equations 5 and 6 by definition. By Equation 13, $\hat{\theta}([\mathbf{w}^*, \mathbf{A}^*], [\mathbf{w}^*, \mathbf{A}^*]) = 0$. Equations 5 and 6 imply that $\hat{\theta}([\mathbf{w}, \mathbf{A}], [\mathbf{w}^*, \mathbf{A}^*])$ is positive and $\hat{\theta}([\mathbf{w}^*, \mathbf{A}^*], [\mathbf{w}, \mathbf{A}])$ is negative for $[\mathbf{w}, \mathbf{A}] \neq [\mathbf{w}^*, \mathbf{A}^*]$. When $\hat{\theta}$ is convex in $[\mathbf{w}_1, \mathbf{A}_1]$ and concave in $[\mathbf{w}_2, \mathbf{A}_2]$, this means that $(\mathbf{w}^*, \mathbf{A}^*)$ is an equilibrium point if and only if $\hat{\theta}$ has a saddle point at position $[\mathbf{w}^*, \mathbf{A}^*], [\mathbf{w}^*, \mathbf{A}^*]$.

Saddle points of convex-concave functions can be located with the Arrow-Hurwicz-Uzawa method [1]. We implement the method as an iterative procedure with a constant stepsize t [20]. In each iteration j, the method computes the gradient of $\hat{\theta}$ with respect to $\mathbf{w}_1$, $\mathbf{w}_2$, $\mathbf{A}_1$ and $\mathbf{A}_2$, and performs a descent by updating previous estimates:

$$(\mathbf{w}_1, \mathbf{A}_1)^{(j+1)} = (\mathbf{w}_1, \mathbf{A}_1)^{(j)} - t\nabla_{(\mathbf{w}_1, \mathbf{A}_1)}\hat{\theta}([\mathbf{w}_1, \mathbf{A}_1]^{(j)}, [\mathbf{w}_2, \mathbf{A}_2]^{(j)})$$
$$(\mathbf{w}_2, \mathbf{A}_2)^{(j+1)} = (\mathbf{w}_2, \mathbf{A}_2)^{(j)} + t\nabla_{(\mathbf{w}_2, \mathbf{A}_2)}\hat{\theta}([\mathbf{w}_1, \mathbf{A}_1]^{(j)}, [\mathbf{w}_2, \mathbf{A}_2]^{(j)}).$$

The final estimator of the equilibrium point after T iterations is the average of all iterates: $(\hat{\mathbf{w}}^*, \hat{\mathbf{A}}^*) = \sum_{j=1}^{T}(\mathbf{w}_1, \mathbf{A}_1)^{(j)}$. For any convex-concave $\hat{\theta}$, this method converges towards a saddle-point.

4.3   Parallelized Methods

Both the inexact line search method sketched in Section 4.1 and the Arrow-Hurwicz-Uzawa method derived in Section 4.2 can be implemented in a batch-parallel manner. To this end, the data is randomly partitioned into k batches $(\mathbf{X}_i, \mathbf{y}_i)$, where $i = 1, \ldots, k$. In practice, rather than splitting the data into k disjoint partitions, it is advisable to split the data into a larger number of portions but have some overlap between the portions. In the *map* step, k parallel nodes perform gradient descent on their respective batch of training examples; in the final *reduce step*, the k parameter vectors are averaged [26]. The execution time of averaging k parameter vectors $\mathbf{w}^i \in \mathbb{R}^m$ is vanishingly small in comparison to the execution time of the parallel gradient descent.

When $\mathbf{w}_1, \ldots, \mathbf{w}_k$ are equilibrium points of the games given by the respective partitions of the sample, then the averaged vector $\mathbf{w} = \frac{1}{k}\sum_{j=1}^{k} \mathbf{w}_j$ still cannot be guaranteed to be an equilibrium point of the game given by the entire sample. In fact, in the experimental study we will find example cases where this is not the case.

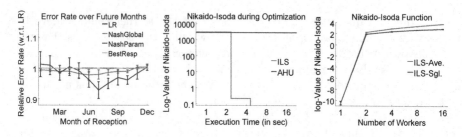

Fig. 1. Relative error (with respect to logistic regression, LR) of classification models evaluated into future (left). Value of Nikaido-Isoda function over time for three different optimization algorithms (center) and parallelized models (right). Error bars show standard errors.

5 Experimental Results

The goal of this section is to explore the robustness and scalability of sequential and parallelized methods that locate equilibrium points. We use a data set of 290,262 emails collected by an email service provider [3]. Each instance contains the term frequency of 226,342 terms. We compute a PCA of the emails and use the first 250 principal components as feature representation for most experiments. The data set is sorted chronologically. Emails that have been received over the final 12 months of the data collection period are held out for evaluation. Emails received in the month before that are used for tuning of the regularization parameters. Training emails are drawn from the remaining set of older emails.

5.1 Reference Methods

We use the logistic loss for all methods and for both learner and data generator. This makes logistic regression (LR) our natural *iid* baseline learning method. In the first experiment, we compare the transformation model derived in Section 2 (*NashParam*) that uses a parameterized function of individual instances to the global transformation model [3] that allows arbitrary dependencies between perturbations of multiple instances (*NashGlobal*). Additionally, we use the following simple game-theoretic reference method (*BestResp*): The data generator chooses the perturbation that is the best response to the standard logistic regression, and the learner chooses the model parameters that are the best response to this perturbation. That is, *BestResp* chooses parameters $\mathbf{w}^*$ according to:

1. $\mathbf{w}' = \arg\min_{\mathbf{w}} \hat{\theta}_f(\mathbf{w}, \mathbf{0}_{m \times m})$
2. $\mathbf{A}' = \arg\min_{\mathbf{A}} \hat{\theta}_g(\mathbf{w}', \mathbf{A})$
3. $\mathbf{w}^* = \arg\min_{\mathbf{w}} \hat{\theta}_f(\mathbf{w}, \mathbf{A})$.

5.2 Performance of the Parameterized Transformation Model

We compare a standard logistic regression approach (LR), game-theoretic heuristic *BestResp*, an equilibrium point with global transformation model (*NashGlobal*), and the equilibrium point with the parameterized transformation model

(*NashParam*). In each iteration, we sample 2500 training instances from the training portion of the data. We tune the free parameters—the regularization parameters of the learner and the data-generator—on the younger tuning portion of the data. All models are then evaluated over the final 12 evaluation months. We repeat this procedure 10 times and average the resulting error rates. For all following experiments, we keep all regularization parameters fixed, using the values that the parameters have been tuned to here.

Figure 1 (left) show the average relative error of all models with respect to logistic regression (*LR*); error bars show the standard error. Both *NashGlobal* and *NashParam* achieve significant improvements over *LR*. The parameterized transformation model *NashParam* reduces the error rate over *NashGlobal* by up to eight percent. The heuristic *BestResp* does not perform better than *LR*.

5.3   Optimization Algorithms

This section compares the convergence rates of the inexact line search (*ILS*) and Arrow-Hurwicz-Uzawa (*AHU*) approaches to finding equilibrium points, discussed in Sections 4.1 and 4.2, respectively.

In each repetition of the experiment, we sample 10,000 instances from the training portion of the data. Here, we use the 500 first principal components as feature representation. In each iteration of the optimization procedures, we measure he Nikaido-Isoda function of the current pair of parameters and the best possible reactions to these parameters—this function reaches zero at an equilibrium point. Figure 1 (center) shows that the *ILS* procedure converges very quickly. By contrast, *AHU* requires several orders of magnitude more time before the Nikaido-Isoda function drops noticably (not visible in the diagram); we have not been able to observe convergence. Increasing the regularization parameters by a factor of 100—which should make the optimization criterion more convex—did not change these findings. We therefore excluded *AHU* from further investigation.

5.4   Parallelized Models

In this section, we study parallel batch gradient descent, as discussed in Section 4.3, based on *ILS*. In each repetition, we sample 3200 instances from the training portion; we average 10 trials. The baseline model *LR-1-Sgl* is trained on all training data. Each of 8 nodes then processes a batch of data and returns a model *LR-8-Sgl*; these parameter vectors are averaged into *LR-8-Avg*. Likewise, *ILS-1-Sgl* is trained on all training data. Each node returns a model *ILS-8-Sgl*; these models are averaged into *ILS-8-Avg*.

For logistic regression, Figures 2 (left diagram, each node processes $\frac{1}{8}$ of the data), 3 (left, each node processes $\frac{1}{4}$ of the data), and 4 (left, each node processes $\frac{1}{\sqrt{8}}$ of the data), the averaged models *LR-8-Avg* consistently outperform the individual models *LR-8-Sgl*. Model *LR-1-Sgl* that has been trained sequentially on all available data outperforms the averaged models—this is consistent with

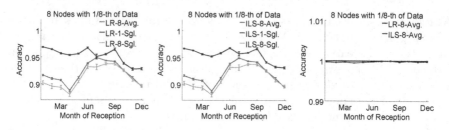

Fig. 2. Accuracy of logistic regression (left) and equilibrium points (center). Error rate of the aggregated equilibrium point relative to the error rate of aggregated logistic regression models (right). Each of 8 nodes processes $\frac{1}{8}$-th of the data. Error bars show standard errors.

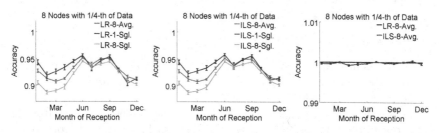

Fig. 3. Accuracy of logistic regression (left) and equilibrium points (center). Error rate of the aggregated equilibrium point relative to the error rate of aggregated logistic regression models (right). Each of 8 nodes processes $\frac{1}{4}$-th of the data. Error bars show standard errors.

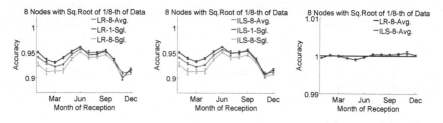

Fig. 4. Accuracy of logistic regression (left) and equilibrium points (center). Error rate of the aggregated equilibrium point relative to the error rate of aggregated logistic regression models (right). Each of 8 nodes processes $\frac{1}{\sqrt{8}}$-th of the data. Error bars show standard errors.

earlier results on parallel stochastic gradient descent [18,26]. The same is true for the equilibrium models found by *ILS*: Figures 2 (center, $\frac{1}{8}$ of the data per node), 3 (center, $\frac{1}{8}$ of the data per node), and 4 (center, $\frac{1}{\sqrt{8}}$ of the data per node) show that the averaged models *ILS-8-Avg* outperform the parallel models *ILS-8-Sgl*. The sequentially trained model *ILS-1-Sgl* outperforms the averaged models.

Figures 2 (right, $\frac{1}{4}$ of the data), and 3 (right, $\frac{1}{8}$ of the data), and 4 (right, $\frac{1}{\sqrt{8}}$ of the data) show the error rate of *ILS-8-Avg* relative to the error rate of *LR-8-Avg*, in analogy to Figure 1 (left). While in Figure Figure 1 (left) the equilibrium points have outperformed *LR*, the averaged model *ILS-8-Avg* tends to have a similar error rate as *LR-8-Avg*. The averaged equilibrium parameters—while still outperforming the equilibrium parameters trained on parallel batches—are no longer more accurate than the averaged logistic regression models.

We investigate further why this is the case. Figure 1 (right) shows the value of the Nikaido-Isoda function at the end of the batch optimization process for a single model trained on $\frac{1}{k}$ of the data (*ILS-Sgl*), and the corresponding Nikaido-Isoda function value for the average of k models trained on $\frac{1}{k}$ of the data each (*ILS-Avg*). Surprisingly, the averaged parameter vectors have a higher function value which means that they are further away from being equilibrium points than the individual models.

We can conclude that for this application (a) equilibrium points tend to be more accurate than standard logistic regression models; (b) averaging parameter vectors that have been trained on different batches of the data always leads to more accurate models; but (c) averaging equilibrium points tends to lead to model parameters that are no longer equilibrium points, and are therefore not generally more accurate than standard logistic regression models.

6   Conclusion

We have derived a model of adversarial learning in which the data generator gets to choose a parametric perturbation function $g_{\mathbf{A}}(\mathbf{x}) = \mathbf{x} + \mathbf{A}\mathbf{x}$ which is used to transform observations at application time. We have shown that the game between learner and data generator has at least one equilibrium point for convex and continuous loss functions. We have shown that the equilibrium point is unique if the loss function of learner and data generator have identical curvatures (as can be achieved with logistic loss functions) and the relationship between the regularization coefficients of learner and data generator are balanced as required by Assumption 2. Empirically, we observe that for the application of email spam filtering, equilibrium points under the derived data generation model maintain a higher accuracy over an evaluation period of 12 months after training than *iid* learning and reference methods.

The MapReduce programming model offers an unrivaled speed-up potential because it requires all synchronization to be limited to a final aggregation step. We derived batch-parallel stochastic gradient methods that identify a unique equilibrium point and can be implemented using the MapReduce model. Prior work on parallel stochastic gradient descent has established that the aggregate of models that have been trained in parallel on subsets of the data are more accurate than the individual, aggregated models, and that the aggregate converges toward to performance of a single model that has been trained sequentially on all the data [18,26]. We observe that this is also true for the aggregates of equilibrium points that have been located in parallel on batches of the data. However, it turns out that aggregates of equilibrium points are not equilibrium points themselves; the Nikaido-Isoda function increases its value during the aggregation step. Therefore, aggregated logistic regression models are about as accurate as aggregated equilibrium points. From a practical point of view, this implies that searching for equilibrium points is advisable for adversarial applications as long as training data, and not computation time, is the limiting factor. As the sample size increases, the computation time needed to locate equilibrium points on a single node becomes the limiting factor. For intermediate sample sizes, it may still be possible (and advisable) to train a model on a single node using *iid* learning. For even larger sample sizes, this becomes impossible. At this point, aggregated batch-parallel gradient descent outperforms sequential optimization using a subset of the data. At this point, however, aggregated equilibrium points offer no advantage over aggregated models trained under the *iid* assumption.

Acknowledgment. This work was supported by the German Science Foundation DFG under grant SCHE540/12-2.

References

1. Arrow, K., Hurwicz, L., Uzawa, H.: Studies in Linear and Non-Linear Programming. Stanford University Press, Stanford (1958)
2. Basar, T., Olsder, G.J.: Dynamic Noncooperative Game Theory. Academic Press, London/New York (1995)
3. Brückner, M., Kanzow, C., Scheffer, T.: Static prediction games for adversarial learning problems. Journal of Machine Learning Research **12**, 2617–2654 (2012)
4. Brückner, M., Scheffer, T.: Stackelberg games for adversarial learning problems. In: Proceedings of the 17th ACM SIGKDD International Conference on Knowledge Discovery and Data Mining (2011)
5. Bu, Y., Howe, B., Balazinska, M., Ernst, M.: Haloop: efficient iterative data processing on large clusters. In: Proceedings of the VLDB Endowment, vol. 3 (2010)
6. Dean, J., Ghemawat, S.: Mapreduce: simplified data processing on large clusters. In: Proceedings of the 6th Symposium on Operating System Design and Implementation (2004)
7. Dekel, O., Shamir, O.: Learning to classify with missing and corrupted features. In: Proceedings of the International Conference on Machine Learning. ACM Press (2008)
8. Dekel, O., Shamir, O., Xiao, L.: Learning to classify with missing and corrupted features. Machine Learning **81**(2), 149–178 (2010)
9. Flam, S., Ruszczynski, A.: Computing normalized equilibria in convex-concave games. Working Papers Working Papers 2006:9, Lund University, Department of Economics (2006)
10. Geiger, C., Kanzow, C.: Theorie und Numerik restringierter Optimierungsaufgaben. Springer, Heidelberg (2002)
11. Ghaoui, L.E., Lanckriet, G.R.G., Natsoulis, G.: Robust classification with interval data. Tech. Rep. UCB/CSD-03-1279, EECS Department, University of California, Berkeley (2003)
12. Globerson, A., Roweis, S.T.: Nightmare at test time: robust learning by feature deletion. In: Proceedings of the International Conference on Machine Learning. ACM Press (2006)

13. Globerson, A., Teo, C.H., Smola, A.J., Roweis, S.T.: An adversarial view of covariate shift and a minimax approach. In: Dataset Shift in Machine Learning, pp. 179–198. MIT Press (2009)
14. Großhans, M., Sawade, C., Brückner, M., Scheffer, T.: Bayesian games for adversarial regression problems. In: Proceedings of the International Conference on Machine Learning (2013)
15. Hardt, M., Megiddo, N., Papadimitriou, C., Wooters, M.: Strategic classification. Unpublished manuscript
16. von Heusinger, A., Kanzow, C.: Relaxation methods for generalized nash equilibrium problems with inexact line search. Journal of Optimization Theory and Applications 143(1), 159–183 (2009)
17. Lanckriet, G.R.G., Ghaoui, L.E., Bhattacharyya, C., Jordan, M.I.: A robust minimax approach to classification. Journal of Machine Learning Research 3, 555–582 (2002)
18. Mann, G., McDonald, R., Mohri, M., Silberman, N., Walker, D.: Efficient large-scale distributed training of conditional maximum entropy models. In: Advances in Neural Information Processing, vol. 22 (2009)
19. Nedic, A., Bertsekas, D., Borkar, V.: Distributed asynchronous incremental subgradient methods. Studies in Computational Mathematics 8, 381–407 (2001)
20. Nedic, A., Ozdaglar, A.: Subgradient methods for saddle-point problems. Journal of Optimization Theory and Applications 142(1), 205–228 (2009)
21. Nikaido, H., Isoda, K.: Note on noncooperative convex games. Pacific Journal of Mathematics 5, 807–815 (1955)
22. Rosen, J.B.: Existence and uniqueness of equilibrium points for concave n-person games. Econometrica 33(3), 520–534 (1965)
23. Teo, C.H., Globerson, A., Roweis, S.T., Smola, A.J.: Convex learning with invariances. In: Advances in Neural Information Processing Systems. MIT Press (2007)
24. Torkamani, M., Lowd, D.: Convex adversarial collective classification. In: Proceedings of the International Conference on Machine Learning (2013)
25. Weimer, M., Condie, T., Ramakrishnan, R.: Machine learning in scalops, a higher order cloud computing language. In: NIPS 2011 Workshop on Parallel and Large-scale Machine Learning (BigLearn) (2011)
26. Zinkevich, M., Weimer, M., Smola, A., Li, L.: Parallelized stochastic gradient descent. In: Advances in Neural Information Processing, vol. 23 (2010)

Structured Regularizer for Neural Higher-Order Sequence Models

Martin Ratajczak[1]([✉]), Sebastian Tschiatschek[2], and Franz Pernkopf[1]

[1] Signal Processing and Speech Communication Laboratory,
Graz University of Technology, Graz, Austria
{martin.ratajczak,pernkopf}@tugraz.at
[2] Learning and Adaptive Systems Group, ETH Zurich, Zurich, Switzerland
sebastian.tschiatschek@inf.ethz.ch

Abstract. We introduce both joint training of neural higher-order linear-chain conditional random fields (NHO-LC-CRFs) and a new structured regularizer for sequence modelling. We show that this regularizer can be derived as lower bound from a mixture of models sharing parts, e.g. neural sub-networks, and relate it to ensemble learning. Furthermore, it can be expressed explicitly as regularization term in the training objective.

We exemplify its effectiveness by exploring the introduced NHO-LC-CRFs for sequence labeling. Higher-order LC-CRFs with linear factors are well-established for that task, but they lack the ability to model non-linear dependencies. These non-linear dependencies, however, can be efficiently modeled by neural higher-order input-dependent factors. Experimental results for phoneme classification with NHO-LC-CRFs confirm this fact and we achieve state-of-the-art phoneme error rate of 16.7% on TIMIT using the new structured regularizer.

Keywords: Structured regularization · Ensemble learning · Additive mixture of experts · Neural higher-order conditional random field

1 Introduction

Overfitting is a common and challenging problem in machine learning. It occurs when a learning algorithm overspecializes to training samples, i.e. irrelevant or noisy information for prediction is learned or even memorized. Consequently, the learning algorithm does not generalize to unseen data samples. This results in large test error, while obtaining small training error. A common assumption is that complex models are prone to overfitting, while simple models have limited predictive expressiveness. Therefore a trade-off between model complexity and predictive expressiveness needs to be found. Usually, a penalty term for

This work was supported by the Austrian Science Fund (FWF) under the project number P25244-N15. Furthermore, we acknowledge NVIDIA for providing GPU computing resources.

A. Appice et al. (Eds.): ECML PKDD 2015, Part I, LNAI 9284, pp. 168–183, 2015.
DOI: 10.1007/978-3-319-23528-8_11

the model complexity is added to the training objective. This penalty term is called *regularization*. Many regularization techniques have been proposed, e.g. in parameterized model priors on individual weights or priors on groups of weights like the l_1-norm and l_2-norm are commonly used.

Recently, dropout [12] and dropconnect [33] have been proposed as regularization techniques for neural networks. During dropout training, input and hidden units are randomly canceled. The cancellation of input units can be interpreted as a special form of input noising and, therefore, as a special type of data augmentation [18,32]. During dropconnect training, the connections between the neurons are dropped [33]. Dropout and dropconnect can be interpreted as mixtures of neural networks with different structures. In this sense, dropout and dropconnect have been interpreted as ensemble learning techniques. In ensemble learning, many different classifiers are trained independently to make the same predictions, i.e. ensembles of different base classifiers. For testing, the predictions of the different classifiers are combined. In the dropout and dropconnect approaches, the mixture of models is trained and utilized for testing. Recently, *pseudo-ensemble learning* [1] has been suggested as a generalization of dropout and dropconnect. In pseudo-ensemble learning, a mixture of child models generated by perturbing a parent model is considered.

We propose a generalization of pseudo-ensemble learning. We introduce a mixture of models which share parts, e.g. neural sub-networks, called *shared-ensemble learning*. The difference is that in shared-ensemble learning, there is no parent model from which we generate child models. The models in the shared-ensemble can be different, but share parts. This is in contrast to traditional ensembles which typically do no share parts. Based on shared-ensembles, we derive a new regularizer as lower bound of the conditional likelihood of the mixture of models. Furthermore, this regularizer can be expressed explicitly as regularization term in the training objective. In this paper, we apply shared-ensemble learning to *linear-chain conditional random fields (LC-CRFs)* [13] in a sequence labeling task, derive a structured regularizer and demonstrate its advantage in experiments. LC-CRFs are established models for sequence labeling [7,35], i.e. we assign some given input sequence **x**, e.g. a time series, to an output label sequence **y**.

First-order LC-CRFs typically consist of transition factors, modeling the relationship between two consecutive output labels, and local factors, modeling the relationship between input observations (usually a sliding window over input frames) and one output label. But CRFs are not limited to such types of factors: *Higher-order LC-CRFs (HO-LC-CRFs)* allow for arbitrary input-independent (such factors depend on the output labels only) [35] and input-dependent (such factors depend on both the input and output variables) higher-order factors [16,23]. That means both types of factors can include more than two output labels. Clearly, the Markov order of the largest factor (on the output side) dictates the order of LC-CRFs.

It is common practice to represent the higher-order factors by linear functions which can reduce the model's expressiveness. To overcome this limitation, a widely used approach is to represent non-linear dependencies by parametrized models and to learn these models from data. Mainly kernel methods [14] and *neural models* [15,19,20,22,24] have been suggested to parametrize *first-order* factors in LC-CRFs, i.e. mapping several input frames to one output label. In summary, most work in the past focused either on (a) higher-order factors represented by simple linear models, or (b) first-order input-dependent factors represented by neural networks. In this work, we explore joint-training of neural *and* higher-order input-dependent factors in LC-CRFs.

Unfortunately, higher-order CRFs significantly increase the model complexity and, therefore, are prone to overfitting. To avoid overfitting, the amount of training data has to be sufficiently large or, alternatively, regularizers can be utilized. In this work, we apply the structured regularizer derived from the shared-ensemble framework to higher-order CRFs and demonstrate its effectiveness.

Our main contributions are:

1. We propose *shared-ensemble learning* as a generalization of pseudo-ensemble learning, i.e. a mixture of models which share parts, e.g. neural sub-networks.
2. From this framework we derive a new regularizer for higher-order sequence models. By lower-bounding the conditional likelihood of the mixture of models, we can explicitly express the regularization term in the training objective.
3. Furthermore, we introduce *joint-training* of *neural higher-order input-dependent factors* in LC-CRFs depending on both sub-sequences of the input and the output labels. These factors are represented by *individual multi-layer perceptron (MLP) networks*.
4. We present experimental results for phoneme classification. NHO-LC-CRFs with the proposed regularizer achieve state-of-the-art performance of 16.7% phone error rate on the TIMIT phoneme classification task.

The remainder of this paper is structured as follows: In Section 2 we briefly review related work. In Section 3 we introduce the NHO-LC-CRF model. Section 4 provides details on the structured regularizer. In Section 5 we evaluate our model on the TIMIT phoneme classification task. Section 6 concludes the paper.

2   Related Work

Dropout applied to the input has been formalized for some linear and log-linear models [18,32]. Assuming a distribution of the dropout noise, an analytical form of the dropout technique has been presented. The training objective has been formulated as the expectation of the loss function under this distribution. Furthermore, this objective has been reformulated as the original objective and an additive explicit regularization.

As mentioned before, HO-LC-CRFs have been applied to sequence labeling in tagging tasks, in handwriting recognition [23,35] and large-scale machine translation [16]. In these works, higher-order factors have not been modeled by neural networks which is the gap we fill. However, first-order factors have been already modeled by several types of neural networks. Conditional neural fields (CNFs) [20] and multi-layer CRFs [22], propose a direct method to optimize MLP networks and LC-CRFs under the conditional likelihood criterion based on error back-propagation. Another approach is to pre-train an unsupervised representation with a deep belief network, transform it into an MLP network and finally fine-tune the network in conjunction with the LC-CRF [5]. *Hidden-unit conditional random fields* (HU-CRFs) [19] replace local factors by *discriminative RBMs* (DRBMs) [15], CNN triangular CRFs [34] by convolutional neural networks (CNNs) and *context-specific deep CRFs* [24] by sum-product networks [6,21] which can be interpreted as *discriminative deep Gaussian mixture models* generalizing discriminative Gaussian mixture models to multiple layers of hidden variables.

In more detail, we contrast our work from [5]: First, although formulated quite general that work focused on neural first-order factors in contrast to neural higher-order factors in our work. Second, they used one shared neural network for all factors in contrast to individual neural networks as in our case. Third, that work utilized unsupervised pre-training as initialization of their MLP network. We jointly train the NHO-LC-CRF and we improved the classification performance using the new structured regularizer. This work is an extension of [25] which focused on discriminative pre-training of neural higher-order factors to produce rich non-linear features. A linear higher-order LC-CRFs is subsequently trained on these features. In contrast, here we train jointly the NHO-LC-CRF. Furthermore, we introduce the new structured regularizer and show its relation to mixture models and ensemble learning.

Finally, in computer vision higher-order factors in Markov random fields [26] and conditional random fields [9,17,30] are much more common than in sequence labeling. Most of that work focus on higher-order factors represented by products of experts [11]. Typically, approximate inference such as belief propagation or a sampling method is utilized.

3   Higher-Order Conditional Random Fields

We consider HO-LC-CRFs for sequence labeling. The HO-LC-CRF defines a conditional distribution

$$p^{CRF}(\mathbf{y}|\mathbf{x}) = \frac{1}{Z(\mathbf{x})} \prod_{t=1}^{T} \prod_{n=1}^{N} \Phi_t(\mathbf{y}_{t-n+1:t}; \mathbf{x}), \tag{1}$$

for an output sequence $\mathbf{y}$ of length T given an input sequence $\mathbf{x}$ of length T where $\Phi_t(\mathbf{y}_{t-n+1:t}; \mathbf{x})$ are non-negative factors that can depend on the label sub-sequence $\mathbf{y}_{t-n+1:t}$ and the whole input sequence $\mathbf{x}$, and where $Z(\mathbf{x})$ is an input-dependent normalization computed as

$$Z(\mathbf{x}) = \sum_{\mathbf{y}} \prod_{t=1}^{T} \prod_{n=1}^{N} \Phi_t(\mathbf{y}_{t-n+1:t}; \mathbf{x}). \tag{2}$$

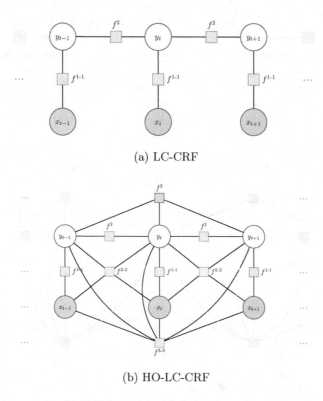

(a) LC-CRF

(b) HO-LC-CRF

Fig. 1. Factor graph of LC-CRFs using (a) input-dependent uni-gram features $f^{1\text{-}1}$ and bi-gram transition features f^2 (typical) and (b) additionally 3-gram features f^3 as well as input-dependent features $f^{2\text{-}2}$ and $f^{3\text{-}3}$.

An $(N-1)^{\text{th}}$-order CRF models label sub-sequences of maximal span N in the corresponding factors. The factors $\Phi_t(\mathbf{y}_{t-n+1:t}; \mathbf{x})$ are assumed to be given in log-linear form, i.e.

$$\Phi_t(\mathbf{y}_{t-n+1:t}; \mathbf{x}) = \exp\left(\sum_k \mathbf{w}_k^{t,n} \mathbf{f}_k(\mathbf{y}_{t-n+1:t}; t, \mathbf{x})\right), \qquad (3)$$

where $\mathbf{f}_k(\mathbf{y}_{t-n+1:t}; t, \mathbf{x})$ are arbitrary vector-valued and (possibly) position-dependent feature functions and $\mathbf{w}_k^{t,n}$ are the weights. These functions can be any functions ranging from simple indicator functions, linear functions, up to functions computed using neural networks as we have in this work. We distinguish the following types of feature functions:

- n-**Gram Input-Independent Features.** These features are observation-independent, i.e. $\mathbf{f}_k(\mathbf{y}_{t-n+1:t}; t, \mathbf{x}) = \mathbf{f}^n(\mathbf{y}_{t-n+1:t})$. Each entry of the vectors corresponds to the indicator function of a certain label sub-sequence $\mathbf{a}_i$, i.e. $\mathbf{f}^n(\mathbf{y}_{t-n+1:t}) = [\mathbf{1}(\mathbf{y}_{t-n+1:t} = \mathbf{a}_1), \mathbf{1}(\mathbf{y}_{t-n+1:t} = \mathbf{a}_2), \ldots]^T$. Typically $\mathbf{a}_1, \mathbf{a}_2, \ldots$ enumerate all possible label sub-sequences of length n. These transition functions are denoted as f^n in Figure 1.

- m-n-**Gram Input-Dependent MLP Features.** These features generalize local factors to longer label sub-sequences. In this way, these feature functions can depend on the label sub-sequence $\mathbf{y}_{t-n+1:t}$ and an input sub-sequence of $\mathbf{x}$, i.e. $\mathbf{f}^{m-n}(\mathbf{y}_{t-n+1:t}; t, \mathbf{x}) = [\mathbf{1}(\mathbf{y}_{t-n+1:t} = \mathbf{a}_1) \mathbf{g}^m(\mathbf{x}, t), \ldots]^T$ where $\mathbf{g}^m(\mathbf{x}, t)$ is an arbitrary function. This function maps an input sub-sequence into a new feature space. In this work, we choose to use MLP networks for this function being able to model complex interactions among the variables. More specific, the hidden activations of the last layer $\mathbf{h}^m(\mathbf{x}, t)$ of the MLP network are used, i.e. $\mathbf{g}^m(\mathbf{x}, t) = \mathbf{h}^m(\mathbf{x}, t)$. We call these features m-n-gram MLP features. They are denoted as f^{m-n} in Figure 1, assuming that they only depend on input-output sub-sequences. Although possible, we do not consider the full input sequence $\mathbf{x}$ to counteract overfitting, but only use a symmetric and centered contextual window of length m around position t or time interval $t - n + 1 : t$. Exemplary, in case of two labels and four input sub-sequences we include the inputs from time interval $t-2 : t+1$. An important extension to prior work [5] is that the m-n-gram MLP features are modeled by individual networks to represent different non-linear interdependences between input and output sub-sequences.

Figure 1 show two LC-CRFs as factor graph. A typical LC-CRF as shown in the top of the Figure 1 consists of input-dependent uni-gram features f^{1-1} and input-independent bi-gram features f^2. In this work, we consider a rarely used extension using higher-order input-dependent m-n-gram features, for example f^{3-3}, shown in the bottom of Figure 1.

The benefit of input-dependent higher-order factors for phoneme classification is substantiated by the fact that the spectral properties of phonemes are strongly influenced by neighboring phonemes. This is illustrated in Figure 2. In conventional speech recognition systems, this well-known fact is tackled by introducing meta-labels in form of tri-phone models [10]. Input-dependent higher-order factors in HO-LC-CRF support this by mapping an input sub-sequence to an output sub-sequence, i.e. several output labels, without introducing meta-labels. Further in HO-LC-CRF, we are able to model arbitrary mappings from input sub-sequences of length m to output sub-sequences of length n, i.e we can also model mono-phones, bi-phones and tri-phones within the same model.

3.1 Parameter Learning

Parameters $\mathbf{w} = \{\mathbf{w}_k^{t,n} \mid \forall k, t, n\}$ are optimized to maximize the conditional log-likelihood of the training-data, i.e.

$$\mathcal{F}(\mathbf{w}) = \sum_{j=1}^{J} \log p(\mathbf{y}^{(j)} | \mathbf{x}^{(j)}), \qquad (4)$$

where $((\mathbf{x}^{(1)}, \mathbf{y}^{(1)}), \ldots, (\mathbf{x}^{(J)}, \mathbf{y}^{(J)}))$ is a collection of J input-output sequence pairs drawn i.i.d. from some unknown data distribution. The partial derivatives of (4) with respect to the weights $\mathbf{w}_k^{t,n}$ can be computed as described in [23, 35].

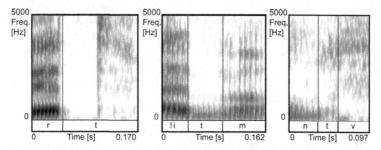

Fig. 2. Three realizations of word-final /t/ in spontaneous Dutch. Left panel: Realization of /rt/ in *gestudeerd* 'studied'. Middle panel: Realization of /'εitm/ in *leeftijd mag* 'age is allowed'. Right panel: Realization of /ntv/ in *want volgens* 'because according' [28].

The weights are shared over time $\mathbf{w}_k^{t,n} = \mathbf{w}_k^n$ as we use time-homogeneous features. To perform parameter learning using gradient ascent all marginal posteriors of the form $p(\mathbf{y}_{t-n+1:t}|t, \mathbf{x}^{(j)})$ are required. These marginals can be efficiently computed using the *forward-backward algorithm* [23,35]. The algorithm can be easily extended to CRFs of order $(N-1) > 2$. However, for simplicity and as we are targeting GPU platforms, we choose another approach. As we describe in more detail in Section 3.2, we compute the conditional log-likelihood by computing just the forward recursion. Then we utilize back-propagation [27] as common in typical neural networks to compute their gradients. The conditional likelihood, the forward recursion and the corresponding gradients are computed using Theano [2], a mathematical expression compiler for GPUs and automatic differentiation toolbox.

3.2   Forward Algorithm for 2nd-Order CRFs

The main ingredient needed for applying the back-propagation algorithm is the forward recursion and the computation of the normalization constant. For a given input-output sequence pair $(\mathbf{x}, \mathbf{y})$, the forward recursion is given in terms of quantities $\alpha_t(\mathbf{y}_{t-1:t})$ that are updated according to

$$\alpha_t(\mathbf{y}_{t-1:t}) = \Phi_t(y_t; \mathbf{x})\Phi_t(\mathbf{y}_{t-1:t}; \mathbf{x}) \sum_{y_{t-2}} \Phi_t(\mathbf{y}_{t-2:t}; \mathbf{x})\alpha_{t-1}(\mathbf{y}_{t-2:t-1}). \quad (5)$$

The recursion is initialized as

$$\alpha_2(\mathbf{y}_{1:2}) = \Phi_2(y_2; \mathbf{x})\Phi_1(\mathbf{y}_{1:2}; \mathbf{x})\Phi_1(y_1; \mathbf{x}). \quad (6)$$

Finally, the normalization constant can be computed as

$$Z(\mathbf{x}) = \sum_{\mathbf{y}_{T-1:T}} \alpha_T(\mathbf{y}_{T-1:T}). \quad (7)$$

The most probable label sequence can be found by the Viterbi algorithm generalized to HO-LC-CRFs: The summation in the forward recursion is exchanged by the maximum operation, i.e. quantities $\hat{\alpha}_t(\mathbf{y}_{t-1:t})$ are computed as

$$\hat{\alpha}_t(\mathbf{y}_{t-1:t}) = \Phi_t(y_t; \mathbf{x})\Phi_t(\mathbf{y}_{t-1:t}; \mathbf{x}) \max_{y_{t-2}} \Phi_t(\mathbf{y}_{t-2:t}; \mathbf{x})\hat{\alpha}_{t-1}(\mathbf{y}_{t-2:t-1}). \tag{8}$$

At the end of the recursion, we identify the most probable state at the last position and apply back-tracking. For details and for time complexities we refer to [23,35].

4   Structured Regularizer

As mentioned in the introduction, NHO-LC-CRFs have high predictive expressiveness but are prone to overfitting. To fully exploit the potential of these models, special regularization techniques must be applied. Therefore, on top of the NHO-LC-CRF, we add a new structured regularizer. In the following, we derive this regularizer in a quite general form based on additive mixture of experts [3]. Our derivation is based on a single training sample $(\mathbf{x}, \mathbf{y})$, the generalization to multiple samples is straightforward.

A mixture of models, i.e. additive mixture of experts, is defined as

$$\log\ p(\mathbf{y}|\mathbf{x}) = \log \sum_{M \in \mathcal{M}} p(\mathbf{y}, M|\mathbf{x}), \tag{9}$$

where $\mathcal{M}$ is the set of models. We assume that the models in $\mathcal{M}$ have K *shared parts* $S_1, \ldots, S_K$, e.g. neural sub-networks. We consider the special case $\mathcal{M} = \{M_{S_1,\ldots,S_K}, M_{S_1}, \ldots, M_{S_K}\}$, where $M_{S_1,\ldots,S_K}$ is the *combination model* which contains all shared parts and M_{S_i} are *sub-models* containing the corresponding parts S_i. The intuition behind this model choice is the following: Shared parts in the combination model should not rely on the predictions of the other parts. Therefore, the sub-models should produce good predictions by itself. This approach improves robustness by counteracting co-adaptation comparable to dropout training in neural networks.

Expanding Equation 9 yields

$$\log\ p(\mathbf{y}|\mathbf{x}) = \log \Big(p(\mathbf{y}, M_{S_1,\ldots,S_K}|\mathbf{x}) + \sum_{M_S \in \mathcal{M}_S} p(\mathbf{y}, M_S|\mathbf{x}) \Big), \tag{10}$$

where the first term in the logarithm is the conditional joint probability of $\mathbf{y}$ and the combination model $M_{S_1,\ldots,S_K}$ and the sum is over the conditional joint probabilities of $\mathbf{y}$ and the sub-models in $\mathcal{M}_S = \{M_{S_1}, \ldots, M_{S_K}\}$. By the chain-rule $p(\mathbf{y}, M|\mathbf{x}) = p(\mathbf{y}|\mathbf{x}, M)\, p(M|\mathbf{x})$ and Jensen's inequality, we obtain a lower bound to the log-likelihood as

$$\log \sum_{M \in \mathcal{M}} p(\mathbf{y}|\mathbf{x}, M)\, p(M|\mathbf{x}) \geq \sum_{M \in \mathcal{M}} p(M|\mathbf{x})\ \log\ p(\mathbf{y}|\mathbf{x}, M), \tag{11}$$

where $\sum_{M \in \mathcal{M}} p(M|\mathbf{x}) = 1$. By lower-bounding the log-likelihood we reformulated the additive mixture of experts as a product of experts, i.e. summation in log-space. By assuming that the model prior satisfies $p(M|\mathbf{x}) = p(M)$, i.e. the prior is independent of the sample $\mathbf{x}$, we obtain $\sum_{M \in \mathcal{M}} p(M) \log p(\mathbf{y}|\mathbf{x}, M)$. To this end, we can rewrite our lower bound as

$$p(M_{S_1,\dots,S_K}) \log p(\mathbf{y}|\mathbf{x}, M_{S_1,\dots,S_K}) + \sum_{M_S \in \mathcal{M}_S} p(M_S) \log p(\mathbf{y}|\mathbf{x}, M_S). \qquad (12)$$

We apply this result to our sequence labeling model introduced in Section 3. We utilize the MLP networks for the different sub-sequences as sub-models M_S and the NHO-LC-CRF as the combination model $M_{S_1,\dots,S_K}$. Assuming a prior probability of λ for the combination model, i.e. $p(M_{S_1,\dots,S_K}) = \lambda$, and a uniform model prior $p(M_S) = (1 - \lambda)/|\mathcal{M}_S|$ for the label sub-sequence models, the final training objective over sequences including the structured regularizer is

$$\mathcal{F}(\mathbf{w}) = \sum_{j=1}^{J} \left(\lambda \log p^{CRF}(\mathbf{y}^{(j)}|\mathbf{x}^{(j)}) + (1 - \lambda) \frac{1}{|\mathcal{M}_S|} \sum_{n} \log R^n(\mathbf{y}^{(j)}|\mathbf{x}^{(j)}) \right), \qquad (13)$$

where $(\mathbf{x}^{(j)}, \mathbf{y}^{(j)})$ is the jth training sample. The regularizers $R^n(\mathbf{y}|\mathbf{x})$ for the corresponding label sub-sequences are defined as

$$\log R^n(\mathbf{y}|\mathbf{x}) = \sum_{t=n:T} \log p^{\text{MLP}}(\mathbf{y}_{t-n+1:t}|t, m, \mathbf{x}), \qquad (14)$$

where the conditional probabilities of the corresponding label sub-sequences are

$$p^{\text{MLP}}(\mathbf{y}_{t-n+1:t}|t, m, \mathbf{x}) = \frac{\exp\left(\mathbf{w}_n^T \mathbf{f}^{m-n}(\mathbf{y}_{t-n+1:t}; t, \mathbf{x})\right)}{Z_n^{\text{MLP}}(\mathbf{x})} \qquad (15)$$

and the normalization constants of the MLP networks are

$$Z_n^{\text{MLP}}(\mathbf{x}) = \sum_{\mathbf{y}_{t-n+1:t}} \exp\left(\mathbf{w}_n^T \mathbf{f}^{m-n}(\mathbf{y}_{t-n+1:t}; t, \mathbf{x})\right). \qquad (16)$$

This means the sub-models are MLP networks sharing the MLP features f^{m-n} with the NHO-LC-CRF, the combination model. The trade-off parameter λ balances the importance of the sequence model against the other sub-models.

For testing we drop the regularizer and find the most probable sequence by utilizing the Viterbi algorithm for NHO-LC-CRFs as described in Section 3.2.

5   Experiments

We evaluated the performance of the proposed models on the TIMIT phoneme classification task. We compared isolated phone classification (without label context information) with MLP networks to phone labeling with neural HO-LC-CRFs. This comparison substantiates the effectiveness of joint-training of neural higher-order factors. Furthermore, we show performance improvements using our introduced structured regularizer during joint-training.

5.1 TIMIT Data Set

The TIMIT data set [36] contains recordings of 5.4 hours of English speech from 8 major dialect regions of the United States. The recordings were manually segmented at phone level. We use this segmentation for phone classification. Note that phone classification should not be confused with phone recognition [10] where no segmentation is provided. We collapsed the original 61 phones into 39 phones. All frames of Mel-frequency cepstral coefficients (MFCC), delta and double-delta coefficients of a phonetic segment are mapped into one feature vector. Details are presented in [8]. The task is, given an utterance and a corresponding segmentation, to infer the phoneme within every segment. The data set consists of a training set, a development set (dev) and a test set (test), containing 140.173, 50.735 and 7.211 phonetic segments, respectively. The development set is used for parameter tuning.

5.2 Experimental Setup

In all experiments, input features were normalized to zero mean and unit variance. Optimization of our models was in all cases performed using stochastic gradient ascent using a batch-size of one sample. An ℓ_2-norm regularizer on the model weights was used. We utilized early stopping determined on the development data set. We trained for up to 500 epochs. However, in most cases less iterations have been required. The proposed model was entirely trained on NVIDIA GPUs using Theano [2], a mathematical expression compiler for GPUs and automatic differentiation toolbox. The classification performance is measured by phone error rate (PER), i.e. Hamming distance between the reference and predicted label sequence for all test samples.

5.3 Labeling Results Using Only MLP Networks

In the first experiment, we trained MLP networks with a single hidden layer to predict the phone label of each segment. We tuned the number of hidden units $H \in \{100, 150, 200, 300, 400, 500\}$ and their activation functions, i.e. rectifier and tanh. Furthermore, we analyzed the effect of the number of input segments, i.e. we used the current segment and three or five surrounding segments centered at the current position index as input. Results in Table 1 (only a sub-set is reported) show that more hidden units result in better performance. For tanh activations, the best performance of 21.0% is achieved with $m = 3$ input segments and using $H = 500$ hidden units. More input segments reduced the performance. In preliminary experiments, we found that more than one hidden layer decreased the performance. Therefore, we used in the following experiments tanh activations and one hidden layer.

5.4 Labeling Results Using LC-CRFs with Linear or Neural Higher-Order Factors

Experiments with linear HO-LC-CRFs as shown in Table 2 reveal that classification performance degrades with linear 3-3 gram factor. The best performance

Table 1. Isolated Phone Classification using MLP networks ($n = 1$) with different number of hidden units H and lengths of the contextual input window m. The classification performance is measured by phone error rate (PER).

		rectifier		tanh				rectifier		tanh	
H	m	dev	test	dev	test	H	m	dev	test	dev	test
100	1	23.2	23.7	23.3	24.1	200	1	22.5	23.2	22.4	22.6
100	3	22.0	23.4	22.3	22.9	200	3	21.3	21.8	21.4	22.6
100	5	22.6	22.9	23.8	24.4	200	5	22.3	22.7	22.7	22.9
150	1	22.6	23.0	22.6	23.0	500	1	22.1	22.6	22.1	22.9
150	3	21.4	22.2	21.8	22.3	500	3	**20.9**	22.1	**20.6**	**21.0**
150	5	22.4	22.9	23.2	23.9	500	5	22.3	22.7	21.9	22.7

Table 2. Linear higher-order CRFs. All results with $m = 1$ and $n = 1$ already include input-independent 2-gram factors.

m=n	1	+ 2	+ 3
dev	25.8	**20.4**	20.7
test	25.9	**20.5**	21.6

of 20.5% is achieved with factors up to order $n = m = 2$. The plus sign indicates additional higher-order factors on top to the ones from previous columns in the table, i.e. the model of column +2 includes the linear factors $\{f^1, f^{1-1}, f^2, f^{2-2}\}$.

In the next set of experiments, we consider LC-CRFs with neural input-dependent higher-order factors and we will show their effectiveness in contrast to their linear counterparts in Table 2. In Table 3, we explore the combination of higher-order factors up to the order $n = m = 3$ as described in Section 3. By including more higher-order factors in the first column of Table 3, the classification performance improves for the baseline using only l_2 regularization only. For the baseline, we tuned the learning rate 0.01, 0.001, 0.0001 and $l2$ regularizer trade-off-parameter 0.1, 0.01, 0.001, 0.0001 and report the best observed performance. The best result of 17.7% is achieved with 150 hidden units and factors up to order $n = m = 3$ which is significantly better than the best performance of 20.5% with linear factors.

Furthermore, we tested our new structured regularizer with factors up to order $n = m = 3$ for various trade-off parameters $\lambda \in \{0.01, 0.3, 0.6\}$ as shown in Table 3. We used fixed a learning rate of 0.001 and l_2 regularizer of 0.001. We achieved the best performance of 16.8% with factors up to order $m = n = 3$ and a trade-off parameter of 0.3. We further tuned the trade-off parameter $\lambda \in \{0.01, 0.05, 0.1, 0.2, 0.3, 0.4, 0.6, 0.95, 1.0\}$ for different number of hidden units $H \in \{100, 150, 200\}$. This is shown in the Figure 3. For different network sizes, we observe a clear optimum with respect to the trade-off parameter

Table 3. Neural higher-order LC-CRF without and with our structured regularizer. The order $m = n$ of the higher order factors is examined. Furthermore, the effectiveness of the new structured regularizer is demonstrated for factors up to order $m = n = 3$. All results with $m = 1$ and $n = 1$ already include input-independent 2-gram factors. In all experiments, the number of hidden units is $H = 150$ and one hidden layer.

		no reg.		$\lambda = 0.6$		$\lambda = 0.3$		$\lambda = 0.01$	
$m = n$		dev	test	dev	test	dev	test	dev	test
	1	21.2	21.5	21.7	22.4	21.7	21.9	25.7	26.4
+	2	17.6	18.3	17.9	18.3	17.5	18.0	19.8	20.4
+	3	**17.9**	**17.7**	16.9	17.5	**16.6**	**16.8**	18.5	19.2

λ and a margin to the baseline results without the regularizer, i.e. $\lambda = 1.0$, which we indicated by a dotted line in Figure 3. By this additional tuning, we improved the result further and achieved the best overall performance of 16.7% with factors up to order $m = n = 3$ and a trade-off parameter of 0.1.

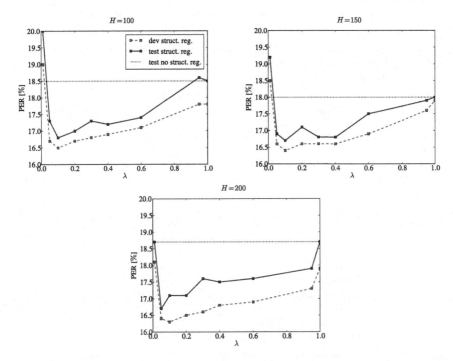

Fig. 3. Performance using the structured regularizer for various trade-off parameters λ. The number of hidden units $H \in \{100, 150, 200\}$ in the neural higher-order factors varies in the different plots. Baseline results without the regularizer $\lambda = 1.0$ are indicated by a dotted line.

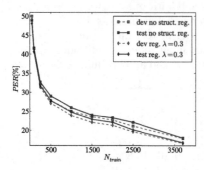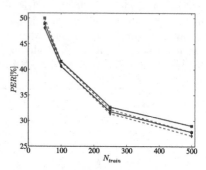

Fig. 4. Test performance for (left) various training set sample sizes with and without our structured regularizer and (right) its zoomed presentation to illustrate the effectiveness for small training set sample sizes.

In additional experiments, we explored the performance for varying numbers of training samples N_{train}. We fixed the number of hidden units $H = 150$, trade-off parameter $\lambda = 0.3$, learning rate of 0.001 and l_2 regularizer of 0.001. Figure 4 shows the effectiveness of our structured regularizer for small and full training sample set, i.e. we are able to outperform the same model without the structured regularizer by a large margin over the range of used training samples. For small sample sizes, the margin between the baseline performance results and the one with the structured regularizer decreased slightly.

Finally, we compare our best result to other state-of-the-art methods based on MFCC features as shown in Table 4. Using the code of [20] we tested CNFs

Table 4. Summary of labeling results. Results marked by ([†]) are from [31], by ([††]) are from [29], by ([†††]) are from [8], by ([††††]) are from [24], and by (*) are from [4]. Performance measure: Phone error rate (PER) in percent.

Model	PER [%]
GMMs ML[††]	25.9
HCRFs[†]	21.5
Large-Margin GMM[††]	21.1
Heterogeneous Measurements[†††]	21.0
CNF	20.67
Linear HO-LC-CRF	20.5
GMM+LC-CRF (1st order)[††††]	22.10
CS-DCRF+MEMM (8th order)[††††]	22.15
CS-DCRF+LC-CRF (1st order)[††††]	19.95
Hierarchical Large-Margin GMMs[*]	**16.7**
NHO-LC-CRF	**17.7**
+ structured regularizer	**16.7**

with 50, 100 and 200 hidden units as well as one and three input segments. We achieved the best result with 100 hidden units and one segment as input. Furthermore, hierarchical large-margin GMMs achieve a performance of 16.7% and outperform most other referenced methods but exploit human-knowledge and committee techniques. However, our best model, the HO-LC-CRF augmented by m-n-gram MLP factors using our new structured regularizer achieves a performance of 16.7% without human-knowledge and is outperforming most of the state-of-the-art methods.

6   Conclusion

We considered NHO-LC-CRFs for sequence labeling. While these models have high predictive expressiveness, they are prone to overfitting due to their high model complexity. To avoid overfitting, we applied a novel structured regularizer derived from the proposed shared-ensemble framework. We show that this regularizer can be derived as lower bound from a mixture of models sharing parts of each other, e.g. neural sub-networks. We demonstrated the effectiveness of this structured regularizer in phoneme classification experiments. Furthermore, we experimentally confirmed the importance of non-linear representations in form of neural higher-order factors in LC-CRFs in contrast to linear ones. In TIMIT phoneme classification, we achieved state-of-the-art phoneme error rate of 16.7% using the NHO-LC-CRFs equipped with our proposed structured regularizer. Future work includes testing of different types of neural networks.

References

1. Bachman, P., Alsharif, O., Precup, D.: Learning with pseudo-ensembles. In: Advances in Neural Information Processing Systems, pp. 3365–3373 (2014)
2. Bergstra, J., Breuleux, O., Bastien, F., Lamblin, P., Pascanu, R., Desjardins, G., Turian, J., Warde-Farley, D., Bengio, Y.: Theano: a CPU and GPU math expression compiler. In: Proceedings of the Python for Scientific Computing Conference (SciPy) (2010). Oral Presentation
3. Bishop, C.M.: Pattern Recognition and Machine Learning. Springer (2006)
4. Chang, H.A., Glass, J.R.: Hierarchical large-margin Gaussian mixture models for phonetic classification. In: Workshop on Automatic Speech Recognition and Understanding (ASRU), pp. 272–277 (2007)
5. Do, T.M.T., Artières, T.: Neural conditional random fields. In: International Conference on Artificial Intelligence and Statistics (AISTATS), pp. 177–184 (2010)
6. Gens, R., Domingos, P.: Learning the structure of sum-product networks. In: International Conference on Machine Learning (ICML), vol. 28, pp. 873–880 (2013)
7. Gunawardana, A., Mahajan, M., Acero, A., Platt, J.C.: Hidden conditional random fields for phone classification. In: Interspeech, pp. 1117–1120 (2005)
8. Halberstadt, A.K., Glass, J.R.: Heterogeneous acoustic measurements for phonetic classification. In: EUROSPEECH, pp. 401–404 (1997)
9. He, X., Zemel, R.S., Carreira-Perpiñán, M.A.: Multiscale conditional random fields for image labeling. In: Conference on Computer Vision and Pattern Recognition (CVPR), pp. 695–703 (2004)

10. Hinton, G., Deng, L., Yu, D., Dahl, G., Mohamed, A., Jaitly, N., Senior, A., Vanhoucke, V., Nguyen, P., Sainath, T., Kingsbury, B.: Deep neural networks for acoustic modeling in speech recognition. Signal Processing Magazine, pp. 82–97 (2012)
11. Hinton, G.E.: Products of experts. In: International Conference on Artificial Neural Networks (ICANN), pp. 1–6 (1999)
12. Hinton, G.E., Srivastava, N., Krizhevsky, A., Sutskever, I., Salakhutdinov, R.: Improving neural networks by preventing co-adaptation of feature detectors. CoRR abs/1207.0580 (2012)
13. Lafferty, J., McCallum, A., Pereira, F.: Conditional random fields: probabilistic models for segmenting and labeling sequence data. In: International Conference on Machine Learning (ICML), pp. 282–289 (2001)
14. Lafferty, J., Zhu, X., Liu, Y.: Kernel conditional random fields: representation and clique selection. In: International Conference on Machine Learning (ICML), p. 64 (2004)
15. Larochelle, H., Bengio, Y.: Classification using discriminative restricted Boltzmann machines. In: International Conference on Machine Learning (ICML), pp. 536–543 (2008)
16. Lavergne, T., Allauzen, A., Crego, J.M., Yvon, F.: From n-gram-based to crf-based translation models. In: Workshop on Statistical Machine Translation, pp. 542–553 (2011)
17. Li, Y., Tarlow, D., Zemel, R.S.: Exploring compositional high order pattern potentials for structured output learning. In: Conference on Computer Vision and Pattern Recognition (CVPR), pp. 49–56 (2013)
18. Maaten, L., Chen, M., Tyree, S., Weinberger, K.Q. In: International Conference on Machine Learning (ICML), pp. 410–418 (2013)
19. van der Maaten, L., Welling, M., Saul, L.K.: Hidden-unit conditional random fields. In: International Conference on Artificial Intelligence and Statistics (AISTATS), pp. 479–488 (2011)
20. Peng, J., Bo, L., Xu, J.: Conditional neural fields. In: Neural Information Processing Systems (NIPS), pp. 1419–1427 (2009)
21. Poon, H., Domingos, P.: Sum-product networks: a new deep architecture. In: Uncertainty in Artificial Intelligence (UAI), pp. 337–346 (2011)
22. Prabhavalkar, R., Fosler-Lussier, E.: Backpropagation training for multilayer conditional random field based phone recognition. In: International Conference on Acoustics Speech and Signal Processing (ICASSP), pp. 5534–5537 (2010)
23. Qian, X., Jiang, X., Zhang, Q., Huang, X., Wu, L.: Sparse higher order conditional random fields for improved sequence labeling. In: International Conference on Machine Learning (ICML), pp. 849–856 (2009)
24. Ratajczak, M., Tschiatschek, S., Pernkopf, F.: Sum-product networks for structured prediction: context-specific deep conditional random fields. In: International Conference on Machine Learning (ICML): Workshop on Learning Tractable Probabilistic Models Workshop (2014)
25. Ratajczak, M., Tschiatschek, S., Pernkopf, F.: Neural higher-order factors in conditional random fields for phoneme classification. In: Interspeech (2015)
26. Roth, S., Black, M.J.: Fields of experts: a framework for learning image priors. In: Conference on Computer Vision and Pattern Recognition (CVPR), pp. 860–867 (2005)
27. Rumelhart, D.E., Hinton, G.E., Williams, R.J.: Parallel distributed processing: Explorations in the microstructure of cognition, vol. 1. chap. Learning Internal Representations by Error Propagation, pp. 318–362 (1986)

28. Schuppler, B.: Automatic Analysis of Acoustic Reduction in Spontaneous Speech. Ph.D. thesis, PhD Thesis, Radboud University Nijmegen, Netherlands (2011)
29. Sha, F., Saul, L.: Large margin Gaussian mixture modeling for phonetic classification and recognition. In: International Conference on Acoustics, Speech and Signal Processing (ICASSP), pp. 265–268 (2006)
30. Stewart, L., He, X., Zemel, R.S.: Learning flexible features for conditional random fields. IEEE Trans. Pattern Anal. Mach. Intell. **30**(8), 1415–1426 (2008)
31. Sung, Y.H., Boulis, C., Manning, C., Jurafsky, D.: Regularization, adaptation, and non-independent features improve hidden conditional random fields for phone classification. In: Workshop on Automatic Speech Recognition and Understanding (ASRU), pp. 347–352 (2007)
32. Wager, S., Wang, S., Liang, P.: Dropout training as adaptive regularization. In: Neural Information Processing Systems (NIPS), pp. 351–359 (2013)
33. Wan, L., Zeiler, M., Zhang, S., Cun, Y.L., Fergus, R.: Regularization of neural networks using dropconnect. In: International Conference on Machine Learning (ICML), pp. 1058–1066 (2013)
34. Xu, P., Sarikaya, R.: Convolutional neural network based triangular crf for joint intent detection and slot filling. In: Workshop on Automatic Speech Recognition and Understanding (ASRU), pp. 78–83 (2013)
35. Ye, N., Lee, W.S., Chieu, H.L., Wu, D.: Conditional random fields with high-order features for sequence labeling. In: Neural Information Processing Systems (NIPS), pp. 2196–2204 (2009)
36. Zue, V., Seneff, S., Glass, J.R.: Speech database development at MIT: Timit and beyond. Speech Communication **9**(4), 351–356 (1990)

Versatile Decision Trees for Learning Over Multiple Contexts

Reem Al-Otaibi[1,2]([⊠]), Ricardo B.C. Prudêncio[3], Meelis Kull[1],
and Peter Flach[1]

[1] Intelligent System Laboratory, Computer Science, University of Bristol, Bristol, UK
{ra12404,meelis.kull,peter.flach}@bristol.ac.uk
[2] King Abdulaziz University, Jeddah, Saudi Arabia
[3] Informatics Center, Universidade Federal de Pernambuco, Recife, Brazil
rbcp@cin.ufpe.br

Abstract. Discriminative models for classification assume that training
and deployment data are drawn from the same distribution. The performance of these models can vary significantly when they are learned and
deployed in different contexts with different data distributions. In the literature, this phenomenon is called dataset shift. In this paper, we address
several important issues in the dataset shift problem. First, how can we
automatically detect that there is a significant difference between training and deployment data to take action or adjust the model appropriately? Secondly, different shifts can occur in real applications (e.g., linear
and non-linear), which require the use of diverse solutions. Thirdly, how
should we combine the original model of the training data with other
models to achieve better performance? This work offers two main contributions towards these issues. We propose a Versatile Model that is rich
enough to handle different kinds of shift without making strong assumptions such as linearity, and furthermore does not require labelled data to
identify the data shift at deployment. Empirical results on both synthetic
shift and real datasets shift show strong performance gains by achieved
the proposed model.

Keywords: Versatile model · Decision Trees · Dataset shift ·
Percentile · Kolmogorov-Smirnov test

1 Introduction

Supervised machine learning is typically concerned with learning a model using
training data and applying this model to new test data. An implicit assumption
made for successfully deploying a model is that both training and test data
follow the same distribution. However, the distribution of the attributes can
change, especially when the training data is gathered in one context, but the
model is deployed in a different context (e.g., the training data is collected in
one country but the predictions are required for another country). The presence
of such *dataset shifts* can harm the performance of a learned model. Different

© Springer International Publishing Switzerland 2015
A. Appice et al. (Eds.): ECML PKDD 2015, Part I, LNAI 9284, pp. 184–199, 2015.
DOI: 10.1007/978-3-319-23528-8_12

kinds of dataset shift have been investigated in the literature [10]. In this work we focus on shifts in continuous attributes caused by hidden transformations from context to another. For instance, a diagnostic test may have different resolutions when produced by different laboratories, or the average temperature may change from city to city. In such cases, the distribution of one or more of the covariates in X changes. This problem is referred as a covariate observation shift [7].

We address this problem in two steps. In the first step, we build Decision Trees (DTs) using percentiles for each attribute to deal with covariate observation shifts. In this proposal, if a certain percentage of training data reaches a child node after applying a decision test, the decision thresholds in deployment are redefined in order to preserve the same percentage (60%) of deployment instances in that node. In the original learned DT, the learned threshold in a decision node corresponds to the 60th percentile of the *training* data. The updated threshold in deployment will be the 60th percentile of the *deployment* data.

The percentile approach assumes that the shift is caused by a monotonic function preserving the ordering of attribute values but ignoring the scale. For some shifts it may be more appropriate to assume a transformation from one linear scale to another. We therefore develop a more general and versatile DT that can choose between different strategies (percentiles, linear shifts or no shift) to update the DT thresholds for each deployment context, according to the shifts observed in the data.

The rest of the paper is organised as follows. Section 2 presents the dataset shift problem and the existing approaches addressing it. In Section 3 we introduce the use of percentiles and the versatile model proposed in our work. Section 4 presents the experiments performed to evaluate the versatile model on both synthetic and non-synthetic dataset shifts, and Section 5 concludes the paper.

2   Dataset Shift

We start by making a distinction between the *training* and *deployment* contexts. In the training context, a set of labelled instances is available for learning a model. The deployment context is where the learned model is actually used for predictions. These contexts are often different in some non-trivial way. For instance, a model may be built using training data collected in a certain period of time and in a particular country, and deployed to data in a future time and/or in a different country. A model built in a training context may fail in a deployment context due to different reasons: in the current paper we focus on performance degradation caused by *dataset shifts* across contexts.

A simple solution to deal with shifts would be to train a new classifier for each new deployment context. However, if there are not enough available labelled instances in the new context, training a new model specific for that context is then unfeasible as the model would not sufficiently generalise. Alternative solutions have to be applied to reuse or adapt existing models, which will depend on the kind of shift observed.

Shifts can occur in the input attributes, in the output or both. Dataset shift happens when training and deployment joint distributions are different [10], i.e.:

$$P_{tr}(Y, X) \neq P_{dep}(Y, X) \tag{1}$$

A shift can occur in the output, i.e., $P_{tr}(Y) \neq P_{dep}(Y)$, while the conditional probability distribution remains the same $P_{tr}(X|Y) = P_{dep}(X|Y)$. This is referred in the literature as the prior probability shift and can be addressed in different ways (e.g., [5]).

In our work we are mainly concerned with situations where the marginal distribution of a covariate changes across contexts, i.e.: $P_{tr}(X) \neq P_{dep}(X)$. Given a change in the marginal distribution, we can further define two different kinds of shifts depending on whether the conditional distribution of the target also changes between training and deployment. In the first case, the marginal distribution of X changes, while the conditional probability of the target Y given X remains the same:

$$P_{tr}(X) \neq P_{dep}(X)$$
$$P_{tr}(Y|X) = P_{dep}(Y|X) \tag{2}$$

For instance, the smoking habits of a population may change over time due to public initiatives but the probability of lung cancer given smoking is expected to remain the same [12]. In the same problem, a labelled training set may be collected biased to patients with bad smoking habits. Again, the marginal distribution in deployment may be different from training while the conditional probability is the same. The above shift is referred in the literature by different terms such as simple covariate shift [12] or sample selection bias [15]. A common solution to deal with simple covariate shifts is to modify the training data distribution by considering the deployment data distribution. A new model can then be learned using the shift-corrected training data distribution. This strategy is adopted by different authors using importance sampling which corrects the training distribution using instance weights proportional to $P_{dep}(X)/P_{tr}(X)$. Examples of such solutions include Integrated Optimisation Problem IOP [3], Kernel Mean Matching [6], Importance Weighted Cross Validation IWCV [14] and others.

In this paper we focus on the second kind of shift in which both the marginal and the conditional distributions can change across contexts, i.e.:

$$P_{tr}(X) \neq P_{dep}(X)$$
$$P_{tr}(Y|X) \neq P_{dep}(Y|X) \tag{3}$$

This is a more difficult situation that can be hard to solve and requires additional assumptions. A suitable assumption in many situations is that there is a hidden transformation of the covariates $\Phi(X)$ for which the conditional probability is unchanged across contexts, i.e.:

$$P_{tr}(X) \neq P_{dep}(X)$$
$$P_{tr}(Y|X) = P_{dep}(Y|\Phi(X)) \tag{4}$$

This is defined in [7] as a *covariate observation shift*. For instance, in prostate cancer detection, shifts can be observed in data from different laboratories due to differences in their equipments and resolution of diagnostic tests [9]. A mapping between attributes can be performed to correct the existing differences in data [9]. As another example [7], suppose that in an image recognition problem, pictures are taken by a camera with two different colour adjustments settings, thus representing two different contexts. This can result in a shift in the covariates. The conditional probability, however, may be the same given an invariant hidden raw camera representation. Finally, a sensor used to detect an event may degrade over time. Such degradation can be seen as a transformation function in the sensor outputs that causes a covariate observation shift.

Previous authors dealt with covariate observation shifts by finding a transformation function Φ to correct the deployment data [1]. Once transformed or 'unshifted' using Φ, the deployment data is given as input to the model learned in the training context. Finding a linear transformation is a natural choice in this approach. In [1], for instance, the authors adopted Stochastic Hill Climbing to find the best linear transformation to apply in the given deployment data. In that work, labelled deployment instances are required in order to evaluate the suitability of a candidate linear map. The parameters of the linear transformation are then optimised to maximise the accuracy obtained by the learned model on the labelled deployment instances (once transformed). A similar idea was proposed in [9], using genetic programming techniques to find more complex transformation functions (both linear and non-linear). As [1], it requires that labelled instances are available in the deployment data to evaluate the adequacy of the transformation functions.

In summary, we emphasise that it can be difficult to recognise or distinguish between the different kinds of shifts that may occur in a dataset. It can be simple in some cases to identify a shift in the covariates, relying on a sufficient number of unlabelled instances in the deployment context. On the other hand, verifying a shift in the conditional probabilities $P(Y|X)$ is not possible if there are only unlabelled instances in deployment or may be unreliable if the number of labelled instances in deployment is low. Additionally, suppose that we have evidence that a change is caused by a transformation in a covariate. Trying to detect a linear transformation to apply in the deployment data may be counter-productive if the true transformation is non-linear instead. Finally, applying a shift-aware method in a deployment context that did not actually change compared to the training context may be detrimental as well. These considerations motivated our proposal of a more sophisticated approach that can adapt to different kinds of dataset shifts under different assumptions.

3   Versatile Decision Trees

In this work, we propose different strategies to build Decision Trees (DTs) in the presence of covariate observation shifts. We make two main contributions. First, we propose a novel approach to build DTs based on percentiles (see Sections 3.1

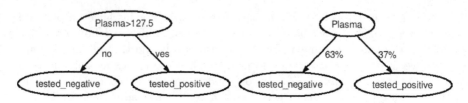

Fig. 1. Two types of models; on the left is the model using a fixed threshold while on the right is the model using percentiles. For each deployment context, the decision tree is deployed in such a way that the deployment instances are split to the leaves in the same percentile amounts of 63% and 37%.

and 3.2). The basic idea is to learn a conventional DT and then to replace the internal decision thresholds by percentiles, which can deal with monotonic shifts. Secondly, we propose a more general Versatile Model (VM) that deploys different strategies (including the percentiles) to update the DT thresholds for each deployment context, according to the shifts observed in the data (see Section 3.3). The shifts are identified by applying a non-parametric statistical test.

3.1 Constructing Splits Using Percentiles

We consider an example using the diabetes dataset from the UCI repository, which has 8 input attributes and 768 instances. Suppose we train a decision stump and the discriminative attribute is the Plasma glucose concentration attribute, which is a numerical attribute. Suppose the decision threshold is 127.5, meaning that any patient with plasma concentration above 127.5 will be classified as diabetic, otherwise classified as non-diabetic as seen in Figure 1 (left). If there is no shift in the attribute from training to deployment, the decision node can be directly applied, i.e., the threshold 127.5 is maintained to split data in deployment. However, if the attribute is shifted in deployment, the original threshold may not work well.

In the current work, we propose to adopt the percentiles[1] of continuous attributes to update the decision thresholds for each deployment context. Back to the example, instead of interpreting the data split in an absolute sense, we will interpret it in terms of ranks: 37% of the training examples with the *highest* values of Plasma reach the right child, while 63% of the training examples with the *lowest* values of Plasma in turn reach the left child (see the right side of Figure 1). We can say that the data was split at the 63th percentile in training. Given a batch set of instances in deployment to classify, the DT can apply the same split rule: the 37% of the examples in deployment with the highest values of Plasma are associated to the right child, while 63% of the examples with the lowest values of Plasma in deployment are associated to the left child. The decision threshold in deployment is updated in such a way that the percentage

---

[1] Percentile is the value below which a given percentage of observations in a group is observed.

of instances split to each child is maintained. In this proposal, it is assumed that the shift is caused by a *monotonic* transformation Φ. Such functions when applied to an input attribute preserve the order of its original values. Different from the previous work [1] the transformation function in the versatile DT is not explicitly estimated, but it is implicitly treated by deploying the percentiles.

Formally, let $\mathcal{L} = \{c_1, \ldots, c_L\}$ be the set of class labels in a problem. Let th_{tr} be the threshold value applied on a numerical attribute X in a decision node. In the previous example $th_{tr} = 127.5$ for the Plasma attribute. Let $n_{left}^{c_l}$ be the number of training instances belonging to class c_l that are associated to the left child node after applying the decision test, i.e., for which $X \leq th_{tr}$. The total number of instances n_{left} associated to this node is:

$$n_{left} = \sum_{c_l \in \mathcal{L}} n_{left}^{c_l} \tag{5}$$

Let $R_{tr}(th_{tr}) = 100 * n_{left}/n$ be the percentage of training instances in the left node, where n is the total number of training instances. Then, th_{tr} is the percentile associated to $R_{tr}(th_{tr})$ for the attribute X. In the above example: $R_{th}(127.5) = 63\%$ and th_{tr} is the 63th percentile of Plasma in the training data. Then the threshold adopted in deployment is defined as:

$$th_{dep} = R_{dep}^{-1}(R_{tr}(th_{tr})) \tag{6}$$

In the above equation, the threshold th_{dep} is the attribute value in deployment that, once adopted to split the deployment data, maintains the percentage of instances in each child node: $R_{dep}(th_{dep}) = R_{tr}(th_{tr})$.

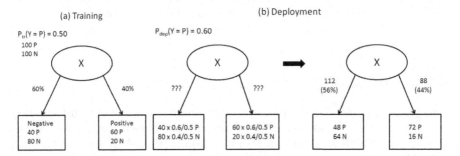

Fig. 2. Example of DT with percentiles when a shift is identified in the class distribution. Part (a) illustrates the percentiles of each leaf for the training context, with prior probability equal to 0.5. Part (b) illustrates the correction of the percentiles for a new deployment context where the prior probability is 0.6. The correction of performed using the ratios of 0.6/0.5 and 0.4/0.5 respectively for the positive and negative instances (left side - (b)). Corrected number of instances expected at each leaf resulted in new estimated percentiles (right side - (b))

3.2   Adapting for Output Shifts

The percentile rule can be adapted to additionally deal with shifts in the class distribution across contexts. Figure 2 illustrates a situation where the prior probability of the positive class was 0.5 in training and then shifted to 0.6 in deployment. In Figure 2(a) we observe a certain number of positives and negatives internally in each child node, which is used to derive the percentiles. If a shift is expected in the target, the percentage of instances expected in deployment for each child node may change as well. For instance, a higher percentage of instances may be observed in the right node in deployment because the probability of positives has increased and the proportion of positives related to negatives in this node is high. In our work, we estimate the percentage of instances in each child node according to the class ratios between training and deployment.

Let $P_{tr}^{c_l}$ and $P_{dep}^{c_l}$ be the probability of class c_l, respectively in training and deployment. $P_{dep}^{c_l}$ can be estimated using available labelled data in deployment or just provided as input. There is a prior shift related to this class label when $P_{tr}^{c_l} \neq P_{dep}^{c_l}$. For each instance belonging to c_l observed in training we expect to observed $P_{dep}^{c_l}/P_{tr}^{c_l}$ instances of c_l in deployment. The number of instances associated to the left child node in deployment is then estimated by the following equation:

$$\hat{n}_{left} = \sum_{c_l \in \mathcal{L}} n_{left}^{c_l} \frac{P_{dep}^{c_l}}{P_{tr}^{c_l}} \tag{7}$$

The percentile is then computed using the corresponding percentage: $R_{tr}(th_{tr}) = 100 * \hat{n}_{left}/n$. In Figure 2(b), the class ratios of 0.6/0.5 and 0.4/0.5 are respectively adopted to correct the number of positive and negative instances in each node. In the left node, for instance, the expected number of positive and negative instances is respectively 48 and 64, resulting in 112 instances. The percentage to be adopted in deployment is now 56%, instead of 60% if no correction is performed. The 56th percentile in the deployment data is then adopted as the decision threshold.

3.3   Versatile Model for Decision Trees

By adopting percentiles in the DT, we are assuming a monotonic transformation Φ across contexts. In this sense, our work is more general compared to the previous work [1] that assumes a linear transformation. Monotonic shifts can not only cover the linear case but also a broad range of non-linear monotonic transformations (e.g., piecewise linear transformations). Even the case where there is no shift can be seen as a monotonic transformation when Φ is the identity function. Despite this generality, the use of percentiles has limitations too. First, percentile estimates (either in training or deployment) can be inaccurate when there is few or sparse data for estimation. Also, it may be worth trying alternative methods if the assumptions made by these methods about the context shifts are actually met. For instance, if we expect the shift to be linear we might be better off fitting an explicit linear transformation between training and deployment.

Algorithm 1. Versatile Model Threshold Selection Algorithm

Input: training attribute vector $X_{tr} = (x_1, \ldots, x_n)$ with n the number of training instances (i.e., a column of the data matrix),; deployment attribute vector $X_{dep} = (x'_1, \ldots, x'_m)$; decision threshold in training th_{tr} for attribute X_{tr}.
Output: deployment decision threshold th_{dep}.
/* Test for no shift. Null hypothesis $H_0 : F(X_{tr}) = F(X_{dep})$ */
p_{value}=**Kolmogorov-Smirnov**(X_{tr},X_{dep})
if p_{value} <0.05 **then**
  /* Reject H_0, X_{dep} is shifted. Try a linear transformation */
  $(X^u_{dep}, \alpha, \beta)$= **Linear Transformation**(X_{tr},X_{dep})
  /* Test corrected shift. Null hypothesis $H_0 : F(X_{tr}) = F(X^u_{dep})$ * /
  p_{value}=**Kolmogorov-Smirnov**(X_{tr},X^u_{dep})
  if p_{value} <0.05 **then**
    /* Reject H_0, X^u_{dep} is still shifted. Use the percentile. */
    $th_{dep} = R^{-1}_{dep}(R_{tr}(th_{tr}))$
  else
    /* Accept H_0, X^u_{dep} is not shifted. Use the linearly corrected threshold */
    $th_{dep} = \alpha \cdot th_{tr} + \beta$
  end if
else
  /* Accept H_0, X_{dep} is not shifted. Use training threshold. */
  $th_{dep} = th_{tr}$
end if
Return th_{dep}

In this section, we propose a versatile decision tree model that employs different strategies to choose the decision threshold according to the shift observed in deployment. Algorithm 1 presents the proposed versatile model, which receives as input the original threshold applied on an attribute, the training and the deployment data of that attribute and returns the appropriate threshold to adopt in deployment. This *versatile model* (VM) can be described in three steps:

1. Initially a statistical test is applied to verify whether the distribution of the attribute differs between training and deployment. In this step, we aim to avoid dealing with shifts when they do not really exist, which could lead to overfitting. In our implementation, the non-parametric Kolgomorov-Smirnoff (KS) test was adopted[2]. If there is no shift in the attribute, the versatile DT is applied using the original thresholds learned in the training context, i.e. $th_{dep} = th_{tr}$.

2. If a shift is detected by the previous test, a linear transformation is fitted and applied to the attribute in deployment, aiming to correct a potential linear shift. In our implementation, α and β parameters were estimated based on

---

[2] We employed the KS test on the values of the attribute: training X_{tr} and deployment X_{dep}. The KS test tests the null hypothesis that the empirical cumulative distribution functions of X_{tr} and X_{dep} are identical against the alternative hypothesis that the two distributions are different.

Algorithm 2. Linear Transformation

Input: training attribute vector $X_{tr} = (x_1, \ldots, x_n)$; deployment attribute vector $X_{dep} = (x'_1, \ldots, x'_m)$.

Output: 'Unshifted' deployment attribute vector X^u_{dep} and corresponding parameters α, β.

/* Estimate the mean and standard deviation of X in training and deployment */

$\mu_{tr} = \frac{1}{n} \sum_{i=1}^{n} x_i$, $\sigma_{tr} = \sqrt{\frac{1}{n} \sum_{i=1}^{n} (x_i - \mu_{tr})^2}$

$\mu_{dep} = \frac{1}{m} \sum_{i=1}^{m} x'_i$, $\sigma_{dep} = \sqrt{\frac{1}{m} \sum_{i=1}^{m} (x'_i - \mu_{dep})^2}$

/* Estimate α and β considering that: $\sigma_{dep} = \alpha \sigma_{tr}$ and $\mu_{dep} = \alpha \mu_{tr} + \beta$ */

$\alpha = \frac{\sigma_{dep}}{\sigma_{tr}}$

$\beta = \mu_{dep} - \alpha \cdot \mu_{tr}$

/* Produce unshifted deployment data X^u_{dep} according to α and β */

$X^u_{dep} = \emptyset$

for $i = 1$ **to** m **do**

$\quad x^u_i = \frac{(x'_i - \beta)}{\alpha}$

$\quad X^u_{dep} = X^u_{dep} \cup x^u_i$

end for

Return X^u_{dep}, α, β

---

the change in mean and standard deviation of the attribute in training and deployment (see Algorithm 2). We then apply the KS test again to compare the distribution of the transformed attribute in deployment and the attribute in the training data. If no shift is observed now, we assume that the linear transformation applied was adequate. The versatile DT is then deployed with a threshold $th_{dep} = \alpha \cdot th_{tr} + \beta$.

3. Finally, if the second test indicates that there is still a shift in the attribute (i.e., the shift was not corrected using the linear transformation), then the percentiles are deployed, assuming a non-linear monotonic shift. In this case the adopted threshold is: $th_{dep} = R^{-1}_{dep}(R_{tr}(th_{tr}))$.

4   Experimental Results

The VM combines 3 strategies for defining the DT thresholds in deployment: original thresholds, linear transformations, and monotonic transformations using percentiles. In the experiments, each strategy was individually compared to the VM, respectively named as Original Model (OM), (α, β) and Perc. Additionally, (α, β) and the Percentile methods were combined with the KS test, referred in the experiments as KS+(α, β) and KS+Perc, respectively. In the former, linear transformation is applied to all shifted attributes, whereas, in the latter, Percentiles are utilised. In both approaches, the original model was applied if there is no shift detected by the KS test.

The first set of experiment applies synthetic shifts to UCI datasets to analyse the performance of the shift detection approach adopted by the VM. We inject two types of shifts into the deployment data to test the VM: a non-linear monotonic transformation and linear shifts with different degrees (see Sections 4.1

Table 1. Values used in the experiments for φ and γ in order to generate the synthetic linear shifts.

φ	γ	Effect
0	0	unshifted data (original)
0	1	μ_{dep} shifted to right
0	-1	μ_{dep} shifted to left
1	0	stretch data
-1	0	compress data
1	1	μ_{dep} shifted to right and stretch the data
1	-1	μ_{dep} shifted to left and stretch the data
-1	1	μ_{dep} shifted to right and compress the data
-1	-1	μ_{dep} shifted to left and compress the data

and 4.2). In Section 4.3 we report on experiments with actual context changes occurring in real-world datasets.

4.1 Generating Synthetic Shifts

In these experiments, linear transformations were applied to numerical attributes in order to simulate shifts between two contexts. Two parameters, α and β, were adopted in each simulation to perform the linear transformation $X_{dep} = \alpha \cdot X_{tr} + \beta$. Let μ_{tr} and σ_{tr} be the mean and standard deviation of attribute X in training. When X is shifted using the parameters α and β, the mean and standard deviation of the transformed variable become

$$\mu_{dep} = \alpha \cdot \mu_{tr} + \beta$$
$$\sigma_{dep} = \alpha \cdot \sigma_{tr}$$

It is useful to re-parametrise α and β as follows.

$$\alpha = 2^{\varphi}$$
$$\beta = (1 - 2^{\varphi}) \cdot \mu_{tr} + \gamma \cdot \sigma_{tr}$$

If φ is negative the attribute values are compressed across contexts, and if φ is positive the values are stretched. The mean is shifted by γ times the standard deviation in training: $\mu_{dep} = \mu_{tr} + \gamma \cdot \sigma_{tr}$. Table 1 shows the values used in the experiments for φ and γ.

To create non-linear monotonic shifts we use the following transformation:

$$X_{dep} = \sigma_{tr} \cdot \left(\frac{X_{tr} - \mu_{tr}}{\sigma_{tr}} \right)^3 + \mu_{tr} \tag{8}$$

We use a cubic rather than a square transformation to ensure monotonicity. In order to preserve the mean and standard deviation of the data we first convert the attribute values to z-scores, apply the cubic transformation and then restore the mean and standard deviation.

4.2 Results of the Synthetic Shifts

We selected 10 datasets from UCI [8] and KEEL [2] with all numerical (real-valued as well as integer-valued) attributes. Each dataset was randomly partitioned into 5 folds. Using 4 folds for training and the 5th fold for deployment, after shifting according to each set of parameters in Table 1. The same shift is applied to all attributes in each dataset. Results are averaged over 5 cross-validation runs for each dataset. Table 2 reports the average accuracy of 5 different runs for all used methods in 4 cases: unshifted, linear shift, non-linear and mixture shift data. Performance of these methods applied to linear shifts is the average of 8 degrees of linear shift as reported in Table 1. We conducted the Friedman test based on the average ranks for all datasets in order to verify whether the differences between algorithms are statistically significant [4]. At significance level 0.05 the Friedman test gives significance for all experiments except the non-linear shifts, so we show critical difference diagrams based on the Nemenyi post-hoc test for the former in Figure 3. We proceed to discuss the results of each experiment.

Unshifted Data. Unsurprisingly, the original model is the best when there is no shift from training to deployment, but the CD diagram demonstrates that the Versatile Model is not significantly worse. Percentiles don't work well in this case, confirming the need for a multi-strategy approach.

Linear Shifts. Estimating a linear shift is the right thing to do here so it is not surprising that $\langle \alpha, \beta \rangle$ performs strongest, with KS+$\langle \alpha, \beta \rangle$ trailing slightly behind as the KS test may sometimes fail to detect the shift. The original model is significantly outperformed by all context-sensitive models except the percentiles. The Versatile Model is a good representative of the context-sensitive models.

Non-Linear Shift. Here the Versatile Model outperforms all other methods in terms of the average ranks, but not significantly. Notice that, while the original model performs worst, there are 2 datasets where the original model performs best: in these datasets many attribute values are in the range $[-1, 1]$ where the cubic transformation has less effect.

Mixture Shift. The aim of this experiment was to test how well the Versatile Model deals with a mixture of different shifts: one-third of the attributes was shifted linearly, one-third non-linearly, and one-third remained unchanged. The results demonstrate that the Versatile Model derives a clear advantage from the ability of being able to distinguish between these different kinds of shift and adapt its strategy.

4.3 Results on Non-synthetic Shifts

The aim of this experiment is to evaluate the Versatile Model on real dataset shift and compared it the with state-of-art covariate shift solvers Integrated Optimisation Problem (IOP) [3] and Kernel Mean Matching (KMM) [6]. IOP and KMM algorithms were retrieved from [11] and run using default parameters.

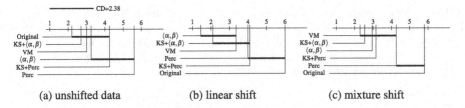

(a) unshifted data   (b) linear shift   (c) mixture shift

Fig. 3. Critical Difference diagrams using pairwise comparisons for those experiments where the Friedman test gives significance at 0.05.

Diabetes. Our first benchmark is a dataset of 4 different ethnic groups of diabetes patients [13]. The original dataset consists of 47 attributes and 101 766 instances. Each instance corresponds to a unique patient diagnosed with diabetes. The features describe the diabetic encounters such as diagnoses, medications, and number of visits in the year preceding the encounter. We rank features using information gain ratio then we select the best 8 numerical features. The classification task is whether the patient was re-admitted to the hospital. The values of the readmission attribute are two: "yes" or "no". In the original dataset, the classes are: readmitted within 30 days "< 30", readmitted after 30 days "> 30" or no.

In our experiment, we split the dataset in 4 subsets according to the "ethnic group" the patient belongs to. There are 4 different groups: Caucasian, African American, Asian, and Hispanic. We evaluate the performance of models trained on ethnic group X and deployed on ethnic group Y, denoted by X-Y. Table 3 shows the performance of the Versatile Model against the original model, IOP and KMM. We also report the number of shifted attributes according to the KS test. The Versatile Model wins most often, followed by the original model.

Heart. Our next benchmark is the heart disease dataset. We split it into two subsets according to gender: male and female. In this dataset there are 5 continuous attributes, 3 of them are indicated as shifted between gender according to KS test, which are age, heart rate and serum cholesterol. Table 4 shows the performance of versatile method against the original model, IOP and KMM. In both contexts the VM has the best accuracy among all three methods including the original model.

Bike Sharing. This dataset [8] contains the hourly and daily count of rental bikes between years 2011 and 2012 in addition to weather information. It contains 4 continuous attributes: actual and apparent temperature in Celsius, humidity and wind speed. The classification task is whether there is a demand in this period of time or not. In order to evaluate the shift effects, we split the dataset as proposed in [1] to obtain the 4 seasons datasets. According to KS, all these 4 attributes are detected as shifted except in 3 cases. First, between Summer-Spring, wind speed is not shifted. Second, in both Summer-Autumn and Autumn-Winter, humidity is not shifted. The performance of Versatile Model and others are shown in Table 5. Again we note the solid performance of the Versatile Model.

Table 2. Cross-validated classification accuracy for both unshifted, linear shift, non-linear shift and mixture shift. The numbers between brackets are ranks. VM is the Versatile Model, OM is the original model, $\langle \alpha, \beta \rangle$ corresponds to a linear shift, Perc corresponds to a percentile shift, and KS+... indicates that the Kolmogorov-Smirnov test is used for detecting the shift.

	VM	OM	$\langle \alpha, \beta \rangle$	KS+$\langle \alpha, \beta \rangle$	Perc	KS+Perc
			unshifted data			
Phoneme	0.851(3)	**0.856(1)**	0.846(5)	0.854(2)	0.819(6)	0.850(4)
Bupa	0.631(3)	**0.632(1.5)**	0.619(5)	**0.632(1.5)**	0.578(6)	0.625(4)
Appendicitis	0.846(5.5)	0.849(4)	**0.853(1)**	0.846(5.5)	0.851(2.5)	0.851(2.5)
Pima	**0.728(1.5)**	0.725(4)	**0.728(1.5)**	0.727(3)	0.711(6)	0.721(5)
Breast-w	**0.947(3)**	**0.947(3)**	**0.947(3)**	**0.947(3)**	0.820(6)	**0.947(3)**
Magic	0.821(4)	**0.834(1)**	0.830(3)	0.833(2)	0.773(6)	0.814(5)
Threenorm	0.674(2.5)	**0.682(1)**	0.673(4)	0.674(2.5)	0.635(6)	0.671(5)
Ringnorm	0.735(2.5)	0.731(4.5)	**0.744(1)**	0.735(2.5)	0.678(6)	0.731(4.5)
Ionosphere	0.893(2.5)	**0.894(1)**	0.851(4)	0.893(2.5)	0.825(5.5)	0.825(5.5)
Sonar	0.752(2.5)	**0.754(1)**	0.739(5)	0.752(2.5)	0.716(6)	0.746(4)
Average	0.787(3)	**0.790(2.2)**	0.783(3.25)	0.789(2.7)	0.740(5.6)	0.778(4.25)
			linear shift			
Phoneme	0.825(3)	0.660(6)	**0.846(1.5)**	**0.846(1.5)**	0.819(4.5)	0.819(4.5)
Bupa	0.601(3)	0.558(6)	**0.619(1.5)**	**0.619(1.5)**	0.578(4.5)	0.578(4.5)
Appendicitis	0.844(4.5)	0.776(6)	**0.853(1)**	0.844(4.5)	0.851(2)	0.846(3)
Pima	0.726(3)	0.624(6)	**0.728(1.5)**	**0.728(1.5)**	0.711(4.5)	0.711(4.5)
Breast-w	0.820(4)	0.782(6)	**0.947(1.5)**	**0.947(1.5)**	0.820(4)	0.820(4)
Magic	0.761(5)	0.579(6)	**0.830(1.5)**	**0.830(1.5)**	0.773(3.5)	0.773(3.5)
Threenorm	0.672(2.5)	0.606(6)	**0.673(1)**	0.672(2.5)	0.635(4.5)	0.635(4.5)
Ringnorm	**0.744(1.5)**	0.608(6)	**0.744(1.5)**	0.743(3)	0.678(4.5)	0.678(4.5)
Ionosphere	0.810(5)	0.694(6)	**0.851(1.5)**	**0.851(1.5)**	0.825(3.5)	0.825(3.5)
Sonar	**0.739(2)**	0.624(6)	**0.739(2)**	**0.739(2)**	0.716(4.5)	0.716(4.5)
Average	0.754(3.35)	0.651(6)	**0.783(1.45)**	0.781(2.1)	0.740(4)	0.740(4.1)
			non-linear shift			
Phoneme	**0.819(2)**	0.746(4)	0.720(5.5)	0.720(5.5)	**0.819(2)**	**0.819(2)**
Bupa	**0.594(1)**	0.506(6)	0.571(4.5)	0.571(4.5)	0.578(2.5)	0.578(2.5)
Appendicitis	0.847(4.5)	0.240(6)	0.849(3)	0.847(4.5)	**0.851(1.5)**	**0.851(1.5)**
Pima	0.715(3)	0.478(6)	**0.728(1.5)**	**0.728(1.5)**	0.711(4.5)	0.711(4.5)
Breast-w	0.820(4)	0.464(6)	**0.916(1.5)**	**0.916(1.5)**	0.820(4)	0.820(4)
Magic	0.761(3)	0.398(6)	0.744(4.5)	0.744(4.5)	**0.773(1.5)**	**0.773(1.5)**
Threenorm	0.651(2.5)	**0.671(1)**	0.607(6)	0.635(4.5)	0.635(4.5)	0.651(2.5)
Ringnorm	0.698(2.5)	**0.731(1)**	0.667(6)	0.680(4)	0.678(5)	0.698(2.5)
Ionosphere	**0.825(2)**	0.820(4)	0.781(5.5)	0.781(5.5)	**0.825(2)**	**0.825(2)**
Sonar	**0.744(2)**	0.478(6)	**0.744(2)**	**0.744(2)**	0.716(4.5)	0.716(4.5)
Average	**0.747(2.65)**	0.553(4.6)	0.732(4)	0.736(3.8)	0.740(3.2)	0.744(2.75)
			mixture shift (unshifted, linear shift, non-linear)			
Phoneme	**0.828(1)**	0.749(6)	0.787(5)	0.789(4)	0.819(3)	0.823(2)
Bupa	**0.605(1)**	0.551(6)	0.595(2)	0.594(3)	0.578(5)	0.592(4)
Appendicitis	0.843(4.5)	0.718(6)	0.847(2.5)	0.843(4.5)	**0.851(1)**	0.847(2.5)
Pima	0.710(5)	0.512(6)	**0.727(1)**	0.724(2)	0.711(4)	0.712(3)
Breast-w	0.935(3.5)	0.797(6)	**0.947(1.5)**	**0.947(1.5)**	0.819(5)	0.935(3.5)
Magic	**0.805(1.5)**	0.510(6)	0.802(3)	**0.805(1.5)**	0.773(5)	0.785(4)
Threenorm	**0.672(1)**	0.647(4)	0.635(5.5)	0.653(2)	0.635(5.5)	0.649(3)
Ringnorm	**0.739(1)**	0.674(6)	0.728(2.5)	0.720(4)	0.678(5)	0.728(2.5)
Ionosphere	0.843(3.5)	0.792(6)	0.843(3.5)	**0.865(1)**	0.825(5)	0.848(2)
Sonar	**0.740(1)**	0.631(6)	0.737(3)	0.738(2)	0.716(4)	0.712(5)
Average	**0.772(2.3)**	0.658(5.8)	0.764(2.95)	0.767(2.55)	0.740(4.25)	0.763(3.15)

AutoMPG. AutoMPG dataset [8] concerns the consumption in miles per gallon of vehicle from 3 different regions: USA, Europe and Japan. It contains 4 numerical attributes: displacement, horsepower, weight and acceleration. All these input attributes have been detected as shifted between regions using KS test. This dataset has been binarised according to the mean value of the target.

Table 3. Classification accuracy for Diabetes dataset. Symbols denote ethnic groups as follows: African-American (AA), Asian (A), Caucasian (C), Hispanic (H). X-Y denotes trained on X, deployed on Y.

	A-AA	A-C	A-H	AA-A	AA-C	AA-H	C-A	C-AA	C-H	H-A	H-AA	H-C
# shifted	6	5	4	6	5	4	5	5	4	4	4	4
VM	0.569	0.529	**0.576**	**0.653**	**0.530**	**0.590**	0.645	0.546	0.588	0.624	**0.565**	**0.564**
OM	**0.574**	**0.538**	0.554	0.642	0.526	0.587	0.641	**0.566**	**0.595**	0.642	0.562	0.563
IOP	0.526	0.499	0.547	0.500	0.494	0.463	0.520	0.488	0.469	0.519	0.509	0.452
KMM	0.467	0.499	0.419	0.352	**0.530**	0.474	**0.647**	0.557	0.580	0.400	0.442	0.507

Table 4. Classification accuracy for Heart dataset, with contexts by gender (F: Female, M: Male).

	M-F	F-M
# shifted	3	3
VM	**0.735**	**0.568**
OM	0.712	0.557
IOP	0.703	0.500
KMM	0.724	0.540

Table 5. Classification accuracy for Bike Sharing dataset, with contexts by season (Sp: Spring, S: Summer, A: Autumn, W: Winter).

	Sp-S	Sp-A	Sp-W	S-Sp	S-A	S-W	A-Sp	A-S	A-W	W-Sp	W-S	W-A
# shifted	3	4	4	3	3	4	4	3	3	4	4	3
VM	**0.641**	**0.558**	**0.601**	0.519	**0.579**	0.601	**0.602**	**0.543**	0.556	0.646	0.565	0.526
OM	0.538	0.468	0.544	0.607	0.547	0.612	0.574	0.521	0.528	0.718	**0.657**	**0.558**
IOP	0.489	0.468	0.533	**0.635**	0.510	**0.657**	0.585	0.534	0.522	0.658	0.630	0.510
KMM	0.559	0.468	0.522	**0.635**	0.521	0.651	0.585	0.521	**0.589**	**0.690**	0.521	0.521

Table 6. Classification accuracy for AutoMPG dataset, with contexts by origin (U: USA, E: Europe, J:Japan).

	U-E	U-J	E-U	E-J	J-U	J-E
# shifted	4	4	4	3	4	3
VM	**0.676**	**0.759**	**0.873**	**0.772**	**0.780**	0.647
OM	0.544	0.607	0.670	0.746	0.747	**0.691**
IOP	0.558	0.493	0.400	0.417	0.600	0.441
KMM	0.558	0.582	0.600	0.582	0.400	0.485

We split the dataset as proposed in [1] to obtain the 3 regions datasets. The performance of Versatile Model and others are shown in Table 6. The VM outperforms all three methods and has only one loss against the original model.

Finally, we report the result of a Friedman test and post-hoc analysis on all non-synthetic shifts. Figure 4 demonstrates that the Versatile Model outperforms all others, significantly so except for the original model.

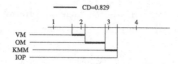

Fig. 4. Critical Difference diagram using pairwise comparisons for non synthetic shift. Average ranks as follows: VM=1.671, OM=2.140, KMM=2.875 and IOP= 3.312. The Friedman test gives significance at 0.05.

5 Conclusion

We proposed a model for adapting to covariate observation shift using unlabelled deployment data. The proposed model is called the Versatile Model and is a Decision Tree model with enhanced splits. The main idea of the VM is that it captures more information about the context during the training process in order to be able to adapt this model for deployment contexts. The VM trains a classifier over the available data and then adapts some of its decisions according to the (usually unlabelled) deployment data. We use a non-parametric test to choose among different strategies to update the decision thresholds in a DT. The VM does not make any strong assumptions such as linear transformation between contexts. Furthermore, it does not need any tuning parameters to adjust the model. Finally, empirical results on both synthetic shift and real dataset shift show strong performance gains by achieved the proposed methods.

This work opens up many avenues for future work. One direction is to adapt the VM to other predictive problems, such as regression. Another direction is to assume that the deployment data is partially labelled and utilise this knowledge in the VM.

Acknowledgments. Reem is a PhD student who is sponsored by King Abdulaziz University, Saudi Arabia. Ricardo Prudencio is financially supported by CNPq (Brazilian Agency) and did this work during a visit to the University of Bristol. This work was partly supported by the REFRAME project granted by the European Coordinated Research on Long-term Challenges in Information and Communication Sciences and Technologies ERA-Net (CHISTERA), and funded by the Engineering and Physical Sciences Research Council in the UK under grant EP/K018728/1.

References

1. Ahmed, C.F., Lachiche, N., Charnay, C., Braud, A.: Reframing continuous input attributes. In: 2014 IEEE 26th International Conference on Tools with Artificial Intelligence (ICTAI), pp. 31–38, November 2014
2. Alcalá-Fdez, J., Fernández, A., Luengo, J., Derrac, J., García, S.: Keel data-mining software tool: Data set repository, integration of algorithms and experimental analysis framework. Multiple-Valued Logic and Soft Computing **17**(2–3), 255–287 (2011)

3. Bickel, S., Brückner, M., Scheffer, T.: Discriminative learning under covariate shift. Journal of Machine Learning Research **10**, 2137–2155 (2009)
4. Demšar, J.: Statistical comparisons of classifiers over multiple data sets. Journal of Machine Learning Research **7**, 1–30 (2006)
5. Elkan, C.: The foundations of cost-sensitive learning. In: Proceedings of the Seventeenth International Joint Conference on Artificial Intelligence, pp. 973–978 (2001)
6. Gretton, A., Smola, A., Huang, J., Schmittfull, M., Borgwardt, K., Schölkopf, B.: Covariate shift by kernel mean matching. In: Quiñonero-Candela, J., Masashi Sugiyama, A.S., Lawrence, N.D. (eds.) Dataset Shift in Machine Learning, pp. 131–160. MIT Press (2009)
7. Kull, M., Flach, P.: Patterns of dataset shift. In: First International Workshop on Learning over Multiple Contexts (LMCE) at ECML-PKDD 2014, Nancy, France, September 2014
8. Lichman, M.: UCI machine learning repository (2013). http://archive.ics.uci.edu/ml
9. Moreno-Torres, J.G., Llorí, X., Goldberg, D.E., Bhargava, R.: Repairing fractures between data using genetic programming-based feature extraction: A case study in cancer diagnosis. Inf. Sci. **222**, 805–823 (2013)
10. Moreno-Torres, J.G., Raeder, T., Alaiz-Rodríguez, R., Chawla, N.V., Herrera, F.: A unifying view on dataset shift in classification. Pattern Recognition **45**(1), 521–530 (2012)
11. Moreno-Torres, J.G., Raeder, T., Aláiz-Rodríguez, R., Chawla, N.V., Herrera, F.: Tackling dataset shift in classification: Benchmarks and methods. http://sci2s.ugr.es/dataset-shift (accessed: March 30, 2015)
12. Storkey, A.J.: When training and test sets are different: characterising learning transfer. In: Quiñonero-Candela, J., Masashi Sugiyama, A.S., Lawrence, N.D. (eds.) Dataset Shift in Machine Learning, chap. 1, pp. 3–28. MIT Press (2009)
13. Strack, B., DeShazo, J.P., Gennings, C., Olmo, J.L., Ventura, S., Cios, K.J., Clore, J.N.: Impact of hba1c measurement on hospital readmission rates: analysis of 70,000 clinical database patient records. BioMed Research International **2014**, 781670 (2014). http://europepmc.org/articles/PMC3996476
14. Sugiyama, M., Krauledat, M., Müller, K.R.: Covariate shift adaptation by importance weighted cross validation. Journal of Machine Learning Research **8**, 985–1005 (2007)
15. Zadrozny, B.: Learning and evaluating classifiers under sample selection bias. In: International Conference on Machine Learning ICML 2004, pp. 903–910 (2004)

When is Undersampling Effective in Unbalanced Classification Tasks?

Andrea Dal Pozzolo[1]([✉]), Olivier Caelen[2], and Gianluca Bontempi[1,3]

[1] Machine Learning Group (MLG), Computer Science Department,
Faculty of Sciences ULB, Université Libre de Bruxelles, Brussels, Belgium
adalpozz@ulb.ac.be
[2] Fraud Risk Management Analytics, Worldline, Brussels, Belgium
[3] Interuniversity Institute of Bioinformatics in Brussels (IB)[2], Brussels, Belgium

Abstract. A well-known rule of thumb in unbalanced classification recommends the rebalancing (typically by resampling) of the classes before proceeding with the learning of the classifier. Though this seems to work for the majority of cases, no detailed analysis exists about the impact of undersampling on the accuracy of the final classifier. This paper aims to fill this gap by proposing an integrated analysis of the two elements which have the largest impact on the effectiveness of an undersampling strategy: the increase of the variance due to the reduction of the number of samples and the warping of the posterior distribution due to the change of priori probabilities. In particular we will propose a theoretical analysis specifying under which conditions undersampling is recommended and expected to be effective. It emerges that the impact of undersampling depends on the number of samples, the variance of the classifier, the degree of imbalance and more specifically on the value of the posterior probability. This makes difficult to predict the average effectiveness of an undersampling strategy since its benefits depend on the distribution of the testing points. Results from several synthetic and real-world unbalanced datasets support and validate our findings.

Keywords: Undersampling · Ranking · Class overlap · Unbalanced classification

1 Introduction

In several binary classification problems, the two classes are not equally represented in the dataset. For example, in fraud detection, fraudulent transactions are normally outnumbered by genuine ones [5]. When one class is underrepresented in a dataset, the data is said to be unbalanced. In such problems, typically, the minority class is the class of interest. Having few instances of one class means that the learning algorithm is often unable to generalize the behavior of the minority class well, hence the algorithm performs poorly in terms of predictive accuracy [14].

When the data is unbalanced, standard machine learning algorithms that maximise overall accuracy tend to classify all observations as majority class

© Springer International Publishing Switzerland 2015
A. Appice et al. (Eds.): ECML PKDD 2015, Part I, LNAI 9284, pp. 200–215, 2015.
DOI: 10.1007/978-3-319-23528-8_13

instances. This translates into poor accuracy on the minority class (low recall), which is typically the class of interest. Degradation of classification performance is not only related to a small number of examples in the minority class in comparison to the number of examples in the majority classes (expressed by the class imbalance ratio), but also to the minority class decomposition into small sub-parts [19] (also known in the literature as *small disjuncts* [15]) and to the overlap between the two classes [16] [3] [11] [10]. In these studies it emerges that performance degradation is strongly caused by the presence of both unbalanced class distributions and a high degree of class overlap. Additionally, in unbalanced classification tasks, the performance of a classifier is also affected by the presence of noisy examples [20] [2].

One possible way to deal with this issue is to adjust the algorithms themselves [14] [23] [7]. Here we will consider instead a data-level strategy known as *undersampling* [13]. Undersampling consists in down-sizing the majority class by removing observations at random until the dataset is balanced. In an unbalanced problem, it is often realistic to assume that many observations of the majority class are redundant and that by removing some of them at random the data distribution will not change significantly. However the risk of removing relevant observations from the dataset is still present, since the removal is performed in an unsupervised manner. In practice, sampling methods are often used to balance datasets with skewed class distributions because several classifiers have empirically shown better performance when trained on balanced dataset [22] [9]. However, these studies do not imply that classifiers cannot learn from unbalanced datasets. For instance, other studies have also shown that some classifiers do not improve their performances when the training dataset is balanced using sampling techniques [4] [14]. As a result, the only way to know if sampling helps the learning process is to run some simulations. Despite the popularity of undersampling, we have to remark that there is not yet a theoretical framework explaining how it can affect the accuracy of the learning process.

In this paper we aim to analyse the role of the two side-effects of undersampling on the final accuracy. The first side-effect is that, by removing majority class instances, we perturb the a priori probability of the training set and we induce a warping in the posterior distribution [8,18]. The second is that the number of samples available for training is reduced with an evident consequence in terms of accuracy of the resulting classifier. We study the interaction between these two effects of undersampling and we analyse their impact on the final ranking of posterior probabilities. In particular we show under which conditions an under sampling strategy is recommended and expected to be effective in terms of final classification accuracy.

2   The Warping Effect of Undersampling on the Posterior Probability

Let us consider a binary classification task $f : R^n \to \{0,1\}$, where $\mathbf{X} \in R^n$ is the input and $\mathbf{Y} \in \{0,1\}$ the output domain. In the following we will also use the

$(\mathcal{X},\mathcal{Y})$ (X,Y)

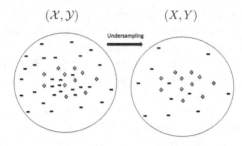

Fig. 1. Undersampling: remove majority class observations until we have the same number of instances in the two classes.

label negative (resp. positive) to denote the label 0 (resp. 1). Suppose that the training set $(\mathcal{X},\mathcal{Y})$ of size N is unbalanced (i.e. the number N^+ of positive cases is small compared to the number N^- of negative ones) and that rebalancing is performed by undersampling. Let $(X,Y) \subset (\mathcal{X},\mathcal{Y})$ be the balanced sample of $(\mathcal{X},\mathcal{Y})$ which contains a subset of the negatives in $(\mathcal{X},\mathcal{Y})$.

Let us introduce a random binary selection variable s associated to each sample in $(\mathcal{X},\mathcal{Y})$, which takes the value 1 if the point is in (X,Y) and 0 otherwise. We now derive how the posterior probability of a model learned on a balanced subset relates to the one learned on the original unbalanced dataset, on the basis of the reasoning presented in [17]. Let us assume that the selection variable s is independent of the input x given the class y (*class-dependent selection*):

$$p(s|y,x) = p(s|y) \tag{1}$$

where $p(s=1|y,x)$ is the probability that a point (x,y) is included in the balanced training sample. Note that the undersampling mechanism has no impact on the class-conditional distribution but that it perturbs the prior probability (i.e. $p(y|s=1) \neq p(y)$).

Let the sign $+$ denote $y=1$ and $-$ denote $y=0$, e.g. $p(+,x) = p(y=1,x)$ and $p(-,x) = p(y=0,x)$. From Bayes' rule we can write:

$$p(+|x,s=1) = \frac{p(s=1|+,x)p(+|x)}{p(s=1|+,x)p(+|x) + p(s=1|-,x)p(-|x)} \tag{2}$$

Using condition (1) in (2) we obtain:

$$p(+|x,s=1) = \frac{p(s=1|+)p(+|x)}{p(s=1|+)p(+|x) + p(s=1|-)p(-|x)} \tag{3}$$

Since undersampling corresponds to set

$$p(s=1|+) = 1 \tag{4}$$

we obtain

$$\frac{p(+)}{p(-)} \leq p(s=1|-) < 1 \tag{5}$$

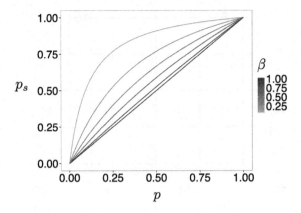

Fig. 2. p and p_s at different β. When β is low, undersampling is strong, which means it is removing a lot of negatives, while for high values the removal is less strong. Low values of β leads to a more balanced problem.

Note that if we set $p(s = 1|-) = \frac{p(+)}{p(-)}$, we obtain a balanced dataset where the number of positive and negative instances is the same. At the same time, if we set $p(s = 1|-) = 1$, no negative instances are removed and no undersampling takes place. Using (4), we can rewrite (3) as

$$p_s = p(+|x, s = 1) = \frac{p(+|x)}{p(+|x) + p(s = 1|-)p(-|x)} = \frac{p}{p + \beta(1 - p)} \qquad (6)$$

where $\beta = p(s = 1|-)$ is the probability of selecting a negative instance with undersampling, $p = p(+|x)$ is the true posterior probability of class $+$ in the original dataset, and $p_s = p(+|x, s = 1)$ is the true posterior probability of class $+$ after sampling. Equation (6) quantifies the amount of warping of the posterior probability due to undersampling. From it, we can derive p as a function of p_s:

$$p = \frac{\beta p_s}{\beta p_s - p_s + 1} \qquad (7)$$

The relation between p and p_s (parametric in β) is illustrated in Figure 2. The top curve of Figure 2 refers to the complete balancing which corresponds to $\beta = \frac{p(+)}{p(-)} \approx \frac{N^+}{N^-}$, assuming that $\frac{N^+}{N^-}$ provides an accurate estimation of the ratio of the prior probabilities.

Figure 3 illustrates the warping effect for two univariate ($n = 1$) classification tasks. In both tasks the two classes are normally distributed ($\mathcal{X}_- \sim N(0, \sigma)$ and $\mathcal{X}_+ \sim N(\mu, \sigma)$), $\sigma = 3$ and $p(+) = 0.1$ but the degree of separability is different (on the left large overlap for $\mu = 3$ and on the right small overlap for $\mu = 15$). It is easy to remark that the warping effect is larger in the low separable case.

As a final remark, consider that when $\beta = \frac{N^+}{N^-}$, the warping due to under-sampling maps two close and low values of p into two values p_s with a larger distance. The opposite occurs for high values of p. In Section 3 we will show how this has an impact on the ranking returned by estimations of p and p_s.

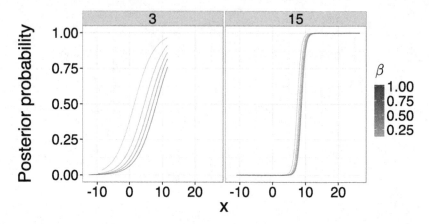

Fig. 3. Posterior probability as a function of β for two univariate binary classification tasks with norm class conditional densities $\mathcal{X}_- \sim N(0, \sigma)$ and $\mathcal{X}_+ \sim N(\mu, \sigma)$ (on the left $\mu = 3$ and on the right $\mu = 15$, in both examples $\sigma = 3$). Note that p corresponds to $\beta = 1$ and p_s to $\beta < 1$.eps

3   The Interaction Between Warping and Variance of the Estimator

The previous section discussed the first consequence of under sampling, i.e. the transformation of the original conditional distribution p into a warped conditional distribution p_s according to equation (6). The second consequence of undersampling is the reduction of the training set size which inevitably leads to an increase of the variance of the classifier. This section discusses how these two effects interact and their impact on the final accuracy of the classifier, by focusing in particular on the accuracy of the ranking of the minority class (typically the class of interest).

Undersampling transforms the original classification task (i.e. estimating the conditional distribution p) into a new classification task (i.e. estimating the conditional distribution p_s). In what follows we aim to assess whether and when under sampling has a beneficial effect by changing the target of the estimation problem.

Let us denote by $\hat{p}$ (resp. $\hat{p}_s$) the estimation of the conditional probability p (resp. p_s). Assume we have two distinct test points having probabilities $p_1 < p_2$ where $\Delta p = p_2 - p_1$ with $\Delta p > 0$. A correct classification aiming to rank the most probable positive samples should rank p_2 before p_1, since the second test sample has an higher probability of belonging to the positive class. Unfortunately the values p_1 and p_2 are not known and the ranking should rely on the estimated values $\hat{p}_1$ and $\hat{p}_2$. For the sake of simplicity we will assume here that the estimator of the conditional probability has the same bias and variance in the two test points. This implies $\hat{p}_1 = p_1 + \epsilon_1$ and $\hat{p}_2 = p_2 + \epsilon_2$, where ϵ_1 and ϵ_2 are two realizations of the random variable $\varepsilon \sim N(b, \nu)$ where b and ν are the bias and the variance of the estimator of p. Note that the estimation errors ϵ_1 and ϵ_2 may induce a wrong ranking if $\hat{p}_1 > \hat{p}_2$.

What happens if instead of estimating p we decide to estimate p_s, as in undersampling? Note that because of the monotone transformation (6), $p_1 < p_2 \Rightarrow p_{s,1} < p_{s,2}$. Is the ranking based on the estimations of $p_{s,1}$ and $p_{s,2}$ more accurate than the one based on the estimations of p_1 and p_2?

In order to answer this question let us suppose that also the estimator of p_s is biased but that its variance is larger given the smaller number of samples. Then $\hat{p}_{s,1} = p_{s,1} + \eta_1$ and $\hat{p}_{s,2} = p_{s,2} + \eta_2$, where $\eta \sim N(b_s, \nu_s)$, $\nu_s > \nu$ and $\Delta p_s = p_{s,2} - p_{s,1}$.

Let us now compute the derivative of p_s w.r.t. p. From (6) we have:

$$\frac{dp_s}{dp} = \frac{\beta}{(p + \beta(1-p))^2} \tag{8}$$

corresponding to a concave function. Let λ be the value of p for which $\frac{dp_s}{dp} = 1$:

$$\lambda = \frac{\sqrt{\beta} - \beta}{1 - \beta}$$

It follows that

$$\beta \leq \frac{dp_s}{dp} \leq \frac{1}{\beta} \tag{9}$$

and

$$1 < \frac{dp_s}{dp} < \frac{1}{\beta}, \qquad \text{when } 0 < p < \lambda$$

while

$$\beta < \frac{dp_s}{dp} < 1 \qquad \text{when } \lambda < p < 1.$$

In particular for $p = 0$ we have $dp_s = \frac{1}{\beta}dp$ while for $p = 1$ it holds $dp_s = \beta dp$.

Let us now suppose that the quantity Δp is small enough to have an accurate approximation $\frac{\Delta p_s}{\Delta p} \approx \frac{dp_s}{dp}$. We can define the probability of obtaining a wrong ranking of $\hat{p}_1$ and $\hat{p}_2$ as:

$$P(\hat{p}_2 < \hat{p}_1) = P(p_2 + \epsilon_2 < p_1 + \epsilon_1)$$
$$= P(\epsilon_2 - \epsilon_1 < p_1 - p_2) = P(\epsilon_1 - \epsilon_2 > \Delta p)$$

where $\epsilon_2 - \epsilon_1 \sim N(0, 2\nu)$. By making an hypothesis of normality we have

$$P(\epsilon_1 - \epsilon_2 > \Delta p) = 1 - \Phi\left(\frac{\Delta p}{\sqrt{2\nu}}\right) \tag{10}$$

where Φ is the cumulative distribution function of the standard normal distribution. Similarly, the probability of a ranking error with undersampling is:

$$P(\hat{p}_{s,2} < \hat{p}_{s,1}) = P(\eta_1 - \eta_2 > \Delta p_s)$$

and

$$P(\eta_1 - \eta_2 > \Delta p_s) = 1 - \Phi\left(\frac{\Delta p_s}{\sqrt{2\nu_s}}\right) \tag{11}$$

We can now say that a classifier learned after undersampling has better ranking w.r.t. a classifier learned with unbalanced distribution when

$$P(\epsilon_1 - \epsilon_2 > \Delta p) > P(\eta_1 - \eta_2 > \Delta p_s) \tag{12}$$

or equivalently from (10) and (11) when

$$1 - \Phi\left(\frac{\Delta p}{\sqrt{2\nu}}\right) > 1 - \Phi\left(\frac{\Delta p_s}{\sqrt{2\nu_s}}\right) \Leftrightarrow \Phi\left(\frac{\Delta p}{\sqrt{2\nu}}\right) < \Phi\left(\frac{\Delta p_s}{\sqrt{2\nu_s}}\right)$$

which boils down to

$$\frac{\Delta p}{\sqrt{2\nu}} < \frac{\Delta p_s}{\sqrt{2\nu_s}} \Leftrightarrow \frac{\Delta p_s}{\Delta p} > \sqrt{\frac{\nu_s}{\nu}} > 1 \tag{13}$$

since Φ is monotone non decreasing and we can assume that $\nu_s > \nu$.

Then it follows that undersampling is useful in terms of more accurate ranking when

$$\frac{\beta}{(p + \beta(1-p))^2} > \sqrt{\frac{\nu_s}{\nu}} \tag{14}$$

The value of this inequality depends on several terms: the rate of under-sampling β, the ratio of the variances of the two classifiers and the posteriori probability p of the testing point. Also the nonlinearity of the first left-hand term suggests a complex interaction between the involved terms. For instance if we plot the left-hand term of (14) as a function of the posteriori probability p (Figure 4(a)) and of the value β (Figure 4(b)), it appears that most favorable configurations for undersampling occur for the lowest values of the posteriori probability (e.g. non separable or badly separable configurations) and interme-diate β (neither too unbalanced nor too balanced). However if we modify β, this has an impact on the size of the training set and consequently on the right-hand term (i.e. variance ratio) too. Also, though the β term can be controlled by the designer, the other two terms vary over the input space. This means that the condition (14) does not necessarily hold for all the test points.

In order to illustrate the complexity of the interaction, let us consider two univariate ($n = 1$) classification tasks where the minority class is normally dis-tributed around zero and the majority class is distributed as a mixture of two gaussians. Figure 5 and 6 show the non separable and separable case, respec-tively: on the left side we plot the class conditional distributions (thin lines) and the posterior distribution of the minority class (thicker line), while on the right side we show the left and the right term of the inequality (14) (solid: left-hand term, dotted: right-hand term). What emerges form the figures is that the least separable regions (i.e. the regions where the posteriori of the minority class is low) are also the regions where undersampling helps more. However, the impact

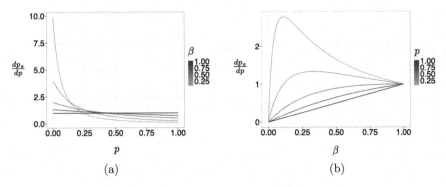

Fig. 4. Left: $\frac{dp_s}{dp}$ as a function of p. Right: $\frac{dp_s}{dp}$ as a function of β

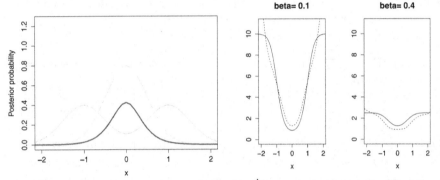

(a) Class conditional distributions (thin (b) $\frac{dp_{s.eps}}{dp}$ (solid lines), $\sqrt{\frac{\nu_s}{\nu}}$ (dotted lines) and the posterior distribution of the lines). minority class (thicker line).

Fig. 5. Non separable case. On the right we plot both terms of inequality 14 (solid: left-hand, dotted: right-hand term) for $\beta = 0.1$ and $\beta = 0.4$

of undersampling on the overall accuracy is difficult to be predicted since the regions where undersampling is beneficial change with the characteristics of the classification task and the rate β of undersampling.

4   Experimental Validation

In this section we assess the validity of the condition (14) by performing a number of tests on synthetic and real datasets.

4.1   Synthetic Datasets

We simulate two unbalanced tasks (5% and 25% of positive samples) with over-lapping classes and generate a testing set and several training sets from the same

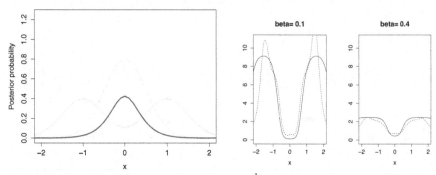

(a) Class conditional distributions (thin lines) and the posterior distribution of the minority class (thicker line).

(b) $\frac{dp_s.eps}{dp}$ (solid lines), $\sqrt{\frac{v_s}{v}}$ (dotted lines).

Fig. 6. Separable case. On the right we plot both terms of inequality 14 (solid: left-hand, dotted: right-hand term) for $\beta = 0.1$ and $\beta = 0.4$

distribution. Figures 7(a) and Figure 9(a) show the distributions of the testing sets for the two tasks.

In order to compute the variance of $\hat{p}$ and $\hat{p}_s$ in each test point, we generate 1000 times a training set ($N = 1000$) and we estimate the conditional probability on the basis of sample mean and covariance.

In Figure 7(b) (first task) we plot $\sqrt{\frac{v_s}{v}}$ (dotted line) and three percentiles $(0.25, 0.5, 0.75)$ of $\frac{dp_s}{dp}$ vs. the rate of undersampling β. It appears that for at least 75% of the testing points, the term $\frac{dp_s}{dp}$ is higher than $\sqrt{\frac{v_s}{v}}$. In Figure 8(a) the points surrounded with a triangle are those one for which $\frac{dp_s}{dp} > \sqrt{\frac{v_s}{v}}$ hold when $\beta = 0.053$ (balanced dataset). For such samples we expect that ranking returned by undersampling (i.e. based on $\hat{p}_s$) is better than the one based on the original data (i.e. based on $\hat{p}$). The plot shows that undersampling is beneficial in the region where the majority class is situated, which is also the area where we expect to have low values of p. Figure 8(b) shows also that this region moves towards the minority class when we do undersampling with $\beta = 0.323$ (90% negatives, 10% positives after undersampling).

In order to measure the quality of the rankings based on $\hat{p}_s$ and $\hat{p}$ we compute the Kendall rank correlation of the two estimates with p, which is the true posterior probability of the testing set that defines the correct ordering. In Table 1 we show the ranking correlations of $\hat{p}_s$ (and $\hat{p}$) with p for the samples where the condition (14) (first five rows) holds and where it does not (last five rows). The results indicate that points for which condition (14) is satisfied have indeed better ranking with $\hat{p}_s$ than $\hat{p}$.

We repeated the experiments for the second task having a larger proportion of positives (25%) (dataset 2 in Figure 9(a)). From the Figure 9(b), plotting $\frac{dp_s}{dp}$

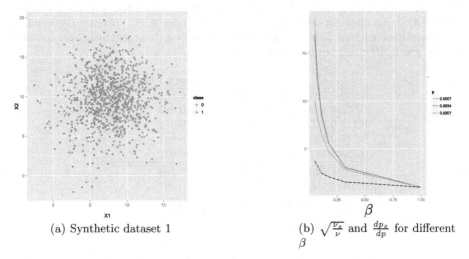

(a) Synthetic dataset 1

(b) $\sqrt{\frac{\nu_s}{\nu}}$ and $\frac{dp_s}{dp}$ for different β

Fig. 7. Left: distribution of the testing set where the positive samples account for 5% of the total. Right: plot of $\frac{dp_s}{dp}$ percentiles (25th, 50th and 75th) and of $\sqrt{\frac{\nu_s}{\nu}}$ (black dashed).

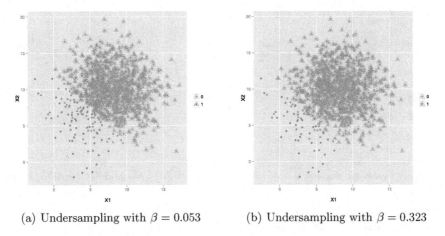

(a) Undersampling with $\beta = 0.053$

(b) Undersampling with $\beta = 0.323$

Fig. 8. Regions where undersampling should work. Triangles indicate the testing samples where the condition (14) holds for the dataset in Figure 7.

and $\sqrt{\frac{\nu_s}{\nu}}$ as a function of β, it appears that only the first two percentiles are over $\sqrt{\frac{\nu_s}{\nu}}$. This means that less points of the testing set satisfy the condition (14). This is confirmed from the results in Table 2 where it appears that the benefit due to undersampling is less significant than for the first task.

Table 1. Classification task in Figure 7: Ranking correlation between the posterior probability $\hat{p}$ ($\hat{p}_s$) and p for different values of β. The value $\mathcal{K}$ ($\mathcal{K}_s$) denotes the Kendall rank correlation without (with) undersampling. The first (last) five lines refer to samples for which the condition (14) is (not) satisfied.

β	$\mathcal{K}$	$\mathcal{K}_s$	$\mathcal{K}_s - \mathcal{K}$	%points satisfying (14)
0.053	0.298	0.749	0.451	0.888
0.076	0.303	0.682	0.379	0.897
0.112	0.315	0.619	0.304	0.912
0.176	0.323	0.555	0.232	0.921
0.323	0.341	0.467	0.126	0.937
0.053	0.749	0.776	0.027	0.888
0.076	0.755	0.773	0.018	0.897
0.112	0.762	0.764	0.001	0.912
0.176	0.767	0.761	-0.007	0.921
0.323	0.768	0.748	-0.020	0.937

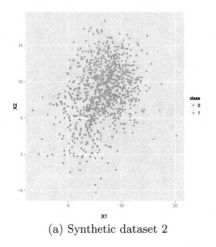

(a) Synthetic dataset 2

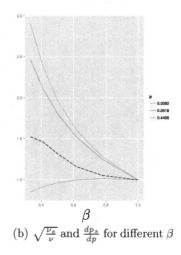

(b) $\sqrt{\frac{\nu_s}{\nu}}$ and $\frac{dp_s}{dp}$ for different β

Fig. 9. Left: distribution of the testing set where the positive samples account for 25% of the total. Right: plot of $\frac{dp_s}{dp}$ percentiles (25th, 50th and 75th) and of $\sqrt{\frac{\nu_s}{\nu}}$ (black dashed).

4.2   Real Datasets

In this section we assess the validity of the condition (14) on a number of real unbalanced binary classification tasks obtained by transforming some datasets from the UCI repository [1] (Table 3)[1].

---

[1] Transformed datasets are available at http://www.ulb.ac.be/di/map/adalpozz/imbalanced-datasets.zip

Table 2. Classification task in Figure 9: Ranking correlation between the posterior probability $\hat{p}$ ($\hat{p}_s$) and p for different values of β. The value $\mathcal{K}$ ($\mathcal{K}_s$) denotes the Kendall rank correlation without (with) undersampling. The first (last) five lines refer to samples for which the condition (14) is (not) satisfied.

β	$\mathcal{K}$	$\mathcal{K}_s$	$\mathcal{K}_s - \mathcal{K}$	% points statifying (14)
0.333	0.586	0.789	0.202	0.664
0.407	0.588	0.761	0.172	0.666
0.500	0.605	0.738	0.133	0.681
0.619	0.628	0.715	0.087	0.703
0.778	0.653	0.693	0.040	0.73
0.333	0.900	0.869	-0.030	0.664
0.407	0.899	0.875	-0.024	0.666
0.500	0.894	0.874	-0.020	0.681
0.619	0.885	0.869	-0.016	0.703
0.778	0.870	0.856	-0.014	0.73

Given the unavailability of the conditional posterior probability function, we first approximate p by fitting a Random Forest over the entire dataset in order to compute the left-hand term of (14). Then we use a boostrap procedure to estimate $\hat{p}$ and apply undersampling to the original dataset to estimate $\hat{p}_s$. We repeat bootstrap and undersampling 100 times to compute the right hand term $\sqrt{\frac{\nu_s}{\nu}}$. This allows us to define the subsets of points for which the condition (14) holds.

Table 3. Selected datasets from the UCI repository [1]

Datasets	N	N^+	N^-	N^+/N
ecoli	336	35	301	0.10
glass	214	17	197	0.08
letter-a	20000	789	19211	0.04
letter-vowel	20000	3878	16122	0.19
ism	11180	260	10920	0.02
letter	20000	789	19211	0.04
oil	937	41	896	0.04
page	5473	560	4913	0.10
pendigits	10992	1142	9850	0.10
PhosS	11411	613	10798	0.05
satimage	6430	625	5805	0.10
segment	2310	330	1980	0.14
boundary	3505	123	3382	0.04
estate	5322	636	4686	0.12
cam	18916	942	17974	0.05
compustat	13657	520	13137	0.04
covtype	38500	2747	35753	0.07

Figure 10 reports the difference between Kendall rank correlation of $\hat{p}_s$ and $\hat{p}$, averaged over different levels of undersampling (proportions of majority vs. minority: $90/10, 80/20, 60/40, 50/50$). Higher difference means that $\hat{p}_s$ returns a better ordering than $\hat{p}$ (assuming that the ranking provided by p is correct). The plot distinguishes between samples for which condition (14) is satisfied and not. In general we see that points with a positive difference corresponds to those having the condition satisfied and the opposite for negative differences. These results seem to confirm the experiments with synthetic data, where a better ordering is given by $\hat{p}_s$ when the condition (14) holds.

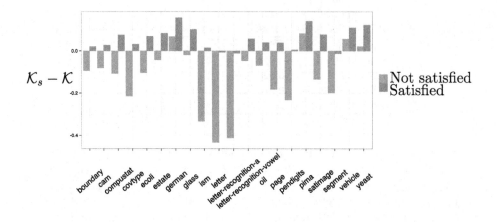

Fig. 10. Difference between the Kendall rank correlation of $\hat{p}_s$ and $\hat{p}$ with p, namely $\mathcal{K}_s$ and $\mathcal{K}$, for points having the conditions (14) satisfied and not. $\mathcal{K}_s$ and $\mathcal{K}$ are calculated as the mean of the correlations over all βs.

In Figure 11 we show the ratio of samples in each dataset satisfying condition 14 averaged over all the (β)s. The proportion of points in which undersampling is useful changes heavily with the dataset considered. For example, in the datasets *vehicle, yeast, german* and *pima*, underdamping returns a better ordering for more than 80% of the samples, while the proportion drops to less than 50% in the *page* dataset.

This seems to confirm our intuition that the right amount of undersampling depends on the classification task (e.g. degree of non separability), the learning algorithm and the targeted test set. It follows that there is no reason to believe that undersampling until the two classes are perfectly balanced is the default strategy to adopt.

It is also worthy to remark that the check of the condition (14) is not easy to be done, since it involves the estimation of $\sqrt{\frac{\nu_s}{\nu}}$ (ratio of the the variance of the classifier before and after undersampling) and of $\frac{dp_s}{dp}$, which demands the knowledge of the true posterior probability p. In practice since p is unknown in real datasets, we can only rely on a data driven approximation of $\frac{dp_s}{dp}$. Also the estimation of $\sqrt{\frac{\nu_s}{\nu}}$ is an hard statistical problem, as known in the statistical literature on ratio estimation [12].

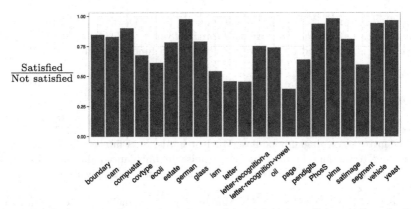

Fig. 11. Ratio between the number of sample satisfying condition (14) and all the instances available in each dataset averaged over all the βs.

5   Conclusion

Undersampling has become the de facto strategy to deal with skewed distributions, but, though easy to be justified, it conceals two major effects: i) it increases the variance of the classifier and ii) it produces warped posterior probabilities. The first effect is typically addressed by the use of averaging strategies (e.g. UnderBagging [21]) to reduce the variability while the second requires the calibration of the probability to the new priors of the testing set [18]. Despite the popularity of undersampling for unbalanced classification tasks, it is not clear how these two effects interact and when undersampling leads to better accuracy in the classification task.

In this paper, we aimed to analyse the interaction between undersampling and the ranking error of the posterior probability. We derive the condition (14) under which undersampling can improve the ranking and we show that when it is satisfied, the posterior probability obtained after sampling returns a more accurate ordering of testing instances. To validate our claim we used first synthetic and then real datasets, and in both cases we registered a better ranking with undersampling when condition (14) was met. It is important to remark how this condition shows that the beneficial impact of undersampling is strongly dependent on the nature of the classification task (degree of unbalancedness and non separability), on the variance of the classifier and as a consequence is extremely dependent on the specific test point. We think that this result sheds light on the reason why several discordant results have been obtained in the literature about the effectiveness of undersampling in unbalanced tasks.

However, the practical use of this condition is not straightforward since it requires the knowledge of the posteriori probability and of the ratio of variances before and after undersampling. It follows that this result should be used mainly as a warning against a naive use of undersampling in unbalanced tasks and should suggest instead the adoption of specific adaptive selection techniques (e.g. racing [6]) to perform a case-by-case use (and calibration) of undersampling.

Acknowledgments. A. Dal Pozzolo is supported by the Doctiris scholarship *Adaptive real-time machine learning for credit card fraud detection* funded by Innoviris, Belgium. G. Bontempi is supported by the project *BridgeIRIS* funded by Innoviris, Belgium.

References

1. Newman, D.J., Asuncion, A.: UCI machine learning repository (2007)
2. Anyfantis, D., Karagiannopoulos, M., Kotsiantis, S., Pintelas, P.: Robustness of learning techniques in handling class noise in imbalanced datasets. In: Artificial intelligence and innovations 2007: From Theory to Applications, pp. 21–28. Springer (2007)
3. Batista, G.E.A.P.A., Prati, R.C., Monard, M.C.: Balancing Strategies and Class Overlapping. In: Famili, A.F., Kok, J.N., Peña, J.M., Siebes, A., Feelders, A. (eds.) IDA 2005. LNCS, vol. 3646, pp. 24–35. Springer, Heidelberg (2005)
4. Batista, G.E.A.P.A., Prati, R.C., Monard, M.C.: A study of the behavior of several methods for balancing machine learning training data. ACM SIGKDD Explorations Newsletter **6**(1), 20–29 (2004)
5. Dal Pozzolo, A., Caelen, O., Borgne, Y.-A.L., Waterschoot, S., Bontempi, G.: Learned lessons in credit card fraud detection from a practitioner perspective. Expert Systems with Applications **41**(10), 4915–4928 (2014)
6. Dal Pozzolo, A., Caelen, O., Waterschoot, S., Bontempi, G.: Racing for Unbalanced Methods Selection. In: Yin, H., Tang, K., Gao, Y., Klawonn, F., Lee, M., Weise, T., Li, B., Yao, X. (eds.) IDEAL 2013. LNCS, vol. 8206, pp. 24–31. Springer, Heidelberg (2013)
7. Domingos, P.: Metacost: a general method for making classifiers cost-sensitive. In: Proceedings of the Fifth ACM SIGKDD International Conference on Knowledge Discovery and Data Mining, pp. 155–164. ACM (1999)
8. Elkan, C.: The foundations of cost-sensitive learning. In: International Joint Conference on Artificial Intelligence, Citeseer, vol. 17, pp. 973–978 (2001)
9. Estabrooks, A., Jo, T., Japkowicz, N.: A multiple resampling method for learning from imbalanced data sets. Computational Intelligence **20**(1), 18–36 (2004)
10. García, V., Mollineda, R.A., Sánchez, J.S.: On the k-nn performance in a challenging scenario of imbalance and overlapping. Pattern Analysis and Applications **11**(3–4), 269–280 (2008)
11. García, V., Sánchez, J., Mollineda, R.A.: An Empirical Study of the Behavior of Classifiers on Imbalanced and Overlapped Data Sets. In: Rueda, L., Mery, D., Kittler, J. (eds.) CIARP 2007. LNCS, vol. 4756, pp. 397–406. Springer, Heidelberg (2007)
12. Hartley, H.O., Ross, A.: Unbiased ratio estimators (1954)
13. He, H., Garcia, E.A.: Learning from imbalanced data. IEEE Transactions on Knowledge and Data Engineering **21**(9), 1263–1284 (2009)
14. Japkowicz, N., Stephen, S.: The class imbalance problem: A systematic study. Intelligent Data Analysis **6**(5), 429–449 (2002)
15. Jo, T., Japkowicz, N.: Class imbalances versus small disjuncts. ACM SIGKDD Explorations Newsletter **6**(1), 40–49 (2004)
16. Prati, R.C., Batista, G.E.A.P.A., Monard, M.C.: Class Imbalances *versus* Class Overlapping: An Analysis of a Learning System Behavior. In: Monroy, R., Arroyo-Figueroa, G., Sucar, L.E., Sossa, H. (eds.) MICAI 2004. LNCS (LNAI), vol. 2972, pp. 312–321. Springer, Heidelberg (2004)

17. Quionero-Candela, J., Sugiyama, M., Schwaighofer, A., Lawrence, N.D.: Dataset shift in machine learning. The MIT Press (2009)
18. Saerens, M., Latinne, P., Decaestecker, C.: Adjusting the outputs of a classifier to new a priori probabilities: a simple procedure. Neural Computation 14(1), 21–41 (2002)
19. Stefanowski, J.: Overlapping, rare examples and class decomposition in learning classifiers from imbalanced data. In: Emerging Paradigms in Machine Learning, pp. 277–306. Springer (2013)
20. Van Hulse, J., Khoshgoftaar, T.: Knowledge discovery from imbalanced and noisy data. Data & Knowledge Engineering 68(12), 1513–1542 (2009)
21. Wang, S., Tang, K., Yao, X.: Diversity exploration and negative correlation learning on imbalanced data sets. In: International Joint Conference on Neural Networks, IJCNN 2009, pp. 3259–3266. IEEE (2009)
22. Weiss, G.M.: Foster Provost. The effect of class distribution on classifier learning: an empirical study. Rutgers Univ (2001)
23. Zadrozny, B., Langford, J., Abe, N.: Cost-sensitive learning by cost-proportionate example weighting. In: Data Mining, ICDM, pp. 435–442. IEEE (2003)

Clustering and Unsupervised Learning

A Kernel-Learning Approach to Semi-supervised Clustering with Relative Distance Comparisons

Ehsan Amid[1]([✉]), Aristides Gionis[1], and Antti Ukkonen[2]

[1] Helsinki Institute for Information Technology,
and Department of Computer Science, Aalto University, Espoo, Finland
{ehsan.amid,aristides.gionis}@aalto.fi
[2] Finnish Institute of Occupational Health, Helsinki, Finland
antti.ukkonen@ttl.fi

Abstract. We consider the problem of clustering a given dataset into k clusters subject to an additional set of constraints on relative distance comparisons between the data items. The additional constraints are meant to reflect side-information that is not expressed in the feature vectors, directly. Relative comparisons can express structures at finer level of detail than must-link (ML) and cannot-link (CL) constraints that are commonly used for semi-supervised clustering. Relative comparisons are particularly useful in settings where giving an ML or a CL constraint is difficult because the granularity of the true clustering is unknown.

Our main contribution is an efficient algorithm for learning a kernel matrix using the log determinant divergence (a variant of the Bregman divergence) subject to a set of relative distance constraints. Given the learned kernel matrix, a clustering can be obtained by any suitable algorithm, such as kernel k-means. We show empirically that kernels found by our algorithm yield clusterings of higher quality than existing approaches that either use ML/CL constraints or a different means to implement the supervision using relative comparisons.

1 Introduction

Clustering is the task of partitioning a set of data items into groups, or clusters. However, the desired grouping of the data may not be sufficiently expressed by the features that are used to describe the data items. For instance, when clustering images it may be necessary to make use of semantic information about the image contents in addition to some standard image features. *Semi-supervised clustering* is a principled framework for combining such external information with features. This information is usually given as labels about the *pair-wise distances* between a few data items. Such labels may be provided by the data analyst, and reflect properties of the data that are hard to express as an easily computable function over the data features.

There are two commonly used ways to formalize such side information. The first are *must-link* (ML) and *cannot-link* (CL) constraints. An ML (CL) constraint between data items i and j suggests that the two items are similar (dissimilar), and should thus be assigned to the same cluster (different clusters).

© Springer International Publishing Switzerland 2015
A. Appice et al. (Eds.): ECML PKDD 2015, Part I, LNAI 9284, pp. 219–234, 2015.
DOI: 10.1007/978-3-319-23528-8_14

The second way to express pair-wise similarities are *relative distance comparisons*. These are statements that specify how the distances between some data items relate to each other. The most common relative distance comparison task asks the data analyst to specify which of the items i and j is closer to a third item k. Note that unlike the ML/CL constraints, the relative comparisons do not as such say anything about the clustering structure.

Given a number of similarity constraints, an efficient technique to implement semi-supervised clustering is *metric learning*. The objective of metric learning is to find a new distance function between the data items that takes both the supplied features as well as the additional distance constraints into account. Metric learning can be based on either ML/CL constraints or relative distance comparisons. Both approaches have been studied extensively in literature, and a lot is known about the problem.

The method we discuss in this paper is *a combination of metric-learning and relative distance comparisons*. We deviate from existing literature by eliciting every constraint with the question

"Which one of the items i, j, and k is the least similar to the other two?"

The labeler should thus select one of the items as an *outlier*. Notably, we also allow the labeler to leave the answer as *unspecified*. The main practical novelty of this approach is in the *capability to gain information also from comparisons where the labeler has not been able to give a concrete solution*. Some sets of three items can be all very similar (or dissimilar) to each other, so that picking one item as an obvious outlier is difficult. In those cases that the labeler gives a "don't know" answer, it is beneficial to use this answer in the metric-learning process as it provides a valuable cue, namely, that the three displayed data items are roughly equidistant.

We cast the metric-learning problem as a *kernel-learning problem*. The learned kernel can be used to easily compute distances between data items, even between data items that did not participate in the metric-learning training phase, and only their feature vectors are available. The use of relative comparisons, instead of hard ML/CL constraints, leads to learning a more accurate metric that captures relations between data items at different scales. The learned metric can be used for multi-level clustering, as well as other data-analysis tasks.

On the technical side, we start with an initial kernel $\mathbf{K}_0$, computed using only the feature vectors of the data items. We then formulate the kernel-learning task as an optimization problem: the goal is to find the kernel matrix $\mathbf{K}$ that is the closest to $\mathbf{K}_0$ and satisfies the constraints induced by the relative-comparison labellings. To solve this optimization task we use known efficient techniques, which we adapt for the case of relative comparisons.

More concretely, we make the following contributions:

1. We design a kernel-learning method that can also use unspecified relative distance comparisons. This is done by extending the method of Anand et al. [1], which works with ML and CL constraints.

2. We perform an extensive experimental validation of our approach and show that the proposed labeling is indeed more flexible, and it can lead to a substantial improvement in the clustering accuracy.

The rest of this paper is organized as follows. We start by reviewing the related literature in Section 2. In Section 3 we introduce our setting and formally define our problem, and in Section 4 we present our solution. In Section 5 we discuss our empirical evaluation, and Section 6 is a short conclusion.

2   Related Work

The idea of semi-supervised clustering was initially introduced by Wagstaff and Cardie [2], and since then a large number of different problem variants and methods have been proposed, the first ones being COP-Kmeans [3] and CCL [4]. Some of the later methods handle the constraints in a probabilistic framework. For instance, the ML and CL constraints can be imposed in the form of a Markov random field prior over the data items [5–7]. Alternatively, Lu [8] generalizes the standard Gaussian process to include the preferences imposed by the ML and CL constraints. Recently, Pei et al. [9] propose a discriminative clustering model that uses relative comparisons and, like our method, can also make use of unspecified comparisons.

The semi-supervising clustering setting has also been studied in the context of spectral clustering, and many spectral clustering algorithms have been extended to incorporate pairwise constraints [10,11]. More generally, these methods employ techniques for semi-supervised graph partitioning and kernel k-means algorithms [12]. For instance, Kulis et al. [13] present a unified framework for semi-supervised vector and graph clustering using spectral clustering and kernel learning.

As stated in the Introduction, our work is based on *metric learning*. Most of the metric-learning literature, starting by the work of Xing et al. [14], aims at finding a Mahalanobis matrix subject to either ML/CL or relative distance constraints. Xing et al. [14] use ML/CL constraints, while Schultz and Joachims [15] present a similar approach to handle relative comparisons. Metric learning often requires solving a semidefinite optimization problem. This becomes easier if Bregman divergences, in particular the log det divergence, is used to formulate the optimization problem. Such an approach was first used for metric learning by Davis et al. [16] with ML/CL constraints, and subsequently by Liu et al. [17] likewise with ML/CL, as well as by Liu et al. [18] with relative comparisons. Our algorithm also uses the log det divergence, and we extend the technique of Davis et al. [16] to handle relative comparisons.

The metric-learning approaches can also be more directly combined with a clustering algorithm. The MPCK-Means algorithm by Bilenko et al. [19] is one of the first to combine metric learning with semi-supervised clustering and ML/CL constraints. Xiang et al. [20] use metric learning, as well, to implement ML/CL constraints in a clustering and classification framework, while Kumar et al. [21] follow a similar approach using relative comparisons. Recently, Anand et al. [1]

use a kernel-transformation approach to adopt the mean-shift algorithm [22] to incorporate ML and CL constraints. This algorithm, called semi-supervised kernel mean shift clustering (SKMS), starts with an initial kernel matrix of the data points and generates a transformed matrix by minimizing the log det divergence using an approach based on the work by Kulis et al. [23]. Our paper is largely inspired by the SKMS algorithm. Our main contribution is to extend the SKMS algorithm so that it handles relative distance comparisons.

3   Kernel Learning with Relative Distances

In this section we introduce the notation used throughout the paper and formally define the problem we address.

3.1   Basic Definitions

Let $\mathcal{D} = \{1, \ldots, n\}$ denote a set of *data items*. These are the data we want to cluster. Let $\mathcal{X} = \{\mathbf{x}_i\}_{i=1}^n$, with $\mathbf{x}_i \in \mathbb{R}^d$, denote a set of vectors in a d dimensional Euclidean space; one vector for every item in $\mathcal{D}$. The vector set $\mathcal{X}$ is the *feature representation* of the items in $\mathcal{D}$. We are also given the set $\mathcal{C}$ of *relative distance comparisons* between data items in $\mathcal{D}$. These distance comparisons are given in terms of some *unknown distance function* $\delta : \mathcal{D} \times \mathcal{D} \to \mathbb{R}$. We assume that δ reflects certain domain knowledge, which is difficult to quantify precisely, and cannot be computed using only the features in $\mathcal{X}$. Thus, the set of distance comparisons $\mathcal{C}$ *augments our knowledge* about the data items in $\mathcal{D}$, in addition to the feature vectors in $\mathcal{X}$. The comparisons in $\mathcal{C}$ are given by human evaluators, or they may come from some other source. We assume that this information is not directly captured by the features.

Given $\mathcal{X}$ and $\mathcal{C}$, our objective is to find a kernel matrix $\mathbf{K}$ that captures more accurately the distance between data items. Such a kernel matrix can be used for a number of different purposes. In this paper, we focus on using the kernel matrix for clustering the data in $\mathcal{D}$. The kernel matrix $\mathbf{K}$ is computed by considering both the similarities between the points in $\mathcal{X}$ as well as the user-supplied constraints induced by the comparisons in $\mathcal{C}$.

In a nutshell, we compute the kernel matrix $\mathbf{K}$ by first computing an initial kernel matrix $\mathbf{K}_0$ using only the vectors in $\mathcal{X}$. The matrix $\mathbf{K}_0$ is computed by applying a Gaussian kernel on the vectors in $\mathcal{X}$. We then solve an optimization problem in order to find the kernel matrix $\mathbf{K}$ that is the closest to $\mathbf{K}_0$ and satisfies the constraints in $\mathcal{C}$.

3.2   Relative Distance Constraints

The constraints in $\mathcal{C}$ express information about distances between items in $\mathcal{D}$ in terms of the distance function δ. However, we do not need to know the absolute distances between any two items $i, j \in \mathcal{D}$. Instead we consider constraints that express information of the type $\delta(i, j) < \delta(i, k)$ for some $i, j, k \in \mathcal{D}$.

In particular, every constraint $C_i \in C$ is a statement about the relative distances between *three* items in $\mathcal{D}$. We consider two types of constraints, i.e., C can be partitioned into two sets $\mathcal{C}_{\text{neq}}$ and $\mathcal{C}_{\text{eq}}$. The set $\mathcal{C}_{\text{neq}}$ contains constraints where one of the three items has been singled out as an "outlier." That is, the distance of the outlying item to the two others is clearly larger than the distance between the two other items. The set $\mathcal{C}_{\text{eq}}$ contains constraints where no item appears to be an obvious outlier. The distances between all three items are then assumed to be approximately the same.

More formally, we define $\mathcal{C}_{\text{neq}}$ to be a set of tuples of the form $(i, j \mid k)$, where every tuple is interpreted as "item k is an outlier among the three items i, j and k." We assume that the item k is an outlier if its distance from i and j is at least γ times larger than the distance $\delta(i, j)$, for some $\gamma > 1$. This is because we assume small differences in the distances to be indistinguishable by the evaluators, and only such cases end up in $\mathcal{C}_{\text{neq}}$ where there is no ambiguity between the distances. As a result each triple $(i, j \mid k)$ in $\mathcal{C}_{\text{neq}}$ implies the following two inequalities

$$(i \leftarrow j \mid k): \quad \gamma\delta(i, j) \leq \delta(i, k) \text{ and} \tag{1}$$
$$(j \leftarrow i \mid k): \quad \gamma\delta(j, i) \leq \delta(j, k), \tag{2}$$

where γ is a parameter that must be set in advance.

Likewise, we define $\mathcal{C}_{\text{eq}}$ to be a set of tuples of the form (i, j, k) that translates to "the distances between items i, j and k are equal." In terms of the distance function δ, each triple (i, j, k) in $\mathcal{C}_{\text{eq}}$ implies

$$\delta(i, j) = \delta(j, k) = \delta(i, k). \tag{3}$$

3.3   Extension to a Kernel Space

As mentioned above, the first step of our approach is forming the initial kernel $\mathbf{K}_0$ using the feature vectors $\mathcal{X}$. We do this using a standard Gaussian kernel. Details are provided in Section 4.2.

Next we show how the constraints implied by the distance comparison sets $\mathcal{C}_{\text{neq}}$ and $\mathcal{C}_{\text{eq}}$ extend to a kernel space, obtained by a mapping $\Phi : \mathcal{D} \to \mathbb{R}^m$. As usual, we assume that an inner product $\Phi(i)^\top \Phi(j)$ between items i and j in $\mathcal{D}$ can be expressed by a symmetric kernel matrix $\mathbf{K}$, that is, $\mathbf{K}_{ij} = \Phi(i)^\top \Phi(j)$. Moreover, we assume that the kernel $\mathbf{K}$ (and the mapping Φ) is connected to the unknown distance function δ via the equation

$$\delta(i, j) = \|\Phi(i) - \Phi(j)\|^2 = \mathbf{K}_{ii} - 2\mathbf{K}_{ij} + \mathbf{K}_{jj}. \tag{4}$$

In other words, we explicitly assume that the distance function δ is in fact the Euclidean distance in some unknown vector space. This is equivalent to assume that the evaluators base their distance-comparison decisions on some implicit features, even if they might not be able to quantify these explicitly.

Next, we discuss the constraint inequalities (Equations (1), (2), and (3)) in the kernel space. Let $\mathbf{e}_i$ denote the vector of all zeros with the value 1 at position i. Equation (4) above can be expressed in matrix form as follows:

$$\mathbf{K}_{ii} - 2\mathbf{K}_{ij} + \mathbf{K}_{jj} = (\mathbf{e}_i - \mathbf{e}_j)^\top \mathbf{K}(\mathbf{e}_i - \mathbf{e}_j) = \mathrm{tr}(\mathbf{K}(\mathbf{e}_i - \mathbf{e}_j)(\mathbf{e}_i - \mathbf{e}_j)^\top), \quad (5)$$

where $\mathrm{tr}(\mathbf{A})$ denotes the trace of the matrix $\mathbf{A}$ and we use the fact that $\mathbf{K} = \mathbf{K}^\top$. Using the previous equation we can write Equation (1) as

$$\gamma\,\mathrm{tr}\left(\mathbf{K}(\mathbf{e}_i - \mathbf{e}_j)(\mathbf{e}_i - \mathbf{e}_j)^\top\right) - \mathrm{tr}\left(\mathbf{K}(\mathbf{e}_i - \mathbf{e}_k)(\mathbf{e}_i - \mathbf{e}_k)^\top\right) \leq 0$$
$$\mathrm{tr}\left(\mathbf{K}\gamma(\mathbf{e}_i - \mathbf{e}_j)(\mathbf{e}_i - \mathbf{e}_j)^\top - \mathbf{K}(\mathbf{e}_i - \mathbf{e}_k)(\mathbf{e}_i - \mathbf{e}_k)^\top\right) \leq 0$$
$$\mathrm{tr}\left(\mathbf{K}(\gamma(\mathbf{e}_i - \mathbf{e}_j)(\mathbf{e}_i - \mathbf{e}_j)^\top - (\mathbf{e}_i - \mathbf{e}_k)(\mathbf{e}_i - \mathbf{e}_k)^\top)\right) \leq 0$$
$$\mathrm{tr}\left(\mathbf{K}\,\mathbf{C}_{(i \leftarrow j|k)}\right) \leq 0,$$

where $\mathbf{C}_{(i \leftarrow j|k)} = \gamma(\mathbf{e}_i - \mathbf{e}_j)(\mathbf{e}_i - \mathbf{e}_j)^\top - (\mathbf{e}_i - \mathbf{e}_k)(\mathbf{e}_i - \mathbf{e}_k)^\top$ is a matrix that represents the corresponding constraint. The constraint matrix $\mathbf{C}_{(j \leftarrow i|k)}$ for Equation (2) can be formed in exactly the same manner. Note that unless we set $\gamma > 1$, the Equations (1) and (2) can be satisfied trivially for a small difference between the longer and the shorter distance and thus, the constraint becomes inactive. Setting $\gamma > 1$ helps avoiding such solutions.

We use a similar technique to represent the constraints in the set $\mathcal{C}_{\mathrm{eq}}$. Recall that the constraint $(i, j, k) \in \mathcal{C}_{\mathrm{eq}}$ implies that i, j, and k are equidistant. This yields three equations on the pairwise distances between the items: $(i \leftrightarrow j, k) :$ $\delta(i, j) = \delta(i, k)$, $(j \leftrightarrow i, k) : \quad \delta(j, i) = \delta(j, k)$, and $(k \leftrightarrow i, j) : \quad \delta(k, i) = \delta(k, j)$. Reasoning as above, we let $\mathbf{C}_{(i \leftrightarrow j,k)} = (\mathbf{e}_i - \mathbf{e}_j)(\mathbf{e}_i - \mathbf{e}_j)^\top - (\mathbf{e}_i - \mathbf{e}_k)(\mathbf{e}_i - \mathbf{e}_k)^\top$, and can thus write the first equation for the constraint $(i, j, k) \in \mathcal{C}_{\mathrm{eq}}$ as

$$\mathrm{tr}(\mathbf{K}\,\mathbf{C}_{(i \leftrightarrow j,k)}) = 0. \qquad (6)$$

The two other equations are defined in a similar manner.

3.4   Log Determinant Divergence for Kernel Learning

Recall that our objective is to find the kernel matrix $\mathbf{K}$ that is close to the initial kernel $\mathbf{K}_0$. Assume that $\mathbf{K}$ and $\mathbf{K}_0$ are both positive semidefinite matrices. We will use the so-called *log determinant divergence* to compute the similarity between $\mathbf{K}_0$ and $\mathbf{K}$. This is a variant of the *Bregman divergence* [24].

The Bregman divergence between two matrices $\mathbf{K}$ and $\mathbf{K}_0$ is defined as

$$\mathrm{D}_\phi(\mathbf{K}, \mathbf{K}_0) = \phi(\mathbf{K}) - \phi(\mathbf{K}_0) - \mathrm{tr}(\nabla\phi(\mathbf{K}_0)^\top(\mathbf{K} - \mathbf{K}_0)), \qquad (7)$$

where ϕ is a strictly-convex real-valued function, and $\nabla\phi(\mathbf{K}_0)$ denotes the gradient evaluated at $\mathbf{K}_0$. Many well-known distance measures are special cases of the Bregman divergence. These can be instantiated by selecting the function ϕ appropriately. For instance, $\phi(\mathbf{K}) = \sum_{ij} K_{ij}^2$ gives the squared Frobenius norm $\mathrm{D}_\phi(\mathbf{K}, \mathbf{K}_0) = \|\mathbf{K} - \mathbf{K}_0\|_F^2$.

For our application in kernel learning, we are interested in one particular case; setting $\phi(\mathbf{K}) = -\log\det(\mathbf{K})$. This yields the so-called log determinant (log det) matrix divergence:

$$D_{ld}(\mathbf{K}, \mathbf{K}_0) = \operatorname{tr}(\mathbf{K}\,\mathbf{K}_0^{-1}) - \log\det(\mathbf{K}\,\mathbf{K}_0^{-1}) - n, \tag{8}$$

The log det divergence has many interesting properties, which make it ideal for kernel learning. As a general result of Bregman divergences, log det divergence is *convex* with respect to the first argument. Moreover, it can be evaluated using the eigenvalues and eigenvectors of the matrices $\mathbf{K}$ and $\mathbf{K}_0$. This property can be used to extend log det divergence to handle rank-deficient matrices [23], and we will make use of this in our algorithm described in Section 4.

3.5   Problem Definition

We now have the necessary ingredients to formulate our semi-supervised kernel learning problem. Given the set of constraints $\mathcal{C} = \mathcal{C}_{neq} \cup \mathcal{C}_{eq}$, the parameter γ, and the initial kernel matrix $\mathbf{K}_0$, we aim to find a new kernel matrix $\mathbf{K}$, which is as close as possible to $\mathbf{K}_0$ while satisfying the constraints in $\mathcal{C}$. This objective can be formulated as the following constrained minimization problem:

$$\underset{\mathbf{K}}{\text{minimize}} \quad D_{ld}(\mathbf{K}, \mathbf{K}_0)$$

subject to

$$\operatorname{tr}\left(\mathbf{K}\,\mathbf{C}_{(i \leftarrow j|k)}\right) \le 0,\ \operatorname{tr}\left(\mathbf{K}\,\mathbf{C}_{(j \leftarrow i|k)}\right) \le 0, \quad \forall(i, j \mid k) \in \mathcal{C}_{neq}$$
$$\operatorname{tr}(\mathbf{K}\,\mathbf{C}_{(i \leftrightarrow j, k)}) = 0,\ \operatorname{tr}(\mathbf{K}\,\mathbf{C}_{(j \leftrightarrow i, k)}) = 0,\ \operatorname{tr}(\mathbf{K}\,\mathbf{C}_{(k \leftrightarrow i, j)}) = 0, \quad \forall(i, j, k) \in \mathcal{C}_{eq}$$
$$\mathbf{K} \succeq 0,$$

$$(9)$$

where $\mathbf{K} \succeq 0$ constrains $\mathbf{K}$ to be a positive semidefinite matrix.

4   Semi-supervised Kernel Learning

We now focus on the optimization problem defined above, Problem (9). It can be shown that in order to have a finite value for the log det divergence, the rank of the matrices must remain equal [23]. This property along with the fact that the domain of the log det divergence is the positive-semidefinite matrices, allow us to perform the optimization without explicitly restraining the solution to the positive-semidefinite cone nor checking for the rank of the solution. This is in contrast with performing the optimization using, say, the Frobenius norm, where the projection to the positive semidefinite cone must be explicitly imposed.

4.1   Bregman Projections for Constrained Optimization

In solving the optimization Problem (9), the aim is to minimize the divergence while satisfying the set of constraints imposed by $\mathcal{C} = \mathcal{C}_{neq} \cup \mathcal{C}_{eq}$. In other words, we seek for a kernel matrix $\mathbf{K}$ by projecting the initial kernel matrix $\mathbf{K}_0$ onto the convex set obtained from the intersection of the set of constraints. The optimization Problem (9) can be solved using the method of *Bregman projections* [23–25]. The idea is to consider one unsatisfied constraint at a time and project the matrix so that the constraint gets satisfied. Note that the projections are not orthogonal and thus, a previously satisfied constraint might become unsatisfied. However, as stated before, the objective function in Problem (9) is convex and the method is guaranteed to converge to the global minimum if all the constraints are met infinitely often (randomly or following a more structured procedure).

Let us consider the update rule for an unsatisfied constraint from $\mathcal{C}_{\mathrm{neq}}$. The procedure for dealing with constraints from $\mathcal{C}_{\mathrm{eq}}$ is similar. We first consider the case of full-rank symmetric positive semidefinite matrices. Let $\mathbf{K}_t$ be the value of the kernel matrix at step t. For an unsatisfied inequality constraint $\mathbf{C}$, the optimization problem becomes[1]

$$\mathbf{K}_{t+1} = \arg\min_{\mathbf{K}} \ D_{\mathrm{ld}}(\mathbf{K}, \mathbf{K}_t),$$
$$\text{subject to } \langle \mathbf{K}, \mathbf{C} \rangle = \mathrm{tr}(\mathbf{K}\mathbf{C}) \leq 0. \tag{10}$$

Using a Lagrange multiplier $\alpha \geq 0$, we can write

$$\mathbf{K}_{t+1} = \arg\min_{\mathbf{K}} \ D_{\mathrm{ld}}(\mathbf{K}, \mathbf{K}_t) + \alpha \, \mathrm{tr}(\mathbf{K}\mathbf{C}). \tag{11}$$

Following standard derivations for computing gradient updates for Bregman projection [25], the solution of Equation (11) can be written as

$$\mathbf{K}_{t+1} = (\mathbf{K}_t^{-1} + \alpha \mathbf{C})^{-1}. \tag{12}$$

Substituting Equation (12) into (10) gives

$$\mathrm{tr}((\mathbf{K}_t^{-1} + \alpha \mathbf{C})^{-1} \mathbf{C}) = 0. \tag{13}$$

Equation (13) does not have a closed form solution for α, in general. However, we exploit the fact that both types of our constraints, the matrix $\mathbf{C}$ has rank 2, i.e., $\mathrm{rank}(\mathbf{C}) = 2$. Let $\mathbf{K}_t = \mathbf{G}\mathbf{G}^\top$ and $\mathbf{W} = \mathbf{G}^\top \mathbf{C}\mathbf{G}$ and therefore $\mathrm{rank}(\mathbf{W}) = 2$, with eigenvalues $\eta_2 \leq 0 \leq \eta_1$ and $|\eta_2| \leq |\eta_1|$. Solving Equation (13) for α gives

$$\frac{\eta_1}{1 + \alpha\eta_1} + \frac{\eta_2}{1 + \alpha\eta_2} = 0, \tag{14}$$

and

$$\alpha^* = -\frac{1}{2}\frac{\eta_1 + \eta_2}{\eta_1\eta_2} \geq 0. \tag{15}$$

Substituting Equation (15) into (12), gives the following update equation for the kernel matrix

$$\mathbf{K}_{t+1} = (\mathbf{K}_t^{-1} + \alpha^* \mathbf{C})^{-1}. \tag{16}$$

---

[1] We skip the subscript for notational simplicity.

Let $\mathbf{C} = \mathbf{U}\mathbf{V}^\top$ where $\mathbf{U}, \mathbf{V}$ are $n \times 2$ matrices of rank 2. Using Sherman-Morrison-Woodbury formula, we can write (16) as

$$\begin{aligned}
\mathbf{K}_{t+1} &= \mathbf{K}_t - \mathbf{K}_t\, \alpha^* \mathbf{U}\, (\mathbf{I} + \alpha^* \mathbf{V}^\top \mathbf{K}_t\, \mathbf{U})^{-1}\, \mathbf{V}^\top \mathbf{K}_t \\
&= \mathbf{K}_t - \triangle\mathbf{K}_t
\end{aligned} \tag{17}$$

in which, $\triangle\mathbf{K}_t$ is the correction term on the current kernel matrix $\mathbf{K}_t$. Calculation of the update rule (17) is simpler since it only involves inverse of a 2×2 matrix, rather than the $n \times n$ matrix in (16).

For a rank-deficient kernel matrix $\mathbf{K}_0$ with $\mathrm{rank}(\mathbf{K}_0) = r$, we employ the results of Kulis et al. [23], which state that for any column-orthogonal matrix $\mathbf{Q}$ with $\mathrm{range}(\mathbf{K}_0) \subseteq \mathrm{range}(\mathbf{Q})$ (e.g., obtained by singular value decomposition of $\mathbf{K}_0$), we first apply the transformation

$$\mathbf{M} \to \hat{\mathbf{M}} = \mathbf{Q}^\top \mathbf{M}\, \mathbf{Q},$$

on all the matrices, and after finding the kernel matrix $\hat{\mathbf{K}}$ satisfying all the transformed constraints, we can obtain the final kernel matrix using the inverse transformation

$$\mathbf{K} = \mathbf{Q}\hat{\mathbf{K}}\,\mathbf{Q}^\top.$$

Note that since log det preserves the matrix rank, the mapping is one-to-one and invertible.

As the final remark, the kernel matrix learned by minimizing the log det divergence subject to the set of constraints $\mathcal{C}_{\mathrm{neq}} \cup \mathcal{C}_{\mathrm{eq}}$ can be also extended to handle *out of sample* data points, i.e., data points that were not present when learning the kernel matrix. The inner product between a pair of out of sample data points $\mathbf{x}, \mathbf{y} \in \mathbb{R}^d$ in the transformed kernel space can written as

$$k(\mathbf{x}, \mathbf{y}) = k_0(\mathbf{x}, \mathbf{y}) + \mathbf{k_x}^\top (\mathbf{K}_0^\dagger (\mathbf{K} - \mathbf{K}_0)\, \mathbf{K}_0^\dagger)\, \mathbf{k_y} \tag{18}$$

where, $k_0(\mathbf{x}, \mathbf{y})$ and the vectors $\mathbf{k_x} = [k_0(\mathbf{x}, \mathbf{x}_1), \ldots, k_0(\mathbf{x}, \mathbf{x}_n)]^\top$ and $\mathbf{k_y} = [k_0(\mathbf{y}, \mathbf{x}_1), \ldots, k_0(\mathbf{y}, \mathbf{x}_n)]^\top$ are formed using the initial kernel function.

4.2 Semi-supervised Kernel Learning with Relative Comparisons

In this section, we summarize the proposed approach, which we name SKLR, for Semi-supervised Kernel-Learning with Relative comparisons. The pseudo-code of the SKLR method is shown in Algorithm 1. As already discussed, the main ingredients of the method are the following.

Selecting the Bandwidth Parameter. We consider an adaptive approach to select the bandwidth parameter of the Gaussian kernel function. First, we set σ_i equal to the distance between point $\mathbf{x}_i$ and its ℓ-th nearest neighbor. Next, we set the kernel between $\mathbf{x}_i$ and $\mathbf{x}_j$ to

$$k_0(\mathbf{x}_i, \mathbf{x}_j) = \exp\left(-\frac{\|\mathbf{x}_i - \mathbf{x}_j\|^2}{\sigma_{ij}^2}\right), \tag{19}$$

Algorithm 1. (SKLR) Semi-supervised kernel learning with relative comparisons

Input: initial $n \times n$ kernel matrix $\mathbf{K}_0$, set of relative comparisons $\mathcal{C}_{\text{neq}}$ and $\mathcal{C}_{\text{eq}}$, constant distance factor γ
Output: kernel matrix $\mathbf{K}$

- **Find low-rank representation:**
  - Compute the $n \times n$ low-rank kernel matrix $\hat{\mathbf{K}}_0$ such that $\text{rank}(\hat{\mathbf{K}}_0) = r \leq n$ using incomplete Cholesky decomposition such that $\frac{\|\hat{\mathbf{K}}_0\|_F}{\|\mathbf{K}_0\|_F} \geq 0.99$
  - Find $n \times r$ column orthogonal matrix $\mathbf{Q}$ such that $\text{range}(\mathbf{K}_0) \subseteq \text{range}(\mathbf{Q})$
  - Apply the transformation $\hat{\mathbf{M}} \leftarrow \mathbf{Q}^\top \mathbf{M} \mathbf{Q}$ on all matrices
- **Initialize the kernel matrix**
  - Set $\hat{\mathbf{K}} \leftarrow \hat{\mathbf{K}}_0$
- **Repeat**
  - (1) Select an unsatisfied constraint $\hat{\mathbf{C}} \in \mathcal{C}_{\text{neq}} \cup \mathcal{C}_{\text{eq}}$
  - (2) Apply Bregman projection (17)

  Until all the constraints are satisfied
- **Return** $\mathbf{K} \leftarrow \mathbf{Q} \hat{\mathbf{K}} \mathbf{Q}^\top$

---

where, $\sigma_{ij}^2 = \sigma_i \sigma_j$. This process ensures a large bandwidth for sparse regions and a small bandwidth for dense regions.

Semi-Supervised Kernel Learning with Relative Comparisons. After finding the low-rank approximation of the initial kernel matrix $\mathbf{K}_0$ and transforming all the matrices by a proper matrix $\mathbf{Q}$, as discussed in Section 4.1, the algorithm proceeds by randomly considering one unsatisfied constrained at a time and performing the Bregman projections (17) until all the constraints are satisfied.

Clustering Method. After obtaining the kernel matrix $\mathbf{K}$ satisfying the set of all relative and undetermined constraints, we can obtain the final clustering of the points by applying any standard kernelized clustering method. In this paper, we consider the kernel k-means because of its simplicity and good performance. Generalization of the method to other clustering techniques such as kernel mean-shift is straightforward.

5 Experimental Results

In this section, we evaluate the performance of the proposed kernel-learning method, SKLR. As the under-the-hood clustering method required by SKLR, we use the standard kernel k-means with Gaussian kernel and without any supervision (Equation (19) and $\ell = 100$). We compare SKLR to three different semi-supervised metric-learning algorithms, namely, ITML [16], SKkm [1] (a variant of SKMS with kernel k-means in the final stage), and LSML [18]. We select the SKkm variant as Anand et al. [1] have shown that SKkm tends to produce more accurate results than other semi-supervised clustering methods. Two of the baselines, ITML and SKkm, are based on pairwise ML/CL constraints, while LSML uses relative comparisons. For ITML and LSML we apply k-means on the transformed feature vectors to find the final clustering, while for SKkm and SKLR we apply kernel k-means on the transformed kernel matrices.

To assess the quality of the resulting clusterings, we use the Adjusted Rand (AR) index [26]. Each experiment is repeated 20 times and the average over all executions is reported. For the parameter γ required by SKLR we use $\gamma = 2$. Our implementation of SKLR is in MATLAB and the code is publicly available.[2] For the other three methods we use publicly available implementations.[3][4][5]

Finally, we note that in this paper we do not report running-time results, but all tested methods have comparable running times. In particular, the computational overhead of our method can be limited by leveraging the fact that the algorithm has to perform rank-2 matrix updates.

5.1 Datasets

We conduct the experiments on three different real-world datasets.

Vehicle:[6] The dataset contains 846 instances from 4 different classes and is available on the LIBSVM repository.

MIT Scene:[7] The dataset contains 2688 outdoor images, each sized 256 × 256, from 8 different categories: 4 natural and 4 man-made. We use the GIST descriptors [27] as the feature vectors.

USPS Digits:[8] The dataset contains 16 × 16 grayscale images of handwritten digits. It contains 1100 instances from each class. The columns of each images are concatenated to form a 256 dimensional feature vector.

5.2 Relative Constraints vs. Pairwise Constraints

We first demonstrate the performance of the different methods using relative and pairwise constraints. For each dataset, we consider two different experiments: (i) *binary* in which each dataset is clustered into *two groups*, based on some predefined criterion, and (ii) *multi-class* where for each dataset the clustering is performed with number of clusters being equal to *number of classes*. In the binary experiment, we aim to find a crude partitioning of the data, while in the multi-class experiment we seek a clustering at a finer granularity.

---

[2] https://github.com/eamid/sklr
[3] http://www.cs.utexas.edu/~pjain/itml
[4] https://github.com/all-umass/metric_learn
[5] https://www.iiitd.edu.in/~anands/files/code/skms.zip
[6] http://www.csie.ntu.edu.tw/~cjlin/libsvmtools/datasets/
[7] http://people.csail.mit.edu/torralba/code/spatialenvelope/
[8] http://cs.nyu.edu/~roweis/data.html

The 2-class partitionings of our datasets required for the binary experiment are defined as follows: For the **vehicle** dataset, we consider class 4 as one group and the rest of the classes as the second group (an arbitrary choice). For the **MIT Scene** dataset, we perform a partitioning of the data into natural vs. man-made scenes. Finally, for the **USPS Digits**, we divide the data instances into even vs. odd digits.

To generate the pairwise constraints for each dataset, we vary the number of labeled instances from each class (from 5 to 19 with step-size of 2) and form all possible ML constraints. We then consider the same number of CL constraints. Note that for the binary case, we only have two classes for each dataset. To compare with the methods that use relative comparisons, we consider an equal number of relative comparisons and generate them by sampling two random points from the same class and one point (outlier) from one of the other classes. Note that for the relative comparisons, there is no need to restrict the points to the labeled samples, as the comparisons are made in a relative manner.

Finally, in these experiments, we consider a subsample of both **MIT Scene** and **USPS Digits** datasets by randomly selecting 100 data points from each class, yielding 800 and 1000 data points, respectively.

The results for the binary and multi-class experiments are shown in Figures 1(a) and 1(b), respectively. We see that all methods perform equally with no constraints. As constraints or relative comparisons are introduced the accuracy of all methods improves very rapidly. The only surprising behavior is the one of ITML in the multi-class setting, whose accuracy drops as the number of constraints increases. From the figures we see that SKLR outperforms all competing methods by a large margin, for all three datasets and in both settings.

5.3   Multi-resolution Analysis

As discussed earlier, one of the main advantages of kernel learning with relative comparisons is the feasibility of multi-resolution clustering using a single kernel matrix. To validate this claim, we repeat the *binary* and *multi-class* experiments described above. However, this time, we mix the binary and multi-class constraints and use the same set of constraints in both experimental conditions. We evaluate the results by performing binary and multi-class clustering, as before.

Figures 1(c) and 1(d) illustrate the performance of different algorithms using the mixed set of constraints. Again, SKLR produces more accurate clusterings, especially in the multi-class setting. In fact, two of the methods, SKkm and ITML, perform worse than the kernel k-means baseline in the multi-class setting. On the other hand all methods outperform the baseline in the binary setting. The reason is that most of the constraints in the multi-class setting are also relevant to the binary setting, but not the other way around.

Figure 2 shows a visualization of the **USPS Digits** dataset using the SNE method [28] in the original space, and the spaces induced by SKkm and SKLR. We see that SKLR provides an excellent separation of the clusters that correspond to even/odd digits as well as the sub-clusters that correspond to individual digits.

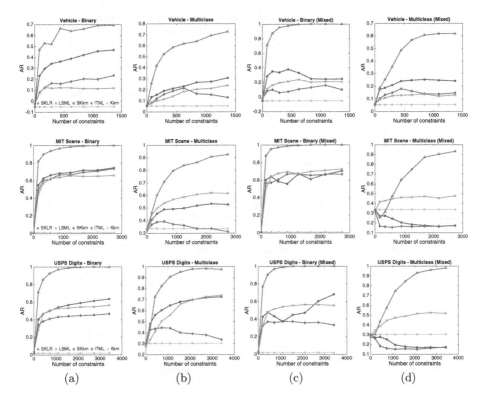

Fig. 1. Clustering accuracy measured with Adjusted Rand index (AR). Rows correspond to different datasets: (1) **Vehicle**; (2) **MIT Scene**; (3) **USPS Digits**. Columns correspond to different experimental settings: (a) binary with separate constraints; (b) multi-class with separate constraints; (c) binary with mixed constraints; (d) multi-class with mixed constraints.

5.4   Generalization Performance

We now evaluate the generalization performance of the different methods to out-of-sample data on the **MIT Scene** and **USPS Digits** datasets (recall that we do not subsample the **Vehicles** dataset). For the baseline kernel k-means algorithm, we run the algorithm on the whole datasets. For ITML and LSML, we apply the learned transformation matrix on the new out-of-sample data points. For SKkm and SKLR, we use Equation (18) to find the transformed kernel matrix of the whole datasets. The results of this experiment are shown in Figure 3. As can be seen from the figure, also in this case, when generalizing to out-of-sample data, SKLR produces significantly more accurate clusterings.

5.5   Effect of Equality Constraints

To evaluate the effect of equality constraints on the clustering, we consider a multi-class clustering scheme. For all datasets, we first generate a fixed number

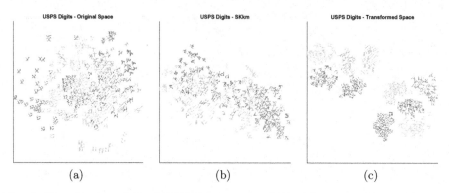

Fig. 2. Visualization of the **USPS Digits** using SNE: (a) original space; (b) space obtained by SK*k*m; (c) space obtained by our method, SKLR.

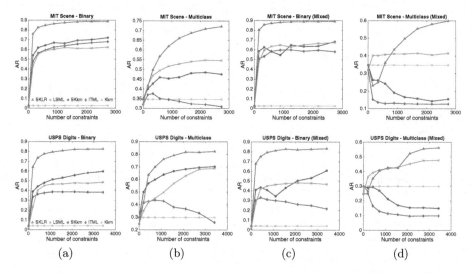

Fig. 3. Clustering accuracy on out-of-sample data (generalization performance). Rows correspond to different datasets: (1) **MIT Scene**; (2) **USPS Digits**. Columns correspond to different experimental settings: (a) binary with separate constraints; (b) multiclass with separate constraints; (c) binary with mixed constraints; (d) multi-class with mixed constraints.

of relative comparisons (360, 720, and 900 relative comparisons for **Vehicle**, **MIT Scene**, and **USPS Digits**, respectively) and then we add some additional equality constraints (up to 200). The equality constraints are generated by randomly selecting three data points, all from the same class, or each from a different class. The results are shown in Figure 4. As can be seen, considering the equality constraint also improves the performance, especially on the **MIT Scene** and **USPS Digits** datasets. Note that none of the other methods can handle these type of constraints.

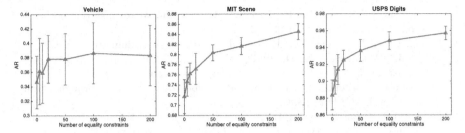

Fig. 4. Effect of equality constraints on the three datasets.

6 Conclusion

We have devised a semi-supervised kernel-learning algorithm that can incorporate various types of relative distance constraints, and used the resulting kernels for clustering. Our experiments show that our method outperforms by a large margin other competing methods, which either use ML/CL constraints or use relative constraints but different metric-learning approaches. Our method is compatible with existing kernel-learning techniques [1] in the sense that if ML and CL constraints are available, they can be used together with relative comparisons. We have also proposed to interpret an "unsolved" distance comparison so that the interpoint distances are roughly equal. Our experiments suggest that incorporating such equality constraints to the kernel learning task can be advantageous, especially in settings where it is costly to collect constraints.

For future work we would like to extend our method to incorporate more robust clustering methods such as spectral clustering and mean-shift. Additionally, the soft formulation of the relative constraints for handling possibly inconsistent constraints is straightforward, however, we leave it for future study.

References

1. Anand, S., Mittal, S., Tuzel, O., Meer, P.: Semi-supervised kernel mean shift clustering. PAMI **36**, 1201–1215 (2014)
2. Wagstaff, K., Cardie, C.: Clustering with instance-level constraints. In: ICML (2000)
3. Wagstaff, K., Cardie, C., Rogers, S., Schroedl, S.: Constrained k-means clustering with background knowledge. In: ICML (2001)
4. Klein, D., Kamvar, S.D., Manning, C.D.: From instance-level constraints to space-level constraints. In: ICML (2002)
5. Basu, S., Bilenko, M., Mooney, R.J.: A probabilistic framework for semi-supervised clustering. In: KDD (2004)
6. Basu, S., Banerjee, A., Mooney, R.J.: Active semi-supervision for pairwise constrained clustering. In: SDM (2004)
7. Lu, Z., Leen, T.K.: Semi-supervised learning with penalized probabilistic clustering. NIPS (2005)
8. Lu, Z.: Semi-supervised clustering with pairwise constraints: A discriminative approach. In: AISTATS (2007)

9. Pei, Y., Fern, X.Z., Rosales, R., Tjahja, T.V.: Discriminative clustering with relative constraints. arXiv:1501.00037 (2014)
10. Lu, Z., Ip, H.H.S.: Constrained Spectral Clustering via Exhaustive and Efficient Constraint Propagation. In: Daniilidis, K., Maragos, P., Paragios, N. (eds.) ECCV 2010, Part VI. LNCS, vol. 6316, pp. 1–14. Springer, Heidelberg (2010)
11. Lu, Z., Carreira-Perpiñán, M.: Constrained spectral clustering through affinity propagation. In: CVPR (2008)
12. Dhillon, I.S., Guan, Y., Kulis, B.: A unified view of kernel k-means, spectral clustering and graph cuts. Technical Report TR-04-25, University of Texas (2005)
13. Kulis, B., Basu, S., Dhillon, I., Mooney, R.: Semi-supervised graph clustering: a kernel approach. Machine Learning **74**, 1–22 (2009)
14. Xing, E.P., Ng, A.Y., Jordan, M.I., Russell, S.J.: Distance metric learning with application to clustering with side-information. In: NIPS (2002)
15. Schultz, M., Joachims, T.: Learning a distance metric from relative comparisons. In: NIPS (2003)
16. Davis, J.V., Kulis, B., Jain, P., Sra, S., Dhillon, I.S.: Information-theoretic metric learning. In: ICML (2007)
17. Liu, W., Ma, S., Tao, D., Liu, J., Liu, P.: Semi-supervised sparse metric learning using alternating linearization optimization. In: KDD (2010)
18. Liu, E.Y., Guo, Z., Zhang, X., Jojic, V., Wang, W.: Metric learning from relative comparisons by minimizing squared residual. In: ICDM (2012)
19. Bilenko, M., Basu, S., Mooney, R.J.: Integrating constraints and metric learning in semi-supervised clustering. In: ICML (2004)
20. Xiang, S., Nie, F., Zhang, C.: Learning a mahalanobis distance metric for data clustering and classification. Pattern Recognition **41**, 3600–3612 (2008)
21. Kumar, N., Kummamuru, K.: Semisupervised clustering with metric learning using relative comparisons. TKDE **20**, 496–503 (2008)
22. Comaniciu, D., Meer, P.: Mean shift: a robust approach toward feature space analysis. PAMI **24**, 603–619 (2002)
23. Kulis, B., Sustik, M.A., Dhillon, I.S.: Low-rank kernel learning with Bregman matrix divergences. JMLR **10**, 341–376 (2009)
24. Bregman, L.: The relaxation method of finding the common point of convex sets and its application to the solution of problems in convex programming. USSR Computational Mathematics and Mathematical Physics **7**, 200–217 (1967)
25. Tsuda, K., Rätsch, G., Warmuth, M.: Matrix exponentiated gradient updates for on-line learning and Bregman projection. JMLR **6**, 995–1018 (2005)
26. Hubert, L., Arabie, P.: Comparing partitions. Journal of Classification **2**, 193–218 (1985)
27. Oliva, A., Torralba, A.: Modeling the shape of the scene: A holistic representation of the spatial envelope. IJCV **42**, 145–175 (2001)
28. Hinton, G., Roweis, S.: Stochastic neighbor embedding. In: NIPS (2003)

Bayesian Active Clustering with Pairwise Constraints

Yuanli Pei[✉], Li-Ping Liu, and Xiaoli Z. Fern

School of EECS, Oregon State University, Corvallis, Oregon 97331, USA
{peiy,liuli,xfern}@eecs.oregonstate.edu

Abstract. Clustering can be improved with *pairwise constraints* that specify similarities between pairs of instances. However, randomly selecting constraints could lead to the waste of labeling effort, or even degrade the clustering performance. Consequently, how to *actively* select effective pairwise constraints to improve clustering becomes an important problem, which is the focus of this paper. In this work, we introduce a Bayesian clustering model that learns from pairwise constraints. With this model, we present an active learning framework that iteratively selects the most informative pair of instances to query an oracle, and updates the model posterior based on the obtained pairwise constraints. We introduce two information-theoretic criteria for selecting informative pairs. One selects the pair with the most uncertainty, and the other chooses the pair that maximizes the marginal information gain about the clustering. Experiments on benchmark datasets demonstrate the effectiveness of the proposed method over state-of-the-art.

1 Introduction

Constraint-based clustering aims to improve clustering using user-provided *pairwise constraints* regarding similarities between pairs of instances. In particular, a must-link constraint states that a pair of instances belong to the same cluster, and a cannot-link constraint implies that two instances are in different clusters. Existing work has shown that such constraints can be effective at improving clustering in many cases [2,4,8,16,19,20,22,24,28]. However, most prior work focus on "passive" learning from constraints, i.e., instance pairs are randomly selected to be labeled by a user. Constraints acquired in this random manner may be redundant and lead to the waste of labeling effort, which is typically limited in real applications. Moreover, when the constraints are not properly selected, they may even be harmful to the clustering performance as has been revealed by Davidson et al. [7]. In this paper, we study the important problem of *actively* selecting effective pairwise constraints for clustering.

Existing work on active learning of pairwise constraints for clustering has mostly focused on neighbourhood-based methods [3,12,14,17,25]. Such methods maintain a neighbourhood structure of the data based on the existing constraints, which represents a partial clustering solution, and they query pairwise constraints to expand such neighborhoods. Other methods that do not rely on

© Springer International Publishing Switzerland 2015
A. Appice et al. (Eds.): ECML PKDD 2015, Part I, LNAI 9284, pp. 235–250, 2015.
DOI: 10.1007/978-3-319-23528-8_15

such structure consider various criteria for measuring the utility of instance pairs. For example, Xu et al. [26] propose to select constraints by examining the spectral eigenvectors of the similarity matrix, and identify data points that are at or close to cluster boundaries. Vu et al. [21] introduce a method that chooses instance pairs involving points on the sparse regions of a k-nearest neighbours graph. As mentioned by Xiong et al. [25], many existing methods often select a batch of pairwise constraints before performing clustering, and they are not designed for iteratively improving clustering by querying new pairs.

In this work, we study Bayesian active clustering with pairwise constraints in an iterative fashion. In particular, we introduce a Bayesian clustering model to find the clustering posterior given a set of pairwise constraints. At every iteration, our task is: a) to select the most informative pair toward improving current clustering, and b) to update the clustering posterior after the query is answered by an oracle/a user. Our goal is to achieve the best possible clustering performance with minimum number of queries.

In our Bayesian clustering model, we use a discriminative logistic model to capture the conditional probability of the cluster assignments given the instances. The likelihood of observed pairwise constraints is computed by marginalizing over all possible cluster assignments using message passing. We adopt a special data-dependent prior that encourages large cluster separations. At every iteration, the clustering posterior is represented by a set of samples ("particles"). After obtaining a new constraint, the posterior is effectively updated with a sequential Markov Chain Monte Carlo (MCMC) method ("particle filter").

We present two information-theoretic criteria for selecting instance pairs to query at each iteration: a) *Uncertain*, which chooses the most uncertain pair based on current posterior, and b) *Info*, which selects the pair that maximizes the information gain regarding current clustering. With the clustering posterior maintained at every iteration, both objectives can be efficiently calculated.

We evaluate our method on benchmark datasets, and the results demonstrate that our Bayesian clustering model is very effective at learning from a small number of pairwise constraints, and our active clustering model outperforms state-of-the-art active clustering methods.

2   Problem Statement

The goal of clustering is to find the underlying cluster structure in a dataset $X = [x_1, \cdots, x_N]$ with $x_i \in \mathbb{R}^d$ where d is the feature dimension. The unknown cluster label vector $Y = [y_1, \cdots, y_N]$, with $y_i \in \{1, \cdots, K\}$ being the cluster label for x_i, denotes the ideal clustering of the dataset, where K is the number of clusters. In the studied *active* clustering, we could acquire some weak supervision, i.e., pairwise constraints, by requesting an oracle to specify whether two instances $(x_a, x_b) \in X \times X$ belong to the same cluster. We represent the response of the oracle as a pair label $z_{a,b} \in \{+1, -1\}$, with $z_{a,b} = +1$ representing that instance x_a and x_b are in the same cluster (a must-link constraint), and $z_{a,b} = -1$ meaning that they are in different clusters (a cannot-link constraint). We assume the cost is uniform for different queries, and the goal of active clustering is to achieve the best possible clustering with the least number of queries.

In this work, we consider sequential active clustering. In each iteration, we select one instance pair to query the oracle. After getting the answer of the query, we update the clustering model to integrate the supervision. With the updated model, we then choose the best possible pair for the next query. So the task of active clustering is an iterative process of posing queries and incorporating new information to clustering.

An active clustering model generally has two key components: the *clustering* component and the *pair selection* component. In every iteration, the task of the clustering component is to identify the cluster structure of the data given the existing constraints. The task of the pair selection component is to score each candidate pair and choose the most informative pair to improve the clustering.

3   Bayesian Active Clustering

3.1   The Bayesian Clustering Model

In our model, we assume that the instance cluster labels y_i's are independent given instance x_i and the model parameter W. Each pair label $z_{a,b}$ only depends on the cluster labels y_a and y_b of the involved instances (x_a, x_b). The proposed Bayesian clustering model consists of three elements: 1) the instance cluster assignment model defined by $P(Y|W, X)$, with parameter W; 2) the conditional distribution of the pair labels given the cluster labels $P(Z|Y)$, where Z contains all pair labels in the constraints; and 3) the data-dependent prior $P(W|X, \theta)$ with parameter θ. The joint distribution of the clustering model is factorized as

$$P(Z, Y, W|X, \theta) = P(Z|Y)P(Y|W, X)P(W|X, \theta) \ . \tag{1}$$

We use the following discriminative logistic model as the clustering assignment model $P(Y|W, X)$:

$$P(y_i = k|W, x_i) = \frac{\exp(W_{\cdot,k}^\top x_i)}{\sum_{k'=1}^{K} \exp(W_{\cdot,k'}^\top x_i)}, \quad \forall 1 \le k \le K, \quad 1 \le i \le N \ , \tag{2}$$

where W is a $d \times K$ matrix, d is the feature dimension, and K is the number of clusters.

Here we use a special prior for W, which combines the Gaussian prior with a data-dependent term that encourages large cluster separations of the data. The logarithmic form of the prior distribution is

$$\log\ P(W|X, \theta) = -\frac{\lambda}{2}\|W\|_F^2 - \frac{\tau}{N}\sum_{i=1}^{N} H(y_i|W, x_i) + constant \ , \tag{3}$$

where the prior parameter $\theta = [\lambda, \tau]$. The first term is the weighted Frobenius norm of W. This term corresponds to the Gaussian prior with zero mean and diagonal covariance matrix with λ as the diagonal elements, and it controls the model complexity. The second term is the average negative entropy of the cluster assignment variable Y. We use this term to encourage large separations among clusters, as similarly utilized by [11] for semi-supervised classification problems. The constant term normalizes the probability. Although it is unknown, inference can be carried out by sampling from the unnormalized distribution (e.g., using slice sampling [18]). We will discuss more details in Sec. 3.3.

With our model assumption, the conditional probability $P(Z|Y)$ is fully factorized based on the pairwise constraints. For a single pair (x_a, x_b), we define the probability of $z_{a,b}$ given cluster labels y_a and y_b as

$$P(z_{a,b} = +1|y_a, y_b) = \begin{cases} \epsilon & \text{if } y_a \neq y_b \\ 1 - \epsilon & \text{if } y_a = y_b \end{cases} ,$$

$$P(z_{a,b} = -1|y_a, y_b) = 1 - P(z_{a,b} = +1|y_a, y_b) ,$$

(4)

where ϵ is a small number to accommodate the (possible) labeling error. In the case where no labeling error exists, ϵ allows for "soft constraints", meaning that the model can make small errors on some pair labels and achieve large cluster separations.

Marginalization of Cluster Labels. In the learning procedure described later, we will need to marginalize some or all cluster labels, for example, in the case of computing the likelihood of the observed pair labels:

$$P(Z|W, X) = \sum_Y P(Z, Y|W, X) = \sum_{Y_{\alpha(Z)}} P(Z|Y_{\alpha(Z)})P(Y_{\alpha(Z)}|W, X_{\alpha(Z)}) , \quad (5)$$

where $\alpha(Z)$ denotes the set of indices for all instances involved in Z.

The marginalization can be solved by performing sum-product message passing [15] on a factor graph defined by all the constraints. Specifically, the set of all instances indexed by $\alpha(Z)$ defines the nodes of the graph, and $P(Y_{\alpha(Z)}|W, X_{\alpha(Z)})$ defines the node potentials. Each queried pair (x_a, x_b) creates an edge, and the edge potential is defined by $P(z_{a,b}|y_a, y_b)$. In this work, we require that the graph formed by the constraints does not contain cycles, and message passing is performed on a tree (or a forest, which is a collection of trees). Since inference on trees are exact, the marginalization is computed exactly. Moreover, due to the simple form of the edge potential (which is a simple modification to the identity matrix as can be seen from (4)), the message passing can be performed very efficiently. In fact, each message propagation only requires $O(K)$ complexity instead of $O(K^2)$ as in the general case. Overall the message passing only takes $O(K|Z|)$, even faster than calculating the node potentials $P(Y_{\alpha(Z)}|W, X_{\alpha(Z)})$, which takes $O(dK|Z|)$.

3.2 Active Query Selection

Now we describe our approach for actively selecting informative pairs at every iteration. Suppose our query budget is T. In each iteration $t, 1 \leq t \leq T$, we need to select a pair (x_a^t, x_b^t) from a pool of unlabeled pairs U^t, and acquire the label $z_{a,b}^t$ from the oracle. We let $U^1 \subseteq X \times X$ be the initial pool of unlabeled pairs. Then $U^t = U^{t-1} \backslash (x_a^{t-1}, x_b^{t-1})$ for $1 \leq t \leq T$. Below we use $Z_t = [z_{a,b}^1, \cdots, z_{a,b}^t]$ to denote all the pair labels obtained up to the t-th iteration.

Selection Criteria. We use two entropy-based criteria to select the best pair at each iteration. The first criterion, which we call *Uncertain*, is to select the pair whose label is the most uncertain. That is, at the t iteration, we choose the pair (x_a^t, x_b^t) that has the largest *marginal* entropy of $z_{a,b}^t$ (over the posterior distribution of W):

$$(x_a^t, x_b^t) = \arg \max_{(x_a, x_b) \in U^t} H(z_{a,b} | Z_{t-1}, X, \theta) . \qquad (6)$$

Similar objective has been considered in prior work on distance metric learning [27] or document clustering [14], where the authors propose different approaches to compute/approximate the entropy objective.

The second criterion is a greedy objective adopted from active learning for classification [6,10,13], which we call *Info*. The idea is to select the query (x_a^t, x_b^t) that maximizes the marginal information gain about the model W:

$$
\begin{aligned}
(x_a^t, x_b^t) &= \arg \max_{(x_a, x_b) \in U^t} I(z_{a,b}, W | Z_{t-1}, X, \theta) \\
&= \arg \max_{(x_a, x_b) \in U^t} H(z_{a,b} | Z_{t-1}, X, \theta) - H(z_{a,b} | W, Z_{t-1}, X, \theta) . \quad (7)
\end{aligned}
$$

Note that here W is a random variable. The *Info* objective is equivalent to maximizing the entropy reduction about W, as can be proved by the chain rule of conditional entropy.

Interestingly, the first entropy term in the *Info* objective (7) is the same with the *Uncertain* objective (6). The additional term to *Info* is the conditional entropy of the pair label $z_{a,b}$ given W, i.e., the second term in (7). Comparing the two objectives, we see that W is marginalized in the *Uncertain* objective and the selected query aims to reduce the maximum uncertainty of the *pair label*. In contrast, the goal of *Info* is to decrease the *model* uncertainty. There is subtle difference between these two types of uncertainties. The additional conditional entropy term in *Info* suggests that it prefers instance pairs whose labels are certain once W is known, yet whose overall uncertainty is high when marginalizing over W. In such sense, *Info* pays more attention to the uncertainty of the model W.

Each of the above selection objectives ranks the candidate pairs from the highest to the lowest. To select a pair to query, we go through the ranking and choose the one that does not create a cycle to the existing graph as described in Sec. 3.1. Since inference on trees are not only exact but also fast, enforcing such acyclic graph structure allows us to compute the selection objectives more effectively and accurately, and select more informative pairs to query.

Computing the Selection Objectives. Now we describe how to compute the two objective values for a candidate instance pair. The two objectives require computing the marginal entropy $H(z_{a,b}|Z_t, X, \theta)$, and the conditional entropy $H(z_{a,b}|W, Z_t, X, \theta)$, for $1 \leq t \leq T$. By definition, the marginal entropy is

$$H(z_{a,b}|Z_t, X, \theta) = -\sum_{z_{a,b}} P(z_{a,b}|Z_t, X, \theta) \log P(z_{a,b}|Z_t, X, \theta) , \qquad (8)$$

where the probability

$$P(z_{a,b}|Z_t, X, \theta) = \int P(\hat{W}|Z_t, X, \theta) P(z_{a,b}|Z_t, \hat{W}, X) d\hat{W} . \qquad (9)$$

The conditional probability is computed as

$$P(z_{a,b}|Z_t, \hat{W}, X) = \frac{P(z_{a,b} \cup Z_t|\hat{W}, X)}{P(Z_t|\hat{W}, X)} , \qquad (10)$$

where calculating both the numerator and the denominator are the same inference problem as (5) and can be solved similarly using message passing. In fact, message propagations for the two calculations are shared except for that a new edge regarding $z_{a,b}$ is introduced to the graph for $P(z_{a,b} \cup Z_t|\hat{W}, X)$. So we can calculate the two values by performing message passing algorithm only once on the graph of $P(z_{a,b} \cup Z_t|\hat{W}, X)$, and record $P(Z_t|\hat{W}, X)$ in the intermediate step.

By definition, the conditional entropy is

$$H(z_{a,b}|W, Z_t, X, \theta) = \int P(\hat{W}|Z_t, X, \theta) H(z_{a,b}|Z_t, \hat{W}, X) d\hat{W} , \qquad (11)$$

where $H(z_{a,b}|\hat{W}, Z_t, X)$ is also easy to compute once we know $P(z_{a,b}|Z_t, \hat{W}, X)$, which has been done in (10).

Now the only obstacle in calculating the two entropies is to take the expectations over the posterior distribution $P(W|Z_t, X, \theta)$ in (9) and (11). Here we use sampling to approximate such expectations. We first sample W's from $P(W|Z_t, X, \theta)$ and then approximate the expectations with the sample means. Directly sampling from the posterior at every iteration is doable but very inefficient. Below we describe a sequential MCMC sampling method ("particle filter") that effectively updates the samples of the posterior.

3.3   The Sequential MCMC Sampling of W

The main idea of the sequential MCMC method is to avoid sampling with random starts at every iteration by utilizing the particles obtained from the previous iteration.[1] Specifically, to obtain particles from distribution $P(W|Z_t, X, \theta)$, the sequential MCMC method first resamples from the particles previously sampled

---

[1] Here we follow the convention of the particle filter field and call samples of W as "particles".

from $P(W|Z_{t-1}, X, \theta)$, and then performs just a few MCMC steps with these particles to prevent degeneration [9].

Here we maintain S particles in each iteration. We denote W_s^t, $1 \le s \le S$, as the s-th particle in the t-th iteration. For initialization, we sample particles $\{W_1^0, \cdots, W_S^0\}$ from the prior distribution $P(W|X, \theta)$ defined in (3) using slice sampling [18] [2], an MCMC method that can uniformly draw samples from an unnormalized density function. Since slice sampling does not require the target distribution to be normalized, the unknown constant in the prior (3) can be neglected here.

At iteration $t, 1 \le t \le T$, after a new pair label $z_{a,b}^t$ is observed, we perform the following two steps to update the particles and get samples from $P(W|Z_t, X, \theta)$.

(1) Resample. The first step is to resample from the particles $\{W_1^{t-1}, \cdots, W_S^{t-1}\}$ obtained from the previous iteration for $P(W|Z_{t-1}, X, \theta)$. We observe that

$$P(W|Z_t, X, \theta) = P(W|z_{a,b}^t, Z_{t-1}, X, \theta)$$
$$\propto P(z_{a,b}^t|Z_{t-1}, W, X)P(W|Z_{t-1}, X, \theta) .$$

So each particle W_s^{t-1} is weighted by $P(z_{a,b}^t|Z_{t-1}, W_s^{t-1}, X)$, which can be calculated the same as (10).

(2) Move. In the second step, we start with each resampled particles, and perform several slice sampling steps for the posterior

$$P(W|Z_t, X, \theta) \propto P(Z_t|W, X)P(W|X, \theta) . \qquad (12)$$

Again $P(Z_t|W, X)$ is calculated by message passing as (5), and the unknown normalizing constant in $P(W|X, \theta)$ can be ignored, since slice sampling does not require the normalization constant.

The *resample-move* method avoids degeneration in the sequence of slice sampling steps. After these two steps, we have updated the particles for $P(W|Z_t, X, \theta)$. Such particles are used to approximate the selection objectives as described in Sec. 3.2, allowing us to select the next informative pair to query.

Note that the distribution $P(W|Z_t, X, \theta)$ is invariant to *label switching*, that is, permuting column vectors of $W = [W_{\cdot,1}, \cdots, W_{\cdot,K}]$ will not change the probability $P(W|Z_t, X, \theta)$. This is because we can not provide any prior of W with label order, nor does the obtained constraints provide any information about the label order. One concern is whether the label switching problem would reduce sampling efficiency and affect the pair selection, since $P(W|Z_t, X, \theta)$ has multiple modes corresponding to different label permutations. Actually it does not cause an issue to the approximation of integrations in (9) and (11), since the term $P(z_{a,b}|Z_t, W, X, \theta)$ is also invariant to label permutations. However, the label switching problem does cause difficulty in getting the Bayesian prediction of clusters labels from distribution $P(Y|Z_t, X, \theta)$, so we will employ the MAP solution W_{map} and predict cluster labels with $P(Y|Z_t, W_{map}, X, \theta)$. We describe this in the following section.

---

[2] Here we use the implementation `slicesample` provided in the MATLAB toolbox.

3.4   Find the MAP Solution

Given a set of constraints with pair labels Z, we first find the MAP estimation W_{map} by maximizing the posterior $P(W|Z, X, \theta)$, or equivalently maximizing the joint distribution $P(W, Z|X, \theta)$ (in the logarithmic form):

$$\max_{W} \quad L = \log P(W, Z|X, \theta) = \log P(Z|W, X) + \log P(W|X, \theta) . \quad (13)$$

The maximization can be solved by off-the-shelf gradient-based optimization approaches. Here we use the quasi-newton method provided in the MATLAB toolbox. The gradient of the objective L with respect to W is

$$\frac{\partial L}{\partial W} = \sum_{i \in \alpha(Z)} x_i(q_i - p_i)^\top - \lambda W - \frac{\tau}{N} \sum_{i=1}^{N} x_i \sum_{k=1}^{K} p_{ik} \log p_{ik} (\mathbf{1}_k - p_i)^\top ,$$

where $p_i = [p_{i1}, \cdots, p_{iK}]^\top$ with $p_{ik} = P(y_i = k|W, x_i)$, $q_i = [q_{i1}, \cdots, q_{iK}]^\top$ with $q_{ik} = P(y_i = k|Z, W, x_i)$, and $\mathbf{1}_k$ is a K dimensional vector that contains 1 on the k-th dimension and 0 elsewhere. Here $\alpha(Z)$ again indexes all the instances involved in the constraints.

With the W_{map} solution to (13), we then find the MAP solution of the cluster labels Y from $P(Y|Z, W_{map}, X)$. This is done in two cases. For the instances that are *not* involved in the constraints, the MAP of Y is simply the most possible assignment of $P(Y|W_{map}, X)$. For the instances involved in the constraints, we need to find

$$\max_{Y_{\alpha(Z)}} \quad P(Y_{\alpha(Z)}|Z, W_{map}, X_{\alpha(Z)}) \propto P(Z|Y_{\alpha(Z)}) P(Y_{\alpha(Z)}|W_{map}, X_{\alpha(Z)}) .$$

The inference can be done by performing max-product algorithm on the same graph as defined for (5), only replacing the "summation" with the "max" operator at every message propagation.

In real applications, we only need to find the MAP solution of Y after the last iteration. In our experiments, we search for the solution at every iteration to show the performance of our method if we stop learning at any iteration. Our overall algorithm is summarized in Algorithm 1.

Note that an alternative of finding the clustering solution is to find the MAP of W and Y at the same time. However, we think our MAP estimation of W which marginalizes Y is more stable, and our calculation method is much simpler compared with the alternative.

4   Experiments

In this section, we empirically examine the effectiveness of the proposed method. In particular, we aim to answer the following questions:

- Is the proposed Bayesian clustering model effective at finding good clustering solutions with a small number of pairwise constraints?
- Is the proposed *active* clustering method more effective than state-of-the-art active clustering approaches?

Algorithm 1. Bayesian Active Clustering

Input: data X, number of clusters K, access to the oracle, initial pool U^1, query budget T, prior parameter θ, number of samples S
Output: a clustering solution of the data

Initialize particles by sampling $\{W_1^0, \cdots, W_S^0\}$ from prior $P(W|X, \theta)$
for $t = 1$ **to** T **do**
   1. Select a pair to query:
       Use particles $\{W_1^{t-1}, \cdots, W_S^{t-1}\}$ to compute the selection objective (6) or (7)
       Choose the best pair (x_a^t, x_b^t) from U^t and acquire $z_{a,b}^t$ from the oracle
   2. Update posterior:
       Resample S particles with weight $P(z_{a,b}^t|Z_{t-1}, W_s^{t-1}, X)$ for W_s^{t-1}
       Perform a few MCMC steps on all particles with distribution $P(W|Z_t, X, \theta)$
   3. Update the pool: $U^{t+1} \leftarrow U^t \backslash (x_a^t, x_b^t)$
end for
Find the MAP solution $W_{map} = \arg\max_W \ \log P(W|Z_T, X, \theta)$
Find the clustering solution $Y_{map} = \arg\max_Y \ \log P(Y|Z_T, W_{map}, X)$

Table 1. Summary of Dataset Information

Dataset	#Inst	#Dim	#Class	#Query
Fertility	100	9	2	60
Parkinsons	195	22	2	60
Crabs	200	5	2	60
Sonar	208	60	2	100
Balance	625	4	3	100
Transfusion	748	4	2	100
Letters-IJ	1502	16	2	100
Digits-389	3165	16	3	100

4.1 Dataset and Setup

We use 8 benchmark UCI datasets to evaluate our method. Table 1 provides a summary of the dataset information. For each dataset, we normalize all features to have zero mean and unit standard deviation.

    We form the pool of unlabeled pairs using all instances in the dataset, and set the query budget to 60 for smaller datasets and to 100 for datasets with large feature dimension (e.g, *Sonar*) or larger dataset size. When a pair of instances is queried, the label is returned based on the ground-truth instance class/cluster labels. We evaluate the clustering results of all methods using pairwise F-Measure [5], which evaluates the harmonic mean of the precision and recall regarding prediction of instance pairwise relations. We repeat all experiments 30 times and average the results.

    For the proposed Bayesian clustering model, we found that its performance is not sensitive to the values of the prior parameter τ or the ϵ used in the pair label distribution (4). Here we set $\tau = 1$ and $\epsilon = 0.05$, where the nonzero value of ϵ

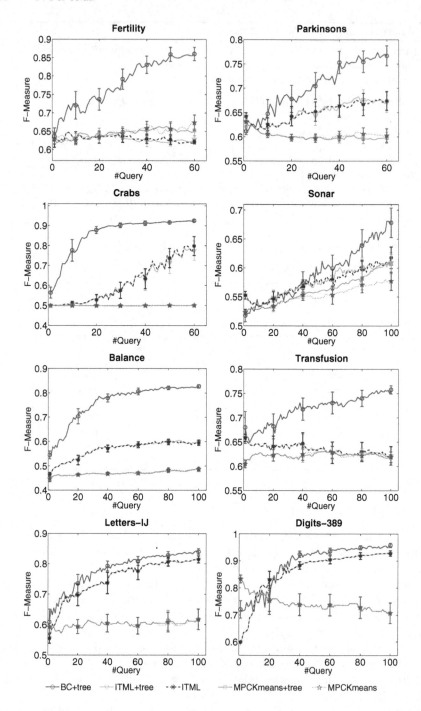

Fig. 1. Pairwise F-Measure clustering results with increasing number of *randomly* selected queries. Results are averaged over 30 runs. Error bars are shown as mean and 95% confidence interval.

allows for "soft constraints". For the parameter λ, which controls the covariance of the Gaussian prior, we experimented with $\lambda \in \{1, 10, 100\}$ and found that $\lambda = 10$ is uniformly good with all datasets, which we fix as the default value. For each dataset, we maintain $S = 2dK$ samples of the posterior at every iteration.

4.2   Effectiveness of the Proposed Clustering Model

To demonstrate the effectiveness of the proposed Bayesian clustering (BC) model, we compare with two well-known methods that learn from pairwise constraints: MPCKmeans [5], and ITML [8][3]. In this set of experiment, we use randomly selected pairwise constraints to evaluate all methods. For our method, we incrementally select random pairs that do not introduce a cycle to the graph formed by existing pairs. To ensure a fair comparison, we evaluate ITML and MPCKmeans with randomly selected pairs with and without the acyclic graph restriction. Thus, all methods in competition are: *BC+tree, ITML, ITML+tree, MPCKmeans, MPCKmeans+tree*, where *BC+tree, ITML+tree*, and *MPCKmeans+tree* use randomly selected constraints that form a tree graph (or a forest), and *ITML* and *MPCKmeans* allow for cycles in the graph.

Figure 1 shows the performance of all methods with increasing number of constraints. We see that our method *BC+tree* outperforms the baselines on most datasets regardless of whether they use constraints with or without the acyclic graph restriction. This demonstrates the effectiveness of our Bayesian clustering model. We also notice that on most datasets we can hardly tell the difference between *ITML* and *ITML+tree*, or *MPCKmeans* and *MPCKmeans+tree*, suggesting that enforcing the acyclic structure in the constraints do not hurt the performance of ITML or MPCKmeans. Interestingly, such enforcement can in some cases produce better performance (e.g, on the *Sonar* dataset). We suspect this is because constraints forming cycles may have larger *incoherence* than those does not.[4] Davidson et al. [7] have shown that constraint sets with large incoherence can potentially degrade the clustering performance.

4.3   Effectiveness of the Overall Active Clustering Model

In this section, we compare our overall active clustering model with existing methods. Our baselines include two recent work on active learning with pairwise constraints: *MinMax* [17], and *NPU* [25]. Both methods provide an active pair selection approach and require a clustering method to learn form the constraints. Here we supply them with MPCKmeans and ITML.[5] So all methods in competition are

---

[3] ITML is a distance metric learning method, and we find the clustering solution by applying Kmeans clustering with the learned metric.

[4] The concept of *incoherence* is formally defined at [7]. Generally, a set of overlapping constraints tends to have higher incoherence than a set of disjoint constraints.

[5] Note that due to our Bayesian clustering model requires the set of constraints to form an acyclic graph, it can not be combined with *MinMax* or *NPU*, as they generally select constraints that form cycles due to their neighbourhood-based approach.

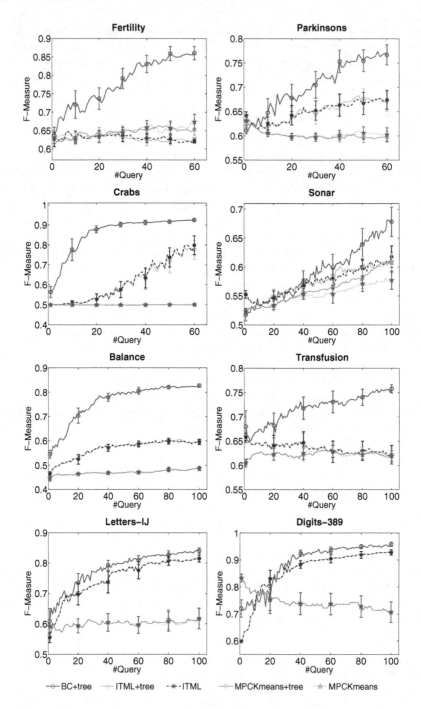

Fig. 2. Pairwise F-Measure clustering results of different *active clustering* methods with increasing number of queries. Results are averaged over 30 runs. Error bars are shown as mean and 95% confidence interval.

- *Info+BC*: The proposed active clustering model with the *Info* criterion (7).
- *Uncertain+BC*: The proposed active clustering model with the *Uncertain* criterion (6).
- *NPU+ITML*: The *NPU* active selection strategy combined with *ITML*.
- *NPU+MPCKmeans*: The *NPU* method with *MPCKmeans*.
- *MinMax+ITML*: The *MinMax* active learning method combined with *ITML*.
- *MinMax+MPCKmeans*: The *MinMax* approach combined with *MPCKmeans*.

Figure 2 reports the performance of all active clustering methods with increasing number of queries. We see that both *Info+BC* and *Uncertain+BC* improve the clustering very quickly as more constraints are obtained, and they outperform all baselines on most datasets. Moreover, *Info+BC* seems to be more effective than *Uncertain+BC* in most cases. We hypothesize this is because *Info* reduces the uncertainty of the model, which might be more appropriate for improving the MAP solution of clustering than decreasing the maximum uncertainty of the pair labels as *Uncertain* does.

To avoid crowding Fig. 2, we did not present the passive learning results of our method *BC+tree* as a baseline in the same figure. The comparison between active learning and passive learning for our method can be done by comparing *Uncertain+BC* and *Info+BC* in Fig. 2 with *BC+tree* in Fig. 1. We see that both our active learning approaches produce better performance than passive learning on most datasets, demonstrating the effectiveness of our pair selection strategies.

We also notice that the performance of *NPU* or *MinMax* highly depends on the clustering method in use. With different clustering methods, their behaviors are very different. In practice, it can be difficult to decide which clustering algorithm should be used in combination with the active selection strategies to ensure good clustering performance. In contrast, our method unifies the clustering and active pair selection model, and the constraints are selected to explicitly reduce the clustering uncertainty and improve the clustering performance.

4.4 Analysis of the Acyclic Graph Restriction

Our method requires the graph formed by the constraints to be a tree (or a forest). Here we show that this restriction will not prevent us from selecting informative pairs. We examine the number of pairs that has been dropped at every iteration in order to find the best pair that does not create a cycle. Table 2 reports the results for the two selection criteria with varied number of queries. We see that for both criteria the number of dropped pairs is very small. For *Uncertain*, there is barely any pair that has been dropped on most datasets, and we see slightly more pairs dropped for the *Info* criteria. Overall, for only less than (often significantly less than) 10% of the number of queries, we encounter the need of dropping a pair. The only exception is the *Fertility* dataset, which is very small in size, making it difficult to avoid cycles with a large number of queries. But from the results in Sec. 4.3, we can see that the active clustering performance was still much better than the competing methods.

Table 2. Number of dropped pairs (*Info/Uncertain*) at different iterations to find the best pair that does not a create cycle. Results are averaged over 30 runs.

Dataset	Query Iteration					
	10	20	30	40	50	60
Fertility	0.4/0.0	0.6/0.1	0.9/0.1	2.7/1.9	4.2/14.3	10.8/32.0
Parkinsons	0.1/0.0	0.0/0.0	0.5/0.0	0.8/0.3	0.9/0.6	1.7/1.7
Crabs	0.6/0.0	0.2/0.0	0.0/0.0	0.1/0.3	0.2/0.6	0.4/1.5
Sonar	0.7/0.0	0.2/0.0	0.4/0.1	0.5/0.2	0.5/0.2	0.6/0.2
Balance	0.0/0.0	0.3/0.0	1.7/0.0	2.6/0.0	3.3/0.1	2.9/0.0
Transfusion	0.3/0.0	1.3/0.0	2.4/0.0	2.3/0.0	4.6/0.0	4.9/0.1
Letters-IJ	0.0/0.0	0.2/0.0	0.3/0.0	0.2/0.0	0.5/0.0	0.7/0.0
Digits-389	0.0/0.0	0.0/0.0	0.1/0.0	0.1/0.0	0.0/0.0	0.3/0.0

In addition, during our experiments, we found that for both criteria the difference between the maximum objective value and objective of the finally selected pair is often negligible. So in the case where some high-ranking pairs are dropped due to the acyclic graph structure restriction, the selected pair is still very informative. Overall, this enforcement does not present any significant negative impact on the final clustering results. It is interesting to note that, the results in Sec. 4.2 suggest that such graph structure restriction can in some cases improve the clustering performance.

5   Related Work

Prior work on active clustering for pairwise constraints has mostly focused on the neighbourhood-based method, where a neighbourhood skeleton is constructed to partially represent the underlying clusters, and constraints are queried to expand such neighbourhoods. Basu et al. [3] first proposed a two-phase method, Explore and Consolidate. The Explore phase incrementally builds K disjoint neighborhoods by querying instance pairwise relations, and the Consolidate phase iteratively queries random points outside the neighborhoods against the existing neighborhoods, until a must-link constraint is found. Mallapragada et al. [17] proposed an improved version, which modifies the Consolidate stage to query the most uncertain points using an MinMax objective. As mentioned by Xiong et al. [25], these methods often select a batch of constraints before performing clustering, and they are not designed for iteratively improving clustering by querying new constraints, as considered in this work.

Wang and Davidson [23], Huang et al. [14] and Xiong et. al. [25] studied active clustering in an iterative manner. Wang and Davidson introduced an active spectral clustering method that iteratively select the pair that maximized the expected error reduction of current model. This method is however restricted to the two-cluster problems. Huang et al. proposed an active document clustering method that iteratively finds probabilistic clustering solution using a language model and they selected the most uncertain pair to query. But this method is limited to the task of document clustering. Xiong et. al. considered a similar iterative framework to Huang et al., and they queried the most uncertain data point against existing neighbourhoods, as apposed to the most uncertain pair in [14]. Xiong et al. only provide a query selection strategy and require a clustering method to learn from the constraints. In contrast, our method is a unified clustering and active pair selection model.

Finally, there are other methods that use various criteria to select pairwise constraints. Xu et al. [26] proposed to select constraints by examining the spectral eigenvectors of the similarity matrix in the two-cluster scenario. Vu et al. [21] proposed to select constraints involving points on the sparse regions of a k-nearest neighbours graph. The work [1,12] used ensemble approaches to select constraints. The scenarios considered in these methods are less similar to what has been studied in this paper.

6   Conclusion

In this work, we studied the problem of active clustering, where the goal is to iteratively improve clustering by querying informative pairwise constraints. We introduced a Bayesian clustering method that adopted a logistic clustering model and a data-dependent prior which controls model complexity and encourages large separations among clusters. Instead of directly computing the posterior of the clustering model at every iteration, our approach maintains a set of samples from the posterior. We presented a sequential MCMC method to efficiently update the posterior samples after obtaining a new pairwise constraint. We introduced two information-theoretic criteria to select the most informative pairs to query at every iteration. Experimental results demonstrated the effectiveness of the proposed Bayesian active clustering method over existing approaches.

Acknowledgments. This material is based upon work supported by the National Science Foundation under Grant No. IIS-1055113. Any opinions, findings, and conclusions or recommendations expressed in this material are those of the author(s) and do not necessarily reflect the views of the National Science Foundation.

References

1. Al-Razgan, M., Domeniconi, C.: Clustering ensembles with active constraints. In: Okun, O., Valentini, G. (eds.) Applications of Supervised and Unsupervised Ensemble Methods. SCI, vol. 245, pp. 175–189. Springer, Heidelberg (2009)
2. Baghshah, M.S., Shouraki, S.B.: Semi-supervised metric learning using pairwise constraints. In: IJCAI, pp. 1217–1222 (2009)
3. Basu, S., Banerjee, A., Mooney, R.J.: Active semi-supervision for pairwise constrained clustering. In: SDM, pp. 333–344 (2004)
4. Basu, S., Bilenko, M., Mooney, R.J.: A probabilistic framework for semi-supervised clustering. In: KDD, pp. 59–68 (2004)
5. Bilenko, M., Basu, S., Mooney, R.J.: Integrating constraints and metric learning in semi-supervised clustering. In: ICML, pp. 81–88 (2004)

6. Dasgupta, S.: Analysis of a greedy active learning strategy. In: NIPS, pp. 337–344 (2005)
7. Davidson, I., Wagstaff, K.L., Basu, S.: Measuring constraint-set utility for partitional clustering algorithms. In: Fürnkranz, J., Scheffer, T., Spiliopoulou, M. (eds.) PKDD 2006. LNCS (LNAI), vol. 4213, pp. 115–126. Springer, Heidelberg (2006)
8. Davis, J.V., Kulis, B., Jain, P., Sra, S., Dhillon, I.S.: Information-theoretic metric learning. In: ICML, pp. 209–216 (2007)
9. Gilks, W.R., Berzuini, C.: Following a moving target - monte carlo inference for dynamic bayesian models. Journal of the Royal Statistical Society: Series B (Statistical Methodology) 63(1), 127–146 (2001)
10. Golovin, D., Krause, A., Ray, D.: Near-optimal bayesian active learning with noisy observations. In: NIPS, pp. 766–774 (2010)
11. Grandvalet, Y., Bengio, Y.: Semi-supervised learning by entropy minimization. In: NIPS, pp. 281–296 (2005)
12. Greene, D., Cunningham, P.: Constraint selection by committee: an ensemble approach to identifying informative constraints for semi-supervised clustering. In: Kok, J.N., Koronacki, J., Lopez de Mantaras, R., Matwin, S., Mladenič, D., Skowron, A. (eds.) ECML 2007. LNCS (LNAI), vol. 4701, pp. 140–151. Springer, Heidelberg (2007)
13. Houlsby, N., Huszár, F., Ghahramani, Z., Lengyel, M.: Bayesian active learning for classification and preference learning. CoRR abs/1112.5745 (2011)
14. Huang, R., Lam, W.: Semi-supervised document clustering via active learning with pairwise constraints. In: ICDM, pp. 517–522 (2007)
15. Kschischang, F.R., Frey, B.J., Loeliger, H.A.: Factor graphs and the sum-product algorithm. IEEE Trans. Inform. Theory 47(2), 498–519 (2001)
16. Lu, Z., Leen, T.K.: Semi-supervised clustering with pairwise constraints: a discriminative approach. In: AISTATS, pp. 299–306 (2007)
17. Mallapragada, P.K., Jin, R., Jain, A.K.: Active query selection for semi-supervised clustering. In: ICPR, pp. 1–4 (2008)
18. Neal, R.M.: Slice sampling. Annals of statistics 31(3), 705–741 (2003)
19. Nelson, B., Cohen, I.: Revisiting probabilistic models for clustering with pairwise constraints. In: ICML, pp. 673–680 (2007)
20. Shental, N., Bar-hillel, A., Hertz, T., Weinshall, D.: Computing gaussian mixture models with em using equivalence constraints. In: NIPS, pp. 465–472 (2003)
21. Vu, V.V., Labroche, N., Bouchon-Meunier, B.: An efficient active constraint selection algorithm for clustering. In: ICPR, pp. 2969–2972 (2010)
22. Wagstaff, K.L., Cardie, C., Rogers, S., Schrödl, S.: Constrained k-means clustering with background knowledge. In: ICML, pp. 577–584 (2001)
23. Wang, X., Davidson, I.: Active spectral clustering. In: ICDM, pp. 561–568 (2010)
24. Xing, E.P., Ng, A.Y., Jordan, M.I., Russell, S.J.: Distance metric learning with application to clustering with side-information. In: NIPS, pp. 505–512 (2002)
25. Xiong, S., Azimi, J., Fern, X.Z.: Active learning of constraints for semi-supervised clustering. IEEE Trans. Knowl. Data Eng. 26(1), 43–54 (2014)
26. Xu, Q., desJardins, M., Wagstaff, K.L.: Active constrained clustering by examining spectral eigenvectors. In: Hoffmann, A., Motoda, H., Scheffer, T. (eds.) DS 2005. LNCS (LNAI), vol. 3735, pp. 294–307. Springer, Heidelberg (2005)
27. Yang, L., Jin, R., Sukthankar, R.: Bayesian active distance metric learning. In: UAI, pp. 442–449 (2007)
28. Yu, S.X., Shi, J.: Segmentation given partial grouping constraints. IEEE Trans. Pattern Anal. Mach. Intell. 26(2), 173–183 (2004)

ConDist: A Context-Driven Categorical Distance Measure

Markus Ring[1]([⊠]), Florian Otto[1], Martin Becker[2], Thomas Niebler[2],
Dieter Landes[1], and Andreas Hotho[2]

[1] Faculty of Electrical Engineering and Informatics, Coburg University of Applied
Sciences and Arts, 96450 Coburg, Germany
{markus.ring,florian.otto,dieter.landes}@hs-coburg.de
[2] Data Mining and Information Retrieval Group, University of Würzburg,
97074 Würzburg, Germany
{becker,niebler,hotho}@informatik.uni-wuerzburg.de

Abstract. A distance measure between objects is a key requirement for many data mining tasks like clustering, classification or outlier detection. However, for objects characterized by categorical attributes, defining meaningful distance measures is a challenging task since the values within such attributes have no inherent order, especially without additional domain knowledge. In this paper, we propose an unsupervised distance measure for objects with categorical attributes based on the idea that categorical attribute values are similar if they appear with similar value distributions on correlated context attributes. Thus, the distance measure is automatically derived from the given data set. We compare our new distance measure to existing categorical distance measures and evaluate on different data sets from the UCI machine-learning repository. The experiments show that our distance measure is recommendable, since it achieves similar or better results in a more robust way than previous approaches.

Keywords: Categorical data · Distance measure · Heterogeneous data · Unsupervised learning

1 Introduction

Distance calculation between objects is a key requirement for many data mining tasks like clustering, classification or outlier detection [13]. Objects are described by a set of attributes. For continuous attributes, the distance calculation is well understood and mostly the Minkowski distance is used [2]. For categorical attributes, defining meaningful distance measures is more challenging since the values within such attributes have no inherent order [4]. The absence of additional domain knowledge further complicates this task.

However, several methods exist to address this issue. Some are based on simple approaches like checking for equality and inequality of categorical values, or create a new binary attribute for each categorical value [2]. An obvious drawback of these two approaches is that they cannot reflect the degree of similarity

A. Appice et al. (Eds.): ECML PKDD 2015, Part I, LNAI 9284, pp. 251–266, 2015.
DOI: 10.1007/978-3-319-23528-8_16

or dissimilarity between two distinct categorical values. Yet, more sophisticated methods incorporate statistical information about the data [6–8].

In this paper, we take the latter approach. In contrast to previous work, we take into account the quality of information that can be extracted from the data, in form of correlation between attributes. The resulting distance measure is called ConDist (Context based Categorical Distance Measure): We first derive a distance measure for each attribute separately. To this end, we take advantage of the fact that categorical attributes are often correlated, as shown in an empirical study [8], or by the fact that entire research fields exist which detect and eliminate such correlations, e.g. feature selection or dimensionality reduction. In order to calculate the distances for the values within a *target attribute*, we first identify the correlated context attributes. The distance measure on target attributes is then based on the idea that attribute values are similar if they appear with similar value distributions on their corresponding set of correlated context attributes. Finally, we combine these distance measures on separate attributes to calculate the distance of objects, again taking into account correlation information. We argue that incorporating the correlation of context attributes and the target attribute itself are important in order to maximize the relevant distance information extracted from the data and mitigate the possibly incorrect influence of uncorrelated attributes.

Table 1 shows a sample data set. Let us assume, we want to calculate the distance between the different values of attribute *height*, i.e., *height* is our target attribute. As mentioned above, our distance measure calculates its distance based on the value distributions of other attributes. For the attributes *weight* and *haircolor* these distributions ($P(X|H = small)$, $P(X|H = medium)$ and $P(X|H = tall)$) are shown in Figure 1. In the case of *weight* the distributions are different. Thus, they will add information to our distance calculations. However, the distributions for *haircolor* are the same for all values of the target attribute. Thus, they will not contribute information to our distance measure. At the same time, we can see that *weight* is correlated to *height*, since higher weight implies greater height with a high probability. For *haircolor* on the other hand, there is no correlation, since *haircolor* does not imply *height*. Since we also take this correlation information into account, we exclude uncorrelated attributes from the distance measure. Therefore, context attribute *haircolor* will not be taken into account when calculating distances between the values of *height*.

Overall, we propose an unsupervised distance measure for objects described by categorical attributes. Our new distance measure *ConDist* calculates distances by identifying and utilizing relevant statistical relationships from the given data set in form of correlations between attributes. This way, *ConDist* tries to compensate for the lack of inherent orders within categorical attribute domains.

The rest of the paper is organized as follows: Related work on categorical distance measures is discussed in Section 2. Section 3 describes the proposed distance measure *ConDist* in detail. Section 4 gives an experimental evaluation and the results are discussed in Section 5. The last section summarizes the paper.

Table 1. Example data set which describes nine people with three categorical attributes. The attributes *height* and *weight* have natural orders. Whereas the attribute *haircolor* has no natural order. *Height* and *weight* are correlated to each other while the attribute *haircolor* is uncorrelated to the other two attributes. *ConDist* uses such correlations between attributes to extract relevant information for distance calculation.

#	height	weight	haircolor
1	small	low	blond
2	small	low	brown
3	small	middle	black
4	medium	low	black
5	medium	middle	brown
6	medium	high	blond
7	tall	middle	blond
8	tall	high	brown
9	tall	high	black

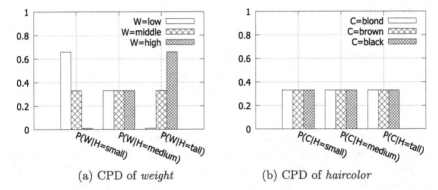

(a) CPD of *weight*   (b) CPD of *haircolor*

Fig. 1. This figure shows the conditional probability distributions (CPDs) of context attributes *weight* and *haircolor*, given the different values of the target attribute *height* based on Table 1. W stands for *weight*, C for *haircolor* and H for *height*. *ConDist* uses the differences between CPDs of context attributes to calculate the distance of target attribute values. Thus, *weight* can be used to calculate a meaningful distance between the values of *height*, while *haircolor* will yield the same distance for all three target attribute values.

2   Related Work

This section reviews related work on categorical distance measures. Distance measures can be divided into supervised and unsupervised. In the supervised setting, the class membership of the objects is provided and this information is exploited by the distance measures. In the unsupervised setting, distance measures are based exclusively on assumptions and statistics of the data. Since the proposed distance measure is unsupervised, the following review considers only unsupervised categorical distance measures. We categorize them into distance calculation (I) without considering context attributes, (II) considering all context attributes, (III) considering a subset of context attributes and (IV) based on entire objects instead of individual attributes.

Boriah et al. [4] give a comprehensive overview of distances measures from category (I). In contrast to *ConDist*, these distance measures ignore available information that could be extracted from context attributes. For example, the distance measure *Eskin* only uses the cardinality of the target attribute domain to calculate distances. [4] evaluated these distance measures for outlier detection and observed that no specific distance measure dominates all others.

Category (II) includes distance measures that employ all context attributes without distinguishing between correlated and uncorrelated. Li and Ho [8] compute the distance between two categorical values as the sum of dissimilarities between the context attributes' conditional probability distribution (CPD) when the target attribute takes these two values. However, they do not recommend their distance measure for data sets with highly independent attributes. Similary, [1] calculates the distance between two values using the co-occurrence probabilities of these two values and the values of the context attributes.

Category (III) selects a subset of context attributes for each target attribute. DILCA [6] is a representative of this category and uses Symmetric Uncertainty (SU) [15] for selecting context attributes. SU calculates the correlation between two attributes. In contrast to our work, all selected context attributes are weighted equally for the distance calculation. Consequently, the potentially differing suitability of the selected context attributes is not reflected in the distance calculation process.

Category (IV) aims to compute distances between entire objects instead of distances between different values within an attribute. Consequently, the distance between different values within an attribute varies in dependence of the whole objects. Recently, Jia and Cheung [7] proposed such a distance measure for cluster analysis. Their basic assumption is that two categorical values with high frequencies should have a higher distance than two categorical values with low frequencies. Therefore, they select and weigh a set of correlated context attributes for each target attribute using the normalized mutual information [3]. Jia and Cheung [7] compared their distance measure with the *Hamming Distance* on four data sets. They conclude that their distance measure performs better than the *Hamming Distance* on the evaluated data sets.

The proposed distance measure *ConDist* neither ignores context attributes (category I) nor simply includes all context attributes (category II). Instead, it follows the approach of the third category but extends the subset selection with a weighting scheme for context attributes. Furthermore, the target attribute itself is included in the distance computation. In contrast to the fourth category, two particular values within an attribute have always the same distance, independent of the corresponding objects. This allows *ConDist* to calculate a distance matrix for each attribute.

3   The Distance Measure ConDist

This section introduces *ConDist*, a new distance measure. Section 3.1 presents the underlying ideas and the core formula. Since *ConDist* first calculates the distance between single attributes before combining them, it requires adjusted distance functions for each attribute. In Section 3.2, we explain how these attribute distance functions are derived. When combining attribute-wise distances, *ConDist* uses a specific weighting scheme which is explained in Section 3.3. Both, the attribute distance functions as well as the weighting scheme use a set of correlated context attributes. Section 3.4 defines how this set is derived and how an impact factor is calculated which accounts for the varying amount of information dependent on different correlation values. Finally, we address the issue of how *ConDist* can be applied to objects characterized by continuous and categorical attributes in Section 3.5.

3.1   Definition of ConDist

This section provides the core formula of *ConDist*, calculating the distance between two objects characterized by attributes.

Let A and B be two objects in the data set D and let each object be characterized by n attributes. Furthermore, let the value of attribute X for object A be denoted by A_X. *ConDist* follows a two-step process: First, it calculates the distance between each of the attributes of the objects A and B and then it combines them using attribute specific weights. Formally, *ConDist* defines the distance for two objects A and B as the weighted sum over all attribute distances:

$$ConDist(A, B) = \sum_X w_X \cdot d_X(A, B), \tag{1}$$

where w_X denotes the weighting factor assigned to attribute X (defined in Section 3.3) and $d_X(A, B)$ denotes the distance of the values A_X and B_X of attribute X in the objects A and B (defined in Section 3.2).

The distance function d_X on the values of each attribute X needs to be calculated beforehand and is based on the idea that attribute values with similar distributions of values in a set of correlated context attributes are similar. The weighting factor w_X accounts for differences in the number of context attributes and the degree of their correlation with the target attribute X. Both, the attribute distance functions d_X as well as the weighting factors w_X incorporate correlation information in order to maximize the relevant information that can be extracted from the data set and mitigate the possibly incorrect influence of uncorrelated attributes. For an example on differently correlated attributes and their influence on distribution based distance measures, please refer to Section 1 as well as Table 1 and Figure 1.

3.2   Attribute Distance d_X

As mentioned in Section 3.1, the distance d_X of values of a single attribute X is based on the idea that attribute values $x \in dom(X)$ are similar if they appear with similar distributions of values in a set of correlated context attributes. Thus, when comparing two objects A and B in attribute X, we first calculate the Euclidean distance between the two conditional probability distributions $P(Y|X = A_X)$ and $P(Y|X = B_X)$ for each attribute $Y \in context_X$ from the set of correlated context attributes $context_X$ of the target attribute X. Then, we weight them using an individual impact factor $impact_X(Y)$ (Section 3.4) and add up these distances for all attributes $Y \in context_X$. The impact factor accounts for the fact that the amount of information about the target attribute X in a context attribute Y decreases with both increasing and decreasing correlation $cor(X|Y)$ as explained in Section 3.4. The resulting formula is:

$$\hat{d}_X(A, B) = \sum_{Y \in context_X} impact_X(Y) \sqrt{\sum_{y \in dom(Y)} \Big(p(y|A_X) - p(y|B_X)\Big)^2}, \quad (2)$$

where $dom(Y)$ is the domain of attribute Y, $p(y|A_X) = p(y|X = A_X)$ denotes the probability that value y of context attribute Y is observed under the condition that value A_X of attribute X is observed in the data set D.

As mentioned above, the attribute distance d_X relies on a set of correlated context attributes $context_X$ as defined in Section 3.4. Because every attribute is correlated to itself, the target attribute is also added to the set of context attributes. The motivation for including the target attribute is two-fold: First, it ensures that the list of context attributes is not empty even if all attributes are independent. Second, the distance between two distinct values is always larger than 0. Thus, if no correlated context attributes can be identified, ConDist calculates the maximum distance for each distinct value-pair in target attribute X. In this case, ConDist reduces to the distance measure Overlap and distinguishes only between equality and inequality of categorical values.

It should be noted that ConDist normalizes the attribute distance by the maximum distance value $d_{X,max}$ between any two values $x, u \in dom(X)$ of attribute X:

$$d_X(A, B) = \frac{\hat{d}_X(A, B)}{d_{X,max}} \quad (3)$$

The proof that ConDist is a distance measure closely follows the proof of the Euclidean metric and exploits the fact that a linear combination of distance measures is also a distance measure. For brevity reasons, we omit the proof.

3.3   Attribute Weighting Function w_X

ConDist compares objects based on the distances between each of the attribute values associated with the objects it compares (see Equation (1)). Each of these attributes is weighted differently by an individual weighting factor w_X. This section explains why these weights w_X are necessary and how they are calculated.

The weight w_X is especially necessary for data sets in which some attributes depend on each other, while others do not: refer back to the example in Table 1.

For attribute *haircolor*, no correlated context attribute can be identified. Consequently, only the attribute *haircolor* itself is used for distance calculation and no additional information can be extracted from context attributes. Therefore, the normalized results of Equation (2) always results in the maximum distance 1 for any pair of non-identical values. In contrast, the attribute *weight* is a correlated context attribute for attribute *height*, and vice versa. Consequently, *ConDist* is able to calculate more meaningful distances for both attributes and these attributes should be weighted higher than *haircolor*.

However, average distances in attribute *haircolor* are larger than in attributes *weight* and *height*. Consequently, distinct values in attribute *haircolor* have implicitly larger relative weight than distinct values in attributes *height* and *weight*.

To solve this issue, the weighting factor w_X assigns a weight to each attribute X based on (I) the amount of identified context attributes and (II) their impact on the target attribute X:

$$w_X = 1 + \frac{\sum\limits_{Y \in context_X} impact_X(Y)}{n \cdot c}, \tag{4}$$

where $context_X$ and $impact_X(Y)$ are defined as in section 3.4, n is the number of attributes in the data set D and c denotes a normalization factor defined as the maximum of the impact function (see Section 3.4) which is independent of the attributes X and Y and amounts to $\frac{8}{27}$.

3.4   Correlation, Context and Impact

The attribute distance measures d_X (Section 3.2) and the weighting scheme w_X (Section 3.3) use the notion of correlation on categorical distance measures as well as a correlation related impact factor. Both are defined here.

Correlation $cor(X|Y)$. A measure of correlation is required to determine an appropriate set of context attributes. For this purpose, we build a correlation measure on the basis of the *Information Gain* (IG) which is widely used in information theory [9]. The IG is calculated as follows:

$$IG(X|Y) = H(X) - H(X|Y), \tag{5}$$

where $H(X)$ is the entropy of an attribute X, and $H(X|Y)$ is the conditional entropy of attribute X given attribute Y. According to this measure, the attribute X is more correlated with attribute Y than attribute W if $IG(X|Y) > IG(X|W)$. The information gain $IG(X|Y)$ is always less than or equal to the entropy $H(X)$. Based on this observation, the function $cor(X|Y)$ is defined as:

$$cor(X|Y) = \frac{IG(X|Y)}{H(X)} \tag{6}$$

and describes a correlation measure which is normalized to the interval $[0,1]$. The quality of possible conclusions from the given attribute Y to the target attribute X can differ from the quality of conclusions from given attribute X to target attribute Y. This aspect is considered in the asymmetric correlation function $cor(X|Y)$ and allows us to always extract the maximum amount of useful information for each target attribute X.

Context $context_X$. For both, the attribute distance d_X (Section 3.2) in *ConDist* as well as for the weighting scheme w_X (Section 3.3), the notion of a set of correlated context attributes $context_X$ is used. This set is defined using the previously defined correlation function $cor(X|Y)$ and a user-defined threshold θ. That is, context attributes Y are included in $context_X$ only if their correlation with target attribute X is equal to or exceeds the threshold θ:

$$context_X = \{Y \mid cor(X|Y) \geq \theta\} \tag{7}$$

Impact $impact_X(Y)$. Again, for both, the attribute distance d_X (Section 3.2) as well as for the weighting scheme w_X (Section 3.3), a so called impact factor $impact_X(Y)$ is used. This factor accounts for the fact that the amount of information about the target attribute X in a context attribute Y decreases with both increasing and decreasing correlation $cor(X|Y)$.

A high correlation value means that a value of a context attribute $Y \in context_X$ implies the value of a target attribute X with a high probability. For example, when we know that someone is heavy, it is more likely that this person is tall than small (see Table 1). Thus, in the extreme case of perfectly correlated attributes, the conditional probability distributions $P(Y|X = A_X)$ and $P(Y|X = B_X)$, for $A_X \neq B_X$ do not overlap. This means that using the Euclidean distance to calculate the similarity of those two CPDs (as in Formula (2)) limits the distance information gained from the context attribute to values of 0 (for $A_X = B_X$) and 1 (for $A_X \neq B_X$) after normalization in Formula (3).

A low correlation value means that a value of a context attribute $Y \in context_X$ implies little to no preference for a particular value of a target attribute X. This means that the similarity between the conditional probability distributions $P(Y|X = A_X)$ and $P(Y|X = B_X)$ may be random, thus, possibly conveying incorrect distance information.

Consequently, non-correlated attributes are excluded to avoid introducing incorrect information. In contrast, perfectly correlated attributes are still used, because they contribute at least no incorrect information. However, since they deliver exclusively high distances for distinct values, their impacts should be reduced. Otherwise, the distances calculated by the other context attributes would be blurred.

Therefore, we choose a weighting function that (I) increases fast at the onset of correlation between attributes, (II) increases more slowly with existing, but partial correlation, and (III) decreases at nearly perfect correlation. In particular, we propose the impact function as depicted in Figure 2 and defined as:

$$impact_X(Y) = cor(X|Y)\Big(1 - 0.5 \cdot cor(X|Y)\Big)^2. \tag{8}$$

In general, this impact function can be replaced by other functions respecting the three properties introduced above.

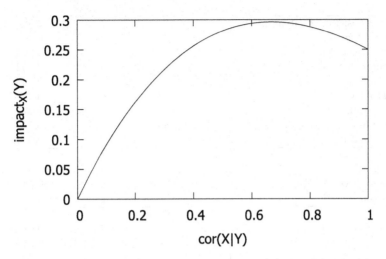

Fig. 2. Graph of the impact function $impact_X(Y)$ as defined in (8).

3.5 Heterogeneous Data Sets

Many real-world data sets contain both continuous and categorical attributes. To apply *ConDist* to such data sets, two situations have to be distinguished: either the target attribute is continuous or the context attribute is continuous.

If the target attribute is continuous, no context attributes are necessary. The Minkowski distance can be used, but should be normalized to the interval $[0, 1]$. Since meaningful distances can be calculated for continuous attributes, the attribute weight w_X (see Section 3.3) should be maximized. If the context attribute is continuous, the continuous value range should be discretized. We propose to use the discretization algorithm TUBE [11], because it does not require any parameters. Other discretization algorithms can be used as well.

4 Experiments

This section presents an experimental evaluation of *ConDist* in the context of classification and clustering. We compared *ConDist* with *DILCA* [6], *JiaCheung* [7], *Quang* [8] and several distance measures presented in [4], namely *Eskin, Gambaryan, Occurrence Frequency (OF)* and *Overlap*. For *DILCA*, we used the non-parametric approach $DILCA_{RR}$ as described in [6] and for *JiaCheung* we set the threshold parameter β to 0.1 as recommended by the authors [7].

Table 2. Characteristics of the data sets. The column *Correlation* contains the average correlation between each pair of attributes in the data set, calculated by the function $cor(X|Y)$, see Equation (6). The value ranges from 0 if no correlation exists to 1 if all attributes are perfectly correlated. The data sets are separated in three subsets from highly correlated to uncorrelated based on their average correlation.

Data Sets	Instances	Attributes	Classes	Correlation
Teaching Assistant Evaluation	151	5	3	0.336
Soybean Large	307	35	19	0.263
Breast Cancer Winconsin	699	10	2	0.216
Mushroom-Extended	8416	22	2	0.162
Mushroom	8124	22	2	0.161
Dermatology	366	34	6	0.098
Lymphography	148	18	4	0.070
Soybean Small	47	35	19	0.070
Breast Cancer	286	9	2	0.054
Audiology-Standard	226	69	24	0.044
Hayes-Roth	160	4	3	0.045
Post-Operative Patient	90	8	3	0.031
TicTacToe	958	9	2	0.012
Monks	432	6	2	0.000
Balance-Scale	625	4	3	0.000
Car	1728	6	4	0.000
Nursey	12960	8	5	0.000

4.1   Evaluation Methodology

Classification. A k-Nearest-Neighbour classifier is used to compare *ConDist* with existing categorical distance measures in the context of classification. For simplification, we fix $k = 7$ in all tests. We evaluate by 10-fold-cross validation and use the classification accuracy as evaluation measure. To reduce confounding effects of the generated subsets, the 10-fold cross-validation is repeated 100 times with different subsets for each data set. We finally compare the averages of the classification accuracies over all executions.

Clustering. The hierarchical WARD algorithm [14] is used to evaluate the performance of *ConDist* in the context of clustering. *ConDist* and its competitors are used to calculate the initial distance matrix as input for WARD. For simplification, the clustering process is terminated when the number of clusters is equal to the number of classes in the data sets. Performance is measured by *Normalized Mutual Information (NMI)* [12] which ranges from 0 for poor clustering to 1 for perfect clustering with respect to the predefined classes.

Data Sets. For the evaluation of *ConDist* the *multivariate categorical data sets for classification* from the UCI machine learning repository [10] are chosen. We exclude data sets with less than 25 objects (e.g., *Balloons*) or mainly binary

Table 3. Classification accuracies for various thresholds θ in *ConDist*. Each column contains the results in percent for particular thresholds θ.

Data Set	Threshold θ								
	0.00	0.01	0.02	0.03	0.05	0.10	0.20	0.50	1.00
Soybean Large	91.74	91.74	91.79	91.80	**91.82**	89.75	89.36	89.63	91.30
Lymphography	**83.36**	**83.36**	83.30	83.24	83.01	81.99	82.01	81.24	81.26
Hayes-Roth	68.11	68.36	68.51	68.60	**69.21**	64.47	64.47	64.47	64.47
TicTacToe	**99.99**	**99.99**	**99.99**	99.98	94.74	94.74	94.74	94.74	94.74
Balance-Scale	77.35	**78.66**	**78.66**	**78.66**	**78.66**	**78.66**	**78.66**	**78.66**	**78.66**
Car	88.98	**90.56**	**90.56**	**90.56**	**90.56**	**90.56**	**90.56**	**90.56**	**90.56**
Average	84.92	85.45	**85.47**	**85.47**	84.67	83.36	83.30	83.22	83.50

attributes (e.g., *Chess*). Furthermore, we include some *multivariate mixed data sets for classification* from the UCI machine learning repository which mainly consist of categorical attributes and some integer attributes with a small set of distinct values (e.g. an integer attribute that contains the number of students in a course): *Teaching Assistant Evaluation, Breast Cancer Winconsin, Dermatology* and *Post-Operative Patient*. Since not all competitors have an explicit way to process integer attributes, we treated all integer attributes as categorical. The final set of data sets is given in Table 2. The data sets are divided in three subgroups: highly-correlated (Correlation $\geq$ 0.05), weakly-correlated (Correlation $>$ 0) and uncorrelated (Correlation $=$ 0).

4.2 Experiment 1 – Context Attribute Selection

Experiment 1 analyzes the effects of varying threshold θ (see Section 3.4) in *ConDist*'s context attribute selection. The threshold θ defines the minimum value of the function $cor(X|Y)$ that a candidate attribute Y has to reach in order to be selected as context attribute for the target attribute X. The higher the threshold θ, the fewer context attributes are used. In the extreme case of $\theta = 0$, all context attributes are used for distance calculation. In the other extreme case $\theta = 1$, only the target attribute itself is used. For this experiment, a representative subset of two highly-correlated (*Soybean Large* and *Lymphography*), two weakly-correlated (*Hayes-Roth* and *TicTacToe*) and two uncorrelated (*Balance-Scale* and *Car*) data sets are used. The results can be seen in Table 3.

The average classification accuracy (I) increases with low thresholds θ, (II) reaches a peak at $\theta = 0.02$ and $\theta = 0.03$, (III) decreases slowly with medium high thresholds, (IV) reaches the minimum at $\theta = 0.5$ and (V) slowly increases with high thresholds again. For nearly all data sets, the classification accuracy stabilizes with increasing thresholds. The lower the attribute correlation within the data set, the faster this effect is reached. For uncorrelated data sets like *Car* and *Balance-Scale*, it can already be observed with thresholds greater than or equal to $\theta = 0.01$. Due to the peak at $\theta = 0.02$, this value is used for the further experiments in this paper.

Table 4. Comparison of categorical distance measures in the context of classification. Each column contains the classification accuracies in percent for a particular distance measure. The data sets are separated in three subsets from highly correlated to uncorrelated based on their average correlation.

Data Set	ConDist	DILCA	Eskin	JiaCheung	Gambaryan	OF	Overlap	Quang
Teaching Assistant. E.	49.85	**50.68**	48.79	49.54	49.44	39.16	45.84	44.48
Soybean Large	91.79	91.48	89.83	89.45	87.18	89.61	91.30	**92.01**
Breast Cancer W.	96.13	95.55	95.67	95.08	92.84	72.47	95.25	**96.28**
Dermatology	96.76	**97.97**	94.91	97.39	91.69	61.12	95.90	96.64
Lymphography	83.30	82.09	79.17	**83.95**	80.72	72.77	81.26	81.53
Breast Cancer	73.85	72.94	73.18	74.30	**74.55**	68.32	74.06	70.45
Audiology Standard	**66.44**	64.80	63.24	60.95	66.16	51.87	61.27	55.56
Hayes-Roth	68.50	67.59	46.71	68.27	60.84	58.71	61.74	**71.19**
Post-Operative Patient	69.62	68.22	68.36	67.28	**69.69**	69.44	68.59	68.69
TicTacToe	**99.99**	90.65	94.74	99.93	98.25	76.80	94.74	99.65
Car	**90.56**	90.25	90.03	90.01	90.25	87.83	**90.56**	88.25
Nursey	**94.94**	92.61	93.29	93.32	93.24	94.65	**94.94**	94.72
Monks	94.50	90.76	87.29	87.34	86.61	**98.67**	94.50	96.66
Balance-Scale	**78.66**	78.43	**78.66**	78.65	77.13	78.54	**78.66**	77.51
Average	**82.49**	81.00	78.85	81.10	79.90	72.85	80.62	80.97

4.3   Experiment 2 – Comparison in the Context of Classification

Experiment 2 compares *ConDist* with several categorical distance measures in the context of classification. All data sets from Table 2 are used. The results are given in Table 4, except for the data sets *Mushroom-Extended*, *Mushroom* and *Soybean Small*. These data sets are omitted in the table since all distance measures reach 100 percent classification accuracy. Consequently, these data sets would only blur the differences between the categorical distance measures.

ConDist achieves the highest average classification accuracy of all distance measures. In the case of highly- and weakly-correlated data sets, context based categorical distance measures (*ConDist, DILCA, JiaCheung* and *Quang*) achieve mostly better results than other distance measures. In the case of uncorrelated data, previous context based categorical distance measures are inferior to *ConDist* and non-context based categorical distance measures.

Statistical Significance Test. In this test, we want to evaluate if the differences in Table 4 are statistically significant. Demšar [5] deals with the statistical comparison of classifiers over multiple data sets. They recommend the Wilcoxon Signed-Ranks Test for the comparison of two classifiers and the Friedman-Test for the comparison of multiple classifiers. Therefore, we use the Friedman-Test to compare all distance measures and the Wilcoxon Signed-Ranks Test for post-hoc tests. The Friedman-Test is significant for $p < 0.001$; thus we can reject the null hypothesis that all distance measures are equivalent. Consequently, we applied

Table 5. Results of the Wilcoxon Signed-Ranks Test comparing the classification accuracies of *ConDist* with each other distance measure. The first row contains the calculated p-value, the second row contains the result of the Wilcoxon Signed-Ranks Test: *yes*, if *ConDist* performs statistically different, *no* otherwise.

	DILCA	Eskin	JiaCheung	Gambaryan	OF	Overlap	Quang
p-value	0.016	0.002	0.045	0.002	0.002	0.008	0.096
significant	yes	yes	yes	yes	yes	yes	no

the Wilcoxon Signed-Ranks Test with $\alpha = 0.05$ on the classification accuracies of Table 4. Table 5 shows that there is a significant difference between *ConDist* and the distance measures *Eskin, JiaCheung, Gambaryan, OF* and *Overlap*. However, the test fails for *ConDist* and *Quang*.

4.4 Experiment 3 – Comparison in the Context of Clustering

Experiment 3 compares *ConDist* with several categorical distance measures in the context of clustering. All data sets from Table 4 are used. The results are given in Table 6.

For some data sets (*Teaching Assistang Evaluation, Lymphography, Breast Cancer, Hayes-Roth, Post-Operative Patient, TicTacToe, Monks, Balance-Scale, Nursey* and *Car*) the clustering fails to reconstruct the predefined classes. For the remaining data sets, no distance measure dominates. However, most distance measures perform poorly on single data sets, whereas *ConDist* achieves more stable results.

Statistical Significance Test. In analogy to Section 4.3, we first apply the Friedman-Test on the results shown in Table 6. Here, the null hypothesis that all distance measures are equivalent cannot be rejected. Nevertheless, we applied the Wilcoxon Signed-Ranks Test ($\alpha = 0.05$) between *ConDist* and the other distance measures. Except for *Eskin* and *Quang*, the results of the Wilcoxon Signed-Ranks Test show no statistically significant differences.

5 Discussion

5.1 Experiment 1 – Context Attribute Selection

Table 3 shows that many useful context attributes are discarded if threshold θ is too high. This is especially the case for weakly correlated data sets, e.g. *Hayes-Roth* and *TicTacToe*. For *Hayes-Roth*, the decrease of classification accuracy is observed for $\theta > 0.05$, and for *TicTacToe* the decrease of classification accuracy is already observed for $\theta > 0.02$. In contrast to this, if the threshold θ is too low, independent context attributes are added which may contribute noise to the distance calculation. This is especially the case for uncorrelated data sets,

Table 6. Comparison of categorical distance measures in the context of clustering. Each column contains the *NMI* of the clustering results found by the *WARD* algorithm where the initial distance matrix is calculated with the particular distance measure. *NMI* assigns low values to poor clusterings and high values to good clusterings with respect to the predefined classes. The data sets are also separated in three subsets based on their average correlation.

Data Set	ConDist	DILCA	Eskin	JiaCheung	Gambaryan	OF	Overlap	Quang
Teaching Assistant Eva.	.078	**.085**	**.085**	**.085**	**.085**	.060	.044	.042
Soybean Large	.803	.785	.758	.735	.772	**.805**	.793	.778
Breast Cancer Winconsin	.785	.557	.749	.656	.601	.217	.621	**.798**
Mushroom Extended	**.597**	**.597**	.317	.223	**.597**	**.597**	**.597**	.245
Mushroom	**.594**	**.594**	.312	**.594**	**.594**	.312	**.594**	.241
Dermatology	.855	**.946**	.832	.879	.863	.292	.847	.859
Lymphography	.209	.303	.165	.207	.163	.243	.226	**.320**
Soybean Small	.687	.690	.687	**.701**	.692	.690	.689	.692
Breast Cancer	.063	**.068**	.031	.074	.001	.002	.100	.001
Audiology-Standard	.661	.612	.623	**.679**	.620	.439	.568	.582
Hayes-Roth	.017	.027	.004	.012	.007	**.166**	.006	.029
Post-Operative Patient	**.043**	.017	.018	.025	.017	.032	.019	.033
TicTacToe	**.087**	.003	.003	.082	.085	.001	.033	.039
Monks	.001	.000	.000	.000	.000	**.081**	.001	.003
Balance-Scale	**.083**	.036	.064	.067	.064	.064	**.083**	.036
Car	.062	.036	**.150**	**.150**	**.150**	.062	.062	.036
Nursey	.048	.006	.037	.037	.037	**.098**	.048	.006
Average	**.334**	.315	.284	.306	.315	.245	.314	.279

e.g. for $\theta = 0$ in *Balance-Scale* and *Car*. However, *ConDist*'s impact function $impact_X(Y)$ accounts for this effect in highly-correlated data sets.

Consequently, the concrete value of the threshold θ is not too crucial as long as two conditions are fulfilled: (I) θ must be large enough to ensure that context attributes are purged whose correlations are caused by too small data sets and (II) θ must be small enough to ensure that context attributes with significant correlations are retained. Therefore, we recommend $\theta = 0.02$ for *ConDist*, because the experiments show that this threshold achieves the best overall results.

5.2   Experiment 2 – Comparison in the Context of Classification

For highly correlated data sets, distance measures using context attributes outperform other distance measures. However, for those data sets no best distance measures can be identified among the context based distance measures.

For uncorrelated data sets, previous context-based distance measures (*DILCA*, *Quang* and *JiaCheung*) achieved inferior results in comparison to *ConDist* and non-context based distance measures. This is because, e.g., *DILCA* and *Quang* use only context attributes for the distance calculation which results in random distances if all context attributes are uncorrelated.

In contrast, *ConDist* achieved acceptable results because not only correlated context attributes, but also the target attributes are considered. This effect is also illustrated by the comparison between *ConDist* and *Overlap*. *ConDist* is equal to *Overlap* if no correlated context attributes can be identified, see uncorrelated data sets (*Monks, Balance-Scale, Nursey* and *Car*) in Table 4. However, for weakly- and highly-correlated data sets, *ConDist's* consideration of context attributes turns into an advantage, leading to better results than *Overlap*. The improvement of *ConDist* can be statistically confirmed by the Wilcoxon Signed-Ranks Test (see Table 5).

5.3   Experiment 3 – Comparison in the Context of Clustering

Table 6 shows that the majority of the different distance measures reach, by and large, similar outcomes for individual data sets. This is because the clustering algorithm and its ability to reconstruct the given classes have much higher impact on the results than the distance measure used to calculate the initial distance matrix. However, it can be seen that the performance of single distance measures strongly decreases for individual data sets. For example, *JiaCheung* which often achieves good results, performs very poorly in the *Mushroom* data set. Similar observations can be made for *OF, Eskin, Quang* and *DILCA*, mainly in the data sets *Breast Cancer Winconsin, Mushroom, Mushroom Extended, Dermatology* and *Audiology-Standard*. In contrast, *ConDist* is almost always among the best results and shows the most stable results for the different data sets.

The Friedman-Test fails for Experiment 3 and the Wilcoxon Signed-Ranks Test shows also no statistically significant differences in the performance of *ConDist* and the compared distance measures, except for *Eskin* and *Quang*. However, the results of Experiment 3 lead to the assumption that *ConDist* may be a more robust distance measure than its competitors.

6   Summary

Categorical distance calculation is a key requirement for many data mining tasks. In this paper, we proposed *ConDist*, an unsupervised categorical distance measure based on the correlation between the target attribute and context attributes. With this approach, we aim to compensate for the lack of inherent orders within categorical attributes by extracting statistical relationships from the data set.

Our experiments show that *ConDist* is a generally usable categorical distance measure. In the case of correlated data sets, *ConDist* is comparable to existing context based categorical distance measures and superior to non-context based categorical distance measures. In the case of weakly and uncorrelated data sets, *ConDist* is comparable to non-context based categorical distance measures and superior to context based categorical distance measures. The overall improvement of *ConDist* can be statistically confirmed in the context of classification. In the context of clustering, this improvement could not be statistically confirmed.

In the future, we want to extend the proposed distance measure so that it can automatically infer the parameter θ from the data sets. Additionally, we want to transform categorical attributes to continuous attributes with aid of the proposed distance measure.

Acknowledgments. This work is funded by the Bavarian Ministry for Economic affairs through the WISENT project (grant no. IUK 452/002) and by the DFG through the PoSTS II project (grant no. STR 1191/3-2). We appreciate the support of our project partners HUK COBURG, Coburg, Germany and Applied Security, Großwallstadt, Germany.

References

1. Ahmad, A., Dey, L.: A method to compute distance between two categorical values of same attribute in unsupervised learning for categorical data set. Pattern Recognition Letters **28**(1), 110–118 (2007)
2. Alamuri, M., Surampudi, B.R., Negi, A.: A survey of distance/similarity measures for categorical data. In: Proc. of IJCNN, pp. 1907–1914. IEEE (2014)
3. Au, W.H., Chan, K.C., Wong, A.K., Wang, Y.: Attribute clustering for grouping, selection, and classification of gene expression data. IEEE/ACM Transactions on Computational Biology and Bioinformatics (TCBB) **2**(2), 83–101 (2005)
4. Boriah, S., Chandola, V., Kumar, V.: Similarity measures for categorical data: A comparative evaluation. In: Proc. SIAM Int. Conference on Data Mining, pp. 243–254 (2008)
5. Demšar, J.: Statistical comparisons of classifiers over multiple data sets. The Journal of Machine Learning Research **7**, 1–30 (2006)
6. Ienco, D., Pensa, R.G., Meo, R.: Context-Based Distance Learning for Categorical Data Clustering. In: Adams, N.M., Robardet, C., Siebes, A., Boulicaut, J.-F. (eds.) IDA 2009. LNCS, vol. 5772, pp. 83–94. Springer, Heidelberg (2009)
7. Jia, H., Cheung, Y.M.: A new distance metric for unsupervised learning of categorical data. In: Proc. of IJCNN, pp. 1893–1899. IEEE (2014)
8. Le, S.Q., Ho, T.B.: An association-based dissimilarity measure for categorical data. Pattern Recognition Letters **26**(16), 2549–2557 (2005)
9. Lehmann, E., Romano, J.: Testing Statistical Hypotheses, Springer Texts in Statistics. Springer (2005)
10. Lichman, M.: Uci machine learning repository (2013). http://archive.ics.uci.edu/ml
11. Schmidberger, G., Frank, E.: Unsupervised Discretization Using Tree-Based Density Estimation. In: Jorge, A.M., Torgo, L., Brazdil, P.B., Camacho, R., Gama, J. (eds.) PKDD 2005. LNCS (LNAI), vol. 3721, pp. 240–251. Springer, Heidelberg (2005)
12. Strehl, A., Ghosh, J.: Cluster ensembles–a knowledge reuse framework for combining multiple partitions. The Journal of Machine Learning Research 3 (2003)
13. Tan, P.N., Steinbach, M., Kumar, V.: Introduction to data mining. Pearson Addison Wesley Boston (2006)
14. Ward Jr., J.H.: Hierarchical grouping to optimize an objective function. Journal of the American statistical association **58**(301), 236–244 (1963)
15. Yu, L., Liu, H.: Feature selection for high-dimensional data: A fast correlation-based filter solution. ICML **3**, 856–863 (2003)

Discovering Opinion Spammer Groups
by Network Footprints

Junting Ye[(✉)] and Leman Akoglu

Department of Computer Science, Stony Brook University, Stony Brook, USA
{juyye,leman}@cs.stonybrook.edu

Abstract. Online reviews are an important source for consumers to evaluate products/services on the Internet (e.g. Amazon, Yelp, etc.). However, more and more fraudulent reviewers write fake reviews to mislead users. To maximize their impact and share effort, many spam attacks are organized as campaigns, by a *group* of spammers. In this paper, we propose a new two-step method to discover spammer groups and their targeted products. First, we introduce NFS (Network Footprint Score), a new measure that quantifies the likelihood of products being spam campaign targets. Second, we carefully devise *GroupStrainer* to cluster spammers on a 2-hop subgraph induced by top ranking products. We demonstrate the efficiency and effectiveness of our approach on both synthetic and real-world datasets from two different domains with millions of products and reviewers. Moreover, we discover interesting strategies that spammers employ through case studies of our detected groups.

Keywords: Opinion spam · Spammer groups · Spam detection · Graph anomaly detection · Efficient hierarchical clustering · Network footprints

1 Introduction

Online reviews of products and services are an increasingly important source of information for consumers. They are valuable since, unlike advertisements, they reflect the testimonials of other, "real" consumers. While many positive reviews can increase the revenue of a business, negative reviews can cause substantial loss. As a result of such financial incentives, opinion spam has become a critical issue [17], where fraudulent reviewers fabricate spam reviews to unjustly promote or demote (e.g., under competition) certain products and businesses.

Opinion spam is surprisingly prevalent; one-third of consumer reviews on the Internet[1], and more than 20% of reviews on Yelp[2] are estimated to be fake. Despite being widespread, opinion spam remains a mostly open and challenging problem for at least two main reasons; (1) humans are incapable of distinguishing fake reviews based on text [25], which renders manual labeling extremely difficult

---

[1] http://www.nytimes.com/2012/08/26/business/book-reviewers-for-hire-meet-a-demand-for-online-raves.html

[2] http://www.businessinsider.com/20-percent-of-yelp-reviews-fake-2013-9

© Springer International Publishing Switzerland 2015
A. Appice et al. (Eds.): ECML PKDD 2015, Part I, LNAI 9284, pp. 267–282, 2015.
DOI: 10.1007/978-3-319-23528-8_17

and hence supervised methods inapplicable, and (2) fraudulent reviewers are often professionals, paid by businesses to write detailed and genuine-looking reviews.

Since the seminal work by Jindal and Liu [17], opinion spam has been the focus of research for the last 7-8 years (Section 5). Most existing work aim to detect *individual* spam reviews [12,17,20,21,25,26] or spammers [1,11,13, 18,22,26]. However, fraud/spam is often a *collective* act, where the involved individuals cooperate in groups to execute spam *campaigns*. This way, they can increase total impact (i.e., dominate the sentiments towards target products via flooding deceptive opinions), split total effort, and camouflage (i.e., hide their suspicious behaviors by balancing workload so that no single individual stands out). Surprisingly, however, only a few efforts aim to detect group-level opinion spam [23,27,28]. Moreover, most existing work employ supervised techniques [12,17,20,25] and/or utilize side information, such as behavioral [17,22,23,27,28] or linguistic [12,25] clues of spammers. The former is inadmissible, due to the difficulty in obtaining ground truth labels. The latter, on the other hand, is not adversarially robust; the spammers can fine-tune their language (e.g., usage of superlatives, self-references, etc.) and behavior (e.g., login times, IPs, etc.) to mimic genuine users as closely as possible and evade detection.

In this work, we propose a new unsupervised and scalable approach for detecting opinion spammer groups solely based on their network footprints. At its heart, our method consists of two key components:

- NFS (**Network Footprint Score**). We introduce a new graph-based measure that quantifies the statistical distortions caused by spamming activities in well-understood network characteristics. NFS is fast to compute and more robust to evasion than linguistic and behavioral measures, provided that spammers have only a partial view of the review network (Section 2).
- *GroupStrainer*. We devise a fast method to group spammers on a carefully induced subnetwork of highly suspicious products and reviewers. *GroupStrainer* employs a hierarchical clustering algorithm that leverages similarity-sensitive hashing to speed up the merging steps. The output is a set of spammer groups and their nested hierarchy, which facilitates sensemaking of their organizational structure, as well as verification by end analysts (Section 3).

In the experiments, we compare our method to various graph anomaly detection methods on synthetic datasets and study its performance on two large real-world datasets (Amazon and iTunes). Our results show that we effectively spot spammer groups with high accuracy, which is superior to existing methods.

2   Measuring Network Footprints

A perfect form of an opinion spam campaign would certainly reflect a near-replica of the characteristics that genuine reviewers exhibit on online review sites. Those characteristics include linguistic, behavioral, and relational (network-level) patterns. One can argue that language and behavior patterns are relatively easier to mimic by the spammers as compared to network-based patterns. For example, spammers could adjust their usage of certain language constructs (e.g., superlatives, self-references, etc.) that have been found to be associated with deception [25], so as to evade classifiers [8].

On the other hand, spoofing network-level characteristics in general is adversarially harder for various reasons. First, spammers (adversaries) often do not have a complete view of the entire review network, due to its sheer scale and access properties. Moreover, the reviewers that belong to a spam campaign have to act as a group, which would create different dynamics in the review network than the independent actions of genuine reviewers. Finally, spammers do not replicate "unimportant" structures in the network due to limited budget.

Therefore, in this work we focus on the network-level characteristics of opinion spammers in the user–product bipartite review network. We develop a network-based method that is able to distinguish the network footprints of spammer groups from those of individual genuine users. In particular, we propose a new measure called the NFS (for Network Footprint Score), that quantifies the extent to which statistical network characteristics of reviewers are distorted by spamming activities.

To design our NFS measure, we leverage two key observations associated with real-world networks:

1. **Neighbor Diversity:** The neighbors of a node in a real network are expected to consist of nodes with varying behavior and levels of activity. As such, the neighbors should not be overly dependent on one another; rather, they should be spread across sources of varying quality or importance. For example, in social networks a person has friends with varying levels of "popularity".
2. **Self-Similarity:** Real-world networks are self-similar [4,5]; that is, portions of the network often have properties similar to the entire network. In particular, the importances of the neighbors of a node follow a skewed, power-law-like distribution, just as the case for all of the nodes in the entire network.

In a nutshell, while the former observation implies (i) local diversity of node importances in the neighborhood of a node, the latter implies (ii) distributional similarity between node importances at local (i.e., neighborhood) and global level (i.e., at large). Importance of nodes in a network can be captured by their centrality. There exist a long list of work on developing various measures for node centralities [10]. In this work, we consider two different measures; degree and Pagerank centrality, which respectively use local and global information to quantify node importances.[3]

In the following, we describe how we utilize each of the above insights to measure the network footprints of spammer groups in review networks. More precisely, a review network is a bipartite graph G that consists of n reviewer nodes connected to m product nodes through review relations.

---

[3] We compute these measures based on the reviewer–product bipartite review network.

2.1   Neighbor Diversity of Nodes

We can translate Observation 1 above to the domain of review networks as follows. An honest set of reviewers for a product arises by independent actions of individuals with varying behavior and levels of activity. As a result, the centrality of the reviewers of a product is expected to vary to a large extent. In analogy to social networks where a person has friends with varying level of "popularity", a product would have reviewers with varying level of network centrality.

In other words, a set of reviewers all with centrality (degree or Pagerank) values falling into a narrow interval is suspicious. Such a large set of highly similar reviewers (i.e., "clones") raises the suspicion that they have emerged through certain means of a *cooperation*, e.g., under a spam campaign.

To quantify the diversity of neighbor centralities of a given product, we first split the centrality values of its reviewers into buckets to create a non-parametric estimation of their density through a histogram. We then compute the skewness of the histogram through entropy. More specifically, given the (degree of Pagerank) centralities $\{c_1^{(i)}, \dots, c_{deg(i)}^{(i)}\}$ of the reviewers of a product i with degree $deg(i)$, we create a list of buckets $k = \{0, 1, \dots\}$. We let the bucket boundary values grow exponentially as $a \cdot b^k$, as both degree and Pagerank values of nodes in real-world networks have been observed to follow skewed distributions [6,9].

For degree centrality we use $a = 3$ and $b = 3$ such that the bucket boundaries are $\{1, 3, 9, 27, \dots\}$ and place each reviewer j with degree $c_j^{(i)}$ to bucket k with $a \cdot b^{k-1} \leq c_j^{(i)} < a \cdot b^k$. On the other hand, we use $a = 0.3$ and $b = 0.3$ for Pagerank where bucket boundaries become $\{1, 0.3, 0.09, 0.027, \dots\}$ as it takes values in $[0, 1]$, and place each reviewer j with Pagerank $c_j^{(i)}$ to bucket k with $a \cdot b^{k-1} \geq c_j^{(i)} > a \cdot b^k$. The choice of a and b has little effect on our results as long as we use a logarithmic binning [24] so as to capture the skewness of data.

Given the placement of reviewers into K buckets by their centrality, we next count the reviewers in each bucket and normalize the counts by the total count $deg(i)$ to obtain a discrete probability distribution $P^{(i)}$ with values $[p_1^{(i)}, \dots, p_K^{(i)}]$. We then compute the Shannon entropy of $P^{(i)}$ as $H_c(i) = -\sum_{k=1}^{K} p_k^{(i)} \log p_k^{(i)}$ for centrality c. As such, a product i receives two neighbor-diversity scores, $H_{deg}(i)$ and $H_{pr}(i)$, for degree and Pagerank respectively. The lower these scores are, the more likely the product is the target of a spam campaign and hence the more suspicious are the reviewers of the product as they appear near-replicas of one another in the network—cooperating around the same goal, leaving similar network footprints.

2.2   Self-Similarity in Real-World Graphs

Similarly, we can leverage Observation 2 to measure the distributional distortions caused by spam activities. Specifically, self-similarity implies that the centrality of reviewers for a particular product should follow a similar distribution as to the centrality of all reviewers in the network. Note that while Observation 1 enforces the neighbor centralities to be diverse, Observation 2 requires them to also closely follow global distributional patterns. It is well-known that degree and Pagerank distributions of nodes in real-world graphs follow power-law-like distributions [6,9]. As such, a diverse but e.g., Gaussian-distributed set of neighbor centralities would still raise a red flag in terms of self-similarity, while considered normal in terms of neighbor diversity.

Therefore, we define a second type of score for each product i as the KL-divergence between the histogram density of the centralities of its reviewers $P^{(i)} = [p_1^{(i)}, \ldots, p_K^{(i)}]$ and that of all reviewers in the network denoted by Q; $KL_c(P^{(i)}\|Q) = \sum_k p_k^{(i)} \log \frac{p_k^{(i)}}{q_k}$. We compute Q in the same way we computed P's as before, where this time we split into buckets the centrality values of all the reviewers in the network.[4] As a result, a product i receives two scores for divergence from self-similarity, $KL_{deg}(i)$ and $KL_{pr}(i)$, for degree and Pagerank respectively. The larger these scores are, the more likely the product is the target of a spam campaign.

2.3   NFS Measure

To quantify the extent to which a product is under attack by a spam campaign, we combine the scores derived from the network footprints into a final score. Overall, we have four suspiciousness scores for a product, two based on neighbor-diversity; H_{deg} and H_{pr}, and two based on self-similarity; KL_{deg} and KL_{pr}. These capture different semantics; a product is likely a target the lower the H and the higher the KL scores. Moreover, they are not normalized within a standard range.

To unify the scores into a single score with a standard scale, we leverage the cumulative distribution function (CDF). In particular, let us denote by $\mathcal{H} = \{H(1), H(2), \ldots\}$ the list of entropy scores (based on degree or Pagerank) we computed for a set of products. To quantify the extremity of a particular $H(i)$, we use the empirical CDF over $\mathcal{H}$ and estimate the probability that the set contains a value as *low* as $H(i)$ as

$$f(H(i)) = P(H \leq H(i)),$$

which is equal to the fraction of scores in $\mathcal{H}$ that are less than or equal to $H(i)$. On the other hand, for KL scores, we estimate the probability that the set $\mathcal{KL} = \{KL(1), KL(2), \ldots\}$ contains a value as *high* as $KL(i)$ by

$$f(KL(i)) = 1 - P(KL \leq KL(i)).$$

As such, $f(\cdot)$ takes low values for low $H(i)$ values and high $KL(i)$ values. Finally, we combine the f values to compute the NFS of a product i as given in Equ. (1), such that $NFS(i) \in [0,1]$ where high values are suspicious.

$$NFS(i) = 1 - \sqrt{\frac{f(H_{deg}(i))^2 + f(H_{pr}(i))^2 + f(KL_{deg}(i))^2 + f(KL_{pr}(i))^2}{4}} \quad (1)$$

---

[4] We use Laplace smoothing for empty buckets.

Algorithm 1. *Compute*NFS

---

1 **Input:** Reviewer–Product graph $G = (V, E)$, degree threshold η
2 **Output:** Network Footprint Score (NFS) of products with degree $\geq \eta$
3 Compute centrality c of each reviewer in G, $c = \{degree, Pagerank\}$
4 Create a list of buckets $k = \{0, 1, \dots\}$
5 **foreach** reviewer j in G, $1 \leq j \leq n$ **do** //*Compute global histogram Q*
6   **if** $c =$ degree **then**
7   place j to bucket$_k$ with $a \cdot b^{k-1} \leq c_j^{(i)} < a \cdot b^k$, $(a = 3, b = 3)$
8   **else** place j to bucket$_k$ with $a \cdot b^{k-1} \geq c_j^{(i)} > a \cdot b^k$, $(a = .3, b = .3)$
9   **forall the** non-empty buckets $k = \{0, 1, \dots\}$ **do**
10         $q_k = |bucket_k|/n$

11 **foreach** product i with $deg(i) \geq \eta$ **do**
12   Create a list of buckets $k = \{0, 1, \dots\}$
13   **foreach** neighbor (reviewer) j of product i **do**
14   **if** $c =$ degree **then**
15   place j to bucket$_k$ with $a \cdot b^{k-1} \leq c_j^{(i)} < a \cdot b^k$, $(a = 3, b = 3)$
16   **else** place j to bucket$_k$ with $a \cdot b^{k-1} \geq c_j^{(i)} > a \cdot b^k$, $(a = .3, b = .3)$
17   **forall the** non-empty buckets $k = \{0, 1, \dots\}$ **do** //*local histogram* $P^{(i)}$
18         $p_k^{(i)} = |bucket_k|/deg(i)$
19   Compute entropy $H_c(i)$ based on $p_k^{(i)}$'s
20      $K' =$ number of uncommon buckets where $q_k \neq 0$ and $p_k^{(i)} = 0$
21   **forall the** buckets k where $q_k \neq 0$ **do** //*local smoothed histogram* $P^{(i)}$
22   **if** $p_k^{(i)} = 0$ **then** $p_k^{(i)} = 1/(deg(i) + K')$ //*Laplace smoothing*
23   **else** $p_k^{(i)} = (p_k^{(i)} \cdot deg(i))/(deg(i) + K')$ //*Re-normalize*
24   Compute divergence $KL_c(P^{(i)}\|Q)$ based on $p_k^{(i)}$'s and q_k's
25   Compute NFS of i by Equ. (1) based on $H_c(i)$ and $KL_c(P^{(i)}\|Q)$

The complete list of steps to compute NFS is given in Algorithm 1. Note that we utilize centrality density distributions P_{deg} and P_{pr} over neighbors to compute the NFS of a product. These distributions are meaningful only when a product has a large number of review(er)s, since only a few data points cannot constitute a reliable distribution (in this work products with less than 20 reviews are ignored, i.e., we set $\eta = 20$). This, however, is not a severe limitation of our approach. The reason is that spam campaigns often involve reasonably large number of reviewers in order to (1) increase the total impact on a target product and (2) share the overall effort. Small spam campaigns are of little spamming power and can be overlooked without much risk.

3 Detecting Spammer Groups

We compute the NFS for the products, as a measure of their abnormality of being targeted by suspiciously similar reviewers. Such groups of highly similar reviewers potentially work together under the same spam campaigns. Our end goal is to identify all such spammer groups.

To achieve this goal, we construct a subnetwork consisting of the top products with the highest NFS values[5] denoted as P_1, all the reviewers of these products R, and all the products that these reviewers reviewed $P_2 \supseteq P_1$. In other words, the subnetwork is the induced subgraph of our original graph G on the nodes within 2-hops away from the top products in P_1, i.e. $G[P_2 \cup R]$. We represent this subgraph with a $p \times u$ adjacency matrix A, where $|P_2| = p$ and $|R| = u$.

An example of A can be seen in Figure 3 (top). This matrix contains highly similar users, i.e., columns, since the subgraph is biased toward products with high NFS values. However, it is clear that the reviewer groups are not immediately evident from the figure. To fully automate the group identification process, we propose a fast algorithm called *GroupStrainer* that finds clusters of highly similar columns of A, which carefully re-organizes/shuffles the columns to better reveal the reviewer groups. The output of *GroupStrainer* on the example matrix can be seen in Figure 3 (bottom), and will be discussed further in Section 4.

Note that the goal here is *not* to cluster all the columns of A (notice the last several in Figure 3 (bottom) that do not belong to any group), but to chip off groups where columns within each group are strongly similar. Moreover, we ideally do not want to pre-specify the number of groups a priori, which is a challenging parameter to set. To achieve both of these goals, *GroupStrainer* adopts an agglomerative hierarchical clustering scheme, where columns are iteratively merged to form larger groups. Such a scheme also reveals the nested, hierarchical structure of the groups that provides further insights to the end analyst.

A naive agglomerative clustering has $O(u^3)$ complexity, where in each step similarities of all-pairs are compared. Moreover, clusters are merged two at a time in each step. In our approach, we leverage Locality-Sensitive Hashing (LSH) [15] to speed up the process of finding similar set of clusters. Provided a set of similar clusters, we can then merge two or more clusters at a time which speeds up the hierarchy construction.

In a nutshell, LSH is a randomized algorithm for similarity search, which ensures similar points are hashed to the same hash bucket with high probability, while non-similar points are rarely hashed to the same bucket. As a result, it quickens the similarity search by narrowing down the search to points that are hashed only to the same buckets. In order to systematically reduce the probability of highly similar points hashing to different buckets, it employs multiple hash functions. At its heart, LSH generates signatures for each data point, where it ensures that the similarity between the original points is proportional to the probability of their signatures to be equal. As a consequence, the more similar two points are, the more likely their signatures match, and the more probable that they hash to the same bucket (hence locality-, or similarity-sensitive hashing). LSH uses different, suitable signature generation functions with respect to

---

[5] Rather than an arbitrary top k, we adopt the mixture-modeling approach in [14], where NFS values are modeled as drawn from two distributions; Gaussian for normal values and exponential for outliers. Model parameters and assignment of points to distributions are learned by the EM algorithm. Top products are then the ones whose NFS values belong to the exponential distribution.

Algorithm 2. *GroupStrainer*

---

1 **Input:** $p \times u$ adjacency matrix A of 2-hop network of top products with highest NFS values, similarity lower bound s_L (default 0.5)
2 **Output:** User groups U^{out} and their hierarchical structure
3 Create a list of similarity thresholds $S = \{0.95, 0.90, \ldots, s_L\}$
4 Set $k^1 = u$, $U^1 := \{1, 2, \ldots, u\} \rightarrow \{1, 2, \ldots, u\}$ // *init reviewer groups*
5 **for** $T = 1$ *to* $|S|$ **do**
6   Estimate LSH parameters r and b for threshold $S(T)$
7   //**Step 1. Generate signatures**
8   Init signature matrix $M[i][j] \in \mathbb{R}^{rb \times k^T}$
9   **if** $T = 1$ **then** // *use Jaccard similarity, generate min-hash signatures*
10   **for** $i = 1$ *to* rb **do**
11              $\pi_i \leftarrow$ generate random permutation $(1 \ldots p)$
12   **for** $j = 1$ *to* k^T **do** $M[i][j] \leftarrow min_{v \in N_j} \pi_i(v)$
13   **else** // *use cosine similarity, generate random-projection signatures*
14   **for** $i = 1$ *to* rb **do**
15              $r_i \leftarrow$ pick a random hyperplane $\in \mathbb{R}^{p \times 1}$
16   **for** $j = 1$ *to* k^T **do** $M[i][j] \leftarrow sign(rsum(A(:, U^T=j)/|U^T=j|) \cdot r_i)$
17   //**Step 2. Generate hash tables**
18   **for** $h = 1$ *to* b **do**
19   **for** $j = 1$ *to* k^T **do** $hash(M[(h-1) \cdot r + 1 : h \cdot r][j])$
20   //**Step 3. Merge clusters from hash tables**
21   Build candidate groups g's: union of clusters that hash to *at least one* same bucket in all hash tables, i.e. $\{c_i, c_j\} \in g$ if $hash_h(c_i) = hash_h(c_j)$ for $\exists h$
22   **foreach** candidate group g **do**
23   **foreach** $c_i, c_j \in g$, $i \neq j$ **do**
24   **if** $sim(\mathbf{v_i}, \mathbf{v_j}) \geq S(T)$ **then** //*merge*
25                  $g = g \backslash \{c_i, c_j\} \cup \{c_i \cup c_j\}$
26                  $k^T = k^T - 1$, $U^T(c_i) = U^T(c_j) = min(U^T(c_i), U^T(c_j))$
27      $k^{T+1} = k^T$, $U^{T+1} = U^T$
28 **return** $U^{out} = U^{|S|+1}$, evolving groups $\{U^1, \ldots, U^{|S|}\}$ to build nested hierarchy

---

different similarity functions. In this paper, we use two: min-hashing for Jaccard similarity and random-projection for cosine similarity.

The details of our *GroupStrainer* is given in Algorithm 2, which consists of three main steps: (1) generate LSH signatures (Lines 7-16), (2) generate hash tables (17-19), and (3) merge clusters using hash buckets (20-24).

In the first iteration of step (1), i.e. $T = 1$, the clusters consist of individual, binary columns of A, represented by $\mathbf{v_j}$'s, $1 \leq j \leq u$. As the similarity measure, we use Jaccard similarity which is high for those columns (i.e., reviewers) with many exclusive common neighbors (i.e., products). Min-hashing is designed to capture the Jaccard similarity between binary vectors; that is, it can be shown that the probability that the min-hash values (signatures) of two binary vectors agree is equal to their Jaccard similarity (Lines 9-12). For $T > 1$, the clusters consist of multiple columns. This time, we represent each cluster j by a length-p real-valued vector $\mathbf{v_j}$ in which the entries denote the density of each row, i.e.,

$\mathbf{v_j} = rsum(A(:, U^T = j)/|U^T = j|$, where U^T is the mapping of columns to clusters at iteration T, $U^T = j$ depicts the indices of columns that belong to cluster j, $rsum$ is the row-sum of entries in the induced adjacency matrix, and $|U^T = j|$ is the size of cluster j. Then, we use the cosine similarity between $\mathbf{v_i}$ and $\mathbf{v_j}$ as the similarity measure of two clusters i and j. In LSH, random-projection based signature generation captures cosine similarity; that is, it can be shown that the probability that the random-projection values of two real-valued vectors agree is proportional to their cosine similarity [19] (Lines 13-16).

In step (2), LSH performs multiple hash operations on different subset of signatures to increase the probability that two highly similar items hash to the same bucket in at least one hash table (Lines 18-19).

Step (3) involves the main merging operations. First we construct the groups of candidate clusters to be merged, where all the clusters that hash to the *same* hash bucket in *at least one* hash table are put into the same group g (Line 21). These are called candidate clusters as LSH is a probabilistic algorithm, and can yield false positives. Therefore, rather than merging all the clusters in each group directly, we verify whether their similarity is above the desired threshold (Line 24), before committing the merge (Lines 25-26). Note that at this merge step, more than two clusters can be merged. With respect to complexity *GroupStrainer* is only sub-quadratic; it performs pairwise similarity computation among clusters only within groups (Line 23), rather than among all the clusters. The number of clusters within each group at a given iteration is often much smaller than the total. This contributes to a significant reduction in running time, while enabling us to focus on merging highly similar clusters.

Finally, we describe the process of constructing a nested hierarchy of clusters. A specific iteration finds groups of clusters with similarity above a desired threshold. At the beginning ($T = 1$), we set a large/conservative threshold of 0.95 such that only extremely similar columns are grouped. As T increases, we gradually lower the similarity threshold so as to allow the hierarchy to grow. Here, the user can specify a lower bound s_L for the threshold (default 0.5), so as to prevent clusters with similarity below the bound to be merged. Depending on s_L, our hierarchy may contain a single or multiple tree(s), as well as singleton columns that do not belong to any cluster. We treat all non-singleton trees as candidates of spammer groups, and inspect them in size order. Other ranking measures can also be used to prioritize inspection.

4   Evaluation

We first evaluate the performance of our method on synthetic datasets, as compared to several existing methods. Our method consists of two steps: NFS computation and *GroupStrainer*. The former tries to capture the targeted products, and the latter focuses on detecting spammer groups through their collusion on the target products. As such, we design separate data generators for each step to best simulate these scenarios. In addition, we apply our method on two real-world review datasets and show through detailed case analyses that it detects many suspicious user groups. A summary of the datasets is given in Table 1.

Table 1. Summary of synthetic and real-world datasets used in this work.

	Synthetic Data				Real-world Data	
	Chung-Lu1	Chung-Lu2	RTG1	RTG2	iTunes	Amazon
# of users	532,742	2,133,399	604,520	876,627	966,808	2,146,074
# of products	157,768	665,381	604,805	876,950	15,093	1,230,916
# of edges	1,299,059	5,191,053	3,097,342	4,644,572	1,132,329	5,838,061

4.1 Performance of NFS on Synthetic Data

Synthetic Data Generation. We use two models to create synthetic graphs: Chung-Lu [7] and the RTG [2]. Chung-Lu creates random graphs with a given degree sequence. We draw the degrees of reviewers and products from a power-law distribution with exponent 2.9 and 2.1, respectively, as observed in the real world [1,6,9]. RTG model also creates realistic bipartite graphs that not only follow power-law degree distribution but also contain communities, which are common in real-world graphs. We create two graphs with different sizes using each model (Table 1). Next, we follow the injection process in [16] to simulate and inject spammer groups into our graphs. Specifically, we add 3 spammer groups with 1000, 2000 and 4000 users respectively. Each spammer group targets a set of designated products (100, 200 and 400 in size). Each injected spammer writes 20 reviews to their target products, with σ percent camouflage written to untargeted ones. There exist two strategies to camouflage: writing reviews (1) to top 100 most popular (highest degree) products; and (2) to random untargeted products. This way, we create four injection configurations; σ=10% or 30% camouflage on popular products, and σ=10% or 30% on random ones, where larger σ and random camouflage are relatively harder to detect.

Compared Methods. NFS measures the suspiciousness of products. In order to rank the users, we utilize the FraudEagle method [1]. FraudEagle computes scores for users and products by propagating *unbiased* beliefs in the review network. We assign NFS values of products as their initial beliefs (i.e., priors). Thus users who targeted many products with large NFS values gain high score at convergence. In our setting, review ratings (often from 1 to 5) are not utilized. Thus we ignore the edge signs in FraudEagle to make these methods comparable.[6]

In addition to (1) FraudEagle [1], we also compare to (2) CatchSync [16], designed to spot synchronized behavior among users and (3) Oddball [3], for detecting users whose neighbors are in near-clique or star shapes. Oddball requires unipartite graphs, thus we use the projected review network on users, where two users with at least 5 common neighbors (products) are connected.

Performance Results. In Section 2, we introduced two key observations that we use to design NFS: neighbor diversity and self-similarity. In Fig. 1, we show the entropy H_{deg} vs. KL-divergence KL_{deg} of products on the Chung-Lu1 graph as

---

[6] Accordinly, we use a single edge compatibility table (i.e., [0.9 0.1; 0.1 0.9]) for FraudEagle [1].

Table 2. AUC of Precision-Recall curve on synthetic datasets. Two values in each entry: (former) performance on high degree users (threshold at 20) and performance on all users. (HDP: number of high degree products (≥ 20); FE: FraudEagle.)

Dataset	Camouf.	HDP	Oddball[3]	CatchSync[16]	FE[1]	NFS+FE
Chung-Lu1	10% Pop.	6170	0.990/0.937	1.000/0.009	0.570/0.569	1.000/1.000
	30% Pop.	6172	0.997/0.973	1.000/0.008	0.570/0.570	1.000/1.000
	10% Rand.	6205	0.982/0.886	1.000/0.007	0.552/0.552	1.000/1.000
	30% Rand.	6266	0.881/0.386	0.957/0.007	0.532/0.526	1.000/1.000
Chung-Lu2	10% Pop.	25306	0.977/0.943	1.000/0.002	0.294/0.294	1.000/1.000
	30% Pop.	25302	0.995/0.988	1.000/0.002	0.294/0.294	1.000/1.000
	10% Rand.	25330	0.955/0.887	1.000/0.002	0.280/0.279	1.000/1.000
	30% Rand.	25392	0.711/0.374	0.982/0.002	0.261/0.256	1.000/0.977
RTG1	10% Pop.	17771	0.945/0.852	1.000/0.008	0.176/0.176	1.000/1.000
	30% Pop.	17766	0.929/0.842	0.997/0.007	0.176/0.176	1.000/1.000
	10% Rand.	17780	0.918/0.803	0.995/0.007	0.168/0.168	1.000/1.000
	30% Rand.	17843	0.637/0.367	0.878/0.007	0.163/0.158	0.952/0.950
RTG2	10% Pop.	25658	0.906/0.778	1.000/0.005	0.129/0.129	1.000/1.000
	30% Pop.	25658	0.879/0.746	1.000/0.005	0.129/0.129	1.000/1.000
	10% Rand.	25678	0.877/0.741	0.987/0.005	0.123/0.123	1.000/1.000
	30% Rand.	25716	0.577/0.331	0.778/0.005	0.119/0.115	0.952/0.951

an example. We can see that the products targeted by a group of spammers reside in the bottom-right part of the figure.

Table 2 presents the performance results (AUC of precision recall curve) of different methods. Each method is tested on high-degree users as well as all the users. Note that for all methods, users with high degrees are easier to rank as they exhibit more information for analysis. However, when all users are considered, the rankings become contaminated with low

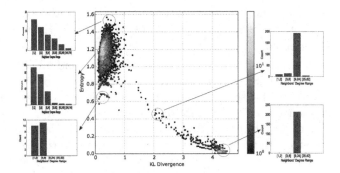

Fig. 1. Degree entropy H_{deg} vs. KL-divergence KL_{deg} of products (dots) in Chung-Lu1 ($\sigma=10\%$ pop. camouflage). Bottom-right depicts suspicious products with neighbors having abnormally similar degrees, whereas upper-left are clustered with normal products whose neighbors have diverse degree distributions that obeys the overall trend.

degree users bubbling up high in the ranking due to errors in scoring, and the performance of the methods drop relatively. This problem is especially evident

Table 3. Group discovery performance of *GroupStrainer* for varying ϵ (camouflage) and SCI (collusion) for spammers. We report NMI/hierarchy similarity threshold.

Dataset	SCI = 1.0	SCI = 0.9	SCI = 0.8	SCI = 0.7	SCI = 0.6	SCI = 0.5	SCI = 0.4
$\epsilon = 0$	1.000/0.95	1.000/0.70	1.000/0.65	1.000/0.65	0.997/0.65	1.000/0.60	0.948/0.55
$\epsilon = 0.2$	0.994/0.65	0.997/0.55	1.000/0.60	0.995/0.60	0.998/0.55	0.990/0.60	0.980/0.50
$\epsilon = 0.4$	0.993/0.50	1.000/0.55	0.993/0.55	0.998/0.55	0.994/0.55	0.988/0.55	0.980/0.50
$\epsilon = 0.6$	0.989/0.55	0.998/0.50	0.991/0.50	0.991/0.55	0.996/0.50	0.995/0.50	0.984/0.45
$\epsilon = 0.8$	0.984/0.50	0.987/0.50	0.989/0.50	0.993/0.50	0.977/0.45	0.991/0.50	0.976/0.45

for CatchSync. In contrast, our approach outperforms others and achieves near-perfect accuracy on all settings (ranking the spammers on top). These results demonstrate the effectiveness and robustness of NFS.

4.2   Performance of *GroupStrainer* on Synthetic Data

Spammer Group Generation. As *GroupStrainer* operates on a carefully induced subgraph, we inject 20 spammer groups into a subgraph with 800 users and 200 products. Spammers are assigned to groups (of sizes between 10 to 40) at random without replacement (each spammer belongs to only one group), while the targeted products (of sizes between 2 and 12) are randomly chosen with replacement (products can be attacked several times).

From real-world datasets, we observed varying degree of collusion among spammers; in some groups they write reviews to *all* the targeted products, while in other groups they are organized into sub-groups to target different subsets of products. To the best of our knowledge, the underlying motivation is to alleviate their suspiciousness and reduce their workload at the same time. To capture this behavior, we use a *Spammer Collusion Index* (SCI) for each spammer group g defined as $SCI(g) = \sum_{g_i,g_j \subset g, i \neq j} \frac{|t(g_i) \cap t(g_j)|}{|t(g_i) \cup t(g_j)|} / \binom{n}{2}$, where g_i, g_j are subgroups in g, $t(g_i)$ denotes the set of products g_i targets, and n is the number of subgroups in g. As such, SCI is the average Jaccard similarity of subgroups' target sets. We divide groups with more than 5 targets randomly into two subgroups to simulate collusion behavior. In addition, all spammers have ϵ probability to randomly write reviews to untargeted products (i.e., camouflage).

Table 3 shows the group detection performance of *GroupStrainer* on datasets simulated with varying levels of camouflage (i.e., ϵ) and collusion (i.e., SCI) among the spammers. We report NMI $\in [0, 1]$ (Normalized Mutual Information) that measures the clustering quality of *GroupStrainer* w.r.t. ground truth, as well as the corresponding level (i.e., similarity threshold) of the hierarchy that maps to the original groups. We find that *GroupStrainer* recovers the (sub)group hierarchy among spammers effectively, with high accuracy even for large ϵ.

4.3   Results on Real-World Data

Having validated the effectiveness of NFS and *GroupStrainer* through synthetic results, we next employ our method on two real-world review datasets; iTunes app-store reviews, and Amazon product reviews.

Table 4. Summary statistics of detected groups in iTunes and Amazon. #P (#U): number of products (users), t: time stamps, ⋆: ratings, s: scattered (distribution), c: concentrated (distribution), *Dup*: number of duplicate reviews/total count.

			iTunes						Amazon		
ID	#P	#U	t, ⋆	*Dup*	Developer	#P	#U	t, ⋆	*Dup*	Category, Author
1	5	31	s, c	51/154	all same	10	20	c, c	90/138	Books, all same
2	8	38	c, s	29/202	2 same	4	12	s, c	32/47	Books, all same
3	4	61	s, c	34/144	all inaccessible	7	9	c, c	44/60	Books, all same
4	4	17	c, s	0/68	1 inaccessible	7	19	s, c	5/88	Books, all same
5	5	102	c, s	8/326	different	23	42	c, c	2/468	Music, all same
6	6	50	s, c	4/173	all same	8	17	s, c	9/73	Books, 4/8 same
7	2	56	c, c	12/112	different	6	18	s, c	4/94	Movies&TV, all same
8	4	42	c, c	8/112	2 same					
9	6	67	s, c	0/137	all same					

Fig. 2 illustrates the scatter plot of H_{deg} vs. KL_{deg} for products in iTunes, where outliers with large NFS values are circled (H_{pr} vs. KL_{pr} looks similar). The adjacency matrix A of the 2-hop induced subnetwork on the outlier products is shown in Fig. 3 (top). While the groups are not directly evident here, the *GroupStrainer* output clearly reveals various colluding user groups as shown in Fig. 3 (bottom). Statistics and properties of the groups are listed in Table 4.

We note that group#1 is the same 31 users spamming 5 products of the same developer with all-5⋆ reviews, as was previously found in [1]. Our method finds other suspicious user groups; e.g., group#2 consists of 8 products, each receiving all their reviews on the

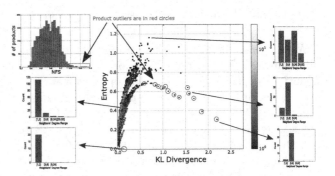

Fig. 2. Degree entropy H_{deg} vs. KL-divergence KL_{deg} of products (dots) in iTunes. Top-left: distribution of NFS scores; circled points: outlier products by NFS.

same day from a subset of 38 users. Interestingly, while the time-stamp is the same (concentrated), the ratings are a bit diversified (not all 5⋆s but 3&4⋆s as well)—potentially for camouflage. This behavior is observed among other groups; while some groups concentrate on both time and ratings (c,c), e.g., all 5⋆ in one day, most groups aim to diversify in one of the aspect. We also note the duplicate reviews across reviewers in the same group.

Similar results are obtained on Amazon, as shown in Fig. 4. We also provide the summary statistics and characteristics of the 7 colluding groups in Table 4.

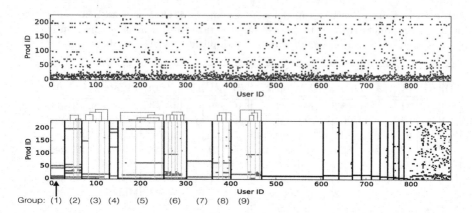

Fig. 3. (top) 2-hop induced network of top-ranking products by NFS in `iTunes`, (bottom) output of *GroupStrainer* with 9 discovered colluding user groups.

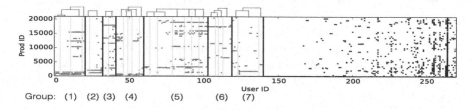

Fig. 4. Output of *GroupStrainer* on `Amazon`, with 7 discovered colluding user groups.

We find that the majority of the targeted products in our groups belong to the Books category. This is not a freak occurrence, as the media has reported that the authors of books get involved in opinion spam to gain popularity (see URL in footnote[1] on pg. 1). For example, group#1 consists of 20 users spamming 10 books. 19/20 users write their reviews on the exact same day. 15/20 has duplicate reviews across products, in total 90/138 reviews have at least one copy. Our dataset listed the same author for 9/10 books. Manual inspection of the authors revealed that the 10th book also belonged to the same author but was mis-indexed by Amazon.

5   Related Work

Opinion spam is one of the new forms of Web-based spam, emerging as a result of review-based sites (TripAdvisor, Yelp, etc.) gaining popularity. We organize related work into two: (*i*) spotting *individual* spam reviewers or fake reviews, and (*ii*) detecting *group* spammers.

Detecting Individual Spam Reviews and Reviewers. To identify individual fake *reviews*, supervised models have been trained based on text and behavioral features [17] or linguistic patterns [12,25]. Another approach [20] employs the semi-supervised co-training method by using review text and reviewer features as two separate views. Relational models have also been explored [21], which correlate reviews written by the same users and from the same IPs. Similarly, behavioral methods have been developed to identify individual spam *reviewers* [13,22]. Other approaches use association based rule mining of rating patterns [18], or temporal analysis [11]. There also exist network-based methods that spot *both* suspicious reviewers and reviews [1,26]. Those infer a suspiciousness score for each node/edge in the user-product or user-review-product network.

Detecting Group Spam. There exist only a few efforts that focus on group-level opinion spam detection [23,27,28]. This is counter-intuitive, since spam/fraud is often an organized act in which the involved individuals cooperate to reduce effort and response time, increase total impact, and camouflage so that no single individual stands out. All related work in this category define group-level or pairwise spam indicators to score reviewer groups by suspiciousness. Their indicators are several behavioral and linguistic features. In contrast, our work solely uses network footprints without relying on side information (e.g., language, time-stamps, etc.) that can be fine-tuned by spammers.

6 Conclusion

We proposed an unsupervised and scalable approach for spotting spammer groups in online review sites solely based on their network footprints. Our method consists of two main components: (1) NFS; a new measure that quantifies the statistical distortions of well-studied properties in the review network, and (2) *GroupStrainer*; a hierarchical clustering method that chips off colluding groups from a subnetwork induced on target products with high NFS values. We validated the effectiveness of our method on both synthetic and real-world datasets, where we detected various groups of users with suspicious colluding behavior.

Acknowledgments. The authors thank the anonymous reviewers for their useful comments. This material is based upon work supported by the ARO Young Investigator Program under Contract No. W911NF-14-1-0029, NSF CAREER 1452425, IIS 1408287 and IIP1069147, a Facebook Faculty Gift, an R&D grant from Northrop Grumman Aerospace Systems, and Stony Brook University Office of Vice President for Research. Any conclusions expressed in this material are of the authors' and do not necessarily reflect the views, either expressed or implied, of the funding parties.

References

1. Akoglu, L., Chandy, R., Faloutsos, C.: Opinion fraud detection in online reviews by network effects. In: ICWSM (2013)
2. Akoglu, L., Faloutsos, C.: RTG: a recursive realistic graph generator using random typing. In: Buntine, W., Grobelnik, M., Mladenić, D., Shawe-Taylor, J. (eds.) ECML PKDD 2009, Part I. LNCS, vol. 5781, pp. 13–28. Springer, Heidelberg (2009)

3. Akoglu, L., McGlohon, M., Faloutsos, C.: Oddball: spotting anomalies in weighted graphs. In: Zaki, M.J., Yu, J.X., Ravindran, B., Pudi, V. (eds.) AKDD 2010, Part II. LNCS, vol. 6119, pp. 410–421. Springer, Heidelberg (2010)

4. Barabási, A.-L., Albert, R., Jeong, H.: Scale-free characteristics of random networks: the topology of the world-wide web. Physica A: Statistical Mechanics and its Applications **281**(1–4), 69–77 (2000)

5. Benczr, A.A., Csalogny, K., Sarls, T., Uher, M.: Spamrank - fully automatic link spam detection. In: AIRWeb, pp. 25–38 (2005)

6. Broder, A.: Graph structure in the web. Computer Networks **33**(1–6), 309–320 (2000)

7. Chung, F.R.K., Lu, L.: The average distance in a random graph with given expected degrees. Internet Mathematics **1**(1), 91–113 (2003)

8. Dalvi, N., Domingos, P., Mausam, Sanghai, S., Verma, D.: Adversarial classification. In: KDD, pp. 99–108 (2004)

9. Faloutsos, M., Faloutsos, P., Faloutsos, C.: On power-law relationships of the internet topology. In: SIGCOMM, pp. 251–262 (1999)

10. Faust, K.: Centrality in affiliation networks. Social Networks **19**(2), 157–191 (1997)

11. Fei, G., Mukherjee, A., Liu, B., Hsu, M., Castellanos, M., Ghosh, R.: Exploiting burstiness in reviews for review spammer detection. In: ICWSM (2013)

12. Feng, S., Banerjee, R., Choi, Y.: Syntactic stylometry for deception detection. In: ACL (2012)

13. Feng, S., Xing, L., Gogar, A., Choi, Y.: Distributional footprints of deceptive product reviews. In: ICWSM (2012)

14. Gao, J., Tan, P.-N.: Converting output scores from outlier detection algorithms to probability estimates. In: ICDM (2006)

15. Gionis, A., Indyk, P., Motwani, R.: Similarity search in high dimensions via hashing. In: VLDB, pp. 518–529 (1999)

16. Jiang, M., Cui, P., Beutel, A., Faloutsos, C., Yang, S.: Catchsync: catching synchronized behavior in large directed graphs. In: KDD, pp. 941–950 (2014)

17. Jindal, N., Liu, B.: Opinion spam and analysis. In: WSDM, pp. 219–230 (2008)

18. Jindal, N., Liu, B., Lim, E.-P.: Finding unusual review patterns using unexpected rules. In: CIKM, pp. 1549–1552. ACM (2010)

19. Johnson, W., Lindenstrauss, J.: Extensions of Lipschitz mappings into a Hilbert space. Contemporary Mathematics, vol. 26, pp. 189–206 (1984)

20. Li, F., Huang, M., Yang, Y., Zhu, X.: Learning to identify review spam. In: IJCAI (2011)

21. Li, H., Chen, Z., Liu, B., Wei, X., Shao, J.: Spotting fake reviews via collective positive-unlabeled learning. In: ICDM (2014)

22. Mukherjee, A., Kumar, A., Liu, B., Wang, J., Hsu, M., Castellanos, M., Ghosh, R.: Spotting opinion spammers using behavioral footprints. In: KDD (2013)

23. Mukherjee, A., Liu, B., Glance, N.S.: Spotting fake reviewer groups in consumer reviews. In: WWW (2012)

24. Newman, M.: Power laws, Pareto distributions and Zipf's law. Contemporary Physics **46**(5), 323–351 (2005)

25. Ott, M., Choi, Y., Cardie, C., Hancock, J.T.: Finding deceptive opinion spam by any stretch of the imagination. In: ACL, pp. 309–319 (2011)

26. Wang, G., Xie, S., Liu, B., Yu, P.S.: Review graph based online store review spammer detection. In: ICDM, pp. 1242–1247 (2011)

27. Xu, C., Zhang, J.: Combating product review spam campaigns via multiple heterogeneous pairwise features. In: SDM. SIAM (2015)

28. Xu, C., Zhang, J., Chang, K., Long, C.: Uncovering collusive spammers in Chinese review websites. In: CIKM, pp. 979–988. ACM (2013)

Gamma Process Poisson Factorization for Joint Modeling of Network and Documents

Ayan Acharya[1]([✉]), Dean Teffer[2], Jette Henderson[2], Marcus Tyler[2],
Mingyuan Zhou[3], and Joydeep Ghosh[1]

[1] Department of ECE, University of Texas at Austin, Austin, USA
{aacharya,jghosh}@utexas.edu
[2] Applied Research Laboratories, University of Texas at Austin, Austin, USA
{dean.teffer,jhende,mtyler}@arlut.utexas.edu
[3] Department of IROM, University of Texas at Austin, Austin, USA
mzhou@utexas.edu

Abstract. Developing models to discover, analyze, and predict clusters within networked entities is an area of active and diverse research. However, many of the existing approaches do not take into consideration pertinent auxiliary information. This paper introduces Joint Gamma Process Poisson Factorization (J-GPPF) to jointly model network and side-information. J-GPPF naturally fits sparse networks, accommodates separately-clustered side information in a principled way, and effectively addresses the computational challenges of analyzing large networks. Evaluated with hold-out link prediction performance on sparse networks (both synthetic and real-world) with side information, J-GPPF is shown to clearly outperform algorithms that only model the network adjacency matrix.

Keywords: Network modeling · Poisson factorization · Gamma process

1 Introduction

Social networks and other relational datasets often involve a large number of nodes N with sparse connections between them. If the relationship is symmetric, it can be represented compactly using a binary symmetric adjacency matrix $\mathbf{B} \in \{0,1\}^{N \times N}$, where $b_{ij} = b_{ji} = 1$ if and only if nodes i and j are linked. Often, the nodes in such datasets are also associated with "side information," such as documents read or written, movies rated, or messages sent by these nodes. Integer-valued side information are commonly observed and can be naturally represented by a count matrix $\mathbf{Y} \in \mathbb{Z}^{D \times V}$, where $\mathbb{Z} = \{0, 1, \ldots\}$. For example, $\mathbf{B}$ may represent a coauthor network and $\mathbf{Y}$ may correspond to a document-by-word count matrix representing the documents written by all these authors. In another example, $\mathbf{B}$ may represent a user-by-user social network and $\mathbf{Y}$ may represent a user-by-item rating matrix that adds nuance and support to the network data. Incorporating such side information can result in better community identification

© Springer International Publishing Switzerland 2015
A. Appice et al. (Eds.): ECML PKDD 2015, Part I, LNAI 9284, pp. 283–299, 2015.
DOI: 10.1007/978-3-319-23528-8_18

and superior link prediction performance as compared to modeling only the network adjacency matrix $\mathbf{B}$, especially for sparse networks.

Many of the popular network models [2,13,18,25,28] are demonstrated to work well for small size networks. However, small networks are often dense, while larger real-world networks tend to be much sparser and hence challenge existing modeling approaches. Incorporating auxiliary information associated with the nodes has the potential to address such challenges, as it may help better identify latent communities and predict missing links. A model that takes advantage of such side information has the potential to outperform network-only models. However, the side information may not necessarily suggest the same community structure as the existing links. Thus a network latent factor model that allows separate factors for side information and network interactions, but at the same time is equipped with a mechanism to capture dependencies between the two types of factors, is desirable.

This paper proposes **J**oint **G**amma **P**rocess **P**oisson **F**actorization (J-GPPF) to jointly factorize $\mathbf{B}$ and $\mathbf{Y}$ in a nonparametric Bayesian manner. The paper makes the following contributions: 1) we present a fast and effective model that uses side information to help discover latent network structures, 2) we perform nonparametric Bayesian modeling for discovering latent structures in both $\mathbf{B}$ and $\mathbf{Y}$, and 3) our model scales with the number of non-zero entries in the network $S_{\mathbf{B}}$ as $\mathrm{O}\,(S_{\mathbf{B}}K_{\mathbf{B}})$, where $K_{\mathbf{B}}$ is the number of network groups inferred from the data.

The remainder of the paper is organized as follows. We present background material and related work in Section 2. J-GPPF and its inference algorithm are explained in Section 3. Experimental results are reported in Section 4, followed by conclusions in Section 5.

2   Background and Related Work

This section presents the related literature and the background materials that are useful for understanding the framework described in Section 3.

2.1   Negative Binomial Distribution

The negative binomial (NB) distribution $m \sim \mathrm{NB}(r,p)$, with probability mass function (PMF) $\Pr(M = m) = \frac{\Gamma(m+r)}{m!\Gamma(r)}p^m(1-p)^r$ for $m \in \mathbb{Z}$, can be augmented into a gamma-Poisson construction as $m \sim \mathrm{Pois}(\lambda)$, $\lambda \sim \mathrm{Gamma}(r,p/(1-p))$, where the gamma distribution is parameterized by its shape r and scale $p/(1-p)$. It can also be augmented under a compound Poisson representation as $m = \sum_{t=1}^{l} u_t, u_t \overset{iid}{\sim} \mathrm{Log}(p), l \sim \mathrm{Pois}(-r\ln(1-p))$, where $u \sim \mathrm{Log}(p)$ is the logarithmic distribution [17]. Consequently, we have the following Lemma.

Lemma 1 ([41]). *If $m \sim \mathrm{NB}(r,p)$ is represented under its compound Poisson representation, then the conditional posterior of l given m and r has PMF:*

$$\Pr(l = j | m, r) = \frac{\Gamma(r)}{\Gamma(m+r)} |s(m,j)| r^j, \; j = 0, 1, \cdots, m, \tag{1}$$

where $|s(m,j)|$ are unsigned Stirling numbers of the first kind. We denote this conditional posterior as $(l|m,r) \sim \mathrm{CRT}(m,r)$, a Chinese restaurant table (CRT) count random variable, which can be generated via $l = \sum_{n=1}^{m} z_n, z_n \sim$ Bernoulli$(r/(n-1+r))$.

Lemma 2. *Let $X = \sum_{k=1}^{K} x_k$, $x_k \sim \mathrm{Pois}(\zeta_k)$ $\forall k$, and $\zeta = \sum_{k=1}^{K} \zeta_k$. If $(y_1, \cdots, y_K | X) \sim \mathrm{Mult}(X, \zeta_1/\zeta, \cdots, \zeta_K/\zeta)$ and $X \sim \mathrm{Pois}(\zeta)$, then the following holds:*

$$P(X, x_1, \cdots, x_K) = P(X, y_1, \cdots, y_K). \tag{2}$$

Lemma 3. *If $x_i \sim \mathrm{Pois}(m_i \lambda)$, $\lambda \sim \mathrm{Gamma}(r, 1/c)$, then $x = \sum_i x_i \sim \mathrm{NB}(r, p)$, where $p = (\sum_i m_i)/(c + \sum_i m_i)$.*

Lemma 4. *If $x_i \sim \mathrm{Pois}(m_i \lambda)$, $\lambda \sim \mathrm{Gamma}(r, 1/c)$, then*

$$(\lambda | \{x_i\}, r, c) \sim \mathrm{Gamma}\left(r + \sum_i x_i, 1/(c + \sum_i m_i)\right). \tag{3}$$

Lemma 5. *If $r_i \sim \mathrm{Gamma}(a_i, 1/b)$ $\forall i$, $b \sim \mathrm{Gamma}(c, 1/d)$, then we have:*

$$(b | \{r_i, a_i\}, c, d) \sim \mathrm{Gamma}\left(\sum_i a_i + c, 1/(\sum_i r_i + d)\right). \tag{4}$$

The proofs of Lemmas 3, 4 and 5 follow from the definitions of Gamma, Poisson and Negative Binomial distributions.

Lemma 6. *If $x_i \sim \mathrm{Pois}(m_i r_2)$, $r_2 \sim \mathrm{Gamma}(r_1, 1/d)$, $r_1 \sim \mathrm{Gamma}(a, 1/b)$, then $(r_1|-) \sim \mathrm{Gamma}(a + \ell, 1/(b - \log(1-p)))$ where $(\ell | x, r_1) \sim \mathrm{CRT}(\sum_i x_i, r_1)$ and $p = \sum_i m_i/(d + \sum_i m_i)$. The proof and illustration can be found in Section 3.3 of [1].*

2.2 Gamma Process

The Gamma Process [12,36] $G \sim \Gamma\mathrm{P}(c, H)$ is a completely random measure defined on the product space $\mathbb{R}_+ \times \Omega$, with concentration parameter c and a finite and continuous base measure H over a complete separable metric space Ω, such that $G(A_i) \sim \mathrm{Gamma}(H(A_i), 1/c)$ are independent gamma random variables for disjoint partition $\{A_i\}_i$ of Ω. The Lévy measure of the Gamma Process can be expressed as $\nu(drd\omega) = r^{-1} e^{-cr} dr H(d\omega)$. Since the Poisson intensity $\nu^+ = \nu(\mathbb{R}_+ \times \Omega) = \infty$ and the value of $\int_{\mathbb{R}_+ \times \Omega} r\nu(drd\omega)$ is finite, a draw from the Gamma Process consists of countably infinite atoms, which can be expressed as follows:

$$G = \sum_{k=1}^{\infty} r_k \delta_{\omega_k}, \; (r_k, \omega_k) \overset{iid}{\sim} \pi(drd\omega), \; \pi(drd\omega)\nu^+ \equiv \nu(drd\omega). \tag{5}$$

A gamma process based model has an inherent shrinkage mechanism, as in the prior the number of atoms with weights greater than $\tau \in \mathbb{R}_+$ follows a Poisson distribution with parameter $H(\Omega) \int_\tau^\infty r^{-1}\exp(-cr)dr$, the value of which decreases as τ increases.

2.3   Network Modeling, Topic Modeling and Count Matrix Factorization

The Infinite Relational Model (IRM [18]) allows for multiple types of relations between entities in a network and an infinite number of clusters, but restricts these entities to belong to only one cluster. The Mixed Membership Stochastic Blockmodel (MMSB [2]) assumes that each node in the network can exhibit a mixture of communities. Though the MMSB has been applied successfully to discover complex network structure in a variety of applications, the computational complexity of the underlying inference mechanism is in the order of N^2, which limits its use to small networks. Computation complexity is also a problem with many other existing latent variable network models, such as the latent feature relational model [25] and its max margin version [44], and the infinite latent attribute model [28]. The Assortative Mixed-Membership Stochastic Blockmodel (a-MMSB [13]) bypasses the quadratic complexity of the MMSB by making certain assumptions about the network structure that might not be true in general. The hierarchical Dirichlet process relational model [19] allows mixed membership with an unbounded number of latent communities; however, it is built on the a-MMSB whose assumptions could be restrictive.

Some of the existing approaches handle sparsity in real-world networks by using some auxiliary information [21,24,39]. For example, in a protein-protein interaction network, the features describing the biological properties of each protein can be used [24]. In an extremely sparse social network, information about each user's profile can be used to better recommend friends [21]. Recommender system and text mining researchers, in contrast, tend to take an orthogonal approach. In recommender systems [10,22], $\mathbf{Y}$ may represent a user-by-item rating matrix and the objective in this setting is to predict the missing entries in $\mathbf{Y}$, and the social network matrix $\mathbf{B}$ plays a secondary role in providing auxiliary information to facilitate this task [22]. Similarly, in the text mining community, many existing models [3,23,26,35] use the network information or other forms of side information to improve the discovery of "topics" from the document-by-word matrix $\mathbf{Y}$. The matrix $\mathbf{B}$ can represent, for example, the interaction network of authors participating in writing the documents. The Relational Topic Model [11] discovers links between documents based on their topic distributions, obtained through unsupervised exploration. The Author-Topic framework (AT [30]) and the Author-Recipient-Topic model (ART [23]) jointly model documents along with the authors of the documents. Block-LDA [3], on the other hand, provides a generative model for the links between authors and recipients in addition to documents. The Group-Topic model [34] addresses the task of modeling events pertaining to pairs of entities with textual attributes that annotate the event. J-GPPF differs from these existing approaches in mathematical formulation, including more effective modeling of both sparsity and the dependence between network interactions and side information.

A large number of discrete latent variable models for count matrix factorization can be united under Poisson factor analysis (PFA) [43], which factorizes a count matrix $\mathbf{Y} \in \mathbb{Z}^{D \times V}$ under the Poisson likelihood as $\mathbf{Y} \sim \mathrm{Pois}(\boldsymbol{\Phi\Theta})$, where $\boldsymbol{\Phi} \in \mathbb{R}_+^{D \times K}$ is the factor loading matrix or dictionary, $\boldsymbol{\Theta} \in \mathbb{R}_+^{K \times V}$ is the factor score matrix. A wide variety of algorithms, although constructed with different motivations and for distinct problems, can all be viewed as PFA with different prior distributions imposed on $\boldsymbol{\Phi}$ and $\boldsymbol{\Theta}$. For example, non-negative matrix factorization [9,20], with the objective to minimize the Kullback-Leibler divergence between $\mathbf{N}$ and its factorization $\boldsymbol{\Phi\Theta}$, is essentially PFA solved with maximum likelihood estimation. LDA [6] is equivalent to PFA, in terms of both block Gibbs sampling and variational inference, if Dirichlet distribution priors are imposed on both $\boldsymbol{\phi}_k \in \mathbb{R}_+^D$, the columns of $\boldsymbol{\Phi}$, and $\boldsymbol{\theta}_k \in \mathbb{R}_+^V$, the columns of $\boldsymbol{\Theta}$. The gamma-Poisson model [8,32] is PFA with gamma priors on $\boldsymbol{\Phi}$ and $\boldsymbol{\Theta}$. A family of negative binomial (NB) processes, such as the beta-NB [7,43] and gamma-NB processes [41,42], impose different gamma priors on $\{\theta_{vk}\}$, the marginalization of which leads to differently parameterized NB distributions to explain the latent counts. Both the beta-NB and gamma-NB process PFAs are nonparametric Bayesian models that allow K to grow without limits [16].

J-GPPF models both $\mathbf{Y}$ and $\mathbf{B}$ using Poisson factorization. As discussed in [1], Poisson factorization has several practical advantages over other factorization methods that use Gaussian assumptions (*e.g.* in [22]). First, zero-valued observations could be efficiently processed during inference, so the model can readily accommodate large, sparse datasets. Second, Poisson factorization is a natural representation of count data. Additionally, the model allows mixed membership across an unbounded number of latent communities using the gamma Process as a prior. The authors in [4] also use Poisson factorization to model a binary interaction matrix. However, this is a parametric model and a KL-divergence based objective is optimized w.r.t. the latent factors without using any prior information. To model the binary observations of the network matrix $\mathbf{B}$, J-GPPF additionally uses a novel Poisson-Bernoulli (PoBe) link, discussed in detail in Section 3, that transforms the count values from the Poisson factorization to binary values. Similar transformation has also been used in the BigCLAM model [37] which builds on the works of [4]. This model was later extended to include non-network information in the form of binary attributes [38]. Neither BigCLAM nor its extension allows non-parametric modeling or imposing prior structure on the latent factors, thereby limiting the flexibility of the models and making the obtained solutions more sensitive to initialization. The collaborative topic Poisson factorization (CTPF) framework proposed in [15] solves a different problem where the objective is to recommend articles to users of similar interest. CTPF is a parametric model and variational approximation is adopted to solve the inference.

3   Joint Gamma Process Poisson Factorization (J-GPPF)

Let there be a network of N users encoded in an $N \times N$ binary matrix $\boldsymbol{B}$. The users in the network participate in writing D documents summarized in a $D \times V$ count matrix $\boldsymbol{Y}$, where V is the size of the vocabulary. Additionally, a binary matrix $\boldsymbol{Z}$ of dimension $D \times N$ can also be maintained, where the unity entries in each column indicate the set of documents in which the corresponding user contributes. In applications where $\boldsymbol{B}$ represents a user-by-user social network and $\boldsymbol{Y}$ represents a user-by-item rating matrix, $\boldsymbol{Z}$ turns out to be an N-dimensional identity matrix. However, in the following model description we consider the more general document-author framework. Also, to make the notations more explicit, the variables associated with the side information have $\boldsymbol{Y}$ as a subscript (e.g., $G_{\boldsymbol{Y}}$) and those associated with the network make similar use of the subscript $\boldsymbol{B}$ (e.g., $G_{\boldsymbol{B}}$). Also, if $\boldsymbol{Y}$ represents a matrix of dimension $D \times V$, then $y_{.w}$ represents the sum over all the rows for the entries in the wth column, and $y_{d.}$ represents the sum over all the columns for the entries in the dth row.

Before providing an explicit description of the model, we introduce two separate Gamma Processes. The first one models the latent factors in the network and also contributes to generate the count matrix. A draw from this Gamma Process $G_{\boldsymbol{B}} \sim \Gamma\mathrm{P}(c_{\boldsymbol{B}}, H_{\boldsymbol{B}})$ is expressed as: $G_{\boldsymbol{B}} = \sum_{k_{\boldsymbol{B}}=1}^{\infty} \rho_{k_{\boldsymbol{B}}} \delta_{\phi_{k_{\boldsymbol{B}}}}$, where $\phi_{k_{\boldsymbol{B}}} \in \Omega_{\boldsymbol{B}}$ is an atom drawn from an N-dimensional base distribution as $\phi_{k_{\boldsymbol{B}}} \sim \prod_{n=1}^{N} \mathrm{Gamma}(a_{\boldsymbol{B}}, 1/\sigma_n)$, $\rho_{k_{\boldsymbol{B}}} = G_{\boldsymbol{B}}(\phi_{k_{\boldsymbol{B}}})$ is the associated weight, and $H_{\boldsymbol{B}}$ is the corresponding base measure. The second Gamma Process models the latent groups of side information. A draw from this gamma process $G_{\boldsymbol{Y}} \sim \Gamma\mathrm{P}(c_{\boldsymbol{Y}}, H_{\boldsymbol{Y}})$ is expressed as: $G_{\boldsymbol{Y}} = \sum_{k_{\boldsymbol{Y}}=1}^{\infty} r_{k_{\boldsymbol{Y}}} \delta_{\beta_{k_{\boldsymbol{Y}}}}$, where $\beta_{k_{\boldsymbol{Y}}} \in \Omega_{\boldsymbol{Y}}$ is an atom drawn from a V-dimensional base distribution as $\beta_{k_{\boldsymbol{Y}}} \sim \mathrm{Dir}(\boldsymbol{\xi}_{\boldsymbol{Y}})$, $r_{k_{\boldsymbol{Y}}} = G_{\boldsymbol{Y}}(\beta_{k_{\boldsymbol{Y}}})$ is the associated weight, and $H_{\boldsymbol{Y}}$ is the corresponding base measure. Also, $\gamma_{\boldsymbol{B}} = H_{\boldsymbol{B}}(\Omega_{\boldsymbol{B}})$ is defined as the mass parameter corresponding to the base measure $H_{\boldsymbol{B}}$ and $\gamma_{\boldsymbol{Y}} = H_{\boldsymbol{Y}}(\Omega_{\boldsymbol{Y}})$ is defined as the mass parameter corresponding to the base measure $H_{\boldsymbol{Y}}$. In the following paragraphs, we explain how these Gamma processes, with the atoms and their associated weights, are used for modeling both $\boldsymbol{B}$ and $\boldsymbol{Y}$.

The $(n, m)^{\text{th}}$ entry in the matrix $\boldsymbol{B}$ is assumed to be derived from a latent count as:

$$b_{nm} = \mathbb{I}_{\{x_{nm} \geq 1\}}, \quad x_{nm} \sim \mathrm{Pois}\left(\lambda_{nm}\right), \lambda_{nm} = \sum_{k_{\boldsymbol{B}}} \lambda_{nmk_{\boldsymbol{B}}}, \tag{6}$$

where $\lambda_{nmk_{\boldsymbol{B}}} = \rho_{k_{\boldsymbol{B}}} \phi_{nk_{\boldsymbol{B}}} \phi_{mk_{\boldsymbol{B}}}$. This is called as the Poisson-Bernoulli (PoBe) link in [1,40]. The distribution of b_{nm} given λ_{nm} is named as the Poisson-Bernoulli distribution, with the PMF: $f(b_{nm}|\lambda_{nm}) = e^{-\lambda_{nm}(1-b_{nm})}(1 - e^{-\lambda_{nm}})^{b_{nm}}$. One may consider $\lambda_{nmk_{\boldsymbol{B}}}$ as the strength of mutual latent community membership between nodes n and m in the network for latent community $k_{\boldsymbol{B}}$, and λ_{nm} as the interaction strength aggregating all possible community membership. Using Lemma 2, one may augment the above representation as

$x_{nm} = \sum_{k_B} x_{nmk_B}$, $x_{nmk_B} \sim \text{Pois}(\lambda_{nmk_B})$. Thus each interaction pattern contributes a count and the total latent count aggregates the countably infinite interaction patters.

Unlike the usual approach that links the binary observations to latent Gaussian random variables with a logistic or probit function, the above approach links the binary observations to Poisson random variables. Thus, this approach transforms the problem of modeling binary network interaction into a count modeling problem, providing several potential advantages. First, it is more interpretable because ρ_{k_B} and ϕ_{k_B} are non-negative and the aggregation of different interaction patterns increases the probability of establishing a link between two nodes. Second, the computational benefit is significant since the computational complexity is approximately linear in the number of non-zeros S_B in the observed binary adjacency matrix $\boldsymbol{B}$. This benefit is especially pertinent in many real-word datasets where S_B is significantly smaller than N^2.

To model the matrix $\boldsymbol{Y}$, its $(d, w)^{\text{th}}$ entry y_{dw} is generated as:

$$y_{dw} \sim \text{Pois}(\zeta_{dw}), \zeta_{dw} = \left(\sum_{k_Y} \zeta_{Y dwk_Y} + \sum_{k_B} \zeta_{B dwk_B} \right),$$

$$\zeta_{Y dwk_Y} = r_{k_Y} \theta_{dk_Y} \beta_{wk_Y}, \zeta_{B dwk_B} = \epsilon \rho_{k_B} \left(\sum_n Z_{nd} \phi_{nk_B} \right) \psi_{wk_B},$$

where $Z_{nd} \in \{0, 1\}$ and $Z_{nd} = 1$ if and only if author n is one of the authors of paper d. One can consider ζ_{dw} as the affinity of document d for word w, This affinity is influenced by two different components, one of which comes from the network modeling. Without the contribution from network modeling, the joint model reduces to a gamma process Poisson matrix factorization model, in which the matrix $\boldsymbol{Y}$ is factorized in such a way that $y_{dw} \sim \text{Pois} \left(\sum_{k_Y} r_{k_Y} \theta_{dk_Y} \beta_{wk_Y} \right)$.

Here, $\boldsymbol{\Theta} \in \mathbb{R}_+^{D \times \infty}$ is the factor score matrix, $\boldsymbol{\beta} \in \mathbb{R}_+^{V \times \infty}$ is the factor loading matrix (or dictionary) and r_{k_Y} signifies the weight of the k_Y^{th} factor. The number of latent factors, possibly smaller than both D and V, would be inferred from the data.

In the proposed joint model, $\boldsymbol{Y}$ is also determined by the users participating in writing the dth document. We assume that the distribution over word counts for a document is a function of both its topic distribution as well as the characteristics of the users associated with it. In the author-document framework, the authors employ different writing styles and have expertise in different domains. For example, an author from machine learning and statistics would use words like "probability", "classifiers", "patterns", "prediction" more often than an author with an economics background. Frameworks such as author-topic model [23, 30] were motivated by a related concept. In the user-rating framework, the entries in $\boldsymbol{Y}$ are also believed to be influenced by the interaction network of the users. Such influence of the authors is modeled by the interaction of the authors in the latent communities via the latent factors $\boldsymbol{\phi} \in \mathbb{R}_+^{N \times \infty}$ and $\boldsymbol{\psi} \in \mathbb{R}_+^{V \times \infty}$, which encodes the writing style of the authors belonging to different latent communities. Since an infinite number of network communities is maintained, each entry y_{dw} is assumed to come from an infinite dimensional interaction. ρ_{k_B} signifies the interaction strength corresponding to the k_B^{th} network community. The contributions of the interaction from all the authors participating in a given document are accumulated to produce the total contribution from the networks in generating y_{dw}. Since $\mathbf{B}$ and $\mathbf{Y}$ might have different levels of sparsity and the range of integers in $\mathbf{Y}$ can be quite large, a parameter ϵ is required to balance the contribution of the network communities in dictating the structure of $\mathbf{Y}$. A low value of ϵ forces disjoint modeling of $\mathbf{B}$ and $\mathbf{Y}$, while a higher value implies joint modeling of $\mathbf{B}$ and $\mathbf{Y}$ where information can flow both ways, from network discovery to topic discovery and vice-versa. We present a thorough discussion of the effect of ϵ in Section 4.1. To complete the generative process, we put Gamma priors over σ_n, ς_d, $c_{\mathbf{B}}$, $c_{\mathbf{Y}}$ and ϵ as:

$$c_{\mathbf{B}} \sim \text{Gamma}(g_{\mathbf{B}}, 1/h_{\mathbf{B}}), c_{\mathbf{Y}} \sim \text{Gamma}(g_{\mathbf{Y}}, 1/h_{\mathbf{Y}}), \epsilon \sim \text{Gamma}(g_0, 1/f_0). \quad (7)$$

$$\sigma_n \sim \text{Gamma}(\alpha_{\mathbf{B}}, 1/\varepsilon_{\mathbf{B}}), \varsigma_d \sim \text{Gamma}(\alpha_{\mathbf{Y}}, 1/\varepsilon_{\mathbf{Y}}). \quad (8)$$

3.1   Inference *via* Gibbs Sampling

Though J-GPPF supports countably infinite number of latent communities for network modeling and infinite number of latent factors for topic modeling, in practice it is impossible to instantiate all of them. Instead of marginalizing out the underlying stochastic process [5,27] or using slice sampling [33] for nonparametric modeling, for simplicity, we consider a finite approximation of the infinite model by truncating the number of graph communities and the latent topics to $K_{\mathbf{B}}$ and $K_{\mathbf{Y}}$ respectively, by letting $\rho_{k_B} \sim \text{Gamma}(\gamma_{\mathbf{B}}/K_{\mathbf{B}}, 1/c_{\mathbf{B}})$ and $r_{k_Y} \sim \text{Gamma}(\gamma_{\mathbf{Y}}/K_{\mathbf{Y}}, 1/c_{\mathbf{Y}})$. Such approximation approaches the original infinite model as both $K_{\mathbf{B}}$ and $K_{\mathbf{Y}}$ approach infinity. With such finite approximation, the generative process of J-GPPF is summarized in Table 1. For notational convenience, we represent the set of documents the nth author contributes to as $\mathcal{Z}_n$ and the set of authors contributing to the dth document as $\mathcal{Z}_d$.

Sampling of $(x_{nmk_B})_{k_B=1}^{K_B}$: We first sample the network links according to the following:

$$(x_{nm}|-) \sim b_{nm}\text{Pois}_+ \left(\sum_{k_B=1}^{K_B} \lambda_{nmk_B} \right). \quad (9)$$

Sampling from a truncated Poisson distribution is described in detail in [40]. Since, one can augment $x_{nm} \sim \text{Pois}\left(\sum_{k_B=1}^{K_B} \lambda_{nmk_B}\right)$ as $x_{nm} = \sum_{k_B=1}^{K_B} x_{nmk_B}$, where $x_{nmk_B} \sim \text{Pois}(\lambda_{nmk_B})$, equivalently, one obtains the following:

$$\left((x_{nmk_B})_{k_B=1}^{K_B}|-\right) \sim \text{Mult}\left(x_{nm}, \left(\lambda_{nmk_B}/\sum_{k_B=1}^{K_B}\lambda_{nmk_B}\right)_{k_B=1}^{K_B}\right). \quad (10)$$

Table 1. Generative Process of J-GPPF

$$b_{nm} = I_{\{x_{nm} \geq 1\}}, x_{nm} \sim \text{Pois}\left(\sum_{k_B} \rho_{k_B} \phi_{nk_B} \phi_{mk_B}\right),$$

$$y_{dw} \sim \text{Pois}\left(\sum_{k_Y} r_{k_Y} \theta_{dk_Y} \beta_{wk_Y} + \epsilon \sum_{k_B} \rho_{k_B} \left(\sum_n Z_{nd} \phi_{nk_B}\right) \psi_{wk_B}\right),$$

$$\phi_{k_B} \sim \prod_{n=1}^N \text{Gamma}(a_B, 1/\sigma_n),\ \psi_{k_B} \sim \text{Dir}(\xi_B),$$

$$\theta_{k_Y} \sim \prod_{d=1}^D \text{Gamma}(a_Y, 1/\varsigma_d),\ \beta_{k_Y} \sim \text{Dir}(\xi_Y),\ \epsilon \sim \text{Gamma}(f_0, 1/g_0),$$

$$\sigma_n \sim \text{Gamma}(\alpha_B, 1/\varepsilon_B),\ \rho_{k_B} \sim \text{Gamma}(\gamma_B/K_B, 1/c_B),$$

$$\gamma_B \sim \text{Gamma}(e_B, 1/f_B),\ c_B \sim \text{Gamma}(g_B, 1/h_B),$$

$$\varsigma_d \sim \text{Gamma}(\alpha_Y, 1/\varepsilon_Y),\ r_{k_Y} \sim \text{Gamma}(\gamma_Y/K_Y, 1/c_Y),$$

$$\gamma_Y \sim \text{Gamma}(e_Y, 1/f_Y),\ c_Y \sim \text{Gamma}(g_Y, 1/h_Y).$$

Sampling of $(y_{dwk})_k$: Since, one can augment $y_{dw} \sim \text{Pois}(\zeta_{dw})$ as $y_{dw} = \sum_{k_Y=1}^{K_Y} y_{dwk_Y} + \sum_{n \in \mathcal{Z}_d} \sum_{k_B=1}^{K_B} y_{dnwk_B},\ y_{dwk_Y} \sim \text{Pois}(\zeta_{dwk_Y}),\ y_{dnwk_B} \sim \text{Pois}(\zeta_{dnwk_B})$, again following Lemma 2, we have:

$$\left((y_{dwk_Y})_{k_Y=1}^{K_Y}, (y_{dnwk_B})_{k_B=1, n \in \mathcal{Z}_d}^{K_B} |-\right) \sim \text{Mult}\left(y_{dw}, \frac{\{\zeta_{dwk_Y}\}_{k_Y}, \{\zeta_{dnwk_B}\}_{n \in \mathcal{Z}_d, k_B}}{\sum_{k_Y} \zeta_{dwk_Y} + \sum_{n \in \mathcal{Z}_d} \sum_{k_B} \zeta_{dnwk_B}}\right). (11)$$

Sampling of ϕ_{nk_B}, ρ_{k_B}, θ_{dk_Y}, r_{k_Y} **and** ϵ : Sampling of these parameters follow from Lemma 4 and are given as follows:

$$(\phi_{nk_B}|-) \sim \text{Gamma}\left(a_B + x_{n \cdot k_B} + y_{\cdot n \cdot k_B}, \frac{1}{\sigma_n + \rho_{k_B}(\phi_{k_B}^{-n} + \epsilon|\mathcal{Z}_n|)}\right), (12)$$

$$(\rho_{k_B}|-) \sim \text{Gamma}\left(\frac{\gamma_B}{K_B} + x_{\cdot \cdot k_B} + y_{\cdot \cdot \cdot k_B}, \frac{1}{c_B + \sum_n \phi_{nk_B} \phi_{k_B}^{-n} + \epsilon \sum_n |\mathcal{Z}_n| \phi_{nk_B}}\right), (13)$$

$$(\theta_{dk_Y}|-) \sim \text{Gamma}\left(a_Y + y_{d \cdot k_Y}, \frac{1}{\varsigma_d + r_{k_Y}}\right), (r_{k_Y}|-) \sim \text{Gamma}\left(\frac{\gamma_Y}{K_Y} + y_{\cdot \cdot k_Y}, \frac{1}{c_Y + \theta_{\cdot k_Y}}\right), (14)$$

$$(\epsilon|-) \sim \text{Gamma}\left(f_0 + \sum_{k=1}^{K_B} y_{\cdot \cdot \cdot k}, \frac{1}{g_0 + q_0}\right), q_0 = \sum_{k=1}^{K_B} \rho_{k_B} \sum_{n=1}^N |\mathcal{Z}_n| \phi_{nk_B}. (15)$$

The sampling of parameters ϕ_{nk_B} and ρ_{k_B} exhibits how information from the count matrix $\mathbf{Y}$ influences the discovery of the latent network structure. The latent counts from $\mathbf{Y}$ impact the shape parameters for both the posterior gamma distribution of ϕ_{nk_B} and ρ_{k_B}, while $\mathbf{Z}$ influences the corresponding scale parameters.

Sampling of ψ_{k_B} : Since $y_{dnwk_B} \sim \text{Pois}(\epsilon \rho_{k_B} \phi_{nk_B} \psi_{wk_B})$, using Lemma 2 we have: $(y_{\cdot \cdot wk_B})_{w=1}^V \sim \text{Mult}(y_{\cdot \cdot \cdot k_B, (\psi_{wk_B})_{w=1}^V})$. Since the Dirichlet distribution is conjugate to the multinomial, the posterior of ψ_{k_B} also becomes a Dirichlet distribution and can be sampled as:

$$(\psi_{k_B}|-) \sim \text{Dir}\left(\xi_{B1} + y_{\cdot \cdot 1k_B}, \cdots, \xi_{BV} + y_{\cdot \cdot Vk_B}\right). (16)$$

Sampling of β_{k_Y} : Since $y_{dwk_Y} \sim \text{Pois}(r_{k_Y}\theta_{dk_Y}\beta_{wk_Y})$, again using Lemma 2, we have:

$$(y_{.wk_Y})_{w=1}^{V} \sim \text{Mult}\left(y_{..k_Y}, (\beta_{wk_Y})_{w=1}^{V}\right).$$

Using conjugacy, the posterior of β_{k_Y} can be sampled as:

$$(\beta_{k_Y}|-) \sim \text{Dir}\left(\boldsymbol{\xi}_{Y1} + y_{.1k_Y}, \cdots, \boldsymbol{\xi}_{YV} + y_{.V+k_Y}\right). \tag{17}$$

Sampling of σ_n, ς_d, c_B and c_Y: Sampling of these parameters follow from Lemma 5 and are given as:

$$(\sigma_n|-) \sim \text{Gamma}\left(\alpha_B + K_B a_B, \frac{1}{\varepsilon_B + \phi_{n.}}\right), (\varsigma_d|-) \sim \text{Gamma}\left(\alpha_Y + K_Y a_Y, \frac{1}{\varepsilon_Y + \theta_{d.}}\right), \tag{18}$$

$$(c_B|-) \sim \text{Gamma}\left(g_B + \gamma_B, \frac{1}{h_B + \sum_{k_B}\rho_{k_B}}\right), (c_Y|-) \sim \text{Gamma}\left(g_Y + \gamma_Y, \frac{1}{h_Y + \sum_{k_Y}r_{k_Y}}\right). \tag{19}$$

Sampling of γ_B : Using Lemma 2, one can show that $x_{..k_B} \sim \text{Pois}(\rho_{k_B})$. Integrating ρ_{k_B} and using Lemma 4, one can have $x_{..k_B} \sim \text{NB}(\gamma_B, p_B)$, where $p_B = 1/(c_B+1)$. Similarly, $y_{..k_B} \sim \text{Pois}(\rho_{k_B})$ and after integrating ρ_{k_B} and using Lemma 4, we have $y_{..k_B} \sim \text{NB}(\gamma_B, p_B)$. We now augment $l_{k_B} \sim \text{CRT}(x_{..k_B} + y_{..k_B}, \gamma_B)$ and then following Lemma 6 sample

$$(\gamma_B|-) \sim \text{Gamma}\left(e_B + \sum_{k_B} l_{k_B}, \frac{1}{f_B - q_B}\right), q_B = \sum_{k_B}\frac{q_{k_B}}{K_B}, q_{k_B} = \log\left(\frac{c_B}{c_B + \sum_n \phi_{nk_B}\phi_{k_B}^{-n}}\right). \tag{20}$$

Sampling of γ_Y : Using Lemma 2, one can show that $y_{..(K_B+k_Y)} \sim \text{Pois}(r_{k_Y})$ and after integrating r_{k_Y} and using Lemma 4, we have $y_{..(K_B+k_Y)} \sim \text{NB}(\gamma_Y, p_Y)$, where $p_Y = 1/(c_Y + 1)$. We now augment $m_{k_Y} \sim \text{CRT}(y_{..(K_B+k_Y)}, \gamma_Y)$ and then following Lemma 6 sample

$$(\gamma_Y|-) \sim \text{Gamma}\left(e_Y + \sum_{k_Y} m_{k_Y}, \frac{1}{f_Y - q_Y}\right), q_Y = \sum_{k_Y}\frac{q_{k_Y}}{K_Y}, q_{k_Y} = \log\left(\frac{c_Y}{c_Y + \theta_{.k_Y}}\right). \tag{21}$$

Table 2. Generative Process of N-GPPF

$b_{nm} = I_{\{x_{nm} \geq 1\}}, \; x_{nm} \sim \text{Pois}\left(\sum_{k_B=1}^{\infty} \lambda_{nmk_B}\right), \; r_{k_B} \sim \text{Gamma}(\gamma_B/K_B, 1/c_B),$
$\phi_{k_B} \sim \prod_{n=1}^{N} \text{Gamma}(a_B, 1/\sigma_n), \; \sigma_n \sim \text{Gamma}(\alpha_B, 1/\varepsilon_B),$
$\gamma_B \sim \text{Gamma}(e_B, 1/f_B), \; c_B \sim \text{Gamma}(g_B, 1/h_B),$

3.2 Special Cases: Network Only GPPF (N-GPPF) and Corpus Only GPPF (C-GPPF)

A special case of J-GPPF appears when only the binary matrix B is modeled without the auxiliary matrix Y. The generative model of N-GPPF is given in Table 2. The update equations of variables corresponding to N-GPPF can be obtained with $Z = 0$ in the corresponding equations. As mentioned in

Table 3. Generative Process of C-GPPF

$$
\begin{aligned}
y_{dw} &\sim \text{Pois}\left(\sum_{k_Y=1}^{\infty} r_{k_Y}\theta_{dk_Y}\beta_{wk_Y}\right), \\
\theta_{k_Y} &\sim \prod_{d=1}^{D}\text{Gamma}(a_Y, 1/\varsigma_d), \quad \beta_{k_Y} \sim \text{Dir}(\xi_Y), \\
\varsigma_d &\sim \text{Gamma}(\eta_Y, 1/\xi_Y), \quad r_{k_Y} \sim \text{Gamma}(\gamma_Y/K_Y, 1/c_Y), \\
\gamma_{Y_Y} &\sim \text{Gamma}(e_Y, 1/f_Y), \quad c_Y \sim \text{Gamma}(g_Y, 1/h_Y).
\end{aligned}
$$

Section 2.3, N-GPPF can be considered as the gamma process infinite edge partition model (EPM) proposed in [40], which is shown to well model assortative networks but not necessarily disassortative ones. Using the techniques developed in [40] to capture community-community interactions, it is relatively straightforward to extend J-GPPF to model disassortative networks Another special case of J-GPPF appears when only the count matrix Y is modeled without using the contribution from the network matrix B. The generative model of C-GPPF is given in Table 3.

3.3 Computation Complexity

The Gibbs sampling updates of J-GPPF can be calculated in $O(K_B S_B + (K_B + K_Y)S_Y + NK_B + DK_Y + V(K_B + K_Y))$ time, where S_B is the number of non-zero entries in B and S_Y is the number of non-zero entries in Y. It is obvious that for large matrices the computation is primarily of the order of $K_B S_B + (K_B + K_Y)S_Y$. Such complexity is a huge saving when compared to other methods like MMSB [2], that only models B and incurs computation cost of $O(N^2 K_B)$; and standard matrix factorization approaches [31] that work with the matrix Y and incur $O(DVK_Y)$ computation cost. Interestingly, the inference in [14] incurs cost $O(K_Y^2 D + K_Y V + K_Y S_Y)$ with K_Y signifying the termination point of stick breaking construction in their model. C-GPPF incurs computation cost $O(DK_Y + K_Y S_Y + VK_Y)$, an apparent improvement over that of [14]. However, one needs to keep in mind that [14] use variational approximation for which the updates are available in closed form solution. Our method does not use any approximation to joint distribution but uses Gibbs sampling, the computation cost of which should also be taken into account. In

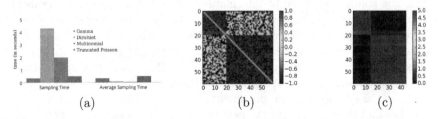

Fig. 1. (a) Time to generate a million of samples, (b) B with held-out data, (c) Y

Fig. 1(a), we show the computation time for generating one million samples from Gamma, Dirichlet (of dimension 50), multinomial (of dimension 50) and truncated Poisson distributions using the samplers available from GNU Scientific Library (GSL) on an Intel 2127U machine with 2 GB of RAM and 1.90 GHz of processor base frequency. To highlight the average complexity of sampling from Dirichlet and multinomial distributions, we further display another plot where the computation time is divided by 50 for these samplers only. One can see that to draw one million samples, our implementation of the sampler for truncated Poisson distribution takes the longest, though the difference from the Gamma sampler in GSL is not that significant.

4   Experimental Results

4.1   Experiments with Synthetic Data

We generate a synthetic network of size 60×60 ($\mathbf{B}$) and a count data matrix of size 60×45 ($\mathbf{Y}$). Each user in the network writes exactly one document and a user and the corresponding document are indexed by the same row-index in $\mathbf{B}$ and $\mathbf{Y}$ respectively. To evaluate the quality of reconstruction in presence of side-information and less of network structure, we hold-out 50% of links and equal number of non-links from $\mathbf{B}$. This is shown in Fig. 1(b) where the links are presented by brown, the non-links by green and the held-out data by deep blue. Clearly, the network consists of two groups. $\mathbf{Y} \in \{0,5\}^{60 \times 45}$, shown in Fig. 1(c), is also assumed to consist of the same structure as $\mathbf{B}$ where the zeros are presented by deep blue and the non-zeros are represented by brown. The performance of N-GPPF is displayed in Fig. 2(a). Evidently, there is not much structure visible in the discovered partition of $\mathbf{B}$ from N-GPPF and that is reflected in the poor value of AUC in Fig. 3(a). The parameter ϵ, when fixed at a given value, plays an important role in determining the quality of reconstruction for J-GPPF. As $\epsilon \to 0$, J-GPPF approaches the performance of N-GPPF on $\mathbf{B}$ and we observe as poor a quality of reconstruction as in Fig. 2(a). When ϵ is increased and set to 1.0, J-GPPF departs from N-GPPF and performs much better in terms of both structure recovery and prediction on held-out data as shown in Fig. 2(e) and Fig. 3(b). With $\epsilon = 10.0$, perfect reconstruction and prediction are recorded as shown in Fig. 2(i) and Fig. 3(c) respectively. In this synthetic example, $\mathbf{Y}$ is purposefully designed to reinforce the structure of $\mathbf{B}$ when most of its links and non-links are held-out. However, in real applications, $\mathbf{Y}$ might not contain as much of information and the Gibbs sampler needs to find a suitable value of ϵ that can carefully glean information from $\mathbf{Y}$.

There are few more interesting observations from the experiment with synthetic data that characterize the behavior of the model and match our intuitions. In our experiment with synthetic data $K_{\mathbf{B}} = K_{\mathbf{Y}} = 20$ is used. Fig. 2(b) demonstrates the assignment of the users in the network communities and Fig. 2(d) illustrates the assignment of the documents to the combined space of network communities and the topics (with the network communities appearing before the topics in the plot). For $\epsilon = 0.001$, we observe disjoint modeling of $\mathbf{B}$ and $\mathbf{Y}$, with

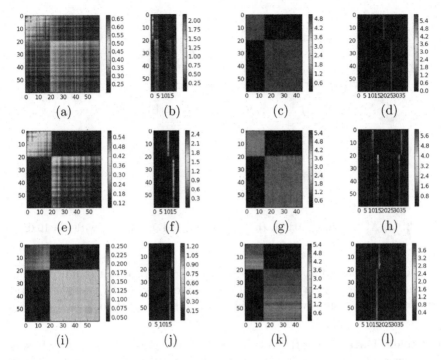

Fig. 2. Performance of J-GPPF: $\epsilon = 10^{-3}$ (top row), $\epsilon = 1$ (middle row), $\epsilon = 10$ (bottom row)

two latent factors modeling $\mathbf{Y}$ and multiple latent factors modeling $\mathbf{B}$ without any clear assignment. As we increase ϵ, we start observing joint modeling of $\mathbf{B}$ and $\mathbf{Y}$. For $\epsilon = 1.0$, as Fig. 2(h) reveals, two of the network latent factors and two of the factors for count data together model $\mathbf{Y}$, the contribution from the network factors being expectedly small. Fig. 2(f) shows how two of the exact same latent factors model $\mathbf{B}$ as well. Fig. 2(j) and Fig. 2(l) show how two of the latent factors corresponding to $\mathbf{B}$ dictate the modeling of both $\mathbf{B}$ and $\mathbf{Y}$ when $\epsilon = 10.0$. This implies that the discovery of latent groups in $\mathbf{B}$ is dictated mostly by the information contained in $\mathbf{Y}$. In all these cases, however, we observe perfect reconstruction of $\mathbf{Y}$ as shown in Fig. 2(c), Fig. 2(g) and Fig. 2(k).

4.2   Experiments with Real World Data

To evaluate the performance of J-GPPF, we consider N-GPPF, the infinite relational model (IRM) of [18] and the Mixed Membership Stochastic Block Model (MMSB) [2] as the baseline algorithms.

NIPS Authorship Network. This dataset contains the papers and authors from NIPS 1988 to 2003. We took the 234 authors who published with the most

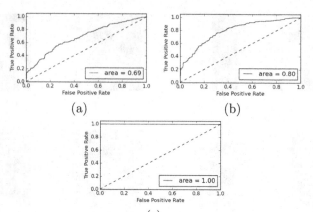

Fig. 3. (a) AUC with $\epsilon = 0.001$, (b) AUC with $\epsilon = 1.0$, (c) AUC with $\epsilon = 10.0$

other people and looked at their co-authors. After pre-processing and removing words that appear less than 50 times, the number of users in the graph is 225 and the total number of unique words is 1354. The total number of documents is 1165.

GoodReads Data. Using the Goodreads API, we collected a base set of users with recent activity on the website. The friends and friends of friends of these users were collected. Up to 200 reviews were saved per user, each consisting of a book ID and a rating from 0 to 5. A similar dataset was used in [10]. After pre-processing and removing words that appear less than 10 times, the number of users in the graph is 84 and the total number of unique words is 189.

Twitter Data. The Twitter dataset is a set of geo-tagged tweets collected by the authors in [29]. We extracted a subset of users located in San Francisco for our analysis. We created a graph within the subset by collecting follower information from the Twitter API. The side information consists of tweets aggregated by user, with one document per user. After pre-processing and removing words that appear less than 25 times, the number of users in the graph is 670 and the total number of unique words is 538.

Experimental Setup and Results. In all the experiments, we initialized ϵ to 2 and let the sampler decide what value works best for joint modeling. We used $K_{\mathbf{B}} = K_{\mathbf{Y}} = 50$ and initialized all the hyper-parameters to 1. For each dataset, we ran 20 different experiments and display the mean AUC and one standard error. Fig. 4 and 5 demonstrate the performances of the models in predicting the held-out data. J-GPPF clearly has advantage over other network-only models when the network is sparse enough and the auxiliary information is sufficiently strong. However, all methods fail when the sparsity increases beyond a certain point. The performance of J-GPPF also drops below the performances

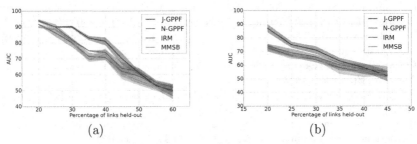

Fig. 4. (a) NIPS Data, (b) GoodReads Data

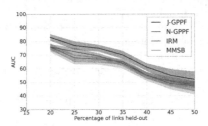

Fig. 5. Twitter Data

of network-only models in highly sparse networks, as the sampler faces additional difficulty in extracting information from both the network and the count matrix.

5 Conclusion

We propose J-GPPF that jointly factorizes the network adjacency matrix and the associated side information that can represented as a count matrix. The model has the advantage of representing true sparsity in adjacency matrix, in latent group membership, and in the side information. We derived an efficient MCMC inference method, and compared our approach to several popular network algorithms that model the network adjacency matrix. Experimental results confirm the efficiency of the proposed approach in utilizing side information to improve the performance of network models.

Acknowledgments. This work is supported by the United States Office of Naval Reseach, grant No. N00014-14-1-0039.

References

1. Acharya, A., Ghosh, J., Zhou, M.: Nonparametric bayesian factor analysis for dynamic count matrices. In: Proc. of AISTATS (to appear, 2015)
2. Airoldi, E.M., Blei, D.M., Fienberg, S.E., Xing, E.P.: Mixed membership stochastic blockmodels. JMLR **9**, 1981–2014 (2008)

3. Balasubramanyan, R., Cohen, W.W.: Block-LDA: jointly modeling entity-annotated text and entity-entity links. In: Proc. of SDM, pp. 450–461 (2011)
4. Ball, B., Karrer, B., Newman, M.: Efficient and principled method for detecting communities in networks. Phys. Rev. E **84**, September 2011
5. Blackwell, D., MacQueen, J.: Ferguson distributions via Pólya urn schemes. The Annals of Statistics (1973)
6. Blei, D.M., Ng, A.Y., Jordan, M.I.: Latent Dirichlet Allocation. JMLR **3**, 993–1022 (2003)
7. Broderick, T., Mackey, L., Paisley, J., Jordan, M.I.: Combinatorial clustering and the beta negative binomial process. arXiv:1111.1802v5 (2013)
8. Canny, J.: Gap: a factor model for discrete data. In: SIGIR (2004)
9. Cemgil, A.T.: Bayesian inference for nonnegative matrix factorisation models. Intell. Neuroscience (2009)
10. Chaney, A., Gopalan, P., Blei, D.: Poisson trust factorization for incorporating social networks into personalized item recommendation. In: NIPS Workshop: What Difference Does Personalization Make? (2013)
11. Chang, J., Blei, D.: Relational topic models for document networks. In: Proc. of AISTATS (2009)
12. Ferguson, T.S.: A Bayesian analysis of some nonparametric problems. Ann. Statist. (1973)
13. Gopalan, P., Mimno, D.M., Gerrish, S., Freedman, M.J., Blei, D.M.: Scalable inference of overlapping communities. In: Proc. of NIPS, pp. 2258–2266 (2012)
14. Gopalan, P., Ruiz, F., Ranganath, R., Blei, D.: Bayesian nonparametric poisson factorization for recommendation systems. In: Proc. of AISTATS (2014)
15. Gopalan, P., Charlin, L., Blei, D.: Content-based recommendations with poisson factorization. In: Proc. of NIPS, pp. 3176–3184 (2014)
16. Hjort, N.L.: Nonparametric Bayes estimators based on beta processes in models for life history data. Ann. Statist. (1990)
17. Johnson, N.L., Kemp, A.W., Kotz, S.: Univariate Discrete Distributions. John Wiley & Sons (2005)
18. Kemp, C., Tenenbaum, J., Griffiths, T., Yamada, T., Ueda, N.: Learning systems of concepts with an infinite relational model. In: Proc. of AAAI, pp. 381–388 (2006)
19. Kim, D.I., Gopalan, P., Blei, D.M., Sudderth, E.B.: Efficient online inference for bayesian nonparametric relational models. In: Proc. of NIPS, pp. 962–970 (2013)
20. Lee, D.D., Seung, H.S.: Algorithms for non-negative matrix factorization. In: NIPS (2001)
21. Leskovec, J., Julian, J.: Learning to discover social circles in ego networks. In: Proc. of NIPS, pp. 539–547 (2012)
22. Ma, H., Yang, H., Lyu, M.R., King, I.: Sorec: social recommendation using probabilistic matrix factorization. In: Proc. of CIKM, pp. 931–940 (2008)
23. McCallum, A., Wang, X., Corrada-Emmanuel, A.: Topic and role discovery in social networks with experiments on enron and academic email. J. Artif. Int. Res. **30**(1), 249–272 (2007)
24. Menon, A.K., Elkan, C.: Link prediction via matrix factorization. In: Gunopulos, D., Hofmann, T., Malerba, D., Vazirgiannis, M. (eds.) ECML PKDD 2011, Part II. LNCS, vol. 6912, pp. 437–452. Springer, Heidelberg (2011)
25. Miller, K.T., Griffiths, T.L., Jordan, M.I.: Nonparametric latent feature models for link prediction. In: Proc. of NIPS, pp. 1276–1284 (2009)
26. Nallapati, R., Ahmed, A., Xing, E., Cohen, W.: Joint latent topic models for text and citations. In: Proc. of KDD, pp. 542–550 (2008)

27. Neal, R.M.: Markov chain sampling methods for Dirichlet process mixture models. Journal of Computational and Graphical Statistics (2000)
28. Palla, K., Ghahramani, Z., Knowles, D.A.: An infinite latent attribute model for network data. In: Proc. of ICML, pp. 1607–1614 (2012)
29. Roller, S., Speriosu, M., Rallapalli, S., Wing, B., Baldridge, J.: Supervised text-based geolocation using language models on an adaptive grid. In: Proc. of EMNLP-CoNLL, pp. 1500–1510 (2012)
30. Rosen-Zvi, M., Griffiths, T., Steyvers, M., Smyth, P.: The author-topic model for authors and documents. In: Proc. of UAI, pp. 487–494 (2004)
31. Salakhutdinov, R., Mnih, A.: Probabilistic matrix factorization. In: Proc. of NIPS (2007)
32. Titsias, M.K.: The infinite gamma-poisson feature model. In: Proc. of NIPS (2008)
33. Walker, S.G.: Sampling the Dirichlet mixture model with slices. Communications in Statistics Simulation and Computation (2007)
34. Wang, X., Mohanty, N., Mccallum, A.: Group and topic discovery from relations and their attributes. In: Proc. of NIPS, pp. 1449–1456 (2006)
35. Wen, Z., Lin, C.: Towards finding valuable topics. In: Proc. of SDM, pp. 720–731 (2010)
36. Wolpert, R.L., Clyde, M.A., Tu, C.: Stochastic expansions using continuous dictionaries: Lévy Adaptive Regression Kernels. Annals of Statistics (2011)
37. Yang, J., Leskovec, J.: Overlapping community detection at scale: a nonnegative matrix factorization approach. In: Proc. of WSDM, pp. 587–596 (2013)
38. Yang, J., McAuley, J.J., Leskovec, J.: Community detection in networks with node attributes. In: Proc. of ICDM, pp. 1151–1156 (2013)
39. Yoshida, T.: Toward finding hidden communities based on user profile. J. Intell. Inf. Syst. **40**(2), 189–209 (2013)
40. Zhou, M.: Infinite edge partition models for overlapping community detection and link prediction. In: Proc. of AISTATS (to appear, 2015)
41. Zhou, M., Carin, L.: Augment-and-conquer negative binomial processes. In: Proc. of NIPS (2012)
42. Zhou, M., Carin, L.: Negative binomial process count and mixture modeling. IEEE Trans. Pattern Analysis and Machine Intelligence (2015)
43. Zhou, M., Hannah, L., Dunson, D., Carin, L.: Beta-negative binomial process and poisson factor analysis. In: Proc. of AISTATS (2012)
44. Zhu, J.: Max-margin nonparametric latent feature models for link prediction. In: Proc. of ICML (2012)

Generalization in Unsupervised Learning

Karim T. Abou-Moustafa$^{(\boxtimes)}$ and Dale Schuurmans

Department of Computing Science, University of Alberta,
Edmonton, AB T6G 2E8, Canada
{aboumous,daes}@ualberta.ca

Abstract. We are interested in the following questions. Given a finite data set $\mathcal{S}$, with neither labels nor side information, and an unsupervised learning algorithm A, can the generalization of A be assessed on $\mathcal{S}$? Similarly, given two unsupervised learning algorithms, A_1 and A_2, for the same learning task, can one assess whether one will generalize "better" on future data drawn from the same source as $\mathcal{S}$? In this paper, we develop a general approach to answering these questions in a reliable and efficient manner using mild assumptions on A. We first propose a concrete generalization criterion for unsupervised learning that is analogous to prediction error in supervised learning. Then, we develop a computationally efficient procedure that realizes the generalization criterion on finite data sets, and propose and extension for comparing the generalization of two algorithms on the same data set. We validate the overall framework on algorithms for clustering and dimensionality reduction (linear and nonlinear).

1 Introduction

The goal of unsupervised learning is to autonomously capture and model latent relations among the variables of a data set. Such latent relations are usually in the form of regularities and statistical dependencies known as *the underlying structure of the data distribution*. Unlike supervised learning, there are no desired target answers to guide and correct the learning process. However, similar to supervised learning, unsupervised learning algorithms generate estimates that are functions of sample data drawn from an unknown distribution $\mathscr{P}$. As such, it is natural to ask questions related to the generalization capability of these estimates, as well as questions on the choice of these estimates (model selection) [11].

In supervised learning, questions of generalization have been scrutinized, equally, in theory and in practice; see for instance [5,6,8,9,14,15,17,20,22,24]. In unsupervised learning, however, few efforts have acknowledged and addressed the problem in general. For instance, [11] approximates the expected loss of finite parametric models such as principle component analysis (PCA) and k-Means clustering based on asymptotic analysis and central limit results.

One possible reason for the scarcity of such efforts is the subjective nature of unsupervised learning, the diversity of tasks covered (such as clustering, density

© Springer International Publishing Switzerland 2015
A. Appice et al. (Eds.): ECML PKDD 2015, Part I, LNAI 9284, pp. 300–317, 2015.
DOI: 10.1007/978-3-319-23528-8_19

estimation, dimensionality reduction, feature learning, etc.), and the lack of a unified framework that incorporates a significant subset of these tasks. Another reason is that the principles underlying supervised learning are often distinct from those underlying unsupervised learning. In supervised learning, the final result of a learning algorithm is a function f^* that minimizes the expected loss (possibly plus a regularizer) under the unknown true distribution $\mathscr{P}$, which can be applied to new points not included during training. Since $\mathscr{P}$ is unknown, the learning algorithm selects f^* that minimizes an empirical average of the loss as a surrogate for the expected loss. Therefore, since the loss measures the difference between the estimated and expected outputs, its average provides an indicator of generalization error. The validity of this mechanism, however, rests on (i) the existence of target outputs, and (ii) consistency of the empirical average of the loss [22].

In unsupervised learning, the characterization is different. First, the target output is not available. Second, an unsupervised learning algorithm A produces an output that is a re-representation of the input; hence loss functions in this setting usually assess a reconstruction error between the output and input [25]. Third, there are various unsupervised learning algorithms that do not minimize a reconstruction error yet still produce an output that is a re-representation of the input: see for example the recent literature on moments-based methods for latent variable models and finite automata [1,2,12,21].

These observations motivate us to deal with unsupervised learning algorithms in an abstract form. In particular, we consider an unsupervised learning algorithm A as an abstract function – a black box – that maps an input x to an output y. The advantage of this view is that (i) it is independent of the learning task, and (ii) it provides a simple unified view for these algorithms without being overly dependent on internal details.

Based on this perspective, we propose a general definition for generalization of an unsupervised learning algorithm on a data set S. The framework is based on a general loss function ℓ that measures the reconstruction error between the input and output of A, which is not necessarily the loss minimized by A (if any). To study the generalization of A under the black box assumption and an external loss ℓ, we will assume that A satisfies a certain notion of algorithmic stability under some mild assumptions on ℓ. Given this notion of stability, we derive a finite useful upper bound on A's expected loss, which naturally lends itself to a generalization criterion for unsupervised learning. As a second contribution, we develop an efficient procedure to realize this generalization criterion on finite data sets, which can be extended to comparing the generalization of two different unsupervised learning algorithms on a common data source. Finally, we apply this generalization analysis framework and evaluation procedure to two unsupervised learning problems; clustering and dimensionality reduction.

1.1 Preliminaries and Setup

Let $\mathcal{X} \subseteq \mathbb{R}^d$ and $\mathcal{Y} \subseteq \mathbb{R}^k$ be the input and output spaces, respectively.[1] Let $S \in \mathcal{X}^n$ be a training set of size n drawn IID from an unknown distribution $\mathscr{P}_{\mathbf{x}}$ defined on a measurable space $(\mathcal{X}, \Sigma)$ with domain $\mathcal{X}$ and σ-algebra Σ. We denote this as $S \sim \mathscr{P}_{\mathbf{x}}$ where $S = \{\mathbf{x}_i\}_{i=1}^n$. For each $\mathbf{x}_i \in S$ there is a corresponding output $\mathbf{y}_i$, $1 \leq i \leq n$, with appropriate dimension k. For convenience, S can be represented as a matrix $\mathbf{X}_{n \times d}$, while the output can also be represented as a matrix $\mathbf{Y}_{n \times k}$.

An unsupervised learning algorithm A is a mapping from $\mathcal{X}^n$ to the class of functions $\mathcal{F}$ s.t. for $f \in \mathcal{F}$, $f : \mathcal{X} \to \mathcal{Y}$. Thus, A takes as input S, selects a particular f^* from $\mathcal{F}$, and estimates an $n \times k$ output matrix $\widehat{\mathbf{Y}} \equiv \mathsf{A}_S(\mathbf{X})$, or $\widehat{\mathbf{y}} \equiv \mathsf{A}_S(\mathbf{x})$,[2] where A_S denotes the output of A (i.e. $f^* \in \mathcal{F}$) after training on S. The algorithm A could also have certain parameters, denoted $\boldsymbol{\theta}_{\mathsf{A}}$, that the user can tune to optimize its performance. We assume that A and its output functions in $\mathcal{F}$ are all measurable maps.

2 A General Learning Framework

The problem of unsupervised learning is that of selecting a function $f^* \in \mathcal{F}$ that transforms input $\mathbf{x}$ into an output $\widehat{\mathbf{y}} \equiv \mathsf{A}_S(\mathbf{x})$ in some desired way. Here we assume that A is a black box that takes S and produces a map f^* from $\mathbf{x}$ to $\widehat{\mathbf{y}}$. Since we are ignoring A's internal details, assessing its generalization requires us to consider an additive *external loss* function $\ell : \mathcal{X} \times \mathcal{Y} \to \mathbb{R}^+$ that measures the reconstruction error between $\mathbf{x}$ and $\widehat{\mathbf{y}}$. Thus, the expected loss for A_S with respect to ℓ is defined as:

$$R(\mathsf{A}_S) \equiv \mathbb{E}\left[\ell(\mathbf{x}, \mathsf{A}_S(\mathbf{x})\right] = \int \ell(\mathbf{x}, \mathsf{A}_S(\mathbf{x}))d\mathscr{P}_{\mathbf{x}}. \tag{1}$$

Unfortunately $R(\mathsf{A}_S)$ cannot be computed since $\mathscr{P}_{\mathbf{x}}$ is unknown, and thus it has to be estimated from $S \in \mathcal{X}^n$. A simple estimator for $R(\mathsf{A}_S)$ is the empirical estimate:

$$\widehat{R}_{\mathrm{EMP}}(\mathsf{A}_S) = \frac{1}{n} \sum_{i=1}^n \ell(\mathbf{x}_i, \mathsf{A}_S(\mathbf{x}_i)). \tag{2}$$

To obtain a practical assessment of the generalization of A, we need to derive an upper bound for the quantity $\widehat{R}_{\mathrm{EMP}}(\mathsf{A}_S) - R(\mathsf{A}_S)$. Given the generality of

---

[1] Notation: Lower case letters x, m, i denote scalars and indices. Upper case letters X, Y denote random variables. Bold lower case letters $\mathbf{x}, \mathbf{y}$ denote vectors. Bold upper case letters $\mathbf{A}, \mathbf{B}$ are matrices. Distributions $\mathscr{P}, \mathscr{G}$ will be written in script. Calligraphic letters $\mathcal{X}, \mathcal{Y}$ denote sets.

[2] For example, in k-Means clustering, the elements of $\widehat{\mathbf{Y}}$ could be the corresponding cluster centers assigned to each $\mathbf{x}_i$ from a set of k such centers. In nonlinear dimensionality reduction, the output could be the $n \times n$ low rank matrix $\widehat{\mathbf{Y}}$. In density estimation using a mixture model, A could output the $n \times 1$ matrix $\mathbf{Y}$ with the density value of each $\mathbf{x}_i$.

this setting, one needs to resort to worst case bounds. However, this cannot be done without introducing additional assumptions about the behaviour of A. For example, if one assumes that A chooses its output from a class of functions $\mathcal{F}$ such that the class of loss random variables $\Lambda : \mathcal{X} \times \mathcal{Y} \to \mathbb{R}+$ induced by $\mathcal{F}$, i.e. $\Lambda = \ell \circ \mathcal{F}$, is uniformly upper bounded by $c < \infty$ and $\mathrm{VCdim}(\Lambda) = h < \infty$, then with probability at least $1 - \eta$ there is a uniform concentration of $\widehat{R}_{\mathrm{EMP}}(A_{\mathcal{S}})$ around $R(A_{\mathcal{S}})$:

$$R(A_{\mathcal{S}}) \leq \widehat{R}_{\mathrm{EMP}}(A_{\mathcal{S}}) + \frac{\tau c}{2} \left(1 + \sqrt{1 + \frac{4\widehat{R}_{\mathrm{EMP}}(A_{\mathcal{S}})}{\tau c}} \right), \tag{3}$$

where $\tau = 4n^{-1} [h(\ln 2n/h + 1) - \ln \eta]$ [22,23]. Rademacher or Gaussian complexities can also be used to obtain similar concentration inequalities [3]. The caveat is that such an analysis is worst case and the resulting bounds, such as (3), are too loose to be useful in practice. This suggests that we need to make stronger assumptions on A to achieve more useful bounds on the quantity $\widehat{R}_{\mathrm{EMP}}(A_{\mathcal{S}}) - R(A_{\mathcal{S}})$.

2.1   Generalization and Stability

To achieve a more practical criterion and assessment procedure, we need to introduce some form of additional assumptions on A without sacrificing too much generality. To this end, we investigate an assumption that A satisfies a particular notion of algorithmic stability that allows us to derive a more useful and a tighter upper bound on $\widehat{R}_{\mathrm{EMP}}(A_{\mathcal{S}}) - R(A_{\mathcal{S}})$. Algorithmic stability has been successfully applied in learning theory to derive generalization bounds for supervised learning algorithms, but has yet to be formally applied to unsupervised learning. Among the different notions of stability, the uniform stability of [5] is considered to be the strongest since it implies other notions of stability such as: hypothesis stability, error stability, point–wise hypothesis stability, everywhere stability, CVLOO stability, etc. [8,14,16,17,20].

To define uniform stability for A in the unsupervised learning context, we require the following definitions. For any $\mathcal{S} \in \mathcal{X}^n$, we define $\forall i, 1 \leq i \leq n$, the modified training set $\mathcal{S}^{\setminus i}$ by removing from $\mathcal{S}$ the i-th element: $\mathcal{S}^{\setminus i} = \{\mathbf{x}_1, \ldots, \mathbf{x}_{i-1}, \mathbf{x}_{i+1}, \ldots, \mathbf{x}_n\}$. We assume that A is symmetric with respect to $\mathcal{S}$; i.e. it does not depend on the elements' order in $\mathcal{S}$. Further, we require that the external loss ℓ be "well behaved" with respect to slight changes in $\mathcal{S}$; i.e. if $\varepsilon = \ell(\mathbf{x}, A_{\mathcal{S}}(\mathbf{x}))$, $\varepsilon' = \ell(\mathbf{x}, A_{\mathcal{S}'}(\mathbf{x}))$, and $\mathcal{S}'$ is slightly different from $\mathcal{S}$ such that $A_{\mathcal{S}}(\mathbf{x}) \approx A_{\mathcal{S}'}(\mathbf{x})$, then the difference between ε and ε' should be small. The notion of "well behaved" is formally imposed by requiring that ℓ is Lipschitz continuous, and that A is uniformly β–stable with respect to ℓ. This uniform β–stability is defined as follows:

Definition 1 (Uniform β–Stability). *An algorithm* A *is uniformly β–stable with respect to the loss function ℓ if for any $\mathbf{x} \in \mathcal{X}$, the following holds:*

$$\forall \mathcal{S} \in \mathcal{X}^n, \max_{i=1,\ldots,n} |\ell(\mathbf{x}, A_{\mathcal{S}}(\mathbf{x})) - \ell(\mathbf{x}, A_{\mathcal{S}^{\setminus i}}(\mathbf{x}))| \leq \beta.$$

Note that β is a function of n and we assume that stability is non-increasing as a function of n. Hence, in the following, β can be denoted by β_n.

Definition 2 (Stable Algorithm). *Algorithm* A *is stable if* $\beta_n \propto \frac{1}{n}$.[3]

The analogy between our definition of uniform β–stability and the uniform β–stability in supervised learning can be explained as follows. The uniform β–stability in [5] is in terms of $\ell(\mathsf{A}_S, z)$ and $\ell(\mathsf{A}_{S \backslash i}, z)$, where $z = (\mathbf{x}, y)$, $\mathbf{x}$ is the input vector, and y is its expected output (or true label). Note that $\ell(\mathsf{A}_S, z)$ can be written as $\ell(f_S(\mathbf{x}), y)$, where f_S is the hypothesis learned by A using the training set S. Similarly, $\ell(\mathsf{A}_{S \backslash i}, z)$ can be written as $\ell(f_{S \backslash i}(\mathbf{x}), y)$. Observe that the difference between $\ell(f_S(\mathbf{x}), y)$ and $\ell(f_{S \backslash i}(\mathbf{x}), y)$ is in the hypotheses f_S and $f_{S \backslash i}$. Note also that in supervised learning, the loss ℓ measures the discrepancy between the expected output y and the estimated output $\widehat{y} = f_S(\mathbf{x})$. In our unsupervised learning setting, the expected output is not available, and the loss ℓ measures the reconstruction error between $\mathbf{x}$ and $\widehat{\mathbf{y}} \equiv \mathsf{A}_S(\mathbf{x})$. Hence, we replace $\ell(\mathsf{A}_S, z)$ by $\ell(\mathbf{x}, \mathsf{A}_S(\mathbf{x}))$, and $\ell(\mathsf{A}_{S \backslash i}, z)$ by $\ell(\mathbf{x}, \mathsf{A}_{S \backslash i}(\mathbf{x}))$ to finally obtain Definition 1.

Note that the uniform β–stability of A with respect to ℓ is complimentary to the continuous Lipschitz condition on ℓ. If A is uniformly β–stable, then a slight change in the input will result in a slight change in the output, resulting in a change in the loss bounded by β. The following corollary upper bounds the quantity $\widehat{R}_{\mathrm{EMP}}(\mathsf{A}_S) - R(\mathsf{A}_S)$ using the uniform β–stability of A.

Corollary 1. *Let* A *be a uniformly β–stable algorithm with respect to ℓ, $\forall\ \mathbf{x} \in \mathcal{X}$, and $\forall\ S \in \mathcal{X}^n$. Then, for any $n \geq 1$, and any $\delta \in (0, 1)$, the following bounds hold (separately) with probability at least $1 - \delta$ over any $S \sim \mathscr{P}_{\mathbf{x}}$:*

$$(i) \quad R(\mathsf{A}_S) \leq \widehat{R}_{EMP}(\mathsf{A}_S) + 2\beta + (4n\beta + c)\sqrt{\frac{\log(1/\delta)}{2n}}, \qquad (4)$$

$$(ii) \quad R(\mathsf{A}_S) \leq \widehat{R}_{LOO}(\mathsf{A}_S) + \beta + (4n\beta + c)\sqrt{\frac{\log(1/\delta)}{2n}}, \quad where \qquad (5)$$

$\widehat{R}_{LOO}(\mathsf{A}_S) = \frac{1}{n}\sum_{i=1}^{n} \ell(\mathbf{x}_i, \mathsf{A}_{S \backslash i}(\mathbf{x}_i))$, *is the leave-one-out (LOO) error estimate.*

Discussion. The generalization bounds in (4) and (5) directly follow from Theorem 12 in [5] for the regression case. The reason we consider A under the regression framework is due to our characterization of unsupervised learning algorithms in which we consider the output $\widehat{y} \in \mathbb{R}^k$ is a re-representation of the input $\mathbf{x} \in \mathbb{R}^d$. This, in turn, defined the form of the external loss ℓ as $\ell : \mathcal{X} \times \mathcal{Y} \to \mathbb{R}^+$. This characterization is very similar to the multivariate regression setting, and hence our reliance on Theorem 12 in [5]. Note that if $\beta \propto \frac{1}{n}$, then the bounds in (4) and (5) will be tight.

Corollary 1 is interesting in our context for a few reasons. First, it defines a generalization criterion for unsupervised learning algorithms in general: if A

---

[3] $\beta_n \propto \frac{1}{n} \implies \beta_n = \frac{\kappa}{n}$, for some constant $\kappa > 0$.

is uniformly β–stable with respect to ℓ on $\mathcal{S}$, then the bounds in (4) and (5) hold with high probability. Note that the bound in (4) is tighter than the one in (3). Second, the bounds for $\widehat{R}_{\text{EMP}}$ and $\widehat{R}_{\text{LOO}}$ are very similar. Various works have reported that $\widehat{R}_{\text{EMP}}$ is an optimistically biased estimate for R, while $\widehat{R}_{\text{LOO}}$ is almost an unbiased estimate [5,8,14].[4] Therefore, an advantage of uniform β–stability is that this discrepancy is mitigated. This also shows that stability based bounds are more suitable for studying algorithms whose empirical error remains close to the LOO error.

Second, this result also shows that to be uniformly stable, a learning algorithm needs to significantly depart from the empirical risk minimization principle that emphasizes the minimization of $\widehat{R}_{\text{EMP}}$. That is, a stable algorithm A might exhibit a larger error during training but this would be compensated by a decrease in complexity of the learned function. This characteristic is exactly what defines the effects of regularization. Therefore, the choice for uniform stability allows one to consider a large class of unsupervised learning algorithms, including those formulated as regularized minimization of an internal loss.

3   Empirical Generalization Analysis

Although the previous section defines a general criterion for generalization in unsupervised learning, in practice this criterion requires assessing the uniform stability of A on a finite data set $\mathcal{S}$. The quantity of interest in the uniform stability criterion is $|\ell(\mathbf{x}, A_{\mathcal{S}}(\mathbf{x})) - \ell(\mathbf{x}, A_{\mathcal{S} \setminus i}(\mathbf{x}))|$, which is the amount of change in the loss with respect to the exclusion of one data point $\mathbf{x}_i$ from $\mathcal{S}$. Taking expectations with respect to $\mathscr{P}_{\mathbf{x}}$ and replacing the expected loss with the empirical estimator, we have that:

$$\forall \, \mathcal{S} \in \mathcal{X}^n \quad \max_{i=1,\ldots,n} |\widehat{R}_{\text{EMP}}(A_{\mathcal{S}}) - \widehat{R}_{\text{EMP}}(A_{\mathcal{S} \setminus i})| \leq \beta_n. \tag{6}$$

This states that for a uniformly β_n–stable algorithm with respect to ℓ on $\mathcal{S}$, the change in the empirical loss due to the exclusion of one sample from $\mathcal{S}$ is at most β_n. In the finite sample setting, this will be:

$$\max_{i=1,\ldots,n} \left| \frac{1}{n} \sum_{j=1}^{n} \ell(\mathbf{x}_j, A_{\mathcal{S}}(\mathbf{x}_j)) - \frac{1}{n-1} \sum_{\substack{j=1 \\ j \neq i}}^{n} \ell(\mathbf{x}_j, A_{\mathcal{S} \setminus i}(\mathbf{x}_j)) \right| \leq \beta_n. \tag{7}$$

Inequality (7) contains an unknown parameter β_n which cannot be upper bounded without any further knowledge on A. In fact, given the black box assumption on A and the absence of information on $\mathscr{P}_{\mathbf{x}}$, we cannot obtain a uniform upper bound on β_n. This suggests that β_n needs to be estimated from

---

[4] The LOO error estimate over n samples, $\widehat{R}_{\text{LOO}_n}$, is an unbiased estimate for $\widehat{R}_{\text{LOO}_{n-1}}$. Since in most interesting cases $\widehat{R}_{\text{LOO}_n}$ converges with probability one, the difference between $\widehat{R}_{\text{LOO}_n}$ and $\widehat{R}_{\text{LOO}_{n-1}}$ becomes negligible for large n [7, Ch. 24].

Algorithm 1. Generalization Analysis for Algorithm A.

1: **Require:** Algorithm A and its input parameters $\boldsymbol{\theta}_\mathsf{A}$, data set $\mathcal{S}$, loss function ℓ, number of subsamples m, and the sizes of subsamples, n_t s.t. $n_1 < n_2 < n_3 < \cdots < n_\tau$.
2: **for** $t = 1$ to τ **do**
3:     **for** $j = 1$ to m **do**
4:         $\mathbf{X}_j \leftarrow$ draw n_t samples uniformly from $\mathcal{S}$
5:         $\widehat{\mathbf{Y}}_j \leftarrow \mathsf{A}_\mathcal{S}(\mathbf{X}_j; \boldsymbol{\theta}_\mathsf{A})$
6:         $\boldsymbol{\Phi} \leftarrow$ hold out one random samples from $\mathbf{X}_j$
7:         $\mathbf{X}'_j \leftarrow \mathbf{X}_j \setminus \boldsymbol{\Phi}$
8:         $\widehat{\mathbf{Y}}'_j \leftarrow \mathsf{A}_{\mathcal{S}\setminus i}(\mathbf{X}'_j; \boldsymbol{\theta}_\mathsf{A})$
9:         $R_j \leftarrow \frac{1}{n_1}\ell(\mathbf{X}_j, \widehat{\mathbf{Y}}_j)$
10:         $R'_j \leftarrow \frac{1}{n_1 - 1}\ell(\mathbf{X}'_j, \widehat{\mathbf{Y}}'_j)$
11:         $B_j = |R_j - R'_j|$
12:     **end for**
13:     $\widehat{\beta}_{n_t} = \text{median}\{B_1, \ldots, B_j, \ldots, B_m\}$
14: **end for**
15: **Return:** $\mathcal{B} = \{\widehat{\beta}_{n_1}, \ldots, \widehat{\beta}_{n_t}, \ldots, \widehat{\beta}_{n_\tau}\}$

---

the data set $\mathcal{S}$. Also, recall from Definitions 1 and 2 that if $\beta_n \propto 1/n$, then the generalization bounds in (4) and (5) will hold with high probability. These two requirements raise the need for two procedures; one to estimate β_n at increasing values of n, and another one to model the relation between the estimated β_n's and the values of n. However, to consider these two procedures for assessing A's generalization, we need to introduce a further mild assumption on A. In particular, we need to assume that A does not change its learning mechanism as the sample size is increasing from n to $n + 1$ for any $n \geq 1$. Note that if A changes its learning mechanism based on the sample size, then A can have inconsistent trends of β_n with respect to n which makes it unfeasible to obtain consistent confidence bounds for $\widehat{R}_{\text{EMP}}(\mathsf{A}_\mathcal{S}) - R(\mathsf{A}_\mathcal{S})$. Therefore, we believe that our assumption is an intuitive one, and is naturally satisfied by most learning algorithms.

3.1   Estimating β_n From a Finite Data Set

Inequality (7) might suggest a simple procedure for estimating β_n: (i) Compute $\widehat{\mathbf{Y}} = \mathsf{A}_\mathcal{S}(\mathbf{X})$. (ii) Set $\mathbf{X}' = \mathbf{X}$, hold out sample $\mathbf{x}_i$ from $\mathbf{X}'$, and compute $\widehat{\mathbf{Y}}' = \mathsf{A}_{\mathcal{S}\setminus i}(\mathbf{X}')$, and set $B_i = |n^{-1}\ell(\mathbf{X}, \widehat{\mathbf{Y}}) - (n-1)^{-1}\ell(\mathbf{X}', \widehat{\mathbf{Y}}')|$. (iii) Repeat step (ii) n times to obtain $\{B_1 \ldots, B_n\}$, and then set set $\widehat{\beta}_n = \max\{B_1 \ldots, B_n\}$. The problem with this procedure is three–fold. First, note that in the finite sample setting, Inequality (7) cannot be evaluated for $\forall \mathcal{S} \in \mathcal{X}^n$ as required in Inequality (6). Note also that the sample maximum is a noisy estimate, and hence is not reliable. Second, the LOO estimate suggested above is computationally expensive since it requires invoking A for n times. Third, using all $\mathbf{X}$ to learn $\widehat{\mathbf{Y}}$ will not reflect A's sensitivity to the randomness in the data. If A easily gets stuck in local minima, or A has tendency to overfit the data, learning using all $\mathbf{X}$ will obscure such traits.

Our proposed procedure for estimating β_n, depicted in Algorithm 1, addresses the above issues in the following ways. First it is based on repeated random sub-sampling (with replacement) from the original data set $\mathcal{S}$, similar in spirit to bootstrapping [10]. Second, for each subsample of size n_t, the procedure obtains an estimate for the empirical loss before and after holding out one random sample. After repeating this subsampling process m times, $m \ll n$, the procedure obtains one estimate for β_n, denoted by $\widehat{\beta}_{n_t}$, for sample size n_t. Note that $\widehat{\beta}_{n_t}$ is the median of B_j's to increase the robustness of the estimate. This process is repeated τ times and the final output of Algorithm 1 is the set of $\widehat{\beta}_{n_t}$'s for the increasing values of n_t.

The proposed procedure is computationally intensive, yet it is efficient, scalable, and provides control over the accuracy of the estimates. First, note that the proposed procedure is not affected by the fact that A is an unsupervised learning algorithm. If A is a supervised learning algorithm, then assessing its generalization through uniform β–stability results will still require $2\tau m$ calls for A, as it is the case for the unsupervised setting discussed here. Thus, the procedure does not impose a computational overhead given the absence of the expected output, and the black box assumption on A. Second, considering scalability for large data sets, the procedure can be fully parallelized on multiple core architectures and computing clusters [13]. Note that in each iteration j the processing steps for each subsample are independent from all other iterations, and hence all m subsamples can be processed in an embarrassingly parallel manner. Note also that in each iteration, $A_{\mathcal{S}}(\mathbf{X}_j)$ and $A_{\mathcal{S}\backslash i}(\mathbf{X}'_j)$ can also be executed in parallel.

Parameters m and size of the subsamples, $n_1, n_2, \ldots, n_\tau$, control the tradeoff between computational efficiency and estimation accuracy. These parameters are user–specified and they depend on the data and problem in hand, its size n, A's complexity, and the available computational resources. Parameter m needs to be sufficient to reduce the variance in $\{R_1, \ldots, R_m\}$ and $\{R'_1, \ldots, R'_m\}$. However, increasing m beyond a certain value will not increase the accuracy of the estimated empirical loss. Reducing the variance in $\{R_1, \ldots, R_m\}$ and $\{R'_1, \ldots, R'_m\}$, in turn, encourages reducing the variance in $\{B_1, \ldots, B_m\}$. Note that for any random variable Z with mean μ, median ν, and variance σ^2, then $|\mu - \nu| \leq \sigma$ with probability one. Therefore, in practice, increasing m encourages reducing the variance in B_j's thereby reducing the difference $|\widehat{\beta}_{n_t} - \mathbb{E}(B_j)|$. Observe that the operator $\max_{i=1,\ldots,n_t}$ defined $\forall \mathcal{S} \in \mathcal{X}^{n_t}$ in (6) is now replaced with the estimate $\widehat{\beta}_{n_t}$.

3.2   The Trend of $\widehat{\beta}_n$ and The Stability Line

The output of Algorithm 1 is the set $\mathcal{B}$ of estimated $\widehat{\beta}_{n_t}$'s for the increasing values of n_t. In order to assess the stability of A, we need to observe whether $\widehat{\beta}_{n_t} = \frac{\kappa}{n_t}$, for some constant $\kappa > 0$. As an example, Figure 1 shows the trend of $\widehat{\beta}_n$ for k–Means clustering and principal component analysis (PCA) on two

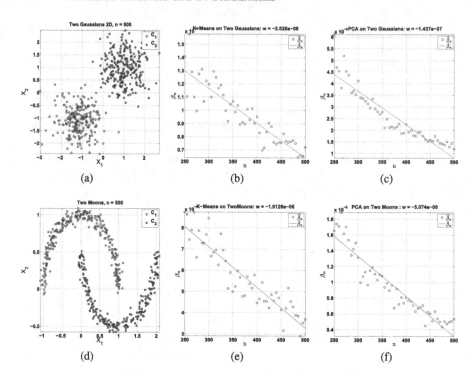

Fig. 1. Left: Two synthetic data sets, (a) two normally distributed clouds of points with equal variance and equal priors, and (d) two moons data points with equal priors. **Middle:** The estimated $\widehat{\beta}_n$ (blue circles) from Algorithm 1 for k–Means clustering on the two synthetic data sets. The fitted stability lines are shown in magenta. The slope of the stability lines is indicated by **w**. **Right:** The estimated $\widehat{\beta}_n$ and stability lines for PCA on the two synthetic data sets. The dispersion of $\widehat{\beta}_n$'s around the stability line is reflected in the norm of the residuals for the stability line (not displayed). Note the difference in the dispersion of points around the stability line for k–Means and PCA. Note also that the more structure in the two moons data set is reflected in a smaller **w** (compared to **w** for the tow Gaussians) for both algorithms.

synthetic toy data sets. The blue circles in the middle and right figures are the estimated $\widehat{\beta}_n$ from Algorithm 1.[5] Observe that $\widehat{\beta}_n$ is decreasing as n is increasing.

To formally detect and quantify this decrease, a line is fitted to the estimated $\widehat{\beta}_n$ (shown in magenta); i.e. $\beta(n_t) = wn_t + \zeta$, where w is the slope of the line, and ζ is the intercept. We call this line, the *Stability Line*. The slope of the stability line indicates its steepness which is an esimtate for the decreasing rate of β_n. For stable algorithms, $w < 0$, and $|w|$ indicates the stability degree of the algorithm. Note that $w = \tan\theta$, where θ is the angle between the stability line and the

---

[5] In these experiments, $m = 100$, and $n_1, n_2, \ldots, n_\tau$ were set to $0.5n, 0.51n, \ldots, 0.99n, n$. The loss ℓ for k–Means is the sum of L_1 distances between each point and its nearest centre, and for PCA, $\ell = \text{tr}(\mathbf{C})$, where $\mathbf{C}$ is the data's sample covariance matrix.

abscissa, and $-\frac{\pi}{2} < \theta < \frac{\pi}{2}$. For $0 \leq \theta < \frac{\pi}{2}$, A is not stable. For $-\frac{\pi}{2} < \theta < 0$, if θ is approaching 0, then A is a less stable algorithm, while if θ is approaching $-\frac{\pi}{2}$, then A is a more stable algorithm. Observe that in this setting, β is a function of n and w, and hence it can be denoted by $\beta(n,w)$. Plugging $\beta(n,w)$ in the inequalities of Corollary 1, we get that:

$$(i) \quad R(A_{\mathcal{S}}) \leq \widehat{R}_{\mathrm{EMP}}(A_{\mathcal{S}}) + 2(wn + \zeta) + [4n(wn + \zeta) + c]\sqrt{\frac{\log(1/\delta)}{2n}}, \quad (8)$$

$$(ii) \quad R(A_{\mathcal{S}}) \leq \widehat{R}_{\mathrm{LOO}}(A_{\mathcal{S}}) + (wn + \zeta) + [4n(wn + \zeta) + c]\sqrt{\frac{\log(1/\delta)}{2n}}. \quad (9)$$

That is, the steeper is the stability line ($w < 0$), the more tight is the confidence bound. Figure 2 shows other examples for stability lines on the synthetic data sets (Gaussians and Moons) using Laplacian eigenmaps (LEM) [4], and Local Linear Embedding (LLE) [19]. The generalization assessment is based on the Laplacian matrix $\mathbf{L}$ for LEM, and the weighted affinity matrix $\mathbf{W}$ for LLE. In particular, the loss for LEM is $\ell = \mathrm{tr}(\mathbf{L}\mathbf{L}^{\top})$, while for LLE, $\ell = \mathrm{tr}(\mathbf{W}\mathbf{W}^{\top})$.

3.3   Comparing Two Algorithms: A_1 vs. A_2

The previous generalization assessment procedure only considered one algorithm. Here we propose an extension for the above procedure to compare the generalization of two unsupervised learning algorithms, A_1 and A_2, under the same loss ℓ, on a given data source. More specifically, the comparative setting addresses the following questions: if A_1 is stable with respect to ℓ on $\mathcal{S}$ (according to Definition 2), and if A_2 is stable with respect to ℓ on $\mathcal{S}$ (according to Definition 2), which algorithm has better generalization on $\mathcal{S}$? The following definition gives a formal answer to these questions.

Definition 3 (Comparing A_1 vs. A_2). *Let A_1 be a stable algorithm with respect to ℓ on $\mathcal{S}$ with slope $w_1 < 0$ for its stability line. Let A_2 be a stable algorithm with respect to ℓ on $\mathcal{S}$ with slope $w_2 < 0$ for its stability line. We say that:*

1. A_1 *is similar to* A_2, *denoted* $A_1 = A_2$, *if* $w_1 \approx w_2$.[6]
2. A_1 *is better than* A_2, *denoted by* $A_1 \succ A_2$, *if* $w_1 < w_2$.
3. A_1 *is worse than* A_2, *denoted* $A_1 \prec A_2$, *if* $w_1 > w_2$.

To develop a formal procedure for such an assessment we proceed by letting Algorithm 1 invoke the two algorithms A_1 and A_2 on the same subsamples $\{\mathbf{X}_1, \ldots, \mathbf{X}_m\}$ and $\{\mathbf{X}'_1, \ldots, \mathbf{X}'_m\}$. The final output of Algorithm 1 will be two sets $\mathcal{B}^1 = \{\widehat{\beta}^1_{n_1}, \ldots, \widehat{\beta}^1_{n_\tau}\}$, and $\mathcal{B}^2 = \{\widehat{\beta}^2_{n_1}, \ldots, \widehat{\beta}^2_{n_\tau}\}$. The analysis then proceeds by fitting the stability line for each algorithm, plotting the curves shown in Figures 1 and 2, and then comparing the slopes of both algorithms. Formal

---

[6] This is done using hypothesis testing for the equality of slopes – See Appendix for details.

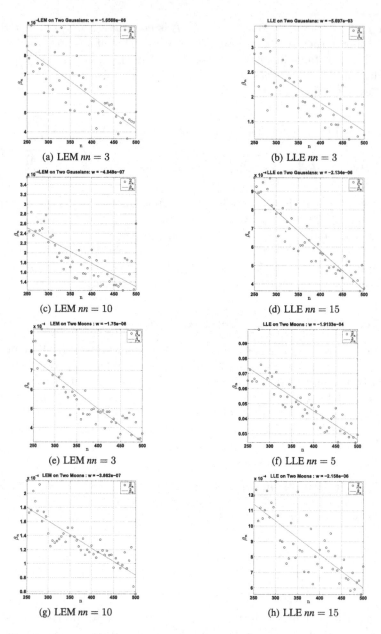

Fig. 2. First Column: Generalization assessment for LEM on two Gaussians (a,c) and two moons (e,g), with different number of nearest neighbours (nn) for constructing the data's neighbourhood graph. Compare the slopes (w) for the stability lines and the dispersion of points around it, and note the sensitivity of LEM to the number of nn. The same follows for the two moons case (e,g). Note also the difference in the stability lines (slope, and dispersion of estimated $\widehat{\beta}_n$'s) for LEM and PCA on the same data sets. **Second Column:** Generalization assessment for LLE on two Gaussians (b,d) and two moons (f,h) data sets, with different number of nn.

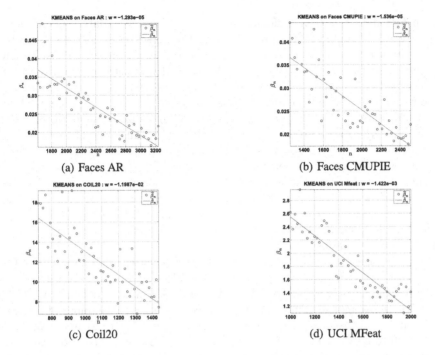

Fig. 3. Generalization assessment for k–Means clustering using stability analysis on four real data sets: (a) Faces AR, (b) Faces CMUPIE, (c) Coil20, and (d) UCI MFeat.

comparison for the slopes w_1 and w_2 is done using hypothesis testing for the equality of slopes:

$$H_0 : w_1 = w_2 \quad \text{vs.} \quad H_1 : w_1 \neq w_2.$$

If H_0 is rejected at a significance level α (usually 0.05 or 0.01), then deciding which algorithm has better generalization can be done using rules 2 and 3 in the above definition. If H_0 cannot be rejected at the desired significance level, then both algorithms have a similar generalization capability. Further insight can be gained through the norm of the residuals, and the spread of the estimated $\widehat{\beta}_n$'s around the stability line.

4   Empirical Validation on Real Data Sets

We have conducted some initial validation tests for the proposed generalization assessment framework. In these experiments, we considered two different unsupervised learning problems: clustering and dimensionality reduction (linear and nonlinear). In particular, we considered the following algorithms: k–Means for clustering, PCA for linear dimensionality reduction, and LEM and LLE for nonlinear dimensionality reduction (NLDR). The four algorithms were run on four data sets from different domains: (i) two faces data sets, AR and CMUPIE with (samples $\times$ features) 3236×2900, and 2509×2900, respectively. (ii) two image features data sets, Coil20 and Multiple Features (MFeat) with (samples $\times$ features) 1440×1024, and 2000×649, respectively, from the UCI Repository for Machine Learning [18].[7] In all these experiments, the number of bootstraps m was set to 100, and the values for $n_1, n_2, \ldots, n_\tau$ were set to $0.5n, 0.51n, 0.52n, \ldots, 0.99n, n$, where n is the original size of the data set.

To apply the proposed generalization assessment, an external loss ℓ needs to be defined for each problem. k–Means minimizes the sum of L_2 distances between each point and its nearest cluster centre. Thus, a suitable loss can be the sum of L_1 distances. Note that the number of clusters k is assumed to be known. Note also that in this setting, for each iteration j in Algorithm 1, the initial k centres are randomly chosen and they remain unchanged after holding out the random sample. That is, k–Means starts from the same initial centres before and after holding out one sample.

For PCA, LEM and LLE, the loss functions are chosen as follows: $\ell = \mathrm{tr}(\mathbf{C})$ for PCA, $\ell = \mathrm{tr}(\mathbf{LL}^\top)$ for LEM, and $\ell = \mathrm{tr}(\mathbf{WW}^\top)$ for LLE, where $\mathbf{C}$ is the data's sample covariance matrix, $\mathbf{L}$ is the Laplacian matrix defined by LEM, and $\mathbf{W}$ is the weighted affinity matrix defined by LLE. The number of nearest neighbours for constructing the neighbourhood graph for LEM and LLE was fixed to 30 to ensure that the neighbourhood graph is connected. Note that we did not perform any model selection for the number of nearest neighbours to simplify the experiments and the demonstrations.

4.1   Generalization Assessment of k–Means Clustering

Figure 3 shows the stability lines for k–Means clustering on the four real data sets used in our experiments. For both faces data sets, AR and CMUPIE, the stability lines have similar slopes despite the different sample size. However, the dispersion of points around the stability line is bigger for CMUPIE than it is for AR. Hypothesis testing for the equality of slopes (at significance level $\alpha = 0.05$) did not reject H_0 (p–value $= 0.92$). For Coil20 and UCI Mfeat, the slopes of stability lines differ by one order of magnitude (despite the different sample size). Indeed, the hypothesis test in this case rejected H_0 with a very small p–value. Note that the estimated $\widehat{\beta}_n$'s for the four data sets do not show a clear trend as is the case for the two Gaussians and the two moons data sets in Figure 1. This behaviour is expected from k–Means on real high dimensional data sets, and is in agreement with what is known about its sensitivity to the initial centres and its convergence to local minima. For a better comparison, observe the stability lines for PCA on the same data sets in Figures 4 and 5.

---

[7] The AR and CMUPIE face data sets were obtained from http://www.face-rec.org/databases/.

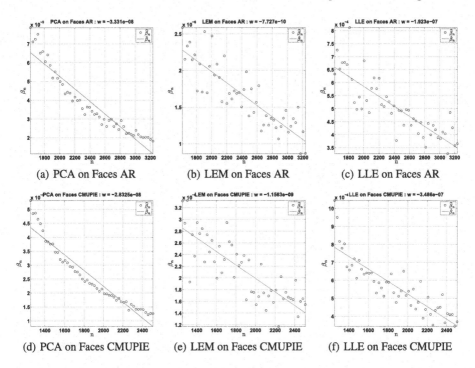

(a) PCA on Faces AR (b) LEM on Faces AR (c) LLE on Faces AR

(d) PCA on Faces CMUPIE (e) LEM on Faces CMUPIE (f) LLE on Faces CMUPIE

Fig. 4. Generalization assessment for PCA, LEM and LLE using stability analysis on two faces data sets: AR (a,b,c), and CMUPIE (d,e,f).

4.2 Generalization Assessment of PCA, LEM, and LLE

Figures 4 and 5 show the stability lines for the three dimensionality reduction algorithms; PCA, LEM and LLE, on the four real data sets used in our experiments. Note that the magnitude of w for these experiments should not be surprising given the scale of n and $\widehat{\beta}_{n_t}$. It can be seen that PCA shows a better trend of the estimated $\widehat{\beta}_n$'s than LEM and LLE (for our choice of fixed neighbourhood size). This trend shows that PCA has better stability (and hence better generalization) than LEM and LLE on these data sets. Note that in this setting, the slope for PCA stability line cannot be compared to that of LEM (nor LLE) since the loss functions are different. However, we can compare the slopes for each algorithm stability lines (separately) on the face data sets and on the features data sets.

Hypothesis testing ($\alpha = 0.05$) for PCA stability lines on AR and CMUPIE rejects H_0 in favour of H_1 with a p–value = 0.0124. For Coil20 and UCI Mfeat, the test did not reject H_0 and the p–value = 0.9. For LEM, the test did not reject H_0 for the slopes of AR and CMUPIE, while it did reject H_0 in favour of H_1 for Coil20 and UCI MFeat. A similar behaviour was observed for LLE.

In these experiments and the previous ones on k–Means clustering, note that no comparison of two algorithms were carried on the same data set. In these

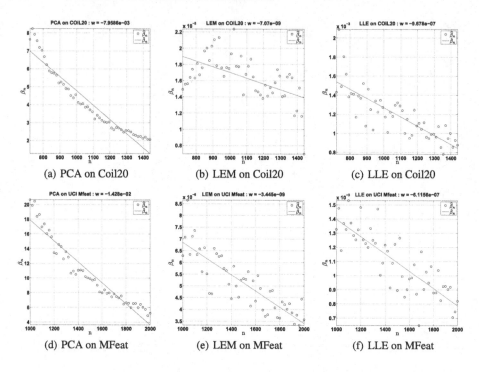

Fig. 5. Generalization assessment for PCA, LEM and LLE using stability analysis on Coil20 (a,b,c), and UCI MFeat (d,e,f).

illustrative examples, the generalization of one algorithm was assessed on two different data sets, following the examples on the synthetic data sets in Figures 1. Note that this scenario is different from the one described in § 3.3. In the above experiments, the trend of $\widehat{\beta}_{n_t}$, the stability line, the slope w, and the scatter of points around the stability line, provided a quantitative and a qualitative evaluation for the generalization capability of k–Means and PCA. However, our experience suggests that when analyzing the generalization of one algorithm on two different data sets, hypothesis testing can give more accurate insight if the sample sizes n_t are equal for both data sets since β_n is known to decrease as κ/n, and $\kappa > 0$.

5   Concluding Remarks

In this paper we proposed a general criterion for generalization in unsupervised learning that is analogous to the prediction error in supervised learning. We also proposed a computationally intensive, yet efficient procedure to realize this criterion on finite data sets, and extended it for comparing two different algorithms on a common data source. Our preliminary experiments showed that, for algorithms from three different unsupervised learning problems, the proposed framework provided a unified mechanism and a unified interface to assess their generalization capability. This type of analysis suggests further rigorous assessment of these algorithms, and is encouraging to conduct similar analysis for other unsupervised learning problems such as density estimation, subspace clustering, feature learning, and layer wise analysis of deep architectures. Further, our framework can be extended to answer model selection questions for unsupervised learning, or it can be complimentary to exiting methods for model selection.

Acknowledgments. We would like to thank our Reviewers for their helpful comments and suggestions, the Alberta Innovates Centre for Machine Learning and NSERC for their support, and Frank Ferrie for additional computational support at McGill's Centre for Intelligent Machines.

Appendix

Hypothesis testing for the equality of slopes w_1 and w_2 for two regression lines $Y_1 = w_1 X_1 + \zeta_1$ and $Y_2 = w_2 X_2 + \zeta_2$, respectively, proceeds as follows. Let $S_1 = \{(x_1^i, y_1^i)\}_{i=1}^{n_1}$ and $S_2 = \{(x_2^j, y_2^j)\}_{j=1}^{n_2}$, be the two data sets to be used for estimating the lines defined by w_1 and w_2, respectively. Let $\{\widehat{y}_1^1, \ldots, \widehat{y}_1^{n_1}\}$ and $\{\widehat{y}_2^1, \ldots, \widehat{y}_2^{n_2}\}$ be the estimated predictions from each regression line. The null and alternative hypotheses of the test are:

$$H_0 : w_1 = w_2 \quad \text{vs.} \quad H_1 : w_1 \neq w_2.$$

That is, $H_0 : w_1 - w_2 = 0$. If H_0 is true, then $w_1 - w_2 \sim \mathscr{G}(0, s_{w_1 w_2})$, where $s_{w_1 w_2}$ is the pooled error variance. Using a t test, we construct the statistic t:

$$t = \frac{w_1 - w_2}{s_{w_1 w_2}} \sim \mathscr{T}_r \, ,$$

where $\mathscr{T}_r$ is the Student's t distribution with r degrees of freedom, and $r = n_1 + n_2 - 4$. The pooled error variance is defined as:

$$s_{w_1 w_2} = \sqrt{s_{w_1}^2 + s_{w_2}^2},$$

where

$$s_{w_k}^2 = \frac{e_k}{\sigma_k^2 (n_k - 1)},$$

$e_k = \sum_{i=1}^{n_k} (y_k^i - \widehat{y}_k^i)^2 / (n_k - 2)$, and $\sigma_k^2 = Var(X_k)$, which can be replaced by the sample variance. For significance level α, we compute the probability of observing the statistic t from $\mathscr{T}_r$ given that H_0 is true; this is the P value of the test. If $P > \alpha$, then H_0 cannot be rejected. Otherwise, reject H_0 in favour of H_1.

References

1. Anandkumar, A., Hsu, D., Kakade, S.: A method of moments for mixture models and hidden Markov models. CoRR abs/1203.0683 (2012)
2. Balle, B., Quattoni, A., Carreras, X.: Local loss optimization in operator models: a new insight into spectral learning. In: Proceedings of the 29th International Conference on Machine Learning, pp. 1879–1886 (2012)
3. Bartlett, P., Mendelson, S.: Rademacher and Gaussian complexities: Risk bounds and structural results. Journal of Machine Learning Research **3**, 463–482 (2003)
4. Belkin, M., Niyogi, P.: Laplacian eigenmaps and spectral techniques for data representation. Neural Computation **15**, 1373–1396 (2003)
5. Bousquet, O., Elisseeff, A.: Stability and generalization. Journal of Machine Learning Research **2**, 499–526 (2002)
6. Demšar, J.: Statistical comparisons of classifiers over multiple data sets. Journal of Machine Learning Research **7**, 1–30 (2006)
7. Devroye, L., Györfi, L., Lugosi, G.: A Probabilistic Theory of Pattern Recognition. Springer, New York (1996)
8. Devroye, L., Wagner, T.: Distribution-free inequalities for the deleted and holdout error estimates. IEEE Transactions on Information Theory **25**(2), 202–207 (1979)
9. Dietterich, T.: Approximate statistical tests for comparing supervised classification learning algorithms. Neural Computation **10**(7), 1895–1923 (1998)
10. Efron, B.: Bootstrap methods: another look at the jackknife. Annals of Statistics **7**, 1–26 (1979)
11. Hansen, L., Larsen, J.: Unsupervised learning and generalization. In: Proceedings of the IEEE International Conference on Neural Networks, pp. 25–30 (1996)
12. Hsu, D., Kakade, S., Zhang, T.: A spectral algorithm for learning hidden Markov models. In: Proceedings of the 22nd Conference on Learning Theory (2009)
13. Jordan, M.: On statistics, computation and scalability. Bernoulli **19**(4), 1378–1390 (2013)
14. Kearns, M., Ron, D.: Algorithmic stability and sanity-check bounds for leave-one-out cross-validation. In: Proceedings of the Conference on Learning Theory, pp. 152–162. ACM (1999)
15. Kohavi, R.: A study of cross-validation and bootstrap for accuracy estimation and model selection. In: Proceedings of the 14th International Joint Conference on Artificial Intelligence, pp. 1137–1143 (1995)
16. Kutin, S., Niyogi, P.: Almost-everywhere algorithmic stability and generalization error. In: Proceedings of the Conference on Uncertainty in Artificial Intelligence, pp. 275–282 (2002)
17. Mukherjee, S., Niyogi, P., Poggio, T., Rifkin, R.: Learning theory: stability is sufficient for generalization and necessary and sufficient for consistency of empirical risk minimization. Advances in Computational Mathemathics **25**, 161–193 (2006)
18. Newman, D., Hettich, S., Blake, C., Merz, C.: UCI Repository of Machine Learning Databases (1998). http://www.ics.uci.edu/~mlearn/MLRepository.html
19. Saul, L., Roweis, S.: Think globally, fit locally: Unsupervised learning of low dimensional manifolds. Journal of Machine Learning Research **4**, 119–155 (2003)
20. Shalev-Shwartz, S., Shamir, O., Srebro, N., Sridharan, K.: Learnability, stability and uniform convergence. Journal of Machine Learning Research **11**, 2635–2670 (2010)
21. Song, L., Boots, B., Siddiqi, S., Gordon, G., Smola, A.: Hilbert space embeddings of hidden Markov models. In: Proceedings of the 27th International Conference on Machine Learning, pp. 991–998 (2010)

22. Vapnik, V.N.: Statistical Learning Theory. John Wiley & Sons, Sussex (1998)
23. Vapnik, V.N.: An overview of statistical learning theory. IEEE Transactions on Neural Networks **10**(5), 988–999 (1999)
24. Xu, H., Mannor, S.: Robustness and generalization. Machine Learning **86**(3), 391–423 (2012)
25. Xu, L., White, M., Schuurmans, D.: Optimal reverse prediction: a unified perspective on supervised, unsupervised and semi-supervised learning. In: Proceedings of the International Conference on Machine Learning, vol. 382 (2009)

Multiple Incomplete Views Clustering via Weighted Nonnegative Matrix Factorization with $L_{2,1}$ Regularization

Weixiang Shao[1], Lifang He[2]([✉]), and Philip S. Yu[1,3]

[1] Department of Computer Science,
University of Illinois at Chicago, Chicago, IL, USA
[2] Institute for Computer Vision, Shenzhen University, Shenzhen, China
lifanghescut@gmail.com
[3] Institute for Data Science, Tsinghua University, Beijing, China

Abstract. With the advance of technology, data are often with multiple modalities or coming from multiple sources. Multi-view clustering provides a natural way for generating clusters from such data. Although multi-view clustering has been successfully applied in many applications, most of the previous methods assumed the completeness of each view (*i.e.*, each instance appears in all views). However, in real-world applications, it is often the case that a number of views are available for learning but none of them is complete. The incompleteness of all the views and the number of available views make it difficult to integrate all the incomplete views and get a better clustering solution. In this paper, we propose MIC (Multi-Incomplete-view Clustering), an algorithm based on weighted nonnegative matrix factorization with $L_{2,1}$ regularization. The proposed MIC works by learning the latent feature matrices for all the views and generating a consensus matrix so that the difference between each view and the consensus is minimized. MIC has several advantages comparing with other existing methods. First, MIC incorporates weighted nonnegative matrix factorization, which handles the missing instances in each incomplete view. Second, MIC uses a co-regularized approach, which pushes the learned latent feature matrices of all the views towards a common consensus. By regularizing the disagreement between the latent feature matrices and the consensus, MIC can be easily extended to more than two incomplete views. Third, MIC incorporates $L_{2,1}$ regularization into the weighted nonnegative matrix factorization, which makes it robust to noises and outliers. Forth, an iterative optimization framework is used in MIC, which is scalable and proved to converge. Experiments on real datasets demonstrate the advantages of MIC.

1 Introduction

With the advance of technology, real data are often with multiple modalities or coming from multiple sources. Such data is called multi-view data. Different views may emphasize different aspects of the data. Integrating multiple views may help

© Springer International Publishing Switzerland 2015
A. Appice et al. (Eds.): ECML PKDD 2015, Part I, LNAI 9284, pp. 318–334, 2015.
DOI: 10.1007/978-3-319-23528-8_20

improve the clustering performance. For example, one news story may be reported by different news sources, user group can be formed based on users' profiles, user's online social connections, users' transaction history or users' credit score in online shopping recommendation system, one patient can be diagnosed with a certain disease based on different measures, including clinical, imaging, immunologic, serological and cognitive measures. Different from traditional data with a single view, these multi-view data commonly have the following properties:

1. Each view can have its own feature sets, and each view may emphasize different aspects. Different views share some consistency and complementary properties. For example, in online shopping recommendation system, user's credit score has numerical features while users' online social connections provide graph relational features. The credit score emphasizes the creditworthiness of the user, while the social connection emphasizes the social life of the user.
2. Each view may suffer from incompleteness. Due to the nature of the data or the cost of data collection, each available view may suffer from incompleteness of information. For example, not all the news stories are covered by all the news sources, $i.e.$, each news source (view) cannot cover all the news stories. Thus, all the views are incomplete.
3. There may be an arbitrary number of sources. In some applications, the number of available views may be small, while in other applications, it may be quite large.

The above properties raise two fundamental challenges for clustering multi-view data:

1. How to combine various number of views to get better clustering solutions by exploring the consistency and complementary properties of different views.
2. How to deal with the incompleteness of the views, $i.e.$, how to effectively and efficiently get better clustering solutions even all of the views are incomplete.

Multi-view clustering [1,7] provides a natural way for generating clusters from such data. A number of approaches have been proposed for multi-view clustering. Existing multi-view clustering algorithms can be classified into two categories according to [28], distributed approaches and centralized approaches. Distributed approaches, such as [4,15,28] first cluster each view independently from the others, using an appropriate single-view algorithm, and then combine the individual clusterings to produce a final clustering result. Centralized approaches, such as [1,5,24,38] make use of multiple representations simultaneously to mine hidden patterns from the data. In this paper, we mainly focus on the centralized approaches.

Most of the previous studies on multi-view clustering focus on the first challenge. They are all based on the assumption that all of the views are complete, $i.e.$, each instance appears in all views. Few of them addresses how to deal with the second challenge. Recently, there are several methods working on the incompleteness of the views [26,32,34]. They either require the completeness of at least one base view or cannot be easily extended to more than two incomplete views. However, in real-world applications, it is often the case that more than two views are available for learning and none of them is complete. For example, in document clustering, we can have documents translated into different languages representing multiple views. However, we may not get all the documents translated into each language. Another example is medical diagnosis. Although multiple measurements from a series of medical examinations may be available for a patient, it is not realistic to have each patient complete all the potential examinations, which may result in the incompleteness of all the views. The incompleteness of all the views and the number of available views make it difficult to directly integrate all the incomplete views and get a better clustering solution.

In this paper, we propose MIC (Multi-Incomplete-view Clustering) to handle the situation of multiple incomplete views by integrating the joint weighted nonnegative matrix factorization and $L_{2,1}$ regularization. Weighted nonnegative matrix factorization [20] is a weighted version of nonnegative matrix factorization [25], and has been successfully used in document clustering [35] and recommendation system [16]. $L_{2,1}$ norm of a matrix was first introduced in [9] as rotational invariant L_1 norm. Because of its robustness to noise and outliers, $L_{2,1}$ has been widely used in many areas [11,13,18,21]. By integrating weighted nonnegative matrix factorization and $L_{2,1}$ norm, MIC tries to learn a latent subspace where the features of the same instance from different views will be co-regularized to a common consensus, while increasing the robustness of the learned latent feature matrices. The proposed MIC method has several advantages comparing with other state-of-art methods:

1. MIC incorporates weighted nonnegative matrix factorization, which will handle the missing instances in each incomplete view. A weight matrix for each incomplete view is introduced to give the missing instances lower weights than the presented instances in each view.
2. By using a co-regularized approach, MIC pushes the learned latent feature matrices to a common consensus. Because MIC only regularizes the difference between the learned latent feature for each view and the consensus, MIC can be easily extended to more than two incomplete views.
3. MIC incorporates $L_{2,1}$ norm into the weighted nonnegative matrix factorization. $L_{2,1}$ regularization added to the objective function will keep the learned latent feature matrices more robust to noises and outliers, which is naturally perfect for the situation of multiple incomplete views.
4. An iterative optimization framework is used in MIC, which is scalable and proved to converge.

The rest of this paper is organized as follows. In the next section, notations and problem formulation are given. The proposed MIC algorithm is then presented in Section 3. Extensive experimental results and analysis are shown in Section 4. Related work is given in Section 5 followed by conclusion in Section 6.

Table 1. Summary of the Notations

Notation	Description
N	Total number of instances.
n_v	Total number of views.
$\mathbf{X}^{(i)}$	Data matrix for the i-th view.
d_i	Dimension of features in the i-th view.
$\mathbf{M}$	The indicator matrix, where $\mathbf{M}_{i,j} = 1$ indicates j-th instance appears in i-th view.
$\mathbf{W}^{(i)}$	The diagonal instance weight matrix for the i-th view.
$\mathbf{U}^{(i)}$	The latent feature matrix for the i-th view.
$\mathbf{V}^{(i)}$	The basis matrix for the i-th view.
$\mathbf{U}^*$	The common consensus, latent feature matrix across all the views.
α_i	Trade-off parameter between reconstruction error and view disagreement for view i.
β_i	Trade-off parameter between reconstruction error and robustness for view i.

2 Problem Formulation and Backgrounds

In this section, we will briefly describe the problem formulation. Then the background knowledge on weighted nonnegative matrix factorization will be introduced.

2.1 Problem Formulation

Before we describe the formulation of the problem, we summarize some notations used in this paper in Table 1. Assume we are given a dataset with N instances and n_v views $\{\mathbf{X}^{(i)}, i = 1, 2, ..., n_v\}$, where $\mathbf{X}^{(i)} \in \mathbb{R}^{N \times d_i}$ represents the dataset in view i. We define an indicator matrix $\mathbf{M} \in \mathbb{R}^{n_v \times N}$ by,

$$\mathbf{M}_{i,j} = \begin{cases} 1 & \text{if } j\text{-th instance is in the } i\text{-th view.} \\ 0 & \text{otherwise.} \end{cases}$$

where each row of $\mathbf{M}$ represent the instance presence for one view. Most of the previous methods on multi-view clustering assume the completeness of all the views. Every view contains all the instances, *i.e.*, $\mathbf{M}$ is an all one matrix, $\sum_{j=1}^{N} \mathbf{M}_{i,j} = N$, $i = 1, 2, ..., n_v$. However, in most real-world situations, one instance may only appear in some of the views, which may result in the incompleteness of all the views. For each view, the data matrix $\mathbf{X}^{(i)}$ will have a number of rows missing, *i.e.*, $\sum_{j=1}^{N} \mathbf{M}_{i,j} < N$, $i = 1, 2, ..., n_v$.

Our goal is to cluster all the N instances into K clusters by integrating all the n_v incomplete views.

2.2 Weighted Nonnegative Matrix Factorization

Let $\mathbf{X} \in \mathbb{R}_+^{N \times M}$ denote the nonnegative data matrix where each row represents an instance and each column represents one attribute. Weighted nonnegative matrix factorization [20] aims to factorize the data matrix $\mathbf{X}$ into two nonnegative matrices, while giving different weights to the reconstruction errors of different entries. We denote the two nonnegative matrices factors as $\mathbf{U} \in \mathbb{R}_+^{N \times K}$ and $\mathbf{V} \in \mathbb{R}_+^{M \times K}$. Here K is the desired reduced dimension. To facilitate discussions, we call $\mathbf{U}$ the *latent feature matrix* and $\mathbf{V}$ the *basis matrix*. The objective function for general weighted nonnegative matrix factorization can be formulated as below:

$$\min_{\mathbf{U},\mathbf{V}} \|\mathbf{W} * (\mathbf{X} - \mathbf{U}\mathbf{V}^T)\|_F^2, \ s.t. \ \mathbf{U} \geq 0, \mathbf{V} \geq 0, \tag{1}$$

where $\|.\|_F$ is the Frobenius norm, $\mathbf{W} \in \mathbb{R}^{N \times M}$ is the weight matrix, $*$ is elementwise production and $\mathbf{U} \geq 0, \mathbf{V} \geq 0$ represent the constraints that all the matrix elements are nonnegative.

3 Multi-Incomplete-View Clustering

In this section, we present the Multi-Incomplete-view Clustering (MIC) framework. We model the multi-incomplete-view clustering as a joint weighted nonnegative matrix factorization problem with $L_{2,1}$ regularization. The proposed MIC learns the latent feature matrices for each view and pushes them towards a consensus matrix. Thus, the consensus matrix can be viewed as the shared latent feature matrix across all the views. In the following, we will first describe the construction of the objective function for the proposed method and derive the solution to the optimization problem. Then the whole MIC framework is presented.

3.1 Objective Function of MIC

Given n_v views $\{\mathbf{X}^{(i)} \in \mathbb{R}^{N \times d_i}, i = 1, 2, ..., n_v\}$, where each of the views suffers from incompleteness, *i.e.*, $\sum_{j=1}^{N} \mathbf{M}_{i,j} < N$. With more than two incomplete views, we cannot directly apply the existing methods to the incomplete data. One simple solution is to fill the missing instances with average features first, and then apply the existing multi-view clustering methods. However, this approach depends on the quality of the filled instances. For small missing percentages, the quality of the information contained in the filled average features may be good. However, when the number of missing instance increase, the quality of information contained in the filling average features may be bad or even misleading. Thus, simply filling the missing instance will not solve this problem.

Borrowing the similar idea from weighted NMF, we introduce a diagonal weight matrix $\mathbf{W}^{(i)} \in \mathbb{R}^{N \times N}$ for each incomplete views i by

$$\mathbf{W}_{j,j}^{(i)} = \begin{cases} 1 & \text{if } i\text{-th view contains } j\text{-th instance, } i.e., \mathbf{M}_{j,i} = 1. \\ w_i & \text{otherwise.} \end{cases}$$

Note that $\mathbf{W}_{j,j}^{(i)}$ indicates the weight of the j-th instance from view i, and w_i is the weight of the filled average feature instances for view i. In our experiment, w_i is defined as the percentage of the available instances for view i:

$$w_i = \frac{\sum_{j=1}^{N} \mathbf{M}_{j,i}}{N}.$$

It can be seen, $\mathbf{W}^{(i)}$ gives lower weights to the missing instances than the presented instances in the i-th view. For different views with different incomplete rates, the weights for missing instances are also different. The diagonal weight matrices give higher weights to the missing instances from views with lower incomplete rate.

A simple objective function to combine multiple incomplete views can be:

$$\min_{\{\mathbf{U}^{(i)}\},\{\mathbf{V}^{(i)}\}} \mathcal{O} = \sum_{i=1}^{n_v}(\|\mathbf{W}^{(i)}(\mathbf{X}^{(i)} - \mathbf{U}^{(i)}\mathbf{V}^{(i)^T})\|_F^2 \ s.t. \ \mathbf{U}^{(i)} \geq 0, \ \mathbf{V}^{(i)} \geq 0, \ i = 1, 2, ..., n_v,$$

(2)

where $\mathbf{U}^{(i)}$ and $\mathbf{V}^{(i)}$ are the latent feature matrix and basis matrix for the i-th view.

However, Eq. (2) only decomposes the different views independently without taking advantages of the relationship between the views. In order to make use of the relation between different views, we push the latent feature matrices for different views towards a common consensus by adding additional term R to Eq. (2) to minimize the disagreement between different views and the common consensus.

$$\min_{\{\mathbf{U}^{(i)}\},\{\mathbf{V}^{(i)}\},\mathbf{U}^*} \sum_{i=1}^{n_v} \left(\|\mathbf{W}^{(i)}(\mathbf{X}^{(i)} - \mathbf{U}^{(i)}\mathbf{V}^{(i)^T})\|_F^2 + \alpha_i R(\mathbf{U}^{(i)}, \mathbf{U}^*) \right)$$

(3)

$$s.t. \ \mathbf{U}^* \geq 0, \mathbf{U}^{(i)} \geq 0, \mathbf{V}^{(i)} \geq 0, \ i = 1, 2, ..., n_v,$$

where $\mathbf{U}^* \in \mathbb{R}^{N*K}$ is the consensus latent feature matrix across all the views, and α_i is the trade-off parameter between reconstruction error and disagreement between view i and the consensus. In this paper we define R as the square of Frobenius norm of the weighted difference between the latent feature matrices:

$$R(\mathbf{U}^{(i)}, \mathbf{U}^*) = \|\mathbf{W}^{(i)}(\mathbf{U}^{(i)} - \mathbf{U}^*)\|_F^2.$$

Additionally, considering the nature of incomplete views, we added $L_{2,1}$ regularization into Eq. 3, which is robust to noises and outliers and widely used in many applications [10,17,37].

Formally, the objective function of MIC is as follows:

$$\min_{\{\mathbf{U}^{(i)}\},\{\mathbf{V}^{(i)}\},\mathbf{U}^*} \mathcal{O} = \sum_{i=1}^{n_v}(\|\mathbf{W}^{(i)}(\mathbf{X}^{(i)} - \mathbf{U}^{(i)}\mathbf{V}^{(i)^T})\|_F^2 + \alpha_i\|\mathbf{W}^{(i)}(\mathbf{U}^{(i)} - \mathbf{U}^*)\|_F^2 + \beta_i\|\mathbf{U}^{(i)}\|_{2,1})$$

$$s.t. \ \mathbf{U}^{(i)} \geq 0, \ \mathbf{V}^{(i)} \geq 0, \ \mathbf{U}^* \geq 0, \ i = 1, 2, ..., n_v.$$

(4)

where β_i is the trade-off between robustness and accuracy of reconstruction for the i-th view, $\| \cdot \|_{2,1}$ is the $L_{2,1}$ norm and defined as:

$$\|\mathbf{U}\|_{2,1} = \sum_{i=1}^{N} \left(\sum_{k=1}^{K} |\mathbf{U}_{i,k}|^2 \right)^{1/2}$$

3.2   Optimization

In the following, we give the solution to Eq. 4. For the sake of convenience, we will see both α_i and β_i as positive in the derivation, and denote $\tilde{\mathbf{W}}^{(i)} = \mathbf{W}^{(i)^T}\mathbf{W}^{(i)}$. As we see, minimizing Eq. 4 is with respect to $\{\mathbf{U}^{(i)}\}$, $\{\mathbf{V}^{(i)}\}$ and $\mathbf{U}^*$, and we cannot give a closed-form solution. We propose an alternating scheme to optimize the objective function. Specifically, the following two steps are repeated until convergence: (1) fixing $\{\mathbf{U}^{(i)}\}$ and $\{\mathbf{V}^{(i)}\}$, minimize $\mathcal{O}$ over $\mathbf{U}^*$, (2) fixing $\mathbf{U}^*$, minimize $\mathcal{O}$ over $\{\mathbf{U}^{(i)}\}$ and $\{\mathbf{V}^{(i)}\}$.

Fixing $\{\mathbf{U}^{(i)}\}$ and $\{\mathbf{V}^{(i)}\}$, minimize $\mathcal{O}$ over $\mathbf{U}^*$. With $\{\mathbf{U}^{(i)}\}$ and $\{\mathbf{V}^{(i)}\}$ fixed, we need to minimize the following objective function:

$$\mathcal{J}(\mathbf{U}^*) = \sum_{i=1}^{n_v} \alpha_i \|\mathbf{W}^{(i)}(\mathbf{U}^{(i)} - \mathbf{U}^*)\|_F^2 \quad s.t. \ \mathbf{U}^* \geq 0 \tag{5}$$

We take the derivative of the objective function $\mathcal{J}$ in Eq. 5 over $\mathbf{U}^*$ and set it to 0:

$$\frac{\partial \mathcal{J}}{\partial \mathbf{U}^*} = \sum_{i=1}^{n_v} 2\alpha_i \tilde{\mathbf{W}}^{(i)}\mathbf{U}^* - 2\alpha_i \tilde{\mathbf{W}}^{(i)}\mathbf{U}^{(i)} = 0 \tag{6}$$

Since $\tilde{\mathbf{W}}^{(i)}$ is a positive diagonal matrix and α_i is a positive constant, $\sum_{i=1}^{n_v} \alpha_i \tilde{\mathbf{W}}^{(i)}$ is invertible. Solving Eq. 6, we have an exact solution for $\mathbf{U}^*$:

$$\mathbf{U}^* = \left(\sum_{i=1}^{n_v} \alpha_i \tilde{\mathbf{W}}^{(i)} \right)^{-1} \left(\sum_{i=1}^{n_v} \alpha_i \tilde{\mathbf{W}}^{(i)}\mathbf{U}^{(i)} \right) \geq 0 \tag{7}$$

Fixing $\mathbf{U}^*$, minimize $\mathcal{O}$ over $\{\mathbf{U}^{(i)}\}$ and $\{\mathbf{V}^{(i)}\}$. With $\mathbf{U}^*$ fixed, the computation of $\mathbf{U}^{(i)}$ and $\mathbf{V}^{(i)}$ does not depend on $\mathbf{U}^{(i')}$ or $\mathbf{V}^{(i')}$, $i' \neq i$. Thus for each view i, we need to minimize the following objective function:

$$\min_{\mathbf{U}^{(i)}, \mathbf{V}^{(i)}} \|\mathbf{W}^{(i)}(\mathbf{X}^{(i)} - \mathbf{U}^{(i)}\mathbf{V}^{(i)^T})\|_F^2 + \alpha_i \|\mathbf{W}^{(i)}(\mathbf{U}^{(i)} - \mathbf{U}^*)\|_F^2 + \beta_i \|\mathbf{U}^{(i)}\|_{2,1}$$
$$s.t. \ \mathbf{U}^{(i)} \geq 0, \mathbf{V}^{(i)} \geq 0 \tag{8}$$

We will iteratively update $\mathbf{U}^{(i)}$ and $\mathbf{V}^{(i)}$ using the following multiplicative updating rules. We repeat the two steps iteratively until the objective function in Eq. 8 converges.

(1) Fixing $\mathbf{U}^*$ and $\mathbf{V}^{(i)}$, minimize $\mathcal{O}$ over $\mathbf{U}^{(i)}$. For each $\mathbf{U}^{(i)}$, we need to minimize the following objective function:

$$\mathcal{J}(\mathbf{U}^{(i)}) = \|\mathbf{W}^{(i)}(\mathbf{X}^{(i)} - \mathbf{U}^{(i)}\mathbf{V}^{(i)^T})\|_F^2 + \alpha_i\|\mathbf{W}^{(i)}(\mathbf{U}^{(i)} - \mathbf{U}^*)\|_F^2 + \beta_i\|\mathbf{U}^{(i)}\|_{2,1} \tag{9}$$
$$s.t. \quad \mathbf{U}^{(i)} \geq 0$$

The derivative of $\mathcal{J}(\mathbf{U}^{(i)})$ with respect to $\mathbf{U}^{(i)}$ is

$$\frac{\partial \mathcal{J}}{\partial \mathbf{U}^{(i)}} = -2\tilde{\mathbf{W}}^{(i)}\mathbf{X}^{(i)}\mathbf{V}^{(i)} + 2\tilde{\mathbf{W}}^{(i)}\mathbf{U}^{(i)}\mathbf{V}^{(i)^T}\mathbf{V}^{(i)} + 2\alpha_i\tilde{\mathbf{W}}^{(i)}\mathbf{U}^{(i)} - 2\alpha_i\tilde{\mathbf{W}}^{(i)}\mathbf{U}^* + \beta_i\mathbf{D}^{(i)}\mathbf{U}^{(i)} \tag{10}$$

Here $\mathbf{D}^{(i)}$ is a diagonal matrix with the j-th diagonal element given by

$$\mathbf{D}_{j,j}^{(i)} = \frac{1}{\|\mathbf{U}_{j,:}^{(i)}\|_2}, \tag{11}$$

where $\mathbf{U}_{j,:}^{(i)}$ is the j-th row of matrix $\mathbf{U}^{(i)}$, and $\|\cdot\|_2$ is the L_2 norm.

Using the Karush-Kuhn-Tucker (KKT) complementary condition [3] for the nonnegativity of $\mathbf{U}^{(i)}$, we get

$$(-2\tilde{\mathbf{W}}^{(i)}\mathbf{X}^{(i)}\mathbf{V}^{(i)} + 2\tilde{\mathbf{W}}^{(i)}\mathbf{U}^{(i)}\mathbf{V}^{(i)^T}\mathbf{V}^{(i)} + 2\alpha_i\tilde{\mathbf{W}}^{(i)}\mathbf{U}^{(i)} - 2\alpha_i\tilde{\mathbf{W}}^{(i)}\mathbf{U}^* + \beta_i\mathbf{D}^{(i)}\mathbf{U}^{(i)})_{j,k}\mathbf{U}_{j,k}^{(i)} = 0 \tag{12}$$

Based on this equation, we can derive the updating rule for $\mathbf{U}^{(i)}$:

$$\mathbf{U}_{j,k}^{(i)} \leftarrow \mathbf{U}_{j,k}^{(i)}\sqrt{\frac{\left(\tilde{\mathbf{W}}^{(i)}\mathbf{X}^{(i)}\mathbf{V}^{(i)} + \alpha_i\tilde{\mathbf{W}}^{(i)}\mathbf{U}^*\right)_{j,k}}{\left(\mathbf{U}^{(i)}\mathbf{V}^{(i)^T}\mathbf{V}^{(i)} + \alpha_i\tilde{\mathbf{W}}^{(i)}\mathbf{U}^{(i)} + 0.5\beta_i\mathbf{D}^{(i)}\mathbf{U}^{(i)}\right)_{j,k}}} \tag{13}$$

(2) Fixing $\mathbf{U}^{(i)}$ and $\mathbf{U}^*$, minimize $\mathcal{O}$ over $\mathbf{V}^{(i)}$. For each $\mathbf{V}^{(i)}$, we need to minimize the following objective function:

$$\mathcal{J}(\mathbf{V}^{(i)}) = \|\mathbf{W}^{(i)}(\mathbf{X}^{(i)} - \mathbf{U}^{(i)}\mathbf{V}^{(i)^T})\|_F^2 \quad s.t. \quad \mathbf{V}^{(i)} \geq 0 \tag{14}$$

The derivative of $\mathcal{J}(\mathbf{V}^{(i)})$ with respect to $\mathbf{V}^{(i)}$ is

$$\frac{\partial \mathcal{L}}{\partial \mathbf{V}^{(i)}} = 2\mathbf{V}^{(i)}\mathbf{U}^{(i)^T}\tilde{\mathbf{W}}^{(i)}\mathbf{U}^{(i)} - 2\mathbf{X}^{(i)^T}\tilde{\mathbf{W}}^{(i)}\mathbf{U}^{(i)} \tag{15}$$

Using the KKT complementary condition for the nonnegativity of $\mathbf{V}^{(i)}$, we get

$$(\mathbf{V}^{(i)}\mathbf{U}^{(i)^T}\tilde{\mathbf{W}}^{(i)}\mathbf{U}^{(i)} - \mathbf{X}^{(i)^T}\tilde{\mathbf{W}}^{(i)}\mathbf{U}^{(i)})_{j,k}\mathbf{V}_{j,k}^{(i)} = 0 \tag{16}$$

Based on this equation, we can derive the updating rule for $\mathbf{V}^{(i)}$:

$$\mathbf{V}_{j,k}^{(i)} \leftarrow \mathbf{V}_{j,k}^{(i)}\sqrt{\frac{(\mathbf{X}^{(i)^T}\tilde{\mathbf{W}}^{(i)}\mathbf{U}^{(i)})_{j,k}}{(\mathbf{V}^{(i)}\mathbf{U}^{(i)^T}\tilde{\mathbf{W}}^{(i)}\mathbf{U}^{(i)})_{j,k}}} \tag{17}$$

Algorithm 1. Multi-Incomplete-view Clustering (MIC)

Input: Nonnegative data matrices for incomplete views $\{\mathbf{X}^{(1)}, \mathbf{X}^{(2)}, ..., \mathbf{X}^{(n_v)}\}$, indicator matrix $\mathbf{M}$, parameters $\{\alpha_1, \alpha_2, ..., \alpha_{n_v}, \beta_1, \beta_2, ..., \beta_{n_v}\}$, number of clusters K.
Output: Basis matrices $\{\mathbf{V}^{(1)}, \mathbf{V}^{(2)}, ..., \mathbf{V}^{(n_v)}\}$, latent feature matrices $\{\mathbf{U}^{(1)}, \mathbf{U}^{(2)}, ..., \mathbf{U}^{(n_v)}\}$, consensus matrix $\mathbf{U}^*$ and clustering results.
1: Fill the missing instances in each incomplete view with average feature values.
2: Normalize each view $\mathbf{X}^{(i)}$ such that $\|\mathbf{X}^{(i)}\|_1 = 1$.
3: Initialize $\mathbf{U}^{(i)}$ and $\mathbf{V}^{(i)}$ for $1 \le i \le n_v$.
4: **repeat**
5:    Fixing $\mathbf{U}^{(i)}s$ and $\mathbf{V}^{(i)}s$, update $\mathbf{U}^*$ by Eq. 7.
6:    **for** $i = 1$ **to** n_v **do**
7:       **repeat**
8:          Fixing $\mathbf{U}^*$ and $\mathbf{V}^{(i)}$, update $\mathbf{U}^{(i)}$ by Eq. 13.
9:          Fixing $\mathbf{U}^{(i)}$ and $\mathbf{U}^*$, update $\mathbf{V}^{(i)}$ by Eq. 17.
10:          Normalize $\mathbf{V}^{(i)}$ and $\mathbf{U}^{(i)}$ by Eq. 18.
11:       **until** Eq. 8 converges.
12:    **end for**
13: **until** Eq. 4 converges.
14: Apply k-means on $\mathbf{U}^*$ to get the clustering result.

It is worth noting that to prevent $\mathbf{V}^{(i)}$ from having arbitrarily large values (which may lead to arbitrarily small values of $\mathbf{U}^{(i)}$), it is common to put a constraint on each basis matrix $\mathbf{V}^{(i)}$ [14], s.t. $\|\mathbf{V}^{(i)}_{:,k}\|_1 = 1$, $\forall\ 1 \le k \le K$. However, the updated $\mathbf{V}^{(i)}$ may not satisfy the constraint. We need to normalize $\mathbf{V}^{(i)}$ and change $\mathbf{U}^{(i)}$ to make the constraint satisfied and keep the accuracy of the approximation $\mathbf{X}^{(i)} \approx \mathbf{U}^{(i)}\mathbf{V}^{(i)T}$:

$$\mathbf{V}^{(i)} \leftarrow \mathbf{V}^{(i)}\mathbf{Q}^{(i)-1}, \mathbf{U}^{(i)} \leftarrow \mathbf{U}^{(i)}\mathbf{Q}^{(i)} \tag{18}$$

Here, $\mathbf{Q}^{(i)}$ is a diagonal matrix with the k-th diagonal element given by $\mathbf{Q}^{(i)}_{k,k} = \sum_j^{d_i} \mathbf{V}^{(i)}_{j,k}$.

The whole procedure is summarized in Algorithm 1. We will first fill the missing instances with average feature values in each incomplete view. Then we normalize the data and initialize the latent feature matrices and basis matrices. We apply the iterative alternating optimization procedure until the objective function converges. k-means is then applied to the learned consensus latent feature matrix to get the clustering solution.

4   Experiments and Results

4.1   Comparison Methods

We compare the proposed MIC method with several state-of-art methods. The details of comparison methods are as follows:

- **MIC:** MIC is the clustering framework proposed in this paper, which applies weighted joint nonnegative matrix with $L_{2,1}$ regularization. If not stated, the co-regularization parameter set $\{\alpha_i\}$ and the robust parameter set $\{\beta_i\}$ are all set to 0.01 for all the views throughout the experiment.
- **Concat:** Feature concatenation is one straightforward way to integrate all the views. We first fill the missing instances with the average features for each view. Then we concatenate the features of all the views, and run k-means directly on this concatenated view representation.
- **MultiNMF:** MultiNMF [27] is one of the most recent multi-view clustering methods based on joint nonnegative matrix factorization. MultiNMF added constraints to original nonnegative matrix factorization that pushes clustering solution of each view towards a common consensus.
- **ConvexSub:** The subspace-based multi-view clustering method developed by [17]. In the experiments, we set $\beta = 1$ for all the views. We run the ConvexSub method using a range of γ values as in the original paper, and present the best results obtained.
- **PVC:** Partial multi-view clustering [26] is one of the state-of-art multi-view clustering methods, which deals with incomplete views. PVC works by establishing a latent subspace where the instances corresponding to the same example in different views are close to each other. In our experiment, we set the parameter λ to 0.01 as in the original paper.
- **CGC:** CGC [6] is the most recent work that deals with many-to-many instance relationship, which can be used in the situation of incomplete views. In order to run the CGC algorithm, for every pair of incomplete views, we generate the mapping between the instances that appears in both views. In the experiment, the parameter λ is set to 1 as in the original paper.

It is worth to note that MultiNMF and ConvexSub are two recent methods for multi-view clustering. Both of them assumes the completeness of all the available views. PVC is among the first works that does not assume the completeness of any view. However, PVC can only works with two incomplete views. For the sake of comparison, all the views are considered with equivalent importance in the evaluation of all the multi-view algorithms. The results evaluated by two metrics, the normalized mutual information (NMI) and the accuracy (AC). Since we use k-means to get the clustering solution at the end of the algorithm, we run k-means 20 times and report the average performance.

4.2   Dataset

In this paper, three different real-world datasets are used to evaluate the proposed method MIC. Among the three datasets, the first one is handwritten digit data, the second one is text data, while the last one is flower image data. The important statistics of them are summarized in Table 2.

- **Handwritten Dutch Digit Recognition (Digit):** This data contains 2000 handwritten numerals ("0"-"9") extracted from a collection of Dutch

Table 2. Statics of the data

Data	size	# views	# clusters
Digit	2000	5	10
3Sources	416	3	6
Flowers	1360	3	17

Table 3. Incomplete rates for 3Sources

Data	V1	V2	V3	size
BBC-Reuters	13.51%	27.76%	-	407
BBC-Guardian	12.87%	25.25%	-	404
Reuters-Guardian	23.44%	21.35%	-	384
3Sources	15.38%	29.33%	27.40%	416

utility maps [12]. The following feature spaces (views) with different vector-based features are available for the numbers: (1) 76 Fourier coefficients of the character shapes, (2) 216 profile correlations, (3) 64 Karhunen-Love coefficients, (4) 240 pixel averages in 2×3 windows, (5) 47 Zernike moments. All these features are conventional vector-based features but in different feature spaces.

- **3-Source Text data (3Sources)**[1] It is collected from three online news sources: BBC, Reuters, and The Guardian, where each news source can be seen as one view for the news stories. In total there are 948 news articles covering 416 distinct news stories from the period February to April 2009. Of these distinct stories, 169 were reported in all three sources, 194 in two sources, and 53 appeared in a single news source. Each story was manually annotated with one of the six topical labels: business, entertainment, health, politics, sport, technology.
- **Oxford Flowers Data (Flowers)**: The Oxford Flower Dataset is composed of 17 flower categories, with 80 images for each category [30]. Each image is described by different visual features using color, shape, and texture. In this paper, we use the χ^2 distance matrices for different flower features (color, shape, texture) as three different views.

Both Digit and Flowers data are complete. We randomly delete instances from each view to make the views incomplete. To simplify the situation, we delete the same number of instances for all the views, and run the experiment under different incomplete percentages from 0% (all the views are complete) to 50% (all the views have 50% instances missing). It is also worth to note that 3Sources is naturally incomplete. Also since PVC can only with with two incomplete views, in order to compare PVC with other methods, we take any two of the three incomplete views and run experiments on them. We also report the results on all the three incomplete views. The statistics of 3Sources data are summarized in Table 3.

4.3   Results

The results for Digit data and Flower data are shown in Figs. 1-4. We report the results for various incomplete rates (from 0% to 50% with 10% as interval). Table 4 contains the results for 3Sources data.

---

[1] http://mlg.ucd.ie/datasets/3sources.html

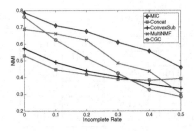

Fig. 1. NMIs for Digit.

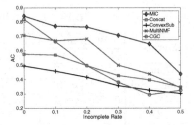

Fig. 2. ACs for Digit.

From Figs. 1 and 2 for Digit data, we can see that the proposed MIC method outperforms all the other methods in all the scenarios, especially for relatively large incomplete rates (about 12% higher than other methods in NMI and about 20% higher in AC for incomplete rates 30% and 40%). It is worth to note that when the incomplete rate is 0, CGC is the second best method in both NMI and AC, which is very close to MIC. However, as the incomplete rate increases, the performance of CGC drops quickly. One of the possible reasons is that CGC works on the similarity matrices/kernels, as the incomplete rate increases, estimated similarity/kernel matrices are not accurate. Also, as the incomplete rate increases, fewer instance mappings between views are available. Combining these two factors, the performance of CGC drops for incomplete views. We can also observe that for incomplete views (incomplete rate > 0), multiNMF gives the second best performance (still at lease 5% lower in NMI and at lease 8% lower in AC).

In Table 4, we can also observe that the proposed method outperforms all the other methods in both NMI and AC. MultiNMF and ConvexSub perform the best among the compared techniques.

From Figs. 3 and 4 for Flowers data, we can observe that in most of the cases, MIC outperforms all the other methods. It is worth to note that when all the views are complete, the performances of ConvexSub and MultiNMF are almost the same as MIC. As the incomplete rate increases, MIC starts to show the advantages over other methods. However, when the incomplete rate is too large (*e.g.*, 50%), the performance of MIC is almost the same as ConvexSub and MultiNMF.

4.4   Parameter Study

There are two sets of parameters in the proposed methods: $\{\alpha_i\}$, trade-off parameter between reconstruction error and view disagreement and $\{\beta_i\}$, trade-off parameter between the reconstruction error and robustness. Here we explore the effects of the view disagreement trade-off parameter and the robust trade-off parameter to clustering performance. We first fix $\{\beta_i\}$ to 0.01, run MIC with various $\{\alpha_i\}$ values (from 10^{-7} to 100). Then fix $\{\alpha_i\}$ to 0.01, run MIC with various $\{\beta_i\}$ values (from 10^{-7} to 100). Due to the limit of space, we only report

Table 4. Results on 3Sources Text Data

Methods	BBC-Reuters		BBC-Guardian		Reuters-Guardian		Three-Source	
	NMI	AC	NMI	AC	NMI	AC	NMI	AC
Concat	0.2591	0.3465	0.2526	0.3599	0.2474	0.3633	0.2757	0.3429
ConvexSub	0.3309	0.3913	0.3576	0.4584	0.3450	0.4370	0.3653	0.4504
PVC	0.2931	0.4252	0.2412	0.4334	0.2488	0.4145	–	–
CGC	0.2336	0.4167	0.2470	0.3857	0.2682	0.4530	0.2875	0.4279
MultiNMF	0.3687	0.4517	0.3647	0.4693	0.3487	0.4281	0.4131	0.4756
MIC	**0.3814**	**0.4912**	**0.3813**	**0.4988**	**0.3800**	**0.4612**	**0.4512**	**0.5631**

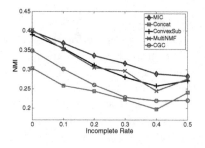

Fig. 3. NMIs for Flowers.   Fig. 4. ACs for Flowers.

the results on 3Souces data with all the three views in Fig. 5. From Fig. 5, we can find that MIC achieves stably good performance when α_i is around 10^{-2} and β_i is from 10^{-5} to 10^{-1}.

4.5   Convergence Study

The three updates rules for $\mathbf{U}^*$, $\{\mathbf{V}^{(i)}\}$ and $\{\mathbf{U}^{(i)}\}$ are iterative. In the supplemental material, we prove that each update will decrease the value objective function and the whole process will converge to a local minima solution. Fig. 6 shows the convergence curve together with its performance for Digit data with

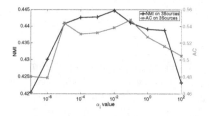

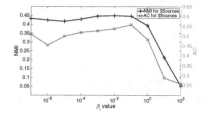

Performance of MIC v.s. α_i Performance of MIC v.s. β_i

Fig. 5. Parameter study on 3Sources.

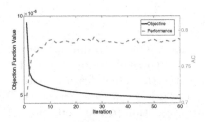
Digit data with 10% incomplete rate.

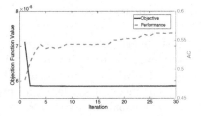
3Sources data with all three views.

Fig. 6. Convergence and corresponding performance curve.

10% incomplete rate and 3Sources data using all the three views. The blue solid line shows the value of the objective function and the red dashed line indicates the accuracy of the method. As can be seen, for Digit data, the algorithm will converge after 30 iterations. For 3Sources data, after less than 10 iterations, the algorithm will converge.

5 Related Work

There are two areas of related works upon which the proposed model is built. Multi-view learning [2,22,29], is proposed to learn from instances which have multiple representations in different feature spaces. Specifically, Multi-view clustering [1,28] is most related to our work. For example, [1] developed and studied partitioning and agglomerative, hierarchical multi-view clustering algorithms for text data. [23,24] are among the first works proposed to solve the multi-view clustering problem via spectral projection. Linked Matrix Factorization [33] is proposed to explore clustering of a set of entities given multiple graphs. Recently, [34] proposed a kernel based approach which allows clustering algorithms to be applicable when there exists at least one complete view with no missing data. As far as we know, [26,32] are the only two works that do not require the completeness of any view. However, both of the methods can only work with two incomplete views.

Nonnegative matrix factorization [25] is the second area that is related to our work. NMF has been successfully used in unsupervised learning [31,36]. Different variations were proposed in the last decade. For example, [8] posed a three factor NMF and added orthogonal constrains for rigorous clustering interpretation. [19] introduced sparsity constraints on the latent feature matrix, which will give more sparse latent representations. [20] proposed a weighted version of NMF, which gives different weights to different entries in the data. Recently, [6,27] propsed to use NMF to clustering data from multiple views/sources. However, they cannot deal with multiple incomplete views. The proposed MIC, which uses weighted joint NMF to handle the incompleteness of the views and maintain the robustness by introducing the $L_{2,1}$ regularization.

6 Conclusion

In this paper, we study the problem of clustering on data with multiple incomplete views, where each view suffers from incompleteness of instances. Based on weighted NMF, the proposed MIC method learns the latent feature matrices for all the incomplete views and pushes them towards a common consensus. To achieve the goal, we use a joint weighted NMF algorithm to learn not only the latent feature matrix for each view but also minimize the disagreement between the latent feature matrices and the consensus matrix. By giving missing instances from each view lower weights, MIC minimizes the negative influences from the missing instances. It also maintains the robustness to noises and outliers by introducing the $L_{2,1}$ regularization. Extensive experiments conducted on three datasets demonstrate the effectiveness of the proposed MIC method on data with multiple incomplete views comparing with other state-of-art methods.

Acknowledgments. This work is supported in part by NSF (CNS-1115234), NSFC (61272050, 61273295, 61472089), Google Research Award, the Pinnacle Lab at Singapore Management University, Huawei grants, and the Science Foundation of Guangdong Province (2014A030313556, 2014A030308008).

References

1. Bickel, S., Scheffer, T.: Multi-view clustering. In: ICDM, pp. 19–26 (2004)
2. Blum, A., Mitchell, T.: Combining labeled and unlabeled data with co-training. In: COLT, New York, NY, USA, pp. 92–100 (1998)
3. Boyd, S., Vandenberghe, L.: Convex Optimization. Cambridge University Press, New York (2004)
4. Bruno, E., Marchand-Maillet, S.: Multiview clustering: a late fusion approach using latent models. In: SIGIR. ACM, New York (2009)
5. Chaudhuri, K., Kakade, S.M., Livescu, K., Sridharan, K.: Multi-view clustering via canonical correlation analysis. In: ICML, New York, NY, USA (2009)
6. Cheng, W., Zhang, X., Guo, Z., Wu, Y., Sullivan, P.F., Wang, W.: Flexible and robust co-regularized multi-domain graph clustering. In: SIGKDD, pp. 320–328. ACM (2013)
7. de Sa, V.R.: Spectral clustering with two views. In: ICML Workshop on Learning with Multiple Views (2005)
8. Ding, C., Li, T., Peng, W., Park, H.: Orthogonal nonnegative matrix T-factorizations for clustering. In: SIGKDD, pp. 126–135. ACM (2006)
9. Ding, C., Zhou, D., He, X., Zha, H.: R1-PCA: rotational invariant L1-norm principal component analysis for robust subspace factorization. In: ICML, pp. 281–288. ACM (2006)
10. Ding, W., Wu, X., Zhang, S., Zhu, X.: Feature selection by joint graph sparse coding. In: SDM, Austin, Texas, pp. 803–811, May 2013
11. L. Du, X. Li, and Y. Shen. Robust nonnegative matrix factorization via half-quadratic minimization. In: ICDM, pp. 201–210 (2012)
12. Duin, R.P.: Handwritten-Numerals-Dataset

13. Evgeniou, A., Pontil, M.: Multi-task Feature Learning. Advances in Neural Information Processing Systems **19**, 41 (2007)
14. Févotte, C.: Majorization-minimization algorithm for smooth itakura-saito nonnegative matrix factorization. In: ICASSP, pp. 1980–1983. IEEE (2011)
15. Greene, D., Cunningham, P.: A matrix factorization approach for integrating multiple data views. In: Buntine, W., Grobelnik, M., Mladenić, D., Shawe-Taylor, J. (eds.) ECML PKDD 2009, Part I. LNCS, vol. 5781, pp. 423–438. Springer, Heidelberg (2009)
16. Gu, Q., Zhou, J., Ding, C.: Collaborative filtering: weighted nonnegative matrix factorization incorporating user and item graphs. In: SDM. SIAM (2010)
17. Guo, Y.: Convex subspace representation learning from multi-view data. In: AAAI, Bellevue, Washington, USA (2013)
18. Huang, H., Ding, C.: Robust tensor factorization using R1 norm. In: CVPR, pp. 1–8. IEEE (2008)
19. Kim, J., Park, H.: Sparse Nonnegative Matrix Factorization for Clustering (2008)
20. Kim, Y., Choi, S.: Weighted nonnegative matrix factorization. In: International Conference on Acoustics, Speech and Signal Processing, pp. 1541–1544 (2009)
21. Kong, D., Ding, C., Huang, H.: Robust nonnegative matrix factorization using L21-norm. In: CIKM, New York, NY, USA, pp. 673–682 (2011)
22. Kriegel, H.P., Kunath, P., Pryakhin, A., Schubert, M.: MUSE: multi-represented similarity estimation. In: ICDE, pp. 1340–1342 (2008)
23. Kumar, A., Daume III, H.: A co-training approach for multi-view spectral clustering. In: ICML, New York, NY, USA, pp. 393–400, June 2011
24. Kumar, A., Rai, P., Daumé III, H.: Co-regularized multi-view spectral clustering. In: NIPS, pp. 1413–1421 (2011)
25. Lee, D., Seung, S.: Learning the Parts of Objects by Nonnegative Matrix Factorization. Nature **401**, 788–791 (1999)
26. Li, S., Jiang, Y., Zhou, Z.: Partial multi-view clustering. In: AAAI, pp. 1968–1974 (2014)
27. Liu, J., Wang, C., Gao, J., Han, J.: Multi-view clustering via joint nonnegative matrix factorization. In: SDM (2013)
28. Long, B., Philip, S.Y., (Mark) Zhang, Z.: A general model for multiple view unsupervised learning. In: SDM, pp. 822–833. SIAM (2008)
29. Nigam, K., Ghani, R.: Analyzing the effectiveness and applicability of co-training. In CIKM, pp. 86–93. ACM, New York (2000)
30. Nilsback, M.-E., Zisserman, A.: A visual vocabulary for flower classification. In: CVPR, vol. 2, pp. 1447–1454 (2006)
31. Shahnaz, F., Berry, M., Pauca, V.P., Plemmons, R.: Document Clustering Using Nonnegative Matrix Factorization. Information Processing & Management **42**(2), 373–386 (2006)
32. Shao, W., Shi, X., Yu, P.: Clustering on multiple incomplete datasets via collective kernel learning. In: ICDM (2013)
33. Tang, W., Lu, Z., Dhillon, I.S.: Clustering with multiple graphs. In: ICDM, Miami, Florida, USA, pp. 1016–1021, December 2009
34. Trivedi, A., Rai, P., Daumé III, H., DuVall, S.L.: Multiview clustering with incomplete views. In: NIPS 2010: Workshop on Machine Learning for Social Computing, Whistler, Canada (2010)
35. Wang, D., Li, T., Ding, C.: Weighted feature subset non-negative matrix factorization and its applications to document understanding. In: ICDM (2010)

36. Xu, W., Liu, X., Gong, Y.: Document clustering based on non-negative matrix factorization. In: SIGIR, pp. 267–273 (2003)
37. Zhang, X., Yu, Y., White, M., Huang, R., Schuurmans, D.: Convex sparse coding, subspace learning, and semi-supervised extensions. In: AAAI (2011)
38. Zhou, D., Burges, C.: Spectral clustering and transductive learning with multiple views. In: ICML, pp. 1159–1166. ACM, New York (2007)

Solving a Hard Cutting Stock Problem
by Machine Learning and Optimisation

Steven D. Prestwich, Adejuyigbe O. Fajemisin[✉], Laura Climent,
and Barry O'Sullivan

Department of Computer Science, Insight Centre for Data Analytics,
University College Cork, Cork, Ireland
{steven.prestwich,ade.fajemisin,laura.climent,
b.osullivan}@insight-centre.org

Abstract. We are working with a company on a hard industrial optimisation problem: a version of the well-known Cutting Stock Problem in
which a paper mill must cut rolls of paper following certain *cutting patterns* to meet customer demands. In our problem each roll to be cut may
have a different size, the cutting patterns are semi-automated so that we
have only indirect control over them via a list of continuous parameters
called a request, and there are multiple mills each able to use only one
request. We solve the problem using a combination of machine learning
and optimisation techniques. First we approximate the distribution of
cutting patterns via Monte Carlo simulation. Secondly we cover the distribution by applying a k-medoids algorithm. Thirdly we use the results
to build an ILP model which is then solved.

1 Introduction

The *Cutting Stock Problem* (CSP) [1] is a well-known NP-complete optimization
problem in Operations Research. It arises from many applications in industry
and a standard application is a paper mill. The mill produces rolls of paper of
a fixed width, but its customers require rolls of a lesser width. The problem is
to decide how many original rolls to make, and how to cut them, in order to
meet customer demands. Typically, the objective is to minimise waste, which
is leftover rolls or pieces of rolls. The problem can be modelled and solved by
Integer Linear Programming (ILP), and for large instances column generation
can be used.

We are working with a company on an industrial project and have encountered a hard optimisation problem. The application is commercially sensitive so
we cannot divulge details, but the problem can be considered as a variant of the
CSP. (We shall refer to "rolls" and "paper mills" but in fact the problem origi-
nates from another industry.) In this CSP variant, the choice of cutting pattern
is semi-automated so the user has only partial control over it via a "request". A
request is a vector of continuous variables so there are infinitely many possibilities, and their effect on the choice is complex. There are multiple paper mills,
and each can use only one request. The rolls made by the mills are of different

© Springer International Publishing Switzerland 2015
A. Appice et al. (Eds.): ECML PKDD 2015, Part I, LNAI 9284, pp. 335–347, 2015.
DOI: 10.1007/978-3-319-23528-8_21

sizes even before they are cut. For each mill, either all or none of its rolls are cut. There are also demands to be met and costs to be minimised. For this paper the interest is in the application of machine learning techniques (multivariate distribution approximation and cluster analysis) to reduce this infinite nonlinear problem to a finite linear problem that can be solved by standard optimisation methods.

This paper is structured as follows. First, in Section 2 the cutting stock problem is described. Second, in Section 3, we define the framework associated with the extra difficult cutting stock problem treated in this paper. We also propose an Integer Linear Program for this problem in Section 4, and give a brief overview of an alternative metaheuristic approach. The machine learning approach presented for the problem is described in Section 5. The approaches are evaluated with a real-life application in Section 6. Finally, the conclusions are commented in Section 7.

2   Cutting Stock Problems

The cutting stock problem is a well-known optimisation problem that is often modelled as an ILP:

$$\text{minimise} \sum_{i=1}^{n} c_i x_i$$

$$\text{s.t.} \sum_{i=1}^{n} a_{ij} x_i \geq d_j \qquad \forall j \in \mathcal{M}, \forall x_i \in \mathbb{N} \qquad (1)$$

where $\mathcal{M}$ is the set of roll types and d_j is the demand for type j. There are n cutting patterns and x is a vector of decision variables which state how many times each pattern is used. The number of rolls of type j generated by pattern i is a_{ij}. The objective function is to minimize the total cost, where c_i is the cost associated with pattern i. The costs depend on the specifications of the problem. For instance, for some problems, such as the model described above, the costs are associated with the patterns used (e.g. some cutting machines incur certain cost), while for other problems the costs are associated with the amount of left-over material (typically called waste if it can not be sold in future orders), etc. For a literature review of cutting stock problems we recommend [2].

Variants of the CSP have been studied. The above problem is one-dimensional but two or three dimensions might be necessary [3]. The problem might be multi-stage, involving further processing after cutting [4], or might be combined with other problems, e.g. [5]. Additional constraints might be imposed because of user requirements. Widths might be continuous though restricted to certain ranges of values.

3   Problem Formalization

As mentioned above, our CSP problem has several extra difficulties from the standard CSP which makes it hard to model and solve. Instead of rolls of a fixed original size which we can generate at will, we have a fixed number $r = 1 \ldots R$ of rolls (possibly several hundred) each with its own dimensions $\sigma_r \in S$; the details of S are confidential and unimportant here, but it involves continuous values. The cutting patterns are vectors $\boldsymbol{p} \in \mathbb{N}^m$ of m integer variables, describing how many rolls of each of the m widths is cut from the original roll. However, we cannot directly choose the cutting pattern because the choice is semi-automated. We have only limited control over it via a *request* $\boldsymbol{v}$ which is a vector of n continuous variables (m and n might be different but are typically less than 20). Each v_i is restricted to the interval $[0, 1]$ and each vector is of length 1:

$$\sum_{i=1}^{n} v_i^2 = 1$$

A request $\boldsymbol{v}$ and a roll r are passed to an algorithm $\mathcal{A}$ which uses $\boldsymbol{v}$ and σ_r to select a cutting pattern $\boldsymbol{p}$. Considering the algorithm as a function $\mathcal{A} : [0, 1]^n \times S \rightarrow \mathbb{N}^m$, experiments reveal it to be quite a complex (nonlinear and discrete) function. We make no assumptions about the form of $\mathcal{A}$ (which is also confidential) and treat it as a black box. Unlike the standard CSP we have several paper mills $j = 1 \ldots J$ (J might be as large as several hundred) each with its own set of R rolls (R might be different for each mill but we ignore this to simplify the description). For each mill, either all its rolls are cut into smaller rolls using the same request, or none of them are. Thus the rolls are partitioned into sets, each of which is treated as a unit and cut using the same request, though not necessarily the same cutting pattern. Finally, each mill's set of rolls has an intrinsic *value* V_j and we would like to satisfy demands using low-value rolls (in order to save the most valuable resources for future demands).

In summary, our problem is as follows. Given customer demand $\boldsymbol{d} \in \mathbb{N}^m$ for the m different roll sizes, we must select a subset of the mills with minimum total value, and a request $\boldsymbol{v}_j$ for each mill j, so that demand is met. This is an infinite nonlinear optimisation problem which we call the Continuous Semi-Automated Weighted Cutting Stock Problem (CSAWCSP):

$$\text{minimize} \sum_{j} V_j b_j$$

s.t.

$$\sum_{j=1}^{J} \sum_{r=1}^{R} b_j \mathcal{A}(\boldsymbol{v}_j, \sigma_r) \geq \boldsymbol{d} \tag{2}$$

$$b_j \in \{0, 1\} \qquad \qquad \forall j \tag{3}$$

$$\boldsymbol{v}_j \in [0, 1]^n \qquad \qquad \forall j \tag{4}$$

where b_j is a binary variable that is set to one iff mill j's rolls are cut.

This problem is very hard to solve because there are infinite number of possible requests $\boldsymbol{v}_j$, and because $\mathcal{A}$ is not a simple function. A metaheuristic approach is possible, searching in the space of b_j and $\boldsymbol{v}_j$ variable assignments with $\sum_j V_j b_j$

as the objective function and penalizing constraint violations, but in experiments this gave poor results. Instead we would like to transform the problem to make it solvable (at least approximately) by a standard method such as ILP.

Another possibility is to combine a metaheuristic approach with the ILP. First, the metaheuristic algorithm searches in the space of v_j for each j roll with the objective function of maximizing the similarity of the percentages of products of the pattern analyzed $(\mathcal{A}(v_j, \sigma_r))$ with respect to the percentages of demanded products (d). When the stopping criterion of the metahuristic has been reached (e.g. cut-off time), the best pattern found for each roll is provided to the ILP model. For implementing such approach, we used the metaheuristic introduced in [6], which is a Simulated Annealing Like Algorithm (SALA) called the Threshold Accepting Algorithm (TA) [7]. This metaheuristic algorithm iteratively generates new requests that are mapped by the non-linear function $\mathcal{A}$ into patterns. The objective function is to maximize the similarity of the percentages of product types obtained by a pattern with respect the percentages demanded. The results obtained with such technique were very poor: the ILP models of most of the instances evaluated in Section 6 did not have solutions (because the sum of all the unique patterns associated to each roll j did not satisfy the demand of at least one type of product); and for those few instances with solutions, their quality was far below that of the solutions found with our technique.

4   ILP Model for the CSAWCSP

To make the problem amenable to an ILP approach we reduce the infinite set of possible requests v_j to a representative finite set of requests u_{jk} ($k = 1 \ldots K$) for each mill j (we assume the same K for each mill to simplify the notation). We then precompute the total number of each roll size obtained for request k and mill j over all its rolls, storing the results in vectors of integer constants $c_{jk} = \sum_{r=1}^{R} \mathcal{A}(u_{jk}, \sigma_r)$ ($\forall j, k$). This eliminates the complexity of $\mathcal{A}$ and the infinite choice of requests, and we can now model the CSAWCSP as an ILP:

$$\text{minimize} \sum_j V_j b_j$$

s.t.

$$\sum_{k=1}^{K} x_{jk} = b_j \qquad \forall j \qquad (5)$$

$$\sum_{j=1}^{J} \sum_{k=1}^{K} c_{jk} x_{jk} \geq d \qquad (6)$$

$$b_j, x_{jk} \in \{0,1\} \qquad \forall j, k \qquad (7)$$

where $b_j = 1$ indicates that all mill j's rolls are cut, and $x_{jk} = 1$ indicates that they are cut using request k. If mill j is not selected then $b_j = 0$ which forces $x_{jk} = 0$ for $k = 1 \ldots K$.

This ILP can be solved by standard optimisation software. But to make this approach practical we must first select a finite set of requests u_{jk} that adequately covers all possible requests. More precisely, the possible sets of cut rolls c_{jk} must be adequately covered. This requires the generation of a finite set of vectors that approximately cover an unknown multivariate probability distribution.

5   Machine Learning Approach for the CSAWCSP

In this section we explain our approach to the problem of covering the unknown multivariate distribution in the CSAWCSP. An illustration of our approach is shown in Figure 1.

In scatter plot (a) the circle represents the hypersphere of possible requests v, with a small random number of them selected shown as dots. Plot (b) shows the result of applying algorithm $\mathcal{A}$ to a mill's rolls using the different v, to obtain a small set of c vectors. The space of c vectors might have a very different shape to that of the v, as shown. As a consequence, a small random set of v might correspond to a very non-random small set of c vectors, showing the inadequacy of merely sampling a few requests.

Instead we sample a large number of v as shown in plot (c), with their corresponding c shown in plot (d): this represents the use of Monte Carlo simulation to approximate the distribution of the c. We then select a small number of c via a k-medoids algorithm to approximately cover the estimated distribution, highlighted in plot (f). Finally we use a record of which v corresponds to which c to derive the non-random set of v highlighted in plot (e). Next we describe these phases of our approach in more detail.

5.1   Distribution Learning

Given a number of samples drawn from a distribution, the goal of distribution learning is to find the distribution from which the samples have been drawn with a high degree of certainty. Let $\mathcal{D}_n$ be a particular distribution class, and let $D \in \mathcal{D}_n$ be a distribution which has a support S, i.e. S is the range over which D is defined. To represent the probability distribution D over S, let G_D be a generator for D [10]. G_D is called a generator because, given a random input y, G_D simulates sampling from the distribution D and outputs an observation $G_D[y] \in S$.

Given independent random samples from D, as well as confidence and approximation parameters δ and ϵ respectively, the goal of any learning algorithm is to output an approximate distribution D' with a probability δ in polynomial time. The distance between this approximate distribution and the original distribution is $d(D, D')$, and can be measured in several ways. These include the Kolmogorov distance (from the Kolmogorov-Smirnoff test) [11], the Total Variation distance [12], and the Kullback-Leibler divergence [13]. When $d(D, D') \leq \epsilon$, G_D is called an ϵ-good generator.

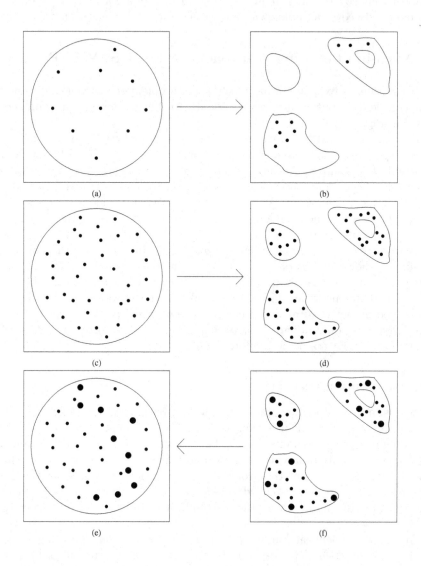

Fig. 1. Illustration of our approach

For our problem it is necessary to understand the relationship between the requests v and the corresponding vectors c which state how many of each roll is produced. Once this is known, given a set of demands for each of the different rolls, the requests required to produce the best cutting patterns for each roll can be determined. As the algorithm $\mathcal{A}$ is complex it is necessary to learn its distribution. $\mathcal{A}$ acts as the generator G_D.

We approximated the distribution of $\mathcal{A}$'s output using Monte Carlo sampling, by generating a large number of random request vectors v and obtaining corresponding vectors c. To avoid bias we generated v by uniform sampling from a vector space, as opposed to simply randomising each component of v.

5.2   Generating Uniformly Distributed Random Vectors

To generate a random vector Y that is uniformly distributed over an irregular n-dimensional region G, an acceptance-rejection method may be used [14].

For a regular region W, where W may be multidimensional in nature, we first of all generate a random vector X, which is uniformly distributed in W. If $X \in G$, we accept $Y = X$ as the random vector uniformly distributed over G. Otherwise, we reject X, and generate a new random vector. In the case when G is an n-dimensional unit ball, i.e.

$$G = \left\{ x : \sum_i x_i^2 \leq 1 \right\} \tag{8}$$

we generate a uniformly distributed random vector $X = (X_1, X_2, \ldots, X_n)^T$, and we accept X if it falls inside the n-ball. If it does not then X is rejected, otherwise it is projected onto the hypersphere. The algorithm used for this is taken from [9] and is described below.

Algorithm 1. Random Vector Generation

---

1. Generate n random variables $U_1, \ldots, U_n$ as iid variables from $U(0, 1)$.
2. Set $X_1 = 1 - 2U_1, \ldots, X_n = 1 - 2U_n$ and $R = \sum_{i=1}^{n} X_i^2$
3. If $R \leq 1$, accept $X = (X_1, \ldots, X_n)^T$ as the desired vector; otherwise go to Step 1.

Using Algorithm 1, we generated a large number of request vectors v which were then passed to $\mathcal{A}$, together with σ_r ($\forall r \in R$). $\mathcal{A}$ then returned the same number of cutting pattern vectors c. For our problem this large number of patterns must be reduced to a smaller number. To do this, we used another well-known machine learning technique called k-medoids clustering.

5.3 k-Medoids Clustering

The k-medoids algorithm is a clustering algorithm used for partitioning a data set X into K homogeneous groups or clusters, i.e. $C = \{C_1, C_2, \ldots, C_K\}$ [15].

Unlike the k-means algorithm, partitioning is done around *medoids* (or *exemplars*) rather than *centroids*. This is vital for our problem because we require a small set of c that are each generated from some known v. A medoid m_k is a data point in a cluster C_k which is most similar to all other points in that cluster.

A k-medoids algorithm seeks to minimize the function

$$\sum_{k=1}^{K} \sum_{i \in C_k} d(x_i, m_k) \qquad (9)$$

where $d(x_i, m_k)$ is a distance metric measuring the dissimilarity between data entity x_i and the medoid m_k of the cluster [15]. Commonly used distance metrics are the Manhattan distance or Euclidean distance [16] and we use the latter. The most common algorithm for k-medoid clustering is the Partitioning Around Medoids (PAM) algorithm [15], presented at a high level in Algorithm 2.

Algorithm 2. Partitioning Around Medoids (PAM)

---

1. Select k out of n data entities as initial medoids $m_1, m_2, ...m_k$.
2. Assign each data entity $x_i \in X$ to cluster C_k of the closest medoid m_k as determined by $d(x_i, m_k)$.
3. For each medoid m_k, select a non-medoid y_i and swap y_i for m_k.
4. Calculate the distance $d(x_i, m_k)$
5. Select the configuration with the lowest value of $d(x_i, m_k)$.
6. Repeat steps 2 to 5 until the lowest possible value of $d(x_i, m_k)$ has been found with final medoids $m_1', m_2', ...m_k'$.

For very large datasets the CLARA algorithm, which is a combination of PAM and random sampling, is commonly used [17,18]. The speedup by CLARA over PAM is achieved by analysing data subsets of fixed size, which has the effect of making both computational and storage complexity linear, as opposed to quadratic [17,19]. The steps for CLARA are outlined in Algorithm 3 below.

Once the large set of cutting patterns c has been reduced to a much smaller representative set c', we can now solve an approximation to the hard cutting stock problem using ILP.

6   Emprirical Study

For empirically studying our approach, we have compared the solutions obtained by our approach for a certain range of k medoids with respect the lower optimality bounds (explained below) of several instances. For such purpose we used real data from our industrial partner. The total volume of the raw material analyzed is $1191.3m^3$ for 8 mills ($J = 8$) with each mill's rolls partitioned into a maximum of 4 different types of products. We generated and solved 20 instances of random demands in a 2.3 GHz Intel Core $i7$ processor. Monte Carlo simulation and

---

Algorithm 3. Clustering LARge Applications (CLARA)

---

1. Randomly choose a number p of sub-datasets of fixed size from the large dataset X.
2. Partition each sub-dataset into k clusters using the PAM algorithm, getting a set $M = \{m_1, m_2, ..., m_k\}$ of k medoids.
3. Associate each data entity x_i of the original large dataset to its nearest medoid m_k.
4. Calculate the mean of the dissimilarities of the observations to their closest medoid using:

$$MeanDistance(M, X) = \frac{\sum_{i=1}^{n} d\left(x_i, rep\left(M, x_i\right)\right)}{|X|} \tag{10}$$

   where:
   $d(x_i, x_j)$ is the dissimilarity between two data points x_i and x_j,
   $rep(M, x_i)$ returns the closest medoid m_k to x_i from the set M, and
   $|X|$ is the number of items in X.
5. Repeat steps 1 to 4 q times, selecting the set of medoids $M' = \{m'_1, m'_2, ...m'_k\}$ with the minimum $MeanDistance(M, X)$.

---

clustering were done in Java and R (using the CLARA algorithm) respectively on a 3.0 GHz Intel Xeon Processor with 8 GB of RAM. For solving the integer linear programming model, we used the CPLEX 12.6 solver with a time cut-off of 1 hour.

To approximate the unknown multivariate distribution, we generated 10,000 random request vectors for each mill. Then, we obtained the same number of corresponding cutting patterns by applying the algorithm $\mathcal{A}$ (see Section 3). For 8 mills, this process resulted in total of 80,000 cutting patterns. The total time required for generating all the cutting patterns was 2 hours and 30 minutes. Note that, in time-sensitive applications, a lower number of random cutting patterns could be generated to reduce this overhead. Next we applied k-medoids clustering to cover the distribution.

For evaluating the effect of the number of medoids used to cover the distribution, we varied k from 25 to 200 in steps of 25 units. The time taken to generate the clusters can be found in Table 1. In addition, Figure 2 shows the total clustering times (for all the mills). Note that the clustering times increase exponentially from 40.54 seconds for $k = 25$, to 2,651.08 seconds ($\sim$ 44 minutes) for $k = 200$. Thus for real-life problems it is very important to make an appropriate selection of the parameter k (especially in on-line applications).

Once all the medoids were obtained we used them as input parameters for the ILP model (see Section 4). Then we solved the 20 instances of random demands. Furthermore, we also applied a relaxed ILP model with the objective of calculating the lower bound of optimality of the instances analyzed. In this variation, we consider feasible any linear combination of cutting patterns. However, it might occur that the combination selected is not feasible for the real life problem. For this reason, this measurement is a lower bound of optimality, since in the lat-

Table 1. Clustering times.

Mill	$k = 25$	$k = 50$	$k = 75$	$k = 100$	$k = 125$	$k = 150$	$k = 175$	$k = 200$
				Time (sec)				
0	5.13	14.47	31.40	56.80	107.00	158.85	229.90	310.47
1	4.94	15.68	35.66	59.88	95.85	156.30	234.19	313.49
2	4.99	14.62	27.59	56.59	94.38	154.96	235.22	313.99
3	5.06	17.59	36.63	62.06	102.29	167.75	266.71	356.86
4	5.17	16.04	33.82	60.64	99.78	164.89	240.40	331.72
5	5.43	16.83	33.04	63.48	95.85	171.37	255.47	351.97
6	4.99	15.77	33.21	61.18	98.10	163.63	241.04	327.87
7	4.83	16.27	34.37	67.03	95.04	168.46	251.94	344.71
Total Time	40.54	127.27	265.72	487.66	788.29	1306.23	1954.87	2651.08

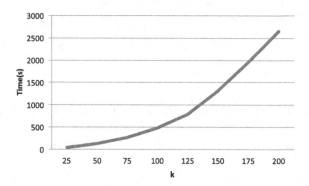

Fig. 2. Total clustering times.

ter case, the optimal solution is greater than this bound. The lower optimality bound is very useful since it allows us to stop the search for a better solution once we have reached such bound (since it can be ensured that this is the optimal solution). For this reason, we incorporated such lower optimality bound in the model solved by CPLEX.

Figure 3 shows the percentage difference between the solutions obtained with our approach and the lower optimality bound. It can be observed that as k increases, the percentage difference decreases, following an exponential inverse function. This suggest that increasing the medoids is more effective when the original medoids are fewer. Furthermore, it can be observed that there is a "saturation" point in which it is not possible to further improve the quality of the solutions. For this instance, the saturation point is located approximately at $k = 125$ since higher values of k provide almost the same results. For this reason, for these instances, the best option is to select $k = 125$, since it is not worth it to spend more time computing higher k values. Note that for this case, the difference in percentage with the lower optimality bound is $\sim 0.4\%$, which indicates that we succeeded in finding optimal and close-optimal solutions.

We also would like to comment how these percentages differences are translated into economical earnings. On average, for these instance, the raw material that was required to satisfy the demands when using $k = 25$ was almost € 800 more expensive than when using $k = 125$. Needless to mention, the benefits in the real life application that will involve to use the approach presented in this paper with a great enough value of k.

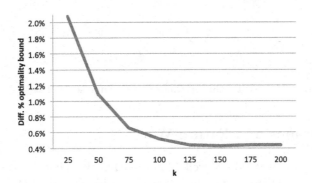

Fig. 3. Percentage Optimality Difference

In Table 2 and Figure 4 we show the mean times for solving the 20 instances for the interval of values of k selected. Note that these times also increase in a non-linear fashion, from a solution time of 1.506 seconds for $k = 25$ to 407.475 seconds ($\sim$ 7 minutes) for $k = 200$. We would like to point out that there is a correlation with the saturation point in the optimality with respect the saturation point in the computation times for solving the ILP model. Note that for higher values of $k = 125$, the increment of computation time is little appreciable.

Table 2. Mean ILP solution times.

k	Average Time (sec)
25	1.506
50	188.454
75	369.16
100	240.120
125	410.091
150	398.914
175	411.180
200	407.475

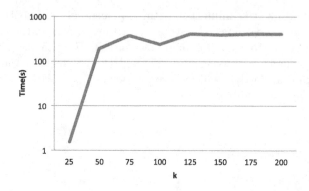

Fig. 4. Mean ILP solution times (logarithmic scale of base 10).

7   Conclusions and Future Work

In this paper, we combined two well-known machine learning techniques to solve a hard optimisation problem. This problem, which we called the CSAWCSP, arose from a real-world application and is a more complicated variant of the traditional CSP. We used Monte Carlo simulation to approximate an unknown multivariate distribution. We generated a large number of random request vectors, which were then provided to an algorithm in order to generate a corresponding number of cutting patterns. Subsequently, we applied k-medoids clustering to cover the distribution.

To study the effect of the number of medoids on the solution, we increased k steadily, and observed that there is a particular value of k (saturation point) above which, no improvements to the solution could be made. In this way, we succeeded in finding optimal and close-optimal solutions for the type of cutting stock problem analyzed in this paper. Regarding the computation time, we observed that linear increases in the value of k led to exponential increases in clustering time, and that for values above the k saturation point, the ILP solution times were relatively constant.

In the future, we aim to improve the sampling approach in order to reduce the number of request vectors needed to be generated. Furthermore, we intend to use an adaptive clustering, which might be more effective in reducing the clustering time. By applying these two new improvements, we expect to reduce computation times. This will be especially beneficial for on-line real-life applications.

Acknowledgments. This publication has emanated from research conducted with the financial support of Science Foundation Ireland (SFI) under Grant Number SFI/12/RC/2289.

References

1. Kantorovich, L.V.: Mathematical methods of organizing and planning production. Management Science **6**(4), 366–422 (1960)
2. Cheng, C.H., Feiring, B.R., Cheng, T.C.E.: The cutting stock problem - a survey. International Journal of Production Economics **36**(3), 291–305 (1994)
3. Gilmore, P.C., Gomory, R.E.: Multistage Cutting Stock Problems of Two and More Dimensions. Operations Research **13**, 94–120 (1965)
4. Furini, F., Malaguti, E.: Models for the two-dimensional two-stage cutting stock problem with multiple stock size. Computers & Operations Research **40**(8), 1953–1962 (2013)
5. Hendry, L.C., Fok, K.K., Shek, K.W.: A cutting stock scheduling problem in the copper industry. Journal of the Operational Research Society **47**, 38–47 (1996)
6. Murphy, G., Marshall, H., Bolding, M.C.: Adaptive control of bucking on harvesters to meet order book constraints. Forest Products Journal and Index **54**(12), 114–121 (2004)
7. Dueck, G., Scheuer, T.: Threshold accepting: a general purpose optimization algorithm appearing superior to simulated annealing. Journal of computational physics **90**(1), 161–175 (1990)
8. Sawilowsky, S.S.: You think you've got trivials? Journal of Modern Applied Statistical Methods **2**(1), 218–225 (2003)
9. Kroese, D.P., Taimre, T., Botev, Z.I.: Handbook of Monte Carlo Methods, Wiley Series in Probability and Statistics. John Wiley and Sons, New York (2011)
10. Kearns, M., Mansour, Y., Ron, D., Rubinfeld, R., Schapire, R., Sellie, L.: On the Learnability of Discrete Distributions. ACM Symposium on Theory of Computing (1994)
11. Chakravarti, I.M., Laha, R.G., Roy, J.: Handbook of Methods of Applied Statistics, vol. I. John Wiley and Sons, pp. 392–394 (1967)
12. Adams, C.R., Clarkson, J.A.: On definitions of bounded variation for functions of two variables. Transactions of the American Mathematical Society **35**, 824–854 (1933)
13. Kullback, S., Leibler, R.A.: On information and sufficiency. Annals of Mathematical Statistics **22**(1), 79–86 (1951). doi:10.1214/aoms/1177729694. MR 39968
14. Rubinstein, R.Y., Kroese, D.P.: Simulation and the Monte Carlo Method, 2nd edn. John Wiley & Sons (2008)
15. de Amorim, R.C., Fenner, T.: Weighting features for partition around medoids using the Minkowski metric. In: Proceedings of the 11th International Symposium in Intelligent Data Analysis, pp. 35–44, October 2012
16. Maechler, M., Rousseeuw, P., Struyf, A., Hubert, M., Hornik, K., Studer, M., Roudier: cluster: Cluster Analysis Extended Rousseeuw et al. R package version 2.0.1, January 2015. http://cran.r-project.org/web/packages/cluster/cluster.pdf
17. Kaufman, L., Rousseeuw, P.J.: Finding Groups in Data: An Introduction to Cluster Analysis. John Wiley & Sons Inc, New York (1990)
18. Wei, C., Lee, Y., Hsu, C.: Empirical comparison of fast clustering algorithms for large data sets. In: Proceedings of the 33rd Hawaii International Conference on System Sciences (2000)
19. Nagpaul, P.S.: 7.1.2 Clustering Large Applications (CLARA). In: Guide to Advanced Data Analysis using IDAMS Software. http://www.unesco.org/webworld/idams/advguide/Chapt7_1_2.htm (access date: October 02, 2015)

Data Preprocessing

Markov Blanket Discovery
in Positive-Unlabelled and Semi-supervised Data

Konstantinos Sechidis$^{(\boxtimes)}$ and Gavin Brown

School of Computer Science, University of Manchester, Manchester M13 9PL, UK
{konstantinos.sechidis,gavin.brown}@manchester.ac.uk

Abstract. The importance of Markov blanket discovery algorithms is twofold: as the main building block in constraint-based structure learning of Bayesian network algorithms and as a technique to derive the optimal set of features in filter feature selection approaches. Equally, learning from partially labelled data is a crucial and demanding area of machine learning, and extending techniques from fully to partially supervised scenarios is a challenging problem. While there are many different algorithms to derive the Markov blanket of fully supervised nodes, the partially-labelled problem is far more challenging, and there is a lack of principled approaches in the literature. Our work derives a generalization of the conditional tests of independence for partially labelled binary target variables, which can handle the two main partially labelled scenarios: *positive-unlabelled* and *semi-supervised*. The result is a significantly deeper understanding of how to control false negative errors in Markov Blanket discovery procedures and how unlabelled data can help.

Keywords: Markov blanket discovery · Partially labelled · Positive unlabelled · Semi supervised · Mutual information

1 Introduction

Markov Blanket (MB) is an important concept that links two of the main activities of machine learning: dimensionality reduction and learning. Using Pellet & Elisseef's [15] wording "Feature selection and causal structure learning are related by a common concept: the Markov blanket."

Koller & Sahami [10] showed that the MB of a target variable is the optimal set of features for prediction. In this context discovering MB can be useful for eliminating irrelevant features or features that are redundant in the context of others, and as a result plays a fundamental role in filter *feature selection*. Furthermore, Markov blankets are important in learning Bayesian networks [14], and can also play an important role in *causal structure learning* [15].

In most real world applications, it is easier and cheaper to collect unlabelled examples than labelled ones, so transferring techniques from fully to *partial-labelled* datasets is a key challenge. Our work shows how we can recover the MB around partially labelled targets. Since the main building block of the MB

© Springer International Publishing Switzerland 2015
A. Appice et al. (Eds.): ECML PKDD 2015, Part I, LNAI 9284, pp. 351–366, 2015.
DOI: 10.1007/978-3-319-23528-8_22

discovery algorithms is the *conditional test of independence*, we will present a method to apply this test despite the partial labelling and how we can use the unlabelled examples in an informative way.

Section 4 explores the scenario of *positive-unlabelled* data. This is a special case of partially-labelling, where we have few labelled examples *only* from the positive class and a vast amount of unlabelled examples. Section 5 extends our work to *semi-supervised* data, where the labelled set contains examples from both classes. Finally, Section 6 presents a semi-supervised scenario that can occur in real world, the *class prior change* scenario, and shows how our approach performs better than the state of the art [1].

Before the formal presentation of the background material (Sections 2 and 3) we will motivate our work with a toy Bayesian network presented in Figure 1. The MB of the target variable Y is the feature set that contains the *parents* (X_4 and X_5), *children* (X_9 and X_{10}) and *spouses* (X_7 and X_8, which are other parents of a child of Y) of the target. There exist many techniques to derive MB by using fully-supervised datasets, Figure 1a. But our work will focus on partially labelled scenarios where we have the values of Y only for a small subset of examples, Figure 1b, while all the other variables are completely observed. We will suggest ways to derive the MB by controlling the two possible errors in the discovery procedure:

Falsely adding variables to the predicted Markov blanket: for example assuming that the variable X_{11} *belongs* to MB.

Falsely not adding variables to the predicted Markov blanket: for example assuming the variable X_4 *does not belong* to MB.

2   Background: Markov Blanket

In this section we will introduce the notation and the background material on Markov blanket discovery algorithms. Assuming that we have a binary classification dataset $\mathcal{D} = \{(\mathbf{x}^i, y^i) | i = 1, ..., N\}$, where the target variable Y takes the value y^+ when the example is positive, and y^- when the example is negative. The feature vector $\mathbf{x} = [x_1...x_d]$ is a realization of the d-dimensional joint random variable $\mathbf{X} = X_1...X_d$. With a slight abuse of notation, in the rest of our work, we interchange the symbol for a set of variables and for their joint random variable. Following Pearl [14] we have the following definitions.

Definition 1 (Markov blanket — Markov boundary).
The Markov blanket of the target Y is a set of features $\mathbf{X}_{\mathrm{MB}}$ with the property $Y \perp\!\!\!\perp \mathbf{Z} | \mathbf{X}_{\mathrm{MB}}$ for every $\mathbf{Z} \subseteq \mathbf{X} \backslash \mathbf{X}_{\mathrm{MB}}$. A set is called Markov boundary if it is a minimal Markov blanket, i.e. non of its subsets is a Markov blanket.

In probabilistic graphical models terminology, the target variable Y becomes conditionally independent from the rest of the graph $\mathbf{X} \backslash \mathbf{X}_{\mathrm{MB}}$ given its Markov blanket $\mathbf{X}_{\mathrm{MB}}$.

---

[1] Matlab code and the supplementary material with all the proofs are available in www.cs.man.ac.uk/~gbrown/partiallylabelled/.

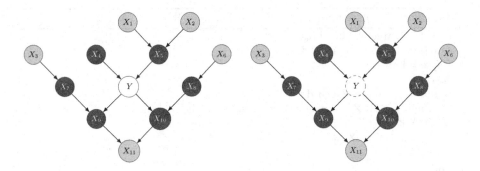

(a) MB in fully supervised target (b) MB in partially labelled target

Fig. 1. Toy Markov blanket example where: white nodes represent the target variable, black ones the features that *belong* to the MB of the target and grey ones the features that *do not belong* to the MB. In 1a we know the value of the target over all examples, while in 1b the target is partially observed (dashed circle) meaning that we know its value only in a small subset of the examples.

Learning the Markov blanket for each variable of the dataset, or in other words inferring the local structure, can naturally lead to *causal structure learning* [15]. Apart from playing a huge role in structure learning of a Bayesian network, Markov blanket is also related to another important machine learning activity: *feature selection.*

Koller & Sahami [10] published the first work about the optimality of Markov blanket in the context of feature selection. Recently, Brown et al. [5] introduced a unifying probabilistic framework and showed that many heuristically suggested feature selection criteria, including Markov blanket discovery algorithms, can be seen as iterative maximizers of a clearly specified objective function: the conditional likelihood of the training examples.

2.1 Markov Blanket Discovery Algorithms

Margaritis & Thrun [12] introduced the first theoretically sound algorithm for Markov blanket discovery, the Grow-Shrink (GS) algorithm. This algorithm consists of two-stages: *growing* where we add features to the Candidate Markov Blanket (CMB) set until the point that the remaining features are independent with the target given the candidate blanket, and *shrinkage*, where we remove potential false positives from the CMB. Tsamardinos & Aliferis [21] suggested an improved version to this approach, the Incremental Association Markov Blanket (IAMB), which can be seen in Algorithm 1. Many measures of association have been used to decide which feature will be added in the candidate blanket during the growing phase (Alg. 1 - Line 4), with the main being the *conditional mutual information* [17]. But, Yaramakala & Margaritis [23] suggested the use of the *significance of the conditional test of independence*, which is more appropriate in statistical terms than the raw conditional mutual information value.

Algorithm 1. Incremental Association Markov Blanket (IAMB)

Input : Target Y, Features $\mathbf{X} = X_1...X_d$, Significance level α
Output: Markov Blanket: $\mathbf{X}_{\mathrm{CMB}}$
1 Phase I: forward — growing
2 $\mathbf{X}_{\mathrm{CMB}} = \emptyset$
3 **while** $\mathbf{X}_{\mathrm{CMB}}$ has changed **do**
4 $\quad$ Find $X \in \mathbf{X} \backslash \mathbf{X}_{\mathrm{CMB}}$ most strongly related with Y given $\mathbf{X}_{\mathrm{CMB}}$
5 $\quad$ **if** $X \not\perp Y | \mathbf{X}_{\mathrm{CMB}}$ using significance level α **then**
6 $\quad\quad$ | Add X to $\mathbf{X}_{\mathrm{CMB}}$
7 $\quad$ **end**
8 **end**
9 Phase II: backward — shrinkage
10 **foreach** $X \in \mathbf{X}_{\mathrm{CMB}}$ **do**
11 $\quad$ **if** $X \perp\!\!\!\perp Y | \mathbf{X}_{\mathrm{CMB}} \backslash X$ using significance level α **then**
12 $\quad\quad$ | Remove X from $\mathbf{X}_{\mathrm{CMB}}$
13 $\quad$ **end**
14 **end**

Finally, there is another class of algorithms that try to control the size of conditioning test in a two-phase procedure: firstly identify parent-children, then identify spouses. The most representative algorithms are the HITON [2] and the Max-Min Markov Blanket (MMMB) [22]. All of these algorithms assume faithfulness of the data distribution. As we already saw, in all Markov blanket discovery algorithms, the conditional test of independence (Alg. 1 - Line 5 and 11) plays a crucial role, and this is the focus of the next paragraph.

2.2 Testing Conditional Independence in Categorical Data

IAMB needs to test the conditional independence of X and Y given a subset of features $\mathbf{Z}$, where in Line 5 $\mathbf{Z} = \mathbf{X}_{\mathrm{CMB}}$ while in Line 11 $\mathbf{Z} = \mathbf{X}_{\mathrm{CMB}} \backslash X$. In fully observed categorical data we can use the G-test, a generalised likelihood ratio test, where the test statistic can be calculated from sample data counts arranged in a contingency table [1].

G-**statistic:** We denote by $O_{x,y,\mathbf{z}}$ the observed count of the number of times the random variable X takes on the value x from its alphabet $\mathcal{X}$, Y takes on $y \in \mathcal{Y}$ and $\mathbf{Z}$ takes on $\mathbf{z} \in \mathcal{Z}$, where $\mathbf{z}$ is a vector of values when we condition on more than one variable. Furthermore denote by $O_{x,.,\mathbf{z}}$, $O_{.,y,\mathbf{z}}$ and $O_{.,.,\mathbf{z}}$ the marginal counts. The estimated expected frequency of $(x, y, \mathbf{z})$, assuming X, Y are conditional independent given $\mathbf{Z}$, is given by $E_{x,y,\mathbf{z}} = \frac{O_{x,.,\mathbf{z}} O_{.,y,\mathbf{z}}}{O_{.,.,\mathbf{z}}} = \widehat{p}(x|\mathbf{z})\widehat{p}(y|\mathbf{z})O_{.,.,\mathbf{z}}$. To calculate the G-statistic we use the following formula:

$$\widehat{G}\text{-statistic} = 2 \sum_{x,y,\mathbf{z}} O_{x,y,\mathbf{z}} \ln \frac{O_{x,y,\mathbf{z}}}{E_{x,y,\mathbf{z}}} = 2 \sum_{x,y,\mathbf{z}} O_{x,y,\mathbf{z}} \ln \frac{O_{.,.,\mathbf{z}} O_{x,y,\mathbf{z}}}{O_{x,.,\mathbf{z}} O_{.,y,\mathbf{z}}} = \tag{1}$$

$$= 2N \sum_{x,y,\mathbf{z}} \widehat{p}(x,y,\mathbf{z}) \ln \frac{\widehat{p}(x,y|\mathbf{z})}{\widehat{p}(x|\mathbf{z})\widehat{p}(y|\mathbf{z})} = 2N\widehat{I}(X;Y|\mathbf{Z}), \tag{2}$$

where $\widehat{I}(X;Y|\mathbf{Z})$ is the maximum likelihood estimator of the conditional mutual information between X and Y given $\mathbf{Z}$ [8].

Hypothesis Testing Procedure: Under the null hypothesis that X and Y are statistically independent given $\mathbf{Z}$, the G-statistic is known to be asymptotically χ^2-distributed, with $\nu = (|\mathcal{X}| - 1)(|\mathcal{Y}| - 1)|\mathcal{Z}|$ degrees of freedom [1]. Knowing that and using (2) we can calculate the $\widehat{p}_{XY|\mathbf{Z}}$ value as $1 - F(\widehat{G})$, where F is the CDF of the χ^2-distribution and $\widehat{G}$ the observed value of the G-statistic. The p-value represents the probability of obtaining a test statistic equal or more extreme than the observed one, given that the null hypothesis holds. After calculating this value, we check to see whether it exceeds a significance level α. If $p_{XY|\mathbf{Z}} \leq \alpha$, we reject the null hypothesis, otherwise we fail to reject it. This is the procedure that we follow to take the decision in Lines 5 and 11 of the IAMB algorithm 1. Furthermore, to choose the most strongly related feature in Line 4, we evaluate the p-values and choose the feature with the smaller one.

Different Types of Error: Following this testing procedure, two possible types of error can occur. The significance level α defines the probability of *Type I error* or False Positive rate, that the test will reject the null hypothesis when the null hypothesis is in fact true. While the probability of *Type II error* or False Negative rate, which is denoted by β, is the probability that the test will fail to reject the null hypothesis when the alternative hypothesis is true and there is an actual effect in our data. Type II error is closely related with the concept of statistical *power* of a test, which is the probability that the test will reject the null hypothesis when the alternative hypothesis is true, i.e. *power* $= 1 - \beta$.

Power Analysis: With such a test, it is common to perform an *a-priori power analysis* [7], where we would take a given sample size N, a required significance level α, an effect size ω, and would then compute the power of the statistical test to detect the given effect size. In order to do this we need a test statistic with a known distribution under the alternative hypothesis. Under the alternative hypothesis (i.e. when X and Y are dependent given $\mathbf{Z}$), the G-statistic has a large-sample *non-central* χ^2 distribution [1, Section16.3.5]. The non-centrality parameter (λ) of this distribution has the same form as the G-statistic, but with sample values replaced by population values, $\lambda = 2NI(X;Y|\mathbf{Z})$. The effect size of the G-test can be naturally expressed as a function of the *conditional mutual information*, since according to Cohen [7] the effect size (ω) is the square root of the non-centrality parameter divided by the sample, thus we have $\omega = \sqrt{2I(X;Y|\mathbf{Z})}$.

Sample Size Determination: One important usage of a-priori power analysis is *sample size determination*. In this prospective procedure we specify the probability of Type I error (e.g. $\alpha = 0.05$), the desired probability of Type II error (e.g. $\beta = 0.01$ or *power* $= 0.99$) and the desired effect size that we want to observe, and we can determine the minimum number of examples (N) that we need to detect that effect.

2.3   Suggested Approach for Semi-supervised MB Discovery

To the best of our knowledge, there is only one algorithm to derive the MB of semi-supervised targets: BASSUM (BAyesian Semi-SUpervised Method) [6]. BASSUM follows the HITON approach, finding firstly the parent-child nodes and then the spouses, and tries to take into account both labelled and unlabelled data. BASSUM makes the "traditional semi-supervised" assumption that the labelled set is an unbiased sample of the overall population, and it uses the unlabelled examples in order to improve the reliability of the conditional independence tests. For example to estimate the G-statistic, in equation (1), it uses both labelled and unlabelled data for the observed counts $O_{.,.,\mathbf{z}}$ and $O_{x,.,\mathbf{z}}$. This technique is known in statistics as *available case analysis* or *pairwise deletion*, and is affected by the ambiguity over the definition of the overall sample size, which is crucial for deriving standard errors and the sampling distributions (the reader can find more details on this issue in Allison [3, page8]). This can lead to unpredictable results, for example there are no guarantees that the G-statistic will follow χ^2 distribution after this substitution. Another weakness of BASSUM is that it cannot be applied in partially labelled environments where we have the restriction that the labelled examples come only from one class, such as the positive-unlabelled data. In order to explore the Markov blanket of this type of data we should explore how to test conditional independence in this scenario and this is the focus of Section 4. Before that, we will formally introduce the partially-labelled data in the following section.

3   Background: Partially-Labelled Data

In this section we will give the background for the two partially-labelled problems on which we will focus: positive-unlabelled and semi-supervised.

3.1   Positive-Unlabelled Data

Positive-Unlabelled (PU) data refers to situations where we have a small number of labelled examples from the positive class, and a large number of entirely unlabelled examples, which could be either positive *or* negative. For reasoning over PU data we will follow the formal framework of Elkan & Noto [9]. Assume that a dataset $\mathcal{D}$ is drawn i.i.d. from the joint distribution $p(\mathbf{X}, Y, S_P)$, where $\mathbf{X}$ and Y are random variables describing the feature set and the target variable, while S_P is a further random variable with possible values 's_P^+' and 's_P^-', indicating if the positive example is labelled (s_P^+) or not (s_P^-). We sample a total number of N examples out of which $N_{S_P^+}$ are labelled as positives. Thus $p(\mathbf{x}|s_P^+)$ is the probability of $\mathbf{X}$ taking the value $\mathbf{x}$ from its alphabet $\mathcal{X}$ conditioned on the labelled set. In this context, Elkan & Noto formalise the *selected completely at random* assumption, stating that the examples for the labelled set are selected completely at random from all the positive examples:

$$p(s_P^+|\mathbf{x}, y^+) = p(s_P^+|y^+) \quad \forall \ \mathbf{x} \in \mathcal{X}. \tag{3}$$

Building upon this assumption, Sechidis et al. [19] proved that we can test independence between a feature X and the unobservable variable Y, by simply testing the independence between X and the observable variable S_P, which can be seen as a *surrogate* version of Y. While this assumption is sufficient for testing independence and guarantees the same probability of false positives, it leads to a less powerful test, and the probability of committing a false negative error is increased by a factor which can be calculated using prior knowledge over $p(y^+)$. With our current work we extend this approach to test conditional independence.

3.2 Semi-supervised Data

Semi-Supervised (SS) data refer to situations where we have a small number of labelled examples from *both* classes and a large number of unlabelled examples. For reasoning over semi-supervised data we will follow the formal framework of Smith & Elkan [20]. Assuming that the dataset is drawn i.i.d. from the joint distribution $p(\mathbf{X}, Y, S)$, where S describes whether an example is labelled (s^+) or not (s^-). We sample a total number of N examples out of which N_{S^+} are labelled as positive or negative. Smith & Elkan [20] presented the "traditional semi-supervised" scenario, where the labels are missing completely at random (MCAR), so the labelled set is an unbiased sample of the population. But apart from this traditional scenario, there are many alternative scenarios that can lead to semi-supervised data [20]. In our work, we will focus on the scenario where labelling an example is conditionally independent of the features given the class:

$$p(s^+|\mathbf{x}, y) = p(s^+|y) \quad \forall \; \mathbf{x} \in \mathcal{X}, y \in \mathcal{Y}. \tag{4}$$

This assumption can be seen as a straightforward extension of the selected completely at random assumption in the semi-supervised scenario, and it is followed in numerous semi-supervised works [11,16,18]. A practical application where we can use this assumption is in *class-prior-change* scenario [16], which occurs when the class balance in the labelled set does not reflect the population class balance. This sampling bias is created because the labels are missing not at random (MNAR), and the missingness mechanism depends directly on the class. The "traditional semi-supervised" assumption is as a restricted version of the assumption described in equation (4), when we furthermore assume $p(s^+|y) = p(s^+) \; \forall \; y \in \mathcal{Y}$.

4 Markov Blanket Discovery in Positive-Unlabelled Data

In this section we present a novel methodology for testing conditional independence in PU data. We will then see how we can use this methodology to derive Markov blanket despite the labelling restriction.

4.1 Testing Conditional Independence in PU Data

With the following theorem we prove that a valid approach to test conditional independence is to assume all unlabelled examples to be negative and as a result use the surrogate variable S_P instead of the unobservable Y.

Theorem 1 (Testing conditional independence in PU data).
In the positive unlabelled scenario, under the selected completely at random assumption, a variable X is independent of the class label Y given a subset of features $\mathbf{Z}$ if and only if X is independent of S_P given $\mathbf{Z}$, so it holds:

$$X \perp\!\!\!\perp Y|\mathbf{Z} \Leftrightarrow X \perp\!\!\!\perp S_P|\mathbf{Z}.$$

The proof of the theorem is available in the supplementary material. Now we will verify the consequences of this theorem in the context of Markov blanket discovery. We use four widely used networks; Appendix A contains all details on data generation and on the experimental protocol. For these networks we know the true Markov blankets and we compare them with the discovered blankets through the IAMB algorithm. As we observe from Figure 2 using S_P instead of Y in the IAMB algorithm does not result to a statistical significant difference in the false positive rate, or in Markov blanket terminology the blankets derived from these two approaches are similar in terms of the variables that were *falsely added to the blanket.*

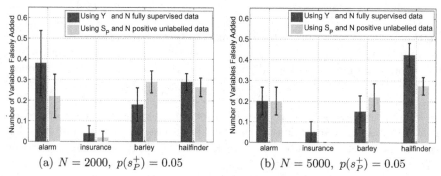

(a) $N = 2000$, $p(s_P^+) = 0.05$ (b) $N = 5000$, $p(s_P^+) = 0.05$

Fig. 2. Verification of Theorem 1. This illustrates the average number of variables falsely added in MB and the 95% confidence intervals over 10 trials when we use IAMB with Y and S_P. 2a for total sample size $N = 2000$ out of which we label only 100 positive examples and 2b for total sample size $N = 5000$ out of which we label only 250 positives.

But, while Theorem 1 tells us that the probability of committing an error is the same for the two tests $G(X; Y|\mathbf{Z})$ and $G(X; S_P|\mathbf{Z})$ when $X \perp\!\!\!\perp Y|\mathbf{Z}$, it does not say anything about the performance of these tests when the variables are conditionally dependent. In this case, we should compare the *power* of the tests, and in order to do so we should explore the non-centrality parameters of the two conditional G-tests of independence.

Theorem 2 (Power of PU conditional test of independence).
In the positive unlabelled scenario, under the selected completely at random assumption, when a variable X is dependent on the class label Y given a subset of features $\mathbf{Z}$, $X \not\perp\!\!\!\perp Y|\mathbf{Z}$, we have: $I(X; Y|\mathbf{Z}) > I(X; S_P|\mathbf{Z})$.

While with the following theorem we will quantify the amount of power that we are loosing with the naive assumption of all unlabelled examples being negative.

Theorem 3 (Correction factor for PU test).
The non-centrality parameter of the conditional G-test between X and S_P given a subset of features $\mathbf{Z}$ takes the form:

$$\lambda_{G(X;S_P|\mathbf{Z})} = \kappa_P \lambda_{G(X;Y|\mathbf{Z})} = \kappa_P 2NI(X;Y|\mathbf{Z}),$$

where $\kappa_P = \dfrac{1-p(y^+)}{p(y^+)} \dfrac{p(s_P^+)}{1-p(s_P^+)} = \dfrac{1-p(y^+)}{p(y^+)} \dfrac{N_{s_P^+}}{N-N_{s_P^+}}.$

The proofs of the last two theorems are also available in the supplementary material. So, by using prior knowledge over the $p(y^+)$ we can use the naive test for sample size determination, and decide the amount of data that we need in order to have similar performance with the unobservable fully supervised test. Now we will illustrate the last theorems again in the context of MB discovery. A direct consequence of Theorem 2 is that using S_P instead of Y results in a higher number of *false negative* errors. In the MB discovery context this will result in a larger number of variables falsely not added to the predicted blanket, since we assumed that the variables were independent when in fact they were dependent. In order to verify experimentally this conclusion we will compare again the discovered blankets by using S_P instead of Y. As we see in Figure 3, the number of variables that were falsely not added is higher when we are using S_P. This Figure also verifies Theorem 3, where we see that the number of variables falsely removed when using the naive test $G(X;S_P|\mathbf{Z})$ with increased sample size N/κ_P is the same as when using the unobservable test $G(X;Y|\mathbf{Z})$ with N data.

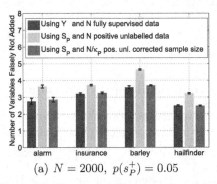

(a) $N = 2000$, $p(s_P^+) = 0.05$

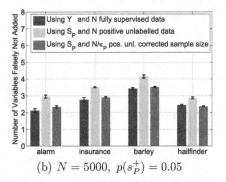

(b) $N = 5000$, $p(s_P^+) = 0.05$

Fig. 3. Verification of Theorems 2 and 3. This illustrates the average number of variables falsely not added to the MB and the 95% confidence intervals over 10 trials when we use IAMB with Y and S_P. 3a for total sample size $N = 2000$ and 3b for total sample size $N = 5000$. In all the scenarios we label 5% of the total examples as positives.

4.2   Evaluation of Markov Blanket Discovery in PU Data

For an overall evaluation of the derived blankets using S_P instead of Y we will use the F-measure, which is the harmonic mean of precision and recall, against the ground truth [17]. In Figure 4, we observe that the assumption of all unlabelled examples to be negative gives worse results than the fully-supervised scenario, and that the difference between the two approaches gets smaller as we increase sample size. Furthermore, using the correction factor κ_P to increase the sample size of the naive approach makes the two techniques perform similar.

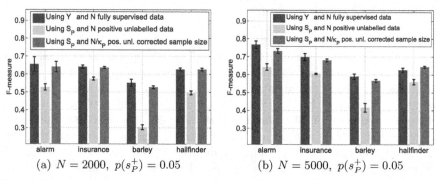

(a) $N = 2000$, $p(s_P^+) = 0.05$ (b) $N = 5000$, $p(s_P^+) = 0.05$

Fig. 4. Comparing the performance in terms of F-measure when we use IAMB with Y and S_P. 4a for total sample size $N = 2000$ and 4b for total sample size $N = 5000$. In all the scenarios we label 5% of the total examples as positives.

5   Markov Blanket Discovery in Semi-supervised Data

In this section we will present two informative ways, in terms of power, to test conditional independence in semi-supervised data. Then we will suggest an algorithm for Markov blanket discovery where we will incorporate prior knowledge to choose the optimal way for testing conditional independence.

5.1   Testing Conditional Independence in Semi-supervised Data

We will introduce two variables in the semi-supervised scenario, which can be seen as noisy versions of the unobservable random variable Y. The first one is S_P, which we already used in the PU scenario, and is a binary random variable that takes the value s_P^+ when a positive example is labelled, and s_P^- in any other case. The second variable is S_N, which is also a binary random variable that takes the value s_N^+ when a negative example is labelled and s_N^- otherwise. Using these two variables, the selected completely at random assumptions described in equation (4) can be written as:

$$p(s_P^+|\mathbf{x}, y^+) = p(s_P^+|y^+) \text{ and } p(s_N^+|\mathbf{x}, y^-) = p(s_N^+|y^-) \quad \forall \ \mathbf{x} \in \mathcal{X}.$$

So, using S_P instead of Y is equivalent to making the assumption that all unlabelled examples are negative, as we did in the positive-unlabelled scenario, while using S_N instead of Y is equivalent to assuming all unlabelled examples being positive. In this section we will prove the versions of the three theorems we presented earlier for both variables S_P and S_N in the semi-supervised scenario.

Firstly we will show that testing conditional independence by assuming the unlabelled examples to be either positive or negative is a valid approach.

Theorem 4 (Testing conditional independence in SS data).
In the semi-supervised scenario, under the selected completely at random assumption, a variable X is independent of the class label Y given a subset of features $\mathbf{Z}$ if and only if X is independent of S_P given $\mathbf{Z}$ and the same result holds for S_N: $X \perp\!\!\!\perp Y|\mathbf{Z} \Leftrightarrow X \perp\!\!\!\perp S_P|\mathbf{Z}$ and $X \perp\!\!\!\perp Y|\mathbf{Z} \Leftrightarrow X \perp\!\!\!\perp S_N|\mathbf{Z}$.

Proof. Since the selected completely at random assumption holds for both classes, this theorem is a direct consequence of Theorem 1.

The consequence of this assumption is that the derived conditional tests of independence are less powerful that the unobservable fully supervised test, as we prove with the following theorem.

Theorem 5 (Power of the SS conditional tests of independence).
In the semi-supervised scenario, under the selected completely at random assumption, when a variable X is dependent of the class label Y given a subset of features $\mathbf{Z}$, $X \not\perp\!\!\!\perp Y|\mathbf{Z}$, we have: $I(X;Y|\mathbf{Z}) > I(X;S_P|\mathbf{Z})$ and $I(X;Y|\mathbf{Z}) > I(X;S_N|\mathbf{Z})$.

Proof. Since the selected completely at random assumption holds for both classes, this theorem is a direct consequence of Theorem 2.

Finally, with the following theorem we can quantify the amount of power that we are loosing by assuming that the unlabelled examples are negative (i.e. using S_P) or positive (i.e. using S_N).

Theorem 6 (Correction factors for SS tests).
The non-centrality parameter of the conditional G-test can take the form:

$$\lambda_{G(X;S_P|\mathbf{Z})} = \kappa_P \lambda_{G(X;Y|\mathbf{Z})} = \kappa_P 2NI(X;Y|\mathbf{Z}) \quad and$$
$$\lambda_{G(X;S_N|\mathbf{Z})} = \kappa_N \lambda_{G(X;Y|\mathbf{Z})} = \kappa_N 2NI(X;Y|\mathbf{Z}),$$

where $\kappa_P = \dfrac{1-p(y^+)}{p(y^+)} \dfrac{p(s_P^+)}{1-p(s_P^+)}$ and $\kappa_N = \dfrac{p(y^+)}{1-p(y^+)} \dfrac{p(s_N^+)}{1-p(s_N^+)}$.

Proof. Since the selected completely at random assumption holds for both classes, this theorem is a direct consequence of Theorem 3.

5.2   Incorporating Prior Knowledge on Markov Blanket Discovery

Since using S_P or S_N are both valid approaches it is preferable to use the most powerful test. In order to do so, we can use some "soft" prior knowledge over the probability $p(y^+)$[2]. We call it "soft" because there is no need to know the exact value, but we only need to know if it is greater or smaller than a quantity calculated from the observed dataset. The following corollary gives more details.

Corollary 1 (Incorporating prior knowledge).
In order to have the smallest number of falsely missing variables from the Markov Blanket we should use S_P instead of S_N, when the following inequality holds

$$\kappa_P > \kappa_N \Leftrightarrow p(y^+) < \cfrac{1}{1 + \sqrt{\dfrac{(1-p(s_P^+))p(s_N^+)}{p(s_P^+)(1-p(s_N^+))}}}.$$

When the opposing inequality holds the most powerful choice is S_N. When equality holds, both approaches are equivalent.

We can estimate $p(s_P^+)$ and $p(s_N^+)$ from the observed data, and, using some prior knowledge over $p(y^+)$, we can decide the most powerful option. In Figure 5 we compare in terms of F-measure the derived Markov blankets when we use the most powerful and the least powerful choice. As we observe by incorporating prior knowledge as Corollary 1 describes, choosing to test with the most powerful option, results in remarkably better performance than with the least powerful option.

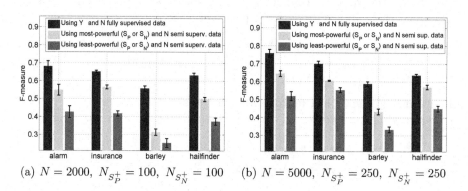

(a) $N = 2000$, $N_{S_P^+} = 100$, $N_{S_N^+} = 100$ (b) $N = 5000$, $N_{S_P^+} = 250$, $N_{S_N^+} = 250$

Fig. 5. Comparing the performance in terms of F-measure when we use the unobservable variable Y and the most and least powerful choice between S_P and S_N. 5a for sample size $N = 2000$ out of which we label only 100 positive and 100 negative examples and 5b for sample size $N = 5000$ out of which we label only 250 positive and 250 negative examples.

---

[2] When the labelling depends directly in the class, eq. (4), we cannot have an unbiased estimator for this probability without further assumptions, more details in [16].

6   Exploring our Framework Under Class Prior Change — When and how the Unlabelled Data Help

In this section, we will present how our approach performs in a real world problem where the class balance in the labelled set does not reflect the balance over the overall population; such situation is known as *class-prior-change* [16]. We compare our framework with the following two approaches: ignoring the unlabelled examples, a procedure known in statistic as *listwise deletion* [3], or using the unlabelled data to have more reliable estimates for the marginal counts of the features, a procedure known in statistics as *available case analysis* or *pairwise deletion* [3]. The latter is followed in BASSUM [6]; Section 2.3 gives more details about this approach and its limitations.

Firstly, let's assume that the semi-supervised data are generated under the "traditional semi-supervised" scenario, where the labelled set is an unbiased sample from the overall population. As a result, the class-ratio in the labelled set is the same to the population class-ratio. In mathematical notation it holds $\frac{p(y^+|s^+)}{p(y^-|s^+)} = \frac{p(y^+)}{p(y^-)}$, where the *lhs* is the class-ratio in the labelled set and in *rhs* the population class-ratio. As we observe in Figure 6, choosing the most powerful option between S_P and S_N performs similarly with ignoring completely the unlabelled examples. As it was expected, using the semi-supervised data with pairwise deletion has unpredictable performance and often performs much worse than using only the labelled examples.

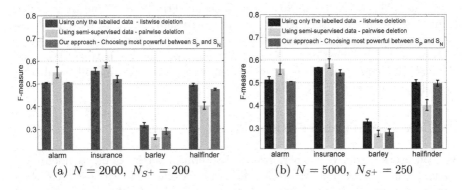

(a) $N = 2000$, $N_{S^+} = 200$ (b) $N = 5000$, $N_{S^+} = 250$

Fig. 6. Traditional semi-supervised scenario. Comparing the performance in terms of F-measure when we have the same class-ratio in the labelled-set as in the overall population. 6a for sample size $N = 2000$ out of which we label only 200 examples and 6b $N = 5000$ out of which we label only 250 examples.

Now, let's assume we have semi-supervised data under the class-prior-change scenario (for more details see Section 3.2). In our simulation we sample the labelled data in order to have a class ratio in the labelled set inverse than the population ratio. In mathematical notation it holds $\frac{p(y^+|s^+)}{p(y^-|s^+)} = \left(\frac{p(y^+)}{p(y^-)}\right)^{-1}$, where the *lhs* is the class-ratio in the labelled set and in *rhs* the inverse of the population class-ratio. As we observe in Figure 7, choosing the most powerful option between S_P and S_N performs statistically better than ignoring the unlabelled examples. Our approach performs better on average than the pairwise deletion, while the latter one performs comparably to the listwise deletion in many settings.

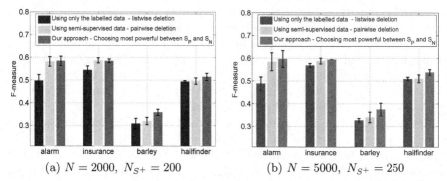

(a) $N = 2000, \ N_{S+} = 200$ (b) $N = 5000, \ N_{S+} = 250$

Fig. 7. Class prior change semi-supervised scenario. Comparing the performance in terms of F-measure when we have inverse class-ratio in the labelled-set than in the overall population. 7a for sample size $N = 2000$ out of which we label only 200 examples and 7b $N = 5000$ out of which we label only 250 examples.

Furthermore, our approach can be applied in scenarios where we have labelled examples only from one class, which cannot be handled with the other two approaches. Also, with our approach, we can control the power of our tests, which is not the case in pairwise deletion procedure. To sum up, in class-prior-change scenarios we can use Corollary 1 and some "soft" prior knowledge over $p(y^+)$ in order to decide which of the following two assumptions is better: to assume all unlabelled examples are negative (i.e. use S_P) or to assume all unlabelled examples are positive (i.e. use S_N).

7 Conclusions and Future Work

With our work we derive a generalization of conditional tests of independence for partially labelled data and we present a framework on how we can use unlabelled data for discovering Markov blankets around partially labelled target nodes.

In positive-unlabelled data, we proved that assuming all unlabelled examples are negative is sufficient for testing conditional independence but it will increase the number of the variables that are falsely missing from the predicted blanket. Furthermore, with a correction factor, we quantified the amount of power we are losing by this assumption, and we present how we can take this into account for adjusting the sample size in order to perform the same as in fully-supervised scenarios.

Then, we extended our methodology to semi-supervised data, where we can make two valid assumptions over the unlabelled examples: assume them either positive or negative. We explored the consequences of these two assumptions again in terms of possible errors in Markov blanket discovery procedures, and we suggested a way to use some "soft" prior knowledge to take the optimal decision. Finally, we presented a practical semi-supervised scenario in which the usage of unlabelled examples under our framework proved to be more beneficial compared to other suggested approaches.

A future research direction could be to explore how we can use our methodology for structure learning of Bayesian networks. Since our techniques are informative in terms of power, they can be used in structure learning approaches that have control over the false negative rate to prevent over constraint structures; for example, our framework generalises the work presented by Bacciu et al. [4] for partially-labelled data. Furthermore, our work for structure learning in partially labelled data can be used in combination with recently suggested methods for parameter learning from incomplete data by Mohan et al. [13].

Table 1. A summary of the networks used in the experimental studies

Network	Number of target nodes	Total number of nodes	Average MB size of target nodes	Average prior prob. $p(y^+)$ of target nodes
alarm	5	37	5.6	0.21
insurance	10	27	6.2	0.32
barley	10	48	5.6	0.31
hailfinder	20	56	4.9	0.31

A   Generation of Network Data and Experimental Protocol

The networks used are standard benchmarks for Markov blanket discovery taken from the Bayesian network repository[3]. For target variables we used nodes that have at least one child, one parent and one spouse in their Markov blanket. Furthermore we chose as positive examples (y^+) those examples with class value 1, while the rest of the examples formed the negative set. We also focused in nodes that had prior probability $p(y^+)$ between 0.15 and 0.50, which is an area of interest for PU data. For the supervised scenarios (i.e. when we used the variable Y) we perform 10 trials of size $N = 2000$ and 5000. For each trial we sample 30 different partially labelled datasets, and the outcome was the most frequently derived Markov blanket. For all experiments we fixed the significance of the tests to be $\alpha = 0.10$. Table 1 presents the summary of the Networks used in the current work.

Acknowledgments. The research leading to these results has received funding from EPSRC Anyscale project EP/L000725/1 and the European Union's Seventh Framework Programme (FP7/2007-2013) under grant agreement n° 318633. This work was supported by EPSRC grant [EP/I028099/1]. Sechidis gratefully acknowledges the support of the Propondis Foundation.

---

[3] Downloaded from http://www.bnlearn.com/bnrepository/

References

1. Agresti, A.: Categorical Data Analysis. Wiley Series in Probability and Statistics, 3rd edn. Wiley-Interscience (2013)
2. Aliferis, C.F., Statnikov, A., Tsamardinos, I., Mani, S., Koutsoukos, X.D.: Local causal and Markov blan. induction for causal discovery and feat. selection for classification part I: Algor. and empirical eval. JMLR **11**, 171–234 (2010)
3. Allison, P.: Missing Data. Sage University Papers Series on Quantitative Applications in the Social Sciences, 07–136 (2001)
4. Bacciu, D., Etchells, T., Lisboa, P., Whittaker, J.: Efficient identification of independence networks using mutual information. Comp. Stats **28**(2), 621–646 (2013)
5. Brown, G., Pocock, A., Zhao, M.J., Luján, M.: Conditional likelihood maximisation: a unifying framework for information theoretic feature selection. The Journal of Machine Learning Research (JMLR) **13**(1), 27–66 (2012)
6. Cai, R., Zhang, Z., Hao, Z.: BASSUM: A Bayesian semi-supervised method for classification feature selection. Pattern Recognition **44**(4), 811–820 (2011)
7. Cohen, J.: Statistical Power Analysis for the Behavioral Sciences, 2nd edn. Routledge Academic (1988)
8. Cover, T.M., Thomas, J.A.: Elements of information theory. J. Wiley & Sons (2006)
9. Elkan, C., Noto, K.: Learning classifiers from only positive and unlabeled data. In: ACM SIGKDD Int. Conf. on Knowledge Discovery and Data Mining (2008)
10. Koller, D., Sahami, M.: Toward optimal feature selection. In: International Conference of Machine Learning (ICML), pp. 284–292 (1996)
11. Lawrence, N.D., Jordan, M.I.: Gaussian processes and the null-category noise model. In: Semi-Supervised Learning, chap. 8, pp. 137–150. MIT Press (2006)
12. Margaritis, D., Thrun, S.: Bayesian network induction via local neighborhoods. In: NIPS, pp. 505–511. MIT Press (1999)
13. Mohan, K., Van den Broeck, G., Choi, A., Pearl, J.: Efficient algorithms for bayesian network parameter learning from incomplete data. In: Conference on Uncertainty in Artificial Intelligence (UAI) (2015)
14. Pearl, J.: Probabilistic Reasoning in Intelligent Systems: Networks of Plausible Inference. Morgan Kaufmann Publishers Inc., San Francisco (1988)
15. Pellet, J.P., Elisseeff, A.: Using Markov blankets for causal structure learning. The Journal of Machine Learning Research (JMLR) **9**, 1295–1342 (2008)
16. Plessis, M.C.d., Sugiyama, M.: Semi-supervised learning of class balance under class-prior change by distribution matching. In: 29th ICML (2012)
17. Pocock, A., Luján, M., Brown, G.: Informative priors for Markov blanket discovery. In: 15th AISTATS (2012)
18. Rosset, S., Zhu, J., Zou, H., Hastie, T.J.: A method for inferring label sampling mechanisms in semi-supervised learning. In: NIPS (2004)
19. Sechidis, K., Calvo, B., Brown, G.: Statistical hypothesis testing in positive unlabelled data. In: Calders, T., Esposito, F., Hüllermeier, E., Meo, R. (eds.) ECML PKDD 2014, Part III. LNCS, vol. 8726, pp. 66–81. Springer, Heidelberg (2014)
20. Smith, A.T., Elkan, C.: Making generative classifiers robust to selection bias. In: 13th ACM SIGKDD Inter. Conf. on Knwl. Disc. and Data Min., pp. 657–666 (2007)
21. Tsamardinos, I., Aliferis, C.F.: Towards principled feature selection: relevancy, filters and wrappers. In: AISTATS (2003)
22. Tsamardinos, I., Aliferis, C.F., Statnikov, A.: Time and sample efficient discovery of Markov blankets and direct causal relations. In: ACM SIGKDD (2003)
23. Yaramakala, S., Margaritis, D.: Speculative Markov blanket discovery for optimal feature selection. In: 5th ICDM. IEEE (2005)

Multi-view Semantic Learning for Data Representation

Peng Luo[1], Jinye Peng[1(✉)], Ziyu Guan[1], and Jianping Fan[2]

[1] College of the Information and Technology,
Northwest University of China, Xi'an, China
luopengpeng@gmail.com, {pjy,ziyuguan}@nwu.edu.cn
[2] Department of Computer Science,
University of North Carolina at Charlotte, Charlotte, NC, USA
jfan@uncc.edu

Abstract. Many real-world datasets are represented by multiple features or modalities which often provide compatible and complementary information to each other. In order to obtain a good data representation that synthesizes multiple features, researchers have proposed different multi-view subspace learning algorithms. Although label information has been exploited for guiding multi-view subspace learning, previous approaches either fail to directly capture the semantic relations between labeled items or unrealistically make Gaussian assumption about data distribution. In this paper, we propose a new multi-view nonnegative subspace learning algorithm called Multi-view Semantic Learning (MvSL). MvSL tries to capture the semantic structure of multi-view data by a novel graph embedding framework. The key idea is to let neighboring intra-class items be near each other while keep nearest inter-class items away from each other in the learned common subspace across multiple views. This nonparametric scheme can better model non-Gaussian data. To assess nearest neighbors in the multi-view context, we develop a multiple kernel learning method for obtaining an optimal kernel combination from multiple features. In addition, we encourage each latent dimension to be associated with a subset of views via sparseness constraints. In this way, MvSL is able to capture flexible conceptual patterns hidden in multi-view features. Experiments on two real-world datasets demonstrate the effectiveness of the proposed algorithm.

Keywords: Multi-view learning · Nonnegative matrix factorization · Graph embedding · Multiple kernel lerning · Structured sparsity

1 Introduction

In many real-world data analytic problems, instances are often described with multiple modalities or views. It becomes natural to integrate multi-view representations to obtain better performance than relying on a single view. A good

© Springer International Publishing Switzerland 2015
A. Appice et al. (Eds.): ECML PKDD 2015, Part I, LNAI 9284, pp. 367–382, 2015.
DOI: 10.1007/978-3-319-23528-8_23

integration of multi-view features can lead to a more comprehensive description of the data items, which could improve performance of many related applications.

An emerging area of multi-view learning is *multi-view latent subspace learning*, which aims to obtain a compact latent representation by taking advantage of inherent structure and relation across multiple views. A pioneering technique in this area is Canonical Correlation Analysis (CCA) [7], which tries to learn the projections of two views so that the correlation between them is maximized. Recently, a lot of methods have been applied to multi-view subspace learning, such as matrix factorization [9], [4], [11], [18], graphical models [3] and spectral embedding [22].

Matrix factorization techniques have received more and more attention as fundamental tools for multi-view latent subspace learning. Since a useful representation acquired by matrix factorization typically makes latent structure in the data explicit (through the basis vectors), and usually reduces the dimensionality of input views, so that further analysis can be effectively and efficiently carried out. Nonnegative Matrix Factorization (NMF) [13] is an attractive matrix factorization method due to its theoretical interpretation and desired performance. NMF aims to find two nonnegative matrices (a basis matrix and an encoding matrix) whose product provides a good approximation to the original matrix. NMF tries to formulate a feasible model for learning object parts, which is closely relevant to human perception mechanism. Recently, variants of NMF have been proposed for multi-view learning [11], [18], [10].

Labeled data has been incorporated to improve NMF's performance in both the single view case [16], [21] and the multi-view case [10]. However, there is still lack of effective methods for learning a common nonnegative latent subspace which captures the semantic structure of multi-view data through label information. Previously, there are mainly two ways to incorporate label information into the NMF framework. The first one is to reconstruct the label indicator matrix through multiplying the encoding matrix by a weight matrix [10], [17],[16]. These methods intrinsically impose *indirect* affinity constraints on encodings of labeled items. Nevertheless, such indirect constraints could be insufficient for capturing the semantic relationships between labeled items. The second one is to regularize the encodings of labeled items by fisher-style discriminative constraints [21], [25]. Although methods of this kind directly penalize distances among labeled items in the latent subspace, they assume the data of each class follows a Gaussian distribution. However, in reality this assumption is too restricted since data often exhibit complex non-Gaussian distribution [2], [24].

In this paper, we propose a novel semi-supervised multi-view representation (i.e. latent subspace) learning algorithm, namely, Multi-view Semantic Learning (MvSL), to better capture the semantic structure of multi-view data. MvSL jointly factorizes data matrices of different views, and each view is factorized into a *basis matrix* and an *encoding matrix* where the encoding matrix is the low dimensional optimal consensus representation shared by multiple views. We regularize the encoding matrix by developing a novel graph embedding framework:

we construct (1) an affinity graph which characterizes the intra-class compactness and connects each data point with its neighboring points of the same class; (2) a discrimination graph which connects the marginal points and characterizes the inter-class separability in the learned subspace. This nonparametric scheme can better capture the complex non-Gaussian distribution of real-world data [24]. A sub-challenge is how to identify nearest neighbors in the multi-view context. To this end, we develop a new multiple kernel learning algorithm to find the optimal kernel combination for multi-view features. The algorithm tries to optimally preserves the semantic relations among labeled items, so that we can assess within-class variance and between-class similarity effectively. Moreover, we impose a $L_{1,2}$ norm regularizer on the basis matrix to encourage some basis vectors to be zero-valued [9]. In this way, each latent dimension has the flexibility to be associated with a subset of views, thus enhancing the expressive power of the model. To solve MvSL, we develop a block coordinate descent [15] optimization algorithm. For empirical evaluation, two real-world multi-view datasets are employed. The encouraging results of MvSL are achieved in comparison with the state-of-the-art algorithms.

2   Related Work

In this section, we present a brief review of related work about NMF-based subspace learning. Firstly, we describe the notations used throughout the paper.

2.1   Common Notations

In this paper, vectors and matrices are denoted by lowercase boldface letters and uppercase boldface letters respectively. For matrix $\mathbf{M}$, we denote its (i,j)-th element by M_{ij}. The i-th element of a vector $\mathbf{b}$ is denoted by b_i. Given a set of N items, we use matrix $\mathbf{X} \in \mathbb{R}_+^{M \times N}$ to denote the nonnegative data matrix where the i-th column vector is the feature vector for the i-th item. In the multi-view setting, we have H views and the data matrix of the v-th view is denoted by $\mathbf{X}^{(v)} \in \mathbb{R}_+^{M_v \times N}$, where M_v is the dimensionality of the v-th view. Throughout this paper, $\|\mathbf{M}\|_F$ denotes the Frobenius norm of matrix $\mathbf{M}$.

2.2   NMF-Based Latent Subspace Learning

NMF is an effective subspace learning method to capture the underlying structure of the data in the parts-based low dimensional representation space, which accords with the cognitive process of human brain from the psychological and physiological studies [13].

Given an input nonnegative data matrix $\mathbf{X} \in \mathbb{R}_+^{M \times N}$ where each column represents a data point and each row represents a feature. NMF aims to find two nonnegative matrices $\mathbf{U} \in \mathbb{R}_+^{M \times K}$ and $\mathbf{V} \in \mathbb{R}_+^{K \times N}$ whose product can well approximate the original data matrix:

$$\mathbf{X} \approx \mathbf{UV}.$$

$K < \min(M, N)$ denotes the desired reduced dimensionality, and to facilitate discussion, we call $\mathbf{U}$ the basis matrix and $\mathbf{V}$ the coefficient matrix.

It is known that the objective function above is not convex in $\mathbf{U}$ and $\mathbf{V}$ together, so it is unrealistic to expect an algorithm to find the global minimum. Lee and Seung [13] presented multiplicative update rules to find the locally optimal solution as follows:

$$U_{ik}^{t+1} = U_{ik}^t \frac{(\mathbf{X}(\mathbf{V}^t)^T)_{ik}}{(\mathbf{U}^t\mathbf{V}^t(\mathbf{V}^t)^T)_{ik}}$$

$$V_{kj}^{t+1} = V_{kj}^t \frac{((\mathbf{U}^{t+1})^T\mathbf{X})_{kj}}{((\mathbf{U}^{t+1})^T\mathbf{U}^{t+1}\mathbf{V}^t)_{kj}}.$$

In recent years, many variants of the basic NMF model have been proposed. We just list a few which are related to our work. One direction related to our work is coupling label information to NMF [21], [25]. These works added discriminative constraints into NMF via regularizing the encoding matrix $\mathbf{V}$ by fisher-style discriminative constraints. Nevertheless, fisher discriminative analysis assumes data of each class is approximately Gaussian distributed, a property that cannot always be satisfied in real-world applications. Our method adopts a nonparametric regularization scheme (i.e. regularization in neighborhoods) and consequently can better model real-life data. Another related direction of NMF is sparse NMF [8]. Sparseness constraints not only encourage local and compact representations, but also improve the stability of the decomposition. Most previous works on sparse NMF employed L_1 norm or ratio between L_1 norm and L_2 norm to achieve sparsity on $\mathbf{U}$ and $\mathbf{V}$. However, the story for our problem is different since we have multiple views and the goal is to allow each latent dimension to be associated with a subset of views. Therefore, $L_{1,2}$ norm [9] is used to achieve this goal.

There are also some extensions of NMF for multi-view data, e.g. clustering [18], image annotation [11], graph regularized multi-view NMF [6] and semi-supervised learning [10],[17]. Although [10] and [17] also exploited label information, they incorporated label information as a factorization constraint on $\mathbf{V}$, i.e. reconstructing the label indicator matrix through multiplying $\mathbf{V}$ by a weight matrix. Hence, those methods intrinsically imposed *indirect* affinity constraints on encodings of labeled items in the latent subspace. On the contrary, our method *directly* captures the semantic relationships between items in the latent subspace through the proposed graph embedding framework. We will compare MvSL with [10] in experiments.

3 Multi-view Semantic Learning

In this section, we present the proposed Multi-view Semantic Learning (MvSL) algorithm for latent representation learning from multi-view data.

3.1   Matrix Factorization with Multi-view Data

The consensus principle is the fundamental principle in multi-view learning [23]. At first, MvSL jointly factorizes $\{\mathbf{X}^{(v)}\}_{v=1}^{H}$ with different basis matrices $\{\mathbf{U}^{(v)}\}_{v=1}^{H}$ and the consensus encoding matrix $\mathbf{V}$ [9], [18], [11]:

$$\min_{\{\mathbf{U}^{(v)}\}_{v=1}^{H},\mathbf{V}} \frac{1}{2}\sum_{v=1}^{H}\|\mathbf{X}^{(v)} - \mathbf{U}^{(v)}\mathbf{V}\|_F^2 \tag{1}$$
$$\text{s.t.} \quad U_{ik}^{(v)} \geq 0, V_{kj} \geq 0, \quad \forall i,j,k,v.$$

However, the standard unsupervised NMF fails to discover the semantic structure in the data. In the next, we introduce our graph embedding framework for multi-view semantic learning.

3.2   Graph Embedding for Multi-view Semantic Learning

Let $\mathbf{V}^l \in \mathbb{R}^{K \times N^l}$, the first N^l columns of $\mathbf{V}$, be the latent representation of the first N^l labeled items and $\mathbf{V}^u \in \mathbb{R}^{K \times N^u}$ be the latent representation of the remaining N^u unlabeled items (i.e. $\mathbf{V} = [\mathbf{V}^l \ \mathbf{V}^u]$). Inspired by [24], we propose a graph embedding framework for capturing the semantic structure of multi-view data. We define an affinity graph G^a and a discrimination graph G^p. The affinity graph $G^a = \{\mathbf{V}^l, \mathbf{W}^a\}$ is an undirected weighted graph with labeled item set $\mathbf{V}^l$ as its vertex set, and similarity matrix $\mathbf{W}^a \in \mathbb{R}^{N^l \times N^l}$ which characterizes the intra-class local similarity structure. The discrimination graph $G^p = \{\mathbf{V}^l, \mathbf{W}^p\}$ characterizes inter-class separability and penalizes the similarity between the most similar inter-class item pairs in the learned subspace. Let $\mathbf{v}_i^l$ be the i-th column of $\mathbf{V}^l$. The graph-preserving criteria are given as follows:

$$\min_{\mathbf{V}^l} \frac{1}{2}\sum_{i=1}^{N^l}\sum_{j=1}^{N^l} W_{ij}^a\|\mathbf{v}_i^l - \mathbf{v}_j^l\|_2^2 = \min_{\mathbf{V}^l}\frac{1}{2}tr\left[\mathbf{V}^l\mathbf{L}^a\left(\mathbf{V}^l\right)^T\right], \tag{2}$$

$$\max_{\mathbf{V}^l} \frac{1}{2}\sum_{i=1}^{N^l}\sum_{j=1}^{N^l} W_{ij}^p\|\mathbf{v}_i^l - \mathbf{v}_j^l\|_2^2 = \max_{\mathbf{V}^l}\frac{1}{2}tr\left[\mathbf{V}^l\mathbf{L}^p\left(\mathbf{V}^l\right)^T\right], \tag{3}$$

where $tr(\cdot)$ denotes the trace of a matrix, and $\mathbf{L}^a = \mathbf{D}^a - \mathbf{W}^a$ is the graph Laplacian matrix for G^a with the (i,i)-th elements of the diagonal matrix $\mathbf{D}^a$ equals $\sum_{j=1}^{N^l} W_{ij}^a$ ($\mathbf{L}^p$ is for G^p). Generally speaking, Eq. (2) means items belonging to the same class should be near each other in the learned latent subspace, while Eq. (3) tries to keep items from different classes as distant as possible. However, only with the nonnegative constraints Eq. (3) would diverge. Note that there is an arbitrary scaling factor in solutions to problem (1): for any invertible $K \times K$ matrix $\mathbf{Q}$, we have $\mathbf{U}^{(v)}\mathbf{V} = (\mathbf{U}^{(v)}\mathbf{Q})(\mathbf{Q}^{-1}\mathbf{V})$. Hence, without loss of generality, we add the constraints $\{V_{kj} \leq 1, \forall k,j\}$ to put an upper bound on (3).

The similarity matrices $\mathbf{W}^a$ and $\mathbf{W}^p$ are defined as follows

$$W_{ij}^a = \begin{cases} 1, & \text{if } i \in N_{k_a}(j) \;\; or \;\; j \in N_{k_a}(i) \\ 0, & \text{otherwise} \end{cases}, \tag{4}$$

where $N_{k_a}(i)$ indicates the index set of the k_a nearest neighbors of item i in the *same* class,

$$W_{ij}^p = \begin{cases} 1, & \text{if } i \in N_{k_p}(j) \;\; or \;\; j \in N_{k_p}(i) \\ 0, & \text{otherwise} \end{cases}, \tag{5}$$

where $N_{k_p}(i)$ indicates the index set of the k_p nearest neighbors of item i in the *distinct* classes. We can see from the definitions of $\mathbf{W}^a$ and $\mathbf{W}^p$ that G^a and G^p intrinsically preserve item semantic relations in local neighborhoods. This nonparametric scheme can better handle real-life datasets which often exhibit non-Gaussian distribution.

The remaining question is how to estimate nearest neighbors, which is a routine function for constructing G^a and G^p. However, since real-life datasets could be diverse and noisy, a single feature may not be sufficient to characterize the affinity relations among items. Hence, we propose to use multiple features for assessing the similarity between data items. In particular, we develop a novel Multiple Kernel Learning (MKL) [20],[5] method for this task.

3.2.1   Multiple Kernel Learning

A kernel function measures the similarity between items in terms of one view. We use $\mathbf{K}_v(i,j)$ to denote the kernel value between items i and j in terms of view v. To make all kernel functions comparable, we normalize each kernel function into $[0,1]$ as follows:

$$\mathbf{K}_v(i,j) \leftarrow \frac{\mathbf{K}_v(i,j)}{\sqrt{\mathbf{K}_v(i,i)\mathbf{K}_v(j,j)}}. \tag{6}$$

To obtain a comprehensive kernel function, we linearly combine multiple kernels as follow:

$$\mathbf{K}(i,j,\boldsymbol{\eta}) = \sum_{v=1}^{H} \eta_v \mathbf{K}_v(i,j), \quad \sum_{v=1}^{H} \eta_v = 1, \eta_v \geq 0, \tag{7}$$

where $\boldsymbol{\eta} = [\eta_1, ..., \eta_H]^T$ is the weight vector to be learned. This combined kernel function can lead to better estimation of similarity among items than any single kernel. For example, only relying on color information could not handel images of concept "zebra" well since the background may change arbitrarily, while adding texture information can better characterize zebra images.

Then we need to design the criterion for learning $\boldsymbol{\eta}$. Since our goal is to model the semantic relations among items, the learned kernel function should be accommodated to the semantic structure among classes. We define an *ideal kernel* to encode the semantic structure:

$$\mathbf{K}_{ideal}(i,j) = \begin{cases} 1, & \text{if } y_i = y_j \\ 0, & \text{otherwise} \end{cases}, \tag{8}$$

where y_i denotes the label of item i. For each pair of items, we require its combined kernel function value to conform to the corresponding ideal kernel value. This leads to the following least square loss

$$l(i, j, \boldsymbol{\eta}) = (\mathbf{K}(i, j, \boldsymbol{\eta}) - \mathbf{K}_{ideal}(i, j))^2 \tag{9}$$

Summing $l(i, j, \boldsymbol{\eta})$ over all pairs of labeled items, we could get the optimization objective. However, in reality we would get imbalanced classes: the numbers of labeled items for different classes can be quite different. The item pairs contributed by classes with much larger number of items will dominate the overall loss. In order to tackle this issue, we normalize the contribution of each pair of classes (including same-class pairs) by its number of item pairs. This is equivalent to multiplying each $l(i, j, \boldsymbol{\eta})$ by a weight t_{ij} which is defined as follows

$$\mathbf{t}_{ij} = \begin{cases} \frac{1}{n_i^2}, & \text{if } y_i = y_j \\ \frac{1}{2n_i n_j}, & \text{otherwise} \end{cases}, \tag{10}$$

where n_i denotes the number of items belonging to the class with label y_i. Therefore, the overall loss becomes $\sum_{i,j} t_{ij} l(i, j, \boldsymbol{\eta})$. To prevent overfitting, a L_2 regularization term is added for $\boldsymbol{\eta}$. The final optimization problem is formulated as

$$\min_{\boldsymbol{\eta}} \sum_{i,j=1}^{N^l} t_{ij} l(i, j, \boldsymbol{\eta}) + \lambda \|\boldsymbol{\eta}\|_2^2$$

$$s.t. \sum_{v=1}^{H} \eta_v = 1, \eta_v \geq 0 \tag{11}$$

where λ is a regularization tradeoff parameter. The optimization problem of (11) is a classical quadratic programming problem which can be solved efficiently using any convex programming software. When $\boldsymbol{\eta}$ is obtained, we could assess the similarity relationship between labeled items in terms of multi-view features according to (7). Then, according to Eqs. (4) and (5) we can construct the affinity matrix $\mathbf{W}^a$ and discriminative matrix $\mathbf{W}^p$, respectively.

3.3   Sparseness Constraint

Since similarities among data items belonging to the same class share the same sparsity pattern, a structured sparseness regularizer is added to objective function to encourage some basis column vectors in $\mathbf{U}^{(v)}$ to become to 0. This makes view v independent of the latent dimensions which correspond to these zeros-valued basis vectors. By employing $L_{1,q}$ norm regularization, the latent factors obtained by NMF can be improved with an additional property of shared sparsity. In this work, we choose $q = 2$. $L_{1,2}$ norm of matrix $\mathbf{U}$ is defined as:

$$\|\mathbf{U}\|_{1,2} = \sum_{k=1}^{K} \|\mathbf{u}_k\|_2, \tag{12}$$

3.4   Objective Function of MvSL

By synthesizing the above objectives, the optimization problem of MvSL is formulated as:

$$
\min_{\{\mathbf{U}^{(v)}\}_{v=1}^{H}, \mathbf{V}} \frac{1}{2} \sum_{v=1}^{H} \|\mathbf{X}^{(v)} - \mathbf{U}^{(v)}\mathbf{V}\|_F^2 + \alpha \sum_{v=1}^{H} \|\mathbf{U}^{(v)}\|_{1,2}
$$
$$
+ \frac{\beta}{2} \left\{ tr\left[\mathbf{V}^l \mathbf{L}^a (\mathbf{V}^l)^T\right] - tr\left[\mathbf{V}^l \mathbf{L}^p (\mathbf{V}^l)^T\right] \right\} \tag{13}
$$
$$
\text{s.t.} \quad U_{ik}^{(v)} \geq 0, 1 \geq V_{kj} \geq 0, \quad \forall i, j, k, v.
$$

4   Optimization

The joint optimization function in (13) is not convex over all variables $\mathbf{U}^{(1)}, ..., \mathbf{U}^{(H)}$ and $\mathbf{V}$ simultaneously. Thus, we propose a block coordinate descent method [15] which optimizes one block of variables while keeping the other block fixed. The procedure is depicted in Algorithm 1. For the ease of representation, we define

$$
\mathcal{O}\{(\mathbf{U}^{(1)}, ..., \mathbf{U}^{(H)}, \mathbf{V})\} = \frac{1}{2} \sum_{v=1}^{H} \|\mathbf{X}^{(v)} - \mathbf{U}^{(v)}\mathbf{V}\|_F^2 + \alpha \sum_{v=1}^{H} \|\mathbf{U}^{(v)}\|_{1,2}
$$
$$
+ \frac{\beta}{2} \left\{ tr\left[\mathbf{V}^l \mathbf{L}^a (\mathbf{V}^l)^T\right] - tr\left[\mathbf{V}^l \mathbf{L}^p (\mathbf{V}^l)^T\right] \right\} \tag{14}
$$

4.1   Optimizing $\{\mathbf{U}^{(v)}\}_{v=1}^{H}$

When $\mathbf{V}$ is fixed, $\mathbf{U}^{(1)}, ..., \mathbf{U}^{(H)}$ are independent with one another. Since the optimization method is the same, here we just focus on an arbitrary view and use $\mathbf{X}$ and $\mathbf{U}$ to denote respectively the data matrix and the basis matrix for the view. The optimization problem involving $\mathbf{U}$ can be written as

$$
\min_{\mathbf{U}} \; \phi(\mathbf{U}) := \frac{1}{2} \|\mathbf{X} - \mathbf{U}\mathbf{V}\|_F^2 + \alpha\|\mathbf{U}\|_{1,2}
$$
$$
\text{s.t.} \quad U_{ik} \geq 0, \; \forall i, k. \tag{15}
$$

Two terms of $\phi(\mathbf{U})$ are convex functions. The first term of $\phi(\mathbf{U})$ is differentiable, and its gradient is Lipschitz continuous. Hence, an efficient convex optimization method can be adopted. [12] presented a variant of Nesterov's first order method, suitable for solving (15). In this paper, we take the optimization method of [12] to update $\mathbf{U}$. Due to the limitation of space, details can be found in [12].

4.2   Optimizing V

When $\{\mathbf{U}^{(v)}\}_{v=1}^{H}$ are fixed, the subproblem for $\mathbf{V}$ can be written as

$$
\begin{aligned}
\min_{\mathbf{V}} \ \psi(\mathbf{V}) := \ & \left(\frac{1}{2}\sum_{v=1}^{H}\|\mathbf{X}^{(v)} - \mathbf{U}^{(v)}\mathbf{V}\|_{F}^{2} \right. \\
& \left. + \frac{\beta}{2}\left\{tr\left[\mathbf{V}^{l}\mathbf{L}^{a}(\mathbf{V}^{l})^{T}\right] - tr\left[\mathbf{V}^{l}\mathbf{L}^{p}(\mathbf{V}^{l})^{T}\right]\right\}\right)
\end{aligned}
\tag{16}
$$
$$
\text{s.t.} \ \ 1 \geq V_{kj} \geq 0, \ \ \forall j, k.
$$

This is a bounded nonnegative quadratic programming problem for $\mathbf{V}$. Sha *et al.* [19] proposed a general multiplicative optimization scheme for this kind of problems. Inspired by their method, we develop a multiplicative update algorithm for optimizing $\mathbf{V}$.

Firstly, recall that $\mathbf{X}^{(v)} = [\mathbf{X}^{(v),l} \ \mathbf{X}^{(v),u}]$ and $\mathbf{V} = [\mathbf{V}^{l} \ \mathbf{V}^{u}]$. We can transform the first term of $\psi(\mathbf{V})$:

$$
\begin{aligned}
& \frac{1}{2}\sum_{v=1}^{H}\|\mathbf{X}^{(v)} - \mathbf{U}^{(v)}\mathbf{V}\|_{F}^{2} \\
=& \frac{1}{2}\sum_{v=1}^{H}\Big(tr[(\mathbf{V}^{l})^{T}(\mathbf{U}^{(v)})^{T}\mathbf{U}^{(v)}\mathbf{V}^{l}] - 2tr[(\mathbf{V}^{l})^{T}(\mathbf{U}^{(v)})^{T}\mathbf{X}^{(v),l}] \\
& + tr[(\mathbf{V}^{u})^{T}(\mathbf{U}^{(v)})^{T}\mathbf{U}^{(v)}\mathbf{V}^{u}] - 2tr[(\mathbf{V}^{u})^{T}(\mathbf{U}^{(v)})^{T}\mathbf{X}^{(v),u}]\Big) + const.
\end{aligned}
$$

For convenience, let $\mathbf{P} = \sum_{v=1}^{H}(\mathbf{U}^{(v)})^{T}\mathbf{U}^{(v)}$ and $\mathbf{Q}^{l} = \sum_{v=1}^{H}(\mathbf{U}^{(v)})^{T}\mathbf{X}^{(v),l}$. $\mathbf{Q}^{u}$ is defined similarly for the unlabeled part. Eq. (16) can be transformed into

$$
\begin{aligned}
\min_{\mathbf{V}} \ & \frac{1}{2}tr[(\mathbf{V}^{l})^{T}\mathbf{P}\mathbf{V}^{l}] - tr[(\mathbf{V}^{l})^{T}\mathbf{Q}^{l}] + \frac{1}{2}tr[(\mathbf{V}^{u})^{T}\mathbf{P}\mathbf{V}^{u}] - tr[(\mathbf{V}^{u})^{T}\mathbf{Q}^{u}] \\
& + \frac{\beta}{2}\left\{tr\left[\mathbf{V}^{l}\mathbf{L}^{a}(\mathbf{V}^{l})^{T}\right] - tr\left[\mathbf{V}^{l}\mathbf{L}^{p}(\mathbf{V}^{l})^{T}\right]\right\}
\end{aligned}
\tag{17}
$$
$$
\text{s.t.} \ \ 1 \geq V_{kj} \geq 0, \ \ \forall j, k.
$$

Since $\mathbf{V}^{l}$ and $\mathbf{V}^{u}$ are independent, we analyze them separately. The objective terms involving $\mathbf{V}^{l}$ can be summarized as

$$
\begin{aligned}
\mathcal{O}^{l}(\mathbf{V}^{l}) =& \frac{1}{2}tr[(\mathbf{V}^{l})^{T}\mathbf{P}\mathbf{V}^{l}] - tr[(\mathbf{V}^{l})^{T}\mathbf{Q}^{l}] \\
& + \frac{\beta}{2}\left\{tr\left[\mathbf{V}^{l}\mathbf{L}^{a}(\mathbf{V}^{l})^{T}\right] - tr\left[\mathbf{V}^{l}\mathbf{L}^{p}(\mathbf{V}^{l})^{T}\right]\right\}
\end{aligned}
\tag{18}
$$

The second term is linear term for $\mathbf{V}^{l}$. We only need to focus on the quadratic terms which can be rewritten as follows

$$\frac{1}{2}tr[(\mathbf{V}^l)^T\mathbf{P}\mathbf{V}^l] = \frac{1}{2}\sum_{j=1}^{N^l}(\mathbf{v}_j^l)^T\mathbf{P}\mathbf{v}_j^l, \tag{19}$$

$$\frac{\beta}{2}\left\{tr\left[\mathbf{V}^l\mathbf{L}^a(\mathbf{V}^l)^T\right] - tr\left[\mathbf{V}^l\mathbf{L}^p(\mathbf{V}^l)^T\right]\right\}$$

$$= \frac{\beta}{2}\sum_{k=1}^{K}\left\{(\bar{\mathbf{v}}_k^l)^T(\mathbf{D}^a + \mathbf{W}^p)\bar{\mathbf{v}}_k^l - (\bar{\mathbf{v}}_k^l)^T(\mathbf{D}^p + \mathbf{W}^a)\bar{\mathbf{v}}_k^l\right\}, \tag{20}$$

where $\mathbf{v}_j^l$ and $\bar{\mathbf{v}}_k^l$ represent the j-th column vector and k-th row vector of $\mathbf{V}^l$, respectively. Each summand in Eq. (19) and (20) is a quadratic function of a vector variable. Therefore, we can provide upper bounds for these summands:

$$(\mathbf{v}_j^l)^T\mathbf{P}\mathbf{v}_j^l \leq \sum_{k=1}^{K}\frac{(\mathbf{P}\mathbf{v}_j^{l,t})_k}{V_{kj}^{l,t}}(V_{kj}^l)^2,$$

$$(\bar{\mathbf{v}}_k^l)^T(\mathbf{D}^a + \mathbf{W}^p)\bar{\mathbf{v}}_k^l \leq \sum_{j=1}^{N^l}\frac{((\mathbf{D}^a + \mathbf{W}^p)\bar{\mathbf{v}}_k^{l,t})_j}{V_{kj}^{l,t}}(V_{kj}^l)^2,$$

$$-(\bar{\mathbf{v}}_k^l)^T(\mathbf{D}^p + \mathbf{W}^a)\bar{\mathbf{v}}_k^l \leq -\sum_{i,j}(\mathbf{D}^p + \mathbf{W}^a)_{ij}V_{ki}^{l,t}V_{kj}^{l,t}\left(1 + \log\frac{V_{ki}^lV_{kj}^l}{V_{ki}^{l,t}V_{kj}^{l,t}}\right),$$

where we use $\mathbf{V}^{l,t}$ to denote the value of $\mathbf{V}^l$ in the t-th iteration of the update algorithm and $\mathbf{v}_j^{l,t}$, $\bar{\mathbf{v}}_k^{l,t}$ are its j-th column vector and k-th row vector, respectively. Note that V_{kj}^l can be viewed both as the k-th element of $\mathbf{v}_j^l$ and as the j-th element of $\bar{\mathbf{v}}_k^l$. The proofs of these bounds follow directly from Lemmas 1 and 2 in [19]. Aggregating the bounds for all the summands, we obtain the auxiliary function for $\mathcal{O}^l(\mathbf{V}^l)$

$$\mathcal{G}^l(\mathbf{V}^{l,t}; \mathbf{V}^l)$$

$$= \frac{1}{2}\sum_{j=1}^{N^l}\sum_{k=1}^{K}\frac{(\mathbf{P}\mathbf{v}_j^{l,t})_k + \beta((\mathbf{D}^a + \mathbf{W}^p)\bar{\mathbf{v}}_k^{l,t})_j}{V_{kj}^{l,t}}(V_{kj}^l)^2$$

$$- \frac{\beta}{2}\sum_{k=1}^{K}\sum_{i,j}(\mathbf{D}^p + \mathbf{W}^a)_{ij}V_{ki}^{l,t}V_{kj}^{l,t}\left(1 + \log\frac{V_{ki}^lV_{kj}^l}{V_{ki}^{l,t}V_{kj}^{l,t}}\right) \tag{21}$$

$$- \sum_{j=1}^{N^l}\sum_{k=1}^{K}Q_{kj}^lV_{kj}^l.$$

The estimate of $\mathbf{V}^l$ in the $(t+1)$-th iteration is then computed as

$$\mathbf{V}^{l,t+1} = \arg\min_{\mathbf{V}^l}\mathcal{G}^l(\mathbf{V}^{l,t}; \mathbf{V}^l). \tag{22}$$

Differentiating $\mathcal{G}^l(\mathbf{V}^{l,t}; \mathbf{V}^l)$ with respect to each V_{kj}^l, we have

$$\frac{\partial \mathcal{G}^l(\mathbf{V}^{l,t}; \mathbf{V}^l)}{\partial V_{kj}^l}$$

$$= \frac{(\mathbf{P}\mathbf{v}_j^{l,t})_k + \beta((\mathbf{D}^a + \mathbf{W}^p)\bar{\mathbf{v}}_k^{l,t})_j}{V_{kj}^{l,t}} V_{kj}^l - \frac{\beta((\mathbf{D}^p + \mathbf{W}^a)\bar{\mathbf{v}}_k^{l,t})_j}{V_{kj}^l} V_{kj}^{l,t} - Q_{kj}^l$$

Setting $\partial \mathcal{G}^l(\mathbf{V}^{l,t}; \mathbf{V}^l)/\partial V_{kj}^l = 0$, we get the update rule for $\mathbf{V}^l$

$$V_{kj}^{l,t+1} = \min\left\{1, V_{kj}^{l,t} \frac{-B_{kj} + \sqrt{B_{kj}^2 + 4A_{kj}C_{kj}}}{2A_{kj}}\right\}, \tag{23}$$

$$A_{kj} = (\mathbf{P}\mathbf{v}_j^{l,t})_k + \beta((\mathbf{D}^a + \mathbf{W}^p)\bar{\mathbf{v}}_k^{l,t})_j,$$
$$B_{kj} = -Q_{kj}^l, C_{kj} = \beta((\mathbf{D}^p + \mathbf{W}^a)\bar{\mathbf{v}}_k^{l,t})_j.$$

Here $\mathbf{v}_j^l$ and $\bar{\mathbf{v}}_k^l$ denote the j-th column vector and the k-th row vector of $\mathbf{V}^l$, respectively. It is easy to verify that $\mathcal{O}^l(\mathbf{V}^{l,t+1}) \leq \mathcal{G}^l(\mathbf{V}^{l,t}; \mathbf{V}^{l,t+1}) \leq \mathcal{G}^l(\mathbf{V}^{l,t}; \mathbf{V}^{l,t}) = \mathcal{O}^l(\mathbf{V}^{l,t})$. Therefore, the update rule for $\mathbf{V}^l$ monotonically decreases Eq. 13. The case for $\mathbf{V}^u$ is simpler since we do not have the graph embedding terms:

$$\mathcal{O}^u(\mathbf{V}^u) = \frac{1}{2}tr[(\mathbf{V}^u)^T\mathbf{P}\mathbf{V}^u] - tr[(\mathbf{V}^u)^T\mathbf{Q}^u] \tag{24}$$

Similarly, the auxiliary function for $\mathcal{O}^u(\mathbf{V}^u)$ can be derived

$$\mathcal{G}^u(\mathbf{V}^{u,t}; \mathbf{V}^u) = \frac{1}{2}\sum_{j=1}^{N^u}\sum_{k=1}^{K} \frac{(\mathbf{P}\mathbf{v}_j^{u,t})_k}{V_{kj}^{u,t}}(V_{kj}^u)^2 - \sum_{j=1}^{N^u}\sum_{k=1}^{K} Q_{kj}^u V_{kj}^u \tag{25}$$

and the update rule can be obtained by setting the partial derivatives to 0:

$$V_{kj}^{u,t+1} = \min\left\{1, V_{kj}^{u,t} \frac{Q_{kj}^u - |Q_{kj}^u|}{2(\mathbf{P}\mathbf{v}_j^{u,t})_k}\right\} \tag{26}$$

5   Experiment

In this section, we conduct the experiments on two real-world data sets to validate the effectiveness of the proposed algorithm MvSL.

Algorithm 1. Optimization of MvSL

    Data: $\{\mathbf{X}^{(v)}\}_{v=1}^{H}, \alpha, \beta$
    Result: $\{\mathbf{U}^{(v)}\}_{v=1}^{H}, \mathbf{V}$

1 **begin**
2   Randomly initialize $U_{ik}^{(v)} \geq 0, 1 \geq V_{kj} \geq 0, \forall i, j, k, v$
3   **repeat**
4   Fix $\mathbf{V}$, update $\mathbf{U}^{(1)}, ..., \mathbf{U}^{(H)}$ as in [12]
5   Fix $\mathbf{U}^{(1)}, ..., \mathbf{U}^{(H)}$, update $\mathbf{V}^{l}$ as in (23) and update $\mathbf{V}^{u}$ as in (26) ;
6   **until** *convergence or max no. iterations reached*
7 **end**

Table 1. Statistics of the datasets.

Dataset	Size	# of categories	Dimensionality of views
Reuters	1800	6	$21,531/15,506/11,547$
MM2.0	5000	25	$64/144/75/128$

5.1 Data Set

We use two real-world datasets to evaluate the proposed factorization method. The first dataset was constructed from the Reuters Multilingual collection [1]. This test collection contains totally 111,740 Reuters news documents written in five different languages. Documents for each language can be divided into a common set of six categories. Each document was translated into the other four languages and represented as TF-IDF vectors. We took documents written in English as the first view and their Italian and Spanish translations as the second and third views. We randomly sampled 1800 English documents, with 300 for each category. The second dataset came from Microsoft Research Asia Internet Multimedia Dataset 2.0 (MSRA-MM 2.0) [14]. MSRA-MM 2.0 consists of about 1 million images which were respective search results for 1165 popular query concepts in Microsoft Live Search. Each concept has approximately 500-1000 images. For each image, its relevance to the corresponding concept was manually labeled with three levels: very relevant, relevant and irrelevant. 7 different low level features were extracted for each image. To form the experimental dataset, we selected 25 query concepts from the Animal, Object and Scene branches, and then randomly sampled 200 images from each concept while discarding irrelevant ones. We took 4 features in MSRA-MM 2.0 as 4 different views: 64D HSV color histogram, 144D color correlogram, 75D edge distribution histogram and 128D wavelet texture. Hereafter, we refer to the two datasets as Reuters and MM2.0, respectively. The statistics of these datasets are summarized in Table 1.

5.2 Baselines

To validate the performance of our method, we compare the proposed MvSL with the following baselines:

- **NMF** [13].
- Feature concatenation (**ConcatNMF**): This method concatenates feature vectors of different views to form a united representation and then applies NMF.
- Multi-view NMF (**MultiNMF**): MultiNMF [18] is an unsupervised multi-view NMF algorithm.
- Semi-supervised Unified Latent Factor method (**SULF**): SULF [10] is a semi-supervised multi-view nonnegative factorization method which models partial label information as a factorization constraint on V^l.
- Graph regularized NMF (**GNMF**): GNMF [2] is a manifold regularized version of NMF. We extended it to the multi-view case and replaced the affinity graph for approximating data manifolds with the within-class affinity graph defined in Eq. (4) to make it a semi-supervised method on multi-view data.

5.3   Evaluation Metric

Accuracy (ACC) is a typical evaluation metric of classification. Let N_u denote the total number of test images to be labeled, the N_r is the number of images that are assigned the right categories or tags by the proposed algorithms according to the ground truth, the ACC is defined as ACC=N_r/N_u.

Table 2. Classification performance of different factorization methods on the Reuters dataset (accuracy±std dev,%).

Labeled Percentage	NMF-b	ConcatNMF	MultiNMF	SULF	GNMF	MvSL
10	61.55±1.08	63.04±1.67	63.69±1.52	67.93±1.92	68.93±1.77	**70.56±1.21**
20	65.71±1.37	66.09±1.08	67.42±1.97	68.40±1.64	70.59±1.65	**72.67±1.02**
30	67.30±0.27	68.40±1.91	69.16±1.52	70.05±1.48	71.80±1.24	**74.78±1.34**
40	68.41±1.96	69.81±1.96	70.28±1.83	71.86±1.38	72.23±1.54	**75.87±1.26**
50	70.44±1.72	70.75±2.03	71.81±1.47	72.78±1.44	73.78±1.75	**77.33±0.79**

Table 3. Classification performance of different factorization methods on the MM2.0 dataset (accuracy±std dev, %).

Labeled Percentage	NMF-b	ConcatNMF	MultiNMF	SULF	GNMF	MvSL
10	24.56±0.98	27.41±0.83	26.26±0.95	27.47±1.03	28.03±1.17	**30.92±0.44**
20	25.37±0.85	31.24±0.93	30.39±1.12	30.94±1.25	31.55±1.14	**33.83±1.52**
30	26.09±0.71	32.47±0.80	31.85±0.87	33.13±0.87	34.15±0.51	**35.80±0.68**
40	28.03±0.46	34.25±0.71	33.48±0.65	34.94±0.65	35.26±0.97	**37.12±0.73**
50	28.06±0.28	35.08±0.48	34.33±0.56	36.32±0.56	36.61±0.57	**38.16±0.65**

5.4   Experiment Results

Table 2 and Table 3 show the classification performance of different factorization methods on MM2.0 and Reuters, respectively. We varied the percentage of training items from 10% to 50%. The observations are revealed as follows. Firstly, Semi-supervised algorithms are superior to unsupervised algorithms in general, which indicated that exploiting label information could lead to latent spaces with better discriminative structures. Secondly, from comparison between multi-view algorithms and single-view algorithm (NMF), it is easy to see that multi-view algorithms are more preferable for multi-view data. This is in accord with the results of previous multi-view learning work. Thirdly, MvSL and GNMF show superior performance over SULF. SULF models partial label information as a factorization constraint on $\mathbf{V}^l$, which can be viewed as indirect affinity constraints on encoding of within-class items. On the contrary, the graph embedding terms in MvSL and GNMF impose direct affinity constraints on item encodings and therefore could lead to more explicit semantic structures in the learned latent spaces. Finally, MvSL outperformed the baseline methods under all cases. The reason should be that MvSL not only directly exploits label information via a graph embedding framework, but also adds regularization by $L_{1,2}$-norm on $\mathbf{U}^{(v)}$ successfully promotes that sparsity pattern is shared among data items or features within classes. These properties could help to learn a clearer semantic latent space.

5.5   Parameter Sensitive Analysis

There are two essential parameters in new methods. β measures the importance of the semi-supervised part of MvSL (i.e. the graph embedding regularization terms), while α controls the degree of sparsity of the basis matrices. We investigate their influence on MvSL's performance by varying one while fixing the other one.

The classification results are shown in Figure 1 for MM2.0 and Reuters. We found the general behavior of the two parameters was the same: when increasing the parameter from 0, the performance curves first went up and then went down. This indicates that when assigned moderate weights, the sparseness and semi-supervised constraints indeed helped learn a better latent subspace. Based observations , we set $\alpha = 10$, $\beta = 0.02$ for experiments.

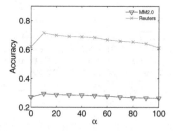
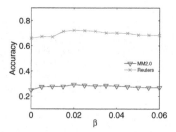

Fig. 1. Influence of different parameter settings on the performance of MvSL: (a) varying α while setting $\beta = 0.02$, (b) varying β while setting $\alpha = 10$

6 Conclusion

We have proposed Multi-view semantic learning (MvSL), a novel nonnegative latent representation learning algorithm for representation learning multi-view data. MvSL tries to learn a semantic latent subspace of items by exploiting both multiple views of items and partial label information. The partial label information was used to construct a graph embedding framework, which encouraged items of the same category to be near with each other and kept items belonging to different categories as distant as possible in the latent subspace. What's more, kernel alignment effectively estimated the items pair similarity among multi-view data, which further extended graph embedding framework. Another novel property of MvSL was that it allowed each latent dimension to be associated with a subset of views by imposing $L_{1,2}$-norm on each basis $\mathbf{U}^{(v)}$. Therefore, MvSL is able to learn flexible latent factor sharing structures which could lead to more meaningful semantic latent subspaces. An efficient multiplicative-based iterative algorithm is developed to solve the proposed optimization problem. The classification experimental results on two real-world data sets have demonstrated the effectiveness of our method.

Graph embedding is a general framework and different definitions of the within class affinity graph G^a and the discriminative graph G^p can be employed. How to propose more suitable similarity criteria with multi-view data is an interesting direction for further study.

Acknowledgments. This research was supported by the National High-tech R&D Program of China (863 Program) (No. 2014AA015201), National Natural Science Foundation of China (No. 61373118).

References

1. Amini, M., Usunier, N., Goutte, C.: Learning from multiple partially observed views-an application to multilingual text categorization. In: Advances in Neural Information Processing Systems, pp. 28–36 (2009)
2. Cai, D., He, X., Han, J., Huang, T.S.: Graph regularized nonnegative matrix factorization for data representation. IEEE Transactions on Pattern Analysis and Machine Intelligence **33**(8), 1548–1560 (2011)
3. Chen, N., Zhu, J., Xing, E.P.: Predictive subspace learning for multi-view data: a large margin approach. In: Advances in Neural Information Processing Systems, pp. 361–369 (2010)
4. Han, Y., Wu, F., Tao, D., Shao, J., Zhuang, Y., Jiang, J.: Sparse unsupervised dimensionality reduction for multiple view data. IEEE Transactions on Circuits and Systems for Video Technology **22**(10), 1485–1496 (2012)
5. He, J., Chang, S.-F., Xie, L.: Fast kernel learning for spatial pyramid matching. In: IEEE Conference on Computer Vision and Pattern Recognition, CVPR 2008, pp. 1–7. IEEE (2008)
6. Hidru, D., Goldenberg, A.: Equinmf: Graph regularized multiview nonnegative matrix factorization (2014). arXiv preprint arXiv:1409.4018
7. Hotelling, H.: Relations between two sets of variates. Biometrika 321–377 (1936)

8. Hoyer, P.O.: Non-negative sparse coding. In: Proceedings of the 2002 12th IEEE Workshop on Neural Networks for Signal Processing, pp. 557–565 (2002)
9. Jia, Y., Salzmann, M., Darrell, T.: Factorized latent spaces with structured sparsity. In: Advances in Neural Information Processing Systems, pp. 982–990 (2010)
10. Jiang, Y., Liu, J., Li, Z., Lu, H.: Semi-supervised unified latent factor learning with multi-view data. Machine Vision and Applications 25(7), 1635–1645 (2014)
11. Kalayeh, M., Idrees, H., Shah, M.: NMF-KNN: Image annotation using weighted multi-view non-negative matrix factorization. In: Proceedings of the IEEE Conference on Computer Vision and Pattern Recognition, pp. 184–191 (2014)
12. Kim, J., Monteiro, R., Park, H.: Group sparsity in nonnegative matrix factorization. In: SDM, pp. 851–862. SIAM (2012)
13. Lee, D.D., Seung, H.S.: Learning the parts of objects by non-negative matrix factorization. Nature 401(6755), 788–791 (1999)
14. Li, H., Wang, M., Hua, X.-S.: Msra-mm 2.0: A large-scale web multimedia dataset. In: IEEE International Conference on Data Mining Workshops, pp. 164–169. IEEE (2009)
15. Lin, C.-J.: Projected gradient methods for nonnegative matrix factorization. Neural computation 19(10), 2756–2779 (2007)
16. Liu, H., Wu, Z., Li, X., Cai, D., Huang, T.S.: Constrained nonnegative matrix factorization for image representation. IEEE Transactions on Pattern Analysis and Machine Intelligence 34(7), 1299–1311 (2012)
17. Liu, J., Jiang, Y., Li, Z., Zhou, Z.-H., Lu, H.: Partially shared latent factor learning with multiview data (2014)
18. Liu, J., Wang, C., Gao, J., Han, J.: Multi-view clustering via joint nonnegative matrix factorization. In: Proc. of SDM, vol. 13, pp. 252–260 (2013)
19. Sha, F., Lin, Y., Saul, L.K., Lee, D.D.: Multiplicative updates for nonnegative quadratic programming. Neural Computation 19(8), 2004–2031 (2007)
20. Shawe-Taylor, N., Kandola, A.: On kernel target alignment. Advances in neural information processing systems 14, 367 (2002)
21. Wang, Y., Jia, Y.: Fisher non-negative matrix factorization for learning local features. In: Proc. Asian Conf. on Comp. Vision. Citeseer (2004)
22. Xia, T., Tao, D., Mei, T., Zhang, Y.: Multiview spectral embedding. IEEE Transactions on Systems, Man, and Cybernetics, Part B: Cybernetics 40(6), 1438–1446 (2010)
23. Xu, C., Tao, D., Xu, C.: A survey on multi-view learning. arXiv preprint arXiv:1304.5634 (2013)
24. Yan, S., Xu, D., Zhang, B., Zhang, H.-J., Yang, Q., Lin, S.: Graph embedding and extensions: a general framework for dimensionality reduction. IEEE Transactions on Pattern Analysis and Machine Intelligence 29(1), 40–51 (2007)
25. Zafeiriou, S., Tefas, A., Buciu, I., Pitas, I.: Exploiting discriminant information in nonnegative matrix factorization with application to frontal face verification. IEEE Transactions on Neural Networks 17(3), 683–695 (2006)

Unsupervised Feature Analysis with Class Margin Optimization

Sen Wang[1], Feiping Nie[2], Xiaojun Chang[3]([✉]), Lina Yao[4], Xue Li[1],
and Quan Z. Sheng[4]

[1] School of ITEE, The University of Queensland, Brisbane, Australia
`sen.wang@uq.edu.au, xueli@itee.uq.edu.au`
[2] Center for OPTIMAL, Northwestern Polytechnical University, Shaanxi, China
`feiping.nie@gmail.com`
[3] Center for QCIS, University of Technology Sydney, Ultimo, Australia
`cxj273@gmail.com`
[4] School of CS, The University of Adelaide, Adelaide, Australia
`lina@cs.adelaide.edu.au, michael.sheng@adelaide.edu.au`

Abstract. Unsupervised feature selection has been attracting research attention in the communities of machine learning and data mining for decades. In this paper, we propose an unsupervised feature selection method seeking a feature coefficient matrix to select the most distinctive features. Specifically, our proposed algorithm integrates the Maximum Margin Criterion with a sparsity-based model into a joint framework, where the class margin and feature correlation are taken into account at the same time. To maximize the total data separability while preserving minimized within-class scatter simultaneously, we propose to embed K-means into the framework generating pseudo class label information in a scenario of unsupervised feature selection. Meanwhile, a sparsity-based model, $\ell_{2,p}$-norm, is imposed to the regularization term to effectively discover the sparse structures of the feature coefficient matrix. In this way, noisy and irrelevant features are removed by ruling out those features whose corresponding coefficients are zeros. To alleviate the local optimum problem that is caused by random initializations of K-means, a convergence guaranteed algorithm with an updating strategy for the clustering indicator matrix, is proposed to iteratively chase the optimal solution. Performance evaluation is extensively conducted over six benchmark data sets. From our comprehensive experimental results, it is demonstrated that our method has superior performance against all other compared approaches.

Keywords: Unsupervised feature selection · Maximum margin criterion · Sparse structure learning · Embedded K-means clustering

1 Introduction

Over the past few years, data are more than often represented by high-dimensional features in a number of research fields, e.g. data mining, computer

© Springer International Publishing Switzerland 2015
A. Appice et al. (Eds.): ECML PKDD 2015, Part I, LNAI 9284, pp. 383–398, 2015.
DOI: 10.1007/978-3-319-23528-8_24

vision, etc. With the inventions of such many sophisticated data representations, a problem has been never a lack of research attention: How to select the most distinctive features from high-dimensional data for subsequent learning tasks, e.g. classification? To answer this question, we take two points into account. First, the number of selected features should be smaller than the one of all features. Due to a lower dimensional representation, the subsequent learning tasks with no doubt can gain benefit in terms of efficiency [31]. Second, the selected features should have more discriminant power than the original all features. Many previous works have proven that removing those noisy and irrelevant features can improve discriminant power in most cases. In light of advantages of feature selection, different new algorithms have been flourished with various types of applications recently [29, 32, 33].

According to the types of supervision, feature selection can be generally divided into three categories, i.e. supervised, semi-supervised, and unsupervised feature selection algorithms. Representative supervised feature selection algorithms include Fisher score [6], Relief[11] and its extension, ReliefF [12], information gain [20], etc [25]. Label information of training data points is utilized to guide the supervised feature selection methods to seek distinctive subsets of features with different search strategies, i.e. *complete search, heuristic search, and non-deterministic search*. In the real world, class information is quite limited, resulting in the development of semi-supervised feature selection methods [3, 4], in which both labeled and unlabeled data are utilized.

In unsupervised scenarios, feature selection is more challenging, since there is no class information to use for selecting features. In the literature, unsupervised feature selection can be roughly categorized into three groups, i.e. *filter, wrapper*, and *embedded methods*. Filter-based unsupervised feature selection methods rank features according to some intrinsic properties of data. Then those features with higher scores are selected for the further learning tasks. The selection is independent to the consequent process. For example, He et al. [8] assume that data from the same class are often close to each other and use the locality preserving power of data, also termed as *Laplacian Score*, to evaluate importance degrees of features. In [30], a unified framework has been proposed for both supervised and unsupervised feature selection schemes using a spectral graph. Tabakhi et al. [23] have proposed an unsupervised feature selection method to select the optimal feature subset in an iterative algorithm, which is based on ant colony optimization. Wrapper-based methods as a more sophisticated way wrap learning algorithms to yield learned results that will be used to select distinctive subsets of features. In [15], for instance, the authors have developed a model that selects relevant features using two backward stepwise selection algorithms without prior knowledges of features. Normally, wrapper-based methods have better performance than filter-based methods, since they use learning algorithms. Unfortunately, the disadvantage is that the computation of wrapper methods is more expensive. Embedded methods are seeking a trade-off between them by integrating feature selection and clustering together into a joint framework. Because clustering algorithms can provide pseudo labels that can reflect the intrinsic information of data, some works [1,14,26] incorporate different clustering algorithms in objective functions to select features.

Most of the existing unsupervised feature selection methods [8,9,14,19,24,30] rely on a graph, e.g. *graph Laplacian*, to reflect intrinsic relationships among data, labeled and unlabeled. When the number of data is extremely large, the computational burden of constructing a graph Laplacian is significantly heavy. Meanwhile, some traditional feature selection algorithms [6,8] neglect correlations among features. The distinctive features are individually selected according to the importance of each feature rather than taking correlations among features into account. Recently, exploiting feature correlations has attracted much research attention [5,17,18,27,28]. It has proven that discovering feature correlation is beneficial to feature selection.

In this paper, we propose a graph-free method to select features by combining Maximum Margin Criterion with feature correlation mining into a joint framework. Specifically, the method, on one hand, aims to learn a feature coefficient matrix that linearly combines features to maximize the class margins. With the increase of the separability of the entire transformed data by maximizing the total scatter, the proposed method also expects distances between data points within the same class to be minimized after the linear transformation by the coefficient matrix. Since there is no class information can be borrowed from, K-means clustering is jointly embedded in the framework to provide pseudo labels. Inspired by recent feature selection works using sparsity-based model on the regularization term [4], on the other hand, the proposed algorithm learns sparse structural information of the coefficient matrix, with the goal of reducing noisy and irrelevant features by removing those features whose coefficients are zeros. The main contributions of this paper can be summarized as follows:

- The proposed method makes efforts to maximize class margins in a framework, where simultaneously considers the separability of the transformed data and distances between the transformed data within the same class. Besides, a sparsity-based regularization model is jointly applied on the feature coefficient matrix to analyze correlations among features in an iterative algorithm;
- K-means clustering is embedded into the framework generating cluster labels, which can be used as pseudo labels. Both maximizing class margins and learning sparse structures can benefit from generated pseudo labels during each iteration;
- Because the performance of K-means is dominated by the initialization, we propose a strategy to avoid our algorithm rapidly converge to a local optimum, which is largely ignored by most of existing approaches using K-means clustering. Theoretical proof of convergence is also provided.
- We have conducted extensive experiments over six benchmark datasets. The experimental results show that our method has better performance than all the compared unsupervised algorithms.

The rest of this paper is organized as follows: Notations and definitions that are used throughout the entire paper will be given in section 2. Our method will be elaborated in section 3, followed by proposing its optimization with an algorithm to guarantee the convergence property in section 4. In section 5, extensive experimental results are reported with related analysis. Lastly, the conclusion of this paper will be given in section 6.

2   Notations and Definitions

To give a better understanding of the proposed method, notations and definitions which are used throughout this paper are summarized in this section. Matrices and vectors are written as boldface uppercase letters and boldface lowercase letters, respectively. Given a data set denoted as $\boldsymbol{X} = [\boldsymbol{x}_1, \ldots, \boldsymbol{x_n}] \in \mathbb{R}^{d \times n}$, where n is the number of training data and d is the feature dimension. The mean of data is denoted as $\bar{\boldsymbol{x}}$. The feature coefficient matrix, $\boldsymbol{W} \in \mathbb{R}^{d \times d'}$, linearly combines data features as $\boldsymbol{W}^T \boldsymbol{X}$, d' is the feature dimension after the linear transformation. Given a cluster centroid matrix for the transformed data, $\boldsymbol{G} = [\boldsymbol{g}_1, \ldots, \boldsymbol{g}_c] \in \mathbb{R}^{d' \times c}$, its cluster indicator of transformed $\boldsymbol{x}_i$ is represented as $\boldsymbol{u}_i = [u_{i1}, \ldots, u_{ic}]$. c is the number of centroids. If transformed $\boldsymbol{x}_i$ belongs to the j-th cluster, $u_{ij} = 1$, $j = 1, \ldots, c$. Otherwise, $u_{ij} = 0$. Correspondingly, the cluster indicator matrix is $\boldsymbol{U} = [\boldsymbol{u}_1^T, \ldots, \boldsymbol{u}_n^T]^T \in \mathbb{R}^{n \times c}$.

For an arbitrary matrix $\boldsymbol{M} \in \mathbb{R}^{r \times l}$, its $\ell_{2,p}$-norm is defined as:

$$\|\boldsymbol{M}\|_{2,p} = \left[\sum_{i=1}^{r} \left(\sum_{j=1}^{l} M_{ij}^2 \right)^{\frac{p}{2}} \right]^{\frac{1}{p}} \tag{1}$$

The i-th row of $\boldsymbol{M}$ is represented by $\boldsymbol{M}^i$. The between-class, within-class and total scatter matrices of data are respectively defined as:

$$\boldsymbol{S_b} = \sum_{i=1}^{c} n_i (\bar{\boldsymbol{x}}_i - \bar{\boldsymbol{x}})(\bar{\boldsymbol{x}}_i - \bar{\boldsymbol{x}})^T,$$

$$\boldsymbol{S_w} = \sum_{i=1}^{c} \sum_{j=1}^{n_i} (\boldsymbol{x}_j - \bar{\boldsymbol{x}}_i)(\boldsymbol{x}_j - \bar{\boldsymbol{x}}_i)^T, \tag{2}$$

$$\boldsymbol{S_t} = \sum_{i=1}^{n} (\boldsymbol{x}_i - \bar{\boldsymbol{x}})(\boldsymbol{x}_i - \bar{\boldsymbol{x}})^T$$

where n_i is the number of data for the c-th class. $\boldsymbol{S_t} = \boldsymbol{S_w} + \boldsymbol{S_b}$. Other notations and definitions will be explained when they are in use.

3   Proposed Method

We now introduce our proposed method for unsupervised feature selection. To exploit distinctive features, an intuitive way is to find a linear transformation matrix which can project the data into a new space where the original data are more separable. PCA is the most popular approach to analyze the separability of features. PCA aims to seek directions on which transformed data have max variances. In other words, PCA is to maximize the separability of linearly transformed data by maximizing the covariance: $\max_{\boldsymbol{W}} \sum_{i=1}^{n} (\boldsymbol{W}^T(\boldsymbol{x}_i - \bar{\boldsymbol{x}}))^T (\boldsymbol{W}^T(\boldsymbol{x}_i - \bar{\boldsymbol{x}}))$. Without losing the generality, we assume the data has zero mean, i.e. $\bar{\boldsymbol{x}} = 0$. Recall the definition of total scatter of data, PCA is equivalent to maximize the total scatter of data. However, if only total scatter is considered as a separability measure, the within-class scatter might be also geometrically maximized with the maximization of the total scatter. This is not helpful to distinctive feature discovery. The representative model, LDA, solves this problem by maximizing Fisher criterion: $\max_{\boldsymbol{W}} \frac{\boldsymbol{W}^T \boldsymbol{S}_b \boldsymbol{W}}{\boldsymbol{W}^T \boldsymbol{S}_w \boldsymbol{W}}$. However, LDA and its variants require class information to construct between-class and within-class scatter matrices [2], which is not suitable for unsupervised feature selection. Before we give the objective that can solve the aforementioned problem, we first look at a supervised feature selection framework:

$$\max_{\boldsymbol{W}} \sum_{i=1}^{n} (\boldsymbol{W}^T \boldsymbol{x}_i)^T (\boldsymbol{W}^T \boldsymbol{x}_i) - \alpha \sum_{i=1}^{c} \sum_{j=1}^{n_i} (\boldsymbol{W}^T(\boldsymbol{x}_j - \bar{\boldsymbol{x}}_i))^T (\boldsymbol{W}^T(\boldsymbol{x}_j - \bar{\boldsymbol{x}}_i)) - \beta \Omega(\boldsymbol{W})$$

$$\text{s.t.} \quad \boldsymbol{W}^T \boldsymbol{W} = \boldsymbol{I},$$

$$(3)$$

where α and β are regularization parameters. In this framework, the first term is to maximize the total scatter, while the second term is to minimize the within-class scatter. The third part is a sparsity-based regularization term which controls the sparsity of $\boldsymbol{W}$. This model is quite similar with the classical LDA-based methods. Due to there is no class information in the unsupervised scenario, we need virtual labels to minimize the distances between data within the same class while maximize the total separability at the same time. To achieve this goal, we apply K-means clustering in our framework to replace the ground truth by generating cluster indicators of data. Given c centroids $\boldsymbol{G} = [\boldsymbol{g}_1, \ldots, \boldsymbol{g}_c] \in \mathbb{R}^{d' \times c}$, the objective function of the traditional K-means algorithm aims to minimize the following function:

$$\sum_{i=1}^{c} \sum_{\boldsymbol{y}_j \in \mathcal{Y}_i} (\boldsymbol{y}_j - \boldsymbol{g}_i)^T (\boldsymbol{y}_j - \boldsymbol{g}_i)$$

$$= \sum_{i=1}^{n} (\boldsymbol{y}_i - \boldsymbol{G}\boldsymbol{u}_i^T)^T (\boldsymbol{y}_i - \boldsymbol{G}\boldsymbol{u}_i^T),$$

$$(4)$$

where $\boldsymbol{y}_i = \boldsymbol{W}^T \boldsymbol{x}_i$. Note that K-means is used to assign cluster labels, which are used as pseudo labels, to minimize the within-class scatter after the linear transformation by $\boldsymbol{W}$. Then, we can substitute (4) into (3):

$$\max_{\boldsymbol{W}} \sum_{i=1}^{n} (\boldsymbol{W}^T \boldsymbol{x}_i)^T (\boldsymbol{W}^T \boldsymbol{x}_i) - \alpha \sum_{i=1}^{n} (\boldsymbol{W}^T \boldsymbol{x}_i - \boldsymbol{G} \boldsymbol{u}_i^T)^T (\boldsymbol{W}^T \boldsymbol{x}_i - \boldsymbol{G} \boldsymbol{u}_i^T) - \beta \Omega(\boldsymbol{W})$$
$$\text{s.t.} \quad \boldsymbol{W}^T \boldsymbol{W} = \boldsymbol{I},$$

$$(5)$$

As mentioned above, the sparsity-based regularization term has been widely used to find out correlated structures among features. The motivation behind this is to exploit sparse structures of the feature coefficient matrix. By imposing the sparse constraint, some of the rows of the feature coefficient matrix shrink to zeros. Those features corresponding to non-zero coefficients are selected as the distinctive subset of features. In this way, noisy and redundant features can be removed. This sparsity-based regularization has been applied in various problems. Inspired by the *"shrinking to zero"* idea, we utilize a sparsity model to uncover the common structures shared by features. To achieve that goal, we propose to minimize the $\ell_{2,p}$-norm of the coefficient matrix, $\|\boldsymbol{W}\|_{2,p}, (0 < p < 2)$. From the definition of $\|\boldsymbol{W}\|_{2,p}$ in (1), outliers or negative impact of the irrelevant $\boldsymbol{w}^i$'s are suppressed by minimizing the $\ell_{2,p}$-norm. Note that p is a parameter that controls the degree of correlated structures among features. The lower p is, the more shared structures among are expected to exploit. After a number of optimization steps, the optimal feature coefficient matrix, $\boldsymbol{W}$, can be obtained. Thus, we impose the $\ell_{2,p}$-norm on the regularization term and re-write the objective function in a matrix representation as follows:

$$\max_{\boldsymbol{W},\boldsymbol{G},\boldsymbol{U}} Tr(\boldsymbol{W}^T \boldsymbol{S}_t \boldsymbol{W}) - \alpha \|\boldsymbol{W}^T \boldsymbol{X} - \boldsymbol{G} \boldsymbol{U}^T\|_F^2 - \beta \|\boldsymbol{W}\|_{2,p}$$
$$\text{s.t.} \quad \boldsymbol{W}^T \boldsymbol{W} = \boldsymbol{I},$$

$$(6)$$

where $\boldsymbol{U}$ is an indicator matrix. $Tr(\cdot)$ is trace operator, while $\|\cdot\|_F^2$ is the Frobenius norm of a matrix. Our proposed method integrates the Maximum Margin Criterion and sparse regularization into a joint framework. Embedding K-means into the framework not only minimizes the distances between within-class data while maximizing total data separability, but also provides cluster labels. The cluster centroids generated by K-means can further guide the sparse structure learning on the feature coefficient matrix in each iterative step of our solution, which will be explained in the next section. We name this method for the unsupervised feature analysis with class margin optimization as **UFCM**.

4   Optimization

In this section, we present our solution to the objective function in (6). Since the $\ell_{2,p}$-norm is used to exploit sparse structures, the objective function cannot be solved in a closed form. Meanwhile, the objective function is not jointly convex with respect to three variables, i.e. $\boldsymbol{W}, \boldsymbol{G}, \boldsymbol{U}$. Thus, we propose to solve the problem as follows.

We define a diagonal matrix $\boldsymbol{D}$ whose diagonal entries are defined as:

$$\boldsymbol{D}^{ii} = \frac{1}{\frac{2}{p} \|\boldsymbol{w}^i\|_2^{2-p}}.$$

$$(7)$$

The objective function in (6) is equivalent to:

$$\max_{W,G,U} Tr(W^T S_t W) - \alpha \| W^T X - G U^T \|_F^2 - \beta Tr(W^T D W)$$
$$\text{s.t.} \quad W^T W = I \tag{8}$$

We propose to optimize the objective function in two steps in each iteration as follows:

(1) Fix W, G and optimize U:

When W is fixed, the first and third terms can be viewed as constants. While the second term can be viewed as the objective function of K-means, assigning cluster labels to each data. Also, the cluster centroid matrix $G = [g_1, \ldots, g_c]$ is also fixed, the optimal U is:

$$U_{ij} = \begin{cases} 1, & j = \underset{k}{\arg\min} \| W^T x_i - g_k \|_F^2, \\ 0, & \text{Otherwise.} \end{cases} \tag{9}$$

This is equivalent to perform K-means on the transformed data, $W^T X$, which means the solution is unique.

(2) Fix U and optimize W, G:

After fixing the indicator matrix, U, we set the derivative of Equation (8) with respect to G equal to 0:

$$-\alpha \frac{\partial Tr(W^T X - G U^T)^T (W^T X - G U^T)}{\partial G} = 0$$
$$\Rightarrow -2\alpha W^T X U + 2\alpha G U^T U = 0 \tag{10}$$
$$\Rightarrow G = W^T X U (U^T U)^{-1}$$

Substituting Equation (10) into Equation (8), we have:

$$Tr(W^T S_t W) - \alpha \| W^T X - W^T X U (U^T U)^{-1} U^T \|_F^2 - \beta Tr(W^T D W)$$
$$= \alpha Tr \left((W^T X U (U^T U)^{-1} U^T - W^T X)(W^T X - W^T X U (U^T U)^{-1} U^T)^T \right)$$
$$+ Tr(W^T S_t W) - \beta Tr(W^T D W)$$
$$= \alpha Tr \left(W^T X U (U^T U)^{-1} U^T X^T W - W^T X X^T W \right)$$
$$+ Tr(W^T S_t W) - \beta Tr(W^T D W)$$
$$= Tr[W^T (S_t + \alpha X U (U^T U)^{-1} U^T X^T - \alpha X X^T - \beta D) W] \tag{11}$$

Thus, the objective function becomes:

$$\max_W Tr[W^T (S_t + \alpha X U (U^T U)^{-1} U^T X^T - \alpha X X^T - \beta D) W]$$
$$\text{s.t.} \quad W^T W = I \tag{12}$$

The objective function can be then solved by performing eigen-decomposition of the following formula:

$$S_t + \alpha X U (U^T U)^{-1} U^T X^T - \alpha X X^T - \beta D \tag{13}$$

Algorithm 1. Unsupervised Feature Analysis with Class Margin Optimization.

Input: Data matrix $\boldsymbol{X} = [\boldsymbol{x}_1, \ldots, \boldsymbol{x}_n] \in \mathbb{R}^{d \times n}$ and parameters α and β.
Output: Feature coefficient matrix $\boldsymbol{W}$ and cluster indicator matrix $\boldsymbol{U}$.
1: Initialize $\boldsymbol{W}$ by PCA on $\boldsymbol{X}$;
2: Initialize $\boldsymbol{U}$ by K-means on $\boldsymbol{W}^T \boldsymbol{X}$;
3: **repeat**
4:     Compute $\boldsymbol{D}$ according to (7);
5:     Update $\boldsymbol{U}$ according to (14);
6:     Update $\boldsymbol{W}$ by eigen-decomposition of (13);
7:     Update $\boldsymbol{G}$ according to (10);
8: **until** Convergence

The optimal $\boldsymbol{W}$ can be determined by choosing d' eigenvectors corresponding to d' largest eigenvalues, $d' \leq d$. Our proposed method can be solved by above steps in an iterative algorithm. Each step can obtain the corresponding optimum. As the cluster indicator matrix $\boldsymbol{U}$ is initialized by K-means, the performance of our algorithm is determined by the initialization of K-means. To alleviate the local optimum problem, an update strategy for $\boldsymbol{U}$ is demanded. Generally speaking, we randomly initialize $\boldsymbol{U}$ a number of times and make comparisons according to the second term in Equation (6). Then we choose how to update the indicator matrix. Specifically, the optimal $\boldsymbol{U}_i^*$ and $\boldsymbol{W}_i^*$ has been derived in the i-th iteration. In the $(i + 1)$-th iteration, we first randomly initialize $\boldsymbol{U}$ r times ($r = 10$ in our experiment) and combine the derived $\boldsymbol{U}_i^*$ in the i-th iteration as an updating candidate set: $\tilde{\boldsymbol{U}}_{i+1} = [\boldsymbol{U}_{i+1}^0, \boldsymbol{U}_{i+1}^1, \ldots, \boldsymbol{U}_{i+1}^r]$, $\boldsymbol{U}_{i+1}^0 = \boldsymbol{U}_i^*$. According to $\|\boldsymbol{W}^T\boldsymbol{X} - \boldsymbol{G}\boldsymbol{U}^T\|_F^2$, the candidate, which yields the smallest value, is chosen to update $\boldsymbol{U}_{i+1}^*$:

$$\boldsymbol{U}_{i+1}^* = \tilde{\boldsymbol{U}}_{i+1}^j, \qquad j = \operatorname*{argmin}_j \|\boldsymbol{W}^T\boldsymbol{X} - \boldsymbol{G}(\tilde{\boldsymbol{U}}_{i+1}^j)^T\|_F^2 \qquad (14)$$

where j is the index of candidate set, $j = 0, 1, \ldots, r$. In this way, we compare the derived cluster indicator matrix with r randomly initialized counterparts to alleviate the local optimum problem. We summarize the solution in Algorithm 1 which outputs the learned feature coefficient matrix $\boldsymbol{W}$ to select distinctive features.

From Algorithm 1, it can be seen that the most computational operation is the eigen-decomposition in Equation (13). The computational complexity is $O(d^3)$. If the dimensionality of the data, d, is very high, dimensionality reduction is desirable. To analyze the convergence of our proposed method, the following proposition and its proof are given.

Proposition 1. *Algorithm 1 monotonically increases the objective function in Equation (6) until convergence.*

Proof. Assuming that, in the i-th iteration, the transformation matrix $\boldsymbol{W}$ and cluster centroid matrix $\boldsymbol{G}$ have been derived as $\boldsymbol{W}_i$ and $\boldsymbol{G}_i$. In the $(i + 1)$-th iteration step, we use $\boldsymbol{W}_i$ and $\boldsymbol{G}_i$ to update $\boldsymbol{U}_{i+1}$ according to the updating strategy in (14). We can have the following inequality:

$$\begin{aligned}
&Tr(\boldsymbol{W}_i^T \boldsymbol{S}_t \boldsymbol{W}_i) - \alpha \|\boldsymbol{W}_i^T \boldsymbol{X} - \boldsymbol{G}_i \boldsymbol{U}_i^T\|_F^2 - \beta \|\boldsymbol{W}_i\|_{2,p} \\
&\leq Tr(\boldsymbol{W}_i^T \boldsymbol{S}_t \boldsymbol{W}_i) - \alpha \|\boldsymbol{W}_i^T \boldsymbol{X} - \boldsymbol{G}_i \boldsymbol{U}_{i+1}^T\|_F^2 - \beta \|\boldsymbol{W}_i\|_{2,p}
\end{aligned} \tag{15}$$

Similarly, when $\boldsymbol{U}_{i+1}$ is fixed to optimize $\boldsymbol{W}$ and $\boldsymbol{G}$ in the $(i+1)$-th iteration, the following inequality can be obtained according to Equation (12):

$$\begin{aligned}
&Tr(\boldsymbol{W}_i^T \boldsymbol{S}_t \boldsymbol{W}_i) - \alpha \|\boldsymbol{W}_i^T \boldsymbol{X} - \boldsymbol{G}_i \boldsymbol{U}_{i+1}^T\|_F^2 - \beta \|\boldsymbol{W}_i\|_{2,p} \\
&\leq Tr(\boldsymbol{W}_{i+1}^T \boldsymbol{S}_t \boldsymbol{W}_{i+1}) - \alpha \|\boldsymbol{W}_{i+1}^T \boldsymbol{X} - \boldsymbol{G}_{i+1} \boldsymbol{U}_{i+1}^T\|_F^2 - \beta \|\boldsymbol{W}_{i+1}\|_{2,p}
\end{aligned} \tag{16}$$

After combining Equation (15) and (16) together, it indicates that the proposed algorithm will monotonically increase the objective function in each iteration. It is worth noting that the algorithm is alleviating the local optimum problem raised by random initializations of K-means, rather than completely solving it. However, our algorithm can avoid to rapidly converge to a local optimum and may converge to the global optimal solution.

5 Experiments

In this section, experimental results will be presented together with related analysis. We compare our method with seven approaches over six benchmark datasets. Besides, we also conduct experiments to evaluate performance variations in different aspects. They are including the impact of different selected feature numbers, the validation of feature correlation analysis, and parameter sensitivity analysis. Lastly, the convergence demonstration is shown.

5.1 Experiment Setup

In the experiments, we have compared our method with seven approaches as follows:

- **All Features:** All original variables are preserved as the baseline in the experiments.
- **Max Variance:** Features are ranked according to the variance magnitude of each feature in a descending order. The highest ranked features are selected.
- **Spectral Feature Selection (SPEC) [30]:** This method employs a unified framework to select features one by one based on spectral graph theory.
- **Multi-Cluster Feature Selection (MCFS) [1]:** This unsupervised approach selects those features who make the multi-cluster structure of the data preserved best. Features are selected using spectral regression with the ℓ_1-norm regularization.
- **Robust Unsupervised Feature Selection (RUFS) [19]:** RUFS jointly performs robust label learning and robust feature learning. To achieve this, robust orthogonal nonnegative matrix factorization is applied to learn labels while the $\ell_{2,1}$-norm minimization is simultaneously utilized to learn the features.

- **Nonnegative Discriminative Feature Selection (NDFS)** [14]: NDFS exploits local discriminative information and feature correlations simultaneously. Besides, the manifold structure information is also considered jointly.
- **Laplacian Score (LapScore)** [8]: This method learns and selects distinctive features by evaluating their powers of locality preserving, which is also called Laplacian Score.

All the parameters (if any) are tuned in the range of $\{10^{-3}, 10^{-1}, 10^1, 10^3\}$ for each algorithm mentioned above and the best results are reported. The size of the neighborhood is set to 5 for any algorithm based on spectral clustering. The number of random initializations required in the update strategy in (14), is set at 10 in the experiment. To measure the performance, two metrics have been used: *Clustering Accuracy (ACC)* and *Normalized Mutual Information (NMI)*.

For a data point x_i, its ground truth label is denoted as p_i and its clustering label that is produced from a clustering algorithm, is represented as q_i. Then, ACC metric over a data set with n data points is defined as follows:

$$ACC = \frac{\sum_{i=1}^{n} \delta(p_i, map(q_i))}{n}, \tag{17}$$

where $\delta(x, y) = 1$ if $x = y$ and $\delta(x, y) = 0$ otherwise. $map(x)$ is the *best mapping function* which permutes clustering labels to match the ground truth labels using the Kuhn-Munkres algorithm. A larger ACC means better performance.

According to the definition in [22], *NMI* is defined as:

$$NMI = \frac{\sum_{l=1}^{c} \sum_{h=1}^{c} t_{l,h} log(\frac{n \times t_{l,h}}{t_l t_h})}{\sqrt{(\sum_{l=1}^{c} t_l log \frac{t_l}{n})(\sum_{h=1}^{c} \tilde{t}_h log \frac{\tilde{t}_h}{n})}}, \tag{18}$$

where t_l is the number of data points in the l-th cluster, $1 \leq l \leq c$, which is generated by a clustering algorithm. While $\tilde{t}_h$ denotes the number of data points in the h-th ground truth cluster. $t_{l,h}$ is the number of data points which are in the intersection of the l-th and h-th clusters. Similarly, a larger *NMI* means better performance.

The performance evaluations are performed over six benchmark datasets as follows:

- **COIL20** [16]: It contains 1,440 gray-scale images of 20 objects (72 images per object) under various poses. The objects are rotated through 360 degrees and taken at the interval of 5 degrees.
- **MNIST** [13]: It is a large-scale dataset of handwritten digits, which has been widely used as a test bed in data mining. The dataset contains 60,000 training images and 10,000 testing images. In this paper, we use its subclass version, MNIST-S, in which one handwritten digit image per ten images, for each class, is randomly sampled from the MNIST database. There are 6,996 handwritten images with a resolution of 28×28.
- **ORL** [21]: This data set which is used as a benchmark for face recognition, consists of 40 different subjects with 10 images each. We also resize each image to 32×32.

Table 1. Summary of data sets.

	COIL20	MNIST	ORL	UMIST	USPS	YaleB
Number of data	1,440	6,996	400	564	9,298	2,414
Number of classes	20	10	40	20	10	38
Feature dimensions	1,024	784	1,024	644	256	1,024

- **UMIST:** UMIST, which is also known as the Sheffield Face Database, consists of 564 images of 20 individuals. Each individual is shown in a variety of poses from profile to frontal views.
- **USPS** [10]: This dataset collects 9,298 images of handwritten digits (0-9) from envelops by the U.S. Postal Service. All images have been normalized to the same size of 16 × 16 pixels in gray scale.
- **YaleB** [7]: It consists of 2,414 frontal face images of 38 subjects. Different lighting conditions have been considered in this dataset. All images are reshaped into 32 × 32 pixels.

The pixel value in data is used as the feature. Details of data sets that are used in this paper are summarized in Table 1.

5.2   Experimental Results

To compare the performance of our proposed algorithm with others, we repeatedly perform the test five times and report the average performance results (ACC and NMI) with standard deviations in Tables 2 and 3. It is observed that our proposed method consistently achieves better performance than all other compared approaches across all the data sets. Besides, it is worth noting that our method is superior to those state-of-the-art counterparts that rely on a graph Laplacian (SPEC, RUFS, NDFS, LapScore).

We study how the number of selected features can affect the performance by conducting an experiment whose results are shown in Figure 1. From the figure, performance variations with respect to the number of selected features using the proposed algorithm over three data sets, including COIL20, MNIST, and USPS, have been illustrated. We only adopt ACC as the metric. Some observations can be obtained: 1) When the number of selected features is small, e.g. 500 on each data set, the accuracy value is relatively small. 2) With the increase of selected features, performance can peak at a certain point. For example, the performance of our algorithm peaks at 0.7475 on COIL20 when the number of selected features increases to 800. Similarly, 0.6392 (800 selected features) and

Table 2. Performance comparison ($ACC \pm STD$).

	COIL20	MNIST	ORL	UMIST	USPS	YaleB
AllFea	0.7051 ± 0.0294	0.6009 ± 0.0063	0.6675 ± 0.0112	0.4800 ± 0.0115	0.7139 ± 0.0272	0.1261 ± 0.0025
MaxVar	0.7124 ± 0.0191	0.6239 ± 0.0100	0.6965 ± 0.0121	0.4984 ± 0.0141	0.7165 ± 0.0186	0.1291 ± 0.0042
SPEC	0.7105 ± 0.0116	0.6254 ± 0.0024	0.6645 ± 0.0065	0.4824 ± 0.0077	0.7037 ± 0.0315	0.1307 ± 0.0049
MCFS	0.7355 ± 0.0050	0.6299 ± 0.0037	0.7055 ± 0.0048	0.5239 ± 0.0038	0.7634 ± 0.0138	0.1355 ± 0.0043
RUFS	0.7365 ± 0.0024	0.6294 ± 0.0028	0.6920 ± 0.0033	0.5110 ± 0.0091	0.7659 ± 0.0076	0.1795 ± 0.0032
NDFS	0.7368 ± 0.0074	0.6291 ± 0.0016	0.7050 ± 0.0031	0.5243 ± 0.0028	0.7630 ± 0.0124	0.1315 ± 0.0034
LapScore	0.7126 ± 0.0249	0.6214 ± 0.0054	0.7100 ± 0.0117	0.5092 ± 0.0062	0.7089 ± 0.0324	0.1255 ± 0.0025
Ours	**0.7475±0.0076**	**0.6392 ± 0.0056**	**0.7210 ± 0.0052**	**0.5343 ± 0.0062**	**0.7813 ± 0.007**	**0.1886 ± 0.0043**

Table 3. Performance comparison (*NMI±STD*).

	COIL20	MNIST	ORL	UMIST	USPS	YaleB
AllFea	0.7884 ± 0.0157	0.5162 ± 0.0027	0.8265 ± 0.0129	0.6715 ± 0.0069	0.6305 ± 0.0029	0.1968 ± 0.0017
MaxVar	0.7932 ± 0.0071	0.5314 ± 0.0063	0.8424 ± 0.0085	0.6825 ± 0.0063	0.6361 ± 0.0021	0.2123 ± 0.0040
SPEC	0.7866 ± 0.0061	0.5367 ± 0.0035	0.8232 ± 0.0021	0.6753 ± 0.0114	0.6215 ± 0.0073	0.2071 ± 0.0027
MCFS	0.8066 ± 0.0025	0.5367 ± 0.0003	0.8460 ± 0.0025	0.7005 ± 0.0053	0.6419 ± 0.0015	0.2024 ± 0.0033
RUFS	0.8045 ± 0.0025	0.5374 ± 0.0021	0.8430 ± 0.0044	0.6898 ± 0.0035	0.6468 ± 0.0027	0.2845 ± 0.0040
NDFS	0.8062 ± 0.0058	0.5376 ± 0.0004	0.8458 ± 0.0026	0.6981 ± 0.0054	0.6452 ± 0.0054	0.2048 ± 0.0041
LapScore	0.7920 ± 0.0101	0.5308 ± 0.0065	0.8421 ± 0.0006	0.6924 ± 0.0027	0.6291 ± 0.0047	0.1945 ± 0.0018
Ours	**0.8119 ± 0.0035**	**0.5422 ± 0.0018**	**0.8518 ± 0.0027**	**0.7112 ± 0.0033**	**0.6535 ± 0.0022**	**0.2959 ± 0.0043**

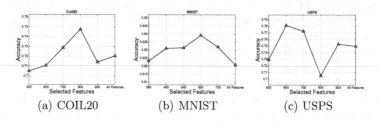

(a) COIL20   (b) MNIST   (c) USPS

Fig. 1. Performance variation results with respect to the number of selected features using the proposed algorithm over three data sets, COIL20, MNIST, and USPS.

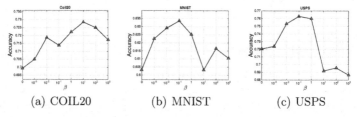

(a) COIL20   (b) MNIST   (c) USPS

Fig. 2. Performance variation results with respect to different values of regularization parameter, βs, over three data sets, COIL20, MNIST, and USPS.

0.7813 (600 selected features) are observed on MNIST and USPS, respectively. 3) When all features are in use, the performance is worse than the best. Similar trends can be also observed on the other data sets. It is concluded that our algorithm can select distinctive features.

To demonstrate exploiting feature correlation is beneficial to the performance, we conduct an experiment in which parameters α and p are both fixed at 1. β varies in a range of $[0, 10^{-3}, 10^{-2}, 10^{-1}, 1, 10^1, 10^2, 10^3]$. The performance variation results with respect to different βs are plotted in Figure 2. The experiment is conducted over three data sets, i.e. COIL20, MNIST, and USPS. From the results, we can observe that the performance is relatively low, when there is no correlation exploiting in the framework, i.e. $\beta = 0$. The performance always peaks at a certain point when a proper degree of sparsity is imposed to the regularization term. For example, the performance is only 0.6993 when $\beta = 0$ on COIL20. The performance increases to 0.7285 when $\beta = 10^1$. Similar observations are also obtained on the other data sets. We can conclude that sparse structure learning on feature coefficient matrix contributes to the performance of our unsupervised feature selection method.

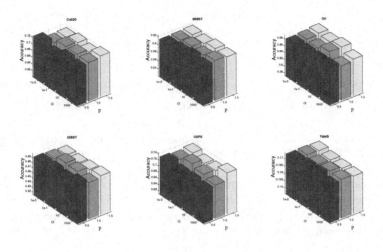

Fig. 3. Performance variation results under different combinations of αs and ps. β is fixed at 10^{-1}.

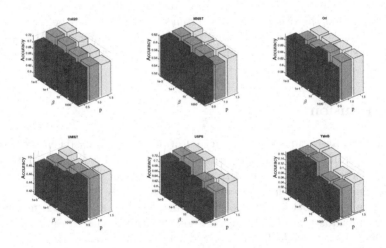

Fig. 4. Performance variation results under different combinations of βs and ps. α is fixed at 10^{-1}.

5.3   Studies on Parameter Sensitivity and Convergence

There are three parameters in our algorithms, which are denoted as α, β and p in (6). α and β are two regularization parameters while p controls the degree of sparsity. To investigate the sensitivity of the parameters, we conduct an experiment to study

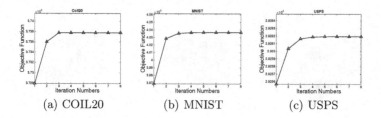

(a) COIL20	(b) MNIST	(c) USPS

Fig. 5. Objective function values of our proposed objective function in (6) over three data sets, COIL20, MNIST, and USPS.

how they exert influences on performance. Firstly, we fix $\beta = 10^{-1}$ and derive the performance variations under different combinations of αs and ps in Figure 3. Secondly, α is fixed at 10^{-1}. The performance variation results with respect to different βs and ps are shown in Figure 4. Both α and β vary in a range of $[10^{-3}, 10^{-1}, 10^{1}, 10^{1}]$. While p changes in $[0.5, 1.0, 1.5]$. We only take ACC as the metric.

To validate that our algorithm will monotonically increase the objective function value in (6), we conduct an experiment to demonstrate this fact. In this experiment, all parameters (α, β, and p) in (6) are fixed at 1. The objective function values and corresponding iteration numbers are drawn in Figure 5. We take COIL20, MNIST, and USPS as examples. Similar observations can be also obtained on the other data sets. From the figure, it can be seen that our algorithm converges to the optimum, usually within eight iteration steps, over three data sets. We can then conclude that the proposed method is efficient and effective.

6   Conclusion

In this paper, an unsupervised feature selection approach has been proposed by using the Maximum Margin Criterion and the sparsity-based model. More specifically, the proposed method seeks to maximize the total scatter on one hand. On the other hand, the within-class scatter is simultaneously considered to minimize. Since there is no label information in an unsupervised scenario, K-means clustering is embedded into the framework jointly. Advantages can be summarized as twofold: First, pseudo labels generated by K-means clustering is beneficial to maximizing class margins in each iteration step. Second, pseudo labels can guide the sparsity-based model to exploit sparse structures of the feature coefficient matrix. Noisy and uncorrelated features can be therefore removed. Since the objective function is non-convex for all variables, we have proposed an algorithm with a guaranteed convergence property. To avoid to rapidly converge to a local optimum which is caused by K-means, we applied an updating strategy to alleviate the problem. In this way, our proposed method might converge to the global optimum. Extensive experimental results have shown that our method has superior performance against all other compared approaches over six benchmark data sets.

Acknowledgments. This work was supported by Australian Research Council Discovery Project (DP140100104). Any opinions, findings, and conclusions or recommendations expressed in this material are those of the author(s) and do not necessarily reflect the views of the Australian Research Council.

References

1. Cai, D., Zhang, C., He, X.: Unsupervised feature selection for multi-cluster data. In: SIGKDD (2010)
2. Chang, X., Nie, F., Wang, S., Yang, Y.: Compound rank-k projections for bilinear analysis. IEEE Trans. Neural Netw. Learning Syst. (2015)
3. Chang, X., Nie, F., Yang, Y., Huang, H.: A convex formulation for semi-supervised multi-label feature selection. In: AAAI (2014)
4. Chang, X., Shen, H., Wang, S., Liu, J., Li, X.: Semi-supervised feature analysis for multimedia annotation by mining label correlation. In: Tseng, V.S., Ho, T.B., Zhou, Z.-H., Chen, A.L.P., Kao, H.-Y. (eds.) PAKDD 2014, Part II. LNCS, vol. 8444, pp. 74–85. Springer, Heidelberg (2014)
5. Chang, X., Yang, Y., Hauptmann, A.G., Xing, E.P., Yu, Y.: Semantic concept discovery for large-scale zero-shot event detection. In: IJCAI (2015)
6. Duda, R.O., Hart, P.E., Stork, D.G.: Pattern classification. John Wiley & Sons (2012)
7. Georghiades, A.S., Belhumeur, P.N., Kriegman, D.: From few to many: Illumination cone models for face recognition under variable lighting and pose. IEEE Transactions on Pattern Analysis and Machine Intelligence (TPAMI) **23**(6), 643–660 (2001)
8. He, X., Cai, D., Niyogi, P.: Laplacian score for feature selection. In: NIPS (2005)
9. Hou, C., Nie, F., Li, X., Yi, D., Wu, Y.: Joint embedding learning and sparse regression: A framework for unsupervised feature selection. IEEE T. Cybernetics **44**(6), 793–804 (2014)
10. Hull, J.J.: A database for handwritten text recognition research. IEEE Transactions on Pattern Analysis and Machine Intelligence (TPAMI) **16**(5), 550–554 (1994)
11. Kira, K., Rendell, L.A.: A practical approach to feature selection. In: IWML (1992)
12. Kononenko, I.: Estimating attributes: analysis and extensions of relief. In: Bergadano, F., De Raedt, L. (eds.) ECML 1994. LNCS, vol. 784, pp. 171–182. Springer, Heidelberg (1994)
13. LeCun, Y., Bottou, L., Bengio, Y., Haffner, P.: Gradient-based learning applied to document recognition. Proceedings of the IEEE **86**(11), 2278–2324 (1998)
14. Li, Z., Yang, Y., Liu, J., Zhou, X., Lu, H.: Unsupervised feature selection using nonnegative spectral analysis. In: AAAI (2012)
15. Maugis, C., Celeux, G., Martin-Magniette, M.L.: Variable selection for clustering with gaussian mixture models. Biometrics **65**(3), 701–709 (2009)
16. Nene, S.A., Nayar, S.K., Murase, H., et al.: Columbia object image library (coil-20). Tech. rep., Technical Report CUCS-005-96 (1996)
17. Nie, F., Huang, H., Cai, X., Ding, C.H.Q.: Efficient and robust feature selection via joint l2, 1-norms minimization. In: NIPS (2010)
18. Nie, F., Huang, H., Cai, X., Ding, C.H.: Efficient and robust feature selection via joint $\ell 2$, 1-norms minimization. In: NIPS, pp. 1813–1821 (2010)
19. Qian, M., Zhai, C.: Robust unsupervised feature selection. In: IJCAI (2013)

20. Raileanu, L.E., Stoffel, K.: Theoretical comparison between the gini index and information gain criteria. AMAI (2004)
21. Samaria, F.S., Harter, A.C.: Parameterisation of a stochastic model for human face identification. In: IEEE Workshop on Applications of Computer Vision (1994)
22. Strehl, A., Ghosh, J.: Cluster ensembles–a knowledge reuse framework for combining multiple partitions. Journal of Machine Learning Research (JMLR) **3**, 583–617 (2003)
23. Tabakhi, S., Moradi, P., Akhlaghian, F.: An unsupervised feature selection algorithm based on ant colony optimization. Engineering Applications of Artificial Intelligence **32**, 112–123 (2014)
24. Wang, D., Nie, F., Huang, H.: Unsupervised feature selection via unified trace ratio formulation and K-means clustering (TRACK). In: Calders, T., Esposito, F., Hüllermeier, E., Meo, R. (eds.) ECML PKDD 2014, Part III. LNCS, vol. 8726, pp. 306–321. Springer, Heidelberg (2014)
25. Wang, S., Chang, X., Li, X., Sheng, Q.Z., Chen, W.: Multi-task support vector machines for feature selection with shared knowledge discovery. Signal Processing (December 2014)
26. Wang, S., Tang, J., Liu, H.: Embedded unsupervised feature selection. AAAI (2015)
27. Yang, Y., Shen, H.T., Ma, Z., Huang, Z., Zhou, X.: l2, 1-norm regularized discriminative feature selection for unsupervised learning. In: IJCAI (2011)
28. Yang, Y., Zhuang, Y., Wu, F., Pan, Y.: Harmonizing hierarchical manifolds for multimedia document semantics understanding and cross-media retrieval. IEEE Transactions on Multimedia **10**(3), 437–446 (2008)
29. Yang, Y., Ma, Z., Hauptmann, A.G., Sebe, N.: Feature Selection for Multimedia Analysis by Sharing Information Among Multiple Tasks. IEEE TMM **15**(3), 661–669 (2013)
30. Zhao, Z., Liu, H.: Spectral feature selection for supervised and unsupervised learning. In: ICML
31. Zhu, X., Huang, Z., Yang, Y., Shen, H.T., Xu, C., Luo, J.: Self-taught dimensionality reduction on the high-dimensional small-sized data. Pattern Recognition **46**(1), 215–229 (2013)
32. Zhu, X., Suk, H.-I., Shen, D.: Matrix-similarity based loss function and feature selection for alzheimer's disease diagnosis. In: IEEE CVPR, pp. 3089–3096 (2014)
33. Zhu, X., Suk, H.-I., Shen, D.: Discriminative feature selection for multi-class alzheimer's disease classification. In: MLMI, pp. 157–164 (2014)

Data Streams and Online Learning

Ageing-Based Multinomial Naive Bayes Classifiers Over Opinionated Data Streams

Sebastian Wagner[1], Max Zimmermann[2], Eirini Ntoutsi[1]([✉]),
and Myra Spiliopoulou[2]

[1] Ludwig-Maximilians University of Munich (LMU), Munich, Germany
sebastian.t.wagner@campus.lmu.de, ntoutsi@dbs.ifi.lmu.de
[2] Otto-von-Guericke-University Magdeburg, Magdeburg, Germany
{max.zimmermann,myra}@iti.cs.uni-magdeburg.de

Abstract. The long-term analysis of opinionated streams requires algorithms that predict the polarity of opinionated documents, while adapting to different forms of concept drift: the class distribution may change but also the vocabulary used by the document authors may change. One of the key properties of a stream classifier is adaptation to concept drifts and shifts; this is typically achieved through ageing of the data. Surprisingly, for one of the most popular classifiers, Multinomial Naive Bayes (MNB), no ageing has been considered thus far. MNB is particularly appropriate for opinionated streams, because it allows the seamless adjustment of word probabilities, as new words appear for the first time. However, to adapt properly to drift, MNB must also be extended to take the age of documents and words into account.

In this study, we incorporate ageing into the learning process of MNB, by introducing the notion of *fading* for words, on the basis of the recency of the documents containing them. We propose two fading versions, gradual fading and aggressive fading, of which the latter discards old data at a faster pace. Our experiments with Twitter data show that the ageing based MNBs outperform the standard accumulative MNB approach and manage to recover very fast in times of change. We experiment with different data granularities in the stream and different data ageing degrees and we show how they "work together" towards adaptation to change.

1 Introduction

Nowadays, we experience an increasing interest on word-of-mouth communication in social media, including opinion sharing [16]. A vast amount of voluntary and bona fide feedback accumulates, referring to products, persons, events etc. Opinionated information is valuable for consumers, who benefit from the experiences of other consumers, in order to make better buying decisions [13], but also for vendors, who can get insights on what customers like and dislike [18]. The extraction of such insights requires a proper analysis of the opinionated data.

In this work, we address the issue of polarity learning over opinionated streams. The accumulating opinionated documents are subject to different forms of drift: the subjects discussed change, the attitude of people towards specific

© Springer International Publishing Switzerland 2015
A. Appice et al. (Eds.): ECML PKDD 2015, Part I, LNAI 9284, pp. 401–416, 2015.
DOI: 10.1007/978-3-319-23528-8_25

products, events or other forms of entities change, the vocabulary changes. As a matter of fact, the impact of the vocabulary is less investigated. It is well-known that the polarity of some words depends on the context they are used in; this subject is investigated e.g. in the context of recurring concepts [8]. However, the impact of the vocabulary in an opinionated stream is much more broad: an opinionated word can appear in more contexts than we can (or want to) trace, since some contexts are rare or do not recur. More importantly, new words emerge and some words are used less. This implies that the polarity learner should be able to cope with an evolving feature space. To deal with concept drift in the opinionated data and their feature space, we propose a *fading* Multinomial Naive Bayes polarity learner. We extend the Multinomial Naive Bayes (MNB) with an ageing function that gradually forgets (fades away) old data and outdated words.

Multinomial Naive Bayes (MNB) classifiers comprise one of the most well-known classifiers and are widely used also for sentiment analysis although most of the approaches cover the non-stream case [17]. For opinionated streams, MNB has a *cardinal advantage*: it allows the seamless adaptation of the vocabulary, by simply requiring the computation of the class probabilities for each word. No other stream classification algorithm can respond so intuitively to an evolving feature space. Surprisingly, so far, MNB is used in opinionated streams mostly without forgetting past data and without extending to a new vocabulary [1]. This is problematic, because the model may overfit to old data and, fore-mostly, to an outdated vocabulary. In this study, we propose two different forgetting mechanisms that differ on how drastically they forget over time.

The rest of the paper is organized as follows: Related work is discussed in Section 2. The basic concepts and motivation are presented in Section 3. The ageing-based MNBs are introduced in Section 4. Experimental results are shown in Section 5. Conclusions and open issues are discussed in Section 6.

2   Related Work

Stream mining algorithms typically assume that the most recent data are the most informative, and thus employ different strategies to downgrade old, obsolete data. In a recent survey [6], Gama et al. discuss two forgetting mechanisms: (i) *abrupt forgetting* where only recent instances, within a sliding window, contribute to the model, and (ii) *gradual forgetting* where all instances contribute to the model but with a weight that is regulated by their age. In the context of our study, the forgetting strategy also affects the vocabulary – the feature space. In particular, if a set of documents is deleted (abrupt forgetting), all words that are in them but in none of the more recent documents are also removed from the feature space. This may harm the classifier, because such words may re-appear soon after their removal. Therefore, we opt for gradual forgetting.

Another approach for selecting features/ words for polarity classification in a dynamic environment is presented in [11]. In this approach, selection does not

---

[1] An exception is our own prior work [24,25].

rely on the age of the words but rather on their *usefulness*, defined as their contribution to the classification task. Usefulness is used in [11] as a selection criterion, when the data volume is high, but is also appropriate for streams with recurring concepts. Concept recurrence is studied e.g. in [8] (where meta-classifiers are trained on data referring to a given concept), and in [12] (where a concept is represented by a data bucket, and recurrence refers to similar buckets). Such methods can be beneficial for opinion stream classification, *if* all encountered opinionated words can be linked to a reasonably small number of concepts that do recur. In our study, we do not pose this requirement; word polarity can be assessed in our MNB model, without linking the words to concepts.

Multinomial Naive Bayes (MNB) [14] is a popular classifier due to its simplicity and good performance, despite its assumption on the class-conditional independence of the words [5,22]. Its simplicity and easy online maintenance constitutes it particularly appealing for data streams. As pointed out in Section 1, MNB is particularly appropriate for adaptation to an evolving vocabulary. In [24,25], we present functions that recompute the class probabilities of each word. In this study, we use different functions, as explained in the last paragraph of this section.

Bermingham et al.   [2] compared the performance of Support Vector Machines (SVM) and MNB classifiers on microblog data and reviews (not streams) and showed that MNB performs well on short-length, opinion-rich microblog messages (rather than on long texts). In [10], popular classification algorithms were studied such as MNBs, Random Forest, Bayesian Logistic Regression and SVMs using sequential minimal optimization for the classification in Twitter streams while building classifiers at different samples. Across the tested classifiers, MNBs showed the best performance for all applied data sets.

In [3], MNB has been compared to Stochastic Gradient Descend (SGD) and Hoeffding Trees for polarity classification on streams. Their MNB approach is incremental, i.e., it accumulates information on class appearances and word-in-class appearances over the stream, however, it did not forget anything. Their experiments showed that MNB had the largest difficulty in dealing with drifts in the stream population, although its performance in times of stability was very good. Regarding runtime, MNB was the fastest model due to its simplicity in predictions but also due to the easy incorporation of new instances in the model. The poor performance of MNB in times of change was also observed in [21], and triggered our ageing-based MNB approach.

Closest to our approach is our earlier work [24]. There, MNB is in the core of a polarity learner that uses two adaptation techniques: i) a forward adaptation technique that selects "useful" instances from the stream for model update and ii) a backward adaptation technique that downgrades the importance of old words from the model based on their age. There are two differences between that earlier method (and our methods that build upon it, e.g. [25]) and the work proposed here. First, the method in [24] is semi-supervised: after receiving an initial seed of labeled documents, it relies solely on the labels it derives from the learner. Backward adaptation is not performing well in that scenario, presumably because the importance of the words in the initial seed of documents diminishes over time. Furthermore, in [24], the ageing of the word-class counts is based directly upon the age of the original documents containing the words. The word-class counts are weighted locally, i.e., within the documents containing the words, and the final word-class counts are aggregations of these local scores. In our current work, we do not monitor the age of the documents. Rather, we use the words as first class objects, which age with time. Therefore the ageing of a word depends solely on the last time the word has been observed in some document from the stream.

3   Basic Concepts

We observe a stream $\mathcal{S}$ of opinionated documents arriving at distinct timepoints $t_0, \ldots, t_i, \ldots$; at each t_i a batch of documents might arrive. The definition of the batch depends on the application per se: i) one can define the batch at the instance level, i.e., a fixed number of instances is received at each timepoint or ii) at the temporal level, i.e., the batch consists of the instances arriving within each time period, e.g. on a daily basis if day is the considered temporal granularity. A document $d \in \mathcal{S}$ is represented by the *bag-of-words* model and for each word $w_i \in d$ its frequency f_i^d is also stored.

Our goal is to build a polarity classifier for the prediction of the polarity of new arriving documents. As it is typical in streams, the underlying population might undergo changes over time, referred in the literature as *concept drift*. The changes are caused by two reasons: i) change in the sentiment of existing words (for example, words have different sentiment for different contexts, e.g. the word "heavy" is negative for a camera, but positive for a solid wood piece of furniture); ii) new words might appear over time and old words might become obsolete (for example, new topics emerge all the time in the news and some topics are not mentioned anymore). The drift in the population might be gradual or drastic; the later is referred also as *concept shift* in the literature. A stream classifier should be able to adapt to drift while maintaining a good predictive power. Except for the *quality of predictions*, another important factor for a stream classifier is *fast adaptation* to the underlying evolving stream population.

3.1   Basic Model: Multinomial Naive Bayes

According to the underpinnings of the Multinomial Naive Bayes (MNB) [14], the probability of a document d belonging to a class c is given by:

$$P(c|d) = P(c) \prod_{i=1}^{|d|} P(w_i|c)^{f_i^d} \tag{1}$$

where $P(c)$ is the prior probability of class c, $P(w_i|c)$ is the conditional probability that word w_i belongs to class c and f_i^d is the number of occurrences of w_i in document d. These probabilities are typically estimated based on a dataset

$\mathcal{D}$ with class labels (training set); we indicate the estimates from now on by a "hat" as in $\hat{P}$.

The class prior $P(c)$ is easily estimated as the fraction of the set of training documents belonging to class c, i.e.,:

$$\hat{P}(c) = \frac{N_c}{|\mathcal{D}|} \tag{2}$$

where N_c is the number of documents in $\mathcal{D}$ belonging to class c and $|\mathcal{D}|$ is the total number of documents in $\mathcal{D}$.

The conditional probability $P(w_i|c)$ is estimated by the relative frequency of the word $w_i \in V$ in documents of class c:

$$\hat{P}(w_i|c) = \frac{N_{ic}}{\sum_{j=1}^{|V|} N_{jc}} \tag{3}$$

where N_{ic} is the number of occurrences of word w_i in documents with label c in $\mathcal{D}$, V is the vocabulary over the training set $\mathcal{D}$. For words that are unknown during prediction, i.e., not in V, we apply the Laplace correction and initialize their probability to $1/|V|$.

From the above formulas, it is clear that the quantities we need in order to estimate the class prior $\hat{P}(c)$ and the class conditional word estimates $\hat{P}(w_i|c)$ are the class prior counts N_c and the word-class counts N_{ic}, where $w_i \in V, c \in C$ are all computed from the training set $\mathcal{D}$. The conventional, static MNB uses the whole training set $\mathcal{D}$ at once to compute these counts.

The typical extension for data streams [3], updates the counts based on new instances from the stream. Let d be a new incoming document from the stream $\mathcal{S}$ with class label c. Updating the MNB model means actually updating the class and word-class counts based on the incoming document. In particular, the number of documents belonging to class c is increased, i.e.,: $N_c + = 1$. Similarly, the class-word counts for each word $w_i \in d$ are updated, i.e., $N_{ic} + = f_i^d$, where f_i^d is the frequency of w_i in d. For existing words in the model, this implies just an update of their counts. For so-far unknown words, this implies that a new entry is created for them in the model. These accumulative counts are used during polarity prediction in Equations 2 and 3. We refer to this model as *accumulativeMNB*.

From the above description it is clear that the typical MNB stream model exhibits a very *long memory* as nothing is forgotten with time, and therefore it cannot respond fast to changes. Next, we present our extensions to MNB: we weight documents and words on their age, ensuring that the model forgets and responds faster to change.

4   Ageing-Based Multinomial Naive Bayes

The reason for the poor performance of an MNB model in times of change is its accumulative nature. As already mentioned, once entering the model nothing is forgotten, neither words nor class prior information. To make MNBs adaptable to change, we introduce the notion of time (Section 4.1) and we show how such a "temporal" model can be used for polarity learning in a stream environment (Sections 4.2, 4.3).

4.1   Ageing-Based MNB Model

In the typical, accumulative MNB classifier there is no notion of time, rather all words are considered equally important independently of their arrival times. We couple the MNB with an ageing mechanism that allows for a differentiation of the words based on their last observation time in the stream.

Ageing is one of the core mechanism in data streams for dealing with concept drifts and shifts in the underlying population. Several ageing mechanisms have been proposed [7] including the landmark window model, the sliding window model and the damped window model. We opt for the damped window model, as already explained, as it comprises a natural choice for temporal applications and data streams [1,4,15]. According to the *damped window model*, the data are subject to ageing based on an ageing function so that recent data are considered more important than older ones. One of the most commonly used ageing functions is the *exponential fading function* that exponentially decays the weights of data instances with time.

Definition 1 (Ageing function). *Let d be a document arriving from the stream S at timepoint t_d. The weight of d at the current timepoint $t \geq t_d$ is given by: $age(d, t) = e^{-\lambda \cdot (t - t_d)}$ where $\lambda > 0$ is the decay rate.*

The weight of the document d decays over time based on the time period elapsing from the arrival of d and the decay factor λ. The higher the value of λ, the lower the impact of historical data comparing to recent data. The ageing factor λ is critical for the ageing process as it determines what is the contribution of old data to the model and how fast old data is forgotten. Another way of thinking of λ is by considering that $\frac{1}{\lambda}$ is the period for an instance to loose half of its original weight. For example, if $\lambda = 0.5$ and timestamps correspond to days, this means that $\frac{1}{0.5} = 2$ days after its observation an instance will loose 50% of its weight. For $\lambda = 0$ there is no ageing and therefore the classifier is equivalent to the accumulative MNB (cf. Section 3.1).

The timestamp of the document from the stream is "transferred" to its component words and finally, to the MNB model. In particular, each word-class pair (w, c) entry in the model is associated with

- *the last observation time*, t_{lo}, which represents the last time that the word w has been observed in the stream in a document of class c.

The t_{lo} entry indicates how recent is the last observation of word w in class c.

Similarly, each class entry in the model is associated with a last observation timestamp indicating the last time that the class was observed in the stream.

Based on the above, the ageing-based MNB model consists of the temporally annotated class prior counts (N_c, t_{lo}^c) and class conditional word counts (N_{ic}, t_{lo}^{ic}).

Hereafter, we focus on how such a temporal model can be used for prediction while being maintained online. We distinguish between a normal fading MNB approach (Section 4.2) and an aggressive/drastic fading MNB approach (Section 4.3).

4.2 Ageing-Based MNB Classification

In order to *predict* the class label of a new document d arriving from the stream S at timepoint t, we employ the ageing-based version of MNB: in particular, the temporal information associated with the class prior counts and the class conditional word counts is incorporated in the class prior estimation $\hat{P}(c)$ and in the class conditional word probability estimation $\hat{P}(w_i|c)$, i.e., in Equations 2 and 3, respectively.

The updated temporal class prior for class $c \in C$ at timepoint t is given by :

$$\hat{P}^t(c) = \frac{N_c^t * e^{-\lambda \cdot (t - t_{lo}^c)}}{|S^t|} \tag{4}$$

where N_c^t is the number of documents in the stream up to timepoint t belonging to class c and $|S^t|$ is the total number of document in the stream thus far, which can be easily derived from the class counts as $|S^t| = \sum_{c' \in C} N_{c'}^t$. Note that the class counts, N_c, are maintained online over the stream as described below. The t_{lo}^c is the last observation of class c in the stream and $(t - t_{lo}^c)$ describes the temporal gap from the last appearance of the class label c in the stream to timepoint t.

The updated temporal class conditional word probability for a word $w_i \in d$ at t is given by:

$$\hat{P}^t(w_i|c) = \frac{N_{ic}^t * e^{-\lambda \cdot (t - t_{lo}^{(w_i,c)})}}{\sum_{j=1}^{|V^t|} N_{jc}^t * e^{-\lambda \cdot (t - t_{lo}^{(w_j,c)})}} \tag{5}$$

where N_{ic}^t is the number of appearances of word w_i in class c in the stream up to timepoint t and V^t is the vocabulary (i.e., distinct words) accumulated from the stream up to timepoint t. It is stressed that the vocabulary changes over time as new words may arrive from the stream.

The word conditional class counts, N_{ic}, are also maintained online over the stream as described hereafter.

Online Model Maintenance. The update of the MNB model consists of updating the class count and word-class count entries and their temporal counterparts. If d with timestamp t is a new document to be included in the MNB model and c is its class label, then the corresponding class count is increased by one, i.e., $N_c + = 1$ and for each word $w_i \in d$, the class conditional word counts

N_{ic} are increased based on the frequency of w_i in d, i.e., $N_{ic}+ = f_{w_i}^d$. The temporal counterpart of N_c is updated w.r.t. arrival time t of d, i.e., $t_{lo}^c = t$.

Similarly the temporal counterparts of any word class combination count N_{ic} in d will be updated, i.e., $t_{lo}^{ic} = t$. We refer to this method as *fadingMNB* hereafter.

4.3   Aggressive Fading MNB Alternative

The *fadingMNB* approach presented in the previous subsection, accumulates evidence about class appearances and word-class appearances in the stream up to the current time and applies the ageing function upon these accumulated counts. The bigger the gap between the last observation of a word in a class and the current timepoint, the more the weight of this word in the specific class would be decayed. However, as soon as we observe the word-class combination again in the stream, the total count is revived. This is because the counts are accumulative and the ageing function is applied *a posteriori*.

An alternative approach to make the exponential fading even more rapid and adapt more quickly to changes is to store in the model the aged-counts instead of the accumulated ones. That is, the ageing is applied over the faded counts. Obviously, such an approach implies a more drastic ageing of the data compared to *fadingMNB*. The decay is not exponential anymore and it depends also on how often a word is observed in the stream. In particular, words that appear constantly in the stream, i.e., at each time point, will not be affected (as with the *fadingMNB* approach) but if some period intervenes between consecutive appearances of a word, the word will be "penalized" for this gap. We refer to this method as *aggressiveFadingMNB*.

5   Experiments

In our experiments, we compare the original MNB stream model (*accumulativeMNB*), to our fading MNB model (*fadingMNB*) and to the aggressive fading MNB model (*aggressiveFadingMNB*). In Section 5.1, we present the Twitter dataset we use for the evaluation, and in Section 5.2 we present the evaluation measures. We present the results on classfication quality in Section 5.3. In Section 5.4, we discuss the role of the fading factor λ. We have run these experiments on different time granularities: hour, day and week. Due to lack of space, we report only on the *hourly-aggregated stream* in Sections 5.3 and 5.4. Then, in Section 5.5, we discuss the role of the stream granularity on the quality of the different models and how it is connected to the fading factor λ.

5.1   Data and Concept Changes

We use the TwitterSentiment dataset [19], introduced in [9]. The dataset was collected by querying the Twitter API for messages between April 6, 2009 and

June 25, 2009. The query terms belong to different categories, such as companies (e.g. query terms "aig", "at&t"), products (e.g. query terms "kindle2", "visa card"), persons (e.g. "warren buffet", "obama"), events (e.g. "indian election", "world cup"). Evidently, the stream is very heterogeneous. It has not been labeled manually. Rather, the authors of [9] derived the labels with a Maximum Entropy classifier that was trained on emoticons.

We preprocessed the dataset as in our previous work [21], including following steps: (i) dealing with data negation (e.g. replacing "not good" with "not_good", "not pretty" with "ugly" etc.), (ii) dealing with colloquial language (e.g. convering "luv" to "love" and "youuuuuuuuu" to "you"), (iii) elimination of superfluous words (e.g. Twitter signs like or #), stopwords (e.g. "the", "and"), special characters and numbers, (iv) stemming (e.g. "fishing", "fisher" were mapped to their root word "fish"). A detailed description of the preprocessing steps is in [20].

The final stream consists of 1,600,000 opinionated tweets, 50% of which are positive and 50% negative (two classes). The class distribution changes over time.

How to choose the temporal granularity of such a stream? On Figure 1, we show the tweets at different levels of temporal granularity: weeks (left), days (center), hours (right). The *weekly-aggregated stream* (Figure 1, left) consists of #12 distinct weeks (the x-axis shows the week of the year). Both classes are present up to week 25, but after that only instances of the negative class appear. In the middle of the figure, we see the same data aggregated at day level: there are #49 days (the horizontal axis denotes the day of the year). Up to day 168, we see positive and negative documents; the positive class (green) is overrepresented. But towards the end of the stream the class distribution changes and the positive class disappears. We see a similar behavior in the *hourly-aggregated stream* (Figure 1, right), where the x-axis depicts the hour (of the year). On Figure 1, we see that independently of the aggregation level, the amount of data received at each timepoint varies: there are high-traffic time points, like day 157 or week 23 and low-traffic ones, like day 96 or week 15. Also, there are "gaps" in the monitoring period. For example, in the daily-aggregated stream there are several 1-day gaps like day 132 but also "bigger gaps" like 10 days of missing observations between day 97 and day 107.

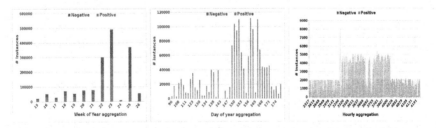

Fig. 1. Class distribution in the stream at different temporal granularities: weekly (left), daily (center), hourly (right).

The time granularity affects the ageing mechanism, since all documents associated with the same time unit (e.g. day) have the same age/weight. In the following, we experiment with all three levels of granularity.

5.2   Evaluation Methods and Evaluation Measures

The two most popular methods for evaluating classification algorithms are hold-out evaluation and prequential evaluation. Their fundamental difference is in the order in which they perform training and testing and the ordering of the dataset [7]. In *hold-out evaluation*, the current model is evaluated over a single independent hold-out test set. The hold-out set is the same over the whole course of the stream. For our experiments, the hold-out set consists of 30% of all instances randomly selected from the stream. In *prequential evaluation*, each instance from the stream is first used for testing and then for training the model. This way, the model is updated continuously based on new instances.

To evaluate the quality of the different classifiers, we employed accuracy and kappa [23] over an *evaluation window, evalW*. *Accuracy* is the percentage of correct classifications in w. Bifet et al. [3] use the kappa statistic defined as $k = \frac{p_0 - p_c}{1 - p_c}$, where p_0 is the accuracy of the studied classifier and p_c is the accuracy of the chance classifier. Kappa lies between -1 and 1.

5.3   Classifier Performance

Accuracy and Kappa in Prequential Evaluation. We compare the performance of our *fadingMNB* and *aggressiveFadingMNB* to the original *accumulativeMNB* algorithm in the hourly-aggregated stream, using prequential evaluation. As criteria for classification performance, we use accuracy (Figure 2) and kappa (Figure 3). In both figures, we see that classification performance has two phases, before and after the arrival of instance 1,341,000; around that time, there has been a drastic change in the class distribution, cf. Figure 1. We discuss these two phases, i.e., the left, respectively right part of the figures, separately.

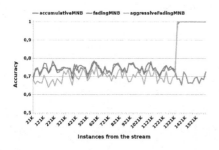
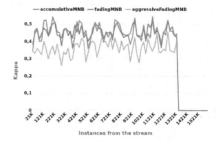

Fig. 2. Prequential evaluation on *accuracy* in the hourly-aggregated stream – fading factor $\lambda = 0.1$, evaluation window *evalW* = 1000

Fig. 3. Prequential evaluation on *kappa* in the hourly-aggregated stream – same parameters as in Figure 2, i.e., $\lambda = 0.1$, *evalW* = 1000

The left part of the accuracy plots on Figure 2 shows that *accumulativeMNB* has the best accuracy, followed closely by *fadingMNB*, while *aggressiveFadingMNB* has slightly inferior performance. In the left part of the plots on kappa (Figure 3), the relative performance is the same, but the performance inferiority of *accumulativeMNB* is more apparent.

In the right part of the accuracy plots on Figure 2, i.e., after the drastic change in the class distribution, we see that *accumulativeMNB* experiences a slight performance drop, while the accuracy of our two algorithms ascends rapidly to 100% (the two curves coincide). The intuitive explanation for the inferior performance of *accumulativeMNB* is that it remembers all past data, so it cannot adapt to the disappearance of the positive class. The proposed *fadingMNB* and *aggressiveFadingMNB*, on the contrary, manage to recover after the change.

The right part of the plots on kappa (Figure 3) gives a different picture: the performance of all three algorithms drops to zero after the concept shift (the three curves coincide). This is owed to the nature of kappa: it juxtaposes the performance of the classifier to that of a random classifier; as soon as there is only one class in the data (here: the negative one), no classifier can be better than the random classifier. Since the accuracy plots and the kappa plots show the same trends before the drastic concept change, and since the accuracy plots reflect the behavior of the classifiers after the change much better than kappa does, we concentrate on accuracy as evaluation measure hereafter.

Accuracy in Hold-Out Evaluation. Under prequential evaluation, each labeled document is used first for testing and, then, immediately for learning. In a more realistic setting, we would expect that the "expert" is not available all the time to deliver fresh labels for each incoming document from the stream. We are therefore interested to study the performance of the algorithms when less labeled data can be exploited. In this experiment, we train the algorithms in a random sample of 70% of the data and test them in the remaining 30%. The results are on Figure 4.

As pointed out in Figure 4, this hold-out evaluation was done after learning on 70% of the complete stream. Thus, the concept drift is incorporated into the learned model and the performance is stable. This allows us to highlight the influence of the hold-out data on the vocabulary. In particular, since the test instances constitute 30% of the dataset, some words belonging to them may be absent from the vocabulary used for training, or be so rare in the training sample that the class probabilities have not been well estimated. This effect cannot be traced under prequential evaluation, because all instances are gradually incorporated into the model. With our variant of a hold-out evaluation, we can observe on Figure 4 how the absence of a part of the vocabulary affects performance.

Figure 4 shows that *aggressiveFadingMNB* performs very poorly. This is expected, because this algorithm forgets instances too fast and thus cannot maintain good estimates of the polarity probabilities of the words. More remarkable is the performance difference between *fadingMNB* and *accumulativeMNB*: the gradual fading of some instances leads to a better model! An explanation may

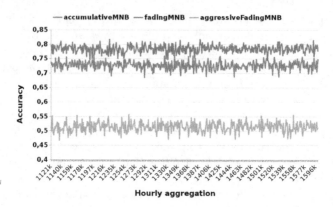

Fig. 4. Hold-out evaluation on *accuracy* in the hourly-aggregated stream – same parameters as in Figure 2, i.e., $\lambda = 0.1$, $evalW = 1000$; 30% of the data are used for testing the stream, after learning on the *complete* training sample (70% of the data).

be that *accumulativeMNB* experiences an overfitting on the large vocabulary of *all* training instances, while *fadingMNB* forgets some training data and thus uses a "smaller vocabulary" that is still adequate to predict the labels of the test instances. We intend to investigate this further by studying the contents of the vocabularies used by the learning algorithms.

5.4 Impact of the Fading Factor λ on the New Algorithms

On Figure 2, we have seen that both *fadingMNB* and *aggressiveFadingMNB* adapted immediately to the drastic change in the class distribution. In this experiment, we investigate how the fading factor λ affects the accuracy of the classifiers. First, we compare the performance of the two algorithms for a very small value of λ, cf. Figure 5.

Figure 5 shows that *aggressiveFadingMNB* manages to adapt to changes, while *fadingMNB* does not. Since small λ values increase the impact of the old data, i.e., enforce a *long memory*, the performance of *fadingMNB* deteriorates, as is the case for *accumulativeMNB*. In contrast, *aggressiveFadingMNB* needs such a small λ to remember some data in the first place.

As an extreme case, $\lambda = 0$ implies that no forgetting takes place and therefore corresponds to *accumulativeMNB*. A high value of λ implies that *aggressiveFadingMNB* forgets all data; a very low value of λ causes *fadingMNB* to remember a lot and therefore it degenerates to *accumulativeMNB*. To visualize these "connections" and to better understand the effect of the fading factor λ on each method, we experiment with different values of λ over the hourly-aggregated stream, cf. Figure 6.

Figure 6 depicts a constant performance of *accumulativeMNB* as it does not depend on λ. The *fadingMNB* and the *aggressiveFadingMNB* have a complementary performance. For small values of λ, *aggressiveFadingMNB* performs best; as

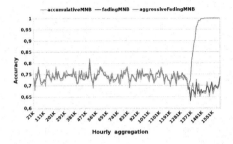

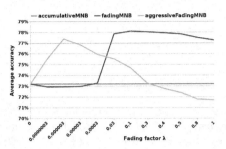

Fig. 5. Prequential evaluation on *accuracy* in the hourly-aggregated stream – same *evalW* = 1000 as in Figure 2, but $\lambda = 0.000003$.

Fig. 6. Effect of fading factor λ on accuracy in the hourly-aggregated stream; prequential evaluation, evaluation window *evalW* = 1000.

the λ increases, its performance drops. This is expected since low values imply that the classifier gradually forgets in a moderate pace, whereas high values mean that the past is forgotten very fast. The performance of *fadingMNB* is the opposite, for very small values of λ there is no actual ageing and therefore the performance is low (and similar to *accumulativeMNB*); whereas as λ increases, *fadingMNB* exploits the ageing of the old data, so its performance improves.

5.5 The Effect of Temporal Granularity and How to Set λ

The temporal granularity of the streams (e.g. hourly, daily, weekly, etc.) and the fading factor λ clearly affect each other. To illustrate this, for a decay degree of $\lambda = 0.1$, the weight of a instance is halved every $\frac{1}{0.1} = 10$ hours, days, weeks for hourly, daily, weekly temporal aggregation of the stream, respectively. The 10 weeks rate might be too low for some fast changing applications, whereas the 10 hours rate might be too high for other applications. Therefore, the choice of the decay factor λ should be done in conjunction with the temporal granularity of the stream. In the subfigures of Figure 7, we show the performance of the ageing-based classifiers for the different temporal-aggregations versions of the stream and different values of λ.

On the left part of Figure 7, we show the accuracy for $\lambda = 0.1$. This value means that an instance *loses half of its weight* after 10 hours, 10 days, 10 weeks for the hourly, daily, weekly aggregated stream respectively. We can see that hourly aggregation deals best with the change in the distributions as it performs best after the distribution changes; closely followed by aggressive fading in a daily aggregation. The worst performance occurs when using weekly aggregation *and* the *fadingMNB*, which indicates that forgetting every 10 weeks is not appropriate in this case; a more frequent forgetting of the data is more appropriate.

We increase the forgetting rate to $\lambda = 0.2$ (mid part of Figure 7), i.e., an instance *loses half its weight* after 5 hours, 5 days, 5 weeks for the hourly, daily, weekly aggregated stream respectively. The results are close to what we observed before; the hourly aggregation has the best performance for this dataset.

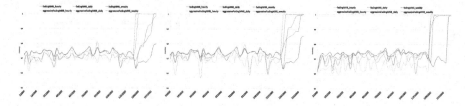

Fig. 7. Prequential accuracy of *fadingMNB*, *aggressiveFadingMNB* in comparison to *accumulativeMNB* (horizontal line) for different levels of granularity in the stream (using $evalW = 1000$): $\lambda = 0.1$ (left), $\lambda = 0.2$ (middle), $\lambda = 1.0$ (right)

The results for a much larger λ, $\lambda = 1.0$, are shown in the right part of Figure 7. A $\lambda = 1.0$ means that an instance *loses half its weight* after 1 hour, 1 day, 1 week for the hourly, daily, weekly aggregated stream respectively. The *aggressiveFadingMNB* performs worse when there is no drift in the stream, because the classifier forgets *too* fast. This fast forgetting though allows the classifier to adapt fast in times of change, i.e., when drift occurs after instance 1,341,000. Among the different streams, in times of stability in the stream, the aggressive fading classifier in the hourly aggregated stream shows the *worst* performance, followed closely by the daily and then the weekly aggregated streams. In times of change however, the behavior is the opposite with the hourly aggregated stream showing the best adaptation rate, because of no memory. Regarding *fadingMNB*, daily and weekly aggregation show best performance in times of stability followed by the hourly aggregated stream. In times of drift *on the other hand*, the hourly aggregation adapts *the fastest*, followed by daily and weekly aggregation streams.

To summarize, the value of λ affects the performance of the classifier over the whole course of the stream. In times of drifts in the stream, a larger λ is preferable as it allows for fast adaptation to the new concepts appearing in the stream. In times of stability though, a smaller λ is preferable as it allows the classifier to exploit already learned concepts in the stream. The selection of λ "works" in collaboration with the stream granularity. Obviously, at a very high granularity (such as a time unit of one hour), lambda can be higher than at a lower granularity (such as a time unit of one week).

6   Conclusions and Outlook

Learning a stream classifier is a challenging task due to the changes in the stream population (at both instance and feature levels) and the necessity for classifier adaptation to change. Adapting to change means that the model should be updated online; this update might mean that existing parts of the model are updated based on new observations, new model parts are added and old model parts are forgotten.

In this work we couple a very popular classifier, Multinomial Naive Bayes, with adaptation-to-change mechanisms. In particular, we introduce the notion of ageing in MNBs and we derive a gradually fading MNB approach (*fadingMNB*) and an aggressive fading MNB approach (*aggressiveFadingMNB*). We compare our methods to the traditional stream MNB approach (*accumulativeMNB*) and we show its superior performance in an evolving stream of tweets. Our ageing-based approaches recover fast after changes in the stream population, while maintaining a good performance in times of stability, i.e., when no drastic changes are observed. We also show how the fading factor, that regulates the ageing of old data, affects the results and its "connection" to the temporal granularity of the stream.

Our ongoing work involves experimenting with different streams from diverse domains and tuning the fading factor λ online based on the stream, instead of having a constant fading factor over time. As we observed in the current experiments, fast forgetting is important for times of change but in times of stability forgetting should be slower. Another direction of future work encompasses bookkeeping of the model at regular time intervals. In particular, one can maintain a more compact model by removing words that do not contribute much in the classification decision due to ageing, or whose observations in the stream are lower than the expected observations based on their age.

References

1. Aggarwal, C.C., Han, J., Wang, J., Yu, P.S.: A framework for projected clustering of high dimensional data streams. In: Proceedings of the 30th International Conference on Very Large Data Bases (VLDB), Toronto, Canada (2004)
2. Bermingham, A., Smeaton, A.F.: Classifying sentiment in microblogs: Is brevity an advantage? In: Proceedings of the 19th ACM International Conference on Information and Knowledge Management, CIKM 2010, pp. 1833–1836. ACM, New York (2010)
3. Bifet, A., Frank, E.: Sentiment knowledge discovery in twitter streaming data. In: Pfahringer, B., Holmes, G., Hoffmann, A. (eds.) DS 2010. LNCS, vol. 6332, pp. 1–15. Springer, Heidelberg (2010)
4. Cao, F., Ester, M., Qian, W., Zhou, A.: Density-based clustering over an evolving data stream with noise. In: Proceedings of the 6th SIAM International Conference on Data Mining (SDM), Bethesda, MD (2006)
5. Domingos, P., Pazzani, M.: On the optimality of the simple bayesian classifier under zero-one loss. Mach. Learn. **29**(2–3), 103–130 (1997)
6. Gama, J.A., Žliobaitė, I., Bifet, A., Pechenizkiy, M., Bouchachia, A.: A survey on concept drift adaptation. ACM Comput. Surv. **46**(4), 44:1–44:37 (2014)
7. Gama, J.: Knowledge Discovery from Data Streams, 1st edn. Chapman & Hall/CRC (2010)
8. Gama, J., Kosina, P.: Recurrent concepts in data streams classification. Knowl. Inf. Syst. **40**(3), 489–507 (2014)
9. Go, A., Bhayani, R., Huang, L.: Twitter sentiment classification using distant supervision. In: Processing, pp. 1–6 (2009). http://www.stanford.edu/alecmgo/papers/TwitterDistantSupervision09.pdf
10. Gokulakrishnan, B., Priyanthan, P., Ragavan, T., Prasath, N., Perera, A.S.: Opinion mining and sentiment analysis on a twitter data stream. In: Proceedings of 2012 International Conference on Advances in ICT for Emerging Regions (ICTer), ICTer 2012, pp. 182–188. IEEE (2012)

11. Guerra, P.C., Meira, Jr., W., Cardie, C.: Sentiment analysis on evolving social streams: how self-report imbalances can help. In: Proceedings of the 7th ACM International Conference on Web Search and Data Mining, WSDM 2014, pp. 443–452. ACM, New York (2014)

12. Lazarescu, M.: A multi-resolution learning approach to tracking concept drift and recurrent concepts. In: Gamboa, H., Fred, A.L.N. (eds.) PRIS, p. 52. INSTICC Press (2005)

13. Liu, Y., Yu, X., An, A., Huang, X.: Riding the tide of sentiment change: Sentiment analysis with evolving online reviews. World Wide Web **16**(4), 477–496 (2013)

14. McCallum, A., Nigam, K.: A comparison of event models for naive bayes text classification. In: AAAI-98 Workshop on Learning for Text Categorization, pp. 41–48. AAAI Press (1998)

15. Ntoutsi, E., Zimek, A., Palpanas, T., Krger, P., peter Kriegel, H.: Density-based projected clustering over high dimensional data streams. In: Proceedings of the 12th SIAM International Conference on Data Mining (SDM), Anaheim, CA, pp. 987–998 (2012)

16. Pang, B., Lee, L.: Opinion mining and sentiment analysis. Found. Trends Inf. Retr. **2**(1–2), 1–135 (2008)

17. Pang, B., Lee, L., Vaithyanathan, S.: Thumbs up?: sentiment classification using machine learning techniques. In: Proceedings of the ACL-02 Conference on Empirical Methods in Natural Language Processing, vol. 10, pp. 79–86. EMNLP, ACL, Stroudsburg (2002)

18. Plaza, L., Carrillo de Albornoz, J.: Sentiment Analysis in Business Intelligence: A survey, pp. 231–252. IGI-Global (2011)

19. Sentiment140: Sentiment140 - a Twitter sentiment analysis tool. http://help.sentiment140.com/

20. Sinelnikova, A.: Sentiment analysis in the Twitter stream. Bachelor thesis, LMU, Munich (2012)

21. Sinelnikova, A., Ntoutsi, E., Kriegel, H.P.: Sentiment analysis in the twitter stream. In: 36th Annual Conf. of the German Classification Society (GfKl 2012), Hildesheim, Germany (2012)

22. Turney, P.D.: Thumbs up or thumbs down?: semantic orientation applied to unsupervised classification of reviews. In: Proceedings of the 40th Annual Meeting on Association for Computational Linguistics, ACL 2002, pp. 417–424. Association for Computational Linguistics, Stroudsburg (2002)

23. Viera, A.J., Garrett, J.M.: Understanding interobserver agreement: The kappa statistic. Family Medicine **37**(5), 360–363 (2005)

24. Zimmermann, M., Ntoutsi, E., Spiliopoulou, M.: Adaptive semi supervised opinion classifier with forgetting mechanism. In: Proceedings of the 29th Annual ACM Symposium on Applied Computing, SAC 2014, pp. 805–812. ACM, New York (2014)

25. Zimmermann, M., Ntoutsi, E., Spiliopoulou, M.: Discovering and monitoring product features and the opinions on them with OPINSTREAM. Neurocomputing **150**, 318–330 (2015)

Drift Detection Using Stream Volatility

David Tse Jung Huang[1]([✉]), Yun Sing Koh[1], Gillian Dobbie[1], and Albert Bifet[2]

[1] Department of Computer Science, University of Auckland, Auckland, New Zealand
{dtjh,ykoh,gill}@cs.auckland.ac.nz
[2] Huawei Noah's Ark Lab, Hong Kong, China
bifet.albert@huawei.com

Abstract. Current methods in data streams that detect concept drifts in the underlying distribution of data look at the distribution difference using statistical measures based on mean and variance. Existing methods are unable to proactively approximate the probability of a concept drift occurring and predict future drift points. We extend the current drift detection design by proposing the use of historical drift trends to estimate the probability of expecting a drift at different points across the stream, which we term the expected drift probability. We offer empirical evidence that applying our expected drift probability with the state-of-the-art drift detector, ADWIN, we can improve the detection performance of ADWIN by significantly reducing the false positive rate. To the best of our knowledge, this is the first work that investigates this idea. We also show that our overall concept can be easily incorporated back onto incremental classifiers such as VFDT and demonstrate that the performance of the classifier is further improved.

Keywords: Data stream · Drift detection · Stream volatility

1 Introduction

Mining data that change over time from fast changing data streams has become a core research problem. Drift detection discovers important distribution changes from labeled classification streams and many drift detectors have been proposed [1,5,8,10]. A drift is signaled when the monitored classification error deviates from its usual value past a certain detection threshold, calculated from a statistical upper bound [6] or a significance technique [9]. The current drift detectors monitor only some form of mean and variance of the classification errors and these errors are used as the only basis for signaling drifts. Currently the detectors do not consider any previous trends in data or drift behaviors. Our proposal incorporates previous drift trends to extend and improve the current drift detection process.

In practice there are many scenarios such as traffic prediction where incorporating previous data trends can improve the accuracy of the prediction process. For example, consider a user using Google Map at home to obtain a fastest route to a specific location. The fastest route given by the system will be based on

© Springer International Publishing Switzerland 2015
A. Appice et al. (Eds.): ECML PKDD 2015, Part I, LNAI 9284, pp. 417–432, 2015.
DOI: 10.1007/978-3-319-23528-8_26

how congested the roads are at the *current time* (prior to leaving home) but is unable to adapt to situations like upcoming peak hour traffic. The user could be directed to take the main road that is not congested at the time of look up, but may later become congested due to peak hour traffic when the user is *en route*. In this example, combining data such as traffic trends throughout the day can help arrive at a better prediction. Similarly, using historical drift trends, we can derive more knowledge from the stream and when this knowledge is used in the drift detection process, it can improve the accuracy of the predictions.

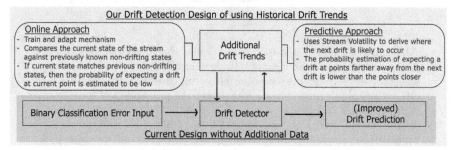

Fig. 1. Comparison of current drift detection process v.s. Our proposed design

The main contribution of this paper is the concept of using historical drift trends to estimate the probability of expecting a drift at each point in the stream, which we term the expected drift probability. We propose two approaches to derive this probability: Predictive approach and Online approach. Figure 1 illustrates the comparison of the current drift detection process against our overall proposed design. The Predictive approach uses Stream Volatility [7] to derive a prediction of where the next drift point is likely to occur. Stream Volatility describes the rate of changes in a stream and using the mean of the rate of the changes, we can make a prediction of where the next drift point is. This prediction from Stream Volatility then indicates periods of time where a drift is less likely to be discovered (*e.g.* if the next drift point is predicted to be 100 steps later, then we can assume that drifts are less likely to occur during steps farther away from the prediction). At these times, the Predictive approach will have a low expected drift probability. The predictive approach is suited for applications where the data have some form of cyclic behavior (*i.e.* occurs daily, weekly, etc.) such as the monitoring of oceanic tides, or daily temperature readings for agricultural structures. The Online approach estimates the expected drift probability by first training a model using previous non-drifting data instances. This model represents the state of the stream when drift is not occurring. We then compare how similar the current state of the stream is against the trained model. If the current state matches the model (*i.e.* current state is similar to previous non-drifting states), then we assume that drift is less likely to occur at this current point and derive a low expected drift probability. The Online approach is better suited for fast changing, less predictive applications such as stock market data. We apply the estimated expected drift probability in the state-of-the-art detector ADWIN [1] by adjusting the detection threshold (*i.e.* the statistical upper bound). When the expected drift probability is low, the detection threshold is adapted and increased to accommodate the estimation. Through experimentation, we offer evidence that using our two new approaches with ADWIN, we achieve a significantly fewer number of false positives.

The paper is structured as follows: in Section 2 we discuss the relevant research. Section 3 details the formal problem definition and preliminaries. In Section 4 our method is presented and we also discuss several key elements and contributions. Section 5 presents our extensive experimental evaluations, and Section 6 concludes the paper.

2 Related Work

Drift Detection: One way of describing a drift is a statistically significant shift in the distribution of a sample of data which initially represents a single homogeneous distribution to a different data distribution. Gama et al. [4] present a comprehensive survey on drift detection methods and points out that techniques generally fall into four categories: sequential analysis, statistical process control (SPC), monitoring two distributions, and contextual.

The Cumulative Sum [9] and the Page-Hinkley Test [9] are sequential analysis based techniques. They are both memoryless but their accuracy heavily depends on the required parameters, which can be difficult to set. Gama et al. [5] adapted the SPC approach and proposed the Drift Detection Method (DDM), which works best on data streams with sudden drift. DDM monitors the error rate and the variance of the classifying model of the stream. When no changes are detected, DDM works like a lossless learner constantly enlarging the number of stored examples, which can lead to memory problems.

More recently Bifet et al. [1] proposed ADaptive WINdowing (ADWIN) based on monitoring distributions of two subwindows. ADWIN is based on the use of the Hoeffding bound to detect concept change. The ADWIN algorithm was shown to outperform the SPC approach and provides rigorous guarantees on false positive and false negative rates. ADWIN maintains a window (W) of instances at a given time and compares the mean difference of any two subwindows $(W_0$ of older instances and W_1 of recent instances) from W. If the mean difference is statistically significant, then ADWIN removes all instances of W_0 considered to represent the old concept and only carries W_1 forward to the next test. ADWIN used a variation of exponential histograms and a memory parameter, to limit the number of hypothesis tests.

Stream Volatility and Volatility Shift: *Stream Volatility* is a concept introduced in [7], which describes the rate of changes in a stream. A high volatility represents a frequent change in data distribution and a low volatility represents an infrequent change in data distribution. Stream Volatility describes the relationship of proximity between consecutive drift points in the data stream. A *Volatility Shift* is when stream volatility changes (*e.g.* from high volatility to low volatility, or vice versa). Stream volatility is a next level knowledge of detected changes in data distribution. In the context of this paper, we employ the idea of stream volatility to help derive a prediction of when the next change point will occur in the Predictive approach.

In [7] the authors describe volatility detection as the discovery of a shift in stream volatility. A volatility detector was developed with a particular focus on finding the shift in stream volatility using a relative variance measure. The proposed volatility detector consists of two components: a buffer and a reservoir. The buffer is used to store recent data and the reservoir is used to store an overall representative sample of the stream. A volatility shift is observed when the variance between the buffer and the reservoir is past a significance threshold.

3   Preliminaries

Let us frame the problem of drift detection and analysis more formally. Let $S_1 = (x_1, x_2, ..., x_m)$ and $S_2 = (x_{m+1}, ..., x_n)$ with $0 < m < n$ represent two samples of instances from a stream with population means μ_1 and μ_2 respectively. The drift detection problem can be expressed as testing the null hypothesis H_0 that $\mu_1 = \mu_2$, i.e. the two samples are drawn from the same distribution against the alternate hypothesis H_1 that they are drawn from different distributions with $\mu_1 \neq \mu_2$. In practice the underlying data distribution is unknown and a test statistic based on sample means is constructed by the drift detector. If the null hypothesis is accepted incorrectly when a change has occurred then a false negative has occurred. On the other hand if the drift detector accepts H_1 when no change has occurred in the data distribution then a false positive has occurred. Since the population mean of the underlying distribution is unknown, sample means need to be used to perform the above hypothesis tests. The hypothesis tests can be restated as the following. We accept hypothesis H_1 whenever $Pr(|\hat{\mu}_1 - \hat{\mu}_2|) \geq \epsilon) \leq \delta$, where the parameter $\delta \in (0, 1)$ and controls the maximum allowable false positive rate, while ϵ is the test statistic used to model the difference between the sample means and is a function of δ.

4   Our Concept and Design

We present how to use historical drift trend to estimate the probability of expecting a drift at every point in the stream using our Predictive approach in Section 4.1 and Online approach in Section 4.2. In Section 4.3 we describe how the expected drift probability is applied onto drift detector ADWIN.

4.1   Predictive Approach

The Predictive approach is based on Stream Volatility. Recall that Stream Volatility describes the rate of changes in the stream. The mean volatility value is the average interval between drift points, denoted $\mu_{volatility}$, and is derived from the history of drift points in the stream. For example, a stream with drift points at times $t = 50$, $t = 100$, and $t = 150$ will have a mean volatility value of 50. The $\mu_{volatility}$ is then used to provide an estimate of a relative position of where the next drift point, denoted $t_{drift.next}$, is likely to occur. In other words, $t_{drift.next} = t_{drift.previous} + \mu_{volatility}$, where $t_{drift.previous}$ is the location of the previous signaled drift point and $t_{drift.previous} < t_{drift.next}$.

The expected drift probability at points t_x in the stream is denoted as ϕ_{t_x}. We use the $t_{drift.next}$ prediction to derive ϕ_{t_x} at each time point t_x in the stream between the previous drift and the next predicted drift point, $t_{drift.previous} < t_x < t_{drift.next}$. When t_x is distant from $t_{drift.next}$, the probability ϕ_{t_x} is smaller and as t_x progresses closer to $t_{drift.next}$, ϕ_{t_x} is progressively increased.

We propose two variations of deriving ϕ_{t_x} based on next drift prediction: one based on the sine function and the other based on sigmoid function. Intuitively the sine function will assume that the midpoint of previous drift and next drift is the least likely point to observe a drift whereas the sigmoid function will assume that immediately after a drift, the probability of observing a drift is low until the stream approaches the next drift prediction.

The calculation of ϕ_{t_x} using the sine and the sigmoid functions at a point t_x where $t_{drift.previous} < t_x < t_{drift.next}$ are defined as:

$$\phi_{t_x}^{\sin} = 1 - \sin\left(\pi \cdot t_r\right) \text{ and } \phi_{t_x}^{sigmoid} = 1 - \frac{1 - t_r}{0.01 + |1 - t_r|}$$

where $t_r = (t_x - t_{drift.previous})/(t_{drift.next} - t_{drift.previous})$

The Predictive approach is applicable when the volatility of the stream is relatively stable. When the volatility of the stream is unstable and highly variable, the Predictive approach will be less reliable at predicting where the next drift point is likely to occur. When this situation arises, the Online approach (described in Section 4.2) should be used.

4.2   Online Adaptation Approach

The Online approach estimates the expected drift probability by first training a model using previous non-drifting data. This model represents the state of the stream when drift is not occurring. We then compare how similar the current state of the stream is against the trained model. If the current state matches the model (i.e. current state is similar to previous non-drifting states), then we assume that drift is less likely to occur at this current point and derive a low expected drift probability. Unlike the Predictive approach, which predicts where a drift point might occur, the Online approach approximates the unlikelihood of a drift occurring by comparing the current state of the stream against the trained model representing non-drifting states.

The Online approach uses a sliding block B of size b that keeps the most recent value of binary inputs. The mean of the binary contents in the block at time t_x is given as $\mu_{B_{t_x}}$ where the block B contains samples with values $v_{x-b}, \cdots, v_x$ of the transactions $t_{x-b}, \cdots, t_x$. The simple moving average of the previous n inputs is also maintained where: $(v_{x-b-n} + \cdots + v_{x-b})/n$. The state of the stream at any given time t_x is represented using the *Running Magnitude Difference*, denoted as γ, and given by: $\gamma_{t_x} = \mu_{B_{tx}} - \text{MovingAverage}$.

We collect a set of γ values $\gamma_1, \gamma_2, \cdots, \gamma_x$ using previous non-drifting data to build a Gaussian training model. The Gaussian model will have the mean μ_γ and the variance σ_γ of the training set of γ values. The set of γ values reflect the stability of the mean of the binary inputs. A stream in its non-drifting states will have a set of γ values that tend to a mean of 0.

To calculate the expected drift probability in the running stream at a point t_x, the estimation ϕ_{t_x} is derived by comparing the Running Magnitude Difference γ_{t_x} against the trained Gaussian model with α as the threshold, the probability calculation is given as:

$$\phi_{t_x}^{online} = 1 \text{ if } f(\gamma_{t_x}, \mu_\gamma, \sigma_\gamma) \leq \alpha$$

and

$$\phi_{t_x}^{online} = 0 \text{ if } f(\gamma_{t_x}, \mu_\gamma, \sigma_\gamma) > \alpha$$

where

$$f(\gamma_{t_x}, \mu_\gamma, \sigma_\gamma) = \frac{1}{\sqrt{2\pi}} e^{-\frac{(\gamma_{t_x} - \mu_\gamma)^2}{2\sigma_\gamma^2}}$$

For example, using previous non-drifting data we build a Gaussian model with $\mu_\gamma = 0.0$ and $\sigma_\gamma = 0.05$, if $\alpha = 0.1$ and we observe that the current state $\gamma_{t_x} = 0.1$, then the expected drift probability is 1 because $f(0.1, 0, 0.05) = 0.054 \leq \alpha$.

In a stream environment, the *i.i.d.* property of incoming random variables in the stream is generally assumed. Although at first glance it may appear that a trained Gaussian model is not suitable to be used in this setting, the central limit theorem provides justification. In drift detection a drift primarily refers to the *real concept drift*, which is a change in the posterior distributions of data $p(y|X)$, where X is the set of attributes and y is the target class label. When the distribution of X changes, the class y might also change affecting the predictive power of the classifier and signals a a drift. Since the Gaussian model is trained using non-drifting data, we assume that the collected γ value originates from the same underlying distribution and remains stable in the non-drifting set of data. Although the underlying distribution of the set of γ values is unknown, the Central Limit Theorem justifies that the mean of a sufficiently large number of random samples will be approximately normally distributed regardless of the underlying distribution. Thus, we can effectively approximate the set of γ values with a Gaussian model. To confirm that the Central Limit Theorem is valid in our scenario, we have generated sets of non-drifting supervised two class labeled streams using the rotating hyperplane generator with the set of numeric attributes X generated from different source distributions such as uniform, binomial, exponential, and Poisson. The sets of data are then run through a Hoeffding Tree Classifier to obtain the binary classification error inputs and the set of γ values are gathered using the Online approach. We plot each set of the γ values and demonstrate that the distribution indeed tends to Gaussian as justified by the Central Limit Theorem in Figure 2.

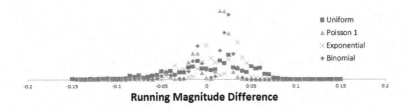

Fig. 2. Demonstration of Central Limit Theorem

4.3   Application onto ADWIN

In Sections 4.1 and 4.2 we have described two approaches at calculating the expected drift probability ϕ and in this section we show how to apply the discovered ϕ in the detection process of ADWIN.

ADWIN relies on using the Hoeffding bound with Bonferroni correction [1] as the detection threshold. The Hoeffding bound provides guarantee that a drift is signalled with probability at most δ (a user defined parameter): $Pr(|\mu_{W_0} - \mu_{W_1}| \geq \epsilon) \leq \delta$ where μ_{W_0} is the sample mean of the reference window of data, W_0, and μ_{W_1} is the sample mean of the current window of data, W_1. The ϵ value is a function of δ parameter and is the test statistic used to model the difference between the sample mean of the two windows. Essentially when the difference in sample means between the two windows is greater than the test statistic ϵ, a drift will be signaled. ϵ is given by the Hoeffding bound with Bonferroni correction as:

$$\epsilon_{hoeffding} = \sqrt{\frac{2}{m} \cdot \sigma_W^2 \cdot ln\frac{2}{\delta'}} + \frac{2}{3m} \, ln\frac{2}{\delta'}$$

where $m = \frac{1}{1/|W_0|+1/|W_1|}$, $\delta' = \frac{\delta}{|W_0|+|W_1|}$.

We incorporate ϕ, expected drift probability, onto ADWIN and propose an Adaptive bound which adjusts the detection threshold of ADWIN in reaction to the probability of seeing a drift at different time t_x across the stream. When ϕ is low, the Adaptive bound (detection threshold) is increased to accommodate the estimation that drift is less likely to occur.

The ϵ of the Adaptive bound is derived as follows:

$$\epsilon_{adaptive} = (1 + \beta \cdot (1 - \phi))\sqrt{\frac{2}{m} \cdot \sigma_W^2 \cdot ln\frac{2}{\delta'}} + \frac{2}{3m} \, ln\frac{2}{\delta'}$$

where β is a tension parameter that controls the maximum allowable adjustment, usually set below 0.5. A comparison of the Adaptive bound using the Predictive approach versus the original Hoeffding bound with Bonferroni correction is shown in Figure 3. In such cases, we can see that by using the Adaptive bound derived from the Predictive approach, we reduce the number of false positives that would have otherwise been signaled by Hoeffding bound.

The Adaptive bound is based on adjusting the Hoeffding bound and maintains similar statistical guarantees as the original Hoeffding bound. We know

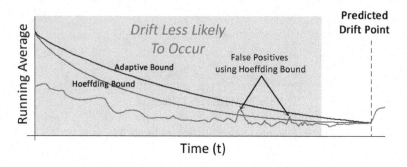

Fig. 3. Demonstration of Adaptive bound vs. Hoeffding bound

that the Hoeffding bound provides guarantee that a drift is signaled with probability at most δ:

$$Pr(|\mu_{W_0} - \mu_{W_1}| \geq \epsilon) \leq \delta$$

and since

$$\epsilon_{adaptive} \geq \epsilon_{hoeffding}$$

therefore,

$$Pr(|\mu_{W_0} - \mu_{W_1}| \geq \epsilon_{adaptive}) \leq Pr(|\mu_{W_0} - \mu_{W_1}| \geq \epsilon_{hoeffding}) \leq \delta$$

As a result, we know that the Adaptive bound is *at least as confident* as the Hoeffding bound and offer the same guarantees as the Hoeffding bound given δ.

The Predictive approach should be used when the input stream allows for an accurate prediction of where the next drift point is likely to occur. A good prediction of next drift point will show major performance increases. The Predictive approach has tolerance to incorrect next drift predictions to a certain margin of error. However, when the stream is too volatile and fast changing to the extent where using volatility to predict the next drift point is unreasonable, the Online approach should be used. The benefits of using the Online approach is that the performance of the approach is not affected irrespective of whether the stream is volatile or not. The Online approach is also better suited for scenarios where a good representative non-drifting training data can be provided.

When using the Predictive approach to estimate ϕ in the Adaptive bound, we note that an extra *warning level* mechanism should be added. The Predictive approach uses the mean volatility value, the average number of transactions between each consecutive drift points in the past, to derive where the next drift point is likely to occur. The mean volatility value is calculated based on previous detection results of the drift detector. The Adaptive bound with the Predictive approach works by increasing the detection threshold at points before the prediction. In cases where the true drift point arrives before the prediction, the drift detector will experience a higher detection delay due to the higher detection threshold of the Adaptive bound. This may affect future predictions based on the mean volatility value. Therefore, we note the addition of the *warning level*, which is when the Hoeffding bound is passed but not the Adaptive bound. A real drift will pass the Hoeffding bound, then pass the Adaptive bound, while a false positive might pass Hoeffding bound but not the Adaptive bound. When a drift is signaled by the Adaptive bound, the mean volatility value is updated using the drift point found when Hoeffding bound was first passed. The addition of the Hoeffding as the warning level resolves the issue that drift points found by the Adaptive bound might influence future volatility predictions.

The β value is a tension parameter used to control the degree at which the statistical bound is adjusted based on drift expectation probability estimation ϕ. Setting a higher β value will increase the magnitude of adjustment of the bound. One sensible way to set the β parameter is to base its value on the confidence of the ϕ estimation. If the user is confident in the ϕ estimation, then setting a high β value (e.g. 0.5) will significantly reduce the number of false positives while still detecting all the real drifts. If the user is less confident in the ϕ estimation, then β can be set low (e.g. 0.1) to make sure drifts of significant value are picked up. An approach for determining the β value is: $\beta = Pr(a) - Pr(e)/2 \cdot (1 - Pr(e))$ where $Pr(a)$ is the confidence of the ϕ estimation and $Pr(e)$ is the confidence of estimation by chance. In most instances $Pr(e)$ should be set at 0.6 as any estimation with confidence lower than 0.6 we can consider as a poor estimation. In practice by setting β at 0.1 reduces the number of false positives found by 50-70% when incorporating our design into ADWIN while maintaining similar true positive rates and detection delay compared to without using our design.

5   Experimental Evaluation

In this section we present the experimental evaluation of applying our expected drift probability ϕ onto ADWIN with the Adaptive bound. We test with both the Predictive approach and the Online approach. Our algorithms are coded in Java and all of the experiments are run on an Intel Core i5-2400 CPU @ 3.10 GHz with 8GB of RAM running Windows 7.

We evaluate with different sets of standard experiments in drift detection: false positive test, true positive test, and false negative test using various β (tension parameter) values from 0.1 to 0.5. The detection delay is also a major indicator of performance therefore reported and compared in the experiments. Our proposed algorithms have a run-time of $< 2ms$. Further supplementary materials and codes can be found online

5.1   False Positive Test

In this test we compare the false positives found between using only ADWIN detector against using our Predictive and Online approaches on top of ADWIN. For this test we replicate the standard false positive test used in [1]. A stream of 100,000 bits with no drifts generated from a Bernoulli distribution with $\mu = 0.2$ is used. We vary the δ parameter and the β tension parameter and run 100 iterations for all experiments. The Online approach is run with a 10-fold cross validation. We use $\alpha = 0.1$ for the Online approach.

Table 1. False Positive Rate Comparison

Predictive Approach - Sine Function						
			Adaptive Bound			
δ	Hoeffding	$\beta = 0.1$	$\beta = 0.2$	$\beta = 0.3$	$\beta = 0.4$	$\beta = 0.5$
0.05	0.0014	0.0006	0.0003	0.0002	0.0001	0.0001
0.1	0.0031	0.0015	0.0008	0.0005	0.0003	0.0002
0.3	0.0129	0.0072	0.0042	0.0027	0.0019	0.0015
Predictive Approach - Sigmoid Function						
			Adaptive Bound			
δ	Hoeffding	$\beta = 0.1$	$\beta = 0.2$	$\beta = 0.3$	$\beta = 0.4$	$\beta = 0.5$
0.05	0.0014	0.0004	0.0002	0.0001	0.0001	0.0001
0.1	0.0031	0.0011	0.0004	0.0003	0.0002	0.0002
0.3	0.0129	0.0057	0.0030	0.0018	0.0015	0.0014
Online Approach						
δ	Hoeffding	$\beta = 0.1$	$\beta = 0.2$	$\beta = 0.3$	$\beta = 0.4$	$\beta = 0.5$
0.05	0.0014	0.0006	0.0005	0.0005	0.0005	0.0005
0.1	0.0031	0.0014	0.0012	0.0011	0.0011	0.0011
0.3	0.0129	0.0073	0.0063	0.0059	0.0058	0.0058

The results are shown in Table 1 and we observe that both the sine function and sigmoid function with the Predictive approach are effective at reducing the number of false positives in the stream. In the best case scenario the number of false positive was reduced by 93%. Even with a small β value of 0.1, we still observe an approximately 50-70% reduction in the number of false positives. For the Online approach we observe around a 65% reduction.

5.2   True Positive Test

In the true positive test we test the accuracy of the three different setting at detecting true drift points. In addition, we look at the detection delay associated with the detection of the true positives. For this test we replicate the true positive test used in [1]. Each stream consists of 1 drift at different points of volatility with varying magnitudes of drift and drift is induced with different slope values over a period of 2000 steps. For each set of parameter values, the experiments are run over 100 iterations using ADWIN as the drift detector with $\delta = 0.05$. The Online approach was run with a 10-fold cross validation.

We observed that for all slopes, the true positive rate for using all different settings (ADWIN only, Predictive_Sine, and Predictive_Sigmoid, Online) is 100%. There was no notable difference between using ADWIN only, Predictive

Table 2. True Positive Test: Detection Delay

Predictive Approach - Sigmoid Function						
Slope	Hoeffding	Adaptive Bound				
		$\beta = 0.1$	$\beta = 0.2$	$\beta = 0.3$	$\beta = 0.4$	$\beta = 0.5$
0.0001	882±(181)	882±(181)	880±(181)	874±(191)	874±(191)	874±(191)
0.0002	571±(113)	569±(112)	564±(114)	562±(119)	562±(119)	562±(119)
0.0003	441±(83)	440±(82)	438±(82)	436±(86)	436±(86)	436±(86)
0.0004	377±(71)	375±(68)	373±(69)	371±(72)	371±(72)	371±(72)

Online Approach						
Slope	Hoeffding	$\beta = 0.1$	$\beta = 0.2$	$\beta = 0.3$	$\beta = 0.4$	$\beta = 0.5$
0.0001	882±(181)	941±(180)	975±(187)	1001±(200)	1015±(211)	1033±(220)
0.0002	571±(113)	597±(116)	611±(123)	620±(130)	629±(136)	632±(140)
0.0003	441±(83)	460±(90)	469±(94)	472±(96)	475±(100)	476±(101)
0.0004	377±(71)	389±(73)	394±(74)	398±(79)	398±(79)	399±(80)

approach, and Online approach in terms of accuracy. The results for the associated detection delay on gradual drift stream are shown in Table 2. We note that the Sine and Sigmoid functions yielded similar results and we only present one of them here. We see that the detection delays remain stable between ADWIN only and the Predictive approach as this test assumes an accurate next drift prediction from volatility and does not have any significant variations. The Online approach observed a slightly higher delay due to the nature of the approach (within one standard deviation of Hoeffding bound delay). There is an increase in delay when β was varied only in the Online approach.

5.3   False Negative Test

The false negative test experiments if predictions are correct. Hence, the experiments are carried out on the Predictive approach and not the Online approach.

For this experiment we generate streams with 100,000 bits containing exactly one drift at a pre-specified location before the presumed drift location (100,000). We experiment with 3 locations at steps 25,000 (the 1/4 point), 50,000 (the 1/2 point), and 75,000 (the 3/4 point). The streams are generated with different drift slopes modelling both gradual and abrupt drift types. We feed the Predictive approach a drift prediction at 100,000. We use ADWIN with $\delta = 0.05$.

In Table 3 we show the detection delay results for varying β and drift types/slopes when the drift is located at the 1/4 point. We observe from the table that as we increase the drift slope, the delay decreases. This is because a drift of a larger magnitude is easier to detect and thus found faster. As we increase β we can see a positive correlation with delay. This is an apparent tradeoff with adapting to a tougher bound. In most cases the increase in delay associated with an unpredicted drift is still acceptable taking into account the

Table 3. Delay Comparison: 1/4 drift point

			Sine Function			
				Adaptive Bound		
Slope	Hoeffding	$\beta = 0.1$	$\beta = 0.2$	$\beta = 0.3$	$\beta = 0.4$	$\beta = 0.5$
0.4	107±(37)	116±(39)	129±(42)	139±(43)	154±(47)	167±(51)
0.6	52±(12)	54±(11)	57±(11)	61±(12)	64±(14)	67±(15)
0.8	27±(10)	28±(11)	32±(14)	39±(16)	44±(15)	50±(12)
0.0001	869±(203)	923±(193)	972±(195)	1026±(200)	1090±(202)	1151±(211)
0.0002	556±(121)	593±(117)	634±(106)	664±(109)	692±(116)	727±(105)
0.0003	434±(89)	463±(91)	488±(84)	514±(83)	531±(80)	557±(75)
0.0004	367±(71)	384±(76)	403±(73)	420±(70)	439±(71)	457±(69)

			Sigmoid Function			
				Adaptive Bound		
Slope	Hoeffding	$\beta = 0.1$	$\beta = 0.2$	$\beta = 0.3$	$\beta = 0.4$	$\beta = 0.5$
0.4	108±(37)	121±(40)	136±(42)	156±(46)	176±(51)	196±(56)
0.6	52±(12)	54±(12)	57±(11)	61±(12)	64±(14)	67±(15)
0.8	27±(10)	28±(11)	32±(14)	39±(16)	44±(15)	50±(12)
0.0001	869±(203)	937±(200)	1013±(198)	1091±(201)	1177±(216)	1259±(221)
0.0002	556±(121)	605±(110)	657±(110)	695±(115)	738±(103)	776±(102)
0.0003	434±(89)	474±(87)	508±(87)	535±(78)	567±(76)	593±(76)
0.0004	367±(71)	390±(75)	415±(71)	442±(70)	471±(65)	492±(61)

Table 4. Delay Comparison: Varying drift point

		Gradual Drift 0.0001				
		Sine			Sigmoid	
β	1/4 point	1/2 point	3/4 point	1/4 point	1/2 point	3/4 point
Hoeffding	869±(203)	885±(183)	872±(183)	869±(203)	885±(183)	872±(183)
0.1	923±(192)	956±(187)	930±(178)	937±(200)	955±(185)	948±(176)
0.2	972±(195)	1026±(197)	982±(173)	1013±(198)	1022±(195)	1020±(183)
0.3	1026±(200)	1095±(203)	1029±(182)	1091±(201)	1091±(202)	1096±(197)
0.4	1090±(202)	1182±(212)	1087±(192)	1177±(216)	1176±(207)	1167±(204)
0.5	1151±(211)	1253±(217)	1133±(198)	1259±(221)	1243±(215)	1245±(214)

magnitude of false positive reductions and the assumption that unexpected drifts should be less likely to occur when volatility predictions are relatively accurate.

Table 4 compares the delay when the drift is thrown at different points during the stream. It can be seen that the sine function does have a slightly higher delay when the drift is at the 1/2 point. This can be traced back to the sine function where the mid-point is the peak of the offset. In general the variations are within reasonable variation and the differences are not significant.

5.4   Real-World Data: Power Supply Dataset

This dataset is obtained from the Stream Data Mining Repository[1]. It contains the hourly power supply from an Italian electricity company from two sources. The measurements from the sources form the attributes of the data and the class label is the hour of the day from which the measurement is taken. The drifts in this dataset are primarily the differences in power usage between different seasons where the hours of daylight vary. We note that because real-world datasets do not have ground truths for drift points, we are unable to report true positive rates, false positive rates, and detection delay. The main objective is to compare the behavior of ADWIN only versus our designs on real-world data and show that they do find drifts at similar locations.

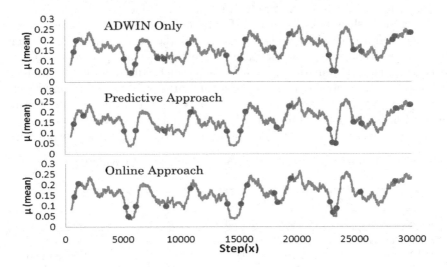

Fig. 4. Power Supply Dataset (drift points are shown with red dots)

Figure 4 shows the comparison using the drifts found between ADWIN Only and our Predictive and Online approaches. We observe that the drift points are fired at similar locations to the ADWIN only approach.

5.5   Case Study: Incremental Classifier

In this study we apply our design into incremental classifiers in data streams. We experiment with both synthetically generated data and real-world datasets. First, we generate synthetic data with the commonly used SEA concepts generator introduced in [11]. Second, we use Forest Covertype and Airlines real-world datasets from MOA website[2]. Note that the ground truth for drifts is not available for the real-world datasets. In all of the experiments we run VFDT [3], an

---

[1] http://www.cse.fau.edu/~xqzhu/stream.html
[2] moa.cms.waikato.ac.nz/datasets/

incremental tree classifier, and compare the prediction accuracy and learning time of the tree using three settings: VFDT without drift detection, VFDT with ADWIN, and VFDT with our approaches and Adaptive bound. In the no drift detection setting, the tree learns throughout the entire stream. In the other settings, the tree is rebuilt using the next n instances when a drift is detected. For SEA data $n = 10000$ and for real-world $n = 300$. We used a smaller n for the real-world datasets because they contain fewer instances and more drift points.

The synthetic data stream is generated from MOA [2] using the SEA concepts with 3 drifts evenly distributed at 250k, 500k, and 750k in a 1M stream. Each section of the stream is generated from one of the four SEA concept functions. We use $\delta = 0.05$ for ADWIN, Sine function for Predictive approach and $\beta = 0.5$ for both approaches and $\alpha = 0.1$ for Online approach. Each setting is run over 30 iterations and the observed results are shown in Table 5.

The results show that by using ADWIN only, the overall accuracy of the classifier is improved. There is also a reduction in the learning time because only parts of the stream are used for learning the classifier as opposed to no drift detection where the full stream is used. Using our Predictive approach and the Online approach showed a further reduction in learning time and an improvement in accuracy. An important observation is the reduction in the number of drifts detected in the stream and an example of drift points is shown in Table 6. We discovered that using Predictive and Online approaches found less false positives.

Table 5. Incremental Classifier Performance Comparisons

SEA Concept Generator (3 actual drifts)			
Setting	Learning Time (ms)	Accuracy	Drifts Detected
No Drift Detection	2763.67±(347.34)	85.71±(0.06)%	-
ADWIN Only	279.07±(65.06)	87.36±(0.15)%	13.67±(3.66)
Predictive Approach	178.63±(41.06)	87.44±(0.22)%	8.67±(1.72)
Online Approach	**161.10±(38.93)**	**87.49±(0.22)%**	**6.60±(1.56)**

Real-World Dataset: Forest Covertype			
Setting	Learning Time (ms)	Accuracy	Drifts Detected
No Drift Detection	44918.47±(149.44)	83.13%	-
ADWIN Only	45474.57±(226.40)	**89.37%**	1719
Predictive Approach	45710.87±(226.40)	89.30%	1701
Online Approach	45143.07±(212.40)	89.09%	**1602**

Real-World Dataset: Airlines			
Setting	Learning Time (ms)	Accuracy	Drifts Detected
No Drift Detection	2051.87±(141.42)	67.44%	-
ADWIN Only	1654.37±(134.79)	**75.96%**	396
Predictive Approach	**1602.07±(120.24)**	75.80%	352
Online Approach	1637.27±(124.16)	75.79%	**320**

Table 6. SEA Dataset Drift Point Comparison

SEA Sample: induced drifts at 250k, 500k, and 750k (false positives colored)											
ADWIN	19167	102463	106367	250399	407807	413535	432415	483519	489407	500223	739423	750143
Predict.	19167	102463	106399	250367						500255		750239
Online	19167			251327						500255		750143

The reduction in the number of drifts detected means that the user does not need to react to unnecessary drift signals. In the real-world dataset experiments, we generally observe a similar trend to the synthetic experiments. Overall the classifier's accuracy is improved when our approaches are applied. Using ADWIN only yields the highest accuracy, however, it is only marginally higher than our approaches while using our approaches the number of drifts detected is reduced. With real-world datasets, we unfortunately do not have the ground truths and cannot produce variance in the accuracy and number of drifts detected, but the eliminated drifts using our approaches did not have apparent effects on the accuracy of the classifier and thus are more likely to be false positives or less significant drifts. Although the accuracy results are not statistically worse or better, we observe a reduction in the number of drifts detected. In scenarios where drift signals incur high costs of action, having a lower number of detected drifts while maintaining similar accuracy is in general more favorable.

6   Conclusion and Future Work

We have described a novel concept of estimating the probability of expecting a drift at each point in the stream based on historical drift trends such as Stream Volatility. To the best of our knowledge this work is the first that investigate this idea. We proposed two approaches to derive the expected drift probability: Predictive approach and Online approach. The Predictive approach uses Stream Volatility [7] to derive a prediction of where the next drift point is likely to occur and based on that prediction the expected drift probability is determined using the proximity of the points to the next drift prediction. The Online approach estimates the expected drift probability by first training a model using previous non-drifting data instances and compare the current state of the stream against the trained model. If the current state matches the model then we assume that drift is less likely to occur at this current point and derive a low expected drift probability. We incorporate the derived expected drift probability in the state-of-the-art detector ADWIN by adjusting the statistical upper bound. When the expected drift probability is low, the bound is increased to accommodate the estimation. Through experimentation, we offer evidence that using our design in ADWIN, we can achieve significantly fewer number of false positives.

Our future work includes applying the Adaptive bound onto other drift detection techniques that utilize similar statistical upper bounds such as SEED [7]. We also want to look at using other stream characteristics such as the types of drifts (*e.g.* gradual and abrupt) to derive the expected drift probability.

References

1. Bifet, A., Gavaldá, R.: Learning from time-changing data with adaptive windowing. In: SIAM International Conference on Data Mining (2007)
2. Bifet, A., Holmes, G., Pfahringer, B., Read, J., Kranen, P., Kremer, H., Jansen, T., Seidl, T.: MOA: a real-time analytics open source framework. In: Gunopulos, D., Hofmann, T., Malerba, D., Vazirgiannis, M. (eds.) ECML PKDD 2011, Part III. LNCS, vol. 6913, pp. 617–620. Springer, Heidelberg (2011)
3. Domingos, P., Hulten, G.: Mining high-speed data streams. In: Proceedings of the 6th ACM SIGKDD International Conference on Knowledge Discovery and Data Mining, pp. 71–80 (2000)
4. Gama, J.A., Žliobaitė, I., Bifet, A., Pechenizkiy, M., Bouchachia, A.: A survey on concept drift adaptation. ACM Computing Surveys 46(4), 44:1–44:37 (2014)
5. Gama, J., Medas, P., Castillo, G., Rodrigues, P.: Learning with drift detection. In: Bazzan, A.L.C., Labidi, S. (eds.) SBIA 2004. LNCS (LNAI), vol. 3171, pp. 286–295. Springer, Heidelberg (2004)
6. Hoeffding, W.: Probability inequalities for sums of bounded random variables. Journal of the American Statistical Association 58, 13–29 (1963)
7. Huang, D.T.J., Koh, Y.S., Dobbie, G., Pears, R.: Detecting volatility shift in data streams. In: 2014 IEEE International Conference on Data Mining (ICDM), pp. 863–868 (2014)
8. Kifer, D., Ben-David, S., Gehrke, J.: Detecting change in data streams. In: Proceedings of the 30th International Conference on VLDB, pp. 180–191. VLDB Endowment (2004)
9. Page, E.: Continuous inspection schemes. Biometrika, 100–115 (1954)
10. Pears, R., Sakthithasan, S., Koh, Y.S.: Detecting concept change in dynamic data streams - A sequential approach based on reservoir sampling. Machine Learning 97(3), 259–293 (2014)
11. Street, W.N., Kim, Y.: A streaming ensemble algorithm (SEA) for large-scale classification. In: Proceedings of the 7th ACM SIGKDD International Conference on Knowledge Discovery and Data Mining, KDD 2001, pp. 377–382. ACM, New York (2001)

Early Classification of Time Series as a Non Myopic Sequential Decision Making Problem

Asma Dachraoui[1,2], Alexis Bondu[1], and Antoine Cornuéjols[2]([✉])

[1] EDF R & D, 1, Avenue du Général de Gaulle, 92141 Clamart Cedex, France
[2] Département MMIP Et INRA UMR-518, AgroParisTech, 16, Rue Claude Bernard, 75231 Paris Cedex 5, France
antoine.cornuejols@agroparistech.fr
http://www.agroparistech.fr/mia/equipes:membres:page:antoine

Abstract. Classification of time series as early as possible is a valuable goal. Indeed, in many application domains, the earliest the decision, the more rewarding it can be. Yet, often, gathering more information allows one to get a better decision. The optimization of this time vs. accuracy tradeoff must generally be solved online and is a complex problem.

This paper presents a formal criterion that expresses this trade-off in all generality together with a generic sequential meta algorithm to solve it. This meta algorithm is interesting in two ways. First, it pinpoints where choices can (*have to*) be made to obtain a computable algorithm. As a result a wealth of algorithmic solutions can be found. Second, it seeks online the earliest time in the future where a minimization of the criterion can be expected. It thus goes beyond the classical approaches that myopically decide at each time step whether to make a decision or to postpone the call one more time step.

After this general setting has been expounded, we study one simple declination of the meta-algorithm, and we show the results obtained on synthetic and real time series data sets chosen for their ability to test the robustness and properties of the technique. The general approach is vindicated by the experimental results, which allows us to point to promising perspectives.

Keywords: Early classification of time series · Sequential decision making

1 Introduction

In many applications, it is natural to acquire the description of an object incrementally, with new measurements arriving sequentially. This is the case in medicine, when a patient undergoes successive examinations until it is determined that enough evidence has been acquired to decide with sufficient certainty the disease he/she is suffering from. Sometimes, the measurements are not controlled and just arrive over time, as when the behavior of a consumer on a web site is monitored on-line in order to predict what add to put on his/her screen.

© Springer International Publishing Switzerland 2015
A. Appice et al. (Eds.): ECML PKDD 2015, Part I, LNAI 9284, pp. 433–447, 2015.
DOI: 10.1007/978-3-319-23528-8_27

In these situations, one is interested in making a prediction as soon as possible, either because each measurement is costly or because it is critical to act as early as possible in order to yield higher returns. However, this generally induces a tradeoff as less measurements commonly entail more prediction errors that can be expensive. The question is therefore how to decide on-line that now is the optimal time to make a prediction.

The problem of deciding when enough information has been gathered to make a reliable decision has historically been studied under the name of *sequential decision making* or *Optimal statistical decisions* [1,2]. One foremost technique being Wald's Sequential Probability Ratio Test [3] which applies to two-classes classification problems and uses the likelihood ratio:

$$R_t \;=\; \frac{P(\langle x_1^i, \ldots, x_t^i \rangle \mid y = -1)}{P(\langle x_1^i, \ldots, x_t^i \rangle \mid y = +1)}$$

where $\langle x_1^i, \ldots, x_t^i \rangle$ is the sequence of t measurements so far that must be classified to either class -1 or class $+1$. As the number of measurements t increases, this ratio is compared to two thresholds set according to the required error of the first kind α (*false positive error*) and error of the second kind β (*false negative error*). The main difficulty lies in the estimation of the conditional probabilities $P(\langle x_1^i, \ldots, x_t^i \rangle \mid y)$. (See also [4], a modern implantation of this idea).

A prominent limitation of this general approach is that the cost of delaying the decision is not taken into account. More recent techniques include the two components of the cost of early classification problems: the cost associated with *the quality* of the prediction and the cost of *the delay* before a prediction is made about the incoming sequence. However, most of them compute an optimal decision time from the learning set, which is then applied to any incoming example whatever their characteristics are. The decision is therefore not adaptive since the delay before making a prediction is independent on the input sequence.

The originality of the method presented here is threefold. *First*, the problem of early classification of time series is formalized as a sequential decision problem involving the two costs: quality and delay of the prediction. *Second*, the method is adaptive, in that the properties of the incoming sequence are taken into account to decide what is the optimal time to make a prediction. And *third*, in contrast to the usual sequential decision making techniques, the algorithm presented is not myopic. At each time step, it computes what is the optimal expected time for a decision in the future, and it is only if this expected time is the current time that a decision is made. A myopic procedure would only look at the current situation and decide whether it is time to stop asking for more data and make a decision or not. It would never try to estimate in advance the best time to make the prediction. The capacity of conjecturing when in the future an optimal prediction should be made with regard to the quality and delay of the prediction is however important and offers valuable opportunities compared to myopic sequential decisions. Indeed, when the prediction is about the breakdown of an equipment or about the possible failure of an organ in a patient, this forecast capacity allows one to make preparations for thwarting as best as possible the breakdown or failure, rather than reacting in haste at the last moment.

The paper is organized as follows. We first review some related work in Section (2). The formal statement of the early classification problem (Section (3)) leads to a generic sequential decision making meta algorithm. Our early decision making proposed approach and its optimal decision rule are formalized in Section (4). In Section (5), we propose one simple implementation of this meta algorithm to illustrate our approach. Experiments and results on synthetic data as well as on real data are presented and discussed in Section (6). The conclusion, in Section (7), underlines the promising features of the approach presented and discusses future works.

2   A Generic Framework and Positions of Related Works

In the following, we will assume that we have a set $\mathcal{S}$ of m training sequences with each training sequence being a couple $(\mathbf{x}_T^i, y_i) \in \mathbb{R}^T \times \mathcal{Y}$, meaning that it is composed of T real valued measurements $\langle x_1^i, \ldots, x_T^i \rangle$, and an associated label $y_i \in \mathcal{Y}$, where $\mathcal{Y}$ is a finite set of classes. The question is to choose the earliest time t^* at which a new incoming and still incomplete sequence $\mathbf{x}_{t^*} = \langle x_1, x_2, \ldots, x_{t^*} \rangle$ can be optimally labeled. Algorithm (1) provides a generic description of early classification methods.

Algorithm 1. Framework of early classification methods

Input:

    – $\mathbf{x}_t \in \mathbb{R}^t$, $t \in \{1, \ldots, T\}$, an incoming time series;
    – $\{h_t\}_{t \in \{1 \ldots, T\}} : \mathbb{R}^t \longrightarrow \mathcal{Y}$, a set of predictive functions h_t learned from the training set;
    – $x_t \in \mathbb{R}$, a new incoming real measurement;
    – $Trigger : \mathbb{R}^t \times h_t \longrightarrow \mathcal{B}, t \in \{1, \ldots, T\}, \mathcal{B} \in \{true, false\}$, a boolean decision function that decides whether it is time or not to output the prediction $h_t(\mathbf{x}_t)$ on the class of $\mathbf{x}_t$;

1: $\mathbf{x}_t \longleftarrow \varnothing$
2: $t \longleftarrow 0$
3: **while** $(\neg Trigger(\mathbf{x}_t, h_t))$ **do**   /* wait for an additional measurement
4:    $\mathbf{x}_t \longleftarrow Concat(\mathbf{x}_t, x_t)$ /* a new measurement is added at the end of $\mathbf{x}_t$
5:    $t \longleftarrow t + 1$
6:    **if** $(Trigger(\mathbf{x}_t, h_t)$ $||$ $t = T)$ **then**
7:        $\hat{y} \longleftarrow h_t(\mathbf{x}_t)$ /* predict the class of $\mathbf{x}_t$ and exit the loop
8:    **end if**
9: **end while**

In the framework outlined above, we suppose that the training set $\mathcal{S}$ has been used in order to learn a series of hypotheses $h_t(t \in \{1, \ldots, T\})$, each hypothesis h_t being able to classify examples of length t: $\mathbf{x}_t = \langle x_1, x_2, \ldots, x_t \rangle$.

Then, the various existing methods for early classification of time series can be categorized according to the $\mathcal{T}rigger$ function which decides when to stop measuring additional information and output a prediction $h_t(\mathbf{x}_t)$ for the class of $\mathbf{x}_t$.

Several papers that are openly motivated by the problem of early classification turn out indeed to be concerned with the problem of classifying from incomplete sequences rather than with the problem of optimizing a tradeoff between the precision of the prediction and the time it is performed. (see for instance [5] where clever classification schemes are presented, but no explicit cost for delaying the decision is taken into account). Therefore there is stricto sensu no $\mathcal{T}rigger$ function used in these algorithms.

In [6], the $\mathcal{T}rigger$ function relies on an estimate of the earliest time at which the prediction $h_t(\mathbf{x}_t)$ should be equal to the one that would be made if the complete example $\mathbf{x}_T$ was known: $h_T(\mathbf{x}_T)$. The so-called *minimum prediction length* (MPL) is introduced, and is estimated using a one nearest neighbor classifier.

In a related work [7,8], the $\mathcal{T}rigger$ function is based on a very similar idea. It outputs *true* when the probability that the assigned label $h_t(\mathbf{x}_t)$ will match the one that would be assigned using the complete time series $h_T(\mathbf{x}_T)$ exceeds some given threshold. To do so, the authors developed a *quadratic discriminant analysis* that estimates a reliability bound on the classifier's prediction at each time step.

In [9], the $\mathcal{T}rigger$ function outputs *true* if the classification function h_t has a sufficient confidence in its prediction. In order to estimate this confidence level, the authors use an ensemble method whereby the level of agreement is translated into a confidence level.

In [10], an early classification approach relying on uncertainty estimations is presented. It extends the *early distinctive shapelet classification* (EDGC) [11] method to provide an uncertainty estimation for each class at each time step. Thus, an incoming time series is labeled at each time step with the class that has the minimum uncertainty at that time. The prediction is triggered once a user-specified uncertainty threshold is met.

It is remarkable that even if the earliness of the decision is mentioned as a motivation in these papers, the decision procedures themselves do not take it explicitly into account. They instead evaluate the confidence or reliability of the current prediction(s) in order to decide if the time is ripe for prediction, or if it seems better to wait one more time step. In addition, the procedures are myopic in that they do not look further than the current time to decide if it a prediction should be made.

In this paper, we present a method that explicitly balance the expected gain in the precision of the decision *at all future time steps* with the cost of delaying the decision. In that way, the optimizing criterion is explicitly a function of both aspects of the early decision problem, and, furthermore, it allows one to estimate, and update if necessary, the future optimal time step for the decision.

3   A Formal Analysis and a Naïve Approach

The question is to learn a decision procedure in order to determine the earliest time t^* at which a new incoming sequence $\mathbf{x}_{t^*} = \langle x_1, x_2, \ldots, x_{t^*} \rangle$ can be optimally labeled. To do so we associate a cost with the prediction quality of the decision procedure and a cost with the time step when the prediction is finally made:

- We assume that a *misclassification cost function* $C_t(\hat{y}|y) : \mathcal{Y} \times \mathcal{Y} \longrightarrow \mathbb{R}$ is given, providing the cost at time t of predicting $\hat{y}$ when the true class is y.
- Each time step t is associated with a real valued *time cost function* $C(t)$ which is non decreasing over time, which means that it is always more costly to wait for making a prediction. Note that, in contrast to most other approaches, this function can be different from a linear one, reflecting the peculiarities of the domain. For instance, if the task is to decide if an electrical power plant must be started or not, the waiting cost rises sharply as the last possible time approaches.

We can now define a cost function f associated with the decision problem.

$$f(\mathbf{x}_t) = \sum_{y \in \mathcal{Y}} P(y|\mathbf{x}_t) \sum_{\hat{y} \in \mathcal{Y}} P(\hat{y}|y, \mathbf{x}_t) \, C_t(\hat{y}|y) \; + \; C(t) \tag{1}$$

This equation corresponds to the expectation of the cost of misclassification after t measurements have been made, added to the cost of having delaying the decision until time t. The optimal time t^* for the decision problem is then defined as :

$$t^* = \operatorname*{ArgMin}_{t \in \{1, \ldots, T\}} f(\mathbf{x}_t) \tag{2}$$

However, this formulation of the decision problem requires that one be able to compute the conditional probabilities $P(y|\mathbf{x}_t)$ and $P(\hat{y}|y, \mathbf{x}_t)$. The first one is unknown, otherwise there would be no learning problem in the first place. The second one is associated with a given classifier, and is equally difficult to estimate.

Short of being able to estimate these terms, one can fall back on the expectation of the cost for *any sequence* (hence the function now denoted $f(t)$):

$$f(t) = \sum_{y \in \mathcal{Y}} P(y) \sum_{\hat{y} \in \mathcal{Y}} P(\hat{y}|y) \, C_t(\hat{y}|y) \; + \; C(t) \tag{3}$$

From the training set $\mathcal{S}$, it is indeed easy to compute the a priori probabilities $P(y)$ and the conditional probabilities $P(\hat{y}|y)$ which are nothing else that the confusion matrix associated with the considered classifier. One gets then the optimal time for prediction as:

$$t^* = \operatorname*{ArgMin}_{t \in \{1, \ldots, T\}} f(t)$$

This can be computed before any new incoming sequence, and, indeed, t^* is independent on the input sequence. Of course, this is intuitively unsatisfactory as one could feel, regarding a new sequence, very confident (resp. not confident) in his/her prediction way before (resp. after) the prescribed time t^*. If such is the case, it seems foolish to make the prediction exactly at time t^*. This is why we propose an adaptive approach.

4   The Proposed Approach

The goal is to estimate the conditional probability $P(\hat{y}|y, \mathbf{x}_t)$ in Equation (1) by taking into account the incoming time series $\mathbf{x}_t$ in order to determine the optimal time t^*. There are several possibilities for this.

In this paper, the idea is to identify a set $\mathcal{C}$ of K clusters c_k ($k \in \{1, \ldots, K\}$) of complete sequences using a training set so that, later, an (incomplete) input sequence $\mathbf{x}_t = \langle x_1, \ldots, x_t \rangle$ can have a membership probability assigned to each of these clusters: $P(c_k \mid \mathbf{x}_t)$, and therefore will be recognized as more or less close to each of the prototype sequences corresponding to the clusters. A complete explanation is given below in Section 5.

The set $\mathcal{C}$ of clusters should obey two constraints as well as possible.

1. Different clusters should correspond to different confusion matrices. Otherwise, Equation (1) will not be able to discriminate the cost between clusters.
2. Clusters should contain similar time series, and be dissimilar to other clusters, so that an incoming sequence will generally be assigned markedly to one of the clusters.

For each time step $t \in [1, \ldots, T]$, a classifier h_t is trained using a learning set $\mathcal{S}'$. One can then estimate the associated confusion matrix for each cluster and classifier h_t: c_k: $P_t(\hat{y}|y, c_k)$ over a distinct learning set $\mathcal{S}''$.

When a new input sequence $\mathbf{x}_t$ of length t is considered, it is compared to each cluster c_k (of complete time series) and is given a probability membership $P_t(\hat{y}|y, c_k)$ for each of them (as detailed in Section (5)). In a way, this compares the input sequence to all families of its possible continuations.

Given that, at time t, $T - t$ measurements are still missing on the incoming sequence, it is possible to compute the expected cost of classifying $\mathbf{x}_t$ at each future time step $\tau \in \{0, \ldots, T - t\}$:

$$f_\tau(\mathbf{x}_t) = \sum_{c_k \in \mathcal{C}} P(c_k|\mathbf{x}_t) \sum_{y \in \mathcal{Y}} \sum_{\hat{y} \in \mathcal{Y}} P_{t+\tau}(\hat{y}|y, c_k)\, C(\hat{y}|y) + C(t + \tau) \qquad (4)$$

Perhaps not apparent at first, this equation expresses two remarkable properties.

First, it is computable, which was not the case of Equation (1). Indeed, each of the terms $P(c_k|\mathbf{x}_t)$ and $P_{t+\tau}(\hat{y}|y, c_k)$ can now be estimated through frequencies observed in the training data (see Figure (1)). Second, the cost depends on the incoming sequence because of the use of the probability memberships $P(c_k|\mathbf{x}_t)$. It is therefore not computed beforehand, once for all.

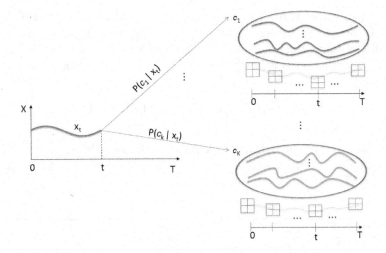

Fig. 1. An incoming (incomplete) sequence is compared to each cluster c_k obtained from the training set of complete time series. The confusion matrices for each time step t and each cluster c_k are computed as explained in the text.

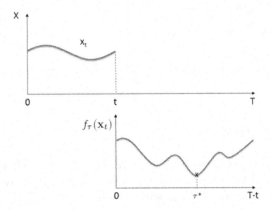

Fig. 2. The first curve represents an incoming time series $\mathbf{x}_t$. The second curve represents the expected cost $f_\tau(\mathbf{x}_t)$ given $\mathbf{x}_t$, $\forall \tau \in \{0, \ldots, T - t\}$. It shows the balance between the gain in the expected precision of the prediction and the cost of waiting before deciding. The minimum of this tradeoff is expected to occur at time τ^*. New measurements can modify the curve of the expected cost and the estimated τ^*.

In addition, the fact that the expected cost $f_\tau(\mathbf{x}_t)$ can be computed for each of the remaining τ time steps allows one to forecast what should be the optimal horizon τ^* for the classification of the input sequence (see Figure (2)):

$$\tau^* = \underset{\tau \in \{0, \ldots, T-t\}}{\text{ArgMin}} \; f_\tau(\mathbf{x}_t) \tag{5}$$

A. Dachraoui et al.

Of course, these costs, and the expected optimal horizon $\tau^\star$, can be re-evaluated when a new measurement is made on the incoming sequence. At any time step t, if the optimal horizon $\tau^* = 0$, then the sequential decision process stops and a prediction is made about the class of the input sequence $\mathbf{x}_t$ using the classifier h_t^k:

$$\hat{y} = h_t^k(\mathbf{x}_t), \quad \text{where } k = \underset{k \in \{1,\dots,K\}}{\text{ArgMax}} \ P(\mathfrak{c}_k | \mathbf{x}_t)$$

Returning to the general framework outlined for the early classification problem in Section (3), the proposed function that triggers a prediction for the incoming sequence is given in Algorithm (2):

Algorithm 2. Proposed $\mathcal{T}rigger(\mathbf{x}_t, h_t)$ function.

Input: $\mathbf{x}_t$, $t \in \{1,\dots,T\}$, an incoming time series;
1: $\mathcal{T}rigger \longleftarrow false$
2: **for all** $\tau \in \{0,\dots,T-t\}$ **do**
3:     compute $f_\tau(\mathbf{x}_t)$ /* see Equation (4)*/
4: **end for**
5: $\tau^* = \underset{\tau \in \{0,\dots,T-t\}}{\text{ArgMin}} \ f_\tau(\mathbf{x}_t)$
6: **if** $(\tau^* = 0)$ **then**
7:     $\mathcal{T}rigger \longleftarrow true$
8: **end if**

---

5 Implementation

Section (3) has outlined the general framework for the early classification problem while Section (4) has presented our proposed approach where the problem is cast as a sequential decision problem with three properties: (i) both the quality of the prediction and the delay before prediction are taken into account in the total criterion to be optimized, (ii) the criterion is adaptive in that it depends upon the incoming sequence $\mathbf{x}_t$, and (iii) the proposed solution leads to a non myopic scheme where the system forecasts the expected optimal horizon τ^* instead of just deciding that now is, or is not, the time to make a prediction.

In order to implement the proposed approach, choices have to be made about:

1. The type of *classifiers* used. For each time step $t \in \{1,\dots,T\}$, the input dimension of the classifier is t.
2. The *clustering method*, which includes the technique (e.g. k-means), the distance used (e.g. the euclidean distance, the time warping distance, ...), and the number of clusters that are looked for.
3. The method for computing the membership probabilities $P(\mathfrak{c}_k | \mathbf{x}_t)$.

In this paper, we have chosen to use simple, direct, techniques to implement each of the choices above, so as to clearly single out the properties of the approach through "baseline results". Better results can certainly be obtained with more sophisticated techniques.

Accordingly, (1) for the classifiers, we have used Naïve Bayes classifiers and Multi-layer Perceptrons with one hidden layer of $\lfloor t + 2/2 \rfloor$ neurons. In Section (6), we only show results obtained using the Multi-Layer Perceptron since both classifiers give similar results. (2) The clustering over complete time series is performed using k-means with euclidean distance. The number K_y of clusters for each of the target classes $y = -1$ and $y = +1$ corresponds to the maximum *silhouettes* factor [12]. (3) The membership probabilities $P(\mathfrak{c}_k|\mathbf{x}_t)$ are computed using the following equation:

$$P(\mathfrak{c}_k|\mathbf{x}_t) \;=\; \frac{s_k}{\sum_i^K s_i}, \quad \text{where } s_k \;=\; \frac{1}{1 + \exp^{-\lambda \Delta_k}} \tag{6}$$

The constant λ used in the sigmoid function s_k is empirically learned from the training set, while $\Delta_k \;=\; \overline{D} - d_k$ is the difference between the average of the distances between $\mathbf{x}_t$ and all the clusters, and the distance between $\mathbf{x}_t$ and the cluster $\mathfrak{c}_k$. The distance between an incomplete incoming time series $\mathbf{x}'_t = \langle x_1, \ldots, x_t \rangle$ and a complete one $\mathbf{x}"_T = \langle x_1, \ldots, x_T \rangle$ is done here using the Euclidian distance between the first t components of the two series.

6   Experiments

Our experiments aimed at checking the validity of the proposed method and at exploring its capacities for various conditions. To this end, we devised controlled experiments with artificial data sets for which we could vary the control parameters: difference between the two target classes, noise level, number of different time series shapes in each class and the cost of waiting before decision $C(t)$. We also applied the method to the real data set TwoLeadECG from UCR Time Series Classification/Clustering repository [13].

6.1   Controlled Experiments

We devised our experiments so that there should be a gain, that we can control, in the prediction accuracy if more measurements are made (increasing t). We have also devised the target classes so that they are composed of several families of time sequences, with, possibly, families that share a strong resemblance between different target classes.

In the reported experiments, the time series in the training set and the testing set are generated according to the following equations:

$$\mathbf{x}_t \;=\; a \, \sin(\omega_i t + phase) + b \, t \;+\; \varepsilon(t) \tag{7}$$

The constant b is used to set a general trend, for instance either ascending ($b > 0$) or descending ($b < 0$), while the first term $a \, \sin(\omega_i t + phase)$ provides

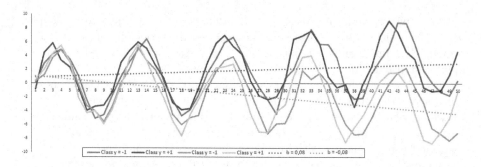

Fig. 3. Subgroups of sequences generated for classes $y = +1$ and $y = -1$, when the trend parameter $b = -0.08$ or $b = +0.08$, and the noise level $\varepsilon(t) = 0.5$.

a shape for this particular family of time series. The last term is a noise factor that makes the overall prediction task more or less difficult.

For instance, Figure (3) shows a set of time series (one for each shape) where:

- $b = -0.08$ or $b = +0.08$
- $a = 5$ and $phase = 0$
- $\omega_1 = 10$ or $\omega_2 = 10.3$ (here, there are 2 groups of time sequences per class)
- $\varepsilon(t)$ is a gaussian term of mean $= 0$ and standard deviation $= 0.5$
- $T = 50$

In this particular setting, it is apparent that it is easy to mix up the two classes $y = -1$ and $y = +1$ until intermediate values of t. However, the waiting cost $C(t)$ may force the system to make a decision before there is enough measurements to make a reasonably sure guess on the class y.

In our experiments, the training set S contained 2,500 examples, and the testing set T contained 1000 examples, equally divided into the two classes $y = -1$ and $y = +1$. (Nota: In case of imbalanced classes, it is easy to compensate this by modifying the misclassification cost function $C_t(\hat{y}|y)$). Each class was made of several subgroups: K_{-1} ones for class -1 and K_{+1} ones for class $+1$. The misclassification costs were set as: $C(\hat{y}|y) = 1$, $\forall~\hat{y}, y$, and the *time cost function* $C(t) = d \times t$, where $d \in \{0.01, 0.05, 0.1\}$.

We varied:

- The level of *distinction between the classes* controlled by b
- The *number of subgroups* in each class and their shape (given by the term $a \sin(\omega_i t + phase)$)
- The *noise* level $\varepsilon(t)$
- The *cost of waiting* before decision $C(t)$

The results for various combinations of these parameters are shown in Table (1) as obtained on the time series of the testing set. It reports $\bar{\tau}^\star$, the average of computed optimal times of decision and its associated standard deviation $\sigma(\tau^\star)$.

Table 1. Experimental results in function of the waiting cost $C(t) = \{0.01, 0.05, 0.1\} \times t$, the noise level $\varepsilon(t)$ and the trend parameter b.

$C(t)$	$\pm b$ $\varepsilon(t)$	0.02 $\bar{\tau}^\star$	$\sigma(\tau^\star)$	AUC	0.05 $\bar{\tau}^\star$	$\sigma(\tau^\star)$	AUC	0.07 $\bar{\tau}^\star$	$\sigma(\tau^\star)$	AUC
	0.2	9.0	2.40	0.99	9.0	2.40	0.99	10.0	0.0	1.00
	0.5	13.0	4.40	0.98	13.0	4.40	0.98	15.0	0.18	1.00
0.01	1.5	24.0	10.02	0.98	32.0	2.56	1.00	30.0	12.79	0.99
	5.0	26.0	7.78	0.84	30.0	18.91	0.87	30.0	19.14	0.88
	10.0	38.0	18.89	0.70	48.0	1.79	0.74	46.0	5.27	0.75
	15.0	23.0	15.88	0.61	32.0	13.88	0.64	29.0	17.80	0.62
	20.0	7.0	8.99	0.52	11.0	11.38	0.55	4.0	1.22	0.52
	0.2	8.0	2.00	0.98	8.0	2.00	0.98	9.0	0.0	1.00
	0.5	10.0	2.80	0.96	8.0	4.0	0.98	14.0	0.41	0.99
0.05	1.5	5.0	0.40	0.68	20.0	0.42	0.95	14.0	4.80	0.88
	5.0	8.0	3.87	0.68	6.0	1.36	0.64	5.0	0.50	0.65
	10.0	4.0	0.29	0.56	4.0	0.25	0.56	4.0	0.34	0.57
	15.0	4.0	0.0	0.54	4.0	0.25	0.56	4.0	0.0	0.55
	20.0	4.0	0.0	0.52	4.0	0.0	0.52	4.0	0.0	0.52
	0.2	6.0	0.80	0.95	7.0	1.60	0.94	8.0	0.40	0.96
	0.5	6.0	0.80	0.84	9.0	2.40	0.93	10.0	0.0	0.95
0.10	1.5	4.0	0.0	0.67	5.0	0.43	0.68	6.0	0.80	0.74
	5.0	4.0	0.07	0.64	4.0	0.05	0.64	4.0	0.11	0.64
	10.0	4.0	0.0	0.56	48.0	1.79	0.74	4.0	0.22	0.56
	15.0	4.0	0.0	0.55	4.0	0.0	0.55	4.0	0.0	0.55
	20.0	4.0	0.0	0.52	11.0	11.38	0.55	4.0	0.0	0.52

Additionally, the Area Under the ROC Curve AUC evaluates the quality of the prediction at the optimal decision time $\tau^\star$ computed by the system.

Globally, one can see that when the noise level is low ($\varepsilon \leq 1.5$) and the waiting cost is low too ($C(t) = c_t \times t$, with $c_t \leq 0.05$), the system is able to reach a high level of performance by waiting increasingly as the noise level augments. When the waiting cost is high ($C(t) = 0.1 \times t$), on the other hand, the system takes a decision earlier at the cost of a somewhat lower prediction performance. Indeed, with rising levels of noise, the system decides that it is not worth waiting and makes a prediction early on, often at the earliest possible moment, which was set to 4 in our experiments[1].

More specifically:

- **Impact of the noise level** $\varepsilon(t)$: As expected, up to a certain value, rising levels of noise $\varepsilon(t)$ entails increasing delays before a decision is decided upon by the system. Then, a decrease of $\bar{\tau}^\star$ is observed, which corresponds to the

---

[1] Below 4 measurements, the classifiers are not effective.

fact that there is no gain to be expected by waiting further. Accordingly, the performance, as measured with the AUC, decreases as well when $\varepsilon(t)$ rises.

– **Impact of the waiting cost $C(t)$**: The role of the waiting cost $C(t)$ appears clearly. When $C(t)$ is very low, the algorithm tends to wait longer before making a decision, often waiting the last possible time. On the other hand, with rising $C(t)$, the optimal decision time $\overline{\tau}^\star$ decreases sharply, converging to the minimal possible value of 4.

– **Impact of the trend parameter b**: While the value of b, which controls the level of distinction of the classes $y = +1$ and $y = -1$, is not striking on the average time of decision $\overline{\tau}^\star$, one can notice however the decrease of the standard deviation when b increases from $b = 0.02$ to $b = 0.05$. At the same time, the AUC increases as well. For small values of the noise level, the decrease of the standard deviation is further observed when $b = 0.07$.

– **Impact of the number of subgroups in each class**: In order to measure the *effect of the complexity of each class* on the decision problem, we changed the number of shapes in each class as well. This is easily done in our setting by using sets of different values of the parameters in Equation (7). For instance, Table (2) reports the results obtained when the number of subgroups of class $y = -1$ was set to $K_{-1} = 3$ while it was set to $K_{+1} = 5$ for class $y = +1$. When the waiting cost is very low $(C(t) = 0.01)$, the number of subgroups in each class, and hence the complexity of the classes, does not influence the results. However, when the waiting cost increases $(C(t) = 0.05 \times t)$, the decision task becomes harder, and the decision time increases while the AUC decreases.

The above results, in Table (1) and Table (2), aggregate the measures on the whole testing set. It is interesting to look as well at individual behaviors. For instance, Figure (4) shows the expected costs $f_\tau(\mathbf{x}_t^1)$ and $f_\tau(\mathbf{x}_t^2)$ for two different incoming sequences $\mathbf{x}_t^1$ and $\mathbf{x}_t^2$, for each of the potentially remaining τ time steps. First, one can notice the overall shape of the cost function $f_\tau(\mathbf{x}_t)$ with a decrease followed by a rise. Second, the *dependence on the incoming sequence* appears clearly, with different optimal times $t^\star$. This confirms that the algorithm takes into account the peculiarities of the incoming sequence.

6.2   Experiments on a Real Data Set

In order to test the ability of the method to solve real problems, we have realized experiments using the real data set TwoLeadECG from the UCR repository. This data set contains 1162 ECG signals all together, that we randomly and disjointedly re-sampled and split into a training set of 70% of examples and the remainder for the test set. Each signal is composed of 81 data point representing the electrical activity of the heart from two different leads. The goal is to detect an abnormal activity in the heart. Our experiments show that it is indeed possible to make an informed decision before all measurements are made.

Table 2. Experimental results in function of the noise level $\varepsilon(t)$, the trend parameter b, and the number of subgroups k_{+1} and k_{-1} in each class. The waiting cost $C(t)$ is fixed to 0.01.

(K_{-1}, K_{+1}) $\pm b$ $\varepsilon(t)$		0.02			0.05			0.07		
		$\overline{\tau}^\star$	$\sigma(\tau^\star)$	AUC	$\overline{\tau}^\star$	$\sigma(\tau^\star)$	AUC	$\overline{\tau}^\star$	$\sigma(\tau^\star)$	AUC
	0.2	9.0	2.40	0.99	9.0	2.40	0.99	10.0	0.0	1.00
	0.5	13.0	4.40	0.98	13.0	4.40	0.98	15.0	0.18	1.00
(3,2)	1.5	24.0	10.02	0.98	32.0	2.56	1.00	30.0	12.79	1.00
	5.0	26.0	7.78	0.84	30.0	18.90	0.87	30.0	19.14	0.88
	10.0	38.0	18.89	0.70	48.0	1.79	0.74	46.0	5.27	0.75
	15.0	23.0	15.88	0.61	32.0	13.88	0.64	29.0	17.80	0.62
	20.0	7.0	8.99	0.52	11.0	11.38	0.55	4.0	1.22	0.52
	0.2	7.0	2.47	0.86	7.0	2.15	0.89	7.0	3.00	0.85
	0.5	11.0	5.10	0.87	10.0	4.87	0.88	14.0	7.07	0.91
(3,5)	1.5	20.0	12.69	0.85	18.0	11.80	0.87	26.0	16.33	0.89
	5.0	44.0	4.75	0.83	46.0	2.81	0.87	38.0	11.49	0.81
	10.0	42.0	6.34	0.67	39.0	7.59	0.68	25.0	8.57	0.61
	15.0	28.0	5.99	0.58	32.0	6.51	0.59	19.0	10.12	0.58
	20.0	17.0	11.72	0.50	13.0	10.72	0.56	17.0	5.93	0.55

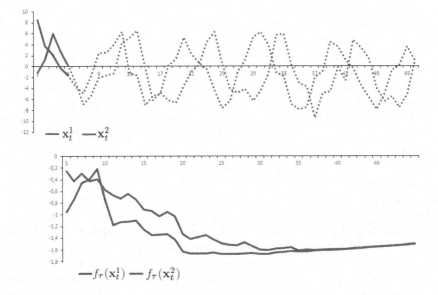

Fig. 4. For two different incoming sequences (top figure), the expected costs (bottom figure) are different. The minima have different values and occur at different instants. These differences confirm that deciding to make a prediction depends on the incoming sequence. (Here, $b = 0.05$, $C(t) = 0.01 \times t$ and $\varepsilon = 1.5$).

Table 3. Experimental results on real data in function of the waiting cost $C(t)$.

$C(t)$	0.01	0.05	0.1
$\bar{\tau}^\star$	22.0	24.0	10.0
$\sigma(\tau^\star)$	6.1214	15.7063	9.7506
AUC	0.9895	0.9918	0.9061

Since the costs involving quality and delay of decision are not provided with this data set, we arbitrarily set these costs to $C(\hat{y}|y) = 1$, $\forall\, \hat{y}, y$, and $C(t) = d \times t$, where $d \in \{0.01, 0.05, 0.1\}$. The question here is whether the method is able to make reliable prediction early and provide reasonable results.

Table (3) reports the average of optimal times of decision $\bar{\tau}^\star$ of test time series, its associated standard deviation $\sigma(\tau^\star)$, and the performance of the prediction AUC. It is remarkable that *a very good performance*, as measured by the AUC, *can be obtained from a limited set of measurements*: E.g. 22 out of 81 if $C(t) = 0.01$, 24 out of 81 if $C(t) = 0.05$, and 10 out of 81 if $C(t) = 0.1$.

We therefore see that the baseline solution proposed here is able to (1) adapt to each incoming sequence and (2) to predict an estimated optimal time of prediction that yields very good prediction performances while controlling the cost of delay.

7   Conclusion and Future Works

The problem of online decision making has been known for decades, but numerous new applications in medicine, electric grid management, automatic transportation, and so on, give a new impetus to research works in this area. In this paper, we have formalized a generic framework for early classification methods that underlines two critical parts: (i) the optimization criterion that governs the *Trigger* boolean function, and (ii) the manner by which the current information about the incoming time sequence is taken into account.

Within this framework, we have proposed an optimization criterion that balances the expected gain in the classification cost in the future with the cost of delaying the decision. One important property of this criterion is that it can be computed at each time step for all future instants. This prediction of the future gains is updated given the current observation and is therefore never certain, but this yields a non myopic sequential decision process.

In this paper, we have sought to determine the baseline properties of our proposed framework. Thus, we have used simple techniques as: (i) clustering of time series in order to compare the incoming time sequence to known shapes from the training set, (ii) a simple formula to estimate the membership probability $P(c_k|\mathbf{x}_t)$, and (iii) not optimized classifiers, here: naïve Bayes or a simple implementation of Multi-Layer Perceptrons.

In this baseline setting, it is a remarkable feat that the experiments exhibit a remarkable fit with desirable properties for an early decision classification algorithm, as stated in Section 6. The system indeed controls the decision time so as to ensure a high level of prediction performance as best as possible given the level of difficulty of the task and the cost of delaying the decision. It is also adaptive by taking into account the peculiarities of the incoming time sequence.

While we have obtained quite satisfying and promising results in the experiments carried out on controlled data and on a real data set, one direction for future work is to boost up this baseline implementation. In particular, we have ideas about how to use training sequences in order to predict the future decision cost of an incoming time sequence without using a clustering approach. Besides, dedicated methods for classifying time sequences should be used rather than naïve Bayes or simple MLP.

Still, even as it is, the method presented here should prove a useful tool for many early classification tasks.

References

1. DeGroot, M.H.: Optimal statistical decisions, vol. 82. John Wiley & Sons (2005)
2. Berger, J.O.: Statistical decision theory and Bayesian analysis. Springer Science & Business Media (1985)
3. Wald, A., Wolfowitz, J.: Optimum character of the sequential probability ratio test. The Annals of Mathematical Statistics, 326–339 (1948)
4. Sochman, J., Matas, J.: Waldboost-learning for time constrained sequential detection. In: IEEE Computer Society Conference on Computer Vision and Pattern Recognition, CVPR 2005, vol. 2, pp. 150–156. IEEE (2005)
5. Ishiguro, K., Sawada, H., Sakano, H.: Multi-class boosting for early classification of sequences. Statistics **28**, 337–407 (2000)
6. Xing, Z., Pei, J., Philip, S.Y.: Early prediction on time series: A nearest neighbor approach. In: IJCAI, pp. 1297–1302. Citeseer (2009)
7. Anderson, H.S., Parrish, N., Tsukida, K., Gupta, M.: Early time-series classification with reliability guarantee. Sandria Report (2012)
8. Parrish, N., Anderson, H.S., Gupta, M.R., Hsiao, D.Y.: Classifying with confidence from incomplete information. J. of Mach. Learning Research **14**, 3561–3589 (2013)
9. Hatami, N., Chira, C.: Classifiers with a reject option for early time-series classification. In: 2013 IEEE Symposium on Computational Intelligence and Ensemble Learning (CIEL), pp. 9–16. IEEE (2013)
10. Ghalwash, M.F., Radosavljevic, V., Obradovic, Z.: Utilizing temporal patterns for estimating uncertainty in interpretable early decision making. In: Proceedings of the 20th ACM SIGKDD International Conference on Knowledge Discovery and Data Mining, pp. 402–411. ACM (2014)
11. Xing, Z., Pei, J., Philip, S.Y., Wang, K.: Extracting interpretable features for early classification on time series. In: SDM, vol. 11, pp. 247–258. SIAM (2011)
12. Rousseeuw, P.J.: Silhouettes: a graphical aid to the interpretation and validation of cluster analysis. J. of computational and applied mathematics **20**, 53–65 (1987)
13. Keogh, E., Xi, X., Wei, L., Ratanamahatana, C.A.: The ucr time series classification/clustering homepage (2006). www.cs.ucr.edu/eamonn/time_series_data/

Ising Bandits with Side Information

Shaona Ghosh[✉] and Adam Prügel-Bennett

University of Southampton, Southampton SO17 1BJ, UK
ghosh.shaona@gmail.com

Abstract. We develop an online learning algorithm for bandits on a graph with side information where there is an underlying Ising distribution over the vertices at low temperatures. We are motivated from practical settings where the graph state in a social or a computer hosts network (potentially) changes at every trial; intrinsically partitioning the graph thus requiring the learning algorithm to play the bandit from the current partition. Our algorithm essentially functions as a two stage process. In the first stage it uses *"minimum-cut"* as the regularity measure to compute the state of the network by using the side label received and acting as a graph classifier. The classifier internally uses a polynomial time linear programming relaxation technique that incorporates the known information to predict the unknown states. The second stage ensures that the bandits are sampled from the appropriate partition of the graph with the potential for exploring the other part. We achieve this by running the adversarial multi armed bandit for the edges in the current partition while exploring the "cut" edges. We empirically evaluate the strength of our approach through synthetic and real world datasets. We also indicate the potential for a linear time exact algorithm for calculating the max-flow as an alternative to the linear programming relaxation, besides promising bounded mistakes/regret in the number of times the "cut" changes.

1 Introduction

Many domains encounter a problem in collection of annotated training data due to the difficulty and costs in requiring efforts of human annotators, while the abundant unlabelled data come for free. What makes the problem more challenging is the data might often exhibit complex interactions that violate the independent and identically distributed assumption of the data generation process. In such domains, it is imperative that learning techniques can learn from unlabelled data and the rich interactions based structure of the data. Learning from unlabelled and a few labelled data falls under the purview of semi-supervised learning. Coupling it with an encoding of the data interdependencies as a graph, results in an attractive problem of learning on graphs.

Often, interesting applications are tied to such problems with rich underlying structure. For example, consider the system of online advertising; serving advertisements on web pages in an incremental fashion. The web pages can be represented as vertices in the graph with the links as edges. At given time t,

© Springer International Publishing Switzerland 2015
A. Appice et al. (Eds.): ECML PKDD 2015, Part I, LNAI 9284, pp. 448–463, 2015.
DOI: 10.1007/978-3-319-23528-8_28

the system receives a request to serve an advertisement on a randomly selected web-page. Moreover, at the same time, the system receives a side information about the state of the web-page: for simplicity we assume the side information to be a rating of 0 or 1. As a consequence, the advertisement pool change with the change in the state of the graph or the ratings, given the already known states and the current advert should be served from the appropriate pool. Once the chosen advertisement is served, the feedback is received and incorporated in serving the next request.

At a deeper level of understanding, the side information can be interpreted as the label of the vertex. There are few available labels at the start; the rest are only incrementally revealed. When a vertex is queried (request for an ad placement made), an action needs to be picked (an advertisement needs to be served) from a set of actions. The algorithm should be able to internally predict what the state of the queried vertex is (how the state of the graph changes) and then select the appropriate action from the action pool that (potentially)changes with the predicted label of the queried vertex.

In this paper, we attempt to tackle this problem by exploiting the knowledge of the non-independence graphical structure of the data in an online setting. We do so by associating a complexity with the labelling. We call this complexity "cut" or "energy" of the labelling on a Markov random field with discrete states (Ising model). The goal of our graph labelling procedure is to minimize the energy while being consistent with the information seen so far when predicting the intrinsic state of the queried vertex at every round. This prediction directs the overall goal towards minimizing the regret of our sequential action selection (bandit) algorithm within the online graph labelling that occurs over the entire sequence.

Related Work. Broadly speaking, there are two central themes that run through our work unified under the common framework of online learning, namely, action selection using bandit feedback and semi-supervised graph labelling. The closest related work that addresses the intersection of these two themes is the work by Claudio et al. [10]. They use bandit feedback to address a multi-class ranking problem. The algorithm outputs a partially ordered subset of classes and receives only bandit feedback (partial information) among the selected classes it observes without any supervised ranking feedback. In contrast, we play the bandit game of sequential action selection, using side information as the class label of the current context. Our feedback for the action selected is still partial (only loss for the selected action is observed). Further, our bandits have a structure associated with the Ising model distribution over the vertices at low temperature. The work of Amin et al.[2], addresses the graphical models for bandit problems to demonstrate the rich interactions between the two worlds in the similar lines of what we try to achieve. Bearing a strong resemblance to our work, they address the similar context-action space. However, in their setting, there is a strong coupling between the context-action space; the algorithm needs to fulfil the entire joint assignment before receiving any feedback. In contrast, our concept-action space is decoupled, labels are revealed gradually determining the current active concept for the learner to choose the action and receive the feedback instantaneously. In their problem formulation under the Ising graph setting, the algorithm tries to pick the action (the label of the concept) that is NP hard. In contrast, we focus on the low temperature setting, where our actions lie on the edges, and are not the labels of the vertices. The computation of the marginal at the vertices is guided by the labels seen so far and the minimal cut. We approximate the labelling of the entire graph rather than predicting the spin configuration of a single vertex using the "cut" as the regularizer that dominates the action selection. The contextual bandits work on online clustering of bandits [9], deals with finding groups or clusters of bandits in the graphs. They have a stochastic assumption of a linear function for reward generation. Similarity is revealed by the parameter vector that is inferred over time. In contrast, we use the similarity over edges to determine the "cut" which in-turns guides the action selection process in adversarial settings. There work extends to running a contextual bandit for every node, whereas ours is a single bandit algorithm, where the context information is captured in the "cut". The work of Castro et al. [7] of edge bandits is similar in the sense that the bandits lie on the edges. However, instead of direct rewards of action selection, rewards are a difference in the values of the vertices. Further, this is the stochastic setting instead of the adversarial one. In Spectral bandits [18], the actions are the nodes, while there is a smooth Laplacian graph function for the rewards. We discuss later the limitations of Laplacian based methods for graph labelling. Further, they do not consider the Ising model that we study. The seminal work of semi supervised graph labelling prediction can be found in [6], where minimum label-separating cut is used for prediction. Laplacian based methods that results neighbouring nodes connected by an edge to share similar values are widely studied in the semi-supervised and manifold learning problems [5,11,12,19,20]. Typically, this information is captured by the semi-norm induced by the Laplacian of the graph. Essentially, the smoothness of the labelling is ensured by the "cut". The "cut" is the number of edges with disagreeing labels. Then, the norm induced by the Laplacian can be considered as the regularizer. However, there are limitations in these methods with increasing unlabelled data [1,16]. Here, we also use "cut" as the regularization measure over an Ising model distribution of the values over the vertices of the graph at low temperatures. We simultaneously find the partition using the "min-cut" and then sample the actions from the relevant partition.

2   Background and Preliminaries

2.1   Semi-supervised Graph Classifier Complexity

The standard approach in semi supervised learning is to construct the graph from the unlabelled and labelled data such that each datum is denoted as a vertex. Traditionally, the norm induced by the graph Laplacian is used to predict the labelling. Typically, either the norm induced by the Laplacian is directly minimized/interpolated with respect to constraints or is used as a regulariser. Both methods help build classifiers on graphs in order to learn sparse labels in

$\mathbb{R}^n$ by incorporating a measure of complexity also called "cut" or energy. For a graph $\mathcal{G} = (V, E)$, where the set of vertices $V = \{v_1, \ldots, v_n\}$ are connected by edges in E. Let a weight of A_{ij} be associated with every edge $(i, j) \in E$, such that $\mathbf{A}$ is the $n \times n$ symmetric adjacency matrix, then the Laplacian $\mathbf{L}$ of the graph is given by $\mathbf{L} = \mathbf{D} - \mathbf{A}$, where $\mathbf{D}$ is the degree matrix with its diagonal values given by $D_{ii} = \sum_j A_{ij}$. We re-state Definition 1 from [14] that relates the quadratic form of the Laplacian with the complexity of the "cut-size" for completeness.

Definition 1 ([14]). *If the labelling of the graph $\mathcal{G}$ is given by $\mathbf{u} \in \mathbb{R}^n$, the "cut size" of $\mathbf{u}$ is given by*

$$\psi_{\mathcal{G}}(\mathbf{u}) = \frac{1}{4}\mathbf{u}^T \mathbf{L} \mathbf{u} = \frac{1}{4} \sum_{(i,j) \in E} A_{ij}(u_i - u_j)^2 \ . \tag{1}$$

When $\mathbf{u} \in \{0, 1\}^n$, the "cut" is on the edge (i, j) where $u_i \neq u_j$, then $\psi_{\mathcal{G}}(\mathbf{u})$ is the number of "cut" edges.

The smoothness functional of $\mathbf{u}^T \mathbf{L} \mathbf{u}$ is generalized in the work of semi-norm interpolation [13] where the Laplacian $p-$seminorm is defined on $\mathbf{u} \in \mathbb{R}^n$ as:

$$||\mathbf{u}||_{\mathcal{G},p} \simeq \psi_{\mathcal{G}}(\mathbf{u}) = \left(\sum_{(i,j) \in E} A_{ij} |u_i - u_j|^p \right)^{\frac{1}{p}} \ . \tag{2}$$

When $p = 2$, this is equivalent to the harmonic energy minimization technique in [20]. Alternatively, this technique is also called the Laplacian interpolated regularization [4]. In [14], the online version of the $p = 2$ case is studied in the context of the already available labels. If $\mathcal{G}$ is a partially labelled graph as in our problem, such that $|V| = N$, and the partial labels $l \leq N$, with the labels given by $\mathbf{y}_l \in \{1, -1\}^l$ on the l vertices, then the minimum semi-norm interpolation gives the labelling:

$$\mathbf{y} = \operatorname{argmin}\{\mathbf{u}^T \mathbf{L} \mathbf{u} : \mathbf{u} \in \mathbb{R}^n, u_r = y_r, r = 1, \ldots, l\} \ .$$

The prediction is made by using $\hat{y}_i = \operatorname{sgn}(y_i)$ [13]. The rationale behind minimizing the cut enables the neighbouring vertices to have similarly valued labels. With $p \to 1$, the prediction problem is reduced to predicting using the label consistent minimum cut.

2.2   Ising Model at Low Temperature

As discussed above, the labelling of the whole graph is obtained by optimizing the objective function constrained on the given labels. From label propagation [20], we saw when $p = 2$, the harmonic energy function $E(\mathbf{u})$ minimized in (1) is quadratic in nature. The technique in (1), chooses the label as a function $u : V \to \mathbb{R}$ and a probability distribution on the function u given by a Gaussian field $P(\mathbf{u}) = \frac{\exp^{-\beta E(\mathbf{u})}}{Z}$, where Z is the partition function and β in the inverse temperature or the uncertainty in the model. There are multiple limitations of the quadratic energy minimization technique. This model is not applicable for $p \rightarrow 1$ in the limit. Not only is the computation slow, the mistake bounds obtained are not the best. Further, in our problem, we relax the values of the labels such that $u : V \rightarrow [0,1]$. With $p \rightarrow 1$, the energy function is equivalent to the the one that finds the minimum cut. Further, when $p \rightarrow 1$ using (2) results in the minimization of a non-strongly convex function per trial that is not differentiable. Also, interesting is that the Laplacian based methods are limited with the abundance of unlabelled data [16]. Hence, we are interested in the Markov random field applicable here with discrete states also known as the Ising model. At low temperatures, the Ising probability distribution over the labellings of a graph $\mathcal{G}$ is defined by:

$$P_T^{\mathcal{G}}(\boldsymbol{u}) \propto \exp\left(-\frac{1}{T}\psi_G(\boldsymbol{u})\right) . \tag{3}$$

where T is the temperature, $\mathbf{u}$ is the labelling over the vertices of $\mathcal{G}$ and $\psi_{\mathcal{G}}(\boldsymbol{u})$ is the complexity of the labelling or the "cut-size". The probabilistic Ising model encodes the uncertainty about the labels of the vertices and at low temperatures favours labellings that minimise the number of edges whose vertices have different labels as shown in (2) with $p = 1$. If the vertex labels pairs seen so far is given by Z_t of vertex label pairs $(j_1, y_1), \ldots, (j_t, y_t)$ such that $(j, y) \in V(\mathcal{G}) \times \{0, 1\}$, then the marginal probability of the label of the vertex v being y conditioned on Z_t is given by: $P_T^{\mathcal{G}}(u_v = y|Z_t) = P_T^{\mathcal{G}}(u_v = y|u_{j_1} = y_1, \ldots, u_{j_t} = y_t)$. At low temperatures and in the limit of zero temperature $T \rightarrow 0$, the marginal favours the labelling that is consistent with the labelling seen so far and the minimum cut. Such label conditioning or label consistency in the context of graph labelling has been extensively studied [11,12,15]. In this paper, we are only interested in the low temperature setting of the Ising model as the environment in which the player functions. However, at low temperatures, the minimum cut is still not unique.

2.3   Multi-Armed Bandit Problem (MAB)

As with any sequential prediction game, the MAB is played between the learner and the environment and proceeds in a series of rounds $t = 1, \ldots, n$. At every time instance t, the forecaster chooses an action I_t from the set of actions or arms $a_t \in \mathcal{A}$, where $\mathcal{A}$ is the action set with K actions. When sampling an arm, the learner suffers a loss l_t that the adversary chooses in a randomized way. The forecaster receives the loss for the selected action only in the bandit setting. The objective of the forecaster is to minimize the regret given by the difference between the incurred cumulative loss on the sequence played and the optimal cumulative loss with respect to the best possible action in hindsight. The decision making process depends on the history of actions sampled and losses received up until time $t - 1$. The notion of regret is expressed as expected (average) regret and pseudo regret, where pseudo regret is the weaker notion because of the comparison with the optimal action in expectation. For the adversarial case, it is given by:

$$\overline{R_n} = \mathbb{E} \sum_{t=1}^{n} l_{I_t,t} - \min_{i=1,...,K} \mathbb{E} \sum_{t=1}^{n} l_{i,t} . \qquad (4)$$

The expectation is with respect to the forecaster's internal randomization and possibly the adversary's randomization. In this work, we consider the adversarial bandit setting with side information (information at queried vertex). Note that unlike in the standard MAB problems where there is no structure defined over the actions, in our setup of the problem, we not only have a structure over the action set but also potentially utilize the associated structural side information that makes the problem more realistic. One more deviation from the standard MAB framework is that at every round, the adversary randomly selects a vertex as the current concept; the value of the concept queried is unknown until after the trial and action selection. Further, our adversary is restricted in that the complexity or "cut-size" of the model of the environment that we have chosen cannot increase across trials. The intuition being, the number of times the learner makes a mistake (predicts the queried state wrong) or does not choose the optimal action, is bounded by the number of times the "cut" changes for the minimum.

2.4   Formulation

We consider an undirected graph $\mathcal{G} = (V(\mathcal{G}), E(\mathcal{G}))$ where the elements of E are called edges that form an unordered pair between the unique elements of V that are called vertices. We assume an unit weight on every edge. The number of vertices in the graph are denoted by N. The vertices of the graph are associated with partially unknown concept values or labels s_i that are gradually revealed, while the bandits lie on the edges in $E(\mathcal{G})$ to form the action set $\mathcal{A}$ with cardinality $|K|$. We assume a κ connected graph, the maximum value of κ such that each vertex has at least κ neighbours. Vertices i and j are neighbours if there is an edge/action connecting them. Note, the number of rounds $n \leq |K|$. In our case, n is equal to number of vertices queried by the environment with unknown labels. A vertex is randomly selected by the environment at every round t, in our case, the queried vertex is given by x_i where $i \in \mathbb{N}_N$. In our example, the queried vertex could represent the request to place an advert on the product website the user currently visits . More specifically, the connections in our graph, not only capture the explicit connections between vertices given by locality, but our bandits or edges also capture the implicit connections between the values of vertices that are possibly differently labelled. In our case, the labels are relaxed such that the label for the i-th vertex is denoted by $s_i = \{-1, 1\}$.

At the start we are given the labels of a small subset of observed vertices, $s^o \in \mathcal{S}^o \subset V(\mathcal{G})$. The labels of the unlabelled vertices $s^u \in \mathcal{S}^u \subset V(\mathcal{G}) \backslash \mathcal{L}$, with $S = S^o \cup S^u$ is revealed one at a time sequentially as at the end of each round as side information. We assume that there are at least two vertices labelled at the start, one in each category. The learning algorithm plays the online bandit game where the adversary at each trial reveals the loss of the selected action and the label of a randomly selected vertex. The goal of the learner is to be able to predict the label of the randomly selected vertex and then sample the appropriate action given the prediction.

3   Maximum Flow Computation

Given a partially labelled graph, the Ising model associates a probability with every labelling that is a consistent completion of the partial labellings. Now, if "cut" of a labelling defines the "energy" of the labelling, then the *low − temperature* Ising is a simplified landscape made up of all such minimum cut (energy) labellings. In a way, the Ising model induces an "energy landscape" over labellings via the "cut." For a n-vertex graph, the energy levels sit inside the n-dimensional hypercube. One can minimize the energy while being consistent with the observations seen so far to achieve the desired goal.

As a first step in the learning process, the learner has to detect the underlying hidden partition in the graph, given the available labels with respect to the currently queried vertex. It can do so by using efficient graph partitioning methods. However, given the partial labelling, the partition detected should respect or be consistent with the labels seen so far. One way to address this is using optimization methods that satisfies the label consistency through constraints. Alternatively, there are very efficient linear time exact methods that can solve this in practise. One such method is "Ford- Fulkerson"[8] algorithm. If one can characterize the labelled vertices in such a way to designate a single source, single sink network, running "Ford Fulkerson"[8] in an online fashion for every round using the side information can be used to efficiently detect the partition. Here, we choose to use a simplified linear programming relaxation to the classic Linear Programming (LP) maximal flow problem (5). Although, the LP formulation we use, can be solved in polynomial time, there is nothing restricting us in using the linear time modified "Ford-Fulkerson" algorithm to achieve the same goal. The objective here is to enable the learner for better predictions and hence lower its regret quicker by detecting the partition early, rather than to illustrate the computational efficiency of the method.

It is known by Menger's theorem of linear programming duality, that maximum flow and the minimum cut are related given a source and a target vertex. Let us introduce the maximum flow or label consistent minimum cut in the graph using the following notation $c^* = \min\{\mathbf{S} \in \{-1,1\}^N : \psi_{\mathcal{G}}(\mathbf{S}|\mathcal{H})\}$ consistent with the trial sequence $\mathcal{H}$ seen so far.

$$E(\mathbf{S}) = \underset{\mathbf{S}\in\{-1,1\}^N}{\arg\min} \sum_{(i,j)\in E(\mathcal{G})} |s_i - s_j| \le c^* . \tag{5}$$

In general. linear programming relaxations are much easier to analyse. Interested readers are referred to the article [17], where LP relaxations are discussed. We use a linear programming relaxation of the above objective as shown in Fig.1

that has auxiliary variables introduced such that there is one variable for every vertex v and one variable for every edge f_{ij}. Since, we have an undirected graph, we assume a directed edge in each direction, for every undirected edge. Hence we have two flow variables per edge in the graph. Essentially, the free variables in the optimization are the unlabelled vertices s_i^u, s_j^u and the flows across every direction f_{ij}. The total flow across all the edges will be our maximum flow for this low temperature Ising model. The formulation in Fig.1 below is what the learner follows to find the minimum cut $\psi_{\mathcal{G}}$. The output from the computation is a directed graph with the value of flow at every edge and the labelling of the vertices consistent with the labels seen so far; $w_{(i,j)}$ is the cost variable of the LP. The sum of the flows is the maximum flow in the Ising model at low temperatures. We fix one of the labelled vertices as a source, and one as target, each with different labels. We assume a unit capacity on every edge. The constraints in Fig. 1 ensure the capacity constraint $f_{(ij)}$ and conservation constraint $s_i - s_j$ are adhered to i.e. the flow in any vertex v other than the source and target, is equal to flow out from v. The largest amount of flow that can pass through any edge is at most 1, as we have unit capacity on every edge. We know that the cost of the maximum flow is equal to the capacity of the minimum cut. The minimum cut obtained as a solution to the optimization problem is an integer.

3.1   Playing Ising Bandits

Figure 2 describes the main algorithm for Ising bandits. It is important to note that `ComputeMaxFlow` can only guide the player towards the active partition with respect to the current context (queried vertex) by detecting the partition early on. $\mathcal{P}$ is a subgraph of $\mathcal{G}$, $\mathcal{P} \subseteq \mathcal{G}$ iff $V(\mathcal{P}) \subseteq V(\mathcal{G})$ and $E(\mathcal{P}) = \{(i,j) : i, j \in V(\mathcal{P}), (i,j) \in E(\mathcal{G})\}$. `SelectPartition` samples the Ising bandits from the best partition with respect to the active concept if the minimum cut changes from previous round. $E(\mathcal{R}), E(\mathcal{J})$ are the partitions of the action set at trial t. Since $\mathbf{S}'$ provides the labelling, it is easy to see which bandits fall in which partition with respect to x_t . The probability distribution r_t over $E(\mathcal{R})$, and j_t over $E(\mathcal{J})$ sum to p_t. Note that if the cut remains the same, player keeps playing the same partition until the cut changes. This has an important implication. Since we assume that the adversary cannot increase the cut at any trial, the cut can only decrease or stay the same. For the rounds it stays the same, the regret that the player suffers is well bounded by the number of times the cut changes. In the best case, the algorithm behaves as a typical Multi-armed bandit (MAB) and in the worst case when the partition changes at every round, the algorithm plays the modified `Ising Bandits`. The algorithm parameter η is the standard MAB value $\eta = \sqrt{\frac{\log |K|}{3n}}$.

4   Experiments

In our experiments, we compare three competitor algorithms with our algorithm `IsingBandits`. The three are `LabProp` [19,20], `Exp3` [3] and `Exp4`[3]. Exp3 and

ComputeMaxFlow(*target vertex:* $s_\sqcup$; *source vertex:* $s_\sqcap$; *trial sequence:* $\mathcal{H} = (x_k, s_k)_{k=1}^t$; *graph:* $\mathcal{G}$)

$$minimize \sum_{(i,j)\in E(\mathcal{G})} w(i,j)f(i,j)$$

subject to:

$$f_{(i,j)} \geq 0 \tag{6}$$
$$s_i - s_j \leq f_{(ij)} \tag{7}$$
$$s_i \geq -1 \tag{8}$$
$$s_i \leq 1$$

Return: *min-cut:* c^*; *flows:* f; *consistent partition:* $\mathbf{S}'$

Fig. 1. Computing the Max-flow

Parameters: Graph: $\mathcal{G}$; $\eta \in \mathbb{R}^+$

Input: Trial Sequence: $\mathcal{H} = \langle (x_1, -1), (x_2, 1), (x_3, s_3), \dots, (x_t, s_t) \rangle$

Initialization: p_1 is the initial distribution over $\mathcal{A}$ such that, $p_1 = (\frac{1}{|K|}, \frac{1}{|K|}, \dots, \frac{1}{|K|})$,

Initial cut-size $c = \infty$; active partition distribution $r_1 = p_1$

for $t = 1, \dots, n$ **do**

    Receive: $x_t \in \mathbb{N}_N$

    $(c^*, f, \mathbf{S}') = $ ComputeMaxFlow$(s_\sqcup, s_\sqcap, \mathcal{H}, \mathcal{G})$

    if $(c \neq c^*)$ **then**   % if cut has changed

      $(E(\mathcal{R}), E(\mathcal{J}), r_t, j_t) = $ SelectPartition$(x_t, p_t, \mathbf{S}', \mathcal{A})$

    Assign: q_t be the distribution over Ising bandits w.r.t x_t, such that,

    $\sum_{i=1}^{|E(\mathcal{R})|} q_{i,t} = r_t$. For any t, $p_t = r_t \cup j_t$

    Play: I_t from q_t

    Receive: Loss z_t; side information s_t

    Compute: Estimated loss $\tilde{z}_{i,t} = \frac{z_{i,t}}{q_{i,t}} \mathbb{1}_{I_t=i}$

    Cumulative estimated loss: $\tilde{Z}_{i,t} = \tilde{Z}_{i,t} + \tilde{z_{i,t}}$

    Update: $q_{i,t+1} = \frac{q_{i,t} \exp(\eta \tilde{Z}_{i,t})}{\sum_{j=1}^{|E(\mathcal{R})|} \exp(\eta \tilde{Z}_{j,t})}$

end

Fig. 2. Ising Bandits Algorithm

Exp4 are from the same family of algorithms for bandits in the adversarial setting. Exp4 is the contextual bandit setting, the close competitor to Ising from the contextual perspective. The experts or contexts in Exp4 for our problem setting are a number of possible labellings. Note that the number of experts selected for prediction have a bearing on the performance of the algorithm. In our experiments, we fixed the number of experts to 10. In reality, even at low temperatures for the model we consider, the set of all possible labellings is exponential in size. LabProp [19,20] is the implementation where the state-of-the-art graph Laplacian based labelling procedure is used to optimize the labelling consistent with the labels seen so far. For all of the above algorithms, we use our own implementation in MATLAB. Since online experiments are extremely time consuming while processing one data point at a time, we have averaged each set of experiments over five trials but for ISOLET, where we average over ten trials. The datasets that we use are the standardized UCI datasets namely the USPS and the ISOLET datasets. All datasets are nearly balanced in our experiments to demonstrate the fairness of the class distribution and for avoiding any majority vote cases where the class with the majority vote wins.

4.1 Dataset Description

The summary of datasets used is captured in Table 1. The USPS handwritten digits is an optical character recognition dataset comprising 16x16 grayscale images of "0" through "9" obtained from scanning handwritten digits. The preprocessed dataset has each image with 256 real valued features without missing values scaled to [-1,1]. We randomly sample the examples for the graph from the 7291 original training points. Each vertex in the graph thus sampled is a digit. We perform several binary graph generation of sampling one digit vs. the other digit to form our underlying graph with edges or connections between the two digits forming our action set.

We use a noisy perceptual dataset for spoken letter recognition called ISOLET consisting of 7797 instances with 617 real valued features. A total of 150 subjects spoke each letter of the English alphabets twice resulting in 52 training examples from each speaker. The total of 150 speakers are split into 30 speakers each into files named as Isolet 1 through to Isolet 5. For the purpose of our experiments here, we build the graph from Isolet 1 comprising 1560 examples from 30 speaker with each letter being spoken twice. Again, we are only interested in binary classified graphs here where we sample the first 13 spoken letters and the last 13 spoken letters as two separate underlying concepts in our graph, the connections between which form our action set.

Fig. 3. Squares image.

Table 1. Datasets used in this paper.

DATA SET	#INSTANCES	#FEATURES	#CLASSES
USPS	7291	256	10
ISOLET	7797	617	26

4.2   Synthetic Dataset

Our synthetic data uses a 2D grid like topology. Figure 3 shows the image used to construct the graph in our experiments. Our interest in using the image for our simulation experiment stems from the natural occurring graph structure in such 2D grids. The image style of `Squares` is chosen based on our interest in smooth and wide regions of similar labels interspersed with dissimilar labelled boundary regions. We use a square image that is constructed using a set of pixels, each with an intensity of 0 or 1. The 0 and 1 intensities are balanced across the pixels i.e. there are equal number of pixels with 0 and 1 intensities. Each pixel in the image corresponds to a vertex in the graph and the intensities correspond to the label or class of the vertex. Here, our graph has 3600 vertices. The neighbourhood system in the graph comprises of edges connecting pairs of neighbouring similar pixels. The connectivity is typically guided by if the pixels are of comparable intensities, if the pixels are structurally close to each other or both. Here, we are only interested in the physical pixel locations that are used to determine connectivity i.e. pixels closer to each other on the grid are connected. The connections eventually form our bandits action set. In this paper, we are only interested in undirected and unweighted graphs. Our grid graph thus generated have a weight of 1 on every edge and there is an edge in either direction. Further, we investigate the type of neighbourhood system, called torus. In the torus grid, each pixel has four neighbours; achieved by connecting the top with the bottom edge pixels and the left with the right edge pixels. Our graph is the same across trials. We randomly sample the available labelled vertices from the graph such that there are equal number of labels from each concept class.

4.3   Graph Generation from Datasets

We design our experiments to test the action selection algorithm under a number of different criteria of graph creation: balanced labels, varying degree of connectedness, varying sizes of initial labels and noise. The parameters that are varied across the experiments are graph size indicated by N, labels available as L, connectivity K, noise levels nse.

In the set of experiments with `ISOLET`, we chose to build the graph from the first 30 speakers in Isolet1 that forms a graph of 1560 vertices of 52 spoken letters (each letter spoken twice) by 30 speakers. The concept classes that are sampled are the first 13 letters of English alphabets as one concept vs. the next 13 letters as the other concept. We build a 3 nearest neighbour graph from the

Euclidean distance matrix constructed using the pairwise distances between the examples (spoken letters). In order to ensure that the graph is connected for such low connectivity, we sample a MST for each graph and always maintain the MST edges in the graph. The MST uses the Euclidean distances as weights. The same underlying graph is used across trials. The edges or connections form the bandits. The available side information is sampled randomly such that the two classes are balanced over the entire graph size.

In the USPS experiments, we randomly sample a different graph for each trial. While sampling the vertices of the graph, we ensure to select vertices equally from each concept class. We use a variety of concept classes 1 vs. 2, 2 vs. 3 and 4 vs. 7. We use the pairwise Euclidean distance as the weights for the MST construction. All the sampled graphs maintain the MST edges. In all the experiments on the datasets, the unweighted minimum spanning tree (MST) and "$K = 3$"-NN graph had their edge sets' "unioned" to create the resultant graph. The motivating reason being that most empirical experiments had shown competitive performance of algorithms at $K = 3$, while the MST guaranteed connectivity in the graph. Besides, MST based graphs are sparse in general, enabling computational efficient completion of the experiments. All the experiments were carried out in a quad-core processor notebooks (@2.30 GHz each) with 8GB RAM and 16 GB RAM.

4.4   Evaluation Criteria

We measure the performance of the algorithms by means of the instantaneous regret or per-round regret of the learning algorithm as compared with the optimal algorithm (lower the better). The instantaneous regret should sub-linearly reduce to zero. The instantaneous regret of the algorithm is measured against time. In our case, time indicates each unlabelled vertex queried in an iterative fashion by the environment, until all unlabelled vertices had been queried. Ideally, the more vertices has been queried and more side information obtained, the lower should be the instantaneous regret of the algorithms. In all the experiments, the hidden concept class distribution in the underlying graph is balanced.

4.5   Results

In the synthetic dataset of concentric squares experiment in Fig. 4, Ising always outperforms Exp3, Exp4 and LabProp. LabProp and Ising are very competitive over uninformed competitors of Exp3, Exp4. Exp3, Exp4 do not use the available side information to sample their action. Note, the overlapping squares create a difficult dataset where closely connected clusters of similar labels white with intensity 1 are surrounded by clusters of opposite labels black with intensity 0 around its boundary. Although, LabProp is good at exploiting connectivity, here we see that Ising captures the opposing boundary side information better than LabProp.

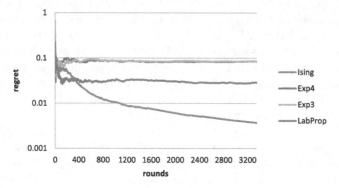

Fig. 4. Results on torus graph generated from `Squares` image with equal number of neighbours $K = 4$, $N = 3600$, $L = 250$.

Our dataset experiments begin with the USPS 2 Vs.3 experiment with connectivity $K = 3$, available labels $L = 8$, and number of data points $N = 1000$. In Fig. 5 below, algorithms `Ising` and `LabProp` are very competitive when side information about more than half of the dataset is obtained. When the side information is very limited at the beginning of the game, `LabProp` outperforms `Ising`.

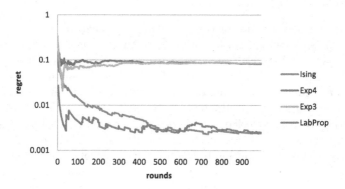

Fig. 5. USPS 2 Vs.3 with $K = 3, N = 1000, L = 8$

In Fig. 6 below, we test the behaviour of the algorithms with varying degree of connectivity. We vary the parameter K over a range to check how well the cluster size affects the performance. It is known from labelling over graph literature that with increasing K the behaviour deteriorates. Here, we see `Ising` outperforms `LabProp` for lower values of K, while `LabProp` wins for higher K.

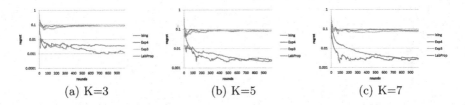

(a) K=3   (b) K=5   (c) K=7

Fig. 6. USPS 4 Vs.7, with varying connectivity $K = 3, K = 5, K = 7$ on randomly sampled graphs with $N = 1000, L = 8$. The color coding is uniform over all the graphs and as indicated in (c) above.

In our experiments over the dataset ISOLET, we sample the graph from ISOLET 1. In Fig. 7, we observe that with $K = 3$ and $L = 128$, Ising outperforms LabProp throughout. The overall regret achieved in ISOLET is higher than the regret achieved in USPS as ISOLET is a noisy dataset.

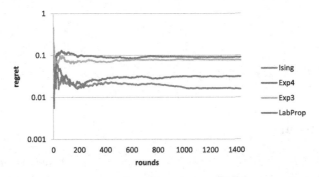

Fig. 7. Experiments on ISOLET with $K = 3, N = 1560, L = 128$

The following set of experiments in Fig. 8 and Fig. 9 test the robustness of our methods in presence of balanced noise. Our noise parameter nse is varied over the percentage range $s = 10, 20, 30, 40$. When noise is say x percent, we randomly eliminate the actions/edges in the graph (from existing connections) for which the noise is less than x percent, and add a balanced equal number of new actions (connections) to the graph. We see that the performance of Ising is the most robust across various noise levels. LabProp suffers with noise as it is heavily dependant on connectivity, and under performs in contrast to Exp4 and Exp3. On the contrary, Ising uses the connectivity for side information, with its action selection unaffected with the introduction of noise. When the noise level increases, the performance of all the algorithms decrease uniformly.

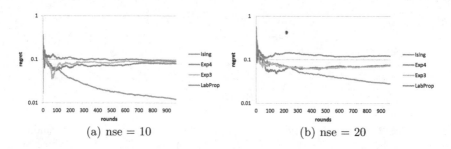

(a) nse = 10   (b) nse = 20

Fig. 8. USPS 1 vs. 2 Robustness Experiments with noise levels 10% and 20%

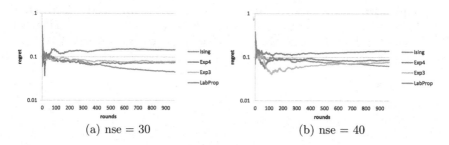

(a) nse = 30   (b) nse = 40

Fig. 9. USPS 1 vs. 2 Robustness Experiments with noise levels 30% and 40%

5 Conclusion

There are real life scenarios where a core minimal subset of connections in a network is responsible for partitioning the graph. Such a core group could be a focus of targeted advertising or content-recommendation as that can have maximum influence on the network with a potential to go viral. Typically, there is a lot of available information in such settings that is potentially usable for detecting the changing partitioning set. We address such advertising and content recommendation challenges by casting the problem as an online Ising graph model of bandits with side information. We use the notion of *cut-size* as a regularity measure in the model to identify the partition and play the bandits game. The best case behaviour of the algorithm when there is a single partition is equivalent to the standard adversarial MAB. We show a polynomial algorithm where the label consistent "cut-size" can guide the sampling procedure. Further, we motivate a linear time exact algorithm for computing the max flow that also respects the label consistency. An interesting effect of the algorithm is that as long as the *cut-size* does not change, the learner keeps playing the same partition on the active action set (size smaller than the actual action set). The regret is then bounded by the number of times the cut changes during the entire game. This can be proven analytically, which we will like to pursue as future work.

References

1. Alamgir, M., von Luxburg, U.: Phase transition in the family of p-resistances. In: Shawe-Taylor, J., Zemel, R.S., Bartlett, P.L., Pereira, F.C.N., Weinberger, K.Q. (eds.) NIPS, pp. 379–387 (2011)
2. Amin, K., Kearns, M., Syed, U.: Graphical models for bandit problems (2012). arXiv preprint arXiv:1202.3782
3. Auer, P., Cesa-Bianchi, N., Freund, Y., Schapire, R.E.: Gambling in a rigged casino: the adversarial multi-armed bandit problem. In: Proceedings of the 36th Annual Symposium on Foundations of Computer Science, 1995, pp. 322–331. IEEE (1995)
4. Belkin, M., Matveeva, I., Niyogi, P.: Regularization and semi-supervised learning on large graphs. In: Shawe-Taylor, J., Singer, Y. (eds.) COLT 2004. LNCS (LNAI), vol. 3120, pp. 624–638. Springer, Heidelberg (2004)
5. Belkin, M., Niyogi, P.: Semi-supervised learning on riemannian manifolds. Mach. Learn. **56**(1–3), 209–239 (2004)
6. Blum, A., Chawla, S.: Learning from labeled and unlabeled data using graph mincuts. In: ICML, pp. 19–26 (2001)
7. Di Castro, D., Gentile, C., Mannor, S.: Bandits with an edge. In: CoRR, abs/1109.2296 (2011)
8. Ford, L.R., Fulkerson, D.R.: Maximal Flow through a Network. Canadian Journal of Mathematics **8**, 399–404 (1956). http://www.rand.org/pubs/papers/P605/
9. Gentile, C., Li, S., Zappella, G.: Online clustering of bandits (2014). arXiv preprint arXiv:1401.8257
10. Gentile, C., Orabona, F.: On multilabel classification and ranking with bandit feedback. The Journal of Machine Learning Research **15**(1), 2451–2487 (2014)
11. Herbster, M.: Exploiting cluster-structure to predict the labeling of a graph. In: Freund, Y., Györfi, L., Turán, G., Zeugmann, T. (eds.) ALT 2008. LNCS (LNAI), vol. 5254, pp. 54–69. Springer, Heidelberg (2008)
12. Herbster, M., Lever, G.: Predicting the labelling of a graph via minimum p-seminorm interpolation. In: Proceedings of the 22nd Annual Conference on Learning Theory (COLT 2009) (2009)
13. Herbster, M., Lever, G.: Predicting the labelling of a graph via minimum p-seminorm interpolation. In: COLT (2009)
14. Herbster, M., Lever, G., Pontil, M.: Online prediction on large diameter graphs. In: Advances in Neural Information Processing Systems, pp. 649–656 (2009)
15. Herbster, M., Pontil, M., Wainer, L.: Online learning over graphs. In: Proceedings of the 22nd international conference on Machine learning ICML 2005, pp. 305–312. ACM, New York (2005)
16. Nadler, B., Srebro, N., Zhou, X.: Statistical analysis of semi-supervised learning: the limit of infinite unlabelled data. In: NIPS, pp. 1330–1338 (2009)
17. Trevisan, L.: Lecture 15:cs261:optimization (2011). http://theory.stanford.edu/trevisan/cs261/lecture15.pdf
18. Valko, M., Munos, R., Kveton, B., Kocák, T.: Spectral bandits for smooth graph functions. In: 31th International Conference on Machine Learning (2014)
19. Zhu, X., Ghahramani, Z.: Towards semi-supervised classification with markov random fields. Tech. Rep. CMU-CALD-02-106, Carnegie Mellon University (2002)
20. Zhu, X., Ghahramani, Z., Lafferty, J.D.: Semi-supervised learning using gaussian fields and harmonic functions. In: ICML, pp. 912–919 (2003)

Refined Algorithms for Infinitely Many-Armed Bandits with Deterministic Rewards

Yahel David$^{(\boxtimes)}$ and Nahum Shimkin

Department of Electrical Engineering,
Technion—Israel Institute of Technology, 32000 Haifa, Israel
yahel83@gmail.com

Abstract. We consider a variant of the Multi-Armed Bandit problem which involves a large pool of a priori identical arms (or items). Each arm is associated with a deterministic value, which is sampled from a probability distribution with unknown maximal value, and is revealed once that arm is chosen. At each time instant the agent may choose a new arm (with unknown value), or a previously-chosen arm whose value is already revealed. The goal is to minimize the cumulative regret relative to the best arm in the pool. Previous work has established a lower bound on the regret for this model, depending on the functional form of the tail of the sample distribution, as well as algorithms that attain this bound up to logarithmic terms. Here, we present a more refined algorithm that attains the same order as the lower bound. We further consider several variants of the basic model, involving an anytime algorithm and the case of non-retainable arms. Numerical experiments demonstrate the superior performance of the suggested algorithms.

Keywords: Many-armed bandits · Regret minimization

1 Introduction

We consider a statistical learning problem in which the learning agent faces a large pool of possible items, or *arms*, each associated with a numerical *value* which is unknown a priori. At each time step the agent chooses an arm, whose exact value is then revealed and considered as the agent's reward at this time step. The goal of the learning agent is to maximize the cumulative reward, or, more specifically, to minimize the cumulative n-step regret (relative to the largest value available in the pool). At every time step, the agent should decide between sampling a new arm (with unknown value) from the pool, or sampling a previously sampled arm with a known value. Clearly, this decision represents the *exploration vs. exploitation* trade-off in the classic multi-armed bandit model. Our model assumes that the number of available arms in the pool is unlimited, and that the value of each newly observed arm is an independent sample from a common probability distribution. We study two variants of the basic model: the *retainable arms* case, in which the learning agent can return to any of the

© Springer International Publishing Switzerland 2015
A. Appice et al. (Eds.): ECML PKDD 2015, Part I, LNAI 9284, pp. 464–479, 2015.
DOI: 10.1007/978-3-319-23528-8_29

previously sampled arms (with known value), and the case of non-retainable arms, where previously sampled arms are lost if not immediately reused.

This model falls within the so-call infinitely-many armed framework, studied in [3,4,6,7,10,11]. In most of these works, which are further elaborated on below, the observed rewards are stochastic and the arms are retainable. Here, we continue the work in [7] that assumes that the potential reward of each arm is fixed and precisely observed once that arm is chosen. This simpler framework allows to obtain sharper bounds which focus on the basic issue of the sample size required to estimate the maximal value in the pool. At the same time, the assumption that the reward is deterministic may be relevant in various applications, such as parts inspection, worker selection, and communication channel selection. For this model, a lower bound on the regret and fixed time horizon algorithms that attain this lower bound up to logarithmic terms were presented in [7]. In the present paper, we propose algorithms that attain the same order as the lower bound (with no additional logarithmic terms) under a fairly general assumption on the tail of the probability distribution of the value. We further demonstrate that these bounds may not be achieved without this assumption. Furthermore, for the case where the time horizon is not specified, we provide an anytime algorithm that also attains the lower bound under similar conditions.

As mentioned above, several papers have studied a similar model with stochastic rewards. A lower bound on the regret was first provided in [3], for the case of Bernoulli arms, with the arm values (namely the expected rewards) distributed uniformly on the interval $[0, 1]$. For a known value distribution, algorithms that attain the same regret order as that lower bound are provided in [3,6,10], and an algorithm which attains that bound exactly under certain conditions is provided in [4]. In [11], the model was analyzed under weaker conditions that involve the form of the tail of the value distribution which is assumed known; however, significantly, the maximal value need not be known a priori. A lower bound and algorithms that achieve it up to logarithmic terms were developed for this case. The assumptions in the present paper are milder, in the sense that the tail distribution is not restricted in its form and only an upper bound on this tail is assumed rather than exact match. Our work also addresses the case of non-retainable arms, which has not been considered in the above-mentioned papers.

In a broader perspective, the present model may be compared to the continuum-armed bandit problem studied in [1,5,9]. In this model the arms are chosen from a continuous set, and the arm values satisfy some continuity properties over this set. In the model discussed here, we do not assume any regularity conditions across arms. The non-retainable arms version of our model is reminiscent of the classical *secretary problem*, see for example [8] and [2] for extensive surveys. In the secretary problem, the learning agent interviews job candidates sequentially, and wishes to maximize the probability of hiring the best candidate in the group. Our model considers the cumulative reward (or regret) as the performance measure.

The paper proceeds as follows. In the next section we present our model and the associated lower bound developed in [7]. Section 3 presents our algorithms and regret bounds for the basic model (with known time horizon and retainable arms). The extensions to anytime algorithms and the case of non-retainable arms are presented in Section 4. Some numerical experiments which compare the performance of the proposed algorithms to previous ones are described in Section 5, followed by concluding remarks.

2   Model Formulation and Lower Bound

We consider an infinite pool of arms, with values that are drawn independently from a common (but unknown) probability distribution with a cumulative distribution function $F(\mu)$, $\mu \in \mathbb{R}$. Let μ^* denote the supremal value, namely, the maximal value in the support of the measure defined by $F(\mu)$. As mentioned, once an arm is sampled its value is revealed, and at each time step $t = 1, ..., n$, a new or a previously sampled arm may be chosen. Our performance measure is the following cumulative regret.

Definition 1. *The regret at time step n is defined as:*

$$regret(n) = E\left[\sum_{t=1}^{n} (\mu^* - r(t))\right], \tag{1}$$

where $r(t)$ is the reward obtained at time t, namely, the value of the arm chosen at time t.

The following notations will be used in this paper:

- μ is a generic random variable with distribution function F.
- For $0 \le \epsilon \le 1$, let

$$D_0(\epsilon) = \inf_{D \ge 0} \{P(\mu \ge \mu^* - D) \ge \epsilon\},$$

Note that $P(\mu \ge \mu^* - D_0(\epsilon)) \ge \epsilon$, with equality if $\mu^* - D_0(\epsilon)$ is a continuity point of F. We refer to $D_0(\epsilon)$ as the *tail function* of F.
- Let $\epsilon_0^*(n)$ be defined as[1]

$$\epsilon_0^*(n) = \sup\left\{\epsilon \in [0,1] : nD_0(\epsilon) \le \frac{1}{\epsilon}\right\}. \tag{2}$$

Note that $nD_0(\epsilon_1) \le \frac{1}{\epsilon_0^*(n)}$ for $\epsilon_1 \le \epsilon_0^*(n)$, and $nD_0(\epsilon_2) \ge \frac{1}{\epsilon_0^*(n)}$ for $\epsilon_2 > \epsilon_0^*(n)$.

For example, when μ is uniform on $[a, b]$, then $D_0(\epsilon) = \frac{\epsilon}{b-a}$, and $\epsilon_0^*(n) = \sqrt{\frac{b-a}{n}}$.

---

[1] If the support of μ is a single interval, then $D_0(\epsilon)$ is continuous. In that case, definition (2) reduced to the equation $nD_0(\epsilon) = \frac{1}{\epsilon}$ which, by monotonicity, has a unique solution for n large enough.

– Furthermore, let $D(\epsilon)$ denote a given upper bound on the tail function $D_0(\epsilon)$, and let $\epsilon^*(n)$ be defined similarly to $\epsilon_0^*(n)$ with $D_0(\epsilon)$ replaced by $D(\epsilon)$, namely,

$$\epsilon^*(n) = \sup\left\{\epsilon \in [0,1] : nD(\epsilon) \le \frac{1}{\epsilon}\right\}. \tag{3}$$

Note that $\epsilon^*(n) \le \epsilon_0^*(n)$. Since $D_0(\epsilon)$ is a non-decreasing function, we assume, without loss of generality, that $D(\epsilon)$ is also a non-decreasing function.

In the following sections, we shall assume that the upper bound $D(\epsilon)$ on the tail function $D_0(\epsilon)$ is known to the learning agent, and that it satisfies the following growth property.

Assumption 1

$$D(\epsilon) \le MD(\epsilon_0)\alpha^{\epsilon/\epsilon_0}$$

for every $0 < \epsilon_0 \le \epsilon \le 1$ *and constants* $M > 1$ *and* $1 \le \alpha < e$.

A general class of distributions that satisfies Assumption 1 is given in the following example, which will further serves us throughout the paper.

Example 1. Suppose that $P\left(\mu \ge \mu^* - \epsilon\right) = \Theta\left(\epsilon^\beta\right)$ for $\epsilon > 0$ small enough, where $\beta > 0$. This is the case considered in [11]. Then $D_0(\epsilon) = \Theta\left(\epsilon^{1/\beta}\right)$, and for $D(\epsilon) = A\epsilon^{1/\beta}$, where $\beta > 0$ and $A > 0$, it can be obtained that $D(\epsilon) \le MD(\epsilon_0)\alpha^{\epsilon/\epsilon_0}$, where $1 < \alpha < e$, $M = \frac{\lambda^{1/\beta}}{\alpha^\lambda}$ and $\lambda = \frac{1}{\beta\ln(\alpha)}$. Hence, in this case Assumption 1 holds. Note that $\beta = 1$ corresponds to a uniform probability distribution which is the case considered in [3] and [4] for $\mu^* = 1$.

Remark 1. Assumption 1 can be extended to any upper bound $\overline{\alpha}$ on the value of α (instead of e). In that case, a proper modification to the algorithms below leads to upper bounds that are larger by a constant multiplicative factor of $\ln(\overline{\alpha})$. However, as the assumption above covers most cases of interest, for simplicity of presentation, we will not go further into this extension. We note that the algorithms presented here do not use the values of α and M.

For the case in which the tail function $D_0(\epsilon)$ itself is known to the learning agent, the following lower bound on the expected regret was established in [7].

Theorem 1. *The n-step regret is lower bounded by*

$$regret(n) \ge (1 - \delta_n)\frac{\mu^* - E[\mu]}{16}\frac{1}{\epsilon_0^*(n)}, \tag{4}$$

where $\epsilon_0^*(n)$ *satisfies* (2), *and* $\delta_n = 1 - 2\exp\left(-\frac{(\mu^* - E[\mu])^2}{8\epsilon_0^*(n)}\right)$.

Note that when $\epsilon_0^*(n) \to 0$ as $n \to \infty$, $\delta_n \to 0$ as $n \to \infty$, so that its effect becomes negligible. Furthermore, this lower bound coincides with the lower bounds presented in [3] and in [11] in the more specific models studied in those papers.

In the following corollary we present a lower bound on the regret for the case in which only a bound on the tail function $D_0(\epsilon)$ is known.

Corollary 1. *Let $D(\epsilon)$ be an upper bound on the tail function $D_0(\epsilon)$ such that*

$$\frac{D(\epsilon)}{D_0(\epsilon)} \leq L < \infty, \quad \forall\, 0 \leq \epsilon \leq 1.$$

Then, the n-step regret is lower bounded by

$$regret(n) \geq (1 - \delta_n) \frac{\mu^* - E[\mu]}{16L} \frac{1}{\epsilon^*(n)}, \tag{5}$$

where $\epsilon^*(n)$ satisfies (3) and δ_n is as defined in Theorem 1.

Proof: Let

$$\epsilon_L^*(n) = \sup\left\{\epsilon \in [0, 1] : n\frac{D(\epsilon)}{L} \leq \frac{1}{\epsilon}\right\}. \tag{6}$$

Then, for every $0 \leq \epsilon_1 \leq 1$ such that $\epsilon_L^*(n) < \epsilon_1$, by (6) and the assumed condition of the Corollary, it follows that $\frac{1}{\epsilon_1} < n\frac{D(\epsilon_1)}{L} \leq nD_0(\epsilon_1)$. Therefore, by Equation (2), $\epsilon_0^*(n) < \epsilon_1$. Thus,

$$\epsilon_0^*(n) \leq \epsilon_L^*(n). \tag{7}$$

Now, we need to compare $\epsilon_L^*(n)$ to $\epsilon^*(n)$. Let $L\epsilon^*(n) < \epsilon_2$. Since the tail function is non-decreasing, it follows that $\frac{L}{\epsilon_2} < nD(\frac{\epsilon_2}{L}) \leq nD(\epsilon_2)$, so that $\frac{1}{\epsilon_2} < n\frac{D(\epsilon_2)}{L}$. Hence, $\epsilon_L^*(n) < \epsilon_2$, and

$$\epsilon_L^*(n) \leq L\epsilon^*(n). \tag{8}$$

Equations (7) and (8) imply that $\epsilon_0^*(n) \leq L\epsilon^*(n)$, or $\frac{1}{L\epsilon^*(n)} \leq \frac{1}{\epsilon_0^*(n)}$. By substituting in Equation (4), the Corollary is obtained.

3   Optimal Sample Size

Here we discuss our most basic model, namely, the retainable arms model for a known time horizon. We present an algorithm that under Assumption 1 achieves a regret of the same order as the lower bound presented in Equation (5). We also present an example for which Assumption 1 does not hold, and show that for this example the lower bound on the regret is larger by a logarithmic factor than the lower bound presented in Equation (4).

The presented algorithm is simple and is based on initially sampling a certain number of new arms, followed by constantly choosing the single best arm found in the initial phase.

The following theorem provides an upper bound on the regret incurred by Algorithm 1.

Theorem 2. *Under Assumption 1, for every $n > 1$, the regret of Algorithm 1 is upper bounded by*

$$regret(n) \leq \left(1 + Me\frac{\alpha}{e - \alpha}\right)\frac{1}{\epsilon^*(n)} + 1, \tag{9}$$

where $\epsilon^*(n)$ is defined in Equation (3), and M and α are as defined in Assumption 1.

Algorithm 1. The Optimal Sampling Algorithm for Retainable Arms – OSR Algorithm

---

1: **Input:** $D(\epsilon)$, an upper bound on the tail function and time horizon $n > 1$.
2: Compute $\epsilon^*(n)$ as defined in (3).
3: Sample $N = \lfloor \frac{1}{\epsilon^*(n)} \rfloor + 1$ arms and keep the best one.
4: Continue by pulling the saved best arm up to the last stage n.

The upper bound obtained in the above Theorem is of the same order as the lower bound in Equation (5). Note that the values of M and α in Assumption 1 are not used in the algorithm, but only appear in the regret bound.

Example 1 (continued). For $\beta = 1$ ($\boldsymbol{\mu}$ is uniform on $[a, b]$), Assumption 1 holds for any $\alpha \in [1, e]$, with $M = \frac{\lambda^{1/\beta}}{\alpha^{\lambda}}$, where $\lambda = \frac{1}{\beta \ln(\alpha)}$ and $\frac{1}{\epsilon^*(n)} = \frac{\sqrt{n}}{\sqrt{b-a}}$. Therefore, for $\beta = 1$, with the optimize choice of $\alpha = 1.47$, we obtain $regret(n) < \frac{4.1\sqrt{n}}{\sqrt{b-a}} + 1$.

Proof of Theorem 2: For $N \geq 1$, we denote by $V_N(1)$ the value of the best arm found by sampling N different arms. Clearly,

$$regret(n) \leq N + (n - N)\Delta(N), \tag{10}$$

where $\Delta(N) = E[\mu^* - V_N(1)]$. Then, for $N = \lfloor \frac{1}{\epsilon^*(n)} \rfloor + 1$, since $D(0) = 0$ we obtain

$$\Delta(N) \leq \Delta^{N,\epsilon}, \tag{11}$$

where

$$\Delta^{N,\epsilon} = \sum_{i=1}^{N} D(i\epsilon) P\left(D(i\epsilon) \geq \mu^* - V_N(1) > D((i-1)\epsilon)\right).$$

Note that if $N\epsilon > 1$ we take $D(N\epsilon) = D(1)$. So, Assumption 1 still holds. Also, for any $0 \leq \epsilon \leq 1$,

$$P\left(\mu^* - V_N(1) > D(\epsilon)\right) \leq (1 - \epsilon)^N.$$

Therefore,

$$\Delta^{N,\epsilon} \leq \sum_{i=1}^{N} D(i\epsilon) P\left(\mu^* - V_N(1) > D((i-1)\epsilon)\right)$$

$$\leq \sum_{i=1}^{N} D(i\epsilon)(1 - (i-1)\epsilon)^N \triangleq \overline{\Delta}^{N,\epsilon}. \tag{12}$$

Observe that $(1 - \epsilon)^{\frac{1}{\epsilon}} \leq e^{-1}$ for $\epsilon \in (0, 1]$. Then, for $\epsilon = \epsilon^*(n)$, since $N \geq \frac{1}{\epsilon^*(n)}$ it follows that $(1 - (i-1)\epsilon^*(n))^N \leq e^{1-i}$. Hence,

$$\overline{\Delta}^{N,\epsilon^*(n)} = \sum_{i=1}^{N} D(i\epsilon^*(n))(1 - (i-1)\epsilon^*(n))^N$$

$$\leq \sum_{i=1}^{N} D(i\epsilon^*(n))e^{1-i} \triangleq \overline{\Delta}_0^{N,\epsilon^*(n)}. \tag{13}$$

Now, by Assumption 1,

$$\overline{\Delta}_0^{N,\epsilon^*(n)} \le \sum_{i=1}^{N} MD(\epsilon^*(n))\alpha^i e^{1-i} < Me\frac{\alpha}{e-\alpha}D(\epsilon^*(n)). \tag{14}$$

Therefore, by (10),

$$regret(n) \le \lfloor\frac{1}{\epsilon^*(n)}\rfloor + 1 + nMe\frac{\alpha}{e-\alpha}D(\epsilon^*(n)) \le (1 + Me\frac{\alpha}{e-\alpha})\frac{1}{\epsilon^*(n)} + 1.$$

Hence, the upper bound on (9) is obtained.

□

For the case that Assumption 1 does not hold, we provide an example for which the regret is larger than the lower bound presented in Equation (4) by a logarithmic term.

Example 2. Suppose that $P(\mu \ge \mu^* - \epsilon) = -\frac{1}{\ln(\epsilon)}$. Then $D_0(\epsilon) = e^{-\frac{1}{\epsilon}}$, and it follows that $\frac{1}{\ln(n)} \le \epsilon_0^*(n) \le \frac{2}{\ln(n)}$.
Take $\epsilon_0 = \frac{1}{2}\epsilon$. Then, for any $\alpha > 1$ and $M > 0$, for ϵ small enough we obtain $\frac{D(\epsilon)}{D(\epsilon_0)} = e^{1/\epsilon_0 - 1/\epsilon} = e^{1/\epsilon} > M\alpha^2 = M\alpha^{\epsilon/\epsilon_0}$. Hence, Assumption 1 does not hold.

Lemma 1. *For the case considered in Example 2, the best regret which can be achieved is larger by multiplicative a logarithmic factor $(\ln(n))$ than the lower bound presented in Equation (4).*

Proof: Let N stand for the number of sampled arms, then, one can find that

$$regret(n) = NE[\boldsymbol{\mu}] + (n-N)\Delta(N), \tag{15}$$

where $\Delta(N) = E[\mu^* - V_N(1)]$. To bound the second term of Equation (15), note that, for any $\overline{N} \le \lfloor\frac{1}{\epsilon}\rfloor$,

$$\Delta(N) \ge \sum_{i=1}^{\overline{N}} D_0(i\epsilon)P(D_0((i+1)\epsilon) \ge \mu^* - V_N(1) > D_0((i)\epsilon))$$

$$= \sum_{i=1}^{\overline{N}} D_0(i\epsilon)(\Delta^{N,\epsilon}(i) - \Delta^{N,\epsilon}(i+1)) \triangleq \tilde{\Delta}(N),$$

where

$$\Delta^{N,\epsilon}(i) = P(\mu^* - V_N(1) > D_0(i\epsilon)).$$

By the fact that $D_0(\epsilon)$ is continuous, it follows that

$$\Delta^{N,\epsilon}(i) = P(\mu^* - V_N(1) > D_0(i\epsilon)) = (1 - i\epsilon)^N,$$

and

$$\Delta^{N,\epsilon}(i+1) = P(\mu^* - V_N(1) > D_0((i+1)\epsilon)) = (1 - (i+1)\epsilon)^N.$$

Noting that $e^{-1} \geq (1-\epsilon)^{\frac{1}{\epsilon}} \geq \exp\left(-1 - \frac{\epsilon}{1-\epsilon}\right)$ for $\epsilon \in (0,1]$, we obtain for the choice of $\epsilon = \frac{1}{N}$ that

$$\Delta^{N,\frac{1}{N}}(i) - \Delta^{N,\frac{1}{N}}(i+1) \geq e^{-i}\beta_N^i ,$$

where $\beta_N^i = e^{\frac{-i}{N-1}} - e^{-1}$.

Now, since $D_0(i\epsilon) = D_0(\epsilon)e^{\frac{i-1}{i\epsilon}}$, again for the choice of $\epsilon = \frac{1}{N}$, it follows that

$$\tilde{\Delta}(N) \geq \sum_{i=1}^{\overline{N}} D_0(\frac{1}{N})e^{N-\frac{N}{i}}e^{-i}\beta_N^i \geq \lfloor\sqrt{N}\rfloor D_0(\frac{1}{N})e^{N-2\sqrt{N}}\beta_N^{\sqrt{N}}.$$

Therefore, since $D_0(\frac{1}{N}) = e^{-\frac{1}{N}}$, for $N \geq 3$ we obtain that

$$regret(n) = NE[\boldsymbol{\mu}] + (n-N)\lfloor\sqrt{N}\rfloor e^{-2\sqrt{N}}\beta_N^{\sqrt{N}},$$

For $N < 3$, noting that $\Delta(N)$ is a non-increasing function of N, we have $\Delta(1), \Delta(2) \geq \Delta(3)$, hence

$$regret(n) = NE[\boldsymbol{\mu}] + (n-2)\lfloor\sqrt{3}\rfloor e^{-2\sqrt{3}}\beta_3^{\sqrt{3}}.$$

By optimizing over N, it can be found that

$$regret(n) \geq A\ln^2(n)$$

where $A = \frac{E[\boldsymbol{\mu}]}{5}$. But, since $\epsilon_0^*(n) \leq \frac{2}{\ln(n)}$, the order of the regret is larger by a logarithmic factor than the lower bound on the regret of Equation (4).

□

4 Extensions

In this section we discuss two extensions of the basic model, the first is the case in which the time horizon is not specified, leading to an anytime algorithm, and the second is the non-reatainable arms model.

4.1 Anytime Algorithm

Consider again the retainable arms model, but assuming now that the time horizon is unspecified. Under Assumption 1 and a mild condition on the tail of the value probability distribution, the proposed algorithm achieves a regret of the same order as the lower bound of Equation (5).

The presented algorithm is a natural extension of Algorithm 1. Here, instead of sampling a certain number of arms at the first phase (as a function of the time horizon) and then sampling the best one among them at the second phase, the algorithm insures that at every time step, the number of sampled arms is larger than a threshold which is a function of time. Since the number of sampled

---

Algorithm 2. The Anytime Optimal Sampling Algorithm for Retainable Arms – AT-OSR Algorithm

---

1: **Input:** $D(\epsilon)$, an upper bound on the tail function.
2: **Initialization:** $m = 0$ the number of sampled arms.
3: Compute $\epsilon^*(t)$ as defined in (3).
4: **if** $m < \lfloor \frac{1}{\epsilon^*(t)} \rfloor + 1$ **then**
5:     Sample a new arm, update $t = t + 1$ and return to step 3.
6: **else**
7:     Pull the best arm so far, update $t = t + 1$ and return to step 3.
8: **end if**

---

arms is increasing gradually, the upper bound on the regret obtained here is worse than that obtained in the case of known time horizon. However, we show in Corollary 2 that it is of the same order, under an additional condition.

We note that applying the standard doubling trick to Algorithm 1 does not serve our purpose here, as it would add a logarithmic factor to the regret bound.

In the following Theorem we provide an upper bound on the regret achieved by the proposed Algorithm.

Theorem 3. *Under Assumption 1, for every $n > 1$, the regret of Algorithm 2 is upper bounded by*

$$regret(n) \leq Me\frac{\alpha}{e-\alpha}\sum_{t=2}^{n}\frac{1}{t\epsilon^*(t)} + \frac{1}{\epsilon^*(n)} + 2 , \tag{16}$$

where $\epsilon^*(n)$ is defined in (3), and M and α are as defined in Assumption 1.

As $\frac{1}{\epsilon^*(n)} \geq \frac{1}{\epsilon^*(t)}$ for $t \leq n$, it is obtained that in the worst case, the bound in Equation (16) is larger than the lower bound in Equation (5) by a logarithmic term. However, as shown in the following corollary, under reasonable conditions on the tail function $D(\epsilon)$, the bound in Equation (16) is of the same order as the lower bound in Equation (5).

Corollary 2. *If $B_1 t^\gamma \leq D(\epsilon) \leq B_2 t^\gamma$ for some constants $0 < B_1 \leq B_2$ and $0 < \gamma$, then*

$$regret(n) \leq \left(2Me\frac{\alpha}{e-\alpha}\left(\frac{B_2}{B_1}\right)^{\frac{1}{1+\gamma}}(1+\gamma)f(n)+1\right)\frac{1}{\epsilon^*(n)} + 2 \tag{17}$$

where $f(n) = \left(\frac{n+1}{n}\right)^{\frac{1}{1+\gamma}}$; note that $f(n) \to 1$ asymptotically.

Example 1 (continued). When $D(\epsilon) = \Theta\left(\epsilon^{1/\beta}\right)$, it follows that $\frac{1}{\epsilon^*(t)} = \Theta\left(n^{\frac{\beta}{1+\beta}}\right)$. Therefore, the condition of Corollary 2 holds.

Proof of Corollary 2: Under the assumed condition $B_1' t^{\gamma'} \leq \frac{1}{\epsilon^*(t)} \leq B_2' t^{\gamma'}$, where $B_1' = B_1^{\frac{1}{1+\gamma}}$, $B_2' = B_2^{\frac{1}{1+\gamma}}$ and $\gamma' = \frac{1}{1+\gamma}$. Therefore,

$$\sum_{t=2}^{n} \frac{1}{t\epsilon^*(t)} \leq \int_{t=2}^{n+1} \frac{1}{(t-1)\epsilon^*(t)} \leq \int_{t=2}^{n+1} \frac{2}{t\epsilon^*(t)} \leq \frac{2B_2'}{\gamma'}(n+1)^{\gamma'} \leq \frac{2B_2' f(n)}{B_1' \gamma'} \frac{1}{\epsilon^*(n)} .$$

Therefore, by Equation (16), Equation (17) is obtained.

□

Proof of Theorem 3: Recall the notation $V_N(1)$ for the value of the best arm found by sampling N different arms. We bound the regret by

$$regret(n) \leq E\left[1 + \sum_{t=2}^{n} I(E_t) + (\mu^* - V_t(1)) I(\overline{E_t})\right] ,$$

where $E_t = \left\{ m_t < \lfloor \frac{1}{\epsilon^*(t)} \rfloor + 1 \right\}$, and $I(\cdot)$ is the indicator function.

Since $\lfloor \frac{1}{\epsilon^*(t)} \rfloor + 1$ is a monotone increasing function it follows that

$$1 + \sum_{t=2}^{n} I(E_t) \leq \lfloor \frac{1}{\epsilon^*(n)} \rfloor + 2 .$$

Recall that $\Delta(t) = E[\mu^* - V_t(1)]$, then, since $V_t(1)$ is non-decreasing, by Equations (11)-(14), we obtain that

$$E\left[\sum_{t=2}^{n} (\mu^* - V_t(1)) I(\overline{E_t})\right] \leq \sum_{t=2}^{n} \Delta\left(\lfloor \frac{1}{\epsilon^*(t)} \rfloor + 1\right) \leq \sum_{t=2}^{n} Me \frac{\alpha}{e - \alpha} D(\epsilon^*(t)) .$$

Also, by Equation (3),

$$\sum_{t=2}^{n} D(\epsilon^*(t)) \leq \sum_{t=2}^{n} \frac{1}{t\epsilon^*(t)} .$$

Combining the above yields the bound in Equation (16).

□

4.2 Non-retainable Arms

Here the learning agent is not allowed to reuse a previously sampled arm, unless this arm was sampled in the last time step. So, Algorithm 1 cannot be applied to this model. However, the values of previously chosen arms can provide a useful information for the agent. In this section we present an algorithm that achieves a regret that is larger by a sublogarithmic term than the lower bound in Equation (5). The following additional assumption will be required here.

Assumption 2

$$D(\epsilon) \leq C\left(\frac{\epsilon}{\epsilon_0}\right)^{\tau} D(\epsilon_0)$$

for every $0 < \epsilon_0 \leq \epsilon \leq 1$ *and constants* $C > 0$ *and* $\tau > 0$.

Algorithm 3. The Optimal Sampling Algorithm for Non-Retainable Arms – OSN Algorithm

---

1: **Input:** An upper bound $D(\epsilon)$ on the tail function, time horizon $n > 1$, and τ under which Assumption 2 holds.
2: Compute $\epsilon^*(n)$ as defined in (3).
3: Sample $N = \lfloor \ln^{-\frac{1}{1+\tau}}(n)\frac{1}{\epsilon^*(n)} \rfloor + 1$ arms and keep the value of the best one.
4: Continue by pulling until observing a value equal or greater than the saved best value. Then, continue by pulling this arm up to the last stage n.

Assumption 2 holds, in particular, in Example 1 for $\tau = \frac{1}{\beta}$.

The proposed algorithm is based on sampling a certain number of arms, such that the value of the best one among them is on one hand large enough, and on the other hand the probability of finding another arm with a larger value is also high enough. Thus, after sampling that number of arms, the algorithm continues by pulling new arms until it finds one with a larger (or equal) value than all previously sampled arms.

In the following theorem, we provide an upper bound on the regret achieved by the presented algorithm.

Theorem 4. *Under Assumptions 1 and 2, for every $n > 1$, the regret of Algorithm 3 is upper bounded by*

$$regret(n) \leq \left(1 + \ln^{\frac{\tau}{1+\tau}}(n)\left(CMe\frac{\alpha}{e-\alpha} + \frac{2}{\ln(2)}\right)\right)\frac{1}{\epsilon^*(n)} + \frac{2\ln^{\frac{\tau}{1+\tau}}(n)}{\ln(2)} + 1 \quad (18)$$

where $\epsilon^*(n)$ is defined in Equation (3), M and α are as defined in Assumption 1 and C and τ are as defined in Assumption 2.

Proof: For $N \geq 1$, recall that $V_N(1)$ stands for the value of the best arm found by sampling N different arms and that $\Delta(N) = E[\mu^* - V_N(1)]$. Clearly,

$$regret(n) \leq N + (n - N)\Delta(N) + E[Y(V_N(1))], \quad (19)$$

where the random variable $Y(V)$ is the number of arms sampled until an arm with a value larger or equal to V is sampled. The second term in Equation (19) can be bounded similarly to the second term in Equation (10). Namely, since $N \geq \ln^{-\frac{1}{1+\tau}}(n)\frac{1}{\epsilon^*(n)}$,

$$\overline{\Delta}^{N,\ln^{\frac{1}{1+\tau}}(n)\epsilon^*(n)} \leq \overline{\Delta}_0^{N,\ln^{\frac{1}{1+\tau}}(n)\epsilon^*(n)},$$

and then, by Assumption 2,

$$\overline{\Delta}_0^{N,\ln^{\frac{1}{1+\tau}}(n)\epsilon^*(n)} \leq \sum_{i=1}^{N} MD(\ln^{\frac{1}{1+\tau}}(n)\epsilon^*(n))\alpha^i e^{1-i} < C\ln^{\frac{\tau}{1+\tau}}(n)Me\frac{\alpha}{e-\alpha}D(\epsilon^*(n)).$$

Thus, as shown in the proof of Theorem 2,

$$(n - N)\Delta(N) < nC \ln^{\frac{\tau}{1+\tau}}(n) Me \frac{\alpha}{e - \alpha} D(\epsilon^*(n)) \leq C \ln^{\frac{\tau}{1+\tau}}(n) Me \frac{\alpha}{e - \alpha} \frac{1}{\epsilon^*(n)}. \tag{20}$$

For bounding the third term, let

$$\hat{\epsilon}_\gamma = \sup\{\epsilon \in [0, 1] | D(\epsilon) \leq \gamma\} ,$$

and note that

$$P\left(\boldsymbol{\mu} \geq \mu^* - \gamma\right) = \hat{\epsilon}_\gamma . \tag{21}$$

Now, let us define:

$$\epsilon_1 = \hat{\epsilon}_{\gamma_1}, \ \gamma_1 = D(\frac{1}{n}) ,$$

as well as the following sequence:

$$\epsilon_{i+1} = \hat{\epsilon}_{\gamma_{i+1}}, \text{ for } \gamma_{i+1} = D(2\epsilon_i), \ \forall i \geq 1 .$$

Let M be such that ϵ_M is the first element in the sequence which is larger or equal to one, and set $\epsilon_M = 1$. Then, since $D(\epsilon_{i+1}) = D(2\epsilon_i) = \gamma_{i+1} \ \forall i \geq 1$, and $E[Y(V)]$ is non-decreasing in V, we obtain that

$$E\left[Y(V_N(1))\right] = E\left[E\left[Y(V_N(1))\right] | V_N(1)\right]$$

$$\leq \sum_{i=1}^{M} E\left[Y(\mu^* - \gamma_i)\right] P\left(\mu^* - \gamma_i \geq V_N(1) > \mu^* - \gamma_{i+1}\right) \tag{22}$$

$$\leq \sum_{i=1}^{M} E\left[Y(\mu^* - \gamma_i)\right] P\left(V_N(1) > \mu^* - \gamma_{i+1}\right) \triangleq \Phi_N .$$

Then, by the expected value of a Geometric distribution, Equation (21), and the fact that $\gamma_i = D(\epsilon_i)$, we obtain that

$$E\left[Y(\mu^* - \gamma_i)\right] = \frac{1}{\epsilon_i} .$$

Also, since $\gamma_{i+1} = D(2\epsilon_i)$, it follows that

$$P\left(V_N(1) > \mu^* - \gamma_{i+1}\right) \leq 2N\epsilon_i .$$

So, since $M \leq \frac{\ln(n)}{\ln(2)}$, we have

$$\Phi_N \leq \sum_{i=1}^{M} 2N \leq 2N \frac{\ln(n)}{\ln(2)} \leq \frac{2 \ln^{\frac{\tau}{1+\tau}}(n)}{\ln(2)} \left(\frac{1}{\epsilon^*(n)} + 1\right). \tag{23}$$

By combining Equations (19), (20), (22) and (23) the claimed bound in Equation (18) is obtained.

□

We note that a combined model which considers the anytime problem for the non-retainable arms case can be analyzed by similar methods. However, we do not consider this variant here.

5   Experiments

We next investigate numerically the algorithms presented in this paper, and compare them to the relevant algorithms from [7,11]. We remind that the present deterministic model was only studied in [7], while the model considered in [11] is similar in its assumptions to the presented one in that only the form of the tail function (rather the exact value distribution) is assumed known. Since the algorithms in [11] are analyzed only for the case of Example 1 (i.e. $D(\epsilon) = \Theta\left(\epsilon^\beta\right)$), we adhere to this model with several values of β for our experiments. The maximal value is taken $\mu^* = 0.99$, but is not known to the learning agent. In addition to that, since the algorithms presented in [11] were planned for the stochastic model, they apply the UCB-V policy on the sampled set of arms. Here, we eliminate this stage which is evidently redundant for the deterministic model considered here.

5.1   Retainable Arms

For the case of retainable arms and a known time horizon, we compare Algorithm 1 with the *KT&RA* Algorithm presented in [7] and the *UCB-F* Algorithm presented in [11]. Since in [11], just an order of the number of arms needed to be sampled is specified (and not exact number), we consider two variations of the *UCB-F* Algorithm, one with a multiplicative factor of 10, and the other with a multiplicative factor of 0.2.

Table 1. Average regret for the retainable arms model for the known time horizon case.

Time Horizon								
$\beta = 0.9$			$\beta = 1$			$\beta = 1.1$		
Algorithm	4×10^4	7×10^4	10×10^4	4×10^4	7×10^4	10×10^4	4×10^4	7×10^4	10×10^4
UCB-F-10	574	740	870	1022	1350	1612	1376	1847	2227
UCB-F-0.2	1043	1410	1778	1043	1410	1778	1129	1445	1764
KT&RA	423	578	705	568	787	970	738	1035	1276
Algorithm 1	242	307	360	287	388	460	381	515	626

In Figure 1, we present the average regret of 200 runs vs. the time horizon for $\beta = 0.9$, $\beta = 1$ and $\beta = 1.1$. The empirical standard deviation is smaller than 5% in all of our results. As shown in Figure 1 and detailed in Table 1, the performance of Algorithm 1 is significantly better than the other algorithms.

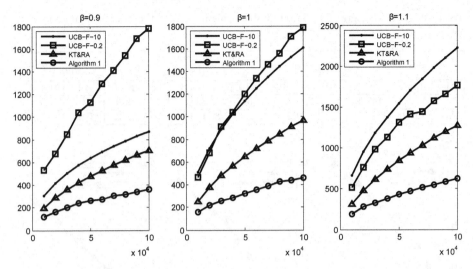

Fig. 1. Average regret (y-axis) vs. the time horizon (x-axis) for $\beta = 0.9$, $\beta = 1$ and $\beta = 1.1$.

5.2 Anytime Algorithm

For the retainable arms model and unspecified time horizon, we compare Algorithm 2 with the *UCB-AIR* Algorithm presented in [11]. Since, these algorithms are identical for $\beta \geq 1$, we run this experiment for $\beta = 0.7$, $\beta = 0.8$ and $\beta = 0.9$. In Figure 2 we present the average regret of 200 runs vs. the time. It is shown in Figure 2 and detailed in Table 2 that the average regret of Algorithm 2 is

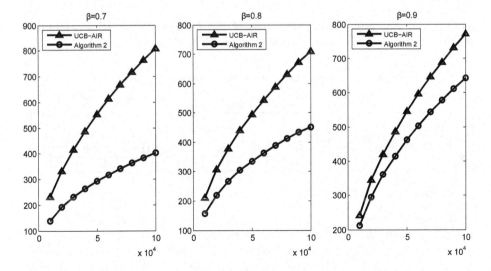

Fig. 2. Average regret (y-axis) vs. time (x-axis) for $\beta = 0.7$, $\beta = 0.8$ and $\beta = 0.9$.

Table 2. Average regret for the retainable arms model for the unknown time horizon case.

	Time Horizon								
	$\beta = 0.7$			$\beta = 0.8$			$\beta = 0.9$		
Algorithm	4×10^4	7×10^4	10×10^4	4×10^4	7×10^4	10×10^4	4×10^4	7×10^4	10×10^4
UCB-AIR	414	667	808	440	589	710	486	646	771
Algorithm 2	264	341	402	305	389	414	542	642	1764

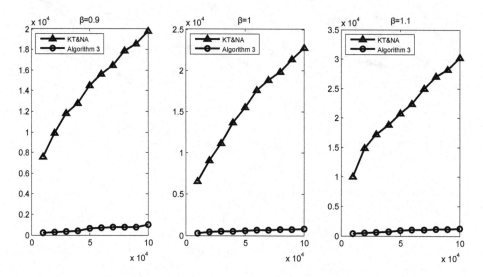

Fig. 3. Average regret (y-axis) vs. the time horizon (x-axis) for $\beta = 0.9$, $\beta = 1$ and $\beta = 1.1$.

smaller and increasing slower than that of the *UCB-AIR* Algorithm. Here the empirical standard deviation is smaller than 7% in all of our results.

Table 3. Average regret for the non-retainable arms model.

	Time Horizon								
	$\beta = 0.7$			$\beta = 0.8$			$\beta = 0.9$		
Algorithm	4×10^4	7×10^4	10×10^4	4×10^4	7×10^4	10×10^4	4×10^4	7×10^4	10×10^4
KT&NA	12800	16460	19760	13670	18800	22760	18850	24950	30170
Algorithm 3	418	741	983	509	646	791	674	1048	1277

5.3   Non-Retainable Arms

In the case of non-retainable arms and a fixed time horizon, we compare Algorithm 3 with the *KT&NA* Algorithm presented in [7]. As in the previous case, we present in Figure 3 the average regret of 200 runs vs. the time horizon for $\beta = 0.9$, $\beta = 1$ and $\beta = 1.1$. Here, the empirical standard deviation is smaller than 10% in all of our results. As shown in Figure 3 and detailed in Table 3, Algorithm 3 outperforms the *KT&NA* Algorithm.

6   Conclusion and Discussion

In this work we provided algorithms with tight bounds on the cumulative regret for the infinitely-many armed problem with deterministic rewards. Our central assumption is that the tail function $D_0(\epsilon)$ is known, to within multiplicative constant. The basic algorithm was extended to the any-time case and to the model with non retainable arms.

A major challenge for future work is further relaxation of the requirement of a known upper bound on the tail function $D_0(\epsilon)$. Initial steps in this direction were presented in [7].

References

1. Auer, P., Ortner, R., Szepesvári, C.: Improved rates for the stochastic continuum-armed bandit problem. In: Bshouty, N.H., Gentile, C. (eds.) COLT. LNCS (LNAI), vol. 4539, pp. 454–468. Springer, Heidelberg (2007)
2. Babaioff, M., Immorlica, N., Kempe, D., Kleinberg, R.: Online auctions and generalized secretary problems. ACM SIGecom Exchanges **7**(2), 1–11 (2008)
3. Berry, D.A., Chen, R.W., Zame, A., Heath, D.C., Shepp, L.A.: Bandit problems with infinitely many arms. The Annals of Statistics, 2103–2116 (1997)
4. Bonald, T., Proutiere, A.: Two-target algorithms for infinite-armed bandits with bernoulli rewards. In: Advances in Neural Information Processing Systems, pp. 2184–2192 (2013)
5. Bubeck, S., Munos, R., Stoltz, G., Szepesvári, C.: X-armed bandits. Journal of Machine Learning Research **12**, 1655–1695 (2011)
6. Chakrabarti, D., Kumar, R., Radlinski, F., Upfal, E.: Mortal multi-armed bandits. In: Advances in Neural Information Processing Systems, pp. 273–280 (2009)
7. David, Y., Shimkin, N.: Infinitely many-armed bandits with unknown value distribution. In: Calders, T., Esposito, F., Hüllermeier, E., Meo, R. (eds.) ECML PKDD 2014, Part I. LNCS, vol. 8724, pp. 307–322. Springer, Heidelberg (2014)
8. Freeman, P.: The secretary problem and its extensions: A review. International Statistical Review, 189–206 (1983)
9. Kleinberg, R., Slivkins, A., Upfal, E.: Multi-armed bandits in metric spaces. In: Proceedings of the 40th Annual ACM Symposium on Theory of Computing, pp. 681–690. ACM (2008)
10. Teytaud, O., Gelly, S., Sebag, M.: Anytime many-armed bandits. In: CAP (2007)
11. Wang, Y., Audibert, J.-Y., Munos, R.: Algorithms for infinitely many-armed bandits. In: Advances in Neural Information Processing Systems, pp. 1729–1736 (2009)

Deep Learning

An Empirical Investigation of Minimum Probability Flow Learning Under Different Connectivity Patterns

Daniel Jiwoong Im[✉], Ethan Buchman, and Graham W. Taylor

School of Engineering, University of Guelph, Guelph, ON, Canada
{imj,ebuchman,gwtaylor}@uoguelph.ca

Abstract. Energy-based models are popular in machine learning due to the elegance of their formulation and their relationship to statistical physics. Among these, the Restricted Boltzmann Machine (RBM), and its staple training algorithm contrastive divergence (CD), have been the prototype for some recent advancements in the unsupervised training of deep neural networks. However, CD has limited theoretical motivation, and can in some cases produce undesirable behaviour. Here, we investigate the performance of Minimum Probability Flow (MPF) learning for training RBMs. Unlike CD, with its focus on approximating an intractable partition function via Gibbs sampling, MPF proposes a tractable, consistent, objective function defined in terms of a Taylor expansion of the KL divergence with respect to sampling dynamics. Here we propose a more general form for the sampling dynamics in MPF, and explore the consequences of different choices for these dynamics for training RBMs. Experimental results show MPF outperforming CD for various RBM configurations.

1 Introduction

A common problem in machine learning is to estimate the parameters of a high-dimensional probabilistic model using gradient descent on the model's negative log likelihood. For exponential models where $p(x)$ is proportional to the exponential of a negative potential function $F(x)$, the gradient of the data negative log-likelihood takes the form

$$\nabla_\theta = \frac{1}{|\mathcal{D}|} \left(\sum_{x \in \mathcal{D}} \frac{\partial F(x)}{\partial \theta} - \sum_x p(x) \frac{\partial F(x)}{\partial \theta} \right) \tag{1}$$

where the sum in the first term is over the dataset, $\mathcal{D}$, and the sum in the second term is over the entire domain of x. The first term has the effect of pushing the parameters in a direction that decreases the energy surface of the model at the training data points, while the second term increases the energy of all possible states. Since the second term is intractable for all but trivial models, we cannot, in practice, accommodate for every state of x, but rather resort to

© Springer International Publishing Switzerland 2015
A. Appice et al. (Eds.): ECML PKDD 2015, Part I, LNAI 9284, pp. 483–497, 2015.
DOI: 10.1007/978-3-319-23528-8_30

sampling. We call states in the sum in the first term *positive particles* and those in the second term *negative particles*, in accordance with their effect on the likelihood (opposite their effect on the energy). Thus, the intractability of the second term becomes a problem of *negative particle selection* (NPS).

The most famous approach to NPS is Contrastive Divergence (CD) [4], which is the centre-piece of unsupervised neural network learning in energy-based models. "CD-k" proposes to sample the negative particles by applying a Markov chain Monte Carlo (MCMC) transition operator k times to each data state. This is in contrast to taking an unbiased sample from the distribution by applying the MCMC operator a large number of times until the distribution reaches equilibrium, which is often prohibitive for practical applications. Much research has attempted to better understand this approach and the reasoning behind its success or failure [6,12], leading to many variations being proposed from the perspective of improving the MCMC chain. Here, we take a more general approach to the problem of NPS, in particular, through the lens of the Minimum Probability Flow (MPF) algorithm [11].

MPF works by introducing a continuous dynamical system over the model's distribution, such that the equilibrium state of the dynamical system is the distribution used to model the data. The objective of learning is to minimize the flow of probability from data states to non-data states after infinitesimal evolution under the model's dynamics. Intuitively, the less a data vector evolves under the dynamics, the closer it is to an equilibrium point; or from our perspective, the closer the equilibrium distribution is to the data. In MPF, NPS is replaced by a more explicit notion of *connectivity* between states. Connected states are ones between which probability can flow under the dynamical system. Thus, rather than attempting to approximate an intractable function (as in CD-k), we run a simple optimization over an explicit, continuous dynamics, and actually never have to run the dynamics themselves.

Interestingly, MPF and CD-k have gradients with remarkably similar form. In fact, the CD-k gradients can be seen as a special case of the MPF gradients - that is, MPF provides a generalized form which reduces to CD-k under a special dynamics. Moreover, MPF provides a consistent estimator for the model parameters, while CD-k as typically formalized is an update heuristic, that can sometimes do bizarre things like go in circles in parameter space [6]. Thus, in one aspect, MPF *solves* the problem of contrastive divergence by re-conceptualizing it as probability flow under an explicit dynamics, rather than the convenient but biased sampling of an intractable function. The challenge thus becomes one of how to design the dynamical system.

This paper makes the following contributions. First, we provide an explanation of MPF that begins from the familiar territory of CD-k, rather than the less familiar grounds of the master equation. While familiar to physicists, the master equation is an apparent obscurity in machine learning, due most likely to its general intractability. Part of the attractiveness of MPF is the way it circumvents that intractability. Second, we derive a generalized form for the MPF transition matrix, which defines the dynamical system. Third, we provide a Theano [1] based implementation of MPF and a number of variants of MPF that run efficiently on GPUs[1]. Finally, we compare and contrast variants of MPF with those of CD-k, and experimentally demonstrate that variants of MPF outperform CD-k for Restricted Boltzmann Machines trained on MNIST and on Caltech-101.

2 Restricted Boltzmann Machines

While the learning methods we discuss apply to undirected probabilistic graphical models in general, we will use the Restricted Boltzmann Machine (RBM) as a canonical example. An RBM is an undirected bipartite graph with visible (observed) variables $\mathbf{v} \in \{0, 1\}^D$ and hidden (latent) variables $\mathbf{h} \in \{0, 1\}^H$ [9]. The RBM is an energy-based model where the energy of state $\mathbf{v}, \mathbf{h}$ is given by

$$E(\mathbf{v}, \mathbf{h}; \theta) = -\sum_i \sum_j W_{ij} v_i h_j - \sum_i b_i v_i - \sum_j c_j h_j \tag{2}$$

where $\theta = \{W, \mathbf{b}, \mathbf{c}\}$ are the parameters of the model. The marginalized probability over visible variables is formulated from the Boltzmann distribution,

$$p(\mathbf{v}; \theta) = \frac{p^*(\mathbf{v}; \theta)}{Z(\theta)} = \frac{1}{Z(\theta)} \sum_\mathbf{h} \exp\left(\frac{-1}{\tau} E(\mathbf{v}, \mathbf{h}; \theta)\right) \tag{3}$$

such that $Z(\theta) = \sum_{\mathbf{v}, \mathbf{h}} \exp\left(\frac{-1}{\tau} E(\mathbf{v}, \mathbf{h}; \theta)\right)$ is a normalizing constant and τ is the thermodynamic temperature. We can marginalize over the binary hidden states in Equation 2 and re-express in terms of a new energy $F(\mathbf{v})$,

$$F(\mathbf{v}; \theta) = -\log \sum_\mathbf{h} \exp\left(\frac{-1}{\tau} E(\mathbf{v}, \mathbf{h})\right) \tag{4}$$

$$= \frac{1}{\tau} \sum_i^D v_i b_i - \frac{1}{\tau} \sum_{j=1}^H \log\left(1 + \exp\left(c_j + \sum_i^D v_i W_{i,j}\right)\right) \tag{5}$$

$$p(\mathbf{v}; \theta) = \frac{\exp\left(-F(\mathbf{v}; \theta)\right)}{Z(\theta)} \tag{6}$$

Following physics, this form of the energy is better known as a free energy, as it expresses the difference between the average energy and the entropy of a distribution, in this case, that of $p(\mathbf{h}|\mathbf{v})$. Defining the distribution in terms of free energy as $p(\mathbf{v}; \theta)$ is convenient since it naturally copes with the presence of latent variables.

---

[1] https://github.com/jiwoongim/minimum_probability_flow_learning

The key characteristic of an RBM is the simplicity of inference due to conditional independence between visible and hidden states:

$$p(\mathbf{h}|\mathbf{v}) = \prod_j p(h_j|\mathbf{v}), \qquad p(h_j = 1|\mathbf{v}) = \sigma(\sum_i W_{ij} v_i + c_j)$$

$$p(\mathbf{v}|\mathbf{h}) = \prod_i p(v_i|\mathbf{h}), \qquad p(v_i = 1|\mathbf{h}) = \sigma(\sum_j W_{ij} h_j + b_i)$$

where $\sigma(z) = 1/(1 + \exp(-z))$.

This leads naturally to a block Gibbs sampling dynamics, used universally for sampling from RBMs. Hence, in an RBM trained by CD-k, the connectivity (NPS) is determined with probability given by k sequential block Gibbs sampling transitions.

We can formalize this by writing the learning updates for CD-k as follows

$$\Delta\theta_{CD-k} \propto -\sum_{j \in \mathcal{D}} \sum_{i \notin \mathcal{D}} \left(\frac{\partial F_j(\theta)}{\partial \theta} - \frac{\partial F_i(\theta)}{\partial \theta}\right) T_{ij} \tag{7}$$

where T_{ij} is the probability of transitioning from state j to state i in k steps of block Gibbs sampling. We can in principle replace T_{ij} by any other transition operator, so long as it preserves the equilibrium distribution. Indeed, this is what alternative methods, like Persistent CD [13], achieve.

3   Minimum Probability Flow

The key intuition behind MPF is that NPS can be reformulated in a firm theoretical context by treating the model distribution as the end point of some explicit continuous dynamics, and seeking to minimize the flow of probability away from the data under those dynamics. In this context then, NPS is no longer a sampling procedure employed to approximate an intractable function, but arises naturally out of the probability flow from data states to non-data states. That is, MPF provides a theoretical environment for the formal treatment of T_{ij} that offers a much more general perspective of that operator than CD-k can. In the same vein, it better formalizes the notion of minimizing divergence between positive and negative particles.

3.1   Dynamics of the Model

The primary mathematical apparatus for MPF is a continuous time Markov chain known as the *master equation*,

$$\dot{p}_i = \sum_{j \neq i} [\Gamma_{ij} p_j^{(t)} - \Gamma_{ji} p_i^{(t)}] \tag{8}$$

where j are the data states and i are the non-data states and Γ_{ij} is the probability flow rate from state j to state i. Note that each state is a full vector of variables, and we are theoretically enumerating all states. $\dot{p}_i$ is the rate of change of the probability of state i, that is, the difference between the probability flowing out of any state j into state i and the probability flowing out of state i to any other state j at time t. We can re-express $\dot{p}_i$ in a simple matrix form as

$$\dot{\mathbf{p}} = \mathbf{\Gamma}\mathbf{p} \qquad (9)$$

by setting $\Gamma_{ii} = -\sum_{i \neq j} \Gamma_{ji} p_i^{(t)}$. We note that if the transition matrix Γ is ergodic, then the model has a unique stationary distribution.

This is a common model for exploring statistical mechanical systems, but it is unwieldy in practice for two reasons, namely, the continuous time dynamics, and exponential size of the state space. For our purposes, we will actually find the former an advantage, and the latter irrelevant.

The objective of MPF is to minimize the KL divergence between the data distribution and the distribution after evolving an infinitesimal amount of time under the dynamics:

$$\theta_{MPF} = \operatorname{argmin}_\theta J(\theta), \ J(\theta) = D_{KL}(p^{(0)}||p^{(\epsilon)}(\theta))$$

Approximating $J(\theta)$ up to a first order Taylor expansion with respect to time t, our objective function reduces to

$$J(\theta) = \frac{\epsilon}{|\mathcal{D}|} \sum_{j \in \mathcal{D}} \sum_{i \notin \mathcal{D}} \Gamma_{ij} \qquad (10)$$

and θ can be optimized by gradient descent on $J(\theta)$. Since Γ_{ij} captures probability flow from state j to state i, this objective function has the quite elegant interpretation of minimizing the probability flow from data states to non-data states [11].

3.2   Form of the Transition Matrix

MPF does not propose to *actually* simulate these dynamics. There is, in fact, no need to, as the problem formulation reduces to a rather simple optimization problem with no intractable component. However, we must provide a means for computing the matrix coefficients Γ_{ij}. Since our target distribution is the distribution defined by the RBM, we require Γ to be a function of the energy, or more particularly, the parameters of the energy function.

A sufficient (but not necessary) means to guarantee that the distribution $\mathbf{p}^\infty(\theta)$ is a fixed point of the dynamics is to choose Γ to satisfy detailed balance, that is

$$\Gamma_{ji} p_i^{(\infty)}(\theta) = \Gamma_{ij} p_j^{(\infty)}(\theta). \qquad (11)$$

The following theorem provides a general form for the transition matrix such that the equilibrium distribution is that of the RBM:

Theorem 1. *1 Suppose $p_j^{(\infty)}$ is the probability of state j and $p_i^{(\infty)}$ is the probability of state i. Let the transition matrix be*

$$\Gamma_{ij} = g_{ij} \exp\left(\frac{o(F_i - F_j) + 1}{2}(F_j - F_i)\right) \tag{12}$$

such that $o(\cdot)$ is any odd function, where g_{ij} is the symmetric connectivity between the states i and j. Then this transition matrix satisfies detailed balance in Equation 11.

The proof is provided in Appendix A.1. The transition matrix proposed by [11] is thus the simplest case of Theorem 1, found by setting $o(\cdot) = 0$ and $g_{ij} = g_{ji}$:

$$\Gamma_{ij} = g_{ij} \exp\left(\frac{1}{2}(F_j(\theta) - F_i(\theta))\right). \tag{13}$$

Given a form for the transition matrix, we can now evaluate the gradient of $J(\theta)$

$$\frac{\partial J(\theta)}{\partial \theta} = \frac{\epsilon}{|\mathcal{D}|} \sum_{j \in \mathcal{D}} \sum_{i \notin \mathcal{D}} \left(\frac{\partial F_j(\theta)}{\partial \theta} - \frac{\partial F_i(\theta)}{\partial \theta}\right) T_{ij}$$

$$T_{ij} = g_{ij} \exp\left(\frac{1}{2}(F_j(\theta) - F_i(\theta))\right)$$

and observe the similarity to the formulation given for the RBM trained by CD-k (Equation 7). Unlike with CD-k, however, this expression was derived through an explicit dynamics and well-formalized minimization objective.

4   Probability Flow Rates Γ

At first glance, MPF might appear doomed, due to the size of Γ, namely $2^D \times 2^D$, and the problem of enumerating all of the states. However, the objective function in Equation 10 summing over the Γ_{ij}'s only considers transitions between data states j (limited in size by our data set) and non-data states i (limited by the sparseness of our design). By specifying Γ to be sparse, the intractability disappears, and complexity is dominated by the size of the dataset.

Using traditional methods, an RBM can be trained in two ways, either with sampled negative particles, like in CD-k or PCD (also known as stochastic maximum likelihood) [4,13], or via an inductive principle, with fixed sets of "fantasy cases", like in general score matching, ratio matching, or pseudo-likelihood [3,5,7]. In a similar manner, we can define Γ by specifying the connectivity function g_{ij} either as a distribution from which to sample or as fixed and deterministic.

In this section, we examine various kinds of connectivity functions and their consequences on the probability flow dynamics.

4.1   1-bit Flip Connections

It can be shown that score matching is a special case of MPF in continuous state spaces, where the connectivity function is set to connect all states within a small Euclidean distance r in the limit of $r \to 0$ [11]. For simplicity, in the case of a discrete state space (Bernoulli RBM), we can fix the Hamming distance to one instead, and consider that data states are connected to all other states 1-bit flip away:

$$g_{ij} = \begin{cases} 1, & \text{if state } i, j \text{ differs by single bit flip} \\ 0, & \text{otherwise} \end{cases} \tag{14}$$

1-bit flip connectivity gives us a sparse Γ with $2^D D$ non-zero terms (rather than a full 2^{2D}), and may be seen as NPS where the only negative particles are those which are 1-bit flip away from data states. Therefore, we only ever evaluate $|\mathcal{D}|D$ terms from this matrix, making the formulation tractable. This was the only connectivity function pursued in [11] and is a natural starting point for the approach.

Algorithm 1. Minimum probability flow learning with single bit-flip connectivity. Note we leave out all g_{ij} since here we are explicit about only connecting states of Hamming distance 1.

- Initialize the parameters θ
- **for** each training example $d \in \mathcal{D}$ **do**
  1. Compute the list of states, L, with Hamming distance 1 from d
  2. Compute the probability flow $\Gamma_{id} = \exp\left(\frac{1}{2}(F_d(\theta) - F_i(\theta))\right)$ for each $i \in L$
  3. The cost function for d is $\sum_{i \in L} \Gamma_{id}$
  4. Compute the gradient of the cost function, $\frac{\partial J(\theta)}{\partial \theta} = \sum_{i \in L} \left(\frac{\partial F_d(\theta)}{\partial \theta} - \frac{\partial F_i(\theta)}{\partial \theta}\right)\Gamma_{id}$
  5. Update parameters via gradient descent with $\theta \leftarrow \theta - \lambda \nabla J(\theta)$
  end for

4.2   Factorized Minimum Probability Flow

Previously, we considered connectivity g_{ij} as a binary indicator function of both states i and j. Instead, we may wish to use a probability distribution, such that g_{ij} is the probability that state j is connected to state i (i.e. $\sum_i g_{ij} = 1$). Following [10], we simplify this approach by letting $g_{ij} = g_i$, yielding an *independence chain* [14]. This means the probability of being connected to state i is independent of j, giving us an alternative way of constructing a transition matrix such that the objective function can be factorized:

$$J(\theta) = \frac{1}{|\mathcal{D}|} \sum_{j \in \mathcal{D}} \sum_{i \notin \mathcal{D}} g_i \left(\frac{g_j}{g_i}\right)^{\frac{1}{2}} \exp\left(\frac{1}{2}(F_j(\mathbf{x}; \theta) - F_i(\mathbf{x}; \theta))\right) \tag{15}$$

$$= \left(\frac{1}{|\mathcal{D}|} \sum_{j \in \mathcal{D}} \exp\left(\frac{1}{2}\left(F_j(\mathbf{x}; \theta) + \log g_j \right) \right) \right) \left(\sum_{i \notin \mathcal{D}} g_i \exp\left(\frac{1}{2}\left(-F_i(\mathbf{x}; \theta) + \log g_i \right) \right) \right)$$
(16)

where $\left(\frac{g_j}{g_i} \right)^{\frac{1}{2}}$ is a scaling term required to counterbalance the difference between g_i and g_j. The independence in the connectivity function allows us to factor all the j terms in 15 out of the inner sum, leaving us with a product of sums, something we could not achieve with 1-bit flip connectivity since the connection to state i depends on it being a neighbor of state j. Note that, intuitively, learning is facilitated by connecting data states to states that are probable under the model (i.e. to contrast the divergence). Therefore, we can use $p(v; \theta)$ to approximate g_i. In practice, for each iteration n of learning, we need the g_i and g_j terms to act as constants with respect to updating θ, and thus we sample them from $p(v; \theta^{n-1})$. We can then rewrite the objective function as $J(\theta) = J_{\mathcal{D}}(\theta) J_{\mathcal{S}}(\theta)$

$$J_{\mathcal{D}}(\theta) = \left(\frac{1}{|\mathcal{D}|} \sum_{\mathbf{x} \in \mathcal{D}} \exp\left[\frac{1}{2}\left(F(\mathbf{x}; \theta) - F(\mathbf{x}; \theta^{n-1}) \right) \right] \right)$$
(17)

$$J_{\mathcal{S}}(\theta) = \left(\frac{1}{|\mathcal{S}|} \sum_{\mathbf{x}' \in \mathcal{S}} \exp\left[\frac{1}{2}\left(-F(\mathbf{x}'; \theta) + F(\mathbf{x}'; \theta^{n-1}) \right) \right] \right)$$
(18)

where $\mathcal{S}$ is the sampled set from $p(\mathbf{v}; \theta^{n-1})$, and the normalization terms in $\log g_j$ and $\log g_i$ cancel out. Note we use the θ^{n-1} notation to refer to the parameters at the previous iteration, and simply θ for the current iteration.

4.3   Persistent Minimum Probability Flow

There are several ways of sampling "fantasy particles" from $p(\mathbf{v}; \theta^{n-1})$. Notice that taking the data distribution with respect to θ^{n-1} is necessary for stable learning.

Previously, persistent contrastive divergence (PCD) was developed to improve CD-k learning [13]. Similarly, persistence can be applied to sampling in MPF connectivity functions. For each update, we pick a new sample based on a MCMC sampler which starts from previous samples. Then we update θ^n, which satisfies $J(\theta^n) \leq J(\theta^{n-1})$ [10]. The pseudo-code for persistent MPF is the same as Factored MPF except for drawing new samples, which is indicated by square brackets in Algorithm 2.

As we will show, using persistence in MPF is important for achieving faster convergence in learning. While the theoretical formulation of MPF guarantees eventual convergence, the focus on minimizing the initial probability flow will have little effect if the sampler mixes too slowly.

Algorithm 2. Factored [Persistent] MPF learning with probabilistic connectivity.

- **for** each epoch n **do**
    1. Draw a new sample S^n based on S^0 $[S^{n-1}]$ using an MCMC sampler.
    2. Compute $J_S(\theta)$
    3. **for** each training example $d \in \mathcal{D}$ **do**
        (a) Compute $J_d(\theta)$. The cost function for d is $J(\theta) = J_d(\theta)J_S(\theta)$
        (b) Compute the gradient of the cost function,
        $$\frac{\partial J(\theta)}{\partial \theta} = J_S(\theta)J_d(\theta)\frac{\partial F_d(\theta)}{\partial \theta}$$
        $$+ \frac{1}{|S|}J_d \sum_{x' \in S}\left(\frac{\partial F(x')}{\partial \theta}\exp\left[\frac{1}{2}\big(F(\mathbf{x}';\theta) - F(\mathbf{x}';\theta^{n-1})\big)\right]\right)$$
        (c) Update parameters via gradient descent with $\theta \leftarrow \theta - \lambda \nabla J(\theta)$
    end for

---

5   Experiments

We conducted the first empirical study of MPF under different types of connectivity as discussed in Section 4. We compared our results to CD-k with varying values for K. We analyzed the MPF variants based on training RBMs and assessed them quantitatively and qualitatively by comparing the log-likelihoods of the test data and samples generated from model. For the experiments, we denote the 1-bit flip, factorized, and persistent methods as MPF-1flip, FMPF, and PMPF, respectively. The goals of these experiments are to

1. Compare among MPF algorithms under different connectivities; and
2. Compare between MPF and CD-k.

In our experiments, we considered the MNIST and CalTech Silhouette datasets. MNIST consists of 60,000 training and 10,000 test images of size 28 × 28 pixels containing handwritten digits from the classes 0 to 9. The pixels in MNIST are binarized based on thresholding. From the 60,000 training examples, we set aside 10,000 as validation examples to tune the hyperparameters in our models. The CalTech Silhouette dataset contains the outlines of objects from the CalTech101 dataset, which are centred and scaled on a 28 × 28 image plane and rendered as filled black regions on a white background creating a silhouette of each object. The training set consists of 4,100 examples, with at least 20 and at most 100 examples in each category. The remaining instances were split evenly between validation and testing[2]. Hyperparameters such as learning rate, number of epochs, and batch size were selected from discrete ranges and chosen based on a held-out validation set. The learning rate for FMPF and PMPF were chosen from the range [0.001, 0.00001] and the learning rate for 1-bit flip was chosen from the range [0.2, 0.001].

---

[2] More details on pre-processing the CalTech Silhouettes can be found in http://people.cs.umass.edu/~marlin/data.shtml

Table 1. Experimental results on MNIST using 11 RBMs with 20 hidden units each. The average training and test log-probabilities over 10 repeated runs with random parameter initializations are reported.

Method	Average log Test	Average log Train	Time(s)	Batchsize
CD1	-145.63 ± 1.30	-146.62 ± 1.72	831	100
PCD	**-136.10 ± 1.21**	**-137.13 ± 1.21**	2620	300
MPF-1flip	-141.13 ± 2.01	-143.02 ± 3.96	2931	75
CD10	-135.40 ± 1.21	-136.46 ± 1.18	17329	100
FMPF10	-136.37 ± 0.17	-137.35 ± 0.19	12533	60
PMPF10	-141.36 ± 0.35	-142.73 ± 0.35	11445	25
FPMPF10	**-134.04 ± 0.12**	**-135.25 ± 0.11**	22201	25
CD15	-134.13 ± 0.82	-135.20 ± 0.84	26723	100
FMPF15	-135.89 ± 0.19	-136.93 ± 0.18	18951	60
PMPF15	-138.53 ± 0.23	-139.71 ± 0.23	13441	25
FPMPF15	**-133.90 ± 0.14**	**-135.13 ± 0.14**	27302	25
CD25	-133.02 ± 0.08	-134.15 ± 0.08	46711	100
FMPF25	-134.50 ± 0.08	-135.63 ± 0.07	25588	60
PMPF25	-135.95 ± 0.13	-137.29 ± 0.13	23115	25
FPMPF25	**-132.74 ± 0.13**	**-133.50 ± 0.11**	50117	25

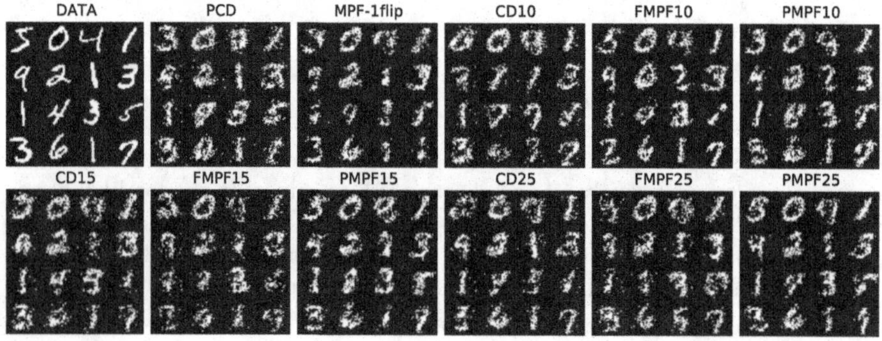

Fig. 1. Samples generated from the training set. Samples in each panel are generated by RBMs trained under different paradigms as noted above each image.

5.1 MNIST - Exact Log Likelihood

In our first experiment, we trained eleven RBMs on the MNIST digits. All RBMs consisted of 20 hidden units and 784 (28×28) visible units. Due to the small number of hidden variables, we calculated the exact value of the partition function by explicitly summing over all visible configurations. Five RBMs were learned by PCD1, CD1, CD10, CD15, and CD25. Seven RBMs were learned by 1 bit flip, FMPF, and FPMPF[3]. Block Gibbs sampling is required for FMPF-k and FPMPF-k similar to CD-k training, where the number of steps is given by k.

The average log test likelihood values of RBMs with 20 hidden units are presented in Table 1. This table gives a sense of the performance under different types of MPF dynamics when the partition function can be calculated exactly. We observed that PMPF consistently achieved a higher log-likelihood than FMPF. MPF with 1 bit flip was very fast but gave poor performance

---

[3] FPMPF is the composition of the FMPF and PMPF connectivities.

Table 2. Experimental results on MNIST using 11 RBMs with 200 hidden units each. The average estimated training and test log-probabilities over 10 repeated runs with random parameter initializations are reported. Likelihood estimates are made with CSL [2] and AIS [8].

| | CSL | | AIS | | | |
Method	Avg. log Test	Avg. log Train	Avg. log Test	Avg. log Train	Time(s)	Batchsize
CD1	-138.63 ± 0.48	-138.70 ± 0.45	-98.75 ± 0.66	-98.61 ± 0.66	1258	100
PCD1	**-114.14 ± 0.26**	**-114.13 ± 0.28**	**-88.82 ± 0.53**	**-89.92 ± 0.54**	2614	100
MPF-1flip	-179.73 ± 0.085	-179.60 ± 0.07	-141.95 ± 0.23	-142.38 ± 0.74	4575	75
CD10	-117.74 ± 0.14	-117.76 ± 0.13	-91.94 ± 0.42	-92.46 ± 0.38	24948	100
FMPF10	-115.11 ± 0.09	-115.10 ± 0.07	-91.21 ± 0.17	-91.39 ± 0.16	24849	25
PMPF10	-114.00 ± 0.08	-113.98 ± 0.09	-89.26 ± 0.13	-89.37 ± 0.13	24179	25
FPMPF10	**-112.45 ± 0.03**	**-112.45 ± 0.03**	**-83.83 ± 0.23**	**-83.26 ± 0.23**	24354	25
CD15	-115.96 ± 0.12	-115.21 ± 0.12	-91.32 ± 0.24	-91.87 ± 0.21	39003	100
FMPF15	-114.05 ± 0.05	-114.06 ± 0.05	-90.72 ± 0.18	-90.93 ± 0.20	26059	25
PMPF15	-114.02 ± 0.11	-114.03 ± 0.09	-89.25 ± 0.17	-89.85 ± 0.19	26272	25
FPMPF15	**-112.58 ± 0.03**	**-112.60 ± 0.02**	**-83.27 ± 0.15**	**-83.84 ± 0.13**	26900	25
CD25	-114.50 ± 0.10	-114.51 ± 0.10	-91.36 ± 0.26	-91.04 ± 0.25	55688	100
FMPF25	-113.07 ± 0.06	-113.07 ± 0.07	-90.43 ± 0.28	-90.63 ± 0.27	40047	25
PMPF25	-113.70 ± 0.04	-113.69 ± 0.04	-89.21 ± 0.14	-89.79 ± 0.13	52638	25
FPMPF25	**-112.38 ± 0.02**	**-112.42 ± 0.02**	**-83.25 ± 0.27**	**-83.81 ± 0.28**	53379	25

compared to FMPF and PMPF. We also observed that MPF-1flip outperformed CD1. FMPF and PMPF always performed slightly worse than CD-k training with the same number of Gibbs steps. However, FPMPF always outperformed CD-k. The advantage of FPMPF in this case may reflect the increased effective number of entries in the transition matrix.

One advantage of FMPF is that it converges much quicker than CD-k or PMPF. This is because we used twice many samples as PMPF as mentioned in Section 4.3. Figure 1 shows initial data and the generated samples after running 100 Gibbs steps from each RBM. PMPF produces samples that are visually more appealing than the other methods.

5.2   MNIST - Estimating Log Likelihood

In our second set of experiments, we trained RBMs with 200 hidden units. We trained them exactly as described in Section 5.1. These RBMs are able to generate much higher-quality samples from the data distribution, however, the partition function can no longer be computed exactly.

In order to evaluate the model quantitatively, we estimated the test log-likelihood using the Conservative Sampling-based Likelihood estimator (CSL) [2] and annealed importance sampling (AIS) [8]. Given well-defined conditional probabilities $P(\mathbf{v}|\mathbf{h})$ of a model and a set of latent variable samples S collected from a Markov chain, CSL computes

$$\log \hat{f}(\mathbf{v}) = \log \mathrm{mean}_{h \in S} P(\mathbf{v}|\mathbf{h}). \qquad (19)$$

The advantage of CSL is that sampling latent variables $\mathbf{h}$ instead of $\mathbf{v}$ has the effect of reducing the variance of the estimator. Also, in contrast to annealed importance sampling (AIS) [8], which tends to overestimate, CSL is much more conservative in its estimates. However, most of the time, CSL is far off from the

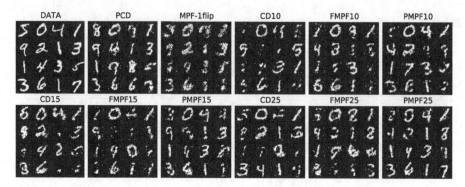

Fig. 2. Samples generated from the training set. Samples in each panel are generated by RBMs trained under different paradigms as noted above each image.

true estimator, so we bound our negative log-likelihood estimate from above and below using both AIS and CSL.

Table 2 demonstrates the test log-likelihood of various RBMs with 200 hidden units. The ranking of the different training paradigms with respect to performance was similar to what we observed in Section 5.1 with PMPF emerging as the winner. However, contrary to the first experiment, we observed that MPF with 1 bit flip did not perform well. Moreover, FMPF and PMPF both tended to give higher test log-likelihoods than CD-k training. Smaller batch sizes worked better with MPF when the number of hiddens was increased. Once again, we observed smaller variances compared to CD-k with both forms of MPF, especially with FMPF. We noted that FMPF and PMPF always have smaller variance compared to CD-k. This implies that FMPF and PMPF are less sensitive to random weight initialization. Figure 2 shows initial data and generated samples after running 100 Gibbs steps for each RBM. PMPF clearly produces samples that look more like digits.

5.3 Caltech 101 Silhouettes - Estimating Log Likelihood

Finally, we evaluated the same set of RBMs on the Caltech-101 Silhouettes dataset. Compared to MNIST, this dataset contains much more diverse structures with richer correlation among the pixels. It has 10 times more categories, contains less training data per category, and each object covers more of the image. For these reasons, we use 500 hidden units per RBM. The estimated average log-likelihood of train and test data is presented in Table 3.

The results for Caltech 101 Silhouettes are consistent with MNIST. In every case, we observed a larger margin between PMPF and CD-k when the number of sampling steps was smaller. In addition, the single bit flip technique was not particularly successful, especially as the number of latent variables grew. We speculate that the reason for this might have to do with the slow rate of convergence for the dynamic system. Moreover, PMPF works better than FMPF for similar reasons. By having persistent samples as the learning progresses, the

Table 3. Experimental results on Caltech-101 Silhouettes using 11 RBMs with 500 hidden units each. The average estimated training and test log-probabilities over 10 repeated runs with random parameter initializations are reported. Likelihood estimates are made with CSL [2] and AIS [8].

| Method | CSL | | AIS | | | |
	Avg. log Test	Avg. log Train	Avg. log Test	Avg. log Train	Time(s)	Batchsize
CD1	-251.30 ± 1.80	-252.04 ± 1.56	-141.87 ± 8.80	-142.88 ± 8.85	300	100
PCD1	**-199.89 ± 1.53**	**-199.95 ± 1.31**	-124.56 ± 0.24	-116.56 ± 2.40	784	100
MPF-1flip	-281.55 ± 1.68	-283.03 ± 0.60	-164.96 ± 0.23	-170.92 ± 0.20	505	100
CD10	-207.77 ± 0.92	-207.16 ± 1.18	-128.17 ± 0.20	-120.65 ± 0.19	4223	100
FMPF10	-211.30 ± 0.84	-211.39 ± 0.90	-135.59 ± 0.16	-135.57 ± 0.18	2698	20
PMPF10	-203.13 ± 0.12	-203.14 ± 0.10	-128.85 ± 0.15	-123.06 ± 0.15	7610	20
FPMPF10	**-200.36 ± 0.16**	**-200.16 ± 0.16**	-123.35 ± 0.16	-108.81 ± 0.15	11973	20
CD15	-205.12 ± 0.87	-204.87 ± 1.13	-125.08 ± 0.24	-117.09 ± 0.21	6611	100
FMPF15	-210.66 ± 0.24	-210.19 ± 0.30	-130.28 ± 0.14	-128.57 ± 0.15	3297	20
PMPF15	-201.47 ± 0.13	-201.67 ± 0.10	-127.09 ± 0.10	-121 ± 0.12	9603	20
FPMPF15	**-198.59 ± 0.17**	**-198.66 ± 0.17**	-122.33 ± 0.13	-107.88 ± 0.14	18170	20
CD25	-201.56 ± 0.11	-201.50 ± 0.13	-124.80 ± 0.13	-117.51 ± 0.23	13745	100
FMPF25	-206.93 ± 0.13	-206.86 ± 0.11	-129.96 ± 0.07	-127.15 ± 0.07	10542	10
PMPF25	-199.53 ± 0.11	-199.51 ± 0.12	-127.81 ± 020	-122.23 ± 0.17	18550	10
FPMPF25	**-198.39 ± 0.0.16**	**-198.39 ± 0.17**	-122.75 ± 0.13	-108.32 ± 0.12	23998	10

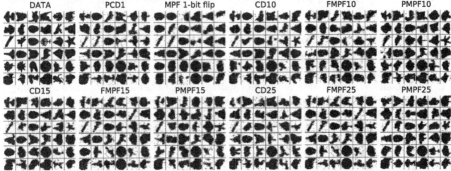

Fig. 3. Random samples generated by RBMs with different training procedures.

dynamics always begin closer to equilibrium, and hence converge more quickly. Figure 3 shows initial data and generated samples after running 100 Gibbs steps for each RBM on Caltech28 dataset.

6 Conclusion

MPF is an unsupervised learning algorithm that can be employed off-the-shelf to any energy-based model. It has a number of favourable properties but has not seen application proportional to its potential. In this paper, we first expounded on MPF and its connections to CD-k training, which allowed us to gain a better understanding and perspective to CD-k. We proved a general form for the transition matrix such that the equilibrium distribution converges to that of an RBM. This may lead to future extensions of MPF based on the choice of $o(\cdot)$ in Equation 12.

One of the merits of MPF is that the choice of designing a dynamic system by defining a connectivity function is left open as long as it satisfies the fixed point equation. Additionally, it should scale similarly to CD-k and its variants when increasing the number of visible and hidden units. We thoroughly explored three different connectivity structures, noting that connectivity can be designed inductively or by sampling. Finally, we showed empirically that MPF, and in particular, PMPF, outperforms CD-k for training generative models. Until now, RBM training was dominated by methods based on CD-k including PCD; however, our results indicate that MPF is a practical and effective alternative.

References

1. Bastien, F., Lamblin, P., Pascanu, R., Bergstra, J., Goodfellow, I.J., Bergeron, A., Bouchard, N., Bengio, Y.: Theano: new features and speed improvements. In: Deep Learning and Unsupervised Feature Learning NIPS 2012 Workshop (2012)
2. Bengio, Y., Yao, L., Cho, K.: Bounding the test log-likelihood of generative models. In: Proceedings of the International Conference on Learning Representations (ICLR) (2013)
3. Besag, J.: Statistical analysis of non-lattice data. The Statistician **24**, 179–195 (1975)
4. Hinton, G.E.: Training products of experts by minimizing contrastive divergence. Neural Computation **14**, 1771–1880 (2002)
5. Hyvärinen, A.: Estimation of non-normalized statistical models by score matching. Journal of Machine Learning Research **6**, 695–709 (2005)
6. MacKay, D.J.C.: Failures of the one-step learning algorithm (2001). http://www.inference.phy.cam.ac.uk/mackay/abstracts/gbm.html, unpublished Technical Report
7. Marlin, B.M., de Freitas, N.: Asymptotic efficiency of deterministic estimators for discrete energy-based models: ratio matching and pseudolikelihood. In: Proceedings of the Uncertainty in Artificial Intelligence (UAI) (2011)
8. Salakhutdinov, R., Murray, I.: On the quantitative analysis of deep belief networks. In: Proceedings of the International Conference of Machine Learning (ICML) (2008)
9. Smolensky, P.: Information processing in dynamical systems: foundations of harmony theory. In: Parallel Distributed Processing: Volume 1: Foundations, pp. 194–281. MIT Press (1986)
10. Sohl-Dickstein, J.: Persistent minimum probability flow. Tech. rep, Redwood Centre for Theoretical Neuroscience (2011)
11. Sohl-Dickstein, J., Battaglino, P., DeWeese, M.R.: Minimum probability flow learning. In: Proceedings of the International Conference of Machine Learning (ICML) (2011)
12. Sutskever, I., Tieleman, T.: On the convergence properties of contrastive divergence. In: Proceedings of the AI & Statistics (AI STAT) (2009)
13. Tieleman, T., Hinton, G.E.: Using fast weights to improve persistent contrastive divergence. In: Proceedings of the International Conference of Machine Learning (ICML) (2009)
14. Tierney, L.: Markov chains for exploring posterior distributions. Annals of Statistics **22**, 1701–1762 (1994)

A Minimum Probability Flow

A.1 Dynamics of The Model

Theorem 2. *1 Suppose $p_j^{(\infty)}$ is the probability of state j and $p_i^{(\infty)}$ is the probability of state i. Let the transition matrix be*

$$\Gamma_{ij} = g_{ij} \exp\left(\frac{o(F_i - F_j) + 1}{2}(F_j - F_i)\right) \tag{20}$$

such that $o(\cdot)$ is any odd function, where g_{ij} is the symmetric connectivity between the states i and j. Then this transition matrix satisfies detailed balance in Equation 11.

Proof. By cancelling out the partition function, the detailed balance Equation 11 can be formulated as

$$\Gamma_{ji} \exp\left(-F_i\right) = \Gamma_{ij} \exp\left(-F_j\right) \tag{21}$$

where $F_i = F(\mathbf{v} = i; \theta)$. By substituting the transition matrix defined in Equation 12, we arrive at the following expression after some straightforward manipulation:

$$\Gamma_{ji} \exp\left(-F_i\right)/\Gamma_{ij} \exp\left(-F_j\right)) = 1$$

$$\exp\left(\frac{o(F_i - F_j) + 1}{2}(F_j - F_i) - F_i\right) / \exp\left(\frac{o(F_j - F_i) + 1}{2}(F_i - F_j) - F_j\right) = 1$$

$$\exp\left(\frac{o(F_i - F_j) + 1}{2}(F_j - F_i) - F_i - \frac{o(F_j - F_i) + 1}{2}(F_i - F_j) + F_j\right) = 1$$

$$\frac{o(F_i - F_j) + 1}{2}(F_j - F_i) - F_i - \frac{o(F_j - F_i) + 1}{2}(F_i - F_j) + F_j = 0$$

$$(F_i - F_j)\left(\frac{o(F_i - F_j) + 1}{2} + \frac{o(F_j - F_i) + 1}{2} - 1\right) = 0$$

$$(F_i - F_j)\left(\frac{o(F_i - F_j)}{2} + \frac{o(F_j - F_i)}{2}\right) = 0$$

Notice that since $o(\cdot)$ is an odd function, this makes the term $\left(\frac{o(F_i - F_j)}{2} + \frac{o(F_j - F_i)}{2}\right) = 0$. Therefore, the detailed balance criterion is satisfied.

Difference Target Propagation

Dong-Hyun Lee[1]([✉]), Saizheng Zhang[1], Asja Fischer[1], and Yoshua Bengio[1,2]

[1] Université de Montréal, Montreal, QC, Canada
[2] CIFAR Senior Fellow, Montreal, Canada
donghyun.lee.dl@gmail.com

Abstract. Back-propagation has been the workhorse of recent successes of deep learning but it relies on infinitesimal effects (partial derivatives) in order to perform credit assignment. This could become a serious issue as one considers deeper and more non-linear functions, e.g., consider the extreme case of non-linearity where the relation between parameters and cost is actually discrete. Inspired by the biological implausibility of back-propagation, a few approaches have been proposed in the past that could play a similar credit assignment role. In this spirit, we explore a novel approach to credit assignment in deep networks that we call target propagation. The main idea is to compute targets rather than gradients, at each layer. Like gradients, they are propagated backwards. In a way that is related but different from previously proposed proxies for back-propagation which rely on a backwards network with symmetric weights, target propagation relies on auto-encoders at each layer. Unlike back-propagation, it can be applied even when units exchange stochastic bits rather than real numbers. We show that a linear correction for the imperfectness of the auto-encoders, called difference target propagation, is very effective to make target propagation actually work, leading to results comparable to back-propagation for deep networks with discrete and continuous units and denoising auto-encoders and achieving state of the art for stochastic networks.

1 Introduction

Recently, deep neural networks have achieved great success in hard AI tasks [2,12,14,19], mostly relying on back-propagation as the main way of performing credit assignment over the different sets of parameters associated with each layer of a deep net. Back-propagation exploits the chain rule of derivatives in order to convert a loss gradient on the activations over layer l (or time t, for recurrent nets) into a loss gradient on the activations over layer $l - 1$ (respectively, time $t - 1$). However, as we consider deeper networks– e.g., consider the recent best ImageNet competition entrants [20] with 19 or 22 layers – longer-term dependencies, or stronger non-linearities, the composition of many non-linear operations becomes more strongly non-linear. To make this concrete, consider the composition of many hyperbolic tangent units. In general, this means that derivatives obtained by back-propagation are becoming either very small (most of the time) or very large (in a few places). In the extreme (very deep computations), one would get discrete functions, whose derivatives are 0 almost everywhere, and

© Springer International Publishing Switzerland 2015
A. Appice et al. (Eds.): ECML PKDD 2015, Part I, LNAI 9284, pp. 498–515, 2015.
DOI: 10.1007/978-3-319-23528-8_31

infinite where the function changes discretely. Clearly, back-propagation would fail in that regime. In addition, from the point of view of low-energy hardware implementation, the ability to train deep networks whose units only communicate via bits would also be interesting.

This limitation of back-propagation to working with precise derivatives and smooth networks is the main machine learning motivation for this paper's exploration into an alternative principle for credit assignment in deep networks. Another motivation arises from the lack of biological plausibility of back-propagation, for the following reasons: (1) the back-propagation computation is purely linear, whereas biological neurons interleave linear and non-linear operations, (2) if the feedback paths were used to propagate credit assignment by back-propagation, they would need precise knowledge of the derivatives of the non-linearities at the operating point used in the corresponding feedforward computation, (3) similarly, these feedback paths would have to use exact symmetric weights (with the same connectivity, transposed) of the feedforward connections, (4) real neurons communicate by (possibly stochastic) binary values (spikes), (5) the computation would have to be precisely clocked to alternate between feedforward and back-propagation phases, and (6) it is not clear where the output targets would come from.

The main idea of target propagation is to associate with each feedforward unit's activation value a *target value* rather than a *loss gradient*. The target value is meant to be close to the activation value while being likely to have provided a smaller loss (if that value had been obtained in the feedforward phase). In the limit where the target is very close to the feedforward value, target propagation should behave like back-propagation. This link was nicely made in [16, 17], which introduced the idea of target propagation and connected it to back-propagation via a Lagrange multipliers formulation (where the constraints require the output of one layer to equal the input of the next layer). A similar idea was recently proposed where the constraints are relaxed into penalties, yielding a different (iterative) way to optimize deep networks [9]. Once a good target is computed, a layer-local training criterion can be defined to update each layer separately, e.g., via the delta-rule (gradient descent update with respect to the cross-entropy loss).

By its nature, target propagation can in principle handle stronger (and even discrete) non-linearities, and it deals with biological plausibility issues (1), (2), (3) and (4) described above. Extensions of the precise scheme proposed here could handle (5) and (6) as well, but this is left for future work.

In this paper, we describe how the general idea of target propagation by using auto-encoders to assign targets to each layer (as introduced in an earlier technical report [4]) can be employed for supervised training of deep neural networks (section 2.1 and 2.2). We continue by introducing a linear correction for the imperfectness of the auto-encoders (2.3) leading to robust training in practice. Furthermore, we show how the same principles can be applied to replace back-propagation in the training of auto-encoders (section 2.4). In section 3 we provide several experimental results on rather deep neural networks as well as discrete and stochastic networks and auto-encoders. The results show that the proposed form of target propagation is comparable to back-propagation with RMSprop [22] - a very popular setting to train deep networks nowadays- and achieves state of the art for training stochastic neural nets on MNIST.

2   Target Propagation

Although many variants of the general principle of target propagation can be devised, this paper focuses on a specific approach, which is based on the ideas presented in an earlier technical report [4] and is described in the following.

2.1   Formulating Targets

Let us consider an ordinary (supervised) deep network learning process, where the training data is drawn from an unknown data distribution $p(\mathbf{x}, \mathbf{y})$. The network structure is defined by

$$\mathbf{h}_i = f_i(\mathbf{h}_{i-1}) = s_i(W_i\mathbf{h}_{i-1}), \quad i = 1, \ldots, M \tag{1}$$

where $\mathbf{h}_i$ is the state of the i-th hidden layer (where $\mathbf{h}_M$ corresponds to the output of the network and $\mathbf{h}_0 = \mathbf{x}$) and f_i is the i-th layer feed-forward mapping, defined by a non-linear activation function s_i (e.g. the hyperbolic tangents or the sigmoid function) and the weights W_i of the i-th layer. Here, for simplicity of notation, the bias term of the i-th layer is included in W_i. We refer to the subset of network parameters defining the mapping between the i-th and the j-th layer $(0 \leq i < j \leq M)$ as $\theta_W^{i,j} = \{W_k, k = i + 1, \ldots, j\}$. Using this notion, we can write $\mathbf{h}_j$ as a function of $\mathbf{h}_i$ depending on parameters $\theta_W^{i,j}$, that is we can write $\mathbf{h}_j = \mathbf{h}_j(\mathbf{h}_i; \theta_W^{i,j})$.

Given a sample $(\mathbf{x}, \mathbf{y})$, let $L(\mathbf{h}_M(\mathbf{x}; \theta_W^{0,M}), \mathbf{y})$ be an arbitrary global loss function measuring the appropriateness of the network output $\mathbf{h}_M(\mathbf{x}; \theta_W^{0,M})$ for the target $\mathbf{y}$, e.g. the MSE or cross-entropy for binomial random variables. Then, the training objective corresponds to adapting the network parameters $\theta_W^{0,M}$ so as to minimize the expected global loss $\mathbb{E}_p\{L(\mathbf{h}_M(\mathbf{x}; \theta_W^{0,M}), \mathbf{y})\}$ under the data distribution $p(\mathbf{x}, \mathbf{y})$. For $i = 1, \ldots, M - 1$ we can write

$$L(\mathbf{h}_M(\mathbf{x}; \theta_W^{0,M}), \mathbf{y}) = L(\mathbf{h}_M(\mathbf{h}_i(\mathbf{x}; \theta_W^{0,i}); \theta_W^{i,M}), \mathbf{y}) \tag{2}$$

to emphasize the dependency of the loss on the state of the i-th layer.

Training a network with back-propagation corresponds to propagating error signals through the network to calculate the derivatives of the global loss with respect to the parameters of each layer. Thus, the error signals indicate how the parameters of the network should be updated to decrease the expected loss. However, in very deep networks with strong non-linearities, error propagation could become useless in lower layers due to exploding or vanishing gradients, as explained above.

To avoid this problem, the basic idea of target propagation is to assign to each $\mathbf{h}_i(\mathbf{x}; \theta_W^{0,i})$ a nearby value $\hat{\mathbf{h}}_i$ which (hopefully) leads to a lower global loss, that is which has the objective to fulfill

$$L(\mathbf{h}_M(\hat{\mathbf{h}}_i; \theta_W^{i,M}), \mathbf{y}) < L(\mathbf{h}_M(\mathbf{h}_i(\mathbf{x}; \theta_W^{0,i}); \theta_W^{i,M}), \mathbf{y}) \ . \tag{3}$$

Such a $\hat{\mathbf{h}}_i$ is called a *target* for the i-th layer.

Given a target $\hat{\mathbf{h}}_i$ we now would like to change the network parameters to make $\mathbf{h}_i$ move a small step towards $\hat{\mathbf{h}}_i$, since – if the path leading from $\mathbf{h}_i$ to $\hat{\mathbf{h}}_i$ is smooth enough – we would expect to yield a decrease of the global loss. To obtain an update direction for W_i based on $\hat{\mathbf{h}}_i$ we can define a layer-local target loss L_i, for example by using the MSE

$$L_i(\hat{\mathbf{h}}_i, \mathbf{h}_i) = ||\hat{\mathbf{h}}_i - \mathbf{h}_i(\mathbf{x}; \theta_W^{0,i})||_2^2 \ . \tag{4}$$

Then, W_i can be updated locally within its layer via stochastic gradient descent, where $\hat{\mathbf{h}}_i$ is considered as a *constant* with respect to W_i. That is

$$W_i^{(t+1)} = W_i^{(t)} - \eta_{f_i} \frac{\partial L_i(\hat{\mathbf{h}}_i, \mathbf{h}_i)}{\partial W_i} = W_i^{(t)} - \eta_{f_i} \frac{\partial L_i(\hat{\mathbf{h}}_i, \mathbf{h}_i)}{\partial \mathbf{h}_i} \frac{\partial \mathbf{h}_i(\mathbf{x}; \theta_W^{0,i})}{\partial W_i} \ , \tag{5}$$

where η_{f_i} is a layer-specific learning rate.

Note, that in this context, derivatives can be used without difficulty, because they correspond to computations performed inside a single layer. Whereas, the problems with the severe non-linearities observed for back-propagation arise when the chain rule is applied through many layers. This motivates target propagation methods to serve as alternative credit assignment in the context of a composition of many non-linearities.

However, it is not directly clear how to compute a target that guarantees a decrease of the global loss (that is how to compute a $\hat{\mathbf{h}}_i$ for which equation (3) holds) or that at least leads to a decrease of the local loss L_i of the next layer, that is

$$L_{i+1}(\hat{\mathbf{h}}_{i+1}, f_{i+1}(\hat{\mathbf{h}}_i)) < L_{i+1}(\hat{\mathbf{h}}_{i+1}, f_{i+1}(\mathbf{h}_i)) \ . \tag{6}$$

Proposing and validating answers to this question is the subject of the rest of this paper.

2.2   How to Assign a Proper Target to Each Layer

Clearly, in a supervised learning setting, the top layer target should be directly driven from the gradient of the global loss

$$\hat{\mathbf{h}}_M = \mathbf{h}_M - \hat{\eta} \frac{\partial L(\mathbf{h}_M, \mathbf{y})}{\partial \mathbf{h}_M} \ , \tag{7}$$

where $\hat{\eta}$ is usually a small step size. Note, that if we use the MSE as global loss and $\hat{\eta} = 0.5$ we get $\hat{\mathbf{h}}_M = \mathbf{y}$.

But how can we define targets for the intermediate layers? In the previous technical report [4], it was suggested to take advantage of an "approximate inverse". To formalize this idea, suppose that for each f_i we have a function g_i such that

$$f_i(g_i(\mathbf{h}_i)) \approx \mathbf{h}_i \quad \text{or} \quad g_i(f_i(\mathbf{h}_{i-1})) \approx \mathbf{h}_{i-1} . \tag{8}$$

Then, choosing

$$\hat{\mathbf{h}}_{i-1} = g_i(\hat{\mathbf{h}}_i) \tag{9}$$

would have the consequence that (under some smoothness assumptions on f and g) minimizing the distance between $\mathbf{h}_{i-1}$ and $\hat{\mathbf{h}}_{i-1}$ should also minimize the loss L_i of the i-th layer. This idea is illustrated in the left of Figure 1. Indeed, if the feed-back mappings were the perfect inverses of the feed-forward mappings $(g_i = f_i^{-1})$, one gets

$$L_i(\hat{\mathbf{h}}_i, f_i(\hat{\mathbf{h}}_{i-1})) = L_i(\hat{\mathbf{h}}_i, f_i(g_i(\hat{\mathbf{h}}_i))) = L_i(\hat{\mathbf{h}}_i, \hat{\mathbf{h}}_i) = 0 . \tag{10}$$

But choosing g to be the perfect inverse of f may need heavy computation and instability, since there is no guarantee that f_i^{-1} applied to a target would yield a value that is in the domain of f_{i-1}. An alternative approach is to learn an approximate inverse g_i, making the f_i / g_i pair look like an *auto-encoder*. This suggests parametrizing g_i as follows:

$$g_i(\mathbf{h}_i) = \bar{s}_i(V_i\mathbf{h}_i), \quad i = 1, ..., M \tag{11}$$

where $\bar{s}_i$ is a non-linearity associated with the decoder and V_i the matrix of feed-back weights of the i-th layer. With such a parametrization, it is unlikely that the auto-encoder will achieve zero reconstruction error. The decoder could be trained via an additional auto-encoder-like loss at each layer

$$L_i^{inv} = ||g_i(f_i(\mathbf{h}_{i-1})) - \mathbf{h}_{i-1}||_2^2 . \tag{12}$$

Changing $\mathbf{V}_i$ based on this loss, makes g closer to f_i^{-1}. By doing so, it also makes $f_i(\hat{\mathbf{h}}_{i-1}) = f_i(g_i(\hat{\mathbf{h}}_i))$ closer to $\hat{\mathbf{h}}_i$, and is thus also contributing to the decrease of $L_i(\hat{\mathbf{h}}_i, f_i(\hat{\mathbf{h}}_{i-1}))$. But we do not want to estimate an inverse mapping only for the concrete values we see in training but for a region around the these values to facilitate the computation of $g_i(\hat{\mathbf{h}}_i)$ for $\hat{\mathbf{h}}_i$ which have never been seen before. For this reason, the loss is modified by noise injection

$$L_i^{inv} = ||g_i(f_i(\mathbf{h}_{i-1} + \epsilon)) - (\mathbf{h}_{i-1} + \epsilon)||_2^2, \quad \epsilon \sim N(0, \sigma) , \tag{13}$$

which makes f_i and g_i approximate inverses not just at $\mathbf{h}_{i-1}$ but also in its *neighborhood*.

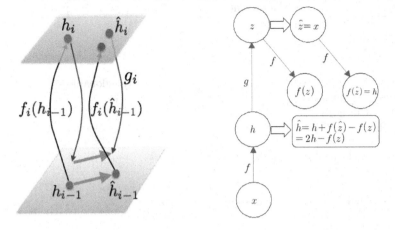

Fig. 1. (left) How to compute a target in the lower layer via difference target propagation. $f_i(\hat{\mathbf{h}}_{i-1})$ should be closer to $\hat{\mathbf{h}}_i$ than $f_i(\mathbf{h}_{i-1})$. (right) Diagram of the back-propagation-free auto-encoder via difference target propagation.

As mentioned above, a required property of target propagation is that the layer-wise parameter updates, each improving a layer-wise loss, also lead to an improvement of the global loss. The following theorem shows that, for the case that g_i is a perfect inverse of f_i and f_i having a certain structure, the update direction of target propagation does not deviate more then 90 degrees from the gradient direction (estimated by back-propagation), which always leads to a decrease of the global loss.

Theorem 1.[1] *Assume that $g_i = f_i^{-1}, i = 1, ..., M$, and f_i satisfies $\mathbf{h}_i = f_i(\mathbf{h}_{i-1}) = W_i s_i(\mathbf{h}_{i-1})$[2] where s_i can be any differentiable monotonically increasing element-wise function. Let δW_i^{tp} and δW_i^{bp} be the target propagation update and the back-propagation update in i-th layer, respectively. If $\hat{\eta}$ in Equation (7) is sufficiently small, then the angle α between δW_i^{tp} and δW_i^{bp} is bounded by*

$$0 < \frac{1 + \Delta_1(\hat{\eta})}{\frac{\lambda_{max}}{\lambda_{min}} + \Delta_2(\hat{\eta})} \le cos(\alpha) \le 1 \tag{14}$$

Here λ_{max} and λ_{min} are the largest and smallest singular values of $(J_{f_M} \ldots J_{f_{i+1}})^T$, where J_{f_k} is the Jacobian matrix of f_k and $\Delta_1(\hat{\eta})$ and $\Delta_2(\hat{\eta})$ are close to 0 if $\hat{\eta}$ is sufficiently small.

2.3 Difference Target Propagation

From our experience, the imperfection of the inverse function leads to severe optimization problems when assigning targets based on equation (9). This brought

---
[1] The proof can be found in the appendix.
[2] This is another way to obtain a non-linear deep network structure.

Algorithm 1. Training deep neural networks via difference target propagation

---

Compute unit values for all layers:
for $i = 1$ to M **do**
   $\mathbf{h}_i \leftarrow f_i(\mathbf{h}_{i-1})$
end for
Making the first target: $\hat{\mathbf{h}}_{M-1} \leftarrow \mathbf{h}_{M-1} - \hat{\eta}\frac{\partial L}{\partial \mathbf{h}_{M-1}}$,   ($L$ is the global loss)
Compute targets for lower layers:
for $i = M - 1$ to 2 **do**
   $\hat{\mathbf{h}}_{i-1} \leftarrow \mathbf{h}_{i-1} - g_i(\mathbf{h}_i) + g_i(\hat{\mathbf{h}}_i)$
end for
Training feedback (inverse) mapping:
for $i = M - 1$ to 2 **do**
   Update parameters for g_i using SGD with following a layer-local loss L_i^{inv}
   $L_i^{inv} = ||g_i(f_i(\mathbf{h}_{i-1} + \epsilon)) - (\mathbf{h}_{i-1} + \epsilon)||_2^2, \quad \epsilon \sim N(0, \sigma)$
end for
Training feedforward mapping:
for $i = 1$ to M **do**
   Update parameters for f_i using SGD with following a layer-local loss L_i
   $L_i = ||f_i(\mathbf{h}_{i-1}) - \hat{\mathbf{h}}_i||_2^2$ if $i < M$, $L_i = L$ (the global loss) if $i = M$.
end for

---

us to propose the following linearly corrected formula for target propagation which we refer to as *"difference target propagation"*

$$\hat{\mathbf{h}}_{i-1} = \mathbf{h}_{i-1} + g_i(\hat{\mathbf{h}}_i) - g_i(\mathbf{h}_i) \ . \tag{15}$$

Note, that if g_i is the inverse of f_i, difference target propagation becomes equivalent to vanilla target propagation as defined in equation (9). The resulting complete training procedure for optimization by difference target propagation is given in Algorithm 1.

In the following, we explain why this linear corrected formula stabilizes the optimization process. In order to achieve stable optimization by target propagation, $\mathbf{h}_{i-1}$ should approach $\hat{\mathbf{h}}_{i-1}$ as $\mathbf{h}_i$ approaches $\hat{\mathbf{h}}_i$. Otherwise, the parameters in lower layers continue to be updated even when an optimum of the global loss is reached already by the upper layers, which then could lead the global loss to increase again. Thus, the condition

$$\mathbf{h}_i = \hat{\mathbf{h}}_i \ \Rightarrow \ \mathbf{h}_{i-1} = \hat{\mathbf{h}}_{i-1} \tag{16}$$

greatly improves the stability of the optimization. This holds for vanilla target propagation if $g_i = f_i^{-1}$, because

$$\mathbf{h}_{i-1} = f_i^{-1}(\mathbf{h}_i) = g_i(\hat{\mathbf{h}}_i) = \hat{\mathbf{h}}_{i-1} \ . \tag{17}$$

Although the condition is not guaranteed to hold for vanilla target propagation if $g_i \neq f_i^{-1}$, for difference target propagation it holds by construction, since

$$\hat{\mathbf{h}}_{i-1} - \mathbf{h}_{i-1} = g_i(\hat{\mathbf{h}}_i) - g_i(\mathbf{h}_i) \ . \tag{18}$$

Furthermore, under weak conditions on f and g and if the difference between $\mathbf{h}_i$ and $\hat{\mathbf{h}}_i$ is small, we can show for difference target propagation that if the input of the i-th layer becomes $\hat{\mathbf{h}}_{i-1}$ (i.e. the $i-1$-th layer reaches its target) the output of the i-th layer also gets closer to $\hat{\mathbf{h}}_i$. This means that the requirement on targets specified by equation (6) is met for difference target propagation, as shown in the following theorem

Theorem 2. [3] *Let the target for layer $i - 1$ be given by Equation (15), i.e.* $\hat{\mathbf{h}}_{i-1} = \mathbf{h}_{i-1} + g_i(\hat{\mathbf{h}}_i) - g_i(\mathbf{h}_i)$. *If $\hat{\mathbf{h}}_i - \mathbf{h}_i$ is sufficiently small, f_i and g_i are differentiable, and the corresponding Jacobian matrices J_{f_i} and J_{g_i} satisfy that the largest eigenvalue of $(I - J_{f_i} J_{g_i})^T (I - J_{f_i} J_{g_i})$ is less than 1, then we have*

$$||\hat{\mathbf{h}}_i - f_i(\hat{\mathbf{h}}_{i-1})||_2^2 < ||\hat{\mathbf{h}}_i - \mathbf{h}_i||_2^2 \ . \tag{19}$$

The third condition in the above theorem is easily satisfied in practice, because g_i is learned to be the inverse of f_i and makes $g_i \circ f_i$ close to the identity mapping, so that $(I - J_{f_i} J_{g_i})$ becomes close to the zero matrix which means that the largest eigenvalue of $(I - J_{f_i} J_{g_i})^T (I - J_{f_i} J_{g_i})$ is also close to 0.

2.4   Training an Auto-Encoder with Difference Target Propagation

Auto-encoders are interesting for learning representations and serve as building blocks for deep neural networks [10]. In addition, as we have seen, training auto-encoders is part of the target propagation approach presented here, where they model the feedback paths used to propagate the targets.

In the following, we show how a regularized auto-encoder can be trained using difference target propagation instead of back-propagation. Like in the work on denoising auto-encoders [23] and generative stochastic networks [6], we consider the denoising auto-encoder as a stochastic network with noise injected in input and hidden units, trained to minimize a reconstruction loss. This is, the hidden units are given by the encoder as

$$\mathbf{h} = f(\mathbf{x}) = sig(W\mathbf{x} + \mathbf{b}) \ , \tag{20}$$

where sig is the element-wise sigmoid function, W the weight matrix and $\mathbf{b}$ the bias vector of the input units. The reconstruction is given by the decoder

$$\mathbf{z} = g(\mathbf{h}) = sig(W^T(\mathbf{h} + \epsilon) + \mathbf{c}), \ \epsilon \sim N(0, \sigma) \ , \tag{21}$$

---

[3] The proof can be found in the appendix.

with $\mathbf{c}$ being the bias vector of the hidden units. And the reconstruction loss is

$$L = ||\mathbf{z} - \mathbf{x}||_2^2 + ||f(\mathbf{x} + \epsilon) - \mathbf{h}||_2^2, \ \epsilon \sim N(0, \sigma) \ , \tag{22}$$

where a regularization term can be added to obtain a contractive mapping. In order to train this network without back-propagation (that is, without using the chain rule), we can use difference target propagation as follows (see Figure 1 (right) for an illustration): at first, the target of $\mathbf{z}$ is just $\mathbf{x}$, so we can train the reconstruction mapping g based on the loss $L_g = ||g(\mathbf{h}) - \mathbf{x}||_2^2$ in which $\mathbf{h}$ is considered as a constant. Then, we compute the target $\hat{\mathbf{h}}$ of the hidden units following difference target propagation where we make use of the fact that f is an approximate inverse of g. That is,

$$\hat{\mathbf{h}} = \mathbf{h} + f(\hat{\mathbf{z}}) - f(\mathbf{z}) = 2\mathbf{h} - f(\mathbf{z}) \ , \tag{23}$$

where the last equality follows from $f(\hat{\mathbf{z}}) = f(\mathbf{x}) = \mathbf{h}$. As a target loss for the hidden layer, we can use $L_f = ||f(\mathbf{x} + \epsilon) - \hat{\mathbf{h}}||_2^2$, where $\hat{\mathbf{h}}$ is considered as a constant and which can be also augmented by a regularization term to yield a contractive mapping.

3   Experiments

In a set of experiments we investigated target propagation for training deep feedforward deterministic neural networks, networks with discrete transmissions between units, stochastic neural networks, and auto-encoders.

For training supervised neural networks, we chose the target of the top hidden layer (number $M-1$) such that it also depends directly on the global loss instead of an inverse mapping. That is, we set $\hat{\mathbf{h}}_{M-1} = \mathbf{h}_{M-1} - \tilde{\eta} \frac{\partial L(\mathbf{h}_M, \mathbf{y})}{\partial \mathbf{h}_{M-1}}$, where L is the global loss (here the multiclass cross entropy). This may be helpful when the number of units in the output layer is much smaller than the number of units in the top hidden layer, which would make the inverse mapping difficult to learn, but future work should validate that.

For discrete stochastic networks in which some form of noise (here Gaussian) is injected, we used a decaying noise level for learning the inverse mapping, in order to stabilize learning, i.e. the standard deviation of the Gaussian is set to $\sigma(e) = \sigma_0/(1 + e/e_0)$ where σ_0 is the initial value, e is the epoch number and e_0 is the half-life of this decay. This seems to help to fine-tune the feedback weights at the end of training.

In all experiments, the weights were initialized with orthogonal random matrices and the bias parameters were initially set to zero. All experiments were repeated 10 times with different random initializations. We put the code of these experiments online (https://github.com/donghyunlee/dtp).

3.1   Deterministic Feedforward Deep Networks

As a primary objective, we investigated training of ordinary deep supervised networks with continuous and deterministic units on the MNIST dataset. We used a held-out validation set of 10000 samples for choosing hyper-parameters. We trained networks with 7 hidden layers each consisting of 240 units (using the hyperbolic tangent as activation function) with difference target propagation and back-propagation.

Training was based on RMSprop [22] where hyper-parameters for the best validation error were found using random search [7]. RMSprop is an adaptive learning rate algorithm known to lead to good results for back-propagation. Furthermore, it is suitable for updating the parameters of each layer based on the layer-wise targets obtained by target propagation. Our experiments suggested that when using a hand-selected learning rate per layer rather than the automatically set one (by RMSprop), the selected learning rates were different for each layer, which is why we decided to use an adaptive method like RMSprop.

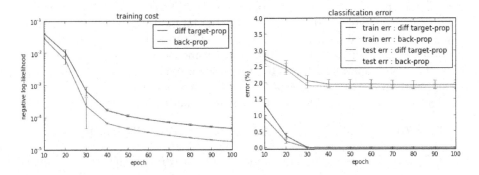

Fig. 2. Mean training cost (left) and train/test classification error (right) with target propagation and back-propagation using continuous deep networks (tanh) on MNIST. Error bars indicate the standard deviation over 10 independent runs with the same optimized hyper-parameters and different initial weights.

The results are shown in Figure 2. We obtained a test error of 1.94% with target propagation and 1.86% with back propagation. The final negative log-likelihood on the training set was 4.584×10^{-5} with target propagation and 1.797×10^{-5} with back propagation. We also trained the same network with rectifier linear units and got a test error of 3.15% whereas 1.62% was obtained with back-propagation. It is well known that this nonlinearity is advantageous for back-propagation [11], while it seemed to be less appropriate for this implementation of target propagation.

In a second experiment we investigated training on CIFAR-10. The experimental setting was the same as for MNIST (using the hyperbolic tangent as activation function) except that the network architecture was 3072-1000-1000-1000-10. We did not use any preprocessing, except for scaling the input values to lay in [0,1], and we tuned the hyper-parameters of RMSprop using a held-out validation set of 1000 samples. We obtained mean test accuracies of 50.71% and 53.72% for target propagation and back-propagation, respectively. It was reported in [15], that a network with 1 hidden layer of 1000 units achieved 49.78% accuracy with back-propagation, and increasing the number of units to 10000 led to 51.53% accuracy. As the current state-of-the-art performance on the permutation invariant CIFAR-10 recognition task, [13] reported 64.1% but when using PCA without whitening as preprocessing and zero-biased auto-encoders for unsupervised pre-training.

3.2 Networks with Discretized Transmission Between Units

To explore target propagation for an extremely non-linear neural network, we investigated training of discrete networks on the MNIST dataset. The network architecture was 784-500-500-10, where only the 1st hidden layer was discretized. Inspired by biological considerations and the objective of reducing the communication cost between neurons, instead of just using the step activation function, we used ordinary neural net layers but with signals being discretized when transported between the first and second layer. The network structure is depicted in the right plot of Figure 3 and the activations of the hidden layers are given by

$$\mathbf{h}_1 = f_1(\mathbf{x}) = \tanh(W_1\mathbf{x}) \quad \text{and} \quad \mathbf{h}_2 = f_2(\mathbf{h}_1) = \tanh(W_2 sign(\mathbf{h}_1)) \quad (24)$$

where $sign(x) = 1$ if $x > 0$, and $sign(x) = 0$ if $x \leq 0$. The network output is given by

$$p(\mathbf{y}|\mathbf{x}) = f_3(\mathbf{h}_2) = softmax(W_3\mathbf{h}_2) \ . \quad (25)$$

The inverse mapping of the second layer and the associated loss are given by

$$g_2(\mathbf{h}_2) = \tanh(V_2 sign(\mathbf{h}_2)) \ , \quad (26)$$

$$L_2^{inv} = ||g_2(f_2(\mathbf{h}_1 + \epsilon)) - (\mathbf{h}_1 + \epsilon)||_2^2, \quad \epsilon \sim N(0,\sigma) \ . \quad (27)$$

If the feed-forward mapping is discrete, back-propagated gradients become 0 and useless when they cross the discretization step. So we compare target propagation to two baselines. As a first baseline, we train the network with back-propagation and the *straight-through estimator* [5], which is biased but was found to work well, and simply ignores the derivative of the step function (which is 0 or infinite) in the back-propagation phase. As a second baseline, we train only the upper layers by back-propagation, while not changing the weight W_1 which are affected by the discretization, i.e., the lower layers do not learn.

The results on the training and test sets are shown in Figure 3. The training error for the first baseline (straight-through estimator) does not converge to zero (which can be explained by the biased gradient) but generalization performance is fairly good. The second baseline (fixed lower layer) surprisingly reached zero training error, but did not perform well on the test set. This can be explained by the fact that it cannot learn any meaningful representation at the first layer. Target propagation however did not suffer from this drawback and can be used to train discrete networks directly (training signals can pass the discrete region successfully). Though the training convergence was slower, the training error did approach zero. In addition, difference target propagation also achieved good results on the test set.

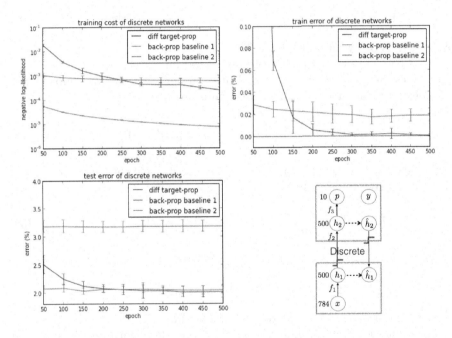

Fig. 3. Mean training cost (top left), mean training error (top right) and mean test error (bottom left) while training discrete networks with difference target propagation and the two baseline versions of back-propagation. Error bars indicate standard deviations over 10 independent runs with the same hyper-parameters and different initial weights. Diagram of the discrete network (bottom right). The output of h_1 is discretized because signals must be communicated from h_1 to h_2 through a long cable, so binary representations are preferred in order to conserve energy. With target propagation, training signals are also discretized through this cable (since feedback paths are computed by bona-fide neurons).

3.3 Stochastic Networks

Another interesting model class which vanilla back-propagation cannot deal with are stochastic networks with discrete units. Recently, stochastic networks have attracted attention [3,5,21] because they are able to learn a multi-modal conditional distribution $P(Y|X)$, which is important for structured output predictions. Training networks of stochastic binary units is also biologically motivated, since they resemble networks of spiking neurons. Here, we investigate whether one can train networks of stochastic binary units on MNIST for classification using target propagation. Following [18], the network architecture was 784-200-200-10 and the hidden units were stochastic binary units with the probability of turning on given by a sigmoid activation

$$\mathbf{h}_i^p = P(\mathbf{H}_i = 1|\mathbf{h}_{i-1}) = sig(W_i\mathbf{h}_{i-1}), \quad \mathbf{h}_i \sim P(\mathbf{H}_i|\mathbf{h}_{i-1}) \ , \tag{28}$$

that is, $\mathbf{h}_i$ is one with probability $\mathbf{h}_i^p$.

As a baseline, we considered training based on the *straight-through* biased gradient estimator [5] in which the derivative through the discrete sampling step is ignored (this method showed the best performance in [18].) That is

$$\delta\mathbf{h}_{i-1}^p = \delta\mathbf{h}_i^p \frac{\partial\mathbf{h}_i^p}{\partial\mathbf{h}_{i-1}^p} \approx sig'(W_i\mathbf{h}_{i-1})W_i^T\delta\mathbf{h}_i^p \ . \tag{29}$$

With difference target propagation the stochastic network can be trained directly, setting the targets to

$$\hat{\mathbf{h}}_2^p = \mathbf{h}_2^p - \eta\frac{\partial L}{\partial\mathbf{h}_2} \quad \text{and} \quad \hat{\mathbf{h}}_1^p = \mathbf{h}_1^p + g_2(\hat{\mathbf{h}}_2^p) - g_2(\mathbf{h}_2^p) \tag{30}$$

where $g_i(\mathbf{h}_i^p) = \tanh(V_i\mathbf{h}_i^p)$ is trained by the loss

$$L_i^{inv} = ||g_i(f_i(\mathbf{h}_{i-1} + \epsilon)) - (\mathbf{h}_{i-1} + \epsilon)||_2^2, \quad \epsilon \sim N(0,\sigma) \ , \tag{31}$$

and layer-local target losses are defined as $L_i = ||\hat{\mathbf{h}}_i^p - \mathbf{h}_i^p||_2^2$.

For evaluation, we averaged the output probabilities for a given input over 100 samples, and classified the example accordingly, following [18]. Results are given in Table 1. We obtained a test error of 1.71% using the baseline method and 1.54% using target propagation, which is – to our knowledge – the best result for stochastic nets on MNIST reported so far. This suggests that target propagation is highly promising for training networks of binary stochastic units.

Table 1. Mean test Error on MNIST for stochastoc networks. The first row shows the results of our experiments averaged over 10 trials. The second row shows the results reported in [18]. M corresponds to the number of samples used for computing output probabilities. We used M=1 during training and M=100 for the test set.

Method	Test Error(%)
Difference Target-Propagation, M=1	1.54%
Straight-through gradient estimator [5] + backprop, M=1 as reported in Raiko et al. [18]	1.71%
as reported in Tang and Salakhutdinov [21], M=20	3.99%
as reported in Raiko et al. [18], M=20	1.63%

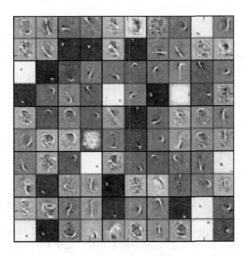

Fig. 4. Filters learned by the back-propagation-free auto-encoder. Each filter corresponds to the hidden weights of one of 100 randomly chosen hidden units. We obtain stroke filters, similar to those usually obtained by regularized auto-encoders.

3.4 Auto-Encoder

We trained a denoising auto-encoder with 1000 hidden units with difference target propagation as described in Section 2.4 on MNIST. As shown in Figure 4 stroke-like filters can be obtained by target propagation. After supervised fine-tuning (using back-propagation), we got a test error of 1.35%. Thus, by training an auto-encoder with target propagation one can learn a good initial representation, which is as good as the one obtained by regularized auto-encoders trained by back-propagation on the reconstruction error.

4 Conclusion

We introduced a novel optimization method for neural networks, called target propagation, which was designed to overcome drawbacks of back-propagation and is biologically more plausible. Target propagation replaces training signals based on partial derivatives by targets which are propagated based on an auto-encoding feedback loop. Difference target propagation is a linear correction for this imperfect inverse mapping which is effective to make target propagation actually work. Our experiments show that target propagation performs comparable to back-propagation on ordinary deep networks and denoising auto-encoders. Moreover, target propagation can be directly used on networks with discretized transmission between units and reaches state of the art performance for stochastic neural networks on MNIST.

Acknowledgments. We would like to thank Junyoung Chung for providing RMSprop code, Caglar Gulcehre and Antoine Biard for general discussion and feedback, Jyri Kivinen for discussion of backprop-free auto-encoder, Mathias Berglund for explanation of his stochastic networks. We thank the developers of Theano [1,8], a Python library which allowed us to easily develop a fast and optimized code for GPU. We are also grateful for funding from NSERC, the Canada Research Chairs, Compute Canada, and CIFAR.

References

1. Bastien, F., Lamblin, P., Pascanu, R., Bergstra, J., Goodfellow, I.J., Bergeron, A., Bouchard, N., Bengio, Y.: Theano: new features and speed improvements. In: Deep Learning and Unsupervised Feature Learning NIPS 2012 Workshop (2012)
2. Bengio, Y.: Learning deep architectures for AI. Now Publishers (2009)
3. Bengio, Y.: Estimating or propagating gradients through stochastic neurons. Tech. Rep. Universite de Montreal (2013). arXiv:1305.2982
4. Bengio, Y.: How auto-encoders could provide credit assignment in deep networks via target propagation. Tech. rep. (2014). arXiv:1407.7906
5. Bengio, Y., Léonard, N., Courville, A.: Estimating or propagating gradients through stochastic neurons for conditional computation (2013). arXiv:1308.3432
6. Bengio, Y., Thibodeau-Laufer, E., Yosinski, J.: Deep generative stochastic networks trainable by backprop. In: ICML 2014 (2014)
7. Bergstra, J., Bengio, Y.: Random search for hyper-parameter optimization. J. Machine Learning Res. **13**, 281–305 (2012)
8. Bergstra, J., Breuleux, O., Bastien, F., Lamblin, P., Pascanu, R., Desjardins, G., Turian, J., Warde-Farley, D., Bengio, Y.: Theano: a CPU and GPU math expression compiler. In: Proceedings of the Python for Scientific Computing Conference (SciPy), oral Presentation, June 2010
9. Carreira-Perpinan, M., Wang, W.: Distributed optimization of deeply nested systems. In: AISTATS 2014, JMLR W&CP, vol. 33, pp. 10–19 (2014)
10. Erhan, D., Courville, A., Bengio, Y., Vincent, P.: Why does unsupervised pre-training help deep learning? In: JMLR W&CP: Proc. AISTATS 2010, vol. 9, pp. 201–208 (2010)

11. Glorot, X., Bordes, A., Bengio, Y.: Deep sparse rectifier neural networks. In: JMLR W&CP: Proceedings of the Fourteenth International Conference on Artificial Intelligence and Statistics (AISTATS 2011), April 2011
12. Hinton, G., Deng, L., Dahl, G.E., Mohamed, A., Jaitly, N., Senior, A., Vanhoucke, V., Nguyen, P., Sainath, T., Kingsbury, B.: Deep neural networks for acoustic modeling in speech recognition. IEEE Signal Processing Magazine **29**(6), 82–97 (2012)
13. Konda, K., Memisevic, R., Krueger, D.: Zero-bias autoencoders and the benefits of co-adapting features. Under review on International Conference on Learning Representations (2015)
14. Krizhevsky, A., Sutskever, I., Hinton, G.: Imagenet classification with deep convolutional neural networks. In: NIPS 2012 (2012)
15. Krizhevsky, A., Hinton, G.: Learning multiple layers of features from tiny images. Master's thesis, University of Toronto (2009)
16. LeCun, Y.: Learning processes in an asymmetric threshold network. In: Fogelman-Soulié, F., Bienenstock, E., Weisbuch, G. (eds.) Disordered Systems and Biological Organization, pp. 233–240. Springer-Verlag, Les Houches (1986)
17. LeCun, Y.: Modèles connexionistes de l'apprentissage. Ph.D. thesis, Université de Paris VI (1987)
18. Raiko, T., Berglund, M., Alain, G., Dinh, L.: Techniques for learning binary stochastic feedforward neural networks. In: NIPS Deep Learning Workshop 2014 (2014)
19. Sutskever, I., Vinyals, O., Le, Q.V.: Sequence to sequence learning with neural networks. Tech. rep. (2014). arXiv:1409.3215
20. Szegedy, C., Liu, W., Jia, Y., Sermanet, P., Reed, S., Anguelov, D., Erhan, D., Vanhoucke, V., Rabinovich, A.: Going deeper with convolutions. Tech. rep. (2014). arXiv:1409.4842
21. Tang, Y., Salakhutdinov, R.: A new learning algorithm for stochastic feedforward neural nets. In: ICML 2013 Workshop on Challenges in Representation Learning (2013)
22. Tieleman, T., Hinton, G.: Lecture 6.5-rmsprop: Divide the gradient by a running average of its recent magnitude. COURSERA: Neural Networks for Machine Learning **4** (2012)
23. Vincent, P., Larochelle, H., Lajoie, I., Bengio, Y., Manzagol, P.A.: Stacked denoising autoencoders: Learning useful representations in a deep network with a local denoising criterion. J. Machine Learning Res. **11** (2010)

Appendix

A   Proof of Theorem 1

Proof. Given a training example $(\mathbf{x}, \mathbf{y})$ the back-propagation update is given by

$$\delta W_i^{bp} = -\frac{\partial L(\mathbf{x}, \mathbf{y}; \theta_W^{0,M})}{\partial W_i} = -J_{f_{i+1}}^T \dots J_{f_M}^T \frac{\partial L}{\partial \mathbf{h}_M} (s_i(\mathbf{h}_{i-1}))^T \ ,$$

where $J_{f_k} = \frac{\partial \mathbf{h}_k}{\partial \mathbf{h}_{k-1}} = W_i \cdot S_i'(\mathbf{h}_{k-1}), k = i+1, \dots, M$. Here $S_i'(\mathbf{h}_{k-1})$ is a diagonal matrix with each diagonal element being element-wise derivatives and J_{f_k} is the Jacobian of $f_k(\mathbf{h}_{k-1})$. In target propagation the target for $\mathbf{h}_M$ is given by $\hat{\mathbf{h}}_M = \mathbf{h}_M - \hat{\eta}\frac{\partial L}{\partial \mathbf{h}_M}$. If all $\mathbf{h}_k$'s are allocated in smooth areas and $\hat{\eta}$ is sufficiently small, we can apply a Taylor expansion to get

$$\hat{\mathbf{h}}_i = g_{i+1}(\ldots g_M(\hat{\mathbf{h}}_M)\ldots) = g_{i+1}(\ldots g_M(\mathbf{h}_M)\ldots) - \hat{\eta}J_{g_{i+1}}\ldots J_{g_M}\frac{\partial L}{\partial \mathbf{h}_M} + \mathbf{o}(\hat{\eta}) \ ,$$

where $\mathbf{o}(\hat{\eta})$ is the remainder satisfying $\lim_{\hat{\eta}\to 0}\mathbf{o}(\hat{\eta})/\hat{\eta} = \mathbf{0}$. Now, for δW_i^{tp} we have

$$\delta W_i^{tp} = -\frac{\partial \|\mathbf{h}_i(\mathbf{h}_{i-1}; W_i) - \hat{\mathbf{h}}_i\|_2^2}{\partial W_i}$$

$$= -(\mathbf{h}_i - (\mathbf{h}_i - \hat{\eta}J_{f_{i+1}}^{-1}\ldots J_{f_M}^{-1}\frac{\partial L}{\partial \mathbf{h}_M} + \mathbf{o}(\hat{\eta})))(s_i(\mathbf{h}_{i-1}))^T$$

$$= -\hat{\eta}J_{f_{i+1}}^{-1}\ldots J_{f_M}^{-1}\frac{\partial L}{\partial \mathbf{h}_M}(s_i(\mathbf{h}_{i-1}))^T + \mathbf{o}(\hat{\eta})(s_i(\mathbf{h}_{i-1}))^T \ .$$

We write $\frac{\partial L}{\partial \mathbf{h}_M}$ as $\boldsymbol{l}$, $s_i(\mathbf{h}_{i-1})$ as $\boldsymbol{v}$ and $J_{f_M}\ldots J_{f_{i+1}}$ as J for short. Then the inner production of vector forms of δW_i^{bp} and δW_i^{tp} is

$$\langle vec(\delta W_i^{bp}), vec(\delta W_i^{tp})\rangle = tr((J^T \boldsymbol{l}\boldsymbol{v}^T)^T(\hat{\eta}J^{-1}\boldsymbol{l}\boldsymbol{v}^T + \mathbf{o}(\hat{\eta})\boldsymbol{v}^T))$$

$$= \hat{\eta}tr(\boldsymbol{v}\boldsymbol{l}^T JJ^{-1}\boldsymbol{l}\boldsymbol{v}^T) - tr(\boldsymbol{v}\boldsymbol{l}^T J\mathbf{o}(\hat{\eta})\boldsymbol{v}^T) = \hat{\eta}\|\boldsymbol{v}\|_2^2\|\boldsymbol{l}\|_2^2 - \langle J^T\boldsymbol{l}, \mathbf{o}(\hat{\eta})\rangle\|\boldsymbol{v}\|_2^2 \ .$$

For $\|vec(\delta W_i^{bp})\|_2$ and $\|vec(\delta W_i^{tp})\|_2$ we have

$$\|vec(\delta W_i^{bp})\|_2 = \sqrt{tr((-J^T \boldsymbol{l}\boldsymbol{v}^T)^T(-J^T \boldsymbol{l}\boldsymbol{v}^T))} = \|\boldsymbol{v}\|_2\|J^T\boldsymbol{l}\|_2 \leq \|\boldsymbol{v}\|_2\|J^T\|_2\|\boldsymbol{l}\|_2$$

and similarly

$$\|vec(\delta W_i^{tp})\|_2 \leq \hat{\eta}\|\boldsymbol{v}\|_2\|J^{-1}\|_2\|\boldsymbol{l}\|_2 + \|\mathbf{o}(\hat{\eta})\|_2\|\boldsymbol{v}\|_2 \ ,$$

where $\|J^T\|_2$ and $\|J^{-1}\|_2$ are matrix Euclidean norms, i.e. the largest singular value of $(J_{f_M}\ldots J_{f_{i+1}})^T$, λ_{max}, and the largest singular value of $(J_{f_M}\ldots J_{f_{i+1}})^{-1}$, $\frac{1}{\lambda_{min}}$ (λ_{min} is the smallest singular value of $(J_{f_M}\ldots J_{f_{i+1}})^T$, because f_k is invertable, so all the smallest singular values of Jacobians are larger than 0). Finally, if $\hat{\eta}$ is sufficiently small, the angle α between $vec(\delta W_i^{bp})$ and $vec(\delta W_i^{tp})$ satisfies:

$$cos(\alpha) = \frac{\langle vec(\delta W_i^{bp}), vec(\delta W_i^{tp})\rangle}{\|vec(\delta W_i^{bp})\|_2 \cdot \|vec(\delta W_i^{tp})\|_2}$$

$$\geq \frac{\hat{\eta}\|\boldsymbol{v}\|_2^2\|\boldsymbol{l}\|_2^2 - \langle J^T\boldsymbol{l}, \mathbf{o}(\hat{\eta})\rangle\|\boldsymbol{v}\|_2^2}{(\|\boldsymbol{v}\|_2\lambda_{max}\|\boldsymbol{l}\|_2)(\hat{\eta}\|\boldsymbol{v}\|_2(\frac{1}{\lambda_{min}})\|\boldsymbol{l}\|_2 + \|\mathbf{o}(\hat{\eta})\|_2\|\boldsymbol{v}\|_2)}$$

$$= \frac{1 + \frac{-\langle J^T\boldsymbol{l}, \mathbf{o}(\hat{\eta})\rangle}{\hat{\eta}\|\boldsymbol{l}\|_2^2}}{\frac{\lambda_{max}}{\lambda_{min}} + \frac{\lambda_{max}\|\mathbf{o}(\hat{\eta})\|_2}{\hat{\eta}\|\boldsymbol{l}\|_2}} = \frac{1 + \Delta_1(\hat{\eta})}{\frac{\lambda_{max}}{\lambda_{min}} + \Delta_2(\hat{\eta})}$$

where the last expression is positive if $\hat{\eta}$ is sufficiently small and $cos(\alpha) \leq 1$ is trivial.

B   Proof of Theorem 2

Proof. Let $\mathbf{e} = \hat{\mathbf{h}}_i - \mathbf{h}_i$. Applying Taylor's theorem twice, we get

$$\hat{\mathbf{h}}_i - f_i(\hat{\mathbf{h}}_{i-1}) = \hat{\mathbf{h}}_i - f_i(\mathbf{h}_{i-1} + g_i(\hat{\mathbf{h}}_i) - g_i(\mathbf{h}_i)) = \hat{\mathbf{h}}_i - f_i(\mathbf{h}_{i-1} + J_{g_i}\mathbf{e} + \mathbf{o}(||\mathbf{e}||_2))$$
$$= \hat{\mathbf{h}}_i - f_i(\mathbf{h}_{i-1}) - J_{f_i}(J_{g_i}\mathbf{e} + \mathbf{o}(||\mathbf{e}||_2)) - \mathbf{o}(||J_{g_i}\mathbf{e} + \mathbf{o}(||\mathbf{e}||_2)||_2)$$
$$= \hat{\mathbf{h}}_i - \mathbf{h}_i - J_{f_i}J_{g_i}\mathbf{e} - \mathbf{o}(||\mathbf{e}||_2) = (I - J_{f_i}J_{g_i})\mathbf{e} - \mathbf{o}(||\mathbf{e}||_2)$$

where the vector $\mathbf{o}(||\mathbf{e}||_2)$ represents the remainder satisfying $\lim_{\mathbf{e}\to 0}\mathbf{o}(||\mathbf{e}||_2)/||\mathbf{e}||_2 = \mathbf{0}$. Then for $||\hat{\mathbf{h}}_i - f_i(\hat{\mathbf{h}}_{i-1})||_2^2$ we have

$$||\hat{\mathbf{h}}_i - f_i(\hat{\mathbf{h}}_{i-1})||_2^2 = ((I - J_{f_i}J_{g_i})\mathbf{e} - \mathbf{o}(||\mathbf{e}||_2))^T((I - J_{f_i}J_{g_i})\mathbf{e} - \mathbf{o}(||\mathbf{e}||_2))$$
$$= \mathbf{e}^T(I - J_{f_i}J_{g_i})^T(I - J_{f_i}J_{g_i})\mathbf{e} - \mathbf{o}(||\mathbf{e}||_2)^T(I - J_{f_i}J_{g_i})\mathbf{e}$$
$$\quad - \mathbf{e}^T(I - J_{f_i}J_{g_i})^T\mathbf{o}(||\mathbf{e}||_2) + \mathbf{o}(||\mathbf{e}||_2)^T\mathbf{o}(||\mathbf{e}||_2))$$
$$= \mathbf{e}^T(I - J_{f_i}J_{g_i})^T(I - J_{f_i}J_{g_i})\mathbf{e} + o(||\mathbf{e}||_2^2)$$
$$\leq \lambda||\mathbf{e}||_2^2 + |o(||\mathbf{e}||_2^2)| \tag{A-1}$$

where $o(||\mathbf{e}||_2^2)$ is the scalar value resulting from all terms depending on $\mathbf{o}(||\mathbf{e}||_2)$ and λ is the largest eigenvalue of $(I - J_{f_i}J_{g_i})^T(I - J_{f_i}J_{g_i})$. If $\mathbf{e}$ is sufficiently small to guarantee $|o(||\mathbf{e}||_2^2)| < (1 - \lambda)||\mathbf{e}||_2^2$, then the left of Equation (A-1) is less than $||\mathbf{e}||_2^2$ which is just $||\hat{\mathbf{h}}_i - \mathbf{h}_i||_2^2$.

Online Learning of Deep Hybrid Architectures for Semi-supervised Categorization

Alexander G. Ororbia II [(⊠)], David Reitter, Jian Wu, and C. Lee Giles

College of Information Sciences and Technology, The Pennsylvania State University,
University Park, State College, PA 16802, USA
{ago109,reitter,jxw394,giles}@psu.edu

Abstract. A hybrid architecture is presented capable of online learning from both labeled and unlabeled samples. It combines both generative and discriminative objectives to derive a new variant of the Deep Belief Network, i.e., the Stacked Boltzmann Experts Network model. The model's training algorithm is built on principles developed from hybrid discriminative Boltzmann machines and composes deep architectures in a greedy fashion. It makes use of its inherent "layer-wise ensemble" nature to perform useful classification work. We (1) compare this architecture against a hybrid denoising autoencoder version of itself as well as several other models and (2) investigate training in the context of an incremental learning procedure. The best-performing hybrid model, the Stacked Boltzmann Experts Network, consistently outperforms all others.

Keywords: Restricted Boltzmann machines · Denoising autoencoders · Semi-supervised learning · Incremental learning · Hybrid architectures

1 Introduction

When it comes to collecting information from unstructured data sources, the challenge is clear for any information harvesting agent: to recognize what is relevant and to categorize what has been found. For applications such as web crawling, models such as the competitive Support Vector Machine are often trained on labeled datasets [6]. However, as the target distribution (such as that of information content from the web) evolves, these models quickly become outdated and require re-training on new datasets. Simply put, while unlabeled data is plentiful, labeled data is not [28]. While incremental approaches such as co-training [15] have been employed to face this challenge, they require careful, time-consuming feature-engineering (to construct multiple views of the data).

To minimize the human effort in gathering data and facilitate scalable learning, a model capable of generalization with only a few labeled examples and vast quantities of easily-acquired unlabeled data is highly desirable. Furthermore, to avoid feature engineering, this model should exploit the representational power afforded by deeper architectures, which have seen success in domains such as computer vision and speech recognition. Such a multi-level model could learn feature abstractions, arguably capturing higher-order feature relationships in an efficient

© Springer International Publishing Switzerland 2015
A. Appice et al. (Eds.): ECML PKDD 2015, Part I, LNAI 9284, pp. 516–532, 2015.
DOI: 10.1007/978-3-319-23528-8_32

manner. In pursuit of this goal, our contribution is the development of a novel hybrid Boltzmann-based architecture and its hybrid denoising autoencoder variant as well as their incremental, semi-supervised learning algorithms and prediction mechanisms. The learning process makes use of compound learning objectives, balancing, in a parametric fashion, the dual goals of generative and discriminative modeling of data. We further experiment with relaxing our approach's strict bottom-up scheme to better handle the online data-stream setting.

The rest of this paper is organized in the following manner. First, we review relevant previous work applying deep models to categorization problems in Section 2. Following this, in Section 3, we describe the algorithmic mechanics of our two incremental, semi-supervised deep architectures. Experimental results of using these deep architectures in a variety of data contexts are presented in Section 4. We sum up our work in Section 5 and consider model limitations and potential algorithmic improvements.

2   Related Work

Our algorithms fall in the realm of representation-learning, designed to learn, "...transformations of the data that make it easier to extract useful information when building classifiers or other predictors" [2]. Shallow learning methods, which require extensive prior human knowledge and large, labeled datasets, have been argued to be limited in terms of learning functions that violate restrictive assumptions such as smoothness and locality [4]. Moreover, architectures with a single unobserved layer require an exponentially increasing number of units to accurately learn complex distributions that deeper architectures, composed of multiple layers of non-linearity, potentially can. Deeper models "exploit the unknown structure" of the input data distribution to generate high-level features that are invariant to most variations in the training examples and yet preserve as much information regarding the input as possible [1].

In both large-scale, image-based [18,39] and language-based problems [14,24,26,31], deep architectures have outperformed popular shallow models. However, these models operate in a multi-stage learning process, where a generative architecture is greedily pre-trained and then used to initialize parameters of a second architecture that is discriminatively fine-tuned. To help deep models deal with potentially uncooperative input distributions or encourage learning of discriminative information earlier in the learning process, some approaches have leveraged auxiliary models in various ways [3,22,41]. A few methods have been proposed for adapting deep architecture construction to incremental learning settings [5,42]. Furthermore, an interesting approach combined the simple idea of *pseudo-labeling* with training deep neural architectures composed of rectified linear activation functions [13], a recent advancement [23].

While fundamentally different compared to purely generative or discriminative ones [21], we hypothesize that deep hybrid models that balance multiple objectives similar to the shallow one in [19] can make good, semi-supervised incremental models for classification. Motivated by this hypothesis, we design two model candidates, building on principles and successes of previous work: the *Stacked Boltzmann Expert Network* (SBEN) and the *Hybrid Stacked Denoising Autoencoders* model (HSDA). Furthermore, we introduce the idea of *layer-wise ensembling*, a simple prediction scheme we shall describe in Section 3.3 to utilize layer-wise information learnt by these models.

3   Deep Hybrid Architectures

In this section, we describe the implementations of our deep hybrid architectures.

3.1   The Stacked Boltzmann Experts Network (SBEN)

Our proposed variant of the Deep Belief Network (DBN), the Stacked Boltzmann Experts Network, follows an approach to construction and training similar to the DBN itself. The key is to, in an efficient, greedy manner, learn a stack of building-block models, and, as a layer is modified, freeze the parameters of lower layers. In practice, this is done by propagating data samples up to the layer targeted for layer-wise training and using the resultant latent representations as observations for constructing a higher level model. In contrast to the DBN, which stacks restricted Boltzmann machines (RBM's) and is often used to initialize a deep multi-layer perceptron (MLP), the SBEN model is constructed by composing hybrid restricted Boltzmann machines and is directly applied to the discriminative task and potentially fine-tuned directly[1].

The hybrid restricted Boltzmann machine (HRBM) [19,20,34], originally referred to as the *ClassRBM*, formalized the RBM extended to handle classification tasks directly. The model has been studied and used in a wide variety of applications [8,12,25,38] including the top of a DBN [36], most of which focus on the directly supervised facet of the model. With defined parameters[2] $\Theta = (\mathbf{W}, \mathbf{U}, \mathbf{b}, \mathbf{c}, \mathbf{d})$, the HRBM is designed to model the joint distribution of a binary pattern vector $\mathbf{x} = (x_1, \cdots, x_D)$ and its corresponding target variable $y \in \{1, \cdots, C\}$ utilizing a set of latent variables $\mathbf{h} = (h_1, \cdots, h_H)$. The HRBM assigns a probability to the triplet $(y, \mathbf{x}, \mathbf{h})$ using:

$$p(y, \mathbf{x}, \mathbf{h}) = \frac{e^{-E(y, \mathbf{x}, \mathbf{h})}}{Z}, \text{ with, } p(y, \mathbf{x}) = \frac{1}{Z} \sum_{\mathbf{h}} e^{-E(y, \mathbf{x}, \mathbf{h})} \tag{1}$$

where $Z = \sum_{(y, \mathbf{x}, \mathbf{h})} e^{-E(y, \mathbf{x}, \mathbf{h})}$ is the partition function meant to ensure that the value assignment is a valid probability distribution. Noting that the $\mathbf{e}_y = (\mathbf{1}_{i=y})_{i=1}^{C}$ is the one-hot vector encoding of y, the model's energy function may be defined as

$$E(y, \mathbf{x}, \mathbf{h}) = -\mathbf{h}^T \mathbf{W} \mathbf{x} - \mathbf{b}^T \mathbf{x} - \mathbf{c}^T \mathbf{h} - \mathbf{d}^T \mathbf{e}_y - \mathbf{h}^T \mathbf{U} \mathbf{e}_y. \tag{2}$$

---

[1] We have developed an algorithm to fine-tune the SBEN jointly, but leave usage and evaluation of this for future work.

[2] $\mathbf{W}$ is the input-to-hidden weight matrix, $\mathbf{U}$ the hidden-to-class weight matrix, $\mathbf{b}$ the visible bias vector, $\mathbf{c}$ the hidden unit bias vector, and $\mathbf{d}$ the class unit bias vector.

It is often not possible to compute $p(y, \mathbf{x}, \mathbf{h})$ or the marginal $p(y, \mathbf{x})$ due to the intractable partition function. However, we may leverage block Gibbs sampling to draw samples of the HRBM's latent variable layer given the current state of the visible layer (composed of $\mathbf{x}$ and $\mathbf{e}_y$) and vice versa, owing to the graphical model's bipartite structure (i.e., no intra-layer connections). This yields implementable equations for conditioning on various layers of the model as follows:

$$p(\mathbf{h}|y, \mathbf{x}) = \prod_j p(h_j|y, \mathbf{x}), \text{ with } p(h_j = 1|y, \mathbf{x}) = \sigma(c_j + U_{jy} + \sum_i W_{ji}x_i) \quad (3)$$

$$p(\mathbf{x}|\mathbf{h}) = \prod_i p(x_i|\mathbf{h}), \text{ with } p(x_i = 1|\mathbf{h}) = \sigma(b_i + \sum_j W_{ji}h_j) \quad (4)$$

$$p(y|\mathbf{h}) = \frac{e^{d_y + \sum_j U_{jy}h_j}}{\sum_{y^\star} e^{d_{y^\star} + \sum_j U_{jy^\star}h_j}} \quad (5)$$

$\sigma(v) = 1/(1 + e^{-v})$. Furthermore, to perform classification directly using the HRBM, one uses the model's free energy function $F(y, \mathbf{x})$ to compute the conditional

$$p(y|\mathbf{x}) = \frac{e^{-F(y,\mathbf{x})}}{\sum_{y^\star \in \{1, \cdots, C\}} e^{-F(y^\star, \mathbf{x})}} \quad (6)$$

where $-F(y, \mathbf{x}) = (d_y + \sum_j \log(1 + \exp(c_j + U_{jy} + \sum W_{ji}x_i)))$.

The hybrid model is trained leveraging a supervised, compound objective loss function that balances a discriminative objective $\mathcal{L}_{disc}$ and generative objective $\mathcal{L}_{gen}$, defined as follows:

$$\mathcal{L}_{disc}(\mathcal{D}_{train}) = -\sum_{t=1}^{|\mathcal{D}_{train}|} \log p(y|\mathbf{x}_t) \quad \mathcal{L}_{gen}(\mathcal{D}_{train}) = -\sum_{t=1}^{|\mathcal{D}_{train}|} \log p(y_t, \mathbf{x}_t)$$

$$(7) \qquad\qquad\qquad\qquad (8)$$

where $\mathcal{D}_{train} = \{(y_t, \mathbf{x}_t)\}$, the labeled training dataset. The gradient for $\mathcal{L}_{disc}$ may be computed directly, following the general form

$$\frac{\partial \log p(y_t|\mathbf{x})}{\partial \theta} = -\mathbb{E}_{\mathbf{h}|y_t, \mathbf{x}_t}\left[\frac{\partial}{\partial \theta}(\mathbb{E}(y_t, \mathbf{x}_t, \mathbf{h}))\right] + \mathbb{E}_{y, \mathbf{h}|\mathbf{x}}\left[\frac{\partial}{\partial \theta}(\mathbb{E}(y, \mathbf{x}, \mathbf{h}))\right] \quad (9)$$

implemented via direct formulation (see [20] for details) or a form of *Dropping*, such as *Drop-Out* or *Drop-Connect* [37]. The generative gradient follows the form

$$\frac{\partial \log p(y_t, \mathbf{x})}{\partial \theta} = -\mathbb{E}_{\mathbf{h}|y_t, \mathbf{x}_t}\left[\frac{\partial}{\partial \theta}(\mathbb{E}(y_t, \mathbf{x}_t, \mathbf{h}))\right] + \mathbb{E}_{y, \mathbf{x}, \mathbf{h}}\left[\frac{\partial}{\partial \theta}(\mathbb{E}(y, \mathbf{x}, \mathbf{h}))\right] \quad (10)$$

Algorithm 1. Contrastive Divergence: Single update for HRBM generative objective.

Input: training sample $(y_t, \mathbf{x}_t)$, HRBM current model parameters Θ
// Note that "$a \leftarrow b$" indicates assignment and "$a \sim b$" indicates a is sampled from b
function COMPUTEGENERATIVEGRADIENT($y_t, \mathbf{x}_t, \Theta$)
    if $y_t = \emptyset$ **then**
        $y_t \sim p(y|\mathbf{x})$ ▷ Obtain a pseudo-label for the unlabeled sample.
    // Conduct Positive Phase
    $y^0 \leftarrow y_t, \mathbf{x}^0 \leftarrow \mathbf{x}_t, \widehat{\mathbf{h}}^0 \leftarrow \sigma(\mathbf{c} + W\mathbf{x}^0 + U\mathbf{e}_{y^0})$
    // Conduct Negative Phase
    $\mathbf{h}^0 \sim p(\mathbf{h}|y^0, \mathbf{x}^0), y^1 \sim p(y|\mathbf{h}^0), x^1 \sim p(\mathbf{x}|\mathbf{h}^0)$
    $\widehat{\mathbf{h}}^1 \leftarrow \sigma(\mathbf{c} + W\mathbf{x}^1 + U\mathbf{e}_{y^1})$
    // Compute Gradient Update
    for $\theta \in \Theta$ **do**
        $\triangledown \leftarrow \frac{\partial}{\partial\theta}\mathbb{E}(y^0, \mathbf{x}^0, \widehat{\mathbf{h}}^0) - \frac{\partial}{\partial\theta}\mathbb{E}(y^1, \mathbf{x}^1, \widehat{\mathbf{h}}^1)$
    return $\triangledown$

---

and, although intractable for any $(y_t, \mathbf{x}_t)$, is approximated via contrastive divergence [17], where the intractable second expectation is replaced by a point estimate using one Gibbs sampling step (after initializing the Markov Chain at the training sample).

In the semi-supervised context, where $\mathcal{D}_{train}$ is small but a large, unlabeled dataset $\mathcal{D}_{unlab}$ is available, the HRBM can be further extended to train with an unsupervised objective $\mathcal{L}_{unsup}$, where negative log-likelihood is optimized according to

$$\mathcal{L}_{unsup}(\mathcal{D}_{unlab}) = -\sum_{t=1}^{|\mathcal{D}_{unlab}|} \log p(\mathbf{x}_t). \tag{11}$$

The gradient for $\mathcal{L}_{unsup}$ can be simply computed using the same contrastive divergence update for $\mathcal{L}_{gen}$ but incorporating an extra step at the beginning by sampling from the model's current estimate of $p(y|\mathbf{u})$ for an unlabeled sample $\mathbf{u}$. This form of the generative update could be viewed as a form of self-training or *Entropy Regularization* [23]. The pseudo-code for the online procedure for computing the generative gradient (either labeled or unlabeled example) for a single HRBM is shown in Algorithm 1.

To train a fully semi-supervised HRBM, one composes the appropriate multi-objective function using a simple weighted summation:

$$\mathcal{L}_{semi}(\mathcal{D}_{train}, \mathcal{D}_{unlab}) = \gamma\mathcal{L}_{disc}(\mathcal{D}_{train}) + \alpha\mathcal{L}_{gen}(\mathcal{D}_{train}) + \beta\mathcal{L}_{unsup}(\mathcal{D}_{unlab}) \tag{12}$$

where α and β are coefficient handles designed to explicitly control the effects that the generative gradients have on the HRBM's learning procedure. We introduced the additional coefficient γ as a means to also directly control the effect of

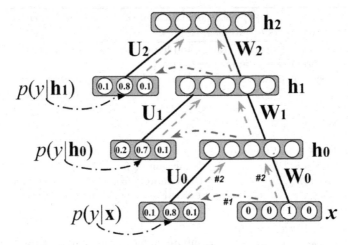

Fig. 1. Architecture of the SBEN model. The flow of data through the system is indicated by the numbered arrows. Given a sample $\mathbf{x}$, the dash-dotted arrow indicates obtaining an estimated label by using the current layer's conditional via Equation 6 (i.e., Step # 1 dash-dotted red arrow). The latent representation is computed using this proxy label and the data vector via Equation 3 (i.e.,Step # 2, dashed green arrow). This procedure is repeated recursively, replacing $\mathbf{x}$ with $\mathbf{h}_n$.

the discriminative gradient in model training. Setting $\gamma = 0$ leads to constructing a purely generative model of $\mathcal{D}_{train}$ and $\mathcal{D}_{unsup}$, and further setting $\beta = 0$ leads to purely supervised generative modeling of labeled dataset $\mathcal{D}_{train}$. If the target task is classification, then γ may be set to any value in $(0, 1]$ (for simplicity, we chose $\gamma = 1$, although future work shall investigate building models with values of this coefficient that shift the balance to models that favor generative features a bit more). These free parameters, though making model selection a bit more challenging, offer an explicit means of controlling the extent to which the final parameters discovered are influenced by generative learning [20], much in contrast to simple generative pre-training of neural architectures.

As mentioned before, to compose our N-layer SBEN (or N-SBEN), one follows the same greedy, layer-wise procedure of a DBN. However, unlike DBN's, where stacking RBM's warrants only a direct feedforward operation (since RBM's contain only a single set of inputs), modifications must be made to account for the architectural design of the HRBM graphical model. In order to unify the SBEN architecture while respecting HRBM building block design, one must combine Equations 3 and 6 to properly compute intermediate data representations during training and prediction. This gives rise to the architecture as depicted in Fig. 1 and its corresponding learning procedure in Algorithm 2, where the representation for the layer above cannot be computed without first obtaining an estimate of the current layer's $p(y|\mathbf{x})$. [3] Each HRBM layer of the SBEN is greedily trained using

---

[3] One may sample from this prediction vector as one would for hidden activations, however, we found that simply using this mean in forward propagation step yielded best results.

Algorithm 2. Greedy, layer-wise construction of an N-SBEN, where N is the desired number of layers of latent variables.

Input: $\mathcal{D}_{train}$, $\mathcal{D}_{unlab}$, learning rate λ and hyper-parameters γ, α, β, $numSteps$, and initial model parameters $\Theta = \{\Theta_1, \Theta_2, ..., \Theta_N\}$

function CONSTRUCTMODEL($\mathcal{D}_{train}$, $\mathcal{D}_{unlab}$, λ, γ, α, β, $numSteps$, Θ)

    $\mathcal{D}_{train}^n \leftarrow \mathcal{D}_{train}$, $\mathcal{D}_{unlab}^n \leftarrow \mathcal{D}_{unlab}$ ▷ Initialize subsets to low-level representations

    for $\Theta_n \in \Theta$ **do**

        $t \leftarrow 0$

        while $t \leq numSteps$ **do**

            $(y_t, \mathbf{x}_t) \sim \mathcal{D}_{train}^n$ ▷ Draw sample from $\mathcal{D}_{train}^n$ without replacement

            $(\mathbf{u}_t) \sim \mathcal{D}_{unlab}^n$ ▷ Draw sample from $\mathcal{D}_{unlab}^n$ without replacement

            $(\nabla_{disc}, \nabla_{gen}, \nabla_{unsup}) \leftarrow$ UPDATELAYER($y_t, \mathbf{x}_t, \mathbf{u}_t, \Theta_n$)

            $\Theta_n \leftarrow \Theta_n + \lambda(-\gamma \nabla_{disc} + \alpha \nabla_{gen} + \beta \nabla_{unsup})$, $t \leftarrow t + 1$

        $\mathcal{D}_{train}^h \leftarrow \emptyset$, $\mathcal{D}_{unlab}^h \leftarrow \emptyset$

        for $(y_t, \mathbf{x}_t) \in \mathcal{D}_{train}^n$ **do**   ▷ Compute latent representation dataset for $\mathcal{D}_{train}^n$

            $\mathcal{D}_{train}^h \leftarrow$ COMPUTELATENTREPRESENTATION($y_t, \mathbf{x}_t, \Theta_n$)

        for $(\emptyset, \mathbf{u}_t) \in \mathcal{D}_{unlab}^n$ **do**   ▷ Compute latent representation dataset for $\mathcal{D}_{unlab}^n$

            $\mathcal{D}_{unlab}^h \leftarrow$ COMPUTELATENTREPRESENTATION($\emptyset, \mathbf{u}_t, \Theta_n$)

        $\mathcal{D}_{train}^n \leftarrow \mathcal{D}_{train}^h$, $\mathcal{D}_{unlab}^n \leftarrow \mathcal{D}_{unlab}^h$

function UPDATELAYER($y_t, \mathbf{x}_t, \mathbf{u}_t, \Theta_n$)

    $\nabla_{disc} \leftarrow$ COMPUTEDISCRIMINATIVEGRADIENT($y_t, \mathbf{x}_t, \Theta_n$)   ▷ See [20] for details

    $\nabla_{gen} \leftarrow$ COMPUTEGENERATIVEGRADIENT($y_t, \mathbf{x}_t, \Theta_n$)   ▷ See Algorithm 1

    $\nabla_{unsup} \leftarrow$ COMPUTEGENERATIVEGRADIENT($\emptyset, \mathbf{u}_t, \Theta_n$)   ▷ See Algorithm 1

    return $(\nabla_{disc}, \nabla_{gen}, \nabla_{unsup})$

function COMPUTELATENTREPRESENTATION($y_t, \mathbf{x}_t, \Theta_n$)

    $y_t^h \leftarrow p(y_t | \mathbf{x}_t, \Theta_n)$ ▷ Equation 6 under the layerwise model

    $\mathbf{h}_t \sim p(\mathbf{h} | y_t^h, \mathbf{x}_t, \Theta_n)$ ▷ Equation 3 under the layerwise model

    return $(y_t, \mathbf{h}_t)$

---

the frozen latent representations of the one below, generated by using lower level expert's inputs and predictions.

The generative objectives (for both unlabeled and labeled samples) of our model can be viewed as a form of data-dependent regularization acting on the discriminative learning gradient of each layer. One key advantage of SBEN training is that each layer's discriminative progress may be tracked directly, since each layer-wise expert is capable of direct classification using Equation 6 to compute the conditional $p(y|\mathbf{h}_{below})$. Note that setting the number of hidden layers equal to 1 recovers the original HRBM architecture (a 1-SEBN). One may notice some similarity with the partially supervised, layer-wise procedure of [3] where a simple softmax classifier was loosely coupled with each RBM of a DBN. However, this only served as a temporary mechanism for pre-training whereas the SBEN leverages the more unified framework of the HRBM during and after training. Note that inputs to the SBEN, like the DBN, can be trivially extended [29,40].

A useful property of the SBEN model is that it also contains a generative model "facet" due to its hybrid nature. One could treat this facet as a directed, top-down generative network and generate fantasy samples for specific, clamped class units. In addition, one could generate class-specific probabilistic scores for input features by adapting the procedure in [37] and, since each layer contains class units, potentially uncover the SBEN hierarchy extracted from the data.

3.2   Hybrid Stacked Denoising Auto-Encoders (HSDA)

The auto-encoder variant of the SBEN is the Hybrid Stacked Denoising Autoencoders model. Instead of building a direct model of the joint distribution of $(y_t, \mathbf{x}_t)$ as in the HRBM, the hybrid denoising autoencoder (hDA) building block, with parameters $\Theta = (\mathbf{W}, \mathbf{W}', \mathbf{U}, \mathbf{b}, \mathbf{b}', \mathbf{d})$, may be viewed as a fusion of a generative model of $p(\mathbf{x})$, or an encoder-decoder reconstruction model, with a conditional model of $p(y|\mathbf{x})$, or a single-layer MLP. The reconstruction model learns a corrupted version of the feature vector $\mathbf{x}_t$, which is created via stochastic mapping $\widehat{\mathbf{x}}_t \sim q_{\mathcal{D}}(\widehat{\mathbf{x}}_t|\mathbf{x})$, where $q_{\mathcal{D}}$ is a function that stochastically corrupts an input vector (i.e., randomly masking entries by setting them to zero under a given probability). The reconstruction model is defined as

$$\mathbf{h} = f_\theta(\widehat{\mathbf{x}}) = \sigma(\mathbf{W}\widehat{\mathbf{x}} + \mathbf{b}) \quad (13) \qquad \mathbf{z} = g_\theta(\mathbf{h}) = \sigma(\mathbf{W}'\mathbf{h} + \mathbf{b}') \quad (14)$$

where parameters $\mathbf{W}$ and $\mathbf{W}'$ may be "tied" by setting $\mathbf{W}' = \mathbf{W}^T$. The encoder deterministically maps an input to a latent representation $\mathbf{h}$ (Equation 13) and the decoder maps this $\mathbf{h}$ back to a reconstruction $\mathbf{z}$ (Equation 14). The coupled neural network is tasked with mapping input $\mathbf{x}$ to y while sharing its latent layer $\mathbf{h}$ with the encoder-decoder pair. To compute the conditional $p(y|\mathbf{x})$, one uses Equation 13 followed by Equation 5 of the HRBM. While the discriminative objective of an hDA is defined similarly to the HRBM (Equation 7), where gradients of the log loss are computed directly (as in an MLP), the generative objective $\mathcal{L}_{gen}$ proceeds a bit differently:

$$\mathcal{L}_{gen}(\mathcal{D}_{train}) = -\sum_{t=1}^{|\mathcal{D}_{train}|} \mathbf{x}_t \log \mathbf{z}_t + (1 - \mathbf{x}_t) \log(1 - \mathbf{z}_t) \qquad (15)$$

This is the cross-entropy of two independent multivariate Bernoulli distributions, or *cross-entropy loss*. Unlike the HRBM, training an hDA under this generative objective is notably simpler since it uses back-propagation combined with the loss as defined in Equaton 15. A full hDA is trained using the weighted, tri-objective framework described in Section 3.1, $\mathcal{L}_{semi}$ (Equation 12), where its unlabeled objective $\mathcal{L}_{unsup}$ uses the same cross-entropy function as $\mathcal{L}_{gen}$ but operates on samples drawn from $\mathcal{D}_{unlab}$. The semi-supervised hDA differs from the HRBM complement in not only gradient calculation but also in that its unsupervised components do not require a corresponding sample of the model's estimate of $p(y_t|\mathbf{u}_t)$ for an unlabeled sample $\mathbf{u}_t$. This is advantageous since generative gradients are computed independently of the existence of a label, saving computational time and avoiding one drawback of self-training schemes: Reinforcement of incorrect predictions through model-generated pseudo-labels.

In the same greedy, layer-wise fashion as the SBEN, the N-layer HSDA (N-HSDA) may be composed by stacking hDA's. By replacing the procedure for generative gradients (Algorithm 1) and the discriminative gradient with the appropriate autoencoder cross-entropy back-propagation alternatives and substituting Equations 6 and 3 with Equation 13 (for computing hidden activities in $COMPUTELATENTREPRESENTATION$ of Algorithm 2), one may build an HSDA using Algorithm 2. The most useful property of the HSDA is that required computation for training and prediction may be reduced since dimensionality of each latent representation in auto-encoder architectures can be gradually decreased for upper levels of the network.

One may notice that the architectures of [42] and [30] may be recovered from our framework by manipulating the coefficients γ, α, and β in the $\mathcal{L}_{semi}$ objective function for the HSDA. Both these studies made use of dual-gradient models, which either focused on a hybrid objective that balanced a discriminative and weighted generative objective on a single sample (where the objective collapsed into a single generative objective when no label was available) [42] or where a generative objective was used as the primary objective and combined with a weighted discriminative objective [30]. Since our HSDA architecture can be viewed as a more general formulation of these original models, it is also amenable to their own particular extensions (such as feature growth/pruning, alternative input units for handling different types of data, etc.).

3.3   Ensembling of Layer-Wise Experts

Both the SBEN and the HSDA models, in addition to unique strengths, possess the interesting property where each layer, or *expert*, of the model is capable of classification given the appropriate latent representation of the data. This implies that the model is ensemble-like in its very nature but differs from standard ensemble methods where many smaller models are horizontally aggregated using well-established schemes such as boosting [33] or majority voting. Traditional feedforward models simply propagate data through the final network to obtain an output prediction from its penultimate layer for a given $\mathbf{x}_t$. In contrast, these hybrid models are capable of a producing a label y_t^n at each level n for $\mathbf{x}_t$, resulting from their layer-wise multi-objective training.

To vertically aggregate layer-wise expert outputs, we experimented with a variety of schemes in development, but found that computing a simple mean predictor, $p(y|\mathbf{x})_{ensemble}$ worked best, defined as:

$$p(y|\mathbf{x})_{ensemble} = \frac{1}{N} \sum_{n=1}^{N} p(y|\mathbf{x})_n \qquad (16)$$

This ensembling scheme allows for all components of the hybrid model to play a role in classification of unseen samples, perhaps leveraging acquired discriminative "knowledge" at their respective level of abstraction in the model hierarchy to ultimately improve final predictive performance. This scheme exploits our model's inherent layer-wise discriminative ability, which stands as an alternative to coupling helper classifiers as in [3] or the "companion objectives" of [22] to solve potential exploding gradients in deep convolution networks for object detection.

4   Experimental Results

We present experimental results on several classification problems in both optical character recognition and document categorization.

Character Recognition. Two experiments were conducted. The first experiment uses the Stanford OCR dataset, which contains 52,152 16×8 binary pixel images labeled as 1 of 26 letters of the English alphabet. Training ($\sim$ 2% of source), validation ($\sim$ 1.9%), unlabeled ($\sim$ 19.2%), and test sets ($\sim$ 77%) are created via a seeded random sampling without replacement procedure that ensured examples from each class appeared in roughly equal quantities in training and validation subsets. The second experiment makes use of a (seeded) stochastic process we implemented that generates 28×28 pixel CAPTCHA images of single characters based on the CAGE model[4], where one of 26 English characters may be generated (26 classes), of either lower or upper-case form in a variety of fonts as well as orientations and scales. We make use of this process in two ways: 1) create a finite dataset of 16,000 samples with ($\sim$ 3.125% in training, $\sim$ 3.125% in validation, $\sim$ 31.125% in unlabeled, and $\sim$ 62.25% in test) and perform an experiment similar to the OCR dataset, and 2) use the process as a controllable data-stream, which allows for compact storage of a complex distribution of image samples. The only pre-processing applied to the CAPTCHA samples was pixel binarization.

Text Categorization. We make use of a pre-processed WEBKB text collection (i.e., font formatting, stop words removed, terms stemmed, and words with length less than 3 removed) [7], which contains pages from a variety of universities (Cornell, Texas, Washington, and Wisconsin and miscellaneous pages from others). The 4-class classification problem as defined by this dataset will be to determine if a webpage can be identified as one belonging to a Student, Faculty, Course, or a Project. The dataset was already pre-partitioned into a training set (2,803 samples) and a test set (1,396 web pages), so using the same sampling scheme as the OCR data, we built from the training split a smaller training ($\sim$ 20.2%) and validation ($\sim$ 14.2%) subset, and put the rest into the unlabeled set ($\sim$ 62.5%), discarding 87 document vectors that contained less than 2 active features. The test set contained 1,344 samples, after discarding 52 samples with less than 2 active features. We simplified the low-level feature representation by using the top 2000 words in the corpus and binarizing the document term vectors.

Models. We compare the HSDA and SBEN models to the non-linear, shallow HRBM (which, as described in Section 3.1, is a *1*-SBEN). For a simpler classifier, we implemented an incremental version of Maximum Entropy (which, as

---

[4] https://akiraly.github.io/cage/index.html

explained in [32], is equivalent to a softmax classifier), or *MaxEnt*. Furthermore, we implemented the Pegasos SVM algorithm (*SVM*) [35] and extended it to follow a proper multi-class scheme [9]. This is the online formulation of the Support Vector Machine, trained via sub-gradient descent on the primal objective followed by a projection step (note that for simplicity we built the linear-kernel version of the model, which is quite fast). Evaluating the Pegasos SVM algorithm in the following experiments allows us to compare our deep semi-supervised models against the incremental version of a strong linear-kernel classifier. To provide some context with previously established deep architectures also learnable in a 1-phase fashion like our own, we present results for a simple sparse Rectifier Network, or *Rect*. [13]. [5] Note that we extended all shallow classifiers and the Rectifier Network to leverage self-training so that they may also learn from unlabeled examples. To do so, we implemented a scheme similar to that of [23] and used a classifier's estimate of $p(y|\mathbf{u})$ for an unlabeled sample. However, a 1-hot proxy encoding using the *argmax* of model's predictor was only created for such a sample if $max[p(y|\mathbf{u})] > \bar{p}$. We found that by explicitly controlling pseudo-labeling through $\bar{p}$ we could more directly improve model performance.

Model Selection. Model selection was conducted using a parallelized multi-setting scheme, where a configuration file for each model was specified, describing a set of hyper-parameter combinations to explore (this is akin to a course-grained grid search, where points of model evaluation are set manually a priori). For the *HSDA, SBEN, HRBM*, and *Rect* we varied model architectures, exploring undercomplete, complete, and over-complete versions, as well as the learning rate, α, and β coefficients (holding γ fixed at 1.0). If a model was trained using its stochastic form (i.e., HRBM, SBEN, or HSDA), to ensure reproducible model behavior, we ran it in feedforward mean-field, where no sampling steps were taken when data vectors were propagated through a network model when collecting layer-wise predictions (we also found that this yielded lowest generalization error). For the SVM, we tuned its slack variable λ. The rectifier network's training also involved using a L2 regularization penalty (0.002), initialization of hidden biases to small positive values ($|N(0, 0.25)|$) [13], and the use of the improved leaky rectifier unit [27].

For all finite dataset experiments, model performance is reported on the test set using the model with lowest validation-set error found during the training step.[6] Generalization performance was evaluated by calculating classification error,

---

[5] Model implementations were computationally verified for correctness when applicable. Since discriminative objectives entailed using an automatic differentiation framework, we checked gradient validity via finite difference approximation.

[6] For the SVM, λ was varied in the interval $[0.0001, 0.5]$ while the learning rate for all other models was varied in $[0.0001, 0.1]$. For HRBM, SBEN, & HSDA, β was explored in the interval $[0.05, 0.1]$, and for HRBM, SBEN, & HSDA, α was explored in $[0.075, 1.025]$. The threshold $\bar{p}$ was varied in $[0.0, 1.0]$ and the number of latent layers N for deeper architectures was explored in $[2, 5]$ where we delineate the optimal number with the prefix "N-".

Table 1. Character identification results on the CAPTCHA simulated dataset. Classification results are reported as 10-trial averages with single standard deviation from the mean.

	Error	Precision	Recall	F1-Score
MaxEnt	0.475 ± 0.010	0.535 ± 0.011	0.524 ± 0.010	0.522 ± 0.010
SVM	0.461 ± 0.011	0.564 ± 0.010	0.537 ± 0.011	0.526 ± 0.011
2-Rect [13, 23]	0.365 ± 0.011	0.651 ± 0.011	0.634 ± 0.011	0.627 ± 0.013
HRBM [20]	0.368 ± 0.009	0.643 ± 0.010	0.631 ± 0.009	0.629 ± 0.009
5-SBEN	$\mathbf{0.324 \pm 0.008}$	$\mathbf{0.681 \pm 0.009}$	$\mathbf{0.675 \pm 0.008}$	$\mathbf{0.671 \pm 0.009}$
5-HSDA	0.359 ± 0.011	0.650 ± 0.011	0.640 ± 0.011	0.633 ± 0.011

Table 2. Character identification results on the Stanford OCR dataset. Classification results are reported as 10-trial averages with single standard deviation from the mean.

	Error	Precision	Recall	F1-Score
MaxEnt	0.425 ± 0.009	0.508 ± 0.006	0.563 ± 0.005	0.512 ± 0.006
SVM	0.428 ± 0.008	0.504 ± 0.004	0.582 ± 0.011	0.510 ± 0.007
3-Rect [13, 23]	0.387 ± 0.009	0.549 ± 0.009	0.592 ± 0.014	0.548 ± 0.011
HRBM [20]	0.399 ± 0.019	0.565 ± 0.009	0.606 ± 0.016	0.552 ± 0.014
3-SBEN	$\mathbf{0.333 \pm 0.009}$	$\mathbf{0.602 \pm 0.009}$	$\mathbf{0.668 \pm 0.009}$	$\mathbf{0.610 \pm 0.012}$
3-HSDA	0.399 ± 0.012	0.546 ± 0.007	0.601 ± 0.012	0.537 ± 0.009

Table 3. Text categorization results on the WEBKB dataset.

	Error	Precision	Recall	F1-Score		Error	Precision	Recall	F1-Score
MaxEnt	0.510	0.386	0.387	0.384	*3-SBEN*	**0.210**	**0.788**	0.770	**0.769**
SVM	0.524	0.404	0.378	0.387	*3-HSDA*	0.219	0.757	**0.780**	0.765

precision, recall, and F-Measure, where F-Measure was chosen to be the harmonic mean of precision and recall, $F1 = 2(\text{precision} \cdot \text{recall})/(\text{precision} + \text{recall})$.

Since the creation of training, validation, and unlabeled subsets was controlled through a seeded random sampling without replacement process, the procedure described above composes a single trial. For the Standford OCR and CAPTCHA datasets, the results we report are 10-trial averages with a single standard deviation from the mean, where each trial used a unique seed value.

4.1 Finite Dataset Learning Performance

On all of the datasets we experimented with, in the case when all samples are available a priori, ranging from vision-based tasks to text classification, we observe that hybrid incremental architectures have, in general, lower error as compared to non-hybrid ones. In the CAPTCHA experiment (Table 1), we observed that both the SBEN and HSDA models reduced prediction error over the SVM by nearly 30% and 22% respectively. Furthermore, both models consistently improved over the error the HRBM, with the SBEN model reducing error by $\sim 12\%$. In the OCR dataset (Table 2), we see the SBEN improving over

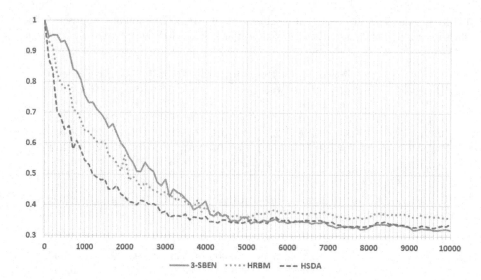

Fig. 2. Online error (y-axis) of 3-SBEN, 3-HSDA, & HRBM (or *1-* SBEN) evaluated every 100 time steps (x-axis). Each curve reported is a 4-trial mean of the lowest validation error model.

the HRBM by more than 16% and the SVM by more than 22%. In this case, the HSDA only marginally improves over the SVM model ($\sim 6\%$) and equal to that of an HRBM, the poor performance we attribute to a coarse search through a meta-parameter space window as opposed to an exhaustive grid search. For WEBKB problem, over the *MaxEnt* model (which slightly outperformed the SVM itself), we see a $\sim 57\%$ improvement in error for the HSDA and $\sim 58\%$ for the SBEN (Table 3). Note that the rectifier network is competitive, however, in both image-based experiments, the SBEN model outperforms it by more than 11% on CAPTCHA and nearly 14% on OCR.

4.2   Incremental Learning Performance

In the online learning setting, samples from $\mathcal{D}_{unlab}$ may not be available at once and instead available at a given rate in a stream for a single time instant (we chose to experiment with one example presented at a given iteration and only constant access to a $|\mathcal{D}_{train}| = 500$). In order to train a deep architecture in this setting, while still exploiting the efficiency of a greedy, layer-wise approach, one may remove the "freezing" step of Algorithm 2 and train all layers disjointly in an incremental fashion as opposed to a purely bottom-up approach. Using the same sub-routines as depicted in Algorithm 2, this procedure may be implemented as shown in Algorithm 3, effectively using a single bottom-up pass to modify model parameters. This approach adapts the training of hybrid architectures, such as the SBEN and HSDA, to the online learning setting.

As evidenced by Fig. 2, it is possible to train the layer-wise experts of a multi-level hybrid architecture simultaneously and still obtain a gain in generalization

Algorithm 3. Online variant of layer-wise construction of a deep hybrid architecture.

Input: $(y_t, \mathbf{x}_t) \in \mathcal{D}_{train}$, $(\mathbf{u}_t) \in \mathcal{D}_{unlab}$, learning rate λ, hyper-parameters γ, α, β, & model parameters $\Theta = \{\Theta_1, \Theta_2, ..., \Theta_N\}$

function CONSTRUCTMODEL($y_t, \mathbf{x}_t, \mathbf{u}_t, \lambda, \gamma, \alpha, \beta, \Theta$)

   $\mathbf{x}_t^h \leftarrow \mathbf{x}_t$, $\mathbf{u}_t^h \leftarrow \mathbf{u}_t$ ▷ Initialize samples to low-level representations

   for $\Theta_n \in \Theta$ **do**

      $(\bigtriangledown_{disc}, \bigtriangledown_{gen}, \bigtriangledown_{unsup}) \leftarrow$ UPDATELAYER($y_t, \mathbf{x}_t^h, \mathbf{u}_t^h, \Theta_n$)

      $\Theta_n \leftarrow \Theta_n + \lambda(-\gamma \bigtriangledown_{disc} + \alpha \bigtriangledown_{gen} + \beta \bigtriangledown_{unsup})$, $t \leftarrow t + 1$

      // Compute latent representation of data samples

      $(y_t, \mathbf{x}_t^h) \leftarrow$ COMPUTELATENTREPRESENTATION($y_t, \mathbf{x}_t^h, \Theta_n$)

      $(\emptyset, \mathbf{u}_t^h) \leftarrow$ COMPUTELATENTREPRESENTATION($\emptyset, \mathbf{u}_t^h, \Theta_n$)

---

performance over a non-linear, shallow model such as the HRBM. The HRBM settles at an online error of 0.356 whereas the 5-HSDA reaches an error of 0.327 and the 5-SBEN an error of 0.319 in a 10,000 iteration sweep. Online error was evaluated by computing classification error on the next 1,000 unseen samples generated by the CAPTCHA process.

While the simultaneous greedy training used in this experiment allows for construction of a deep hybrid model in unity when faced with a data stream, we note that instability may occur in the form of "shifting representations". This is where an upper level model is dynamically trained on a latent representation of a lower-level model that has not yet settled since it has not yet seen enough samples from the data distribution.

5 Conclusions

We developed two hybrid models, the SBEN and the HSDA, and their training algorithms in the context of incremental, semi-supervised learning. They combine efficient greedy, layer-wise construction of deeper architectures with a multi-objective learning approach. We balance the goal of learning a generative model of the data with extracting discriminative regularity to perform useful classification. More importantly, the framework we describe facilitates more explicit control over the multiple objectives involved. Additionally, we presented a vertical aggregation scheme, *layer-wise ensembling*, for generating predictions that exploit discriminative knowledge acquired at all levels of abstraction defined by the architecture's hierarchical form. Our framework allows for explicit control over generative and discriminative objectives as well as a natural scheme for tracking layer-wise learning.

Models were evaluated in two problem settings: optical character recognition and text categorization. We compared results against shallow models and found that our hybrid architectures outperform the others in all datasets investigated. We found that the SBEN performed the best, improving classification error by as much 58% (compared to Maximum Entropy on WEBKB). Furthermore, we found that improvement in performance holds when hybrid learning is adapted to an online setting (relaxing the purely bottom-up framework in Section 3.1). We observe that we are able to improve error while significantly minimizing the number of required labeled samples (as low as 2% of total available data in some cases).

The hybrid deep architectures presented in this paper are not without potential limitations. First, there is the danger of "shifting representations" if using Algorithm 3 for online learning. To combat this, samples could be pooled into mini-batch matrices before computing gradients and minimize some of the noise of online error-surface descent. Alternatively, all layer-wise experts could be extended temporally to Conditional RBM-like structures, potentially improving performance as in [43]. Second, additional free parameters were introduced that require tuning, creating a more challenging model selection process for the human user. This may be alleviated with a parallelized, automated approach, however, a model that adapts its objective weights during the learning process would be better, altering its hyperparameters in response to error progress on data subsets. Our frameworks may be augmented with automatic latent unit growth for both auto-encoder [42] and Boltzmann-like variants [10] or perhaps improved by "tying" all layer-wise expert outputs together in a scheme like that in [11].

The models presented in this paper offer promise in the goal of incrementally building powerful models that reduce expensive labeling and feature engineering effort. They represent a step towards ever-improving models that adapt to "in-the-wild" samples, capable of more fully embracing the "...unreasonable effectiveness of data" [16].

Acknowledgments. The first author acknowledges support from Pennsylvania State University & the National Science Foundation under IGERT award # DGE-1144860, Big Data Social Science. We thank Hugo Larochelle and Roberto Calandra for their correspondence and advice, although all shortcomings of the presented work are ours.

References

1. Bengio, Y.: Deep learning of representations for unsupervised and transfer learning. Journal of Machine Learning Research-Proceedings Track (2012)
2. Bengio, Y., Courville, A.C., Vincent, P.: Unsupervised feature learning and deep learning: Review and new perspectives (2012). CoRR abs/1206.5538
3. Bengio, Y., Lamblin, P., Popovici, D., Larochelle, H.: Greedy layer-wise training of deep networks. Advances in Neural Information Processing Systems (2007)
4. Bengio, Y., LeCun, Y.: Scaling learning algorithms towards AI. Large-Scale Kernel Machines **34**, 1–41 (2007)
5. Calandra, R., Raiko, T., Deisenroth, M.P., Pouzols, F.M.: Learning deep belief networks from non-stationary streams. In: Villa, A.E.P., Duch, W., Érdi, P., Masulli, F., Palm, G. (eds.) ICANN 2012, Part II. LNCS, vol. 7553, pp. 379–386. Springer, Heidelberg (2012)
6. Caragea, C., Wu, J., Williams, K., Das, S., Khabsa, M., Teregowda, P., Giles, C.L.: Automatic identification of research articles from crawled documents. In: Web-Scale Classification: Classifying Big Data from the Web, co-located with WSDM (2014)

7. Cardoso-Cachopo, A.: Improving methods for single-label text categorization. PdD Thesis, Instituto Superior Tecnico, Universidade Tecnica de Lisboa (2007)
8. Chen, G., Srihari, S.H.: Restricted Boltzmann machine for classification with hierarchical correlated prior (2014). arXiv preprint arXiv:1406.3407
9. Crammer, K., Singer, Y.: On the algorithmic implementation of multiclass kernel-based vector machines. Journal of Machine Learning Research **2**, 265–292 (2002)
10. Côté, M.A., Larochelle, H.: An infinite Restricted Boltzmann Machine (2015). arXiv preprint arXiv:1502.02476
11. Elfwing, S., Uchibe, E., Doya, K.: Expected energy-based restricted Boltzmann machine for classification. Neural Networks **64**, 29–38
12. Fiore, U., Palmieri, F., Castiglione, A., De Santis, A.: Network anomaly detection with the Restricted Boltzmann Machine. Neurocomputing **122** (2013)
13. Glorot, X., Bordes, A., Bengio, Y.: Deep sparse rectifier networks. In: Proc. 14th International Conference on Artificial Intelligence and Statistics, vol. 15, pp. 315–323 (2011)
14. Glorot, X., Bordes, A., Bengio, Y.: Domain adaptation for large-scale sentiment classification: a deep learning approach. In: Proc. 28th International Conference on Machine Learning (ICML 2011), pp. 513–520 (2011)
15. Gollapalli, S.D., Caragea, C., Mitra, P., Giles, C.L.: Researcher homepage classification using unlabeled data. In: Proc. 22nd International Conference on World Wide Web, Geneva, Switzerland, pp. 471–482 (2013)
16. Halevy, A., Norvig, P., Pereira, F.: The unreasonable effectiveness of data. IEEE Intelligent Systems **24**(2), 8–12 (2009)
17. Hinton, G.E.: Training products of experts by minimizing contrastive divergence. Neural Computation **14**(8), 1771–1800 (2002)
18. Hinton, G.E.: What kind of graphical model is the brain? In: Proc. 19th International Joint Conference on Artificial Intelligence, Edinburgh, Scotland, vol. 5, pp. 1765–1775 (2005)
19. Larochelle, H., Bengio, Y.: Classification using discriminative Restricted Boltzmann Machines. In: Proc. 25th International Conference on Machine learning, pp. 536–543 (2008)
20. Larochelle, H., Mandel, M., Pascanu, R., Bengio, Y.: Learning algorithms for the classification Restricted Boltzmann Machine. Journal of Machine Learning Research **13**, 643–669 (2012)
21. Lasserre, J.A., Bishop, C.M., Minka, T.P.: Principled hybrids of generative and discriminative models. In: Proc. 2006 IEEE Computer Society Conference on Computer Vision and Pattern Recognition, vol. 1, pp. 87–94. IEEE Computer Society, Washington (2006)
22. Lee, C.Y., Xie, S., Gallagher, P., Zhang, Z., Tu, Z.: Deeply-supervised nets (2014). arXiv:1409.5185 [cs, stat]
23. Lee, D.H.: Pseudo-label: the simple & efficient semi-supervised learning method for deep neural networks. In: Workshop on Challenges in Representation Learning, ICML (2013)
24. Liu, T.: A novel text classification approach based on deep belief network. In: Wong, K.W., Mendis, B.S.U., Bouzerdoum, A. (eds.) ICONIP 2010, Part I. LNCS, vol. 6443, pp. 314–321. Springer, Heidelberg (2010)
25. Louradour, J., Larochelle, H.: Classification of sets using Restricted Boltzmann Machines (2011). arXiv preprint arXiv:1103.4896
26. Lu, Z., Li, H.: A deep architecture for matching short texts. In: Burges, C.J.C., Bottou, L., Welling, M., Ghahramani, Z., Weinberger, K.Q. (eds.) Advances in Neural Information Processing Systems, vol. 26, pp. 1367–1375. Curran Associates, Inc. (2013)

27. Maas, A.L., Hannun, A.Y., Ng, A.Y.: Rectifier nonlinearities improve neural network acoustic models. In: Proc. ICML, vol. 30 (2013)
28. Masud, M.M., Woolam, C., Gao, J., Khan, L., Han, J., Hamlen, K.W., Oza, N.C.: Facing the reality of data stream classification: Coping with scarcity of labeled data. Knowledge and Information Systems **33**(1), 213–244 (2012)
29. Nair, V., Hinton, G.E.: Rectified linear units improve Restricted Boltzmann Machines. In: Proc. 27th International Conference on Machine Learning (ICML 2010), pp. 807–814 (2010)
30. Ranzato, M.A., Szummer, M.: Semi-supervised learning of compact document representations with deep networks. In: Proc. 25th International Conference on Machine Learning, pp. 792–799. ACM (2008)
31. Salakhutdinov, R., Hinton, G.: Semantic Hashing. International Journal of Approximate Reasoning **50**(7), 969–978 (2009)
32. Sarikaya, R., Hinton, G., Deoras, A.: Application of Deep Belief Networks for natural language understanding. IEEE/ACM Transactions on Audio, Speech, and Language Processing **22**(4), 778–784 (2014)
33. Schapire, R.E.: The strength of weak learnability. Machine Learning **5**(2), 197–227 (1990)
34. Schmah, T., Hinton, G.E., Small, S.L., Strother, S., Zemel, R.S.: Generative versus discriminative training of RBMs for classification of fMRI images. In: Advances in Neural Information Processing Systems, pp. 1409–1416 (2008)
35. Shalev-Shwartz, S., Singer, Y., Srebro, N., Cotter, A.: Pegasos: Primal estimated subgradient solver for SVM. Mathematical Programming **127**(1), 3–30 (2011)
36. Sun, X., Li, C., Xu, W., Ren, F.: Chinese microblog sentiment classification based on deep belief nets with extended multi-modality features. In: 2014 IEEE International Conference on Data Mining Workshop (ICDMW), pp. 928–935 (2014)
37. Tomczak, J.M.: Prediction of breast cancer recurrence using classification Restricted Boltzmann Machine with dropping (2013). arXiv preprint arXiv:1308.6324
38. Tomczak, J.M., Ziba, M.: Classification restricted Boltzmann machine for comprehensible credit scoring model. Expert Systems with Applications **42**(4), 1789–1796 (2015)
39. Vincent, P., Larochelle, H., Lajoie, I., Bengio, Y., Manzagol, P.A.: Stacked denoising autoencoders: Learning useful representations in a deep network with a local denoising criterion. Journal of Machine Learning Research **11**, 3371–3408 (2010)
40. Welling, M., Rosen-zvi, M., Hinton, G.E.: Exponential family harmoniums with an application to information retrieval. In: Saul, L.K., Weiss, Y., Bottou, L. (eds.) Advances in Neural Information Processing Systems, vol. 17, pp. 1481–1488. MIT Press (2005)
41. Zhang, J., Tian, G., Mu, Y., Fan, W.: Supervised deep learning with auxiliary networks. In: Proc. 20th ACM SIGKDD International Conference on Knowledge Discovery and Data Mining pp. 353–361. ACM (2014)
42. Zhou, G., Sohn, K., Lee, H.: Online incremental feature learning with denoising autoencoders. In: Proc. 15th International Conference on Artificial Intelligence and Statistics, pp. 1453–1461 (2012)
43. Zhou, J., Luo, H., Luo, Q., Shen, L.: Attentiveness detection using continuous restricted Boltzmann machine in e-learning environment. In: Wang, F.L., Fong, J., Zhang, L., Lee, V.S.K. (eds.) ICHL 2009. LNCS, vol. 5685, pp. 24–34. Springer, Heidelberg (2009)

Scoring and Classifying with Gated Auto-Encoders

Daniel Jiwoong Im$^{(\boxtimes)}$ and Graham W. Taylor

School of Engineering, University of Guelph, Guelph, ON, Canada
{gwtaylor,imj}@uoguelph.ca

Abstract. Auto-encoders are perhaps the best-known non-probabilistic methods for representation learning. They are conceptually simple and easy to train. Recent theoretical work has shed light on their ability to capture manifold structure, and drawn connections to density modeling. This has motivated researchers to seek ways of auto-encoder scoring, which has furthered their use in classification. Gated auto-encoders (GAEs) are an interesting and flexible extension of auto-encoders which can learn transformations among different images or pixel covariances within images. However, they have been much less studied, theoretically or empirically. In this work, we apply a dynamical systems view to GAEs, deriving a scoring function, and drawing connections to Restricted Boltzmann Machines. On a set of deep learning benchmarks, we also demonstrate their effectiveness for single and multi-label classification.

1 Introduction

Representation learning algorithms are machine learning algorithms which involve the learning of features or explanatory factors. Deep learning techniques, which employ several layers of representation learning, have achieved much recent success in machine learning benchmarks and competitions, however, most of these successes have been achieved with purely supervised learning methods and have relied on large amounts of labeled data [10,22]. Though progress has been slower, it is likely that unsupervised learning will be important to future advances in deep learning [1].

The most successful and well-known example of non-probabilistic unsupervised learning is the auto-encoder. Conceptually simple and easy to train via backpropagation, various regularized variants of the model have recently been proposed [20,21,25] as well as theoretical insights into their operation [6,24].

In practice, the latent representation learned by auto-encoders has typically been used to solve a secondary problem, often classification. The most common setup is to train a single auto-encoder on data from all classes and then a classifier is tasked to discriminate among classes. However, this contrasts with the way probabilistic models have typically been used in the past: in that literature, it is more common to train one model per class and use Bayes' rule for classification. There are two challenges to classifying using per-class auto-encoders. First, up until very recently, it was not known how to obtain the score of data under

© Springer International Publishing Switzerland 2015
A. Appice et al. (Eds.): ECML PKDD 2015, Part I, LNAI 9284, pp. 533–545, 2015.
DOI: 10.1007/978-3-319-23528-8_33

an auto-encoder, meaning how much the model "likes" an input. Second, auto-encoders are non-probabilistic, so even if they can be scored, the scores do not integrate to 1 and therefore the per-class models need to be calibrated.

Kamyshanska and Memisevic have recently shown how scores can be computed from an auto-encoder by interpreting it as a dynamical system [7]. Although the scores do not integrate to 1, they show how one can combine the unnormalized scores into a generative classifier by learning class-specific normalizing constants from labeled data.

In this paper we turn our interest towards a variant of auto-encoders which are capable of learning higher-order features from data [15]. The main idea is to learn relations between pixel intensities rather than the pixel intensities themselves by structuring the model as a tri-partite graph which connects hidden units to pairs of images. If the images are different, the hidden units learn how the images transform. If the images are the same, the hidden units encode within-image pixel covariances. Learning such higher-order features can yield improved results on recognition and generative tasks.

We adopt a dynamical systems view of gated auto-encoders, demonstrating that they can be scored similarly to the classical auto-encoder. We adopt the framework of [7] both conceptually and formally in developing a theory which yields insights into the operation of gated auto-encoders. In addition to the theory, we show in our experiments that a classification model based on gated auto-encoder scoring can outperform a number of other representation learning architectures, including classical auto-encoder scoring. We also demonstrate that scoring can be useful for the structured output task of multi-label classification.

2 Gated Auto-Encoders

In this section, we review the gated auto-encoder (GAE). Due to space constraints, we will not review the classical auto-encoder. Instead, we direct the reader to the reviews in [8,15] with which we share notation. Similar to the classical auto-encoder, the GAE consists of an encoder $h(\cdot)$ and decoder $r(\cdot)$. While the standard auto-encoder processes a datapoint $\mathbf{x}$, the GAE processes input-output pairs $(\mathbf{x}, \mathbf{y})$. The GAE is usually trained to reconstruct $\mathbf{y}$ given $\mathbf{x}$, though it can also be trained symmetrically, that is, to reconstruct both $\mathbf{y}$ from $\mathbf{x}$ and $\mathbf{x}$ from $\mathbf{y}$. Intuitively, the GAE learns *relations* between the inputs, rather than representations of the inputs themselves[1]. If $\mathbf{x} \neq \mathbf{y}$, for example, they represent sequential frames of a video, intuitively, the mapping units $\mathbf{h}$ learn *transformations*. In the case that $\mathbf{x} = \mathbf{y}$ (i.e. the input is copied), the mapping units learn pixel covariances.

In the simplest form of the GAE, the M hidden (mapping) units are given by a basis expansion of $\mathbf{x}$ and $\mathbf{y}$. However, this leads to a parameterization that it is at least quadratic in the number of inputs and thus, prohibitively large. Therefore, in practice, $\mathbf{x}$, $\mathbf{y}$, and $\mathbf{h}$ are projected onto matrices or ("latent

---

[1] Relational features can be mixed with standard features by simply adding connections that are not gated.

factors"), W^X, W^Y, and W^H, respectively. The number of factors, F, must be the same for X, Y, and H. Thus, the model is completely parameterized by $\theta = \{W^X, W^Y, W^H\}$ such that W^X and W^Y are $F \times D$ matrices (assuming both $\mathbf{x}$ and $\mathbf{y}$ are D-dimensional) and W^H is an $M \times F$ matrix. The encoder function is defined by

$$h(\mathbf{x}, \mathbf{y}) = \sigma(W^H((W^X \mathbf{x}) \odot (W^Y \mathbf{y}))) \tag{1}$$

where $\odot$ is element-wise multiplication and $\sigma(\cdot)$ is an activation function. The decoder function is defined by

$$r(\mathbf{y}|\mathbf{x}, h) = (W^Y)^T((W^X \mathbf{x}) \odot (W^H)^T h(\mathbf{x}, \mathbf{y})). \tag{2}$$

$$r(\mathbf{x}|\mathbf{y}, h) = (W^X)^T((W^Y \mathbf{y}) \odot (W^H)^T h(\mathbf{x}, \mathbf{y})), \tag{3}$$

Note that the parameters are usually shared between the encoder and decoder. The choice of whether to apply a nonlinearity to the output, and the specific form of objective function will depend on the nature of the inputs, for example, binary, categorical, or real-valued. Here, we have assumed real-valued inputs for simplicity of presentation, therefore, Eqs. 2 and 3 are bi-linear functions of $\mathbf{h}$ and we use a squared-error objective:

$$J = \frac{1}{2}\|r(\mathbf{y}|\mathbf{x}) - \mathbf{y}\|^2. \tag{4}$$

We can also constrain the GAE to be a symmetric model by training it to reconstruct both $\mathbf{x}$ given $\mathbf{y}$ and $\mathbf{y}$ given $\mathbf{x}$ [15]:

$$J = \frac{1}{2}\|r(\mathbf{y}|\mathbf{x}) - \mathbf{y}\|^2 + \frac{1}{2}\|r(\mathbf{x}|\mathbf{y}) - \mathbf{x}\|^2. \tag{5}$$

The symmetric objective can be thought of as the non-probabilistic analogue of modeling a *joint* distribution over $\mathbf{x}$ and $\mathbf{y}$ as opposed to a conditional [15].

3   Gated Auto-Encoder Scoring

In [7], the authors showed that data could be scored under an auto-encoder by interpreting the model as a *dynamical system*. In contrast to the probabilistic views based on score matching [6,21,24] and regularization, the dynamical systems approach permits scoring under models with either linear (real-valued data) or sigmoid (binary data) outputs, as well as arbitrary hidden unit activation functions. The method is also agnostic to the learning procedure used to train the model, meaning that it is suitable for the various types of regularized auto-encoders which have been proposed recently. In this section, we demonstrate how the dynamical systems view can be extended to the GAE.

3.1   Vector Field Representation

Similar to [7], we will view the GAE as a dynamical system with the vector field defined by

$$F(\mathbf{y}|\mathbf{x}) = r(\mathbf{y}|\mathbf{x}) - \mathbf{y}.$$

The vector field represents the local transformation that $\mathbf{y}|\mathbf{x}$ undergoes as a result of applying the reconstruction function $r(\mathbf{y}|\mathbf{x})$. Repeatedly applying the reconstruction function to an input $\mathbf{y}|\mathbf{x} \rightarrow r(\mathbf{y}|\mathbf{x}) \rightarrow r(r(\mathbf{y}|\mathbf{x})|\mathbf{x}) \rightarrow \cdots \rightarrow r(r \cdots r(\mathbf{y}|\mathbf{x})|\mathbf{x})$ yields a trajectory whose dynamics, from a physics perspective, can be viewed as a force field. At any point, the potential force acting on a point is the gradient of some potential energy (negative goodness) at that point. In this light, the GAE reconstruction may be viewed as pushing pairs of inputs $\mathbf{x}, \mathbf{y}$ in the direction of lower energy.

Our goal is to derive the energy function, which we call a scoring function, and which measures how much a GAE "likes" a given pair of inputs $(\mathbf{x}, \mathbf{y})$ up to normalizing constant. In order to find an expression for the potential energy, the vector field must be able to be written as the derivative of a scalar field [7]. To check this, we can submit to Poincaré's integrability criterion: For some open, simple connected set $\mathcal{U}$, a continuously differentiable function $F : \mathcal{U} \rightarrow \Re^m$ defines a gradient field if and only if

$$\frac{\partial F_i(\mathbf{y})}{\partial y_j} = \frac{\partial F_j(\mathbf{y})}{\partial y_i}, \ \forall i, j = 1 \cdots n.$$

The vector field defined by the GAE indeed satisfies Poincaré's integrability criterion; therefore it can be written as the derivative of a scalar field. A derivation is given in the Supplementary Material, Section 1.1. This also applies to the GAE with a symmetric objective function (Eq. 5) by setting the input as $\boldsymbol{\xi}|\boldsymbol{\gamma}$ such that $\boldsymbol{\xi} = [\mathbf{y}; \mathbf{x}]$ and $\boldsymbol{\gamma} = [\mathbf{x}; \mathbf{y}]$ and following the exact same procedure.

3.2   Scoring the GAE

As mentioned in Section 3.1, our goal is to find an energy surface, so that we can express the energy for a specific pair $(\mathbf{x}, \mathbf{y})$. From the previous section, we showed that Poincaré's criterion is satisfied and this implies that we can write the vector field as the derivative of a scalar field. Moreover, it illustrates that this vector field is a conservative field and this means that the vector field is a gradient of some scalar function, which in this case is the energy function of a GAE:

$$r(\mathbf{y}|\mathbf{x}) - \mathbf{y} = \nabla E.$$

Hence, by integrating out the trajectory of the GAE $(\mathbf{x}, \mathbf{y})$, we can measure the energy along a path. Moreover, the line integral of a conservative vector field is path independent, which allows us to take the anti-derivative of the scalar function:

$$E(\mathbf{y}|\mathbf{x}) = \int (r(\mathbf{y}|\mathbf{x}) - \mathbf{y})d\mathbf{y} = \int W^Y \left((W^X \mathbf{x}) \odot W^H h(\mathbf{u}) \right) d\mathbf{y} - \int \mathbf{y} d\mathbf{y}$$

$$= W^Y \left((W^X \mathbf{x}) \odot W^H \int h(\mathbf{u}) \, d\mathbf{y} \right) - \int \mathbf{y} d\mathbf{y}, \tag{6}$$

where $\mathbf{u}$ is an auxiliary variable such that $\mathbf{u} = W^H((W^Y \mathbf{y}) \odot (W^X \mathbf{x}))$ and $\frac{d\mathbf{u}}{d\mathbf{y}} = W^H(W^Y \odot (W^X \mathbf{x} \otimes \mathbf{1}_D))$, and $\otimes$ is the Kronecker product. Moreover, the decoder can be re-formulated as

$$r(\boldsymbol{y}|\boldsymbol{x}) = (W^Y)^T (W^X \boldsymbol{x} \odot (W^H)^T h(\boldsymbol{y}, \boldsymbol{x}))$$
$$= \left((W^Y)^T \odot (W^X \boldsymbol{x} \otimes \mathbf{1}_D) \right) (W^H)^T h(\boldsymbol{y}, \boldsymbol{x}).$$

Re-writing Eq. 6 in terms of the auxiliary variable $\mathbf{u}$, we get

$$E(\mathbf{y}|\mathbf{x}) = \left((W^Y)^T \odot (W^Y \mathbf{x} \otimes \mathbf{1}_D) \right) (W^H)^T \tag{7}$$

$$\int h(\mathbf{u}) \left(W^H \left(W^Y \odot (W^X \mathbf{x} \otimes \mathbf{1}_D) \right) \right)^{-1} d\mathbf{u} - \int \mathbf{y} d\mathbf{y}$$

$$= \int h(\mathbf{u}) d\mathbf{u} - \frac{1}{2} \mathbf{y}^2 + \text{const.} \tag{8}$$

A more detailed derivation from Eq. 6 to Eq. 8 is provided in the Supplementary Material, Section 1.2. Identical to [7], if $h(\mathbf{u})$ is an element-wise activation function and we know its anti-derivative, then it is very simple to compute $E(\mathbf{x}, \mathbf{y})$.

4   Relationship to Restricted Boltzmann Machines

In this section, we relate GAEs through the scoring function to other types of Restricted Boltzmann Machines, such as the Factored Gated Conditional RBM [23] and the Mean-covariance RBM [19].

4.1   Gated Auto-Encoder and Factored Gated Conditional Restricted Boltzmann Machines

Kamyshanska and Memisevic showed that several hidden activation functions defined gradient fields, including sigmoid, softmax, tanh, linear, rectified linear function (ReLU), modulus, and squaring. These activation functions are applicable to GAEs as well.

In the case of the sigmoid activation function, $\sigma = h(\mathbf{u}) = \frac{1}{1+\exp(-\mathbf{u})}$, our energy function becomes

$$E_\sigma = 2 \int (1 + \exp -(\mathbf{u}))^{-1} d\mathbf{u} - \frac{1}{2}(\mathbf{x}^2 + \mathbf{y}^2) + \text{const},$$

$$= 2 \sum_k \log \left(1 + \exp \left(W^H_{k\cdot}(W^X \mathbf{x} \odot W^X \mathbf{y}) \right) \right) - \frac{1}{2}(\mathbf{x}^2 + \mathbf{y}^2) + \text{const}.$$

Note that if we consider the conditional GAE we reconstruct $\mathbf{x}$ given $\mathbf{y}$ only, this yields

$$E_\sigma(\mathbf{y}|\mathbf{x}) = \sum_k \log\left(1 + \exp\left(W^H(W_{k\cdot}^Y \mathbf{y} \odot W_{k\cdot}^X \mathbf{x}))\right)\right) - \frac{\mathbf{y}^2}{2} + \text{const.} \qquad (9)$$

This expression is identical, up to a constant, to the free energy in a Factored Gated Conditional Restricted Boltzmann Machine (FCRBM) with Gaussian visible units and Bernoulli hidden units. We have ignored biases for simplicity. A derivation including biases is provided in the Supplementary Material, Section 2.1.

4.2   Mean-Covariance Auto-Encoder and Mean-Covariance Restricted Boltzmann Machines

The Covariance auto-encoder (cAE) was introduced in [15]. It is a specific form of symmetrically trained auto-encoder with identical inputs: $\mathbf{x} = \mathbf{y}$, and tied input weights: $W^X = W^Y$. It maintains a set of relational mapping units to model covariance between pixels. One can introduce a separate set of mapping units connected pairwise to only one of the inputs which model the mean intensity. In this case, the model becomes a Mean-covariance auto-encoder (mcAE).

Theorem 1. *Consider a cAE with encoder and decoder:*

$$h(\mathbf{x}) = h(W^H((W^X\mathbf{x})^2) + \mathbf{b})$$
$$r(\mathbf{x}|h) = (W^X)^T(W^X\mathbf{x} \odot (W^H)^T h(\mathbf{x})) + \mathbf{a},$$

where $\theta = \{W^X, W^H, \mathbf{a}, \mathbf{b}\}$ *are the parameters of the model, and* $h(\mathbf{z}) = \frac{1}{1+\exp(-\mathbf{z})}$ *is a sigmoid. Moreover, consider a Covariance RBM [19] with Gaussian-distributed visibles and Bernoulli-distributed hiddens, with an energy function defined by*

$$E^c(\mathbf{x}, \mathbf{h}) = \frac{(\mathbf{a}-\mathbf{x})^2}{\sigma^2} - \sum_f P\mathbf{h}(C\mathbf{x})^2 - \mathbf{bh}.$$

Then the energy function of the cAE with dynamics $r(\mathbf{x}|\mathbf{y}) - \mathbf{x}$ *is equivalent to the free energy of Covariance RBM up to a constant:*

$$E(\mathbf{x}, \mathbf{x}) = \sum_k \log\left(1 + \exp\left(W^H(W^X\mathbf{x})^2 + \mathbf{b}\right)\right) - \frac{\mathbf{x}^2}{2} + \text{const.} \qquad (10)$$

The proof is given in the Supplementary Material, Section 2.2. We can extend this analysis to the mcAE by using the above theorem and the results from [7].

Corollary 1. *The energy function of a mcAE and the free energy of a Mean-covariance RBM (mcRBM) with Gaussian-distributed visibles and Bernoulli-distributed hiddens are equivalent up to a constant. The energy of the mcAE is:*

$$E = \sum_k \log\left(1 + \exp\left(-W^H(W^X\mathbf{x})^2 - \mathbf{b}\right)\right) + \sum_k \log\left(1 + \exp(W\mathbf{x} + \mathbf{c})\right) - \mathbf{x}^2 + \text{const}$$

$$(11)$$

where $\theta^m = \{W, \mathbf{c}\}$ *parameterizes the mean mapping units and* $\theta^c = \{W^X, W^H,$ $\mathbf{a}, \mathbf{b}\}$ *parameterizes the covariance mapping units.*

Proof. The proof is very simple. Let $E_{mc} = E_m + E_c$, where E_m is the energy of the mean auto-encoder, E_c is the energy of the covariance auto-encoder, and E_{mc} is the energy of the mcAE. We know from Theorem 1 that E_c is equivalent to the free energy of a covariance RBM, and the results from [7] show that that E_m is equivalent to the free energy of mean (classical) RBM. As shown in [19], the free energy of a mcRBM is equal to summing the free energies of a mean RBM and a covariance RBM. $\qquad\square$

5   Classification with Gated Auto-Encoders

Kamyshanska and Memisevic demonstrated that one application of the ability to assign energy or scores to auto-encoders was in constructing a classifier from class-specific auto-encoders. In this section, we explore two different paradigms for classification. Similar to that work, we consider the usual multi-class problem by first training class-specific auto-encoders, and using their energy functions as confidence scores. We also consider the more challenging structured output problem, specifically, the case of multi-label prediction where a data point may have more than one associated label, and there may be correlations among the labels.

5.1   Classification Using Class-Specific Gated Auto-Encoders

One approach to classification is to take several class-specific models and assemble them into a classifier. The best-known example of this approach is to fit several directed graphical models and use Bayes' rule to combine them. The process is simple because the models are normalized, or calibrated. While it is possible to apply a similar technique to undirected or non-normalized models such as auto-encoders, one must take care to calibrate them.

The approach proposed in [7] is to train K class-specific auto-encoders, each of which assigns a non-normalized energy to the data $E_i(\mathbf{x}), i = 1 \dots, K$, and then define the conditional distribution over classes z_i as

$$P(z_i|\mathbf{x}) = \frac{\exp\left(E_i\left(\mathbf{x}\right) + B_i\right)}{\sum_j \exp\left(E_j\left(\mathbf{x}\right) + B_j\right)}, \tag{12}$$

where B_i is a learned bias for class i. The bias terms take the role of calibrating the unnormalized energies. Note that we can similarly combine the energies from a symmetric gated auto-encoder where $\mathbf{x} = \mathbf{y}$ (i.e. a covariance auto-encoder) and apply Eq. 12. If, for each class, we train both a covariance auto-encoder and a classical auto-encoder (i.e. a "mean" auto-encoder) then we can combine both sets of unnormalized energies as follows

$$P_{mcAE}(z_i|\mathbf{x}) = \frac{\exp(E_i^M(\mathbf{x}) + E_i^C(\mathbf{x}) + B_i)}{\sum_j \exp(E_j^M(\mathbf{x}) + E_j^C(\mathbf{x}) + B_j)}, \tag{13}$$

where $E_i^M(\mathbf{x})$ is the energy which comes from the "mean" (standard) auto-encoder trained on class i and $E_i^C(\mathbf{x})$ the energy which comes from the "covariance" (gated) auto-encoder trained on class i. We call the classifiers in Eq. 12 and Eq. 13 "Covariance Auto-encoder Scoring" (cAES) and "Mean-Covariance Auto-encoder Scoring" (mcAES), respectively.

The training procedure is summarized as follows:

1. Train a (mean)-covariance auto-encoder individually for each class. Both the mean and covariance auto-encoder have tied weights in the encoder and decoder. The covariance auto-encoder is a gated auto-encoder with tied inputs.
2. Learn the B_i calibration terms using maximum likelihood, and backpropagate to the GAE parameters.

Experimental Results. We followed the same experimental setup as [16] where we used a standard set of "Deep Learning Benchmarks" [11]. We used mini-batch stochastic gradient descent to optimize parameters during training. The hyperparameters: number of hiddens, number of factors, corruption level, learning rate, weight-decay, momentum rate, and batch sizes were chosen based on a held-out validation set. Corruption levels and weight-decay were selected from $\{0, 0.1, 0.2, 0.3, 0.4, 0.5\}$, and number of hidden and factors were selected from 100,300,500. We selected the learning rate and weight-decay from the range $(0.001, 0.0001)$.

Classification error results are shown in Table 1. First, the error rates of auto-encoder scoring variant methods illustrate that across all datasets AES outperforms cAES and mcAES outperforms both AES and cAES. AE models pixel means and cAE models pixel covariance, while mcAE models both mean and covariance, making it naturally more expressive. We observe that cAES and mcAES achieve lower error rates by a large margin on rotated MNIST with backgrounds (final row). On the other hand, both cAES and mcAES perform poorly on MNIST with random white noise background (second row from bottom). We believe this phenomenon is due to the inability to model covariance in this dataset. In MNIST with random white noise the pixels are typically uncorrelated, where in rotated MNIST with backgrounds the correlations are present and consistent.

5.2 Multi-label Classification via Optimization in Label Space

The dominant application of deep learning approaches to vision has been the assignment of images to discrete classes (e.g. object recognition). Many applications, however, involve "structured outputs" where the output variable is high-dimensional and has a complex, multi-modal joint distribution. Structured output prediction may include tasks such as multi-label classification where there are regularities to be learned in the output, and segmentation, where the output is as high-dimensional as the input. A key challenge to such approaches lies in

Table 1. Classification error rates on the Deep Learning Benchmark dataset. SAA$_3$ stands for three-layer Stacked Auto-encoder. SVM and RBM results are from [24], DEEP and GSM are results from [15], and AES is from [7].

DATA	SVM RBF	RBM SAA$_3$	DEEP	GSM	AES	cAES	mcAES
RECT	2.15	4.71	2.14	0.56	0.84	0.61	**0.54**
RECT$_{IMG}$	24.04	23.69	24.05	22.51	21.45	22.85	**21.41**
CONVEX	19.13	19.92	18.41	**17.08**	21.52	21.6	20.63
MNIST$_{SMALL}$	3.03	3.94	3.46	3.70	**2.61**	3.65	3.65
MNIST$_{ROT}$	11.11	14.69	**10.30**	11.75	11.25	16.5	13.42
MNIST$_{RAND}$	14.58	9.80	11.28	10.48	**9.70**	18.65	16.73
MNIST$_{ROTIM}$	55.18	52.21	51.93	55.16	47.14	39.98	**35.52**

developing models that are able to capture complex, high level structure like shape, while still remaining tractable.

Though our proposed work is based on a deterministic model, we have shown that the energy, or scoring function of the GAE is equivalent, up to a constant, to that of a conditional RBM, a model that has already seen some use in structured prediction problems [12,18].

GAE scoring can be applied to structured output problems as a type of "post-classification" [17]. The idea is to let a naïve, non-structured classifier make an initial prediction of the outputs in a fast, feed-forward manner, and then allow a second model (in our case, a GAE) clean up the outputs of the first model. Since GAEs can model the relationship between input $\mathbf{x}$ and structured output $\mathbf{y}$, we can initialize the output with the output of the naïve model, and then optimize its energy function with respect to the outputs. Input $\mathbf{x}$ is held constant throughout the optimization.

Li *et al* recently proposed Compositional High Order Pattern Potentials, a hybrid of Conditional Random Fields (CRF) and Restricted Boltzmann Machines. The RBM provides a global shape information prior to the locally-connected CRF. Adopting the idea of *learning* structured relationships between outputs, we propose an alternate approach which the inputs of the GAE are not $(\mathbf{x}, \mathbf{y})$ but $(\mathbf{y}, \mathbf{y})$. In other words, the post-classification model is a covariance auto-encoder. The intuition behind the first approach is to use a GAE to learn the relationship between the input $\mathbf{x}$ and the output $\mathbf{y}$, whereas the second method aims to learn the correlations between the outputs $\mathbf{y}$.

We denote our two proposed methods GAE$_{XY}$ and GAE$_{Y^2}$. GAE$_{XY}$ corresponds to a GAE, trained conditionally, whose mapping units directly model the relationship between input and output and GAE$_{Y^2}$ corresponds to a GAE which models correlations between output dimensions. GAE$_{XY}$ defines $E(\mathbf{y}|\mathbf{x})$, while GAE$_{Y^2}$ defines $E(\mathbf{y}|\mathbf{y}) = E(\mathbf{y})$. They differ only in terms of the data vectors that they consume. The training and test procedures are detailed in Algorithm 1.

Algorithm 1. Structured Output Prediction with GAE scoring

1: **procedure** MULTI-LABEL CLASSIFICATION($\mathcal{D} = \{(\mathbf{x}_i, \mathbf{y}_i) \in \mathcal{X}_{train} \times \mathcal{Y}_{train}\}$)

2:     Train a Multi-layer Perceptron (MLP) to learn an input-output mapping $f(\cdot)$:

$$\operatorname*{argmin}_{\theta_1} l(\mathbf{x}, \mathbf{y}; \theta_1) = \sum_i \text{loss}_1 \left((f(\mathbf{x}_i; \theta_1) - \mathbf{y}_i) \right) \tag{14}$$

  where loss$_1$ is an appropriate loss function for the MLP.[2]

3:     Train a Gated Auto-encoder with inputs $(\mathbf{x}_i, \mathbf{y}_i)$; For the case of GAE$_{Y2}$, set $\mathbf{x}_i = \mathbf{y}_i$.

$$\operatorname*{argmin}_{\theta_2} l(\mathbf{x}, \mathbf{y}; \theta_2) = \sum_i \text{loss}_2 \left(r(\mathbf{y}_i | \mathbf{x}_i, \theta_2) - \mathbf{y}_i \right) \tag{15}$$

  where loss$_2$ is an appropriate reconstructive loss for the auto-encoder.

4:     **for** each test data point $\mathbf{x}_i \in \mathcal{X}_{test}$ **do**

5:         Initialize the output using the MLP.

$$\mathbf{y}_0 = f(\mathbf{x}_{test}) \tag{16}$$

6:         **while** $\|E(\mathbf{y}_{t+1}|\mathbf{x}) - E(\mathbf{y}_t|\mathbf{x})\| > \epsilon$ or $\leq$ max. iter. **do**

7:             Compute $\nabla_{\mathbf{y}_t} E$

8:             Update $\mathbf{y}_{t+1} = \mathbf{y}_t - \lambda \nabla_{\mathbf{y}_t} E$

9:             where ϵ is the tolerance rate with respect to the convergence of the optimization.

Experimental Results. We consider multi-label classification, where the problem is to classify instances which can take on more than one label at a time. We followed the same experimental set up as [18]. Four multi-labeled datasets were considered: Yeast [5] consists of biological attributes, Scene [2] is image-based, and MTurk [13] and MajMin [14] are targeted towards tagging music. Yeast consists of 103 biological attributes and has 14 possible labels, Scene consists of 294 image pixels with 6 possible labels, and MTurk and MajMin each consist of 389 audio features extracted from music and have 92 and 96 possible tags, respectively. Figure 1 visualizes the covariance matrix for the label dimensions in each dataset. We can see from this that there are correlations present in the labels which suggests that a structured approach may improve on a non-structured predictor.

We compared our proposed approaches to logistic regression, a standard MLP, and the two structured CRBM training algorithms presented in [18]. To permit a fair comparison, we followed the same procedure for training and reporting errors as in that paper, where we cross validated over 10 folds and training, validation, test examples are randomly separated into 80%, 10%, and 10% in each fold. The error rate was measured by averaging the errors on each label dimension.

---

[2] In our experiments, we used the cross-entropy loss function for loss$_1$ and loss$_2$.

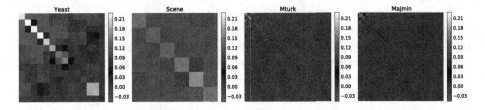

Fig. 1. Covariance matrices for the multi-label datasets: Yeast, Scene, MTurk, and MajMin.

Table 2. Error rate on multi-label datasets. As in previous work, we report the mean across 10 repeated runs with different random weight initializations.

Method	Yeast	Scene	MTurk	MajMin
LogReg	20.16	10.11	8.10	4.34
HashCRBM*	20.02	8.80	7.24	4.24
MLP	19.79	8.99	7.13	4.23
GAES$_{XY}$	**19.27**	6.83	**6.59**	**3.96**
GAES$_{Y^2}$	19.58	**6.81**	**6.59**	4.29

The performance on four multi-label datasets is shown in Table 2. We observed that adding a small amount of Gaussian noise to the input y improved the performance for GAE$_{XY}$. However, adding noise to the input x did not have as much of an effect. We suspect that adding noise makes the GAE more robust to the input provided by the MLP. Interestingly, we found that the performance of GAE$_{Y^2}$ was negatively affected by adding noise. Both of our proposed methods, GAES$_{XY}$ and GAES$_{Y^2}$ generally outperformed the other methods except for GAES$_{Y^2}$ on the MajMin dataset. At least for these datasets, there is no clear winner between the two. GAES$_{XY}$ achieved lower error than GAES$_{Y^2}$ for Yeast and MajMin, and the same error rate on the MTurk dataset. However, GAES$_{Y^2}$ outperforms GAES$_{XY}$ on the Scene dataset. Overall, the results show that GAE scoring may be a promising means of post-classification in structured output prediction.

6   Conclusion

There have been many theoretical and empirical studies on auto-encoders [6,7,20,21,24,25], however, the theoretical study of gated auto-encoders is limited apart from [4,15]. The GAE has several intriguing properties that a classical auto-encoder does not, based on its ability to model relations among pixel intensities rather than just the intensities themselves. This opens up a broader set of applications. In this paper, we derive some theoretical results for the GAE that enable us to gain more insight and understanding of its operation.

We cast the GAE as a dynamical system driven by a vector field in order to analyze the model. In the first part of the paper, by following the same procedure as [7], we showed that the GAE could be scored according to an energy function. From this perspective, we demonstrated the equivalency of the GAE energy to the free energy of a FCRBM with Gaussian visible units, Bernoulli hidden units, and sigmoid hidden activations. In the same manner, we also showed that the covariance auto-encoder can be formulated in a way such that its energy function is the same as the free energy of a covariance RBM, and this naturally led to a connection between the mean-covariance auto-encoder and mean-covariance RBM. One interesting observation is that Gaussian-Bernoulli RBMs have been reported to be difficult to train [3,9], and the success of training RBMs is highly dependent on the training setup [26]. Auto-encoders are an attractive alternative, even when an energy function is required.

Structured output prediction is a natural next step for representation learning. The main advantage of our approach compared to other popular approaches such as Markov Random Fields, is that inference is extremely fast, using a gradient-based optimization of the auto-encoder scoring function. In the future, we plan on tackling more challenging structured output prediction problems.

References

1. Bengio, Y., Thibodeau-Laufer, É.: Deep generative stochastic networks trainable by backprop (2013). arXiv preprint arXiv:1306.1091
2. Boutell, M.R., Luob, J., Shen, X., Brown, C.M.: Learning multi-label scene classification. Pattern Recognition **37**, 1757–1771 (2004)
3. Cho, K.H., Ilin, A., Raiko, T.: Improved learning of Gaussian-Bernoulli restricted Boltzmann machines. In: Honkela, T. (ed.) ICANN 2011, Part I. LNCS, vol. 6791, pp. 10–17. Springer, Heidelberg (2011)
4. Droniou, A., Sigaud, O.: Gated autoencoders with tied input weights. In: ICML (2013)
5. Elisseeff, A., Weston, J.: A kernel method for multi-labelled classification. In: NIPS (2002)
6. Guillaume, A., Bengio, Y.: What regularized auto-encoders learn from the data generating distribution. In: ICLR (2013)
7. Kamyshanska, H., Memisevic, R.: On autoencoder scoring. In: ICML, pp. 720–728 (2013)
8. Kamyshanska, H., Memisevic, R.: The potential energy of an auto-encoder. IEEE Transactions on Pattern Analysis and Machine Intelligence **37**(6), 1261–1273 (2014)
9. Krizhevsky, A.: Learning multiple layers of features from tiny images. Tech. rep., Department of Computer Science, University of Toronto (2009)
10. Krizhevsky, A., Sutskever, I., Hinton, G.E.: Imagenet classification with deep convolutional neural networks. In: NIPS (2012)
11. Larochelle, H., Erhan, D., Courville, A., Bergstra, J., Bengio, Y.: An empirical evaluation of deep architectures on problems with many factors of variation. In: ICML (2007)
12. Li, Y., Tarlow, D., Zemel, R.: Exploring compositional high order pattern potentials for structured output learning. In: CVPR (2013)
13. Mandel, M.I., Eck, D., Bengio, Y.: Learning tags that vary within a song. In: ISMIR (2010)

14. Mandel, M.I., Ellis, D.P.W.: A web-based game for collecting music metadata. Journal New of Music Research **37**, 151–165 (2008)
15. Memisevic, R.: Gradient-based learning of higher-order image features. In: ICCV (2011)
16. Memisevic, R., Zach, C., Hinton, G., Pollefeys, M.: Gated softmax classification. In: NIPS (2010)
17. Mnih, V., Hinton, G.E.: Learning to detect roads in high-resolution aerial images. In: Daniilidis, K., Maragos, P., Paragios, N. (eds.) ECCV 2010, Part VI. LNCS, vol. 6316, pp. 210–223. Springer, Heidelberg (2010)
18. Mnih, V., Larochelle, H., Hinton, G.E.: Conditional restricted Boltzmann machines for structured output prediction. In: UAI (2011)
19. Ranzato, M., Hinton, G.E.: Modeling pixel means and covariances using factorized third-order Boltzmann machines. In: CVPR (2010)
20. Rifai, S.: Contractive auto-encoders: explicit invariance during feature extraction. In: ICML (2011)
21. Swersky, K., Ranzato, M., Buchman, D., Freitas, N.D., Marlin, B.M.: On autoencoders and score matching for energy based models. In: ICML, pp. 1201–1208 (2011)
22. Szegedy, C., Liu, W., Jia, Y., Sermanet, P., Reed, S., Anguelov, D., Erhan, D., Vanhoucke, V., Rabinovich, A.: Going deeper with convolutions (2014). arXiv preprint arXiv:1409.4842
23. Taylor, G.W., Hinton, G.E.: Factored conditional restricted Boltzmann machines for modeling motion style. In: ICML, pp. 1025–1032 (2009)
24. Vincent, P.: A connection between score matching and denoising auto-encoders. Neural Computation **23**(7), 1661–1674 (2010)
25. Vincent, P., Larochelle, H., Bengio, Y., Manzagol, P.: Extracting and composing robust features with denoising autoencoders. In: ICML (2008)
26. Wang, N., Melchior, J., Wiskott, L.: Gaussian-binary restricted Boltzmann machines on modeling natural image statistics. Tech. rep., Institut fur Neuroinformatik Ruhr-Universitat Bochum, Bochum, 44780, Germany (2014)

Sign Constrained Rectifier Networks
with Applications to Pattern Decompositions

Senjian An[1]([✉]), Qiuhong Ke[1], Mohammed Bennamoun[1], Farid Boussaid[2],
and Ferdous Sohel[1]

[1] School of Computer Science and Software Engineering,
The University of Western Australia, Crawley, WA 6009, Australia
{senjian.an,mohammed.bennamoun,ferdous.sohel}@uwa.edu.au,
qiuhong.ke@research.uwa.edu.au
[2] School of Electrical, Electronic and Computer Engineering,
The University of Western Australia, Crawley, WA 6009, Australia
farid.boussaid@uwa.edu.au

Abstract. In this paper we introduce sign constrained rectifier networks
(SCRN), demonstrate their universal classification power and illustrate
their applications to pattern decompositions. We prove that the proposed
two-hidden-layer SCRN, with sign constraints on the weights of the output
layer and on those of the top hidden layer, are capable of separating any two
disjoint pattern sets. Furthermore, a two-hidden-layer SCRN of a pair of
disjoint pattern sets can be used to decompose one of the pattern sets into
several subsets so that each subset is convexly separable from the entire
other pattern set; and a single-hidden-layer SCRN of a pair of convexly
separable pattern sets can be used to decompose one of the pattern sets
into several subsets so that each subset is linearly separable from the entire
other pattern set. SCRN can thus be used to learn the pattern structures
from the decomposed subsets of patterns and to analyse the discriminant
factors of different patterns from the linear classifiers of the linearly separable subsets in the decompositions. With such pattern decompositions
exhibiting convex separability or linear separability, users can also analyse
the complexity of the classification problem, remove the outliers and the
non-crucial points to improve the training of the traditional unconstrained
rectifier networks in terms of both performance and efficiency.

Keywords: Rectifier neural network · Pattern decomposition

1 Introduction

Deep rectifier networks have achieved great success in object recognition
[4,8,10,18], face verification [14,15], speech recognition ([3,6,12] and handwritten digit recognition [2]. However, the lack of understanding of the roles of the
hidden layers makes the deep learning network difficult to interpret for tasks
of discriminant factor analysis and pattern structure analysis. Towards a clear
understanding of the success of the deep rectifier networks, a recent work [1] provides a constructive proof for the universal classification power of two-hidden-
layer rectifier networks. For binary classification, the proof uses the first hidden

© Springer International Publishing Switzerland 2015
A. Appice et al. (Eds.): ECML PKDD 2015, Part I, LNAI 9284, pp. 546–559, 2015.
DOI: 10.1007/978-3-319-23528-8_34

layer to make the pattern sets convexly separable. The second hidden layer is then used to achieve linear separability, and finally a linear classifier is used to separate the patterns. Although this strategy can be used in constructive proofs, it cannot be used to analyse the learnt rectifier network since it might not be verified in the empirical learning from data. Fortunately, this paper will show that such a strategy can be verified if additional sign constraints are imposed on the weights of the output layer and on those of the top hidden layer. A fundamental result of this paper is that a pair of pattern sets can be separated by a single-hidden-layer rectifier network with non-negative output layer weights and non-positive bias *if and only if* one of the pattern sets is disjoint to the *convex hull* of the other. With this fundamental result, this paper introduces sign constrained rectifier networks (SCRN) and proves that the two-hidden-layer SCRNs are capable of separating any two disjoint pattern sets. SCRN can automatically learn a rectifier network classifier which achieves convex separability and linear separability in the first and second hidden layers respectively. For any pair of disjoint pattern sets, a two-hidden-layer SCRN can be used to decompose one of the pattern sets into several subsets each convexly separable from the entire other pattern set; and for any pair of convexly separable pattern sets, a single-hidden-layer SCRN can be used to decompose one of the pattern sets into several subsets each linearly separable from the entire other pattern set. SCRN thus can be used to analyse the pattern structures and the discriminant factors of different patterns.

Compared to traditional unconstrained rectifier networks, SCRN is more interpretable and convenient for tasks of discriminant factor analysis and pattern structure analysis. It can help in initializations or refining of the traditional rectifier networks. The outliers and the non-crucial points of the decomposed subsets can be identified. Classification accuracy can thus be improved by training after removal of outliers, while training efficiency can be improved by removing the non-crucial training patterns, especially when the original training size is large.

Notations: Throughout the paper, we use capital letters to denote matrices, lower case letters for scalar terms, and bold lower letters for vectors. For instance, we use $\mathbf{w}_i$ to denote the ith column of a matrix W, and use b_i to denote the ith element of a vector $\mathbf{b}$. For any integer m, we use $[m]$ to denote the integer set from 1 to m, i.e., $[m] \triangleq \{1, 2, \cdots, m\}$. We use I to denote the identity matrix with proper dimensions, $\mathbf{0}$ a vector with all elements being 0, and $\mathbf{1}$ a vector with all elements being 1. $W \succeq 0$ and $\mathbf{b} \succeq 0$ denote that all elements of W and $\mathbf{b}$ are non-negative while $W \preceq 0$ and $\mathbf{b} \preceq 0$ denote that all elements of W and $\mathbf{b}$ are non-positive. Given a finite number of points $\mathbf{x}_i$ ($i \in [m]$) in $\mathbb{R}^n$, a convex combination $\mathbf{x}$ of these points is a linear combination of these points, in which all coefficients are non-negative and sum to 1. The convex hull of a set $\mathcal{X}$, denoted by $\mathrm{CH}(\mathcal{X})$, is a set of all convex combinations of the points in $\mathcal{X}$.

The rest of this paper is organised as follows. In Section 2, the categories of separable pattern sets are described with a brief review on the disjoint convex hull decomposition models of patterns [1]. In Section 3, we address the formulation of binary classifiers with rectifier networks and introduce the sign-constrained single-hidden-layer and two-hidden-layer binary classifiers. Section 4 presents the fundamental result of this paper, that is, a pair of pattern sets can be separated by a sign-constrained single-hidden-layer classifier *if and only if* they are convexly separable. Section 5 addresses two-hidden-layer sign-constrained rectifier networks, their universal classification power and their capacity for pattern decompositions. In Section 6, we conclude the paper with a discussion on related works and future research directions.

2   The Categories of Separable Pattern Sets

In [1], a disjoint convex hull model of pattern sets is introduced for theoretical analysis of rectifier networks. We will use this model to address the universal classification power of the proposed sign-constrained rectifier networks. This section gives a brief review on this model and describes the categories of separable pattern sets.

Let $\mathcal{X}_1$ and $\mathcal{X}_2$ be two finite pattern sets in $\mathbb{R}^n$. Each of them can be modelled by a union of several subsets under the condition that the convex hulls of every two subsets from distinct classes are disjoint. A decomposition of $\mathcal{X}_1$ and $\mathcal{X}_2$, namely, $\mathcal{X}_k = \bigcup_{i=1}^{L_k} \mathcal{X}_k^i$ where $k = 1, 2$ and L_k are the numbers of subsets, is called a disjoint convex hull decomposition if the unions of the convex hulls of $\mathcal{X}_k^i$, denoted by $\hat{\mathcal{X}}_k \triangleq \bigcup_{i=1}^{L_k} \mathrm{CH}(\mathcal{X}_k^i)$, are still disjoint, i.e.,

$$\hat{\mathcal{X}}_1 \cap \hat{\mathcal{X}}_2 = \emptyset, \tag{1}$$

or equivalently, for all $i \in [L_1], j \in [L_2]$,

$$\mathrm{CH}(\mathcal{X}_1^i) \cap \mathrm{CH}(\mathcal{X}_2^j) = \emptyset. \tag{2}$$

A pair of pattern sets, namely $\mathcal{X}_1$ and $\mathcal{X}_2$, are called linearly separable if there exists $\mathbf{w}$ and b such that

$$\begin{aligned} \mathbf{w}^T\mathbf{x} + b > 0, \ \forall \, \mathbf{x} \in \mathcal{X}_1 \\ \mathbf{w}^T\mathbf{x} + b < 0, \ \forall \, \mathbf{x} \in \mathcal{X}_2. \end{aligned} \tag{3}$$

It is known that two pattern sets are linearly separable if and only if their convex hulls are disjoint, i.e., $\mathrm{CH}(\mathcal{X}_1) \cap \mathrm{CH}(\mathcal{X}_2) = \emptyset$. The disjoint convex hull decomposition of two pattern sets decomposes each pattern set into several subsets so that every pair of subsets from distinct classes are linearly separable.

For finite pattern sets $\mathcal{X}_k, k = 1, 2$, a trivial disjoint convex hull decomposition is to select each point as a subset. Hence, any disjoint pattern sets have at least one disjoint convex hull decomposition. According to the minimal numbers of subsets in their disjoint convex hull decompositions, a pair of pattern sets can be categorized into the following three categories:

1) *Linearly Separable Pattern Sets*: The two pattern sets are linearly separable if they have a disjoint decomposition convex model with $L_1 = L_2 = 1$, i.e., $\mathrm{CH}(\mathcal{X}_1) \cap \mathrm{CH}(\mathcal{X}_2) = \emptyset$;

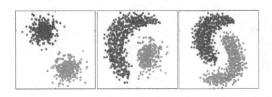

Fig. 1. Illustration of the categories of pattern sets (Left: linear separable; Middle: convexly separable; Right: convexly inseparable). Best viewed in color.

2) *Convexly Separable Pattern Sets*: The two pattern sets have a disjoint decomposition convex hull model with $\min(L_1, L_2) = 1$, i.e., $\mathrm{CH}(\mathcal{X}_1) \cap \mathcal{X}_2 = \emptyset$ or $\mathrm{CH}(\mathcal{X}_2) \cap \mathcal{X}_1 = \emptyset$. These pattern sets are referred to as convexly separable because there exists a convex region which can separate one class from the other;

3) *Disjoint but Convexly Inseparable Pattern Sets*: $\mathcal{X}_1$ and $\mathcal{X}_2$ have no common points, $\mathcal{X}_1 \cap \mathcal{X}_2 = \emptyset$, and all their disjoint convex hull decompositions satisfy $\min(L_1, L_2) > 1$.

Figure 1 demonstrates the three categories of pattern sets. There exists a hyperplane to separate the linearly separable patterns, and the discriminant factor can be characterized by the geometrically interpretable linear classifiers (i.e., the separating hyperplanes). However, patterns are rarely linearly separable in practice, and nonlinear classifiers are required to separate the patterns. Existing nonlinear classification methods such as kernel methods [13] and deep rectifier network methods [4,8,10,18] are not geometrically interpretable due to the nonlinear transformations induced by kernels or hidden layers. In this paper, we investigate the methods to decompose the convexly inseparable pattern sets into convexly separable pattern subsets, and the methods to decompose the convexly separable pattern sets into linearly separable subsets, so that pattern structures and discriminant factors can be analysed through the decomposed pattern subsets by linear classifiers.

3 Binary Classification with Rectifier Networks

With rectifier activation $\max(0, x)$, a single-hidden-layer binary classifier can be described as

$$f(\mathbf{x}) \triangleq \mathbf{a}^T \max(\mathbf{0}, W^T \mathbf{x} + \mathbf{b}) + \beta \tag{4}$$

where $W \in \mathbb{R}^{n \times m}$, $\mathbf{b} \in \mathbb{R}^m$, $\mathbf{a} \in \mathbb{R}^m$, β is a real number, m is the number of hidden nodes and n is the dimension of the inputs.

For pattern sets $\mathcal{X}_+$ and $\mathcal{X}_-$ labelled positive and negative respectively, a single-hidden-layer binary classifier $f(\mathbf{x})$, as defined in (4), is called a single hidden layer separator of $\mathcal{X}_+$ and $\mathcal{X}_-$ if it satisfies

$$\begin{aligned} f(\mathbf{x}) > 0, \ \forall \, \mathbf{x} \in \mathcal{X}_+ \\ f(\mathbf{x}) < 0, \ \forall \, \mathbf{x} \in \mathcal{X}_-. \end{aligned} \tag{5}$$

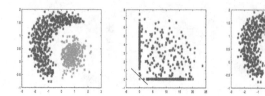

Fig. 2. Illustration of a single-hidden-layer SCRN for convexly separable pattern sets (Left: patterns in the input space; Middle: outputs of the hidden layer with two hidden nodes, and the linear separating boundary; Right: the nonlinear separating boundary in the input space). Best viewed in color.

If it further satisfies $\mathbf{a} \succeq 0, \beta \leq 0$, we call it a *sign-constrained single-hidden-layer separator* of $\mathcal{X}_+$ and $\mathcal{X}_-$.

Similarly, a two-hidden-layer binary classifier, with n dimensional input, l bottom hidden nodes, m top hidden nodes and a single output, can be described by

$$f\{G(\mathbf{x})\} = \mathbf{a}^T \max(\mathbf{0}, W^T G(\mathbf{x}) + \mathbf{b}) + \beta$$
$$G(\mathbf{x}) = \max(\mathbf{0}, V^T \mathbf{x} + \mathbf{c}) \tag{6}$$

where β is a scalar number, $\mathbf{a} \in \mathbb{R}^m, \mathbf{b} \in \mathbb{R}^m, \mathbf{c} \in \mathbb{R}^l, \mathbf{W} \in \mathbb{R}^{l \times m}$ and $V \in \mathbb{R}^{n \times l}$.

We say that a two-hidden-layer binary classifier $f\{G(\mathbf{x})\}$, as defined in (6), is a two-hidden-layer separator of $\mathcal{X}_+$ and $\mathcal{X}_-$ if it satisfies

$$f\{G(\mathbf{x})\} > 0, \ \forall\, \mathbf{x} \in \mathcal{X}_+$$
$$f\{G(\mathbf{x})\} < 0, \ \forall\, \mathbf{x} \in \mathcal{X}_-. \tag{7}$$

If it further satisfies $\mathbf{a} \succeq 0, \beta \leq 0$ and $\mathbf{W} \preceq 0, \mathbf{b} \succeq 0$, we call it a *sign-constrained two-hidden-layer separator* of $\mathcal{X}_+$ and $\mathcal{X}_-$.

Remarks: For sign-constrained single-hidden-layer and two-hidden-layer binary classifiers, due to the non-negativeness of $\mathbf{a}$, β must be non-positive if $f(\mathbf{x})$ is a separator of a pair of pattern sets. If $\beta < 0$, one can always scale it to be -1 and scale $\mathbf{a}$ accordingly so that the sign of $f(\mathbf{x})$ remains unchanged, and thus β can be constrained to be in $\{0, -1\}$ without loss of generality.

4   Single-Hidden-Layer Sign Constrained Rectifier Networks

This section investigates the relationship between convex separability of pattern sets and the sign constraints of single-hidden-layer classifiers. Figure 2 is provided to help understand the ideas of the proofs and the separating boundaries obtained by a single-hidden-layer SCRN.

Lemma 1. *Let $\mathcal{X}_+, \mathcal{X}_-$ be a pair of finite pattern sets in $\mathbb{R}^n$ and be labelled positive and negative respectively. Then $\mathcal{X}_+, \mathcal{X}_-$ can be separated by a sign-constrained single-hidden-layer classifier, as defined in (4) and satisfying $\beta \leq 0$*

and $\mathbf{a} \succeq 0$, *if and only if the convex hull of* $\mathcal{X}_-$ *is disjoint to* $\mathcal{X}_+$, *i.e.,* $\mathcal{X}_+ \cap \mathrm{CH}(\mathcal{X}_-) = \emptyset$.

Proof: (Sufficiency). Suppose $\mathrm{CH}(\mathcal{X}_-) \cap \mathcal{X}_+ = \emptyset$. Let n_+ be the number of training patterns in $\mathcal{X}_+$ and $\mathbf{x}_i^+$ be the ith member of $\mathcal{X}_+$. Since $\mathbf{x}_i^+ \notin \mathrm{CH}(\mathcal{X}_-)$ for any $i \in [n_+]$, there exists $\mathbf{w}_i, b_i, i \in [n_+]$ such that

$$\begin{aligned} \mathbf{w}_i^T \mathbf{x}_i^+ + b_i &> 0 \\ \mathbf{w}_i^T \mathbf{x} + b_i &< 0, \ \forall\, \mathbf{x} \in \mathcal{X}_-. \end{aligned} \tag{8}$$

Denote

$$\begin{aligned} W &= [\mathbf{w}_1, \mathbf{w}_2, \cdots, \mathbf{w}_{n_+}] \\ \mathbf{b} &= [b_1, b_2, \cdots, b_{n_+}]^T \\ \mathbf{z} &= \max(\mathbf{0}, W^T \mathbf{x} + \mathbf{b}). \end{aligned} \tag{9}$$

Then we have

$$\begin{aligned} \mathcal{Z}_- &\triangleq \{\mathbf{z} = \max(0, W^T \mathbf{x} + \mathbf{b}) : \mathbf{x} \in \mathcal{X}_-\} \\ &= \{\mathbf{0}\} \\ \mathcal{Z}_+ &\triangleq \{\mathbf{z} = \max(0, W^T \mathbf{x} + \mathbf{b}) : \mathbf{x} \in \mathcal{X}_+\} \\ &\subset \{\mathbf{z} : \mathbf{1}^T \mathbf{z} > \gamma_{min}, z_i \geq 0, \forall\, i \in [n_+]; \ z_j > 0, \exists\, j \in [n_+]\} \end{aligned} \tag{10}$$

where

$$\begin{aligned} \gamma_{min} &\triangleq \min_{\mathbf{x} \in \mathcal{X}_+} \mathbf{1}^T \max(0, W^T \mathbf{x} + \mathbf{b}) \\ &> 0. \end{aligned} \tag{11}$$

The middle subfigure of Fig 2 provides an example of $\mathcal{Z}_+$ (in red color) and $\mathcal{Z}_-$ (in green color), the transformed pattern sets of a convexly separable pattern sets as shown in the left subfigure of Figure 2.

For a single-hidden-layer binary classifier $f(\mathbf{x})$, as described in (4), if we choose $\beta = -1$ and $\mathbf{a} = \frac{2}{\gamma_{min}}\mathbf{1} \succeq 0$, then

$$f(\mathbf{x}) = \frac{2}{\gamma_{min}}\mathbf{1}^T \max(0, W^T \mathbf{x} + \mathbf{b}) - 1$$

satisfies

$$\begin{aligned} f(\mathbf{x}) &\geq 1 \ > 0, \ \forall\, \mathbf{x} \in \mathcal{X}_+, \\ f(\mathbf{x}) &= -1 < 0, \ \forall\, \mathbf{x} \in \mathcal{X}_- \end{aligned} \tag{12}$$

which imply that $\mathcal{X}_+$ and $\mathcal{X}_-$ can be separated by a sign-constrained single-hidden-layer binary classifier.

(Necessity). Suppose that $\mathcal{X}_+, \mathcal{X}_-$ can be separated by a sign-constrained single-hidden-layer binary classifier, that is, there exist $\mathbf{a} \succeq 0, \beta \leq 0$, and $W, \mathbf{b}$, such that $f(\mathbf{x})$, as defined in (4), satisfies (5). Next, we will prove the convexity of the set $\{\mathbf{x} : f(\mathbf{x}) < 0\}$ and show that $f(\mathbf{x}) < 0$ holds for all $\mathbf{x}$ in the convex hull of $\mathcal{X}_-$.

Let z_0, z_1 be two arbitrary real numbers and let $z_\lambda = \lambda z_1 + (1 - \lambda) z_0$ be their linear combination. Since

$$\max(0, z_\lambda) \leq \lambda \max(0, z_1) + (1 - \lambda) \max(0, z_0), \ \forall\, \lambda \in [0, 1], \tag{13}$$

we have

$$\mathbf{a}^T \max(0, \mathbf{z}_\lambda) \le \lambda \mathbf{a}^T \max(0, \mathbf{z}_0) + (1-\lambda)\mathbf{a}^T \max(0, \mathbf{z}_1), \forall\, \lambda \in [0,1] \qquad (14)$$

for any $\mathbf{a} \succeq 0$ and $\mathbf{z}_0, \mathbf{z}_1$ with same dimensions, where $\mathbf{z}_\lambda \triangleq \lambda \mathbf{z}_1 + (1-\lambda)\mathbf{z}_0$.
In particular, let $\mathbf{z}_\lambda = W^T \mathbf{x}_\lambda + \mathbf{b}$ with $\mathbf{x}_\lambda = \lambda \mathbf{x}_1 + (1-\lambda)\mathbf{x}_0$. Then we have

$$\begin{aligned}
f(\mathbf{x}_\lambda) &= \mathbf{a}^T \max(0, \mathbf{z}_\lambda) + \beta \\
&\le \lambda \left[\mathbf{a}^T \max(0, \mathbf{z}_0) + \beta\right] + (1-\lambda)\left[\mathbf{a}^T \max(0, \mathbf{z}_1) + \beta\right] \qquad (15)\\
&= \lambda f(\mathbf{x}_0) + (1-\lambda) f(\mathbf{x}_1), \forall\, \lambda \in [0,1]
\end{aligned}$$

and therefore

$$f(\mathbf{x}_\lambda) < 0, \ \forall\, \lambda \in [0,1] \qquad (16)$$

if and only if

$$f(\mathbf{x}_\lambda) < 0, \ \forall\, \lambda = 0, 1. \qquad (17)$$

Hence $\{\mathbf{x} : f(\mathbf{x}) < 0\}$ is a convex set, and thus

$$f(\mathbf{x}) < 0, \forall\, \mathbf{x} \in \mathrm{CH}(\mathcal{X}_-) \qquad (18)$$

follows from $f(\mathbf{x}) < 0, \forall\, \mathbf{x} \in \mathcal{X}_-$. Note that $f(\mathbf{x}) > 0$ for all $\mathbf{x} \in \mathcal{X}_+$ (from (5)). So $\mathcal{X}_+$ and $\mathrm{CH}(\mathcal{X}_-)$ are separable and thus $\mathrm{CH}(\mathcal{X}_-) \cap \mathcal{X}_+ = \emptyset$, which completes the proof.

$\square$

The following Lemma shows the capacity of sign-constrained single-hidden-layer classifiers in decomposing the positive pattern set into several subsets so that each subset is linearly separable from the negative pattern set.

Lemma 2. *Let $\mathcal{X}_+$ and $\mathcal{X}_-$ be two convexly separable pattern sets with $\mathcal{X}_+ \cap \mathrm{CH}(\mathcal{X}_-) = \emptyset$, and let $f(\mathbf{x})$, as defined in (4) with m hidden nodes and satisfying $\mathbf{a} \succeq 0, \beta \le 0$, be one of their sign-constrained single-hidden-layer separators. Define*

$$f_\mathcal{I}(\mathbf{x}) \triangleq \left(\sum_{i \in \mathcal{I}} a_i(\mathbf{w}_i^T \mathbf{x} + b_i)\right) + \beta \qquad (19)$$

and

$$\mathcal{X}_+^\mathcal{I} \triangleq \{\mathbf{x} : f_\mathcal{I}(\mathbf{x}) > 0, \mathbf{x} \in \mathcal{X}_+\}. \qquad (20)$$

for any subset $\mathcal{I} \subset [m]$. Then we have

$$X_+ = \bigcup_{\mathcal{I} \subset [m]} \mathcal{X}_+^\mathcal{I} \qquad (21)$$

and

$$\mathrm{CH}\left(\mathcal{X}_+^\mathcal{I}\right) \cap \mathrm{CH}(\mathcal{X}_-) = \emptyset, \qquad (22)$$

i.e, $\mathcal{X}_+^\mathcal{I}$ and $\mathcal{X}_-$ are linearly separable, and furthermore, $f_\mathcal{I}(\mathbf{x})$ is their linear separator satisfying

$$\begin{aligned}
f_\mathcal{I}(\mathbf{x}) &> 0, \ \forall\, \mathbf{x} \in \mathcal{X}_+^\mathcal{I} \\
f_\mathcal{I}(\mathbf{x}) &< 0, \ \forall\, \mathbf{x} \in \mathcal{X}_-.
\end{aligned} \qquad (23)$$

Before proving this Lemma, we give an example about the subsets of $[m]$ and explain the decomposed subsets of the positive pattern set with Fig 2. When $m = 2$, $[m]$ has three subsets: $\mathcal{I}_1 = \{1\}, \mathcal{I}_2 = \{2\}$ and $\mathcal{I}_3 = \{1, 2\}$. In the example of Fig 2, two hidden nodes are used and the positive pattern set (in red color) can be decomposed into three subsets, each associated with one of the three (extended) lines of the separating boundary in the right subfigure of Fig 2, and the middle line of the boundary is associated with $\mathcal{I}_3$. Note that the decomposed subsets have overlaps. The number of the subsets is determined by the number of hidden nodes. For compact decompositions and meaningful discriminate factor analysis, small numbers of hidden nodes are preferable. The significance of this Lemma is in the discovery that single-hidden-layer SCRN can decompose the convexly separable pattern sets into linearly separable subsets so that the discriminate factors of convexly separable patterns can be analysed through the linear classifies between one pattern set (labelled negative) and the subsets of the other pattern set (labelled positive).

Proof: From $\mathbf{a} \succeq 0$, it follows that

$$f_{\mathcal{I}}(\mathbf{x}) \le f(\mathbf{x}), \; \forall \, \mathcal{I} \subset [m], \mathbf{x} \in \mathbb{R}^n \tag{24}$$

and consequently

$$f_{\mathcal{I}}(\mathbf{x}) < 0, \; \forall \, \mathcal{I} \subset [m], \mathbf{x} \in \mathcal{X}_-. \tag{25}$$

Then (23) follows straightforward from the above inequality and the definition of $\mathcal{X}_+^{\mathcal{I}}$. Note that $f_{\mathcal{I}}(\mathbf{x})$ is a linear classifier satisfying (23), $f_{\mathcal{I}}(\mathbf{x})$ is a linear separator of $\mathcal{X}_+^{\mathcal{I}}$ and $\mathcal{X}_-$, and (22) holds consequently.

To complete the proof, it remains to prove (21). Let $\mathbf{x} \in \mathcal{X}_+$ be any pattern with positive label and let $\mathcal{I} \subset [m]$ be the index set so that $\mathbf{w}_i^T \mathbf{x} + b_i > 0$ for all $i \in \mathcal{I}$ and $\mathbf{w}_i^T \mathbf{x} + b_i \le 0$ for all $i \notin \mathcal{I}$. Then $f_{\mathcal{I}}(\mathbf{x}) = f(\mathbf{x}) > 0$ and thus $\mathbf{x} \in \mathcal{X}_+^{\mathcal{I}}$. This proves that any element in $\mathcal{X}_+$ is in $\mathcal{X}_+^{\mathcal{I}}$ for some $\mathcal{I} \subset [m]$. Hence (21) is true and the proof is completed. □

5   Two-Hidden-Layer Sign Constrained Rectifier Networks

This section investigates the universal classification power of sign-constrained two-hidden-layer binary classifiers and their capacity to decompose one pattern set into smaller subsets so that each subset is convexly separable from the other pattern set. Figure 3 is provided to help understand the universal classification power of SCRN by achieving convex separability through the first hidden layer and then achieving the linear separability in the second hidden layer. In this example, three hidden nodes are used in the first hidden layer to achieve convex separability (i.e., the set of the red points and the convex hull of the green points are disjoint) and two hidden nodes are used in the second hidden layer to achieve linear separability.

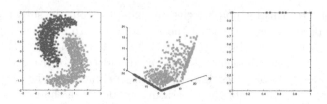

Fig. 3. Illustration of a two-hidden-layer SCRN (Left: patterns in the input space; Middle: outputs of the first hidden layer with three hidden nodes; Right: outputs of the second hidden layer with two hidden nodes). Best viewed in color.

Theorem 3. *For any two disjoint pattern sets, namely $\mathcal{X}_+$ and $\mathcal{X}_-$, in $\mathbb{R}^n$, there exists a sign-constrained two-hidden-layer binary classifier $f\{G(\mathbf{x})\}$, as defined in (6) and satisfying $\mathbf{a} \succeq 0, \beta \leq 0, W \preceq 0, \mathbf{b} \succeq 0$, such that $f\{G(\mathbf{x})\} > 0$ for all $\mathbf{x} \in \mathcal{X}_+$ and $f\{G(\mathbf{x})\} < 0$ for all $\mathbf{x} \in \mathcal{X}_-$.*

Proof: Let

$$\mathcal{X}_+ = \bigcup_{i=1}^{L_1} \mathcal{X}_+^i, \mathcal{X}_- = \bigcup_{j=1}^{L_2} \mathcal{X}_-^j \tag{26}$$

be the disjoint convex hull decomposition of $\mathcal{X}_+$ and $\mathcal{X}_-$. Then we have

$$\mathrm{CH}(\mathcal{X}_+^i) \cap \mathrm{CH}(\mathcal{X}_-^j) \neq \emptyset, \ \forall\, i \in [L_1], j \in [L_2] \tag{27}$$

which implies that

$$\mathcal{X}_- \cap \mathrm{CH}(\mathcal{X}_+^i) \neq \emptyset, \ \forall\, i \in [L_1]. \tag{28}$$

Apply Lemma 1 on the pair of pattern sets, $\mathcal{X}_-$ and $\mathcal{X}_+^i$ (corresponding to the positive pattern set and the negative pattern set respectively), for each i separately, there exists a sign-constrained single-hidden-layer separator between $\mathcal{X}_-$ and $\mathcal{X}_+^i$. More precisely, there exist $\mathbf{w}_i \succeq 0, b_i \leq 0, V, \mathbf{c}$ such that

$$\begin{aligned} g_i(\mathbf{x}) > 0, \ \forall \mathbf{x} \in \mathcal{X}_- \\ g_i(\mathbf{x}) < 0, \ \forall \mathbf{x} \in \mathcal{X}_+^i \end{aligned} \tag{29}$$

where

$$g_i(\mathbf{x}) \triangleq \mathbf{w}_i^T \max(0, V^T \mathbf{x} + \mathbf{c}) + b_i. \tag{30}$$

Note that $\mathcal{X}_-$ is treated as the pattern set with positive labels when applying Lemma 1, and $\mathcal{X}_+^i$ is treated as the pattern set with negative labels correspondingly.

Let $W = [\mathbf{w}_1, \mathbf{w}_2, \cdots, \mathbf{w}_{L_1}], \mathbf{b} = [b_1, b_2, \cdots, b_{L_1}]^T$ and consider the transformation

$$\mathbf{z} = G(\mathbf{x}) \triangleq \max(0, -W^T \max(0, V^T \mathbf{x} + \mathbf{c}) - \mathbf{b}) \tag{31}$$

where $-W$ and $-\mathbf{b}$, instead of W and $\mathbf{b}$, are used in the above transformation so that the responses of the negative patterns in $\mathcal{X}_-$ are $\mathbf{0}$.

Denote

$$
\begin{aligned}
\mathcal{Z}_- &\triangleq \{\mathbf{z} : \mathbf{z} = G(\mathbf{x}), \mathbf{x} \in \mathcal{X}_-\} \\
&= \{\mathbf{0}\} \\
\mathcal{Z}_+ &\triangleq \{\mathbf{z} : \mathbf{z} = G(\mathbf{x}), \mathbf{x} \in \mathcal{X}_+\} \\
&\subset \{\mathbf{z} : \mathbf{1}^T \mathbf{z} > \gamma_{min}, z_i \geq 0, \forall i \in [L_1]; z_j > 0, \exists\, j \in [L_1]\}.
\end{aligned}
\tag{32}
$$

where

$$
\begin{aligned}
\gamma_{min} &\triangleq \min_{\mathbf{x} \in \mathcal{X}_+} \mathbf{1}^T \max(0, -W^T G(\mathbf{x}) - \mathbf{b}) \\
&> 0.
\end{aligned}
\tag{33}
$$

Let $\mathbf{a} = \frac{2}{\gamma_{min}}\mathbf{1} \succeq 0, \beta = -1$ and $f(\mathbf{z}) \triangleq \mathbf{a}^T \mathbf{z} + \beta$. Then $f(\mathbf{z}) \geq 1 > 0$ for $\mathbf{z} \in \mathcal{Z}_+$ and $f(\mathbf{z}) = -1 < 0$ for $\mathbf{z} \in \mathcal{Z}_-$, or equivalently

$$
f\{G(\mathbf{x})\} \triangleq \mathbf{a}^T \max(0, -W^T G(\mathbf{x}) - \mathbf{b}) + \beta
\tag{34}
$$

satisfies $f\{G(\mathbf{x})\} > 0$ for $\mathbf{x} \in \mathcal{X}_+$ and $f\{G(\mathbf{x})\} < 0$ for $\mathbf{x} \in \mathcal{X}_-$. Note that $-W \preceq 0, -\mathbf{b} \succeq 0, \mathbf{a} \succeq 0, \beta \leq 0$, $f\{G(\mathbf{x})\}$ is a sign-constrained two-hidden-layer binary classifier, and the proof is completed.

□

Next, we investigate the applications of the proposed two-hidden-layer SCRN to decompose one pattern set (labelled positive) into several subsets so that each subset is convexly separable from the other pattern set.

Theorem 4. *Let $\mathcal{X}_+, \mathcal{X}_-$ be two disjoint pattern sets and let $f\{G(\mathbf{x})\}$, as defined in (6) and satisfying $\mathbf{a} \succeq 0, \beta \leq 0, W \preceq 0, \mathbf{b} \succeq 0$, be one of their sign-constrained two-hidden-layer binary separators with m top hidden nodes and satisfying (7). Define*

$$
\begin{aligned}
f_{\mathcal{I}}\{G(\mathbf{x})\} &\triangleq \left(\sum_{i \in \mathcal{I}} a_i[\mathbf{w}_i^T G(\mathbf{x}) + b_i] \right) + \beta \\
\mathcal{X}_+^{\mathcal{I}} &\triangleq \{\mathbf{x} : f_{\mathcal{I}}\{G(\mathbf{x})\} > 0, \mathbf{x} \in \mathcal{X}_+\}
\end{aligned}
\tag{35}
$$

for any subset $\mathcal{I}$ in $[m]$. Then we have

$$
X_+ = \bigcup_{\mathcal{I} \subset [m]} \mathcal{X}_+^{\mathcal{I}}
\tag{36}
$$

and

$$
\mathrm{CH}\left(\mathcal{X}_+^{\mathcal{I}}\right) \cap \mathcal{X}_- = \emptyset,
\tag{37}
$$

i.e, $\mathcal{X}_+^{\mathcal{I}}$ and $\mathcal{X}_-$ are convexly separable, and furthermore, $f_{\mathcal{I}}\{G(\mathbf{x})\}$ is their single-hidden-layer separator satisfying

$$
\begin{aligned}
f_{\mathcal{I}}\{G(\mathbf{x})\} &> 0, \ \forall\, \mathbf{x} \in \mathcal{X}_+^{\mathcal{I}} \\
f_{\mathcal{I}}\{G(\mathbf{x})\} &< 0, \ \forall\, \mathbf{x} \in \mathcal{X}_-.
\end{aligned}
\tag{38}
$$

Proof: Note that $a_i \geq 0$, we have

$$f_{\mathcal{I}}\{G(\mathbf{x})\} \leq f\{G(\mathbf{x})\}, \ \forall \, \mathbf{x} \in \mathbb{R}^n \tag{39}$$

and therefore

$$f_{\mathcal{I}}\{G(\mathbf{x})\} < 0, \ \forall \, \mathbf{x} \in \mathcal{X}_- \tag{40}$$

which implies (38), together with the definition of $\mathcal{X}_+^{\mathcal{I}}$ in (35). Hence, $f_{\mathcal{I}}\{G(\mathbf{x})\}$ is a single-hidden-layer separator of $\mathcal{X}_-$ and $\mathcal{X}_+^{\mathcal{I}}$. Next, we show that $-f_{\mathcal{I}}\{G(\mathbf{x})\}$ can be described as a *sign-constrained* single-hidden-layer separator of $\mathcal{X}_-$ and $\mathcal{X}_+^{\mathcal{I}}$ if they are labelled positive and negative respectively.

Let

$$\begin{aligned} g_{\mathcal{I}}(x) &\triangleq -f_{\mathcal{I}}\{G(\mathbf{x})\} \\ &= \hat{\mathbf{a}}^T G(\mathbf{x}) + \hat{\beta}_{\mathcal{I}} \\ &= \hat{\mathbf{a}}^T \max(0, V^T \mathbf{x} + \mathbf{c}) + \hat{\beta}_{\mathcal{I}} \end{aligned} \tag{41}$$

where

$$\begin{aligned} \hat{\mathbf{a}} &= -\sum_{i \in \mathcal{I}} a_i \mathbf{w}_i \succeq 0 \\ \hat{\beta} &= -\sum_{i \in \mathcal{I}} a_i b_i - \beta. \end{aligned} \tag{42}$$

Then, from (38), it follows that

$$\begin{aligned} g_{\mathcal{I}}(\mathbf{x}) &> 0 \, , \ \forall \, \mathbf{x} \in \mathcal{X}_- \\ g_{\mathcal{I}}(\mathbf{x}) &< 0 \, , \ \forall \, \mathbf{x} \in \mathcal{X}_+^{\mathcal{I}}. \end{aligned} \tag{43}$$

Note that $\hat{\mathbf{a}} \succeq 0$. If $\mathcal{X}_+^{\mathcal{I}}$ is not empty, $\hat{\beta}_{\mathcal{I}}$ must be non-positive, and $g_{\mathcal{I}}(\mathbf{x})$ is thus a sign-constrained single-hidden-layer separator of $\mathcal{X}_-$ and $\mathcal{X}_+^{\mathcal{I}}$. Then by Lemma 1, we have (37). Note that $\mathcal{X}_-$ corresponds to the positive set while $\mathcal{X}_+^{\mathcal{I}}$ corresponds to the negative set when applying Lemma 1, with $g_{\mathcal{I}}$ being one of their sign-constrained single-hidden-layer classifiers.

Now it remains to prove (36). It suffices to prove that, for any $\mathbf{x} \in \mathcal{X}_+$, there exists $\mathcal{I} \subset [m]$ such that $\mathbf{x} \in \mathcal{X}_+^{\mathcal{I}}$. Let $\mathbf{x}$ be a member in $\mathcal{X}_+$ and let $\mathcal{I} \subset [m], \mathcal{I} \neq \emptyset$ be the index set such that $\mathbf{w}_i^T G(\mathbf{x}) + b_i > 0$ for all $i \in \mathcal{I}$ and $\mathbf{w}_i^T G(\mathbf{x}) + b_i \leq 0$ for all $i \notin \mathcal{I}$. Then $f_{\mathcal{I}}\{G(\mathbf{x})\} = f\{G(\mathbf{x})\} > 0$ and thus $\mathbf{x}$ is in $\mathcal{X}_+^{\mathcal{I}}$. $\qquad \square$

Theorem 4 states that the positive pattern set can be decomposed into several subsets by a two-hidden-layer SCRN, namely

$$\mathcal{X}_+ = \bigcup_{i=1}^{t} \mathcal{X}_+^i$$

so that each $\mathcal{X}_+^i$ is convexly separable from $\mathcal{X}_-$. Then by labelling $\mathcal{X}_-$ as positive and $\mathcal{X}_+^i$ as negative, and from Lemma 2, $\mathcal{X}_-$ can be decomposed into a number, namely t_i, of subsets by a single-hidden-layer SCRN, namely,

$$\mathcal{X}_- = \bigcup_{j=1}^{t_i} \mathcal{X}_-^j,$$

so that $\mathcal{X}_+^i$ and $\mathcal{X}_-^j$ are linearly separable. Hence, one can investigate the discriminant factors of the two patterns by using the linear classifiers of these subsets of the patterns. With the decomposed subsets, one can investigate the pattern structures. The numbers of the subsets are determined by the numbers of hidden nodes in the top hidden layers of the two-hidden-layer SCRNs and the numbers of the hidden nodes of the single-hidden-layer SCRNs. To find compact pattern structures and meaningful discriminant factors, small number of hidden nodes is preferable. Since the convex hulls of the subsets can be represented by the convex hull of a small set of boundary points, and the interior points are not crucial for training rectifier networks (similar to the training points other than support vectors in support vector machine (SVM) training [13, 16]), one can use the proposed SCRN to find these non-crucial training patterns, and by removing them to speed up the training of the unconstrained rectifier networks. One can also use SVMs to separate the decomposed subsets of the patterns and identify the outliers, and by removing them to improve classification accuracy.

For multiple classes of pattern sets, multiple times of SCRN training and analysis are required, each time labelling one class positive and all the other classes negative.

6   Discussion

In this paper, we have shown that, with sign constraints on the weights of the output layer and on those of the top hidden layer, two-hidden-layer SCRNs are capable of separating any two finite training patterns as well as decomposing one of them into several subsets so that each subset is convexly separable from the other pattern set; and single-hidden-layer SCRNs are capable of separating any pair of convexly separable pattern sets as well as decomposing one of them into several subsets so that each subset is linearly separable from the other pattern set.

The proposed SCRN not only enables pattern and feature analysis for knowledge discovery but also provides insights on how to improve the efficiency and accuracy of training the rectifier network classifiers. Potential applications include: 1) Feature analysis such as in health and production management of precision livestock farming[17], where one needs to identify the key features associated with diseases (e.g. hock burn of broiler chickens) on commercial farms, using routinely collected farm management data [5]. 2). User-supervised neural network training. Since each decomposed subset of the pattern sets via the proposed SCRN is convexly separable from the other pattern set, one can visualize the clusters of each pattern set, identify the outliers, and check the separating boundaries. By removing the outliers and adjusting the boundaries, users could manage to improve the training and/or validation performance. By removing the non-crucial points (i.e., interior points) of the convex hull of each decomposed subset, users could speed up the training especially for big size training data.

Related Works: This work on the universal classification power of the proposed SCRN is related to [7, 9, 11], which address the universal approximation power of deep neural networks for functions or for probability distributions. This work is also closely related to [1], which proves that any multiple pattern sets can be transformed to be linearly separable by two hidden layers, with additional distance preserving properties. In this paper, we prove that any two disjoint pattern sets can be separated by a two-hidden-layer rectifier network with additional sign constraints on the weights of the output layer and on those of the top hidden layer. The significance of SCRN lies in the fact that, through two hidden layers, it can decompose one of any pair of pattern sets into several subsets so that each subset is convexly separable from the entire other pattern set; and through a single hidden layer, it can decompose one of the convexly separable pair of pattern sets into several subsets so that each subset is *linearly* separable from the entire other pattern set. This decomposition can be used to analyse pattern sets and identify the discriminative features for different patterns. Our effort to make the deep rectifier network interpretable is related to [18], which manages to visualize and understand the learnt convolutional neural network features. The visualization technique developed in [18] has been shown useful to improve classification accuracy as evidenced by the state-of-the-art performance in object classification.

Limitations and Future Works: This paper focuses on the theoretical development of interpretable rectifier networks aiming to understand the learnt hidden layers and pattern structures. Future works are required to develop learning algorithms for SCRN and demonstrate their applications to image classification on popular databases.

Acknowledgements. This work was supported by the ARC grants DP150100294, DP150104251, DE120102960 and a UWA FECM grant.

References

1. An, S., Boussaid, F., Bennamoun, M.: How can deep rectifier networks achieve linear separability and preserve distances? In: Proceedings of The 32nd International Conference on Machine Learning, pp. 514–523 (2015)
2. Ciresan, D., Meier, U., Schmidhuber, J.: Multi-column deep neural networks for image classification. In: 2012 IEEE Conference on Computer Vision and Pattern Recognition (CVPR), pp. 3642–3649. IEEE (2012)
3. Deng, L., Li, J., Huang, J.T., Yao, K., Yu, D., Seide, F., Seltzer, M., Zweig, G., He, X., Williams, J., et al.: Recent advances in deep learning for speech research at Microsoft. In: 2013 IEEE International Conference on Acoustics, Speech and Signal Processing (ICASSP), pp. 8604–8608. IEEE (2013)
4. He, K., Zhang, X., Ren, S., Sun, J.: Delving deep into rectifiers: Surpassing human-level performance on ImageNet classification. arXiv preprint arXiv:1502.01852 (2015)
5. Hepworth, P.J., Nefedov, A.V., Muchnik, I.B., Morgan, K.L.: Broiler chickens can benefit from machine learning: support vector machine analysis of observational epidemiological data. Journal of the Royal Society Interface **9**(73), 1934–1942 (2012)

6. Hinton, G., Deng, L., Yu, D., Dahl, G.E., Mohamed, A.r., Jaitly, N., Senior, A., Vanhoucke, V., Nguyen, P., Sainath, T.N., et al.: Deep neural networks for acoustic modeling in speech recognition: The shared views of four research groups. IEEE Signal Processing Magazine 29(6), 82–97 (2012)

7. Hornik, K., Stinchcombe, M., White, H.: Multilayer feedforward networks are universal approximators. Neural Networks 2(5), 359–366 (1989)

8. Krizhevsky, A., Sutskever, I., Hinton, G.E.: ImageNet classification with deep convolutional neural networks. In: Advances in Neural Information Processing Systems, pp. 1097–1105 (2012)

9. Le Roux, N., Bengio, Y.: Deep belief networks are compact universal approximators. Neural Computation 22(8), 2192–2207 (2010)

10. Lee, C.Y., Xie, S., Gallagher, P., Zhang, Z., Tu, Z.: Deeply-supervised nets. arXiv preprint arXiv:1409.5185 (2014)

11. Montufar, G., Ay, N.: Refinements of universal approximation results for deep belief networks and restricted Boltzmann machines. Neural Computation 23(5), 1306–1319 (2011)

12. Seide, F., Li, G., Yu, D.: Conversational speech transcription using context-dependent deep neural networks. In: Interspeech, pp. 437–440 (2011)

13. Shawe-Taylor, J., Cristianini, N.: Kernel methods for pattern analysis. Cambridge University Press (2004)

14. Sun, Y., Chen, Y., Wang, X., Tang, X.: Deep learning face representation by joint identification-verification. In: Advances in Neural Information Processing Systems, pp. 1988–1996 (2014)

15. Taigman, Y., Yang, M., Ranzato, M., Wolf, L.: Deepface: Closing the gap to human-level performance in face verification. In: 2014 IEEE Conference on Computer Vision and Pattern Recognition (CVPR), pp. 1701–1708. IEEE (2014)

16. Vapnik, V.N., Vapnik, V.: Statistical learning theory, vol. 2. Wiley, New York (1998)

17. Wathes, C., Kristensen, H.H., Aerts, J.M., Berckmans, D.: Is precision livestock farming an engineer's daydream or nightmare, an animal's friend or foe, and a farmer's panacea or pitfall? Computers and Electronics in Agriculture 64(1), 2–10 (2008)

18. Zeiler, M.D., Fergus, R.: Visualizing and understanding convolutional networks. In: Fleet, D., Pajdla, T., Schiele, B., Tuytelaars, T. (eds.) ECCV 2014, Part I. LNCS, vol. 8689, pp. 818–833. Springer, Heidelberg (2014)

Aggregation Under Bias: Rényi Divergence Aggregation and Its Implementation via Machine Learning Markets

Amos J. Storkey[✉], Zhanxing Zhu, and Jinli Hu

Institute of Adaptive Neural Computation, School of Informatics,
The University of Edinburgh, Edinburgh EH8 9AB, UK
a.storkey@ed.ac.uk

Abstract. Trading in information markets, such as machine learning markets, has been shown to be an effective approach for aggregating the beliefs of different agents. In a machine learning context, aggregation commonly uses forms of linear opinion pools, or logarithmic (log) opinion pools. It is interesting to relate information market aggregation to the machine learning setting.

In this paper we introduce a spectrum of compositional methods, Rényi divergence aggregators, that interpolate between log opinion pools and linear opinion pools. We show that these compositional methods are maximum entropy distributions for aggregating information from agents subject to individual biases, with the Rényi divergence parameter dependent on the bias. In the limit of no bias this reduces to the optimal limit of log opinion pools. We demonstrate this relationship practically on both simulated and real datasets.

We then return to information markets and show that Rényi divergence aggregators are directly implemented by machine learning markets with isoelastic utilities, and so can result from autonomous self interested decision making by individuals contributing different predictors. The risk averseness of the isoelastic utility directly relates to the Rényi divergence parameter, and hence encodes how much an agent believes (s)he may be subject to an individual bias that could affect the trading outcome: if an agent believes (s)he might be acting on significantly biased information, a more risk averse isoelastic utility is warranted.

Keywords: Probabilistic model aggregation · Rényi divergence · Machine learning markets

1 Introduction

Aggregation of predictions from different agents or algorithms is becoming increasingly necessary in distributed, large scale or crowdsourced systems. Much previous focus is on aggregation of classifiers or point predictions. However, aggregation of probabilistic predictions is also of particular importance, especially where quantification of risk matters, generative models are required or

© Springer International Publishing Switzerland 2015
A. Appice et al. (Eds.): ECML PKDD 2015, Part I, LNAI 9284, pp. 560–574, 2015.
DOI: 10.1007/978-3-319-23528-8_35

where probabilistic information is critical for downstream analyses. In this paper we focus on aggregation of probability distributions (including conditional distributions).

The problem of probabilistic aggregation in machine learning can be cast as choosing a single aggregate distribution given no (or little) direct data, but given instead the beliefs of a number of independent agents. We have no control over what these agents do, other than that we know they *do* have direct access to data and we expect them to have obtained their beliefs using that data. The data the agents observe is generated from a scenario that is the same as or similar to the target scenario we care about. We wish to choose an aggregate distribution that has high log probability under data drawn from that target scenario.

One recent approach for aggregating probabilistic machine learning predictions uses information markets [18, 27, 28] as an aggregation mechanism via the market price. In a machine learning market, agents make utility maximizing decisions regarding trades in securities. These securities are tied to the random variables of the machine learning problem. For example they could be Arrow-Debreu securities defined on each possible predicted outcome. Given the trading desires of each agent, the equilibrium price in the market then defines a distribution that is an aggregation of the beliefs of different agents. Machine learning markets combine an incentivization mechanism (to ensure agents' actions reflect their beliefs P_i) and a aggregation mechanism (via the trading process).

Understanding the relationship between individual actions and the aggregate market price is an interesting open question for information markets. In addition, finding efficient methods of arriving at market equilibria is key to their practical success.

The main novel contributions of this paper are

- Introducing the class of Rényi divergence based aggregators which interpolate between linear opinion pools and log opinion pools, and showing that they are the maximum entropy estimators for aggregation of beliefs potentially subject to bias. We also demonstrate this relationship practically via simulated and real problems.
- Directly relating Rényi divergence aggregators to machine learning markets with different isoelastic utilities, and showing that the risk averseness of the isoelastic utility relates to the Rényi divergence parameter that is used to control the assumed bias.

2   Background

Aggregation methods have been studied for some time, and have been discussed in a number of contexts. Aggregation methods differ from ensemble approaches (see e.g. [9]), as the latter also involves some control over the form of the individuals within the ensemble: with aggregation, the focus is entirely on the method of combination - there is no control over the individual agent beliefs. In addition, most aggregation methods focus on aggregating hard predictions (classifications, mean predictive values etc.) [4, 10]. Some, but not all of those are suitable for aggregation of probabilistic predictions [7,20], where full predictive distributions are given. This issue has received significant attention in the context of aggregating Bayesian or probabilistic beliefs [8,19,21,27,29]. Full predictive distributions are generally useful for a Bayesian analysis (where the expected loss function is computed from the posterior predictive distribution), in situations where full risk computations must be done, or simply to get the most information from the individual algorithms. Wolpert [30] describes a general framework for aggregation, where an aggregator is trained using the individual predictions on a held out validation set as inputs, and the true validation targets as outputs. This requires specification of the aggregation function. The work in this paper fits within this framework, with Rényi mixtures as the aggregator. In crowdsourcing settings, issues of reliability in different contexts come into play. Log opinion pools have been generalized to weighted log opinion pools using Bayesian approaches with an event-specific prior [17]. This emphasises that expert models can work with aggregators at many different levels, from individual data points to whole datasets within a corpus.

Recently, prediction markets, and methods derived from securities market settings [3,5,7,18,21,27,28], have provided a particular foundation for belief aggregation. That securities markets can perform belief aggregation was first discussed by Rubinstein [23–25]. Belief aggregation of this form is of importance in crowdsourcing settings, or settings combining information from different autonomous agents. In such settings, the beliefs of different agents can be subject to various biases.

One other area that aggregation has shown importance is in machine learning competitions, including the Netflix Challenge [14], the PASCAL Visual Object Classes challenge [11]), and many challenges set in the Kaggle challenge environment [13]. Many workshops (e.g. KDD) also run a variety of machine learning challenges. One of the most consistent take-home messages from all the challenges is that aggregation of individual entries provides a performance benefit. The final winning Netflix submission was itself a large scale aggregation of 107 different methods [22].

3   Problem Statement

We will postpone the discussion of information markets and start by introducing Rényi divergence aggregators and their properties, as Rényi divergence aggregators are new to this paper. We will show that Rényi divergence aggregators are intimately related to the issue of bias in individual agent beliefs.

The problem setting is as follows. We have a prediction problem to solve, in common with a number of agents. These agents have learnt probabilistic predictors on each of their own training datasets, using their own machine learning algorithms, and provide the predictions for the test scenario. We wish to combine the agents' predictions to make the best prediction we can for our setting. We don't have access to the training data the agents see, but are potentially given the held out performance of each agent on their training data, and we may have access to their predictions for a small validation set of our own data which we know relates to our domain of interest (the distribution of which we denote by P_G). We consider the case where it may be possible that the data individual agents see are different in distribution (i.e. biased) with respect to our domain of interest.

Our objective is to minimize the negative log likelihood for a model P for future data generated from an unknown data generating distribution P^G. This can be written as desiring $\arg\min_P KL(P^G||P)$, where KL denotes the KL-Divergence. However in an aggregation scenario, we do not have direct access to data that can be used to choose a model P by a machine learning method. Instead we have access to beliefs P_i from $i = 1, 2, \ldots, N_A$ other agents, which *do* have direct access to some data, and we must use those agent beliefs P_i to form our own belief P.

We have no control over the agents' beliefs P_i, but we can expect that the agents have learnt P_i using some learning algorithm with respect to data drawn from individual data distributions P_i^G. Hence agents will choose P_i with low $KL(P_i||P_i^G)$ with respect to their individual data, drawn from P_i^G. For example agents can choose their own posterior distributions P_i with respect to the data they observe.

We also assume that each P_i^G is 'close' to the distribution P^G we care about. Where we need to be specific, we use the measure $KL(P_i^G||P^G)$ as the measure of closeness, which is appropriate if P_i^G is obtained by sample selection bias [26] from P^G. In this case $KL(P_i^G||P^G)$ gives a standardized expected log acceptance ratio, which is a measure of how the acceptance rate varies across the data distribution. Lower KL divergence means lower variation in acceptance ratio and P_i is closer to P. The simplest case is to assume $KL(P_i^G||P^G) = 0 \ \forall i$, which implies an unbiased data sample.

4   Weighted Divergence Aggregation

Weighted divergence-based aggregation was proposed in [12]. The idea was, given individual distributions P_i, to choose an aggregate distribution P given by

$$P = \arg\min_Q \sum_i w_i D(P_i, Q), \tag{1}$$

where w_i is a weight and $D(P_i, Q)$ represents a choice of divergence between P_i and Q, where $D(A, B) \geq 0$, with equality iff $A = B$. This framework generalizes several popular opinion pooling methods, e.g., linear opinion pooling when $D(P_i, Q) = KL(P_i||Q)$, and log opinion pooling when $D(P_i, Q) = KL(Q||P_i)$. Concretely, a linear opinion pool is given by $P(y|\cdot) = \sum_{j=1}^{N_A} w_j P_j(y|\cdot)$, where $w_j \geq 0 \ \forall j$ and $\sum_{j=1}^{N_A} w_j = 1$. The weight vector $\mathbf{w}$ can be chosen using maximum entropy arguments if we know the test performance of the individual models. Alternatively, w_i can be optimized by maximizing the log likelihood of a validation set with simplex constraints, or via an expectation maximization procedure. By convexity, the solution of both optimization approaches is equivalent.

By contrast, a logarithmic opinion pool is given by $P(y|\cdot) = \frac{1}{Z(\mathbf{w})} \prod_{j=1}^{N_A} P(y|\cdot)^{w_j}$ where $w_j \geq 0 \ \forall j$, where we use the $P(y|\cdot)$ notation to reflect that this applies to both conditional and unconditional distributions. The logarithmic opinion pool is more problematic to work with due to the required computation of the normalization constant, which is linear in the number of states. Again the value of $\mathbf{w}$ can be obtained using a maximum-entropy or a gradient-based optimizer. Others (see e.g. [16]) have used various approximate schemes for log opinion pools when the state space is a product space.

Weighted Divergence aggregation is very general but we need to choose a particular form of divergence. In this paper we analyse the family of Rényi divergences for weighted divergence aggregation. This choice is motivated by two facts:

- Rényi divergence aggregators satisfy maximum entropy arguments for the aggregator class under highly relevant assumptions about the biases of individual agents.
- Rényi divergence aggregators are implemented by machine learning markets, and hence can result from autonomous self interested decision making by the individuals contributing different predictors without centralized imposition. Hence this approach can *incentivize* agents to provide their best information for aggregation.

In much of the analysis that follows we will drop the conditioning (i.e. write $P(y)$ rather than $P(y|x)$) for the sake of clarity, but without loss of generality as all results follow through in the conditional setting.

4.1   Weighted Rényi Divergence Aggregation

Here we introduce the family of weighted Rényi divergence methods.

Definition 1 (Rényi Divergence). *Let* y *be a random variable taking values* $y = 1, 2, \ldots, K$. *The Rényi divergence of order* γ *($\gamma > 0$) from a distribution P to a distribution Q is defined as*

$$D_\gamma^R[P||Q] = \frac{1}{\gamma - 1} \log \left(\sum_{y=1}^K P(y)^\gamma Q(y)^{1-\gamma} \right). \tag{2}$$

The Rényi divergence has two relevant special cases: $\lim_{\gamma \to 1} (1/\gamma) D_\gamma^R(P||Q) = KL(P||Q)$, and $\lim_{\gamma \to 0} (1/\gamma) D_\gamma^R(P||Q) = KL(Q||P)$ (which can be seen via L'hôpital's rule). We assume the value for the Rényi divergence for $\gamma = 1$ is defined by $KL(P||Q)$ via analytical continuation.

Definition 2 (Weighted Rényi Divergence Aggregation). *The* weighted Rényi divergence aggregation *is a weighted divergence aggregation given by (1), where each divergence* $D(P_i, Q) = \gamma_i^{-1} D_{\gamma_i}^R[P_i||Q]$.

Note that each component i in (1) can have a Rényi divergence with an individualized parameter γ_i. Sometimes we will assume that all divergences are the same, and refer to a single $\gamma = \gamma_i \ \forall i$ used by all the components.

Properties. The following propositions outline some properties of weighted Rényi divergence aggregation.

Proposition 1. *Weighted Rényi divergence aggregation satisfies the implicit equation for $P(y)$ of*

$$P(y) = \frac{1}{Z} \sum_i w_i \gamma_i^{-1} \frac{P_i(y)^{\gamma_i} P(y)^{1-\gamma_i}}{\sum_{y'} P_i(y')^{\gamma_i} P(y')^{1-\gamma_i}} \tag{3}$$

where w_i are given non-negative weights, and $Z = Z(\{\gamma_i\}) = \sum_i w_i \gamma_i^{-1}$ is a normalisation constant, and $\{\gamma_i\}$ is the set of Rényi divergence parameters.

Proof. Outline: Use $D(P_i, Q) = \gamma_i^{-1} D_{\gamma_i}^R[P_i \| Q]$ from (2) in Equation (1), and build the Lagrangian incorporating the constraint $\sum_y Q(y) = 1$ with Lagrange multiplier Z. Use calculus of variations w.r.t. $Q(y)$ to get K equations

$$\sum_i w_i \gamma_i^{-1} \frac{P_i(y)^{\gamma_i} P(y)^{-\gamma_i}}{\sum_{y'=1}^K P_i(y')^{\gamma_i} P(y')^{1-\gamma_i}} - Z = 0 \tag{4}$$

for the optimum values of $P(y)$. Multiply each equation with $P(y)$ and find $Z = \sum_j w_j \gamma_j^{-1}$ by summing over all equations. Rearrange to obtain the result.

Proposition 2. *Weighted Rényi divergence aggregation interpolates between linear opinion pooling ($\gamma \to 1$) and log opinion pooling ($\gamma \to 0$).*

Proof. Outline: Set $\gamma_i = 1$ in (3) to obtain a standard linear opinion pool. For log opinion pool, set $\gamma_i = \gamma$, and take $\gamma \to 0$. Note (3) can be written $Z = \sum_i w_i \gamma_i^{-1} \frac{\partial}{\partial Q} D_{\gamma_i}^R[P_i \| Q]$. Using L'Hôpital's rule on each element in the sum and switching the order of differentiation $(\partial/\partial \gamma_i)(\partial/\partial Q) = (\partial/\partial Q)(\partial/\partial \gamma_i)$ gives the result.

In the next section we show that Rényi divergence aggregation provides the maximum entropy distribution for combining together agent distributions where the belief of each agent is subject to a particular form of bias. Two consequences that are worth alerting the reader to ahead of that analysis are:

1. If all agents form beliefs on data drawn from the same (unbiased) distribution then the maximum entropy distribution is of the form of a log opinion pool.
2. If all agents form beliefs on unrelated data then the maximum entropy distribution is of the form of a linear opinion pool.

5 Maximum Entropy Arguments

Consider the problem of choosing an aggregator distribution P to model an unknown target distribution P^G given a number of individual distributions P_i. These individual distributions are assumed to be learnt from data by a number of individual agents. We will assume the individual agents did not (necessarily) have access to data drawn from P^G, but instead the data seen by the individual agents was biased, and instead sampled from distribution P_i^G. In aggregating the agent beliefs, we neither know the target distribution P^G, nor any of the individual bias distributions P_i^G, but model them with P and Q_i respectively.

As far as the individual agents are concerned they train and evaluate their methods on their individual data, unconcerned that their domains were biased with respect to the domain we care about. We can think of this scenario as convergent dataset shift [26], where there is a shift from the individual training to a common test scenario. The result is that we are given information regarding the test log likelihood performance for each P_i in their own domains: $\sum_y P^G(y) \log P_i(y) = a_i$.

The individual agent data is biased, not unrelated, and so we make the assumption that the individual distributions P_i^G are related to P in some way. We assume that $KL(P_i^G \| P^G)$ is subject to some bound (and call this the nearness constraint). As mentioned in the Problem Statement this is a constraint on the standardized expected log acceptance ratio, under an assumption that P_i^G is derived from P^G via a sample selection bias.

Given this scenario, a reasonable ambition is to find maximum entropy distributions Q_i to model P_i^G that capture the performance of the individual distributions P_i, while at the same time being related via an unknown distribution P. As we know the test performance, we write this as the constraints:

$$\sum_y Q_i(y) \log P_i(y) = a_i, \tag{5}$$

The nearness constraints[1] for Q_i are written as

$$KL(Q_i \| P) \le A_i \tag{6}$$

$$\Rightarrow \sum_y Q_i(y) \log \frac{Q_i(y)}{P(y)} \le A_i \text{ for some } P. \tag{7}$$

encoding that our model Q_i for P_i^G must be near to the model P for P^G. That is the KL divergence between the two distributions must be bounded by some value A_i.

Given these constraints, the maximum entropy (minimum negative entropy) Lagrangian optimisation can be written as $\arg\min_{\{Q_i\}, P} L(\{Q_i\}, P)$, where

$$L(\{Q_i\}, P) = \sum_i \sum_y Q_i(y) \log Q_i(y) + \sum_i b_i(1 - \sum_y Q_i(y))$$

$$- \sum_i \lambda_i \left(\left[\sum_y Q_i(y) \log P_i(y) \right] - a_i \right) + c(1 - \sum_y P(y))$$

$$+ \sum_i \rho_i \left(\left[\sum_y Q_i(y) \log \frac{Q_i(y)}{P(y)} \right] - A_i + s_i \right) \tag{8}$$

---

[1] We could work with a nearness penalty of the same form rather than a nearness constraint. The resulting maximum entropy solution would be of the same form.

where s_i are slack variables $s_i \geq 0$, and ρ_i, λ_i, b_i and c are Lagrange multipliers. This minimisation chooses maximum entropy Q_i, while ensuring there is a distribution P for which the nearness constraints are met. The final two terms of (8) are normalisation constraints for Q_i and P.

Taking derivatives with respect to $Q_i(y)$ and setting to zero gives

$$Q_i(y) = \frac{1}{Z_i} P(y)^{\frac{\rho_i}{1+\rho_i}} P_i(y)^{\frac{\lambda_i}{1+\rho_i}} \tag{9}$$

where Z_i is a normalisation constant.

Given these Q_i, we can find also find an optimal, best fitting P. Taking derivatives of the Lagrangian with respect to $P(y)$ and setting to zero gives

$$P(y) = \sum_i \frac{\rho_i}{\sum_{i'} \rho_{i'}} Q_i(y) = \sum_i w_i \frac{(P_i(y)^{\lambda_i})^{\gamma_i} P(y)^{1-\gamma_i}}{Z_i} \tag{10}$$

where $w_i = \rho_i / \sum_i' \rho_{i'}$, and $\gamma_i = 1/(1 + \rho_i)$, and $Z_i = \sum_{y'} (P_i(y')^{\lambda_i})^{\gamma_i} P(y')^{1-\gamma_i}$. Comparing this with (3) we see that this form of maximum entropy distribution is equivalent to the Rényi divergence aggregator of annealed forms of P_i. The maximum entropy parameters of the aggregator could be obtained by solving for the constraints or estimated using test data from $P(y)$. Empirically we find that, if all the P_i are trained on the same data, or on data subject to sample-selection bias (rather than say an annealed form of the required distribution), then $\lambda_i \approx 1$.

Note that the parameter ρ_i controls the level of penalty there is for a mismatch between the biased distributions Q_i and the distribution P. If all the ρ_i are zero for all i then this penalty is removed and the Q_i can bear little resemblance to the P and hence to one another. In this setting (10) becomes a standard mixture and the aggregator is a linear opinion pool. If however ρ_i tends to a large value for all i, then the distributions Q_i are required to be much more similar. In this setting (10) becomes like a log opinion pool.

Interim Summary. We have shown that the Rényi divergence aggregator is not an arbitary choice of aggregating distribution. Rather it is the maximum entropy aggregating distribution when the individual agent distributions are expected to be biased using a sample selection mechanism.

6 Implementation

Renyi divergence aggregators can be implemented with direct optimization, stochastic gradient methods, or using a variational optimization for the sum of weighted divergences, which is described here. The weighted Rényi Divergence objective given by Definition 2 can be lower bounded using

$$\sum_i w_i D(P_i, Q) \geq \sum_{i,y} \frac{w_i \gamma_i}{\gamma_i - 1} Q_i(y) \log \frac{[P_i(y)^{\gamma_i} Q(y)^{1-\gamma_i}]}{Q_i(y)} \tag{11}$$

where we have introduced variational distributions Q_i, and used Jensen's inequality. Note equality is obtained in (11) for $Q_i(y) \propto P_i(y)^{\gamma_i} Q(y)^{1-\gamma_i}$. Optimizing for Q gives $P(y) = Q_{\text{opt}}(y) = \sum_i w_i^* Q_i(y)$ with $w_i^* = w_i \gamma_i^{-1} / \sum_i w_i \gamma_i^{-1}$.

This leads to an iterative variational algorithm that is guaranteed (using the same arguments as EM, and using the convexity of to optimize (11): iteratively set $Q_i(y) \propto P_i(y)^{\gamma_i} Q(y)^{1-\gamma_i}$, and then set $Q(y) \propto \sum_i w_i^* Q_i(y)$. The optimization of the parameters w_i^* also naturally fits within this framework. $Q(y)$ is a simple mixture of $Q_i(y)$. Hence given $Q_i(y)$, the optimal w_i^* are given by the optimal mixture model parameters. These can be determined using a standard inner Expectation Maximization loop. In practice, we get faster convergence if we use a single loop. First set $Q_i(y) \propto P_i(y)^{\gamma_i} Q(y)^{1-\gamma_i}$. Second compute $q_{in} = w_i^* Q_i(y_n) / \sum_i w_i^* Q_i(y_n)$. Third set $w_i^* = \sum_n q_{in} / \sum_{in} q_{in}$. Finally set $Q(y) \propto \sum_i w_i^* \gamma_i Q_i(y)$. This is repeated until convergence. All constants of proportionality are given by normalisation constraints. Note that where computing the optimal Q may be computationally prohibitive, this process also gives rise to an approximate divergence minimization approach, where Q_i is constrained to a tractable family while the optimizations for Q_i are performed.

7   Experiments

To test the practical validity of the maximum entropy arguments, the following three tasks were implemented.

Task 1: Aggregation on Simulated Sata. We aim to test the variation of the aggregator performance as the bias of the agent datasets is gradually changed. This requires that the data does not dramatically change across tests of different biases. We tested this process using a number of bias generation procedures, all with the same implication in terms of results.

The details of the data generation and testing is given in Algorithm 1. We used $N_A = 10$, $K = 64$, $N_{Va} = 100$, P^* was a discretized $N(32, 64/7)$, $f_i(y)$ $U([0,1])$ to generate the artificial data that gave the results displayed here. Equivalent results were found for all (non-trivial) parameter choices we tried, as well as using completely different data generation procedures generating biased agent data.

Task 2: Aggregation on Chords from Bach Chorales. This task aims to accurately predict distributions of chords from Bach chorales [2]. The Bach chorales data was split equally and randomly into training and test distributions. Then training data from half of the chorales was chosen to be shared across all the agents. After that each agent received additional training data from a random half of the remaining chorales. Each agent was trained using a mixture of Bernoulli's with a randomized number of mixture components between 5 and 100, and a random regularisation parameter between 0 and 1. 10 agents were used and after all 10 agents were fully trained, the Rényi mixture weights were optimized using the whole training dataset. Performance results were computed on the held out test data.

Algorithm 1. Generate test data for agents with different biases, and test aggregation methods.

Select a target discrete distribution $P^*(.)$ over K values. Choose N_A, the number of agents.

Sample IID a small number N_{Va} of values from the target distribution to get a validation set $\mathcal{D}_{Va}$

Sample IID a large number N of values $\{y_n; n = 1, 2, 3, \ldots, N\}$ from the target distribution to get the base set $\mathcal{D}$ from which agent data is generated.

Sample bias probabilities $f_i(y)$ for each agent to be used as a rejection sampler.

for annealing parameter $\beta = 0$ TO 4 **do**

    for each agent i **do**

        Anneal f_i to get $f_k^*(y) = f_k(y)^\beta./\max_y f_i(y)^\beta$.

        For each data point y_i, reject it with probability $(1 - f_k^*(y_i))$.

        Collect the first 10000 unrejected points, and set P_i to be the resulting empirical distribution.

        This defines the distribution P_i for agent i given the value of β.

    end for

    Find aggregate $P(.)$ for different aggregators given agent distributions P_i and an additional P_0 corresponding to just the uniform distribution, using the validation dataset $\mathcal{D}_{Va}$ for any parameter estimation.

    Evaluate the performance of each aggregator using the KL Divergence between the target distribution $P^*(.)$ and the aggregate distribution $P(.)$: $KL(P^*\|P)$.

end for

Algorithm 2. Competition Data Preparation

Load image data. Discretize to 64 gray scales. Put in INT8 format. Define stopping criterion ϵ

for j=1 to 140000 **do**

    Pick random image and random pixel at least 40 pixels away from edge of image and find 35×30 patch including that pixel at the bottom-middle of the patch.

    Record $x(j) =$vectorisation of all pixels in patch 'before' that pixel in patch in raster-scan terms, $y(j) =$grayscale value at chosen pixel,$i(j) =$image number

end for

Produce three Matlab datasets. Set 1: x and y and i values in one .mat for 100000 training records. Set 2: x and i values in one .mat file for 40000 test records. Set 3: y values for the corresponding test cases, not publicly available.

Task 3: Aggregation on Kaggle Competition. To analyze the use of combination methods in a realistic competition setting, we need data from an appropriate competitive setup. For this purpose we designed and ran the Kaggle-in-Class competition. The competition consisted of a critical problem in low-level image analysis: the image coding problem, which is fundamental in image compression, infilling, super-resolution and denoising. We used data consisting of images from van Hateren's Natural Image Dataset[2] [15]. The data was preprocessed using Algorithm 2 to put it in a form suitable for a Kaggle competition, and ensure

---

[2] http://bethgelab.org/datasets/vanhateren/

the data sizes were sufficient for use on student machines, and that submission files were suitable for uploading (this is the reason for the 6 bit grayscale representation).

The problem was to provide a probabilistic prediction on the next pixel y given information from previous pixels in a raster scan. The competitor's performance was measured by the perplexity on a public set at submission time, but the final ranked ordering was on a private test set. We chose as agent distributions the 269 submissions that had perplexity greater than that given by a uniform distribution and analysed the performance of a number of aggregation methods for the competition: weighted Rényi divergence aggregators, simple averaging of the top submissions (with an optimized choice of number), and a form of heuristic Bayesian model averaging, via an annealed likelihoood: $P(y|\cdot) \propto \sum_j P_j(y|\cdot) \left(P(j|\mathcal{D}_{\text{tr}})\right)^\alpha$, where α is an aggregation parameter choice. The weighted Rényi divergence aggregators were optimized using stochastic gradient methods, until the change between epochs became negligible. The validation set (20,000 pixels) is used for learning the aggregation parameters. The test set (also 20,000 pixels) is only used for the test results.

Results. For Task 1, Figure 1(a) shows the test performance on different biases for different values of $\log(\gamma_i)$ in (10), where all γ_i are taken to be identical and equal to γ. Figure 1(b) shows how the optimal value of γ changes, as the bias parameter β changes. Parameter optimization was done using a conjugate gradient method. The cost of optimization for Rényi mixtures is comparable to that of log opinion pools. For Task 2, Figure 2(a) shows the performance on the Bach chorales with 10 agents, with the implementation described in the Implementation section. Again in this real data setting, the Rényi mixtures show improved performance.

The two demonstrations show that when agents received a biased subsample of the overall data then Rényi-mixtures perform best as an aggregation method, in that they give the lowest KL divergence. As the bias increases, so the optimal value of γ increases. In the limit that the agents see almost the same data from the target distribution, Rényi-mixtures with small γ perform the best, and are indistinguishable from the $\gamma = 0$ limit. Rényi mixtures are equivalent to log opinion pools for $\gamma \to 0$.

For Task 3, all agents see unbiased data and so we would expect log opinion pools to be optimal. The perplexity values as a function of $\eta = 1/\gamma$ for all the methods tested on the test set can be seen in Figure 2(b). The parameter-based pooling methods perform better than simple averages and all forms of heuristic model averaging as these are inflexible methods. There is a significant performance benefit of using logarithmic opinion pooling over linear pooling, and weighted Rényi divergence aggregators interpolate between the two opinion pooling methods. This figure empirically supports the maximum entropy arguments.

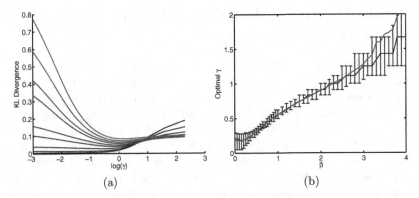

(a)   (b)

Fig. 1. (a) Task 1: Plot of the KL divergence against $\log \gamma$ for one dataset with $\beta = 0$ (lower lines, blue) through to $\beta = 4$ (upper lines, red) in steps of 0.5. Note that, unsurprisingly, more bias reduces performance. However the optimal value of γ (lowest KL), changes as β changes. for low values of β the performance of $\gamma = 0$ (log opinion pools) is barely distinguishable from other low γ values. Note that using a log opinion pool (low γ) when there is bias produces a significant hit on performance. (b) Task 1: Plot of the optimal γ (defining the form of Rényi mixture) for different values of β (determining the bias in the generated datasets for each agent). The red (upper) line is the mean, the blue line the median and the upper and lower bars indicate the 75th and 25th percentiles, all over 100 different datasets. For $\beta = 0$ (no bias) we have optimal aggregation with lower γ values, approximately corresponding to a log opinion pool. As β increases, the optimal γ gets larger, covering the full range of Rényi Mixtures.

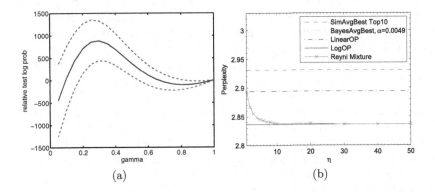

(a)   (b)

Fig. 2. (a) Task 2: test log probability results (relative to the log probability for a mixture) for the Bach chorales data for different values of γ, indicating the benefit of Rényi mixtures over linear ($\gamma = 1$) and log ($\gamma = 0$) opinion pools. Error bars are standard errors over 10 different allocations of chorales to agents prior to training. (b) Task 3: perplexity on the test set of all the compared aggregation methods against $\eta = 1/\gamma$. For each method, the best performance is plotted. Log opinion pools perform best as suggested by the maximum entropy arguments, and is statistically significantly better than the linear opinion pool($p = 8.0 \times 10^{-7}$). All methods perform better than the best individual competition entry (2.963).

8   Machine Learning Markets and Rényi Divergence Aggregation

Machine learning markets with isoelastic utilities [28] are an information market based aggregation method. Independent agents with different beliefs trade in a securities market. The equilibrium prices of the goods in that securities market can then be taken as an aggregate probability distribution, aggregating the individual agent beliefs. Following the notation and formalism in Storkey [28], agents indexed by i with belief $P_i(y)$, wealth W_i and utility function $U_i(.)$ trade in Arrow-Debreu securities derived from each possible outcome of an event. Given the agents maximize expected utility, the market equilibrium price of the securities $c(y)$ is used as an aggregate model $P(y) = c(y)$ of the agent beliefs. When each agent's utility is an isoelastic utility of the form $U_i(W) = W^{1-\eta_i}/(1 - \eta_i)$ with a risk-averseness parameter η_i, the market equilibrium $P(y)$ is implicitly given by

$$P(y) = \sum_i \frac{W_i}{\sum_l W_l} \frac{P_i(y)^{\gamma_i} P(y)^{1-\gamma_i}}{\sum_{y'} P_i(y')^{\gamma_i} P(y')^{1-\gamma_i}} \tag{12}$$

with $\gamma_i = \eta_i^{-1}$ (generalising (10) in [28]). This shows the isoelastic market aggregator linearly mixes together components that are implicitly a weighted product of the agent belief and the final solution. Simple comparison of this market equilibrium with the Rényi Divergence aggregator (3) shows that the market solution and the Rényi divergence aggregator are of exactly the same form.

We conclude that a machine learning market implicitly computes a Rényi divergence aggregation via the actions of individual agents. The process of obtaining the market equilibrium is a process for building the Rényi Divergence aggregator, and hence machine learning markets provide a method of implementation of weighted Rényi divergence aggregators. The benefit of market mechanisms for machine learning is that they are *incentivized*. There is no assumption that the individual agents behave cooperatively, or that there is an overall controller who determines agents' actions. Simply, if agents choose to maximize their utility (under myopic assumptions) then the result is weighted Rényi Divergence aggregation.

In general, equilibrium prices are not necessarily straightforward to compute, but the algorithm in the implementation section provides one such method. As this iterates computing an interim P (corresponding to a market price) and an interim Q_i corresponding to agent positions given that price, the mechanism in this paper can lead to a form of tâtonnement algorithm with a guaranteed market equilibrium – see e.g. [6].

The direct relationship between the risk averseness parameter for the isoelastic utilities and the bias controlling parameter of the Rényi mixtures ($\gamma_i = \eta_i^{-1}$) provides an interpretation of the isoelastic utility parameter: if agents know they are reasoning with respect to a biased belief, then an isoelastic utility is warranted, with a choice of risk averseness that is dependent on the bias.

In [28] the authors show, on a basket of UCI datasets, that market aggregation with agents having isoelastic utilities performs better than simple linear opinion pools (markets with log utilities) and products (markets with exponential utilities) when the data agents see is biased. As such markets implement Rényi mixtures, this provides additional evidence that Rényi mixtures are appropriate when combining biased predictors.

9 Discussion

When agents are training and optimising on different datasets than one another, log opinion pooling is no longer a maximum entropy aggregator. Instead, under certain assumptions, the weighted Rényi divergence aggregator is the maximum entropy solution, and tests confirm this practically. The weighted Rényi divergence aggregator can be implemented using isoelastic machine learning markets.

Though there is some power in providing aggregated prediction mechanisms as part of competition environments, there is the additional question of the competition mechanism itself. With the possibility of using the market-based aggregation mechanisms, it would be possible to run competitions as prediction market or collaborative scenarios [1], instead of as winner takes all competitions. This alternative changes the social dynamics of the system and the player incentives, and so it is an open problem as to the benefits of this. We recognize the importance of such an analysis as an interesting direction for future work.

References

1. Abernethy, J., Frongillo, R.: A collaborative mechanism for crowdsourcing prediction problems. In: Advances in Neural Information Processing Systems 24 (NIPS 2011) (2011)
2. Bache, K., Lichman, M.: UCI machine learning repository (2013). http://archive.ics.uci.edu/ml
3. Barbu, A., Lay, N.: An introduction to artificial prediction markets for classification (2011). arXiv:1102.1465v3
4. Breiman, L.: Bagging predictors. Machine Learning **24**(2), 123–140 (1996)
5. Chen, Y., Wortman Vaughan, J.: A new understanding of prediction markets via no-regret learning. In: Proceedings of the 11th ACM Conference on Electronic Commerce (2010)
6. Cole, R., Fleischer, L.: Fast-converging tatonnement algorithms for the market problem. Tech. rep., Dept. Computer Science. Dartmouth College (2007)
7. Dani, V., Madani, O., Pennock, D., Sanghai, S., Galebach, B.: An empirical comparison of algorithms for aggregating expert predictions. In: Proceedings of the Conference on Uncertainty in Artificial Intelligence (UAI) (2006)
8. Dietrich, F.: Bayesian group belief. Social Choice and Welfare **35**(4), 595–626 (2010)
9. Dietterich, T.G.: Ensemble methods in machine learning. In: Kittler, J., Roli, F. (eds.) MCS 2000. LNCS, vol. 1857, pp. 1–5. Springer, Heidelberg (2000)
10. Domingos, P.: Why does bagging work? a Bayesian account and its implications. In: Proceedings KDD (1997)

11. Everingham, M., et al.: The 2005 PASCAL visual object classes challenge. In: Quiñonero-Candela, J., Dagan, I., Magnini, B., d'Alché-Buc, F. (eds.) MLCW 2005. LNCS (LNAI), vol. 3944, pp. 117–176. Springer, Heidelberg (2006)
12. Garg, A., Jayram, T., Vaithyanathan, S., Zhu, H.: Generalized opinion pooling. AMAI (2004)
13. Goldbloom, A.: Data prediction competitions - far more than just a bit of fun. In: IEEE International Conference on Data Mining Workshops (2010)
14. Green, K.: The $1 million Netflix challenge. Technology Review (2006)
15. van Hateren, J.H., van der Schaaf, A.: Independent component filters of natural images compared with simple cells in primary visual cortex. Proceedings: Biological Sciences 265(1394), 359–366 (1998)
16. Heskes, T.: Selecting weighting factors in logarithmic opinion pools. In: Advances in Neural Information Processing Systems 10 (1998)
17. Kahn, J.M.: A generative Bayesian model for aggregating experts' probabilities. In: Proceedings of the 20th Conference on Uncertainty in Artificial Intelligence, pp. 301–308. AUAI Press (2004)
18. Lay, N., Barbu, A.: Supervised aggregation of classifiers using artificial prediction markets. In: Proceedings of ICML (2010)
19. Maynard-Reid, P., Chajewska, U.: Aggregating learned probabilistic beliefs. In: Proceedings of the Seventeenth Conference on Uncertainty in Artificial Intelligence, pp. 354–361. Morgan Kaufmann Publishers Inc. (2001)
20. Ottaviani, M., Sørensen, P.: Aggregation of information and beliefs in prediction markets, fRU Working Papers (2007)
21. Pennock, D., Wellman, M.: Representing aggregate belief through the competitive equilibrium of a securities market. In: Proceedings of the Thirteenth Conference on Uncertainty in Artificial Intelligence, pp. 392–400 (1997)
22. Bell, R.M., Koren, Y., Volinsky, C.: All together now: A perspective on the NETFLIX PRIZE. Chance 24 (2010)
23. Rubinstein, M.: An aggregation theorem for securities markets. Journal of Financial Economics 1(3), 225–244 (1974)
24. Rubinstein, M.: Securities market efficiency in an Arrow-Debreu economy. American Economic Review 65(5), 812–824 (1975)
25. Rubinstein, M.: The strong case for the generalised logarithmic utility model as the premier model of financial markets. Journal of Finance 31(2), 551–571 (1976)
26. Storkey, A.: When training and test sets are different: Characterising learning transfer. In: Lawrence, C.S.S. (ed.) Dataset Shift in Machine Learning, chap. 1, pp. 3–28. MIT Press (2009). http://mitpress.mit.edu/catalog/item/default.asp?ttype=2&tid=11755
27. Storkey, A.: Machine learning markets. In: Proceedings of Artificial Intelligence and Statistics, vol. 15. Journal of Machine Learning Research W&CP (2011). http://jmlr.csail.mit.edu/proceedings/papers/v15/storkey11a/storkey11a.pdf
28. Storkey, A., Millin, J., Geras, K.: Isoelastic agents and wealth updates in machine learning markets. In: Proceedings of ICML 2012 (2012)
29. West, M.: Bayesian aggregation. Journal of the Royal Statistical Society 147, 600–607 (1984)
30. Wolpert, D.H.: Stacked generalization. Neural Networks 5(2), 241–259 (1992). http://www.sciencedirect.com/science/article/pii/S0893608005800231

Distance and Metric Learning

Higher Order Fused Regularization for Supervised Learning with Grouped Parameters

Koh Takeuchi[1(✉)], Yoshinobu Kawahara[2], and Tomoharu Iwata[1]

[1] NTT Communication Science Laboratories, Kyoto, Japan
{takeuchi.koh,iwata.tomoharu}@lab.ntt.co.jp
[2] The Institute of Scientific and Industrial Research (ISIR),
Osaka University, Osaka, Japan
ykawahara@sanken.osaka-u.ac.jp

Abstract. We often encounter situations in supervised learning where there exist possibly groups that consist of more than two parameters. For example, we might work on parameters that correspond to words expressing the same meaning, music pieces in the same genre, and books released in the same year. Based on such auxiliary information, we could suppose that parameters in a group have similar roles in a problem and similar values. In this paper, we propose the Higher Order Fused (HOF) regularization that can incorporate smoothness among parameters with group structures as prior knowledge in supervised learning. We define the HOF penalty as the Lovász extension of a submodular higher-order potential function, which encourages parameters in a group to take similar estimated values when used as a regularizer. Moreover, we develop an efficient network flow algorithm for calculating the proximity operator for the regularized problem. We investigate the empirical performance of the proposed algorithm by using synthetic and real-world data.

1 Introduction

Various regularizers for supervised learning have been proposed, aiming at preventing a model from overfitting and at making estimated parameters more interpretable [1,3,16,30,31]. Least absolute shrinkage and selection operator (Lasso) [30] is one of the most well-known regularizers that employs the ℓ_1 norm over a parameter vector as a penalty. This penalty enables a sparse estimation of parameters that is robust to noise in situations with high-dimensional data. However, Lasso does not explicitly consider relationships among parameters. Recently, *structured* regularizers have been proposed to incorporate auxiliary information about structures in parameters [3]. For example, the Fused Lasso proposed in [31] can incorporate the smoothness encoded with a similarity graph defined over the parameters into its penalty.

While such a graph representation is useful to incorporate information about pairwise interactions of variables (i.e. the second-order information), we often encounter situations where there exist possibly overlapping groups that consist of more than two parameters. For example, we might work on parameters that

© Springer International Publishing Switzerland 2015
A. Appice et al. (Eds.): ECML PKDD 2015, Part I, LNAI 9284, pp. 577–593, 2015.
DOI: 10.1007/978-3-319-23528-8_36

correspond to words expressing the same meaning, music pieces in the same genre, and books released in the same year. Based on such auxiliary information, we naturally suppose that a group of parameters would provide similar functionality in a supervised learning problem and thus take similar values.

In this paper, we propose Higher Order Fused (HOF) regularization that allows us to employ such prior knowledge about the similarity on groups of parameters as a regularizer. We define the HOF penalty as the Lovász extension of a submodular higher-order potential function, which encourages parameters in a group to take similar estimated values when used as a regularizer. Our penalty has effects not only on such variations of estimated values in a group but also on supports over the groups. That is, it could detect whether a group is effective for a problem, and utilize only effective ones by solving the regularized estimation. Moreover, our penalty is robust to noise of the group structure because it encourages an effective part of parameters within the group to have the same value and allows the rest of the parameters to have different estimated values.

The HOF penalty is defined as a non-smooth convex function. Therefore, a forward-backward splitting algorithm [7] can be applied to solve the regularized problem with the HOF penalty, where the calculation of a proximity operator [22] is a key for the efficiency. Although it is not straightforward to develop an efficient way of solving the proximity operator for the HOF penalty due to its inseparable form of the HOF penalty, we develop an efficient network flow algorithm based on [14] for calculating the proximity operator.

Note that Group Lasso (GL) [34] is also known as a class of regularizers to use explicitly a group structure of parameters. However, while our HOF penalty encourages the smoothness over parameters in a group, GL imposes parameters to be sparse in a group-wise manner.

In this paper, we conduct experiments on regression with both synthetic and real-world data. In the experiments with the synthetic data, we investigate the comparative performance of our method on two settings of overlapping and non-overlapping groups. In the experiments with the real-world data, We first test the predictive performances about the average rating of each item (such as movie and book) from a set of users who watched or read items, given user demographic groups. And then, we confirm the predictive performance on a rating value from a review text given semantic and positive-negative word groups.

The rest of this paper is organized as below. In Section 2, we introduce regularized supervised learning and the forward-backward splitting algorithm. In Section 3, we propose Higher Order Fused regularizer. In Section 4, we derive a efficient flow algorithm for solving the proximity operator of HOF. In Section 5, we review related work of our method. In Section 6, we conduct experiments to compare our methods and existing regularizers. We conclude this paper and discuss future work in Section 6.

2   Regularized Supervised Learning

We denote the number of observations as N and the number of variables as M. An observed sample is denoted as $\{y_n, \boldsymbol{x}_n\}$ where $y_n \in \mathcal{Y}$ is a target value

---

Algorithm 1. Forward-backward splitting algorithm with Nesterov's acceleration

---

Initialize $\boldsymbol{\beta}_0 \in \mathbb{R}^d$, set $\boldsymbol{\zeta}_0 = \boldsymbol{\beta}_0$ and $\eta_0 = 1$.
for $t = 0, 1, \cdots$ **do**
$\quad \hat{\boldsymbol{\beta}}_t = \boldsymbol{\zeta}_t - L^{-1}\nabla l(\boldsymbol{\zeta}_t)$.
$\quad \boldsymbol{\beta}_{t+1} = \text{prox}_{L^{-1}\Omega}\, \hat{\boldsymbol{\beta}}_t$.
$\quad \eta_{t+1} = (1 + \sqrt{4\eta_t^2 + 1})/2$.
$\quad \lambda_t = 1 + (\eta_t - 1)/\eta_{t+1}$.
$\quad \boldsymbol{\zeta}_{t+1} = \boldsymbol{\beta}_t + \lambda_t(\boldsymbol{\beta}_{t+1} - \boldsymbol{\beta}_t)$.
end for

---

and $\boldsymbol{x}_n = (x_1, x_2, \cdots, x_M) \in \mathbb{R}^M$ is an explanatory variable vector. We denote a parameter vector as $\boldsymbol{\beta} = (\beta_1, \beta_2, \cdots, \beta_d) \in \mathbb{R}^d$ where d is the total number of parameters. An object function of regularized supervised learning problem is: $\mathcal{L}(\boldsymbol{\beta}) = \frac{1}{N}\sum_{n=1}^N l(\boldsymbol{\beta}; y_n, \boldsymbol{x}_n) + \gamma\Omega(\boldsymbol{\beta})$, where $l(\boldsymbol{\beta}; y_n, \boldsymbol{x}_n) : \mathbb{R}^d \to \mathbb{R}$ is an empirical risk, $\Omega(\boldsymbol{\beta}) : \mathbb{R}^d \to \mathbb{R}$ is a regularizer, and γ is a hyper parameter of the regularizer. A problem of supervised learning attains a solution: $\arg\min_{\boldsymbol{\beta}} \mathcal{L}(\boldsymbol{\beta})$. This formulation includes well-known regularized supervised learning problems such as Lasso, logistic regression [17], elastic net [36], and SVM [28].

When l is a differentiable convex function where its gradient ∇l is L-Lipschitz continuous , i.e.,

$$\left(\forall(\boldsymbol{\beta}, \hat{\boldsymbol{\beta}}) \in \mathbb{R}^d \times \mathbb{R}^d\right) \; \|\nabla l(\boldsymbol{\beta}) - \nabla l(\hat{\boldsymbol{\beta}})\|_2^2 \leq L\|\boldsymbol{\beta} - \hat{\boldsymbol{\beta}}\|_2^2, \tag{1}$$

where $L \in (0, +\infty)$. And Ω is a lower semicontinuous function whose proximity operator is provided, a minimization problem of $\mathcal{L}$ can be solved by employing the forward-backward splitting algorithm[6,7]. Its solutions are characterized by the fixed point equation.

$$\boldsymbol{\beta} = \text{prox}_{\gamma\Omega}(\boldsymbol{\beta} - \gamma\nabla l(\boldsymbol{\beta})), \tag{2}$$

where $\text{prox}_{\gamma\Omega} : \mathbb{R}^d \to \mathbb{R}^d$ is a proximity operator [6,22] for Ω and $\gamma \in (0, +\infty)$. The proximity operator utilizes the Moreau envelope [22] of the regularizer $^\gamma\Omega :$ $\mathbb{R}^d \to \mathbb{R} : \hat{\boldsymbol{\beta}} \to \min_{\boldsymbol{\beta}} \Omega(\boldsymbol{\beta}) + 1/2\gamma\|\boldsymbol{\beta} - \hat{\boldsymbol{\beta}}\|_2^2$, whose gradient is $1/\gamma$-Lipschitz continuous [5]. The forward-backward splitting algorithm is also known to the proximal gradient method. The convergence of the forward-backward splitting algorithm can achieve $O(1/t^2)$ rate by utilizing Nesterov's acceleration [25,26] (the same idea is also proposed in FISTA [4]), where t is the number of iteration counts, see Algorithm 1.

3   Higher Order Fused Regularizer

In this section, we define Higher Order Fused (HOF) regularizer through the Lovász extension of the higher order potential function, called the robust P^n potential function, and discuss the sparsity property in supervised learning with the HOF penalty.

3.1   Review of Submodular Functions and Robust P^n Potential

Let $V = \{1, 2, \ldots, d\}$. A set function $f : 2^V \to \mathbb{R}$ is called *submodular* if it satisfies:

$$f(S) + f(T) \geq f(S \cup T) + f(S \cap T), \tag{3}$$

for any $S, T \subseteq V$ [8]. A submodular function is known to be a counterpart of a convex function, which is described through a continuous relaxation of a set function called *the Lovász extension*. The Lovász extension $\hat{f} : \mathbb{R}^V \to \mathbb{R}$ of a set function f is defined as:

$$\hat{f}(\boldsymbol{\beta}) = \sum_{i=1}^{d} \beta_{j_i} \left(f(\{j_1, \ldots, j_i\}) - f(\{j_1, \ldots, j_{i-1}\}) \right), \tag{4}$$

where $j_1, j_2, \ldots, j_d \in V$ are the distinct indices corresponding to a permutation that arranges the entries of $\boldsymbol{\beta}$ in non increasing order, i.e., $\beta_{j_1} \geq \beta_{j_2} \geq \cdots \geq \beta_{j_d}$. It is known that a set function f is submodular if and only if its Lovász extension $\hat{f}$ is convex [21]. For a submodular function f with $f(\emptyset) = 0$, *the base polyhedron* is defined as:

$$B(f) = \{\mathbf{x} \in \mathbb{R}^V \mid \mathbf{x}(S) \leq f(S) \ (\forall S \subseteq V), \mathbf{x}(V) = f(V)\}. \tag{5}$$

Many problems in computer vision are formulated as the energy minimization problem, where a graph-cut function is often used as the energy for incorporating *the smoothness* in an image. A graph-cut function is known to be almost equivalent to a second order submodular function [13] (i.e., it represents a relationship between two nodes). Meanwhile, recently several higher order potentials have been considered for taking into account the smoothness among more than two. For example, Kohli et al.[18] propose the robust P^n model, which can be minimized efficiently with a network flow algorithm. Let us denote a group of indices as $g \subset V$ and a set of groups as $\mathcal{G} = \{g_1, g_2, \cdots, g_K\}$, where K is the number of groups. We denote hyper parameters that are weights of parameters in the k-th group as:

$$\mathbf{c}_0^k, \mathbf{c}_1^k \in \mathbb{R}_{\geq 0}^d, \ c_{0,i}^k = \begin{cases} c_{0,i}^k & \text{if } i \in g_k, \\ 0 & \text{otherwise} \end{cases}, \ c_{1,i}^k = \begin{cases} c_{1,i}^k & \text{if } i \in g_k, \\ 0 & \text{otherwise} \end{cases}, \ (i \in V). \tag{6}$$

The potential can be represented in the form of a set function as:

$$f_{\text{ho}}(S) = \sum_{k=1}^{K} \min \left(\theta_0^k + \mathbf{c}_0^k(V \setminus S), \ \theta_1^k + \mathbf{c}_1^k(S), \ \theta_{\max}^k \right), \tag{7}$$

where θ_0^k, θ_1^k and $\theta_{\max}^k \in \mathbb{R}_{\geq 0}$ are hyper parameters for controlling consistency of estimated parameters in the k-th group that satisfy $\theta_{\max}^k \geq \theta_0^k$, $\theta_{\max}^k \geq \theta_1^k$ and, for all $S \subset V$, $(\theta_0^k + \mathbf{c}_0^k(V \setminus S) \geq \theta_{\max}^k) \vee (\theta_1^k + \mathbf{c}_1^k(S) \geq \theta_{\max}^k) = 1$.

3.2 Definition of HOF Penalty

As mentioned in [2,32], Generalized Fused Lasso (GFL) can be obtained as the Lovász extension of a graph-cut function. This penalty, used in supervised learning, prefers parameters that take similar values if a pair of them are adjacent on a given graph, which is a similar structured property to a graph-cut function as an energy function. Now, based on an analogy with this relationship between GFL and a graph-cut function, we define our HOF penalty, which encourages parameters in a groups to take similar values, using the structural property of the higher order potential Eq. (7).

Suppose that a set of groups is given as described in the previous section. Then, we define the HOF penalty as the Lovász extension of the higher order potential Eq. (7), which is described as:

$$
\Omega_{\mathrm{ho}}(\boldsymbol{\beta}) = \sum_{k=1}^{K} \left(\sum_{i \in \{j_1, \ldots, j_{s-1}\}} (\beta_i - \beta_{j_s}) c_{1,i}^k + \beta_{j_s} (\theta_{\max}^k - \theta_1^k) \right.
$$

$$
\left. + \beta_{j_t}(\theta_0^k - \theta_{\max}^k) + \sum_{i \in \{j_{t+1}, \cdots, j_d\}} (\beta_{j_t} - \beta_i) c_{0,i}^k \right), \tag{8}
$$

where $c_0^k, c_1^k, \theta_0^k, \theta_1^k, \theta_{\max}^k$ correspond to the ones in Eq. (7) and,

$$
\begin{aligned}
j_s^k &= \min \left\{ j' \mid \theta_1^k + \textstyle\sum_{i \in \{j_1, \cdots, j'\}} c_{1,i}^k \geq \theta_{\max}^k \right\}, \\
j_t^k &= \min \left\{ j' \mid \theta_0^k + \textstyle\sum_{i \in \{j', \cdots, j_d\}} c_{0,i}^k < \theta_{\max}^k \right\}.
\end{aligned} \tag{9}
$$

The first term in Eq. (8) enforces parameters larger than β_{j_s} to have the same value of β_{j_s}. The second and third terms can be rewritten as $\theta_{\max}^k(\beta_{j_s} - \beta_{j_t}) - \beta_{j_s}\theta_1^k + \beta_{j_t}\theta_0^k$. $\theta_{\max}^k(\beta_{j_s} - \beta_{j_t})$ enforces all of parameters between β_{j_s} and β_{j_t} to have the same value because parameters are sorted by the decreasing order and $\beta_{j_s} = \beta_{j_t}$ can be satisfied if and only if all parameters between β_{j_s} and β_{j_t} have the same estimated value (see an example of parameters between s and t in Figure 1(b)). $-\beta_{j_s}\theta_1^k + \beta_{j_t}\theta_0^k$ encourages β_{j_s} and β_{j_t} to have larger and smaller estimated values, respectively. The fourth term enforces parameters smaller than β_{j_t} to have the same value of β_{j_t}. The HOF penalty is robust to noise of the group structure because it allows parameters outside of $(\beta_{j_s}, \cdots, \beta_{j_t})$ to have different estimated values and then it utilizes only an effective part of the group and discard the others.

Proposition 1. $\Omega_{\mathrm{ho}}(\boldsymbol{\beta})$ *is the Lovász extension of the higher order potential Eq. (7).*

Proof. We denote $U_i = \{j_1, \ldots, j_i\}$ and $f_{\mathrm{ho}}^k(U_i) = \min \left(\theta_0^k + c_0^k(V \setminus S),\ \theta_1^k + c_1^k(S),\ \theta_{\max}^k\right)$, then,

$$
f_{\mathrm{ho}}^k(U_i) = \begin{cases} \theta_1^k + c_1^k(U_i) & (1 \leq i < s) \\ \theta_{\max}^k & (s \leq i < t) \\ \theta_0^k + c_0^k(V \setminus U_i) & (t \leq i \leq d) \end{cases} \tag{10}
$$

and hence,

$$
\beta_{j_i}\left(f_{\mathrm{ho}}^k(U_i) - f_{\mathrm{ho}}^k(U_{i-1})\right) = \begin{cases} \beta_{j_i} c_1^k(\{j_i\}) & (1 \leq i < s) \\ \beta_{j_s}\left(\theta_{\max}^k - (\theta_1^k + c_1^k(U_{s-1}))\right) & (i = s) \\ 0 & (s < i < t) \\ \beta_{j_t}\left(\theta_0^k + c_0^k(V \setminus U_t) - \theta_{\max}^k\right) & (i = t) \\ -\beta_{j_i} c_0^k(\{j_i\}) & (t < i \leq d) \end{cases}, \quad (11)
$$

where $c_1^k(U_i) = \sum_{i \in \{j_1, \cdots, j_i\}} c_{1,i}^k$ and $c_0^k(V \setminus U_i) = \sum_{i \in \{j_{i+1}, \cdots, j_d\}} c_{0,i}^k$. As a result, we have $\Omega_{\mathrm{ho}}(\boldsymbol{\beta})$ by summing all of these from the definition of the Lovász extension Eq. (4).

Although the penalty $\Omega_{\mathrm{ho}}(\boldsymbol{\beta})$ includes many hyper parameters (such as c_0^k, c_1^k, θ_0^k, θ_1^k and $\theta_{\max}^k$), it would be convenient to use the same value for $\theta_0^k, \theta_0^k, \theta_{\max}^k$ for different $g \in \mathcal{G}$ and constant values for non-zero elements in c_0^k and c_1^k, respectively, in practice. We show an example of Eq. (10) in Figure 1(a), and parameters that minimizes the potential in Figure 1(b). As described in [1], the Lovász extension of a submodular function with $f(\emptyset) = f(V) = 0$ has the sparsity effects not only on the support of $\boldsymbol{\beta}$ but also on all sup-level set $\{\boldsymbol{\beta} \geq \alpha\}$ $(\alpha \in \mathbb{R})$.[1] A necessary condition for $S \subseteq V$ to be inseparable for the function $g : A \to f_{\mathrm{ho}}(S \cup A) - f_{\mathrm{ho}}(S)$ is that S is a set included in some unique group g_k. Thus, Ω_{ho} as a regularizer has an effect to encourage the values of parameters in a group to be close.

4   Optimization

4.1   Proximity Operator via Minimum-Norm-Point Problem

From the definition, the HOF penalty belongs to the class of the lower semicontinuous convex function but is non-smooth. To attain a solution of the penalty, we define the proximity operator as:

$$
\mathrm{prox}_{\gamma \Omega_{\mathrm{ho}}} \hat{\boldsymbol{\beta}} = \underset{\boldsymbol{\beta} \in \mathbb{R}^d}{\arg\min}\, \Omega_{\mathrm{ho}}(\boldsymbol{\beta}) + \frac{1}{2\gamma}\|\hat{\boldsymbol{\beta}} - \boldsymbol{\beta}\|_2^2, \quad (12)
$$

and we denote a solution of the proximity operator $\mathrm{prox}_{\gamma \Omega_{\mathrm{ho}}} \hat{\boldsymbol{\beta}}$ as $\boldsymbol{\beta}^*$. By plugging $\Omega_{\mathrm{ho}}(\boldsymbol{\beta}) = \max_{\boldsymbol{s} \in B(f_{\mathrm{ho}})} \boldsymbol{\beta}^{\mathrm{T}} \boldsymbol{s}$ [11] into Eq. (12), the proximity operator can be shown as the following minimization problem on a base polyhedron [32].

$$
\min_{\boldsymbol{\beta} \in \mathbb{R}^d} \Omega_{\mathrm{ho}}(\boldsymbol{\beta}) + \frac{1}{2\gamma}\|\hat{\boldsymbol{\beta}} - \boldsymbol{\beta}\|_2^2 = \min_{\boldsymbol{\beta}} \max_{\boldsymbol{s} \in B(f_{\mathrm{ho}})} \boldsymbol{\beta}^{\mathrm{T}} \boldsymbol{s} + \frac{1}{2\gamma}\|\hat{\boldsymbol{\beta}} - \boldsymbol{\beta}\|_2^2
$$

$$
= \max_{\boldsymbol{s} \in B(f_{\mathrm{ho}})} -\frac{1}{2}\|\boldsymbol{s} - \gamma^{-1}\hat{\boldsymbol{\beta}}\|_2^2 + \frac{1}{2\gamma}\|\hat{\boldsymbol{\beta}}\|_2^2 \quad \left(\because\ \arg\min_{\boldsymbol{\beta}} \boldsymbol{\beta}^{\mathrm{T}} \boldsymbol{s} + \tfrac{1}{2\gamma}\|\hat{\boldsymbol{\beta}} - \boldsymbol{\beta}\|_2^2 = \boldsymbol{\beta} - \gamma \boldsymbol{s}\right)
$$

$$
\leftrightarrow \min_{\boldsymbol{s} \in B(f_{\mathrm{ho}})} \|\boldsymbol{s} - \gamma^{-1}\hat{\boldsymbol{\beta}}\|_2^2. \quad (13)
$$

---

[1] The higher order potential $f_{\mathrm{ho}}(S)$ can be always transformed by excluding the constant terms θ_0 and θ_1 and by accordingly normalizing c_0 and c_1 respectively.

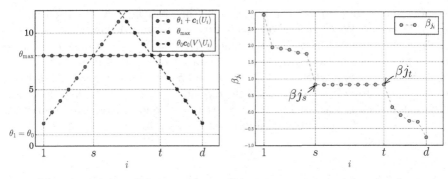

(a) values of the set functions   (b) parameters sorted by the decreasing order

Fig. 1. (a) An example of f_{ho} where $K = 1$, $c_{1,i} = c_{0,i} = 1$ $(i \in V)$, $\theta_1 = \theta_0 = 1$, and $\theta_{\max} = 8$. The horizontal and vertical axes correspond to the index of parameters and values of set functions, respectively. Red, Green, and Blue lines correspond to each lines in Eq. (10), respectively. (b) Parameters $\boldsymbol{\beta}$ sorted by the decreasing order. The horizontal and vertical axes correspond to the sorted index and values of parameters, respectively.

Let $\boldsymbol{t} = \boldsymbol{s} - \gamma^{-1}\hat{\boldsymbol{\beta}}$ and, with the basic property of the base polyhedron of a submodular function, the proximity operator goes equal to a minimal point problem,

$$\min_{\boldsymbol{s} \in B(g)} \|\boldsymbol{s} - \gamma^{-1}\hat{\boldsymbol{\beta}}\|_2^2 = \min_{\boldsymbol{t} \in B(f_{\mathrm{ho}}-\gamma^{-1}\hat{\boldsymbol{\beta}})} \|\boldsymbol{t}\|_2^2. \tag{14}$$

From the derivation, it follows that $\boldsymbol{\beta}^* = -\gamma \boldsymbol{t}^*$ where $\boldsymbol{t}^*$ is the solution of Eq. (14).

In general, the problem in Eq. (14) can be solved with submodular minimization algorithms including Minimum-Norm-Point (MNP) algorithm proposed by [12]. However, the time complexity of the fastest algorithm among existing submodular minimization algorithms is $O(d^5 EO + d^6)$, where EO is a cost for evaluating the function. Therefore, those algorithm are infeasible when the size of parameters d is large.

4.2   Network Flow Algorithm

We utilize a parametric property of MNP problem to solve the problem in Eq. (14). With this property, we can apply a parametric flow algorithm that attains the exact solution of the problem more efficiently than existing submodular minimization algorithms.

The set function $h(S) = f_{\mathrm{ho}}(S) - \hat{\beta}(S)$ in Eq. (14) is submodular because the sum of a submodular and modular functions are submodular [11]. Therefore, Eq. (14) is a special case of a minimization problem of a separable convex function under submodular constraints [23] that can be solved via parametric optimization. We denote a parameter $\alpha \in \mathbb{R}_{\geq 0}$ and define a set function $h_\alpha(S) = h(S) - \alpha \mathbf{1}(S), (\forall S \subset V)$, where $\mathbf{1}(S) = \sum_{i \in S} 1$. When h is nondecreasing submodular function, there exists a set of $r + 1$ $(\leq d)$ subsets: $S^* = \{S_0 \subset S_1 \subset \cdots \subset S_r\}$, where $S_j \subset V$, $S_0 =$, and $S_r = V$, respectively. And there are $r + 1$ subintervals Q_r of α: $Q_0 = [0, \alpha_0), Q_1 = [\alpha_1, \alpha_2), \cdots, Q_r = [\alpha_r, \infty)$, such that, for each $j \in \{0, 1, \cdots, r\}$, S_j is the unique maximal minimizer of $h_\alpha(S), \forall \alpha \in Q_j$ [23]. The optimal minimizer of Eq. (14) $t^* = (t_1^*, t_2^*, \cdots, t_d^*)$ is then determined as:

$$t_i^* = \frac{f_{\text{ho}}(S_{j+1}) - f_{\text{ho}}(S_j)}{\mathbf{1}(S_{j+1} \setminus S_j)}, \ \forall i \in (S_{j+1} \setminus S_j), \ j = (1, \cdots, r). \tag{15}$$

We introduce two lemmas from [24] to ensure that h is a non-decreasing submodular function.

Lemma 1. *For any $\eta \in \mathbb{R}$ and a submodular function h, t^* is an optimal solution to $\min_{t \in \mathcal{B}(h)} \|t\|_2^2$ if and only if $t^* - \eta \mathbf{1}$ is an optimal solution to $\min_{t \in \mathcal{B}(h) + \eta \mathbf{1}} \|t\|_2^2$.*

Lemma 2. *Set $\eta = \max_{i=1,\cdots,d}\{0, h(V \setminus \{i\}) - h(V)\}$, then $h + \eta \mathbf{1}$ is a nondecreasing submodular function.*

With Lemma 2, we solve

$$\min_{S \subset V} f_{\text{ho}}(S) - \hat{\beta}(S) + (\eta - \alpha)\mathbf{1}(S), \tag{16}$$

and then apply Lemma 1 to obtain a solution of the original problem. Because Eq. (16) is a specific form of a min cut problem, we can be solved the problem efficiently.

Theorem 1. *Problem in Eq. (16) is equivalent to a minimum s/t-cut problem defined as in Figure. 2.*

Proof. The cost function in Eq. (16) is a sum of a modular and submodular functions, because the higher order potential can be transformed as a second order submodular function. Therefore, this cost function is a $\mathcal{F}^2$ energy function [19] that is known to be "graph-representative". In Figure. 2, the groups of parameters are represented with hyper nodes u_1^k, u_0^k that correspond to each group, and capacities of edges between hyper nodes and ordinal nodes $v_i \in V$. These structures are not employed in [32]. Edges between source and sink nodes correspond to input parameters like [32]. We can attain a solution of s/t min cut problem via graph cut algorithms. We employ an efficient parametric flow algorithm provided by [14] that run in $O(d|E| \log(d^2/|E|))$ as the worst case, where $|E|$ is the number of edges of the graph in Figure 2.

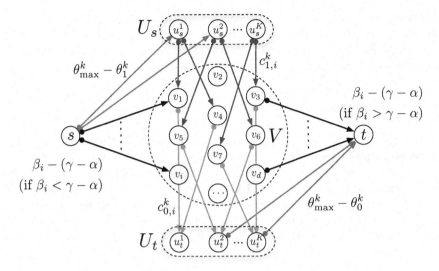

Fig. 2. A minimum s/t-cut problem of Problem 16. Given a graph $G = (V, E)$ for the HOF penalty, capacities of edges are defined as: $c(s, u_1^k) = \theta_{\max} - \theta_1$, $c(u_1^k, v_i) = c_{1,i}^k$, $c(v_i, u_0^k) = c_{0,i}^k$, $c(u_0^k, t) = \theta_{\max}^k - \theta_0^k$, $c(s, v_i) = z_i - (\gamma - \alpha)$ if $z_i > \gamma - \alpha$, and $c(v_i, t) = (\gamma - \alpha) - z_i$ if $z_i < \gamma - \alpha$. Nodes u_1^k and u_0^k, $k = (1, \cdots, K)$ are hyper nodes that correspond to the groups. And s, t, and v_i are source-node, sink-node, and nodes of parameters, respectively.

5   Related Work

Lasso [30] is one of the most well-known sparsity-inducing regularizers, which employs a sum of ℓ_1 norm of parameters as a penalty: $\Omega_{\text{Lasso}}(\boldsymbol{\beta}) = \sum_{i=1}^{d} \|\beta_i\|_1$. The penalty is often minimized by the soft-thresholding that is a proximity operator of ℓ_1 norm. Fused Lasso (FL) is a one of the structured regularizers proposed by [31] to utilize similarities of parameters. FL is also known as the total variation [27] in the field of optimization. Generalized Fused Lasso (GFL) is an extension of FL to adopt a graph structure into the structured norm. We denote a similarity between parameters i and j as $w_{i,j} \in \mathbb{R}_{\geq 0}$. Let us denote a set of edges among parameters, whose similarities are not equal to zero as $E = \{(i, j)|w_{i,j} \neq 0\}$. GFL imposes a fused penalty as: $\Omega_{\text{GFL}}(\boldsymbol{\beta}) = \sum_{(i,j) \in E} w_{i,j} \|\beta_i - \beta_j\|_1$. Because the penalty of GFL is not separable, efficient minimization for this penalty is a challenging problem. A flow algorithm for GFL was proposed by [32] that showed significant improvement on a computational time of proximity operator from existing algorithm. The computational time was reduced by transforming the minimization problem into a separable problem under a submodular constraint [23].

Group Lasso was proposed by [34] to impose a group sparsity as a ℓ_1/ℓ_2 norm on grouped parameters. The Group Lasso imposes a group-wise sparsity penalty as: $\Omega_{\text{GL}}(\boldsymbol{\beta}) = \sum_{k=1}^{K} \|\boldsymbol{\beta}(g_k)\|_2^2$. The penalty works as a group-wise Lasso that selects feature groups effective to a problem. Group Lasso has been extended to groups having overlaps, and efficient calculations of the proximity operator were proposed in [15,33,35]. [9] proposed the Sparse Group Lasso (SGL) which combines both the $\ell 1$ norm and $\ell 1/\ell 2$ norm imposed it as a regularizer. Group Lasso was extended to groups having overlap by [35].

6   Experiments

In this section, we compared our proposed method with existing methods on linear regression problems[2] using both synthetic and real-world data. We employed the ordinary least squares (OLS), Lasso[3], Sparse Group Lasso (SGL) [20], and Generalized Fused Lasso (GFL) as comparison methods. We added the ℓ_1 penalty of Lasso to GFL and our proposed method by utilizing a property: $\text{prox}_{\Omega_{\text{Lasso}}+\Omega} = \text{prox}_{\Omega_{\text{Lasso}}} \circ \text{prox}_{\Omega}$ [10]. With GFL, we encoded groups of parameters by constructing cliques that connect edges between whole pairs of parameters in the group.

6.1   Synthetic Data

We conducted regression experiments with arfically synthesized data. We employed two settings in which parameters had different group structures. In the first setting, parameters had five non-overlapping groups. In the second setting, groups were overlapped.

With the first non-overlapping groups setting, we set the true parameters of features within the group to the same value. With the second overlapping groups setting, we set the true parameters of features having no overlap to the same value, and those of features belonging to two groups to a value of either of the two groups. The explanatory variables $x_{n,i}$ were randomly generated with the Gaussian distribution with mean 0 and variance 1. Then, we obtained target values y from the Gaussian distribution where its mean and variance are $\sum_{i=1}^{d} \beta_i x_{n,i}$ and 5, respectively. The size of the feature dimension D was 100 and the number of observed data points N was $30, 50, 70, 100,$ and 150. Hyper parameters were selected by 10-fold cross validation. The hyper parameters of regularizers γ were selected from $\{0.0001, 0.001, 0.01, 0.1, 1.0, 10.0\}$. θ_{max}^k was selected from $0.01, 0.1,$ and 1.0. $c_{0,i}^k$ and $c_{1,i}^k$ were set to have the same value that was selected from 1.0 and 10.0. θ_0^k and θ_1^k were set to 0. We employed the following Root Mean Squared Error (RMSE) on the test data to evaluate the performances: $\sqrt{\frac{1}{N} \sum_{n=1}^{N} \|y_n - \hat{y}_n\|_2^2}$.

The results are summarized in Table 1. In the first setting with non-overlapping groups, our proposed method and GFL showed superior performances than SGL, Lasso, and OLS. Errors of our proposed method and GFL were almost similar. The SGL and Lasso fell in low performances since these

---

[2] Where the number of variables and features are equal $(m = d)$.

[3] We used matlab built-in codes of OLS and Lasso.

Table 1. Average RMSE and their standard deviations with synthetic data. Hyper parameters were selected from 10-fold cross validation. Values in bold typeface are statistically better ($p < 0.01$) than those in normal typeface as indicated by a paired t-test.

(a) *non-overlapping* groups

N	Proposed	SGL	GFL	Lasso	OLS
30	**0.58 ± 0.32**	174.40 ± 75.60	**0.48 ± 0.29**	189.90 ± 74.50	208.80 ± 119.00
50	**0.56 ± 0.14**	119.70 ± 40.80	**0.57 ± 0.14**	115.30 ± 54.10	260.40 ± 68.80
70	**0.40 ± 0.19**	128.10 ± 39.90	**0.40 ± 0.19**	125.00 ± 48.10	313.20 ± 42.40
100	**0.47 ± 0.13**	120.40 ± 42.00	**0.47 ± 0.13**	112.80 ± 45.60	177.10 ± 68.90
150	**0.51 ± 0.08**	106.80 ± 22.00	**0.51 ± 0.08**	79.40 ± 20.90	1.08 ± 0.13

(b) *overlapping* groups

N	Proposed	SGL	GFL	Lasso	OLS
30	**84.40 ± 76.40**	156.20 ± 64.20	173.50 ± 67.30	162.10 ± 97.10	187.70 ± 108.30
50	**40.90 ± 11.30**	108.60 ± 43.80	103.20 ± 27.10	122.80 ± 57.70	246.40 ± 70.50
70	**9.95 ± 9.22**	119.40 ± 36.20	138.40 ± 54.10	138.80 ± 44.20	317.80 ± 36.60
100	**3.19 ± 6.15**	115.70 ± 38.20	149.20 ± 28.90	101.50 ± 37.70	208.50 ± 76.30
150	**0.53 ± 0.06**	104.50 ± 15.50	135.30 ± 21.00	12.30 ± 4.93	1.08 ± 0.13

methods had no ability to fuse parameters. In the second setting with overlapping groups, our proposed method showed superior performance than SGL, GFL, Lasso, and OLS. When $N < D$, existing methods suffered from overfitting; however, our proposed method showed small errors even if $N = 30$. GFL showed low performance in this setting because the graph cannot represents groups.

Examples of estimated parameters on an experiment ($N = 30$) are shown in Figures 3 and 4. In this situation ($N < D$), the number of observation was less than the number of features; therefore, the problems of parameter estimation became undetermined system problems. From Figure 3, we confirmed that our proposed method and GFL successfully recovered the true parameters by utilizing the group structure. From Figure 4, we confirmed that our proposed methods were able to recover true parameters with overlapping groups. This is because our proposed method can represent overlapping groups appropriately. GFL fell into an imperfect result because it employed the pairwise representation that cannot describe groups.

6.2 Real-World Data

We conducted two settings of experiments with real-world data sets. With the first setting, we predicted the average rating of each item (movie or book) from a set of users who watched or read items. We used publicly available real-world data provided by MovieLens100k, EachMovie, and Book-Crossing[4]. We utilized a group structure of users, for example; age, gender, occupation and country as auxiliary

---

[4] http://grouplens.org

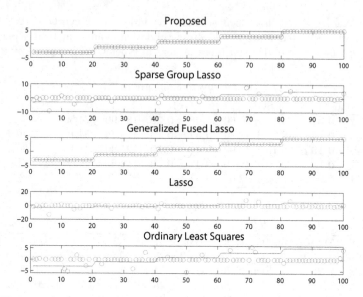

Fig. 3. Estimated parameters from synthetic data with five *non-overlapping* groups. Circles and Blue lines correspond to estimated and true parameter values, respectively.

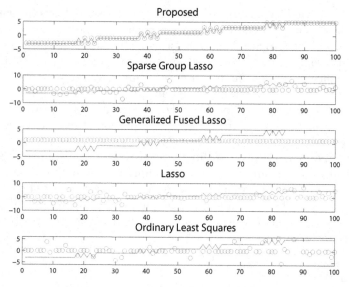

Fig. 4. Estimated parameters from synthetic data with five *overlapping* groups. Circles and Blue lines correspond to estimated and true parameter values, respectively.

information. The MovieLens100k data contained movie rating records with three types groups including ages, genders and occupations. The EachMovie data consisted of movie rating records with two types groups including ages and genders.

Table 2. Summaries of real-world data. N_{all}, D and G correspond to a total number of observations, a dimension of features, and a total number of groups, respectively.

	N_{all}	D	K	types of groups
MovieLens100k	1,620	942	31	8 age, 2 gender, 21 occupation
EachMovie	1,623	1,000	21	11 age, 2 gender
Book-Crossing	1,835	1,275	41	12 age , 29 country

We used the 1,000 most frequently watching users. The Book-Crossing data was made up of book rating records with two types of groups including ages and countries. We eliminated users and books whose total reading counts were less than 30 from the Book-Crossing data. Summaries of real-world data are shown in Table 2. To check the performance of each method, we changed the number of training data N. $c_{0,i}^k$ and $c_{1,i}^k$ were set to have the same value that was 1.0 if the i-th item belonged to the k-th group or 0.0 otherwise. In each experiment, other hyper parameters were selected by 10-fold cross validation in the same manner as previous experiments.

The results are summarized in Table 3. With the MovieLens100k data, our proposed method showed the best performance on whole settings of N because it was able to utilize groups as auxiliary information for parameter estimations. When $N = 1600$, SGL and GFL also showed competitive performance. With the EachMovie and Book-Crossing data, we confirmed that our proposed model showed the best performance. SGL and Lasso showed competitive performance on some settings of N. With the EachMovie and Book-Crossing data sets, estimated parameters were almost sparse therefore SGL and Lasso showed competitive performance.

Next, we conducted another type of an experiment employing the Yelp data[5]. The task of this experiment was to predict a rating value from a review text. We randomly extracted reviews and used the 1,000 most frequently occurred words, where stop words were eliminated by using a list[6]. We employed two types of groups of words. We attained 50 semantic groups of words by applying k-means to semantic vectors of words. The semantic vectors were learned form the GoogleNews data by word2vec[7]. We also utilized a positive-negative word dictionary[8] to construct two positive and negative groups of words [29]. Other settings were set to be the same as the MovieLens100k data.

The results are shown in Table 4. We confirmed that our proposed method showed significant improvements over other existing methods with the Yelp data. GFL also showed competitive performance when the number of training data $N = 1,000$. The semantic groups constructed by k-means have no overlap and overlap was only appeared between semantic and positive-negative groups. When

---

[5] http://www.yelp.com/dataset_challenge

[6] https://code.google.com/p/stop-words

[7] https://code.google.com/p/word2vec/

[8] http://www.lr.pi.titech.ac.jp/~takamura/pndic_en.html

Table 3. Average RMSE and their standard deviations with real-world data sets. Hyper Parameters were selected from 10-fold cross validation. Values in bold typeface are statistically better ($p < 0.01$) than those in normal typeface as indicated by a paired t-test.

(a) MovieLens100k

N	Proposed	SGL	GFL	Lasso	OLS
200	**0.30 ± 0.02**	0.32 ± 0.02	0.33 ± 0.02	1.11 ± 0.71	3.81 ± 1.97
400	**0.28 ± 0.02**	0.32 ± 0.02	0.33 ± 0.02	0.82 ± 0.31	2718.00 ± 6575.00
800	**0.27 ± 0.02**	0.31 ± 0.02	0.33 ± 0.02	0.54 ± 0.21	134144.00 ± 370452.00
1200	**0.27 ± 0.03**	0.32 ± 0.03	0.33 ± 0.03	0.48 ± 0.31	4.19 ± 2.97
1600	**0.27 ± 0.07**	**0.30 ± 0.09**	**0.31 ± 0.09**	0.44 ± 0.45	1.01 ± 0.81

(b) EachMovie

N	Proposed	SGL	GFL	Lasso	OLS
200	**0.86 ± 0.03**	**0.86 ± 0.02**	0.92 ± 0.02	1.24 ± 0.15	2.15 ± 1.17
400	**0.83 ± 0.03**	0.85 ± 0.02	0.90 ± 0.03	1.17 ± 0.09	3.20 ± 1.66
800	**0.81 ± 0.03**	0.84 ± 0.02	0.89 ± 0.03	1.09 ± 0.06	14.30 ± 14.70
1200	**0.80 ± 0.05**	0.84 ± 0.05	0.88 ± 0.05	1.06 ± 0.07	2479.00 ± 9684.00
1500	**0.79 ± 0.09**	**0.83 ± 0.07**	0.87 ± 0.09	1.01 ± 0.12	29.90 ± 29.60

(c) Book-Crossing

N	Proposed	SGL	GFL	Lasso	OLS
200	**0.71 ± 0.02**	0.73 ± 0.02	0.82 ± 0.02	0.92 ± 0.14	3.98 ± 0.83
400	**0.70 ± 0.02**	**0.72 ± 0.02**	0.82 ± 0.02	0.79 ± 0.03	66.60 ± 109.20
800	**0.68 ± 0.02**	0.70 ± 0.02	0.81 ± 0.02	0.71 ± 0.02	34.00 ± 27.70
1200	**0.67 ± 0.04**	0.71 ± 0.04	0.82 ± 0.04	**0.70 ± 0.03**	551.00 ± 1532.00
1700	**0.64 ± 0.07**	**0.68 ± 0.07**	0.78 ± 0.08	**0.66 ± 0.06**	1.18 ± 0.12

N is small, words having overlap scarcely appeared in review texts. Therefore, GFL showed competitive performance.

We show estimated parameters of four semantic groups in Figure 5. Colors of words corresponded to a sign of an estimated parameter value. Blue corresponds to the plus (positive) value and red corresponds to minus (negative) value. The size of words indicates absolute values of an estimated parameter value. As we have explained in Section 3.2, to make our proposed method robust, our proposed method is designed to allow inconsistency of estimated values within a group. This effect was confirmed by those illustrations. In the top two figures, parameters of words attained almost the same values. On the other hand, in the bottom two figures, parameters of words attained different estimated signs and absolute values. We supposed that the first two semantic groups of words were fitted for this regression problem. Therefore, consistency of estimated values was high. Whereas, the second two semantic groups of words were not fitted, and then resulted in low consistency of estimated values. Those results indicated that our proposed method was able to detect effective groups of words from given overlapping groups with Yelp data.

Table 4. Linear regression problems with Yelp data ($D = 1,000$) and $G = 52$ (50 semantic groups and two positive and negative groups). Means and standard deviations of the loss on the test data are shown. Parameters were selected from 10-fold cross validation. Bold font corresponds to significant difference of t-test ($p < 0.01$).

N	Proposed	SGL	GFL	Lasso	OLS
1000	**1.23 $\pm$ 0.02**	3.31 $\pm$ 0.20	**1.24 $\pm$ 0.01**	1.62 $\pm$ 0.11	135.60 $\pm$ 211.00
2000	**1.20 $\pm$ 0.02**	1.58 $\pm$ 0.05	1.23 $\pm$ 0.01	1.27 $\pm$ 0.02	1.61 $\pm$ 0.06
3000	**1.13 $\pm$ 0.02**	1.34 $\pm$ 0.07	1.22 $\pm$ 0.01	1.18 $\pm$ 0.03	1.35 $\pm$ 0.07
5000	**1.10 $\pm$ 0.02**	1.18 $\pm$ 0.03	1.22 $\pm$ 0.01	1.12 $\pm$ 0.02	1.18 $\pm$ 0.03

(a) Positive group.   (b) Negative group.   (c) Positive dominant group.   (d) Negative dominant group.

Fig. 5. Estimated parameters of four semantic groups of words. Blue and Red correspond to plus and minus of estimated parameters, respectively. The size of words correspond to the absolute values of estimated parameters.

7 Conclusion

We proposed a structured regularizer named Higher Order Fused (HOF) regularization in this paper. HOF regularizer exploits groups of parameters as a penalty in regularized supervised learning. We defined the HOF penalty as a Lovástz extension of a robust higher order potential named the robust P^n potential. Because the penalty is non-smooth and non-separable convex function, we provided the proximity operator of the HOF penalty. We also derived a flow algorithm to calculate the proximity operator efficiently, by showing that the robust P^n potential is graph-representative. We examined experiments of linear regression problems with both synthetic and real-world data and confirmed that our proposed method showed significantly higher performance than existing structured regularizers. We also showed that our proposed method can incorporate groups properly by utilizing the robust higher-order representation.

We provided the proximity operator of the HOF penalty but only adopted it to linear regression problems in this paper. We can apply the HOF penalty to other supervised or unsupervised learning problems including classification and learning to rank, and also to other applicational fields including signal processing and relational data analysis.

Acknowledgements. This work was partially supported by JSPS KAKENHI Grant Numbers 14435225 and 14500801.

References

1. Bach, F.R.: Structured sparsity-inducing norms through submodular functions. In: Proc. of NIPS, pp. 118–126 (2010)
2. Bach, F.R.: Shaping level sets with submodular functions. In: Proc. of NIPS, pp. 10–18 (2011)
3. Bach, F.R., Jenatton, R., Mairal, J., Obozinski, G.: Structured sparsity through convex optimization. Statistical Science **27**(4), 450–468 (2012)
4. Beck, A., Teboulle, M.: A fast iterative shrinkage-thresholding algorithm for linear inverse problems. SIAM Journal on Imaging Sciences **2**(1), 183–202 (2009)
5. Chaux, C., Combettes, P.L., Pesquet, J.C., Wajs, V.R.: A variational formulation for frame-based inverse problems. Inverse Problems **23**(4), 1495 (2007)
6. Combettes, P.L., Pesquet, J.C.: Proximal splitting methods in signal processing. In: Fixed-Point Algorithms for Inverse Problems in Science and Engineering, pp. 185–212. Springer (2011)
7. Combettes, P.L., Wajs, V.R.: Signal recovery by proximal forward-backward splitting. Multiscale Modeling & Simulation **4**(4), 1168–1200 (2005)
8. Edmonds, J.: Submodular functions, matroids, and certain polyhedra. In: Combinatorial Structures and their Applications, pp. 69–87 (1970)
9. Friedman, J., Hastie, T., Tibshirani, R.: A note on the group lasso and a sparse group lasso. arXiv preprint arXiv:1001.0736 (2010)
10. Friedman, J., Hastie, T., Höfling, H., Tibshirani, R., et al.: Pathwise coordinate optimization. The Annals of Applied Statistics **1**(2), 302–332 (2007)
11. Fujishige, S.: Submodular functions and optimization, vol. 58. Elsevier (2005)
12. Fujishige, S., Hayashi, T., Isotani, S.: The minimum-norm-point algorithm applied to submodular function minimization and linear programming. Technical report, Research Institute for Mathematical Sciences Preprint RIMS-1571, Kyoto University, Kyoto, Japan (2006)
13. Fujishige, S., Patkar, S.B.: Realization of set functions as cut functions of graphs and hypergraphs. Discrete Mathematics **226**(1), 199–210 (2001)
14. Gallo, G., Grigoriadis, M.D., Tarjan, R.E.: A fast parametric maximum flow algorithm and applications. SIAM Journal on Computing **18**(1), 30–55 (1989)
15. Jacob, L., Obozinski, G., Vert, J.P.: Group lasso with overlap and graph lasso. In: Proc. of ICML, pp. 433–440 (2009)
16. Jenatton, R., Audibert, J.Y., Bach, F.: Structured variable selection with sparsity-inducing norms. The Journal of Machine Learning Research **12**, 2777–2824 (2011)
17. Koh, K., Kim, S.J., Boyd, S.P.: An interior-point method for large-scale l1-regularized logistic regression. Journal of Machine Learning Research **8**(8), 1519–1555 (2007)
18. Kohli, P., Ladicky, L., Torr, P.H.S.: Robust higher order potentials for enforcing label consistency. International Journal of Computer Vision **82**(3), 302–324 (2009)
19. Kolmogorov, V., Zabin, R.: What energy functions can be minimized via graph cuts? IEEE Transactions on Pattern Analysis and Machine Intelligence **26**(2), 147–159 (2004)
20. Liu, J., Ji, S., Ye, J.: SLEP: Sparse Learning with Efficient Projections. Arizona State University (2009)

21. Lovász, L.: Submodular functions and convexity. In: Mathematical Programming the State of the Art, pp. 235–257. Springer (1983)
22. Moreau, J.J.: Fonctions convexes duales et points proximaux dans un espace hilbertien. CR Acad. Sci. Paris Sér. A Math. **255**, 2897–2899 (1962)
23. Nagano, K., Kawahara, Y.: Structured convex optimization under submodular constraints. In: Proc. of UAI, pp. 459–468 (2013)
24. Nagano, K., Kawahara, Y., Aihara, K.: Size-constrained submodular minimization through minimum norm base. In: Proc. of ICML, pp. 977–984 (2011)
25. Nesterov, Y.E.: A method of solving a convex programming problem with convergence rate $O(1/k^2)$. Soviet Mathematics Doklady **27**, 372–376 (1983)
26. Nesterov, Y.E.: Smooth minimization of non-smooth functions. Mathematical Programming **103**(1), 127–152 (2005)
27. Rudin, L.I., Osher, S., Fatemi, E.: Nonlinear total variation based noise removal algorithms. Physica D: Nonlinear Phenomena **60**(1), 259–268 (1992)
28. Suykens, J.A., Vandewalle, J.: Least squares support vector machine classifiers. Neural Processing Letters **9**(3), 293–300 (1999)
29. Takamura, H., Inui, T., Okumura, M.: Extracting semantic orientations of words using spin model. In: Proc. of ACL, pp. 133–140. Association for Computational Linguistics (2005)
30. Tibshirani, R.: Regression shrinkage and selection via the lasso. Journal of the Royal Statistical Society. Series B (Methodological), 267–288 (1996)
31. Tibshirani, R., Saunders, M., Rosset, S., Zhu, J., Knight, K.: Sparsity and smoothness via the fused lasso. Journal of the Royal Statistical Society: Series B (Statistical Methodology) **67**(1), 91–108 (2005)
32. Xin, B., Kawahara, Y., Wang, Y., Gao, W.: Efficient generalized fused lasso with its application to the diagnosis of alzheimers disease. In: Proc. of AAAI, pp. 2163–2169 (2014)
33. Yuan, L., Liu, J., Ye, J.: Efficient methods for overlapping group lasso. In: Proc. of NIPS, pp. 352–360 (2011)
34. Yuan, M., Lin, Y.: Model selection and estimation in regression with grouped variables. Journal of the Royal Statistical Society: Series B (Statistical Methodology) **68**(1), 49–67 (2006)
35. Zhang, X., Burger, M., Osher, S.: A unified primal-dual algorithm framework based on bregman iteration. Journal of Scientific Computing **46**(1), 20–46 (2011)
36. Zou, H., Hastie, T.: Regularization and variable selection via the elastic net. Journal of the Royal Statistical Society: Series B (Statistical Methodology) **67**(2), 301–320 (2005)

Joint Semi-supervised Similarity Learning for Linear Classification

Maria-Irina Nicolae[1,2]([✉]), Éric Gaussier[2], Amaury Habrard[1],
and Marc Sebban[1]

[1] Université Jean Monnet, Laboratoire Hubert Curien, Saint-Étienne, France
{Amaury.Habrard,Marc.Sebban}@univ-st-etienne.fr
[2] Université Grenoble Alpes, CNRS-LIG/AMA, Saint-Martin-d'Héres, France
{Irina.Nicolae,Eric.Gaussier}@imag.fr

Abstract. The importance of metrics in machine learning has attracted a growing interest for distance and similarity learning. We study here this problem in the situation where few labeled data (and potentially few unlabeled data as well) is available, a situation that arises in several practical contexts. We also provide a complete theoretical analysis of the proposed approach. It is indeed worth noting that the metric learning research field lacks theoretical guarantees that can be expected on the generalization capacity of the classifier associated to a learned metric. The theoretical framework of (ϵ, γ, τ)-good similarity functions [1] has been one of the first attempts to draw a link between the properties of a similarity function and those of a linear classifier making use of it. In this paper, we extend this theory to a method where the metric and the separator are jointly learned in a semi-supervised way, setting that has not been explored before, and provide a theoretical analysis of this joint learning via Rademacher complexity. Experiments performed on standard datasets show the benefits of our approach over state-of-the-art methods.

Keywords: Similarity learning · (ϵ, γ, τ)-good similarity · Rademacher complexity

1 Introduction

Many researchers have used the underlying geometry of the data to improve classification algorithms, *e.g.* by learning Mahanalobis distances instead of the standard Euclidean distance, thus paving the way for a new research area termed *metric learning* [5,6]. If most of these studies have based their approaches on distance learning [3,9,10,22,24], similarity learning has also attracted a growing interest [2,12,16,20], the rationale being that the cosine similarity should in some cases be preferred over the Euclidean distance. More recently, [1] have proposed a complete framework to relate similarities with a classification algorithm making use of them. This general framework, that can be applied to any

© Springer International Publishing Switzerland 2015
A. Appice et al. (Eds.): ECML PKDD 2015, Part I, LNAI 9284, pp. 594–609, 2015.
DOI: 10.1007/978-3-319-23528-8_37

bounded similarity function (potentially derived from a distance), provides generalization guarantees on a linear classifier learned from the similarity. Their algorithm does not enforce the positive definiteness constraint of the similarity, like most state-of-the-art methods. However, to enjoy such generalization guarantees, the similarity function is assumed to be known beforehand and to satisfy (ϵ, γ, τ)-goodness properties. Unfortunately, [1] do not provide any algorithm for learning such similarities. In order to overcome these limitations, [4] have explored the possibility of independently learning an (ϵ, γ, τ)-good similarity that they plug into the initial algorithm [1] to learn the linear separator. Yet the similarity learning step is done in a completely supervised way, while the setting in [1] opens the door to the use of unlabeled data.

In this paper, we aim at better exploiting the semi-supervised setting underlying the theoretical framework of [1], which is based on similarities between labeled data and unlabeled reasonable points (roughly speaking, the reasonable points play the same role as that of support vectors in SVMs). Furthermore, and unlike [4], we propose here to jointly learn the metric and the classifier, so that both the metric and the separator are learned in a semi-supervised way. To our knowledge, this approach has not been explored before in metric learning. Enforcing (ϵ, γ, τ)-goodness allows us to preserve the theoretical guarantees from [1] on the classifier in relation to the properties of the similarity. We use the Rademacher complexity to derive new generalization bounds for the joint optimization problem. Lastly, we provide an empirical study on seven datasets and compare our method to different families of supervised and semi-supervised learning algorithms.

The remainder of this paper is organized as follows: Section 2 reviews some previous results in metric and similarity learning. Section 3 reminds the theory of (ϵ, γ, τ)-good similarities and introduces our method that jointly learns the metric and the linear classifier, followed in Section 4 by generalization guarantees for our formulation. Finally, Section 5 presents an experimental study on various standard datasets.

2   Related Work

We denote vectors by lower-case bold symbols $(\mathbf{x})$ and matrices by upper-case bold symbols $(\mathbf{A})$. Consider the following learning problem: we are given access to labeled examples $\mathbf{z} = (\mathbf{x}, y)$ drawn from some unknown distribution P over $S = \mathcal{X} \times \mathcal{Y}$, where $\mathcal{X} \subseteq \mathbb{R}^d$ and $\mathcal{Y} = \{-1, 1\}$ are respectively the instance and the output spaces. A pairwise similarity function over $\mathcal{X}$ is defined as $K : \mathcal{X} \times \mathcal{X} \to [-1, 1]$, and the hinge loss as $[c]_+ = \max(0, 1 - c)$. We denote the L_1 norm by $||\cdot||_1$, the L_2 norm by $||\cdot||_2$ and the Frobenius norm by $||\cdot||_{\mathcal{F}}$.

Metric learning aims at finding the parameters of a distance or similarity function that best account for the underlying geometry of the data. This information is usually expressed as pair-based ($\mathbf{x}$ and $\mathbf{x}'$ should be (dis)similar) or triplet-based constraints ($\mathbf{x}$ should be more similar to $\mathbf{x}'$ than to $\mathbf{x}''$). Typically, the learned metric takes the form of a matrix and is the result of solving an optimization problem. The approaches that have received the most attention in this field involve learning a Mahalanobis distance, defined as $d_{\mathbf{A}}(\mathbf{x}, \mathbf{x}') = \sqrt{(\mathbf{x} - \mathbf{x}')^T \mathbf{A}(\mathbf{x} - \mathbf{x}')}$, in which the distance is parameterized by the symmetric and positive semi-definite (PSD) matrix $\mathbf{A} \in \mathbb{R}^{d \times d}$. One nice feature of this type of approaches is its interpretability: the Mahalanobis distance implicitly corresponds to computing the Euclidean distance after linearly projecting the data to a different (possibly lower) feature space. The PSD constraint on $\mathbf{A}$ ensures $d_{\mathbf{A}}$ is a pseudo metric. Note that the Mahalanobis distance reduces to the Euclidean distant when $\mathbf{A}$ is set to the identity matrix.

Mahalanobis distances were used for the first time in metric learning in [25]. In this study, they aim to learn a PSD matrix $\mathbf{A}$ as to maximize the sum of distances between dissimilar instances, while keeping the sum of distances between similar instances small. Eigenvalue decomposition procedures are used to ensure that the learned matrix is PSD, which makes the computations intractable for high-dimensional spaces. In this context, LMNN [23,24] is one of the most widely-used Mahalanobis distance learning methods. The constraints they use are pair- and triplet-based, derived from each instance's nearest neighbors. The optimization problem they solve is convex and has a special-purpose solver. The algorithm works well in practice, but is sometimes prone to overfitting due to the absence of regularization, especially when dealing with high dimensional data. Another limitation is that enforcing the PSD constraint on $\mathbf{A}$ is computationally expensive. One can partly get around this latter shortcoming by making use of specific solvers or using information-theoretic approaches, such as ITML [9]. This work was the first one to use LogDet divergence for regularization, and thus provides an easy and cheap way for ensuring that $\mathbf{A}$ is a PSD matrix. However, the learned metric $\mathbf{A}$ strongly depends on the initial value $\mathbf{A}_0$, which is an important shortcoming, as $\mathbf{A}_0$ is handpicked.

The following metric learning methods use a semi-supervised setting, in order to improve the performance through the use of unlabeled data. LRML [14,15] learns Mahalanobis distances with manifold regularization using a Laplacian matrix. Their approach is applied to image retrieval and image clustering. LRML performs particularly well compared to fully supervised methods when side information is scarce. M-DML [28] uses a similar formulation to that of LRML, with the distinction that the regularization term is a weighted sum using multiple metrics, learned over auxiliary datasets. SERAPH [19] is a semi-supervised information-theoretic approach that also learns a Mahalanobis distance. The metric is optimized to maximize the entropy over labeled similar and dissimilar pairs, and to minimize it over unlabeled data.

However, learning Mahalanobis distances faces two main limitations. The first one is that enforcing the PSD and symmetry constraints on $\mathbf{A}$, beyond the cost it induces, often rules out natural similarity functions for the problem at hand. Secondly, although one can experimentally notice that state-of-the-art Mahalanobis distance learning methods give better accuracy than using the Euclidean distance, no theoretical guarantees are provided to establish a link between the quality of the metric and the behavior of the learned classifier. In this context, [20,21] propose to learn similarities with theoretical guarantees for the kNN based algorithm making use of them, on the basis of perceptron algorithm presented in [11]. The performance of the classifier obtained is competitive with those of ITML and LMNN. More recently, [1] introduced a theory for learning with so called (ϵ, γ, τ)-good similarity functions based on non PSD matrices. This was the first stone to establish generalization guarantees for a *linear* classifier that would be learned with such similarities. Their formulation is equivalent to a relaxed L_1-norm SVM [29]. The main limitation of this approach is however that the similarity function K is predefined and [1] do not provide any learning algorithm to design (ϵ, γ, τ)-good similarities. This problem has been fixed by [4] who optimize the (ϵ, γ, τ)-goodness of a bilinear similarity function under Frobenius norm regularization. The learned metric is then used to build a good global linear classifier. Moreover, their algorithm comes with a uniform stability proof [8] which allows them to derive a bound on the generalization error of the associated classifier. However, despite good results in practice, one limitation of this framework is that it imposes to deal with strongly convex objective functions.

Recently, [13] extended the theoretical results of [4]. Using the Rademacher complexity (instead of the uniform stability) and Khinchin-type inequalities, they derive generalization bounds for similarity learning formulations that are regularized w.r.t. more general matrix-norms including the L_1 and the mixed $L_{(2,1)}$-norms. Moreover, they show that such bounds for the learned similarities can be used to upper bound the true risk of a linear SVM. The main distinction between this approach and our work is that we propose a method that jointly learns the metric and the linear separator at the same time. This allows us to make use of the semi-supervised setting presented by [1] to learn well with only a small amount of labeled data.

3   Joint Metric and Classifier Learning

In this section, we first briefly recall the (ϵ, γ, τ)-good framework [1] that we are using, prior to presenting our semi-supervised framework for jointly learning a similarity function and a linear separator from data. The (ϵ, γ, τ)-good framework is based on the following definition of a good similarity.

Definition 1. *[1] K is a (ϵ, γ, τ)-good similarity function in hinge loss for a learning problem P if there exists a random indicator function $R(\mathbf{x})$ defining a probabilistic set of "reasonable points" such that the following conditions hold:*

1. *We have*
$$\mathbb{E}_{(\mathbf{x},y)\sim P}\left[[1 - yg(\mathbf{x})/\gamma]_+\right] \leq \epsilon, \qquad (1)$$
   where $g(\mathbf{x}) = \mathbb{E}_{(\mathbf{x}',y'),R(\mathbf{x}')}[y'K(\mathbf{x},\mathbf{x}')|R(\mathbf{x}')]$.
2. *$\Pr_{\mathbf{x}'}(R(\mathbf{x}')) \geq \tau$.*

The first condition can be interpreted as having a $(1 - \epsilon)$ proportion of examples $\mathbf{x}$ on average 2γ more similar to random reasonable examples $\mathbf{x}'$ of their own label than to random reasonable examples $\mathbf{x}'$ of the other label. It also expresses the tolerated margin violations in an averaged way: this allows for more flexibility than pair- or triplet-based constraints. The second condition sets the minimum mass of reasonable points one must consider (greater than τ). Notice that no constraint is imposed on the form of the similarity function. Definition 1 is used to learn a linear separator:

Theorem 1. *[1] Let K be an (ϵ, γ, τ)-good similarity function in hinge loss for a learning problem P. For any $\epsilon_1 > 0$ and $0 < \delta < \gamma\epsilon_1/4$ let $S = \{\mathbf{x}'_1, \mathbf{x}'_2, \ldots, \mathbf{x}'_{d_u}\}$ be a sample of $d_u = \frac{2}{\tau}\left(log(2/\delta) + 16\frac{log(2/\delta)}{(\epsilon_1\gamma)^2}\right)$ landmarks drawn from P. Consider the mapping $\phi^S : \mathcal{X} \to \mathbb{R}^{d_u}$, $\phi_i^S(\mathbf{x}) = K(\mathbf{x}, \mathbf{x}'_i), i \in \{1, \ldots, d_u\}$. With probability $1 - \delta$ over the random sample S, the induced distribution $\phi^S(P)$ in $\mathbb{R}^{d_u}$, has a separator achieving hinge loss at most $\epsilon + \epsilon_1$ at margin γ.*

In other words, if K is (ϵ, γ, τ)-good according to Definition 1 and enough points are available, there exists a linear separator $\boldsymbol{\alpha}$ with error arbitrarily close to ϵ in the space ϕ^S. The procedure for finding the separator involves two steps: first using d_u potentially unlabeled examples as landmarks to construct the feature space, then using a new labeled set of size d_l to estimate $\boldsymbol{\alpha} \in \mathbb{R}^{d_u}$. This is done by solving the following optimization problem:

$$\min_{\boldsymbol{\alpha}}\left\{ \sum_{i=1}^{d_l}\left[1 - \sum_{j=1}^{d_u}\alpha_j y_i K(\mathbf{x}_i, \mathbf{x}_j)\right]_+ : \sum_{j=1}^{d_u}|\alpha_j| \leq 1/\gamma\right\}. \tag{2}$$

This problem can be solved efficiently by linear programming. Furthermore, as it is L_1-constrained, tuning the value of γ will produce a sparse solution. Lastly, the associated classifier takes the following form:

$$y = \operatorname{sgn}\sum_{j=1}^{d_u}\alpha_j K(\mathbf{x}, \mathbf{x}_j). \tag{3}$$

We now extend this framework to jointly learn the similarity and the separator in a semi-supervised way. Let S be a sample set of d_l labeled examples $(\mathbf{x}, y) \in \mathcal{Z} = \mathcal{X} \times \{-1; +1\})$ and d_u unlabeled examples. Furthermore, let $K_\mathbf{A}(\mathbf{x}, \mathbf{x}')$ be a generic (ϵ, γ, τ)-good similarity function, parameterized by the matrix $\mathbf{A} \in \mathbb{R}^{d \times d}$. We assume that $K_\mathbf{A}(\mathbf{x}, \mathbf{x}') \in [-1; +1]$ and that $||\mathbf{x}||_2 \leq 1$, but all our developments and results can directly be extended to any bounded similarities and datasets. Our goal here is to find the matrix $\mathbf{A}$ and the global separator $\boldsymbol{\alpha} \in \mathbb{R}^{d_u}$ that minimize the empirical loss (in our case, the hinge loss) over a finite sample S, with some guarantees on the generalization error of the associated classifier. To this end, we propose a generalization of Problem (2) based on a joint optimization of the metric and the global separator:

$$\min_{\boldsymbol{\alpha}, \mathbf{A}} \sum_{i=1}^{d_l}\left[1 - \sum_{j=1}^{d_u}\alpha_j y_i K_\mathbf{A}(\mathbf{x}_i, \mathbf{x}_j)\right]_+ + \lambda||\mathbf{A} - \mathbf{R}|| \tag{4}$$

$$\text{s.t. } \sum_{j=1}^{d_u} |\alpha_j| \leq 1/\gamma \tag{5}$$

$$\mathbf{A} \text{ diagonal}, \quad |A_{kk}| \leq 1, \quad 1 \leq k \leq d, \tag{6}$$

where $\lambda > 0$ is a regularization parameter, and $\mathbf{R} \in \mathbb{R}^{d \times d}$ is a fixed diagonal matrix such that $\|\mathbf{R}\| \leq d$. Here, the notation $\|\cdot\|$ refers to a generic matrix norm, for instance L_1 or L_2 norms.

The novelty of this formulation is the *joint optimization* over $\mathbf{A}$ and $\boldsymbol{\alpha}$: by solving Problem (4), we are learning the metric and the separator at the same time. One of its significant advantages is that it extends the semi-supervised setting from the separator learning step to the metric learning, and the two problems are solved using the same data. This method can naturally be used in situations where one has access to few labeled examples and some unlabeled ones: the labeled examples are used in this case to select the unlabeled examples that will serve to classify new points. Another important advantage of our technique, coming from [1], is that the constraints on the pair of points do not need to be satisfied entirely, as the loss is averaged on all the reasonable points. In other words, this formulation is less restrictive than pair or triplet-based settings. Constraint (5) takes into account the desired margin γ and is the same as in [1]. Constraint (6) ensures that the learned similarity is bounded in $[-1; +1]$. Note that the diagonality constraint on $\mathbf{A}$ can be relaxed (in which case the bound constraint becomes $\|\mathbf{A}\| \leq 1$ and $\mathbf{R}$ is no longer diagonal); we restrict ourselves to diagonal matrices to simplify the presentation and to limit the number of parameters to be learned.

The matrix $\mathbf{R}$ can be used to encode prior knowledge one has on the problem, in a way similar to what is proposed in [9]. If the non parameterized version of the similarity considered performs well, then a natural choice of $\mathbf{R}$ is the identity matrix I, so that the learned matrix will preserve the good properties of the non parameterized version (and will improve it through learning). Another form of prior knowledge relates to the importance of each feature according to the classes considered. Indeed, one may want to give more weight to features that are more representative of one of the classes $\{-1; +1\}$. One way to capture this importance is to compare the distributions of each feature on the two classes, *e.g.* through Kullback-Leibler (KL) divergence. We assume here that each feature follows a Gaussian distribution in each class, with means μ_1 (class $+1$) and μ_2 (class -1) and standard deviations σ_1 (class $+1$) and σ_2 (class -1). The KL divergence is expressed in that case as:

$$D_{KL}^k = \log\left(\frac{\sigma_1}{\sigma_2}\right) + \frac{1}{2}\left(\frac{\sigma_1^2}{\sigma_2^2} - \frac{\sigma_2^2}{\sigma_1^2} + \frac{(\mu_2 - \mu_1)^2}{\sigma_2^2}\right), \quad 1 \leq k \leq d.$$

and the matrix $\mathbf{R}$ corresponds to $\text{diag}(D_{KL}^1, D_{KL}^2, \cdots, D_{KL}^d)$.

Lastly, once $\mathbf{A}$ and $\boldsymbol{\alpha}$ have been learned, the associated (binary) classifier takes the form given in Eq. (3).

4 Generalization Bound for Joint Similarity Learning

In this section, we establish a generalization bound for our joint similarity learning formulation (4) under constraints (5) and (6). This theoretical analysis is based on the Rademacher complexity and holds for any regularization parameter $\lambda > 0$. Note that when $\lambda = 0$, we can also prove consistency results based on the algorithmic robustness framework [26,27]. In such a case, the proof is similar to the one in [18]. Before stating the generalization bound, we first introduce some notations.

Definition 2. *A pairwise similarity function* $K_\mathbf{A} : \mathcal{X} \times \mathcal{X} \to [-1, 1]$, *parameterized by a matrix* $\mathbf{A} \in \mathbb{R}^{d \times d}$, *is said to be* (β, c)-*admissible if, for any matrix norm* $||\cdot||$, *there exist* $\beta, c \in \mathbb{R}$ *such that* $\forall \mathbf{x}, \mathbf{x}' \in \mathcal{X}, |K_\mathbf{A}(\mathbf{x}, \mathbf{x}')| \leq \beta + c \cdot ||\mathbf{x}'\mathbf{x}^T|| \cdot ||\mathbf{A}||$.

Examples: Using some classical norm properties and the Frobenius inner product, we can show that:

- The bilinear similarity $K_\mathbf{A}^1(\mathbf{x}, \mathbf{x}') = \mathbf{x}^T \mathbf{A} \mathbf{x}'$ is $(0, 1)$-admissible, that is $|K_\mathbf{A}^1(\mathbf{x}, \mathbf{x}')| \leq ||\mathbf{x}'\mathbf{x}^T|| \cdot ||\mathbf{A}||$;
- The similarity derived from the Mahalanobis distance $K_\mathbf{A}^2(\mathbf{x}, \mathbf{x}') = 1 - (\mathbf{x} - \mathbf{x}')^T \mathbf{A}(\mathbf{x} - \mathbf{x}')$ is $(1, 4)$-admissible, that is $|K_\mathbf{A}^2(\mathbf{x}, \mathbf{x}')| \leq 1 + 4 \cdot ||\mathbf{x}'\mathbf{x}^T|| \cdot ||\mathbf{A}||$.

Note that we will make use of these two similarity functions $K_\mathbf{A}^1$ and $K_\mathbf{A}^2$ in our experiments. For any $\mathbf{B}, \mathbf{A} \in \mathbb{R}^{n \times d}$ and any matrix norm $|| \cdot ||$, its dual norm $|| \cdot ||_*$ is defined, for any $\mathbf{B}$, by $||\mathbf{B}||_* = \sup_{||\mathbf{A}|| \leq 1} \mathrm{Tr}(\mathbf{B}^T \mathbf{A})$, where $\mathrm{Tr}(\cdot)$ denotes the trace of a matrix. Denote $X_* = \sup_{\mathbf{x}, \mathbf{x}' \in \mathcal{X}} ||\mathbf{x}'\mathbf{x}^T||_*$.

Let us now rewrite the minimization problem (4) with a more generalized notation of the loss function:

$$\min_{\boldsymbol{\alpha}, \mathbf{A}} \quad \frac{1}{d_l} \sum_{i=1}^{d_l} \ell(\mathbf{A}, \boldsymbol{\alpha}, \mathbf{z}_i = (\mathbf{x}_i, y_i)) + \lambda ||\mathbf{A} - \mathbf{R}||, \tag{7}$$

$$\text{s.t.} \quad \sum_{j=1}^{d_u} |\alpha_j| \leq 1/\gamma \tag{8}$$

$$\mathbf{A} \text{ diagonal}, \quad |A_{kk}| \leq 1, \quad 1 \leq k \leq d, \tag{9}$$

where $\ell(\mathbf{A}, \boldsymbol{\alpha}, \mathbf{z}_i = (\mathbf{x}_i, y_i))) = \left[1 - \sum_{j=1}^{d_u} \alpha_j y_i K_\mathbf{A}(\mathbf{x}_i, \mathbf{x}_j)\right]_+$ is the instantaneous loss estimated at point $(\mathbf{x}_i, y_i)$. Note that from constraints (8) and (9), we deduce that $||\mathbf{A}|| < d$. Let $\mathcal{E}_\mathcal{S}(\mathbf{A}, \boldsymbol{\alpha}) = \frac{1}{d_l} \sum_{i=1}^{d_l} \ell(\mathbf{A}, \boldsymbol{\alpha}, \mathbf{z}_i)$ be the overall empirical loss over the training set $\mathcal{S}$, and let $\mathcal{E}(\mathbf{A}, \boldsymbol{\alpha}) = \mathbb{E}_{\mathbf{z} \sim \mathcal{Z}} \ell(\mathbf{A}, \boldsymbol{\alpha}, \mathbf{z})$ be the true loss w.r.t. the unknown distribution $\mathcal{Z}$. The target of generalization analysis for joint similarity learning is to bound the difference $\mathcal{E}(\mathbf{A}, \boldsymbol{\alpha}) - \mathcal{E}_\mathcal{S}(\mathbf{A}, \boldsymbol{\alpha})$.

Our generalization bound is based on the Rademacher complexity which can be seen as an alternative notion of the complexity of a function class and has the particularity to be (unlike the VC-dimension) data-dependent.

Definition 3. *Let $\mathcal{F}$ be a class of uniformly bounded functions. For every integer n, we call*

$$R_n(\mathcal{F}) := \mathbb{E}_{\mathcal{S}} \mathbb{E}_{\sigma} \left[\sup_{f \in \mathcal{F}} \frac{1}{n} \sum_{i=1}^{n} \sigma_i f(z_i) \right]$$

the Rademacher average over $\mathcal{F}$, where $\mathcal{S} = \{z_i : i \in \{1, \ldots, n\}\}$ are independent random variables distributed according to some probability measure and $\{\sigma_i : i \in \{1, \ldots, n\}\}$ are independent Rademacher random variables, that is, $\Pr(\sigma_i = 1) = \Pr(\sigma_i = -1) = \frac{1}{2}$.

The Rademacher average w.r.t. the dual matrix norm is then defined as:

$$\mathcal{R}_{d_l} := \mathbb{E}_{\mathcal{S}, \sigma} \left[\sup_{\tilde{\mathbf{x}} \in \mathcal{X}} \left\| \frac{1}{d_l} \sum_{i=1}^{d_l} \sigma_i y_i \mathbf{x}_i \tilde{\mathbf{x}}^T \right\|_* \right]$$

We can now state our generalization bound.

Theorem 2. *Let $(\mathbf{A}_{\mathcal{S}}, \boldsymbol{\alpha}_{\mathcal{S}})$ be the solution to the joint problem (7) and $K_{\mathbf{A}_{\mathcal{S}}}$ a (β, c)-admissible similarity function. Then, for any $0 < \delta < 1$, with probability at least $1 - \delta$, the following holds:*

$$\mathcal{E}(\mathbf{A}_{\mathcal{S}}, \boldsymbol{\alpha}_{\mathcal{S}}) - \mathcal{E}_{\mathcal{S}}(\mathbf{A}_{\mathcal{S}}, \boldsymbol{\alpha}_{\mathcal{S}}) \leq 4 \mathcal{R}_{d_l} \left(\frac{cd}{\gamma} \right) + \left(\frac{\beta + cX_* d}{\gamma} \right) \sqrt{\frac{2 \ln \frac{1}{\delta}}{d_l}}.$$

Theorem 2 proves that learning $\mathbf{A}$ and $\boldsymbol{\alpha}$ in a joint manner from a training set minimizes the generalization error, as the latter is bounded by the empirical error of our joint regularized formulation. Its proof makes use of the Rademacher symmetrization theorem and contraction property (Theorem 3 and Lemma 1):

Theorem 3. *[7] Let $R_n(\mathcal{F})$ be the Rademacher average over $\mathcal{F}$ defined as previously. We have:*

$$\mathbb{E} \left[\sup_{f \in \mathcal{F}} \left(\mathbb{E} f(\mathcal{S}) - \frac{1}{n} \sum_{i=1}^{n} f(z_i) \right) \right] \leq 2 R_n(\mathcal{F}).$$

Lemma 1. *[17] Let F be a class of uniformly bounded real-valued functions on (Ω, μ) and $m \in \mathbb{N}$. If for each $i \in \{1, \ldots, m\}, \phi_i : \mathbb{R} \to \mathbb{R}$ is a function having a Lipschitz constant c_i, then for any $\{x_i\}_{i \in \mathbb{N}_m}$,*

$$\mathbb{E}_{\epsilon} \left(\sup_{f \in F} \sum_{i \in \mathbb{N}_m} \epsilon_i \phi_i(f(x_i)) \right) \leq 2 \mathbb{E}_{\epsilon} \left(\sup_{f \in F} \sum_{i \in \mathbb{N}_m} c_i \epsilon_i f(x_i) \right).$$

Proof (Theorem 2).
    Let $\mathbb{E}_{\mathcal{S}}$ denote the expectation with respect to sample $\mathcal{S}$. Observe that $\mathcal{E}_{\mathcal{S}}(\mathbf{A}_{\mathcal{S}}, \boldsymbol{\alpha}_{\mathcal{S}}) - \mathcal{E}(\mathbf{A}_{\mathcal{S}}, \boldsymbol{\alpha}_{\mathcal{S}}) \leq \sup_{\mathbf{A}, \boldsymbol{\alpha}} [\mathcal{E}_{\mathcal{S}}(\mathbf{A}, \boldsymbol{\alpha}) - \mathcal{E}(\mathbf{A}, \boldsymbol{\alpha})]$. Also, for any $\mathcal{S} = (\mathbf{z}_1, \ldots, \mathbf{z}_k, \ldots, \mathbf{z}_{d_l})$ and $\tilde{\mathcal{S}} = (\mathbf{z}_1, \ldots, \tilde{\mathbf{z}}_k, \ldots, \mathbf{z}_{d_l})$, $1 \leq k \leq d_l$:

$$\left| \sup_{\mathbf{A},\boldsymbol{\alpha}} [\mathcal{E}_{\mathcal{S}}(\mathbf{A},\boldsymbol{\alpha}) - \mathcal{E}(\mathbf{A},\boldsymbol{\alpha})] - \sup_{\mathbf{A},\boldsymbol{\alpha}} [\mathcal{E}_{\tilde{\mathcal{S}}}(\mathbf{A},\boldsymbol{\alpha}) - \mathcal{E}(\mathbf{A},\boldsymbol{\alpha})] \right|$$

$$\leq \sup_{\mathbf{A},\boldsymbol{\alpha}} |\mathcal{E}_{\mathcal{S}}(\mathbf{A},\boldsymbol{\alpha}) - \mathcal{E}_{\tilde{\mathcal{S}}}(\mathbf{A},\boldsymbol{\alpha})|$$

$$= \frac{1}{d_l} \sup_{\mathbf{A},\boldsymbol{\alpha}} \left| \sum_{\mathbf{z}=(\mathbf{x},y)\in\mathcal{S}} \left[1 - \sum_{j=1}^{d_u} \alpha_j y K_{\mathbf{A}}(\mathbf{x},\mathbf{x}_j) \right]_+ - \sum_{\tilde{\mathbf{z}}=(\tilde{\mathbf{x}},\tilde{y})\in\tilde{\mathcal{S}}} \left[1 - \sum_{j=1}^{d_u} \alpha_j \tilde{y} K_{\mathbf{A}}(\tilde{\mathbf{x}},\mathbf{x}_j) \right]_+ \right|$$

$$= \frac{1}{d_l} \sup_{\mathbf{A},\boldsymbol{\alpha}} \left| \left[1 - \sum_{j=1}^{d_u} \alpha_j y_k K_{\mathbf{A}}(\mathbf{x}_k,\mathbf{x}_j) \right]_+ - \left[1 - \sum_{j=1}^{d_u} \alpha_j \tilde{y}_k K_{\mathbf{A}}(\tilde{\mathbf{x}}_k,\mathbf{x}_j) \right]_+ \right|$$

$$= \frac{1}{d_l} \sup_{\mathbf{A},\boldsymbol{\alpha}} \left| \sum_{j=1}^{d_u} \alpha_j \tilde{y}_k K_{\mathbf{A}}(\tilde{\mathbf{x}}_k,\mathbf{x}_j) - \sum_{j=1}^{d_u} \alpha_j y_k K_{\mathbf{A}}(\mathbf{x}_k,\mathbf{x}_j) \right| \tag{10}$$

$$\leq \frac{2}{d_l} \sup_{\mathbf{A},\boldsymbol{\alpha}} \left| \sum_{j=1}^{d_u} \alpha_j y_k^{max} K_{\mathbf{A}}(\mathbf{x}_k^{max},\mathbf{x}_j) \right| \quad \text{where } \mathbf{z}_k^{max} = \operatorname*{arg\,max}_{\mathbf{z}=(\mathbf{x},y)\in\{\mathbf{z}_k,\tilde{\mathbf{z}}_k\}} y K_{\mathbf{A}}(\mathbf{x},\mathbf{x}_j)$$

$$\leq \frac{2}{d_l} \sup_{\mathbf{A},\boldsymbol{\alpha}} \left\{ \sum_{j=1}^{d_u} |\alpha_j| \cdot |y_k^{max}| \cdot |K_{\mathbf{A}}(\mathbf{x}_k^{max},\mathbf{x}_j)| \right\}$$

$$\leq \frac{2}{d_l} \left(\frac{\beta + cX_*d}{\gamma} \right) \tag{11}$$

Inequality (10) comes from the 1-lipschitzness of the hinge loss; Inequality (11) comes from Constraint (8), $\|\mathbf{A}\| \leq d$ and the (β,c)-admissibility of $K_{\mathbf{A}}$. Applying McDiarmid's inequality to the term $\sup_{\mathbf{A},\boldsymbol{\alpha}} [\mathcal{E}_{\mathcal{S}}(\mathbf{A},\boldsymbol{\alpha}) - \mathcal{E}(\mathbf{A},\boldsymbol{\alpha})]$, with probability $1 - \delta$, we have:

$$\sup_{\mathbf{A},\boldsymbol{\alpha}} [\mathcal{E}_{\mathcal{S}}(\mathbf{A},\boldsymbol{\alpha}) - \mathcal{E}(\mathbf{A},\boldsymbol{\alpha})] \leq \mathbb{E}_{\mathcal{S}} \sup_{\mathbf{A},\boldsymbol{\alpha}} [\mathcal{E}_{\mathcal{S}}(\mathbf{A},\boldsymbol{\alpha}) - \mathcal{E}(\mathbf{A},\boldsymbol{\alpha})] + \left(\frac{\beta + cX_*d}{\gamma} \right) \sqrt{\frac{2\ln\frac{1}{\delta}}{d_l}}.$$

In order to bound the gap between the true loss and the empirical loss, we now need to bound the expectation term of the right hand side of the above equation.

$$\mathbb{E}_{\mathcal{S}} \sup_{\mathbf{A},\boldsymbol{\alpha}} [\mathcal{E}_{\mathcal{S}}(\mathbf{A},\boldsymbol{\alpha}) - \mathcal{E}(\mathbf{A},\boldsymbol{\alpha})]$$

$$= \mathbb{E}_{\mathcal{S}} \sup_{\mathbf{A},\boldsymbol{\alpha}} \left\{ \frac{1}{d_l} \sum_{i=1}^{d_l} \left[1 - \sum_{j=1}^{d_u} \alpha_j y_i K_{\mathbf{A}}(\mathbf{x}_i,\mathbf{x}_j) \right]_+ - \mathcal{E}(\mathbf{A},\boldsymbol{\alpha}) \right\}$$

$$\leq 2\mathbb{E}_{\mathcal{S},\sigma} \sup_{\mathbf{A},\boldsymbol{\alpha}} \left\{ \frac{1}{d_l} \sum_{i=1}^{d_l} \sigma_i \left[1 - \sum_{j=1}^{d_u} \alpha_j y_i K_{\mathbf{A}}(\mathbf{x}_i,\mathbf{x}_j) \right]_+ \right\} \tag{12}$$

$$\leq 4\mathbb{E}_{\mathcal{S},\sigma} \sup_{\mathbf{A},\boldsymbol{\alpha}} \left| \frac{1}{d_l} \sum_{i=1}^{d_l} \sigma_i y_i \sum_{j=1}^{d_u} \alpha_j K_{\mathbf{A}}(\mathbf{x}_i,\mathbf{x}_j) \right| \tag{13}$$

$$\leq 4 \left(\frac{cd}{\gamma} \right) \mathbb{E}_{\mathcal{S},\sigma} \sup_{\tilde{\mathbf{x}}} \left\| \frac{1}{d_l} \sum_{i=1}^{d_l} \sigma_i y_i \mathbf{x}_i \tilde{\mathbf{x}}^T \right\|_* = 4\mathcal{R}_{d_l} \left(\frac{cd}{\gamma} \right). \tag{14}$$

We obtain Inequality (12) by applying Theorem 3, while Inequality (13) comes from the use of Lemma 1. The Inequality on line (14) makes use of the (β, c)-admissibility of the similarity function $K_{\mathbf{A}}$ (Definition 2). Combining Inequalities (11) and (14) completes the proof of the theorem.   □

After proving the generalization bound of our joint similarity approach, we now move to the experimental validation of the approach proposed.

5 Experiments

The state of the art in metric learning is dominated by algorithms designed to work in a purely supervised setting. Furthermore, most of them optimize a metric adapted to kNN classification (*e.g.* LMNN, ITML), while our work is designed for finding a global linear separator. For these reasons, it is difficult to propose a totally fair comparative study. In this section, we first evaluate the effectiveness of problem (4) with constraints (5) and (6) (JSL, for Joint Similarity Learning) with different settings. Secondly, we extensively compare it with state-of-the-art algorithms from different categories (supervised, kNN-oriented). Lastly, we study the impact of the quantity of available labeled data on our method. We conduct the experimental study on 7 classic datasets taken from the UCI Machine Learning Repository[1], both binary and multi-class. Their characteristics are presented in Table 1. These datasets are widely used for metric learning evaluation.

Table 1. Properties of the datasets used in the experimental study.

	Balance	Ionosphere	Iris	Liver	Pima	Sonar	Wine
# Instances	625	351	150	345	768	208	178
# Dimensions	4	34	4	6	8	60	13
# Classes	3	2	3	2	2	2	3

5.1 Experimental Setting

In order to provide a comparison as complete as possible, we compare two main families of approaches[2]:

1. *Linear classifiers*: in this family, we consider the following methods:
   - BBS, corresponding to Problem (2) and discussed above;
   - SLLC [4], an extension of BBS in which a similarity is learned prior to be used in the BBS framework;
   - JSL, the joint learning framework proposed in this study;
   - Linear SVM with L_2 regularization, which is the standard approach for linear classification;

---

[1] http://archive.ics.uci.edu/ml/.
[2] For all the methods, we used the code provided by the authors.

2. *Nearest neighbor approaches*: in this family, we consider the methods:
   - Standard 3-nearest neighbor classifier (3NN) based on the Euclidean distance;
   - ITML [9], which learns a Mahalanobis distance that is used here in 3NN classification;
   - LMNN with a full matrix and LMNN with a diagonal matrix (LMNN-diag) [23,24], also learning a Mahalanobis distance used here in 3NN classification;
   - LRML [14,15]; LRML also learns a Mahalanobis distance used in 3NN classifier, but in a semi-supervised setting. This method is thus the "most" comparable to JSL (even though one is learning a linear separator and the other only a distance).

All classifiers are used in their binary version, in a one-vs-all setting when the number of classes is greater than two. BBS, SLLC and JSL rely on the same classifier from Eq. (3), even though learned in different ways. We solve BBS and JSL using projected gradient descent. In JSL, we rely on an alternating optimization that consists in fixing $\mathbf{A}$ (resp. $\boldsymbol{\alpha}$) and optimizing for $\boldsymbol{\alpha}$ (resp. $\mathbf{A}$), then changing the variable, until convergence.

Data Processing and Parameter Settings. All features are centered around zero and scaled to ensure $||\mathbf{x}||_2 \leq 1$, as this constraint is necessary for some of the algorithms. We randomly choose 15% of the data for validation purposes, and another 15% as a test set. The training set and the unlabeled data are chosen from the remaining 70% of examples not employed in the previous sets. In order to illustrate the classification using a restricted quantity of labeled data, the number of labeled points is limited to 5, 10 or 20 examples per class, as this is usually a reasonable minimum amount of annotation to rely on. The number of landmarks is either set to 15 points or to all the points in the training set (in which case their label is not taken into account). These two settings correspond to two practical scenarios: one in which a relatively small amount of unlabeled data is available, and one in which a large amount of unlabeled data is available. When only 15 unlabeled points are considered, they are chosen from the training set as the nearest neighbor of the 15 centroids obtained by applying k-means++ clustering with $k = 15$. All of the experimental results are averaged over 10 runs, for which we compute a 95% confidence interval. We tune the following parameters by cross-validation: $\gamma, \lambda \in \{10^{-4}, \ldots, 10^{-1}\}$ for BBS and JSL (λ only for the second), $\lambda_{ITML} \in \{10^{-4}, \ldots, 10^{4}\}$, choosing the value yielding the best accuracy. For SLLC, we tune $\gamma, \beta \in \{10^{-7}, \ldots, 10^{-2}\}$, $\lambda \in \{10^{-3}, \ldots, 10^{2}\}$, as done by the authors, while for LRML we consider $\gamma_s, \gamma_d, \gamma_i \in \{10^{-2}, \ldots, 10^{2}\}$. For LMNN, we set $\mu = 0.5$, as done in [24].

5.2   Experimental Results

Analysis of JSL. We first study here the behavior of the proposed joint learning framework w.r.t. different families of similarities and regularization functions (choice of $\mathbf{R}$ and $|| \cdot ||$). In particular, we consider two types of similarity measures:

Table 2. Average accuracy (%) with confidence interval at 95%, 5 labeled points per class, 15 unlabeled landmarks.

Sim.	Reg.	Balance	Ionosphere	Iris	Liver	Pima	Sonar	Wine
K_A^1	I-L_1	85.2±3.0	85.6±2.4	76.8±3.2	63.3±6.2	71.0±4.1	72.9±3.6	**91.9**±4.2
	I-L_2	85.1±2.9	85.6±2.6	76.8±3.2	63.1±6.3	71.0±4.0	73.2±3.8	91.2±4.5
	KL-L_1	84.9±2.9	85.0±2.6	77.3±2.7	63.9±5.5	71.0±4.0	72.9±3.6	90.8±4.7
	KL-L_2	85.2±3.0	85.8±3.3	76.8±3.2	62.9±6.4	71.3±4.3	74.2±3.8	90.0±5.4
K_A^2	I-L_1	**87.2**±2.9	**87.7**±2.6	78.6±4.6	64.7±5.6	75.1±3.5	73.9±5.7	80.8±9.5
	I-L_2	86.8±3.0	**87.7**±2.8	75.9±5.7	64.3±5.4	**75.6**±3.6	74.8±5.8	80.8±8.6
	KL-L_1	**87.2**±2.9	87.3±2.4	78.6±4.6	62.9±5.6	75.0±3.7	75.5±6.2	79.6±11.8
	KL-L_2	87.1±2.7	85.8±3.3	**79.1**±5.4	**64.9**±5.9	**75.6**±3.4	**77.1**±5.2	79.6±9.7

Table 3. Average accuracy (%) with confidence interval at 95%, all points used as landmarks.

Sim.	Reg.	Balance	Ionosphere	Iris	Liver	Pima	Sonar	Wine
K_A^1	I-L_1	85.8±2.9	88.8±2.5	74.5±3.1	65.5±4.5	71.4±3.8	70.3±6.6	85.8±5.0
	I-L_2	85.8±2.9	87.7±2.7	74.5±3.5	64.7±5.5	71.7±4.1	68.7±6.7	84.6±5.5
	KL-L_1	85.6±3.1	87.9±3.4	75.0±3.5	65.3±4.9	71.6±4.2	70.3±6.8	85.4±5.3
	KL-L_2	85.1±3.1	88.5±3.7	**75.9**±3.4	65.1±4.8	72.1±4.2	71.9±6.7	**86.5**±6.0
K_A^2	I-L_1	85.9±2.3	90.4±2.2	71.8±6.1	67.3±3.5	73.1±3.5	72.9±4.2	81.5±8.4
	I-L_2	**86.2**±2.5	**90.6**±2.2	73.2±6.6	**68.6**±3.3	73.3±3.2	**73.2**±4.2	82.7±9.0
	KL-L_1	85.8±2.6	89.4±2.0	72.7±5.5	67.5±3.8	**73.8**±3.5	71.0±4.1	80.0±7.4
	KL-L_2	85.9±2.4	89.6±2.2	74.5±6.2	68.4±3.6	73.1±3.8	72.3±4.8	80.0±11.5

bilinear (cosine-like) similarities of the form $K_A^1(\mathbf{x}, \mathbf{x}') = \mathbf{x}^T \mathbf{A} \mathbf{x}'$ and similarities derived from the Mahalanobis distance $K_A^2(\mathbf{x}, \mathbf{x}') = 1 - (\mathbf{x} - \mathbf{x}')^T \mathbf{A}(\mathbf{x} - \mathbf{x}')$. For the regularization term, $\mathbf{R}$ is either set to the identity matrix (JSL-I), or to the approximation of the Kullback-Leibler divergence (JSL-KL) discussed in Section 3. As mentioned above, these two settings correspond to different prior knowledge one can have on the problem. In both cases, we consider L_1 and L_2 regularization norms. We thus obtain 8 settings, that we compare in the situation where few labeled points are available (5 points per class), using a small amount (15 instances) of unlabeled data or a large amount (the whole training set) of unlabeled data. The results of the comparisons are reported in Tables 2 and 3.

As one can note from Table 2, when only 15 points are used as landmarks, K_A^2 obtains better results in almost all of the cases, the difference being more important on Iris, Pima and Sonar. The noticeable exception to this better behavior of K_A^2 is Wine, for which cosine-like similarities outperform Mahalanobis-based similarities by more than 10 points. A similar result was also presented in [21]. The difference between the use of the L_1 or L_2 norms is not as marked, and there is no strong preference for one or the other, even though the L_2 norm leads to slightly better results in average than the L_1 norm. Regarding the regularization matrix $\mathbf{R}$, again, the difference is not strongly marked, except maybe on Sonar.

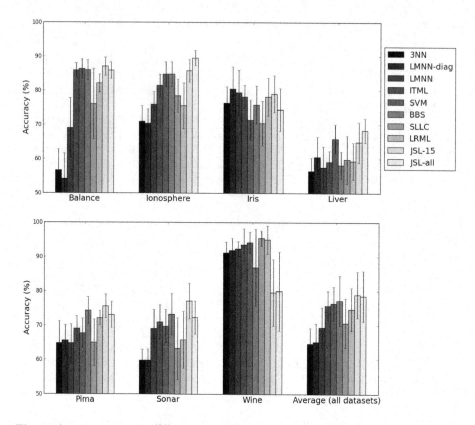

Fig. 1. Average accuracy (%) with confidence interval at 95%, 5 labeled points per class, 15 unlabeled landmarks.

In average, regularizing through the Kullback-Leibler divergence leads to slightly better results than regularizing through the identity matrix.

When all points are used as landmarks (Table 3), similar conclusions can be drawn regarding the similarity functions and the norms used. However, in that case, the regularization based on the identity matrix yields better results than the one based on the KL divergence. It is important to note also that the overall results are in general lower than the ones obtained when only 15 points are used as landmarks. We attribute this effect to the fact that one needs to learn more parameters (via $\boldsymbol{\alpha}$), whereas the amount of available labeled data is the same.

From the above analysis, we focus now on two JSL based methods: JSL-15 with $K_{\mathbf{A}}^2$, L_2 norm and $\mathbf{R} =$KL when 15 points are used as landmarks, and JSL-all with $K_{\mathbf{A}}^2$, L_2 norm and $\mathbf{R} = I$ when all the points are used as landmarks.

Comparison of the Different Methods. We now study the performance of our method, compared to state-of-the-art algorithms. For this, we consider JSL-15 and JSL-all with 5, 10, respectively 20 labeled examples per class. As our

Table 4. Average accuracy (%) over all datasets with confidence interval at 95%.

Method	5 pts./cl.	10 pts./cl.	20 pts./cl.
3NN	64.6±4.6	68.5±5.4	70.4±5.0
LMNN-diag	65.1±5.5	68.2±5.6	71.5±5.2
LMNN	69.4±5.9	70.9±5.3	73.2±5.2
ITML	75.8±4.2	76.5±4.5	76.3±4.8
SVM	76.4±4.9	76.2±7.0	77.7±6.4
BBS	77.2±7.3	77.0±6.2	77.3±6.3
SLLC	70.5±7.2	75.9±4.5	75.8±4.8
LRML	74.7±6.2	75.3±5.9	75.8±5.2
JSL-15	**78.9±6.7**	**77.6±5.5**	77.7±6.4
JSL-all	78.2±7.3	76.6±5.8	**78.4±6.7**

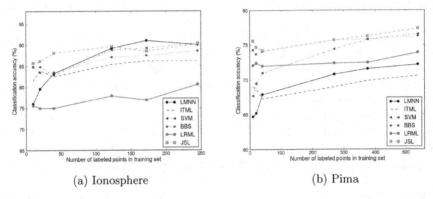

(a) Ionosphere   (b) Pima

Fig. 2. Average accuracy w.r.t. the number of labeled points with 15 landmarks.

methods are tested using the similarity based on the Mahalanobis distance, we use the euclidean distance for BBS to ensure fairness.

Figure 1 presents the average accuracy per dataset obtained with 5 labeled points per class. In this setting, JSL outperforms the other algorithms on 5 out of 7 datasets and has similar performances on one other. The exception is the Wine dataset, where none of the JSL settings yields competitive results. As stated before, this is easily explained by the fact cosine-similarities are more adapted for this dataset. Even though JSL-15 and JSL-all perform the same when averaged over all datasets, the difference between them is marked on some datasets: JSL-15 is considerably better on Iris and Sonar, while JSL-all significantly outperforms JSL-15 on Ionosphere and Liver. Averaged over all datasets (Table 4), JSL obtains the best performance in all configurations with a limited amount of labeled data, which is particularly the setting that our method is designed for. The values in bold are significantly better than the rest of their respective columns, confirmed by a one-sided Student t-test for paired samples with a significance level of 5%.

Impact of the Amount of Labeled Data. As an illustration of the methods'
behavior when the level of supervision varies, Figure 2 presents the accuracies
on two representative datasets, Ionosphere and Pima, with an increasing number
of labeled examples. In both cases, the best results are obtained by JSL (and
more precisely JSL-15) when less than 50% of the training set is used. This
is in agreement with the results reported in Table 4. The results of JSL are
furthermore comparable only to BBS for the Pima dataset. Lastly, the accuracy
of JSL improves slightly when adding more labeled data, and the results on the
whole training set are competitive w.r.t. the other algorithms.

6   Conclusion

In this paper, we have studied the problem of learning similarities in the situation
where few labeled (and potentially few unlabeled) data is available. To do so,
we have developed a semi-supervised framework, extending the (ϵ, γ, τ)-good
of [1], in which the similarity function and the classifier are learned at the same
time. To our knowledge, this is the first time that such a framework is provided.
The joint learning of the similarity and the classifier enables one to benefit
from unlabeled data for both the similarity and the classifier. We have also
showed that the proposed method was theoretically well-founded as we derived a
Rademacher-based bound on the generalization error of the learned parameters.
Lastly, the experiments we have conducted on standard metric learning datasets
show that our approach is indeed well suited for learning with few labeled data,
and outperforms state-of-the-art metric learning approaches in that situation.

Acknowledgments. Funding for this project was provided by a grant from Région
Rhône-Alpes.

References

1. Balcan, M.-F., Blum, A., Srebro, N.: Improved guarantees for learning via similarity
   functions. In: COLT, pp. 287–298. Omnipress (2008)
2. Bao, J.-P., Shen, J.-Y., Liu, X.-D., Liu, H.-Y.: Quick asymmetric text similarity
   measures. ICMLC **1**, 374–379 (2003)
3. Baoli, L., Qin, L., Shiwen, Y.: An adaptive k-nearest neighbor text categorization
   strategy. ACM TALIP (2004)
4. Bellet, A., Habrard, A., Sebban, M.: Similarity learning for provably accurate
   sparse linear classification. In: ICML, pp. 1871–1878 (2012)
5. Bellet, A., Habrard, A., Sebban, M.: A survey on metric learning for feature vectors
   and structured data. arXiv preprint arXiv:1306.6709 (2013)
6. Bellet, A., Habrard, A., Sebban, M.: Metric Learning. Synthesis Lectures on Artificial Intelligence and Machine Learning. Morgan & Claypool Publishers (2015)
7. Boucheron, S., Bousquet, O., Lugosi, G.: Theory of classification : a survey of some
   recent advances. ESAIM: Probability and Statistics **9**, 323–375 (2005)
8. Bousquet, O., Elisseeff, A.: Stability and generalization. JMLR **2**, 499–526 (2002)

9. Davis, J.V., Kulis, B., Jain, P., Sra, S., Dhillon, I.S.: Information-theoretic metric learning. In: ICML, pp. 209–216. ACM, New York (2007)
10. Diligenti, M., Maggini, M., Rigutini, L.: Learning similarities for text documents using neural networks. In: ANNPR (2003)
11. Freund, Y., Schapire, R.E.: Large margin classification using the perceptron algorithm. Machine Learning **37**(3), 277–296 (1999)
12. Grabowski, M., Szałas, A.: A Technique for Learning Similarities on Complex Structures with Applications to Extracting Ontologies. In: Szczepaniak, P.S., Kacprzyk, J., Niewiadomski, A. (eds.) AWIC 2005. LNCS (LNAI), vol. 3528, pp. 183–189. Springer, Heidelberg (2005)
13. Guo, Z.-C., Ying, Y.: Guaranteed classification via regularized similarity learning. CoRR, abs/1306.3108 (2013)
14. Hoi, S.C.H., Liu, W., Chang, S.-F.: Semi-supervised distance metric learning for collaborative image retrieval. In: CVPR (2008)
15. Hoi, S.C.H., Liu, W., Chang, S.-F.: Semi-supervised distance metric learning for collaborative image retrieval and clustering. TOMCCAP **6**(3) (2010)
16. Hust, A.: Learning Similarities for Collaborative Information Retrieval. In: Machine Learning and Interaction for Text-Based Information Retrieval Workshop, TIR 2004, pp. 43–54 (2004)
17. Ledoux, M., Talagrand, M.: Probability in Banach Spaces: Isoperimetry and Processes. Springer, New York (1991)
18. Nicolae, M.-I., Sebban, M., Habrard, A., Gaussier, É., Amini, M.: Algorithmic robustness for learning via (ϵ, γ, τ)-good similarity functions. CoRR, abs/1412.6452 (2014)
19. Niu, G., Dai, B., Yamada, M., Sugiyama, M.: Information-theoretic semi-supervised metric learning via entropy regularization. In: ICML. Omnipress (2012)
20. Qamar, A.M., Gaussier, É.: Online and batch learning of generalized cosine similarities. In: ICDM, pp. 926–931 (2009)
21. Qamar, A.M., Gaussier, É., Chevallet, J., Lim, J.: Similarity learning for nearest neighbor classification. In: ICDM, pp. 983–988 (2008)
22. Shalev-Shwartz, S., Singer, Y., Ng, A.Y.: Online and batch learning of pseudo-metrics. In: ICML. ACM, New York (2004)
23. Weinberger, K., Saul, L.: Fast solvers and efficient implementations for distance metric learning. In: ICML, pp. 1160–1167. ACM (2008)
24. Weinberger, K., Saul, L.: Distance metric learning for large margin nearest neighbor classification. JMLR **10**, 207–244 (2009)
25. Xing, E.P., Ng, A.Y., Jordan, M.I., Russell, S.: Distance metric learning, with application to clustering with side-information. NIPS **15**, 505–512 (2002)
26. Xu, H., Mannor, S.: Robustness and generalization. In: COLT, pp. 503–515 (2010)
27. Xu, H., Mannor, S.: Robustness and generalization. Machine Learning **86**(3), 391–423 (2012)
28. Zha, Z.-J., Mei, T., Wang, M., Wang, Z., Hua, X.-S.: Robust distance metric learning with auxiliary knowledge. In: IJCAI, pp. 1327–1332 (2009)
29. Zhu, J., Rosset, S., Hastie, T., Tibshirani, R.: 1-norm support vector machines. In: NIPS, page 16. MIT Press (2003)

Learning Compact and Effective Distance Metrics with Diversity Regularization

Pengtao Xie[✉]

Machine Learning Department, Carnegie Mellon University,
5000 Forbes Avenue, Pittsburgh, PA 15213, USA
pengtaox@cs.cmu.edu

Abstract. Learning a proper distance metric is of vital importance for many distance based applications. Distance metric learning aims to learn a set of latent factors based on which the distances between data points can be effectively measured. The number of latent factors incurs a trade-off: a small amount of factors are not powerful and expressive enough to measure distances while a large number of factors cause high computational overhead. In this paper, we aim to achieve two seemingly conflicting goals: keeping the number of latent factors to be small for the sake of computational efficiency, meanwhile making them as effective as a large set of factors. The approach we take is to impose a diversity regularizer over the latent factors to encourage them to be uncorrelated, such that each factor can capture some unique information that is hard to be captured by other factors. In this way, a small amount of latent factors can be sufficient to capture a large proportion of information, which retains computational efficiency while preserving the effectiveness in measuring distances. Experiments on retrieval, clustering and classification demonstrate that a small amount of factors learned with diversity regularization can achieve comparable or even better performance compared with a large factor set learned without regularization.

1 Introduction

In data mining and machine learning, learning a proper distance metric is of vital importance for many distance based tasks and applications, such as retrieval [22], clustering [18] and classification [16]. In retrieval, a better distance measure can help find data entries that are more relevant with the query. In k-means based clustering, data points can be grouped into more coherent clusters if the distance metric is properly defined. In k-nearest neighbor (k-NN) based classification, to find better nearest neighbors, the distances between data samples need to be appropriately measured. All these tasks rely heavily on a good distance measure. Distance metric learning (DML) [3,16,18] takes pairs of data points which are labeled either as similar or dissimilar and learns a distance metric such that similar data pairs will be placed close to each other while dissimilar pairs will be separated apart. While formulated in various ways, most DML approaches choose to learn a Mahalanobis distance $(x - y)^{\mathsf{T}} M (x - y)$, where x, y are d-dimensional feature vectors and $M \in \mathbb{R}^{d \times d}$ is a positive semidefinite matrix to be

© Springer International Publishing Switzerland 2015
A. Appice et al. (Eds.): ECML PKDD 2015, Part I, LNAI 9284, pp. 610–624, 2015.
DOI: 10.1007/978-3-319-23528-8_38

learned. DML can be interpreted as a latent space model. By factorizing M into $M = A^{\mathsf{T}}A$, the Mahalanobis distance can be written as $\|Ax - Ay\|^2$, which can be interpreted as first projecting the data from the original feature space to a latent space using the linear projection matrix $A \in \mathbb{R}^{k \times d}$, then measuring the squared Euclidean distance in the latent space. Each row of A corresponds to one latent factor (or one dimension of the latent space). Ax is the latent representation of x and can be used as input of downstream tasks. These latent factors are aimed at capturing the latent features of the observed data. The latent features usually carry high-level semantic meanings and reflect the inherent characteristics of data, thus measuring distance in the latent feature space could be more effective.

In choosing the number k of latent factors (or the dimension of the latent space), there is an inherent tradeoff between the effectiveness of the distance matrix A and computational efficiency. A larger k would bestow A more expressiveness and power in measuring distances. However, the resultant latent representations would be of high dimensionality, which incurs high computational complexity and inefficiency. This is especially true for retrieval where performing nearest neighbor search on high-dimensional representations is largely difficult. On the other hand, a smaller k can reduce the computational cost, but would render the distance matrix less effective.

In this paper, we aim to investigate whether it is possible to achieve the best of both worlds: given a sufficiently small k which facilitates computational efficiency, can the effectiveness of the distance matrix be comparable to that of a large k? In other words, the goal is to learn a compact (with small k) but effective distance matrix. Our way to approach this goal is motivated by Principal Component Analysis (PCA). Similar to DML, PCA also learns a linear projection matrix, where the row vectors are called components. Unlike DML which imposes no constraints on the row vectors (latent factors), PCA requires the components to be orthogonal to each other. Such an orthogonality constraint renders the components to be uncorrelated and each component captures information that cannot be captured by other components. As a result, a small number of components are able to capture a large proportion of information. This inspires us to place an orthogonality constraint over the row vectors of A in DML, with the hope that a small number of latent factors are sufficient to effectively measure distances. However, as verified in our experiments, requiring the latent factors to be strictly orthogonal may be too restrictive, which hurts the quality of distance measurement. Instead, we impose a *diversity* regularizer over the latent factors to encourage them to approach orthogonality, but not necessarily to be orthogonal. We perform experiments on retrieval, clustering and classification to demonstrate that with diversity regularization, a distance matrix with small k can achieve comparable or even better performance in comparison with an unregularized matrix with large k.

The rest of the paper is organized as follows. In Section 2, we review related works. In Section 3, we present how to diversity DML. Section 4 gives experimental results and Section 5 concludes the paper.

2   Related Works

Many works [3,5,8,16,18,21] have been proposed for distance metric learning. Please see [15,19] for an extensive review. There are many problem settings regarding the form of distance supervision, the type of distance metric to be learned and the learning objective. Among them, the most common setting [3,5,18] is given data pairs labeled either as similar or dissimilar, learning a Mahalanobis distance metric, such that similar pairs will be placed close to each other and dissimilar pairs will be separated apart. As first formulated in [18], a Mahalanobis distance metric is learned under similarity and dissimilarity constraints by minimizing the distances of similar pairs while separating dissimilar pairs with a certain margin. Guillaumin *et al* [5] proposed to use logistic discriminant to learn a Mahalanobis metric from a set of labelled data pairs, with the goal that positive pairs have smaller distances than negative pairs. Kostinger *et al* [6] proposed to learn Mahalanobis metrics using likelihood test, which defines the Mahalanobis matrix to be the difference of covariance matrices of two Gaussian distributions used for modeling dissimilar pairs and similar pairs respectively. Ying and Li [21] developed an eigenvalue optimization framework for learning a Mahalanobis metric, which is shown to be equivalent to minimizing the maximal eigenvalue of a symmetric matrix.

Some works take other forms of distance supervision such as class labels [14], rankings [10], triple constraints [13], time series alignments [8] to learn distance metrics for specific tasks, such as k-nearest neighbor classification [16], ranking [10], time series aligning [8]. Globerson and Roweis [4] assumed the class label for each sample is available and proposed to learn a Mahalanobis matrix for classification by collapsing all samples in the same class to a single point and pushing samples in other classes infinitely far away. Weinberger *et al* [16] proposed to learn distance metrics for k-nearest neighbor classification with the goal that the k-nearest neighbors always belong to the same class while samples from different classes are separated by a large margin. This method also requires the presence of class labels of all samples. Trivedi *et al* [14] formulated the problem of metric learning for k nearest neighbor classification as a large margin structured prediction problem, with a latent variable representing the choice of neighbors and the task loss directly corresponding to classification error. In this paper, we focus on the pairwise similarity/dissimilarity constraints which are considered as the most common and natural supervision of distance metric learning. Other forms of distance supervision, together with the corresponding specific-purpose tasks, will be left for future study.

To avoid overfitting, various methods have been proposed to regularize distance metric learning. Davis *et al* [3] imposed a regularization that the Mahalanobis distance matrix should be close to a prior matrix and the Bregman divergence is utilized to measure how close two matrices are. Ying and Li [20] and Niu *et al* [11] utilized a mixed-norm regularization to encourage the sparsity of the projection matrix. Qi *et al* [12] used ℓ_1 regularization to learn sparse metrics for high dimensional problems with small sample size. Qian *et al* [13] applied dropout to regularize distance metric learning. In this paper,

Algorithm 1.. Algorithm for solving DDML.

Input: $\mathcal{S},\mathcal{D},k,\lambda$
repeat
   Fix $\tilde{A}$, optimize g using subgradient method
   Fix g, optimize $\tilde{A}$ using projected subgradient method
until converge

Table 1. Statistics of datasets

	Feature Dim.	#training data	#data pairs
20-News	5000	11.3K	200K
15-Scenes	1000	3.2K	200K
6-Activities	561	7.4K	200K

we investigate a different regularizer for DML: diversity regularization, which has not been studied by previous works.

The problem of diversifying the latent factors in latent variable models has been studied in [7,17], with the goal to reduce the redundancy of latent factors and improve the coverage of infrequent latent features/structures. Zou and Adams [23] used a Determinantal Point Process (DPP) prior to diversify the latent factors and Xie *et al* [17] defined a diversity measure based on pairwise angles between latent factors. In this paper, we study the diversification of distance metric learning, aiming to learn compact distance metrics without compromising their effectiveness.

3   Diversify Distance Metric Learning

In this section, we begin with reviewing the DML problem proposed in [18] and reformulate it as a latent space model using ideas introduced in [16]. Then we present how to diversify DML.

3.1   A Latent Space Modeling View of DML

Distance metric learning represents a family of models and has various formulations regarding the distance metric to learn, the form of distance supervision and how the objective function is defined. Among them, the most popular setting is: (1) distance metric: Mahalanobis distance $(x - y)^{\mathsf{T}} M(x - y)$, where x and y are feature vectors of two data instances and M is a symmetric and positive semidefinite matrix to be learned; (2) the form of distance supervision: pairs of data instances labeled either as similar or dissimilar; (3) learning objective: to learn a distance metric to place similar points as close as possible and separate dissimilar points apart. Given a set of pairs labeled as similar $\mathcal{S} = \{(x_i, y_i)\}_{i=1}^{|\mathcal{S}|}$ and a set of pairs labeled as dissimilar $\mathcal{D} = \{(x_i, y_i)\}_{i=1}^{|\mathcal{D}|}$, DML learns a Mahalanobis

Table 2. Retrieval average precision (%) on 20-News dataset

k	10	100	300	500	700	900
DML	72.4	74.0	74.9	75.4	75.8	76.2
DDML	**76.7**	**81.0**	**81.1**	**79.2**	**78.3**	**77.8**

Table 3. Retrieval average precision (%) on 15-Scenes dataset

k	10	50	100	150	200
DML	79.5	80.2	80.7	80.7	80.8
DDML	**82.4**	**83.6**	**83.3**	**83.1**	**82.8**

distance matrix M by optimizing the following problem

$$\min_M \frac{1}{|\mathcal{S}|} \sum_{(x,y)\in\mathcal{S}} (x-y)^\mathsf{T} M(x-y)$$
$$s.t. \quad (x-y)^\mathsf{T} M(x-y) \geq 1, \forall(x,y)\in\mathcal{D} \quad (1)$$
$$M \succeq 0$$

where $M \succeq 0$ denotes that M is required to be positive semidefinite. This optimization problem aims to minimize the Mahalanobis distances between pairs labeled as similar while separating dissimilar pairs with a margin 1. M is required to be positive semidefinite to ensure that the Mahalanobis distance is a valid distance metric.

By re-parametrizing M with $A^\mathsf{T} A$ [16], where A is a matrix of size $k \times d$ ($k \leq d$) and $A^\mathsf{T} A$ guarantees the positive semi-definiteness of M, the Mahalanobis distance can be written as $\|Ax - Ay\|^2$, which can be interpreted as first projecting the data from the original space to a latent space using the linear projection matrix A, then computing the squared Euclidean distance in the latent space. Each row of A corresponds to a latent factor. Accordingly, the problem defined in Eq.(1) can be written as

$$\min_A \frac{1}{|\mathcal{S}|} \sum_{(x,y)\in\mathcal{S}} \|Ax - Ay\|^2$$
$$s.t. \quad \|Ax - Ay\|^2 \geq 1, \forall(x,y)\in\mathcal{D} \quad (2)$$

To this end, we see that the DML problem can be interpreted as a latent space modeling problem. The goal is to seek a latent space where the squared Euclidean distances of similar data pairs are small and those of dissimilar pairs are large. The latent space is characterized by the projection matrix A.

3.2 Diversify DML

To diversify DML, we use the diversity measure proposed in [17] to regularize the latent factors to encourage them to approach orthogonality. Given k latent factors in $A \in R^{k \times d}$, one can compute the non-obtuse angle $\theta(a_i, a_j)$ between each pair of latent factors a_i and a_j, where a_i is the ith row of A. $\theta(a_i, a_j)$

Table 4. Retrieval average precision (%) on 6-Activities dataset

k	10	50	100	150	200
DML	93.2	94.3	94.5	94.5	94.5
DDML	**96.2**	**95.5**	**95.9**	**95.3**	**95.1**

Table 5. Retrieval average precision (%) on three datasets

	20-News	15-Scenes	6-Activities
EUC	62.8	65.3	85.0
DML [18]	76.2	80.8	94.5
LMNN [16]	67.0	70.3	71.5
ITML [3]	74.7	79.1	94.2
DML-eig [21]	71.2	71.3	86.7
Seraph [11]	75.8	82.0	89.2
DDML	**81.1**	**83.6**	**96.2**

is defined as $\arccos(\frac{|a_i \cdot a_j|}{\|a_i\|\|a_j\|})$. A larger $\theta(a_i, a_j)$ indicates that a_i and a_j are more different from each other. Given the pairwise angles, the diversity measure $\Omega(A)$ is defined as $\Omega(A) = \Psi(A) - \Pi(A)$, where $\Psi(A)$ and $\Pi(A)$ are the mean and variance of all pairwise angles. The mean $\Psi(A)$ measures how these factors are different from each other on the whole and the variance $\Pi(A)$ is intended to encourage these factors to be evenly different from each other. The larger $\Omega(A)$ is, the more diverse these latent factors are. And $\Omega(A)$ attains the global maximum when the factors are orthogonal to each other.

Using this diversity measure to regularize the latent factors, we define a Diversified DML (DDML) problem as:

$$\min_A \frac{1}{|\mathcal{S}|} \sum_{(x,y)\in\mathcal{S}} \|Ax - Ay\|^2 - \lambda\Omega(A)$$
$$s.t. \quad \|Ax - Ay\|^2 \geq 1, \forall(x,y) \in \mathcal{D} \tag{3}$$

where $\lambda > 0$ is a tradeoff parameter between the distance loss and the diversity regularizer. The term $-\lambda\Omega(A)^1$ in the new objective function encourages the latent factors in A to be diverse. λ plays an important role in balancing the fitness of A to the distance loss $\sum_{(x,y)\in\mathcal{S}} \|Ax - Ay\|^2$ and its diversity. Under a small λ, A is learned to best minimize the distance loss and its diversity is ignored. Under a large λ, A is learned with high diversity, but may not be well fitted to the distance loss and hence lose the capability to properly measure distances. A proper λ needs to be chosen to achieve the optimal balance.

3.3 Optimization

In this section, we present an algorithm to solve the problem defined in Eq.(3), which is summarized in Algorithm 1. First, we adopt a strategy similar to [16]

---

[1] Note that a negative sign is used here because the overall objective function is to be minimized but $\Omega(A)$ is intended to be maximized.

Table 6. Clustering accuracy (%) on 20-News dataset

k	10	100	300	500	700	900
DML	23.7	25.1	26.2	26.9	28.1	28.4
DDML	**33.4**	**42.7**	**44.6**	**39.5**	**40.6**	**41.3**

Table 7. Normalized mutual information (%) on 20-News dataset

k	10	100	300	500	700	900
DML	34.1	35.4	36.8	36.9	38.0	38.2
DDML	**42.5**	**49.7**	**51.1**	**47.2**	**47.8**	**48.1**

to remove the constraints. By introducing slack variables ξ to relax the hard constraints, we get

$$\min_A \frac{1}{|\mathcal{S}|} \sum_{(x,y)\in\mathcal{S}} \|Ax - Ay\|^2 - \lambda\Omega(A) + \frac{1}{|\mathcal{D}|} \sum_{(x,y)\in\mathcal{D}} \xi_{x,y} \tag{4}$$
$$s.t. \quad \|Ax - Ay\|^2 \geq 1 - \xi_{x,y}, \xi_{x,y} \geq 0, \forall(x,y) \in \mathcal{D}$$

Using hinge loss, the constraint in Eq.(4) can be further eliminated

$$\min_A \frac{1}{|\mathcal{S}|} \sum_{(x,y)\in\mathcal{S}} \|Ax - Ay\|^2 - \lambda\Omega(A)$$
$$+ \frac{1}{|\mathcal{D}|} \sum_{(x,y)\in\mathcal{D}} \max(0, 1 - \|Ax - Ay\|^2) \tag{5}$$

Since $\Omega(A)$ is non-smooth and non-convex, which is hard to optimize directly, Xie *et al* [17] instead optimized a low bound of $\Omega(A)$, and they proved that maximizing the lower bound čan increase $\Omega(A)$. Factorizing A into $\mathrm{diag}(g)\widetilde{A}$, where g is a vector and g_i denotes the ℓ_2 norm of the ith row of A, then the ℓ_2 norm of each row vector in $\widetilde{A}$ is one. According to the definition of $\Omega(A)$, it is clear that $\Omega(A) = \Omega(\widetilde{A})$. The problem defined in Eq.(5) can be reformulated as

$$\min_{\widetilde{A},g} \frac{1}{|\mathcal{S}|} \sum_{(x,y)\in\mathcal{S}} \|\mathrm{diag}(g)\widetilde{A}(x - y)\|^2 - \lambda\Omega(\widetilde{A})$$
$$+ \frac{1}{|\mathcal{D}|} \sum_{(x,y)\in\mathcal{D}} \max(0, 1 - \|\mathrm{diag}(g)\widetilde{A}(x - y)\|^2) \tag{6}$$
$$s.t. \quad \|\widetilde{A}_i\| = 1, \forall i = 1, \cdots, k$$

where $\widetilde{A}_i$ denotes the ith row of $\widetilde{A}$. This problem can be optimized by alternating between g and $\widetilde{A}$: optimizing g with $\widetilde{A}$ fixed and optimizing $\widetilde{A}$ with g fixed. With $\widetilde{A}$ fixed, the problem defined over g is

$$\min_g \frac{1}{|\mathcal{S}|} \sum_{(x,y)\in\mathcal{S}} \|\mathrm{diag}(g)\widetilde{A}(x - y)\|^2 + \frac{1}{|\mathcal{D}|} \sum_{(x,y)\in\mathcal{D}} \max(0, 1 - \|\mathrm{diag}(g)\widetilde{A}(x - y)\|^2)$$
$$\tag{7}$$

Table 8. Clustering accuracy (%) on 15-Scenes dataset

k	10	50	100	150	200
DML	33.9	36.5	40.1	37.0	37.8
DDML	**46.9**	**51.3**	**46.2**	**46.5**	**49.6**

Table 9. Normalized mutual information (%) on 15-Scenes dataset

k	10	50	100	150	200
DML	41.4	41.0	42.0	41.4	41.6
DDML	**46.7**	**48.9**	**47.3**	**48.8**	**47.1**

which can be optimized with subgradient method. Fixing g, the problem defined over $\widetilde{A}$ is

$$\min_{\widetilde{A}} \frac{1}{|\mathcal{S}|} \sum_{(x,y)\in\mathcal{S}} \|\mathrm{diag}(g)\widetilde{A}(x-y)\|^2 - \lambda\Omega(\widetilde{A})$$
$$+\frac{1}{|\mathcal{D}|} \sum_{(x,y)\in\mathcal{D}} \max(0, 1 - \|\mathrm{diag}(g)\widetilde{A}(x-y)\|^2) \tag{8}$$
$$s.t. \quad \|\widetilde{A}_i\| = 1, \forall i = 1, \cdots, k$$

Since $\Omega(\widetilde{A})$ is non-smooth and non-convex, (sub)gradient method is not applicable. Xie *et al* [17] derived a smooth lower bound of $\Omega(\widetilde{A})$ and instead optimized the low bound with projected gradient descent. Please refer to [17] for details.

4 Experiments

In this section, on three tasks — retrieval, clustering and classification — we corroborate that through diversification it is possible to learn distance metrics that are both compact and effective.

4.1 Datasets

We used three datasets in the experiments: 20 Newsgroups[2] (20-News), 15-Scenes [9] and 6-Activities [1]. The 20-News dataset has 18846 documents from 20 categories, where 60% of the documents were for training and the rest were for testing. Documents were represented with *tfidf* vectors whose dimensionality is 5000. We randomly generated 100K similar pairs and 100K dissimilar pairs from the training set to learn distance metrics. Two documents were labeled as similar if they belong to the same group and dissimilar otherwise. The 15-Scenes dataset contains 4485 images belonging to 15 scene classes. 70% of the images were used for training and 30% were for testing. Images were represented with bag-of-words vectors whose dimensionality is 1000. Similar to 20-News, we generated 100K similar and 100K dissimilar data pairs for distance learning according to whether two images are from the same scene class or not. The 6-Activities

---

[2] http://qwone.com/~jason/20Newsgroups/

Table 10. Clustering accuracy (%) on 6-Activities dataset

k	10	50	100	150	200
DML	75.0	75.6	76.1	75.6	75.7
DDML	**94.9**	**96.3**	**96.6**	**95.1**	**95.7**

Table 11. Normalized mutual information (%) on 6-Activities dataset

k	10	50	100	150	200
DML	83.6	83.5	84.0	83.5	83.5
DDML	**90.3**	**91.9**	**91.3**	**91.4**	**91.1**

dataset is built from recordings of 30 subjects performing six activities of daily living while carrying a waist-mounted smart phone with embedded inertial sensors. The features are 561-dimensional sensory signals. There are 7352 training instances and 2947 testing instances. Similarly, 100K similar pairs and 100K dissimilar pairs were generated to learn distance metrics. Table 1 summarizes the statistics of these three datasets.

4.2   Experimental Settings

Our method DDML contains two key parameters — the number k of latent factors and the tradeoff parameter λ — both of which were tuned using 5-fold cross validation. We compared with 6 baseline methods, which were selected according to their popularity and the state of the art performance. They are: (1) Euclidean distance (EUC); (2) Distance Metric Learning (DML) [18]; (3) Large Margin Nearest Neighbor (LMNN) metric learning [16]; (4) Information Theoretical Metric Learning (ITML) [3]; (5) Distance Metric Learning with Eigenvalue Optimization (DML-eig) [21]; (6) Information-theoretic Semi-supervised Metric Learning via Entropy Regularization (Seraph) [11]. Parameters of the baseline methods were tuned using 5-fold cross validation. Some methods, such as ITML, achieve better performance on lower-dimensional representations which are obtained via Principal Component Analysis. The number of leading principal components were selected via 5-fold cross validation.

4.3   Retrieval

We first applied the learned distance metrics for retrieval. To evaluate the effectiveness of the learned metrics, we randomly sampled 100K similar pairs and 100K dissimilar pairs from 20-News test set, 50K similar pairs and 50K dissimilar pairs from 15-Scenes test set, 100K similar pairs and 100K dissimilar pairs from 6-Activities test set and used the learned metrics to judge whether these pairs were similar or dissimilar. If the distance was greater than some threshold t, the pair was regarded as similar. Otherwise, the pair was regarded as dissimilar. We used average precision (AP) to evaluate the retrieval performance.

Table 12. Clustering accuracy (%) on three datasets

	20-News	15-Scenes	6-Activities
EUC	36.5	29.0	61.6
DML [18]	28.4	40.1	76.1
LMNN [16]	32.9	33.6	56.9
ITML [3]	34.5	38.2	93.4
DML-eig [21]	27.3	26.6	63.3
Seraph [11]	**48.1**	48.2	74.8
DDML	44.6	**51.3**	**96.6**

Table 13. Normalized mutual information (%) on three datasets

	20-News	15-Scenes	6-Activities
EUC	37.9	28.7	59.9
DML [18]	38.2	42.0	83.6
LMNN [16]	33.3	34.3	58.2
ITML [3]	39.2	41.5	87.0
DML-eig [21]	34.0	31.8	58.6
Seraph [11]	49.7	47.5	71.1
DDML	**51.1**	**48.9**	**91.9**

Table 2, 3 and 4 show the average precision under different number k of latent factors on 20-News, 15-Scenes and 6-Activities dataset respectively. As shown in these tables, DDML with a small k can achieve retrieval precision that is comparable to DML with a large k. For example, on the 20-News dataset (Table 2), with 10 latent factors, DDML is able to achieve a precision of 76.7%, which cannot be achieved by DML with even 900 latent factors. As another example, on the 15-Scenes dataset (Table 3), the precision obtained by DDML with $k = 10$ is 82.4%, which is largely better than the 80.8% precision achieved by DML with $k = 200$. Similar behavior is observed on the 6-Activities dataset (Table 4). This demonstrates that, with diversification, DDML is able to learn a distance metric that is as effective as (if not more effective than) DML, but is much more compact than DML. Such a compact distance metric greatly facilitates retrieval efficiency. Performing retrieval on 10-dimensional latent representations is much easier than on representations with hundreds of dimensions. It is worth noting that the retrieval efficiency gain comes without sacrificing the precision, which allows one to perform fast and accurate retrieval. For DML, increasing k consistently increases the precision, which corroborates that a larger k would make the distance metric to be more expressive and powerful. However, k cannot be arbitrarily large, otherwise the distance matrix would have too many parameters that lead to overfitting. This is evidenced by how the precision of DDML varies as k increases.

Table 5 presents the comparison with the state of the art distance metric learning methods. As can be seen from this table, our method achieves the best performances across all three datasets. The Euclidean distance does not

Table 14. 3-NN accuracy (%) on 20-News dataset

k	10	100	300	500	700	900
DML	39.1	48.0	53.0	55.0	56.4	57.5
DDML	**51.3**	**64.1**	**64.5**	**63.3**	**62.9**	**61.4**

Table 15. 10-NN accuracy (%) on 20-News dataset

k	10	100	300	500	700	900
DML	39.4	49.4	54.3	56.2	57.9	58.6
DDML	**54.2**	**66.6**	**66.8**	**66.1**	**65.3**	**64.5**

Table 16. 3-NN accuracy (%) on 15-Scenes dataset

k	10	50	100	150	200
DML	47.7	47.7	50.8	51.7	51.1
DDML	**57.4**	**57.5**	**57.9**	**58.8**	**57.3**

Table 17. 10-NN accuracy (%) on 15-Scenes dataset

k	10	50	100	150	200
DML	51.6	51.7	54.0	54.4	54.9
DDML	**59.2**	**60.9**	**60.5**	**60.6**	**59.6**

incorporate distance supervision provided by human, thus its performance is inferior. DML-eig imposes no regularization over the distance metric, which is thus prone to overfitting. To avoid overfitting, ITML utilized a Bregman divergence regularizer and Seraph used a sparsity regularizer. But the performances of both regularizers are inferior to the diversity regularizer utilized by DDML. LMNN is specifically designed for k-NN classification, thus the learned distance metrics cannot guarantee to be effective in retrieval tasks.

4.4 Clustering

The second task we study is to apply the learned distance metrics for k-means clustering, where the number of clusters was set to the number of categories in each dataset and k-means was run 10 times with random initialization of the centroids. Following [2], we used two metrics to measure the clustering performance: accuracy (AC) and normalized mutual information (NMI). Please refer to [2] for their definitions.

Table 6,8 and 10 show the clustering accuracy on 20-News, 15-Scenes and 6-Activity test set respectively under various number of latent factors k. Table 7, 9 and 11 show the normalized mutual information on 20-News, 15-Scenes and 6-Activity test set respectively. These tables show that the clustering performance achieved by DDML under a small k is much better than DML under a much larger k. For instance, DDML can achieve 33.4% accuracy on the 20-News dataset (Table 6) with 10 latent factors, which is much better than the 28.4% accuracy

Table 18. 3-NN accuracy (%) on 6-Activities dataset

k	10	50	100	150	200
DML	94.9	94.8	94.6	95.1	95.0
DDML	**94.3**	**96.2**	**96.5**	**95.5**	**95.9**

Table 19. 10-NN accuracy (%) on 6-Activities dataset

	10	50	100	150	200
DML	95.3	95.0	95.2	95.2	95.3
DDML	**96.6**	**96.8**	**96.4**	**96.3**	**96.1**

Table 20. 3-NN accuracy (%) on three datasets

	20-News	15-Scenes	6-Activities
EUC	42.6	42.5	88.7
DML [18]	57.5	51.7	95.1
LMNN [16]	60.6	53.5	91.5
ITML [3]	50.9	51.9	93.5
DML-eig [21]	39.2	33.1	82.3
Seraph [11]	**67.9**	55.2	91.4
DDML	64.5	**58.8**	**96.5**

Table 21. 10-NN accuracy (%) on three datasets

k	20-News	15-Scenes	6-Activities
EUC	41.7	44.9	90.2
DML [18]	58.6	54.9	95.3
LMNN [16]	62.7	56.2	91.5
ITML [3]	54.8	54.3	94.0
DML-eig [21]	43.8	34.0	82.8
Seraph [11]	**69.8**	60.3	92.5
DDML	66.8	**60.9**	**96.8**

obtained by DML with 900 latent factors. As another example, the NMI obtained by DDML on the 15-Scenes dataset (Table 9) with $k = 10$ is 46.7%, which is largely better than the 41.6% NMI achieved by DML with $k = 200$. This again corroborates that the diversity regularizer can enable DDML to learn compact and effective distance metrics, which significantly reduce computational complexity while preserving the clustering performance.

Table 12 and 13 present the comparison of DDML with the state of the art methods on clustering accuracy and normalized mutual information. As can be seen from these two tables, our method outperforms the baselines in most cases except that the accuracy on 20-News dataset is worse than the Seraph method. Seraph performs very well on 20-News and 15-Scenes dataset, but its performance is bad on the 6-Activities dataset. DDML achieves consistently good performances across all three datasets.

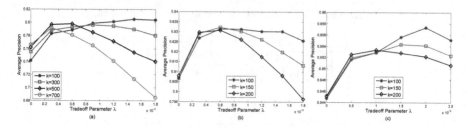

Fig. 1. Sensitivity of DDML to the tradeoff parameter λ on (a) 20-News dataset (b) 15-Scenes dataset (c) 6-Activities dataset

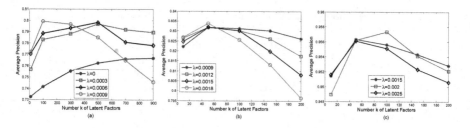

Fig. 2. Sensitivity of DDML to the number of latent factors k on (a) 20-News dataset (b) 15-Scenes dataset (c) 6-Activities dataset

4.5   Classification

We also apply the learned metrics for k-nearest neighbor classification, which is also an algorithm that largely depends on a good distance measure. For each testing sample, we find its k-nearest neighbors in the training set and use the class labels of the nearest neighbors to classify the test sample. Table 14, 16 and 18 show the 3-NN classification accuracy on the 20-News, 15-Scenes and 6-Activities dataset. Table 15, 17 and 19 show the 10-NN classification accuracy on the 20-News, 15-Scenes and 6-Activities dataset. Similar to retrieval and clustering, DDML with a small k can achieve classification accuracy that is comparable to or better than DML with a large k. Table 20 and 21 present the comparison of DDML with the state of the art methods on 3-NN and 10-NN classification accuracy. As can be seen from these two tables, our method outperforms the baselines in most cases except that the accuracy on 20-News dataset is worse than the Seraph method.

4.6   Sensitivity to Parameters

We study the sensitivity of DDML to the two key parameters: tradeoff parameter λ and the number of latent factors k. Figure 1 shows how the retrieval average precision (AP) varies as λ increases on the 20-News, 15-Scenes and 6-Activities dataset respectively. The curves correspond to different k. As can be seen from the figure, initially increasing λ improves AP. The reason is that a larger λ

encourages the latent factors to be more uncorrelated, thus different aspects of the information can be captured more comprehensively. However, continuing to increase λ degrades the precision. This is because if λ is too large, the diversify regularizer dominates the distance loss and the resultant distance metric is not tailored to the distance supervision and loses effectiveness in measuring distances.

Figure 2 shows how AP varies as k increases on the 20-News, 15-Scenes and 6-Activities dataset respectively. The curves correspond to different λ. When k is small, the average precision is low. This is because a small amount of latent factors are insufficient to capture the inherent complex pattern behind data, hence lacking the capability to effectively measure distances. As k increases, the model capacity increases and the AP increases accordingly. However, further increasing k causes performance to drop. This is because a larger k incurs higher risk of overfitting to training data.

5   Conclusions

In this paper, we study the problem of diversifying distance metric learning, with the purpose to learn compact distance metrics without losing their effectiveness in measuring distances. Diversification encourages the latent factors in the distance metric to be different from each other, thus each latent factor is able to capture some unique information that is hard to be captured by other factors. Accordingly, the number of latent factors required to capture the total information can be greatly reduced. Experiments on retrieval, clustering and classification corroborate the effectiveness of the diversity regularizer in learning compact and effective distance metrics.

References

1. Anguita, D., Ghio, A., Oneto, L., Parra, X., Reyes-Ortiz, J.L.: Human activity recognition on smartphones using a multiclass hardware-friendly support vector machine. In: Ambient Assisted Living and Home Care, pp. 216–223. Springer (2012)
2. Cai, D., He, X., Han, J.: Locally consistent concept factorization for document clustering. IEEE Transactions on Knowledge and Data Engineering **23**(6), 902–913 (2011)
3. Davis, J.V., Kulis, B., Jain, P., Sra, S., Dhillon, I.S.: Information-theoretic metric learning. In: Proceedings of the 24th International Conference on Machine Learning, pp. 209–216. ACM (2007)
4. Globerson, A., Roweis, S.T.: Metric learning by collapsing classes. In: Advances in Neural Information Processing Systems, pp. 451–458 (2005)
5. Guillaumin, M., Verbeek, J., Schmid, C.: Is that you? metric learning approaches for face identification. In: IEEE International Conference on Computer Vision, pp. 498–505. IEEE (2009)
6. Kostinger, M., Hirzer, M., Wohlhart, P., Roth, P.M., Bischof, H.: Large scale metric learning from equivalence constraints. In: IEEE Conference on Computer Vision and Pattern Recognition, pp. 2288–2295. IEEE (2012)

7. Kwok, J.T., Adams, R.P.: Priors for diversity in generative latent variable models. In: Advances in Neural Information Processing Systems, pp. 2996–3004 (2012)
8. Lajugie, R., Garreau, D., Bach, F., Arlot, S.: Metric learning for temporal sequence alignment. In: Advances in Neural Information Processing Systems, pp. 1817–1825 (2014)
9. Lazebnik, S., Schmid, C., Ponce, J.: Beyond bags of features: Spatial pyramid matching for recognizing natural scene categories. In: IEEE Conference on Computer Vision and Pattern Recognition, vol. 2, pp. 2169–2178. IEEE (2006)
10. Lim, D., Lanckriet, G., McFee, B.: Robust structural metric learning. In: Proceedings of The 30th International Conference on Machine Learning, pp. 615–623 (2013)
11. Niu, G., Dai, B., Yamada, M., Sugiyama, M.: Information-theoretic semisupervised metric learning via entropy regularization. Neural Computation, 1–46 (2012)
12. Qi, G.-J., Tang, J., Zha, Z.-J., Chua, T.-S., Zhang, H.-J.: An efficient sparse metric learning in high-dimensional space via l1-penalized log-determinant regularization. In: Proceedings of the 26th Annual International Conference on Machine Learning, pp. 841–848. ACM (2009)
13. Qian, Q., Hu, J., Jin, R., Pei, J., Zhu, S.: Distance metric learning using dropout: a structured regularization approach. In: Proceedings of the 20th ACM SIGKDD International Conference on Knowledge Discovery and Data Mining, pp. 323–332. ACM (2014)
14. Trivedi, S., Mcallester, D., Shakhnarovich, G.: Discriminative metric learning by neighborhood gerrymandering. In: Advances in Neural Information Processing Systems, pp. 3392–3400 (2014)
15. Wang, F, Sun, J.: Survey on distance metric learning and dimensionality reduction in data mining. Data Mining and Knowledge Discovery, 1–31 (2014)
16. Weinberger, KQ., Blitzer, J., Saul, L.K.: Distance metric learning for large margin nearest neighbor classification. In: Advances in Neural Information Processing Systems, pp. 1473–1480 (2005)
17. Xie, P., Deng, Y., Xing, E.P.: Diversifying restricted boltzmann machine for document modeling. In: ACM SIGKDD Conference on Knowledge Discovery and Data Mining (2015)
18. Xing, E.P., Jordan, M.I., Russell, S., Ng, A.Y.: Distance metric learning with application to clustering with side-information. In: Advances in Neural Information Processing Systems, pp. 505–512 (2002)
19. Liu, Y., Rong, J.: Distance metric learning: A comprehensive survey. Michigan State Universiy, vol. 2 (2006)
20. Ying, Y., Huang, K., Campbell, C.: Sparse metric learning via smooth optimization. In: Advances in Neural Information Processing Systems, pp. 2214–2222 (2009)
21. Ying, Y., Li, P.: Distance metric learning with eigenvalue optimization. The Journal of Machine Learning Research **13**(1), 1–26 (2012)
22. Zhang, P., Zhang, W., Li, W.-J., Guo, M: Supervised hashing with latent factor models. In: SIGIR (2014)
23. Zou, J.Y., Adams, R.P.: Priors for diversity in generative latent variable models. In: Advances in Neural Information Processing Systems, pp. 2996–3004 (2012)

Scalable Metric Learning for Co-Embedding

Farzaneh Mirzazadeh[1], Martha White[2], András György[1(✉)],
and Dale Schuurmans[1]

[1] Department of Computing Science, University of Alberta, Edmonton, Canada
gyorgy@ualberta.ca
[2] Department of Computer Science and Informatics, Indiana University,
Bloomington, USA

Abstract. We present a general formulation of metric learning for co-embedding, where the goal is to relate objects from different sets. The framework allows metric learning to be applied to a wide range of problems—including link prediction, relation learning, multi-label tagging and ranking—while allowing training to be reformulated as convex optimization. For training we provide a fast iterative algorithm that improves the scalability of existing metric learning approaches. Empirically, we demonstrate that the proposed method converges to a global optimum efficiently, and achieves competitive results in a variety of co-embedding problems such as multi-label classification and multi-relational prediction.

1 Introduction

The goal of *metric learning* is to learn a distance function that is tuned to a target task. For example, a useful distance between person images would be significantly different when the task is pose estimation versus identity verification. Since many machine learning algorithms rely on distances, metric learning provides an important alternative to hand-crafting a distance function for specific problems. For a single modality, metric learning has been well explored (Xing et al. 2002; Globerson and Roweis 2005; Davis et al. 2007; Weinberger and Saul 2008, 2009; Jain et al. 2012). However, for multi-modal data, such as comparing text and images, metric learning has been less explored, consisting primarily of a slow semi-definite programming approach (Zhang et al. 2011) and local alternating descent approaches (Xie and Xing 2013).

Concurrently, there is a growing literature that tackles *co-embedding problems*, where *multiple* sets or modalities are embedded into a common space to improve prediction performance, reveal relationships and enable zero-shot learning. Current approaches to these problems are mainly based on deep neural networks (Ngiam et al. 2011; Srivastava and Salakhutdinov 2012; Socher et al. 2013a, b; Frome et al. 2013) and simpler non-convex objectives (Chopra et al. 2005; Larochelle et al. 2008; Weston et al. 2010; Cheng 2013; Akata et al. 2013). Unlike metric learning, the focus of this previous work has been on exploring heterogeneous data, but without global optimization techniques. This disconnect

© Springer International Publishing Switzerland 2015
A. Appice et al. (Eds.): ECML PKDD 2015, Part I, LNAI 9284, pp. 625–642, 2015.
DOI: 10.1007/978-3-319-23528-8_39

appears to be unnecessary however, since the standard association scores used for co-embedding are related to a Euclidean metric.

In this paper, we demonstrate that co-embedding can be cast as metric learning. Once formalized, this connection allows metric learning methods to be applied to a wider class of problems, including link prediction, multi-label and multi-class tagging, and ranking. Previous formulations of co-embedding as metric learning were either non-convex (Zhai et al. 2013; Duan et al. 2012), introduced approximation (Akata et al. 2013; Huang et al. 2014), dropped positive semi-definiteness (Chechik et al. 2009; Kulis et al. 2011), or required all data to share the same dimensionality (Garreau et al. 2014). Instead, we provide a convex formulation applicable to heterogeneous data.

Once the general framework has been established, the paper then investigates optimization strategies for metric learning that guarantee convergence to a global optimum. Although many metric learning approaches have been based on convex formulations, these typically introduce a semi-definite constraint over a matrix variable, $C \succeq 0$, which hampers scalability. An alternative approach that has been gaining popularity has been to work with a low-rank factorization Q that implicitly maintains positive semi-definiteness through $C = QQ'$ (Burer and Monteiro 2003). This approach allows one to optimize over smaller matrices while avoiding the semi-definite constraint. Recently, Journée et al. (2010) proved that if Q has more columns than the globally optimal rank, a locally optimal Q^* provides a *global* solution $C^* = Q^*Q^{*\prime}$, provided that the objective is smooth and convex *in C*. This result is often neglected in the metric learning literature. However, by using this result, we are able to develop a fast approach to metric learning that improves previous approaches (Journée et al. 2010; Zhang et al. 2012).

The paper then concludes with an empirical investigation of a metric learning task and two co-embedding tasks: multi-label classification and tagging. We demonstrate that the diversity of local minima contracts rapidly in these problems and that local solutions approach global optimality well before the true rank is attained.

2   Metric Learning

The goal of metric learning is to learn a distance function between data instances that helps solve prediction problems. To obtain task-specific distances without extensive manual design, supervised metric learning formulations attempt to exploit task-specific information to guide the learning process. For example, to recognize individual people in images a distance function needs to emphasize certain distinguishing features (such as hair color, etc.), whereas to recognize person-independent facial expressions in the same data, different features should be emphasized (such as mouth shape, etc.).

Suppose one has a sample of t observations, $\mathbf{x}_i \in \mathcal{X}$, and a feature map $\phi : \mathcal{X} \to \mathbb{R}^n$. Then a training matrix $\phi(X) = [\phi(\mathbf{x}_1), \ldots, \phi(\mathbf{x}_t)] \in \mathbb{R}^{n \times t}$ can be obtained by applying ϕ to each of the original data points.[1] A natural distance function between points $\mathbf{x}_1, \mathbf{x}_2 \in \mathcal{X}$ can then be given by a Mahalanobis distance over the feature space

$$d_C(\mathbf{x}_1, \mathbf{x}_2) = (\phi(\mathbf{x}_1) - \phi(\mathbf{x}_2))'C(\phi(\mathbf{x}_1) - \phi(\mathbf{x}_2)) \tag{1}$$

specified by some positive semi-definite inverse covariance matrix $C \in \mathcal{C} \subset \mathbb{R}^{n \times n}$.

Although an inverse covariance in this form can be learned in an unsupervised manner, there is often side information that should influence the learning. As a general framework, Kulis (2013) unifies metric learning problems as learning a positive semi-definite matrix C that minimizes a sum of loss functions plus a regularizer:[2]

$$\min_{C \succeq 0, C \in \mathcal{C}} \sum_i L_i(\phi(X)'C\phi(X)) + \beta \operatorname{reg}(C). \tag{2}$$

For example, in large margin nearest neighbor learning, one might want to minimize

$$L(\phi(X)'C\phi(X)) = \sum_{(i,j) \in \mathcal{S}} d_C(\mathbf{x}_i, \mathbf{x}_j) + \sum_{(i,j,k) \in \mathcal{R}} [1 + d_C(\mathbf{x}_i, \mathbf{x}_j) - d_C(\mathbf{x}_i, \mathbf{x}_k)]_+$$

where $\mathcal{S}$ is a set of "should link" pairs, and $\mathcal{R}$ provides a set of triples (i, j, k) specifying that if $(i, j) \in \mathcal{S}$ then $\mathbf{x}_k$ should have a different label than $\mathbf{x}_i$.

Although supervised metric learning has typically been used for classification, it can also be applied to other settings where distances between data points are useful, such as for kernel regression or ranking. Interestingly, the applicability of metric learning can be extended well beyond the framework (2) by additionally observing that *co-embedding* elements from different sets can also be expressed as a *joint* metric learning problem.

3   Co-embedding as Metric Learning

Co-embedding considers the problem of mapping elements from distinct sets into a common (low dimensional) Euclidean space. Once so embedded, simple Euclidean proximity can be used to determine associations between elements from different sets. This idea underlies many useful formulations in machine learning. For example, in retrieval and recommendation, Bordes et al. (2014) use co-embedding of questions and answers to rank appropriate answers to a query, and Yamanishi (2008) embeds nodes of a heterogeneous graph for link prediction. In natural language processing, Globerson et al. (2007) embed documents, words and authors for semantic document analysis, while Bordes et al. (2012) embed words and senses for word sense disambiguation.

---

[1] Throughout the paper we extend functions $\mathbb{R} \to \mathbb{R}$ to vectors or matrices element-wise.

[2] Kulis (2013) equivalently places the trade-off parameter on the loss rather than the regularizer.

Despite the diversity of these formulations, we show that co-embedding can be unified in a simple metric learning framework. Such a unification is inspired by (Mirzazadeh et al. 2014), who proposed a general framework for bi-linear co-embedding models but did not investigate the extension to metric learning. Here we develop a full formulation of co-embedding as metric learning and develop algorithmic advances.

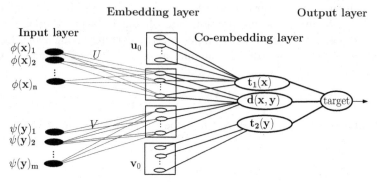

Fig. 1. A neural network view of co-embedding

For co-embedding, assume we are given two sets of data objects $\mathcal{X}$ and $\mathcal{Y}$ with feature maps $\phi(\mathbf{x}) \in \mathbb{R}^n$ and $\psi(\mathbf{y}) \in \mathbb{R}^m$ respectively. Without loss of generality, we assume that the number of samples from $\mathcal{Y}$, t_y, is no more than t, the number of samples from $\mathcal{X}$; that is, $t_y \leq t$. The goal is to map the elements $\mathbf{x} \in \mathcal{X}$ and $\mathbf{y} \in \mathcal{Y}$ from each set into a common Euclidean space.[3]

A standard approach is to consider linear maps into a common d dimensional space where $U \in \mathbb{R}^{d \times n}$ and $V \in \mathbb{R}^{d \times m}$ are parameters. To provide decision thresholds two dummy items can also be embedded from each space, parameterized by $\mathbf{u}_0$ and $\mathbf{v}_0$ respectively. Figure 1 depicts this standard co-embedding set-up as a neural network, where the trainable parameters, U, V, $\mathbf{u}_0$ and $\mathbf{v}_0$, are in the first layer. The inputs to the network are the feature representations $\phi(\mathbf{x}) \in \mathbb{R}^n$ and $\psi(\mathbf{y}) \in \mathbb{R}^m$. The first hidden layer, the *embedding layer*, linearly maps input to embeddings in a common d dimensional space via:

$$\mathbf{u}(\mathbf{x}) = U\phi(\mathbf{x}), \qquad \vec{(\mathbf{y})} = V\psi(\mathbf{y}).$$

The second hidden layer, the *co-embedding layer*, computes the distance function between embeddings, $d(\mathbf{x}, \mathbf{y})$, and decision thresholds, $t_1(\mathbf{x})$ and $t_2(\mathbf{y})$:

$$d(\mathbf{x}, \mathbf{y}) = \|\mathbf{u}(\mathbf{x}) - \vec{(\mathbf{y})}\|^2, \quad t_1(\mathbf{x}) = \|\mathbf{u}(\mathbf{x}) - \mathbf{u}_0\|^2, \quad t_2(\mathbf{y}) = \|\vec{(\mathbf{y})} - \mathbf{v}_0\|^2. \quad (3)$$

The output layer nonlinearly combines the association scores and thresholds to predict targets. For example, in a multi-label classification problem, given an element $\mathbf{x} \in \mathcal{X}$, its association to each $\mathbf{y} \in \mathcal{Y}$ can be determined via:

---

[3] The extension to more than two sets can be achieved by considering tensor representations.

$\text{label}(\mathbf{y}|\mathbf{x}) = \text{sign}(t_1(\mathbf{x}) - d(\mathbf{x}, \mathbf{y}))$. Alternatively, in a symmetric (i.e. undirected) link prediction problem, the association between a pair of elements $\mathbf{x} \in \mathcal{X}$, $\mathbf{y} \in \mathcal{Y}$ can be determined by $\text{label}(\mathbf{x}, \mathbf{y}) = \text{sign}(\min(t_1(\mathbf{x}), t_2(\mathbf{y})) - d(\mathbf{x}, \mathbf{y}))$, and so on.

Although the relationship to metric learning might not be obvious, it is useful to observe that the quantities in (3) can be expressed in terms of underlying covariances:

$$d(\mathbf{x}, \mathbf{y}) = \begin{bmatrix} \phi(\mathbf{x}) \\ -\psi(\mathbf{y}) \end{bmatrix}' \begin{bmatrix} U'U & U'V \\ V'U & V'V \end{bmatrix} \begin{bmatrix} \phi(\mathbf{x}) \\ -\psi(\mathbf{y}) \end{bmatrix} = \begin{bmatrix} \phi(\mathbf{x}) \\ -\psi(\mathbf{y}) \end{bmatrix}' C_1 \begin{bmatrix} \phi(\mathbf{x}) \\ -\psi(\mathbf{y}) \end{bmatrix}$$

$$t_1(\mathbf{x}) = \begin{bmatrix} \phi(\mathbf{x}) \\ -1 \end{bmatrix}' \begin{bmatrix} U'U & U'\mathbf{u}_0 \\ \mathbf{u}_0'U & \mathbf{u}_0'\mathbf{u}_0 \end{bmatrix} \begin{bmatrix} \phi(\mathbf{x}) \\ -1 \end{bmatrix} = \begin{bmatrix} \phi(\mathbf{x}) \\ -1 \end{bmatrix}' C_2 \begin{bmatrix} \phi(\mathbf{x}) \\ -1 \end{bmatrix}$$

$$t_2(\mathbf{y}) = \begin{bmatrix} \psi(\mathbf{y}) \\ -1 \end{bmatrix}' \begin{bmatrix} V'V & V'\mathbf{v}_0 \\ \mathbf{v}_0'V & \mathbf{v}_0'\mathbf{v}_0 \end{bmatrix} \begin{bmatrix} \psi(\mathbf{y}) \\ -1 \end{bmatrix} = \begin{bmatrix} \psi(\mathbf{y}) \\ -1 \end{bmatrix}' C_3 \begin{bmatrix} \psi(\mathbf{y}) \\ -1 \end{bmatrix}$$

where C_1, C_2 and C_3 are symmetric positive semi-definite matrices.

Although our previous work on bi-linear coembedding (Mirzazadeh et al. 2014) did not suggest embedding the thresholds, these turn out to be essential. In fact, to ensure the construction of a common metric space where the inverse covariances are mutually consistent (but without introducing auxiliary equality constraints), one must merge C_1, C_2 and C_3 into a common inverse covariance matrix, $C \in \mathbb{R}^{p \times p}$, $p = n + m + 2$, via:

$$C = \begin{bmatrix} U & V & \mathbf{u}_0 & \mathbf{v}_0 \end{bmatrix}' \begin{bmatrix} U & V & \mathbf{u}_0 & \mathbf{v}_0 \end{bmatrix} \tag{4}$$

From (4), the distance functions d, t_1 and t_2, can then be expressed by

$$\begin{aligned} d(\mathbf{x}, \mathbf{y}) &= [\phi(\mathbf{x}), \ -\psi(\mathbf{y}), \ 0, \ 0] \ C \ [\phi(\mathbf{x}), \ -\psi(\mathbf{y}), \ 0, \ 0]' \\ t_1(\mathbf{x}) &= [\phi(\mathbf{x}), \ 0, \ -1, \ 0] \ C \ [\phi(\mathbf{x}), \ 0, \ -1, \ 0]' \\ t_2(\mathbf{y}) &= [0, \ -\psi(\mathbf{y}), \ 0, \ -1] \ C \ [0, \ -\psi(\mathbf{y}), \ 0, \ -1]'. \end{aligned} \tag{5}$$

This yields a novel distance function representation with mutually consistent thresholds.

Finally, based on this new representation, we can extend the general framework (2) to encompass co-embedding in a novel formulation. Let $Y \in \mathbb{R}^{t_y \times m}$ denote the data matrix from the $\mathcal{Y}$ space and let $\widehat{\psi}(Y) \in \mathbb{R}^{t \times m}$ denote a zero-padded version of $\psi(Y)$; that is, a matrix whose top $t_y \times m$ block is $\psi(Y)$ with the remaining $t - t_y$ rows being all zero. Then, defining $\mathbf{f}(X, Y) = [\phi(X)', -\widehat{\psi}(Y)', -\mathbf{1}, -\mathbf{1}]' \in \mathbb{R}^{t \times (n+m+2)}$, where $\mathbf{1}$ denotes an all-one vector (of dimension t in this case), we propose to find C by solving

$$\min_{C \in \mathbb{R}^{p \times p}, C \succeq 0} \sum_i L_i(\mathbf{f}(X, Y)' C \mathbf{f}(X, Y)) + \beta \operatorname{reg}(C) . \tag{6}$$

Duan et al. (2012) developed a similar algorithm for domain adaptation, which learned a matrix $C \succeq 0$ instead of U and V; however, they approached a less general setting, which, for example, did not include thresholds nor general losses. Furthermore, their formulation leads to a non-convex optimization problem.

Regularization. Regularization is also an important consideration since the risk of over-fitting is ever present. We focus on the most widely used regularizer, the Frobenius norm, which if applied to the factors yields the trace norm regularizer on C:

$$\|U\|_F^2 + \|V\|_F^2 + \|\mathbf{u}_0\|_F^2 + \|\mathbf{v}_0\|_F^2 = \text{tr}(C) = \|C\|_{\text{tr}}.$$

The trace norm (aka nuclear norm) is the sum of the singular values of C. This is a common choice for metric learning since it is the tightest convex lower bound to the rank of a matrix, a widely desired objective for compact learned models and generalization. Moreover, for metric learning, since we have the constraint $C \succeq 0$, the trace norm simplifies to $\|C\|_{\text{tr}} = \text{tr}(C)$, which allows efficient optimization.

4 Algorithm

Given the formulation (6), we consider how to efficiently solve it. First note that the objective can be written, using $L(C) = \sum_i L_i(\mathbf{f}(X,Y)'C\mathbf{f}(X,Y))$, as

$$\min_{C \in \mathbb{R}^{p \times p}, C \succeq 0} f(C) \qquad \text{where } f(C) = L(C) + \beta \, \text{tr}(C). \tag{7}$$

One way to encode the semi-definite constraint is via a change of variable $C = QQ'$:

$$\min_{Q \in \mathbb{R}^{p \times d}} f(QQ') = \min_{Q \in \mathbb{R}^{p \times d}} L(QQ') + \beta \, \text{tr}(QQ'). \tag{8}$$

This optimization, however, becomes non-convex in Q. Recently, however, Journée et al. (2010) showed that local optimization of a related trace constrained problem attains global solutions for rank-deficient local minima $Q \in \mathbb{R}^{p \times d}$; that is, if Q is a local minimum of (8) with $\text{rank}(Q) < d$, then QQ' is a global optimum of (7). In what follows, C^* will denote an optimum of (7) and d^* its rank. Although we have inequality rather than equality constraints, the proof follows easily for our case using the techniques developed in (Bach et al. 2008; Journée et al. 2010; Haeffele et al. 2014), and is an easy consequence of the following, more general result.

Proposition 1. *Consider a local solution of (8), yielding a Q such that $\nabla L(QQ')Q + \beta Q = 0$. Let $\mathbf{u}_1, ..., \mathbf{u}_k$ be the eigenvectors corresponding to the top k **positive** eigenvalues $\lambda_1, ..., \lambda_k$ of $-\nabla L(C) - \beta I$. Then, if C is not a solution to (7), it follows that*

1. *$k > 0$*
2. *$\mathbf{u}_1, ..., \mathbf{u}_k$ are orthogonal to Q, yielding $Q_k = [Q \; \mathbf{u}_1 \; ... \; \mathbf{u}_k]$ such that $C_k = Q_k Q_k' = C + \sum_{i=1}^k \mathbf{u}_i \mathbf{u}_i'$ satisfies $\text{rank}(C_k) = \text{rank}(C) + k$; and*
3. *the descent direction $\sum_{i=1}^k \mathbf{u}_i \mathbf{u}_i'$ is the solution to*

$$\operatorname*{argmin}_{\substack{\|\mathbf{u}_i\| \leq 1, i=1,...,k \\ \mathbf{u}_i' \mathbf{u}_j = 0, i \neq j, \; \mathbf{u}_i \neq 0}} \left\langle -\nabla L(C) - \beta I, \sum_{i=1}^k \mathbf{u}_i \mathbf{u}_i' \right\rangle. \tag{9}$$

Proof. **Part 1:** First, form the Lagrangian of (7), given by $L(C) + \beta \operatorname{tr}(C) - \operatorname{tr}(SC)$ with $S \succeq 0$, and consider the KKT conditions:

$$S = \nabla L(C) + \beta I, \quad S \succeq 0, \quad C \succeq 0, \quad SC = 0. \tag{10}$$

The problem is strictly feasible, since $C = I$ is a strictly feasible point; therefore, Slater's condition holds and (10) is sufficient for optimality. Consequently, an optimal solution is reached when $-S \preceq 0$; that is, the largest eigenvalue of $-\nabla L(C) - \beta I$ is negative or zero. We assumed that C is not optimal, therefore $k > 0$.

Part 2: We know that $0 = \nabla L(QQ')Q + \beta Q = SQ$. Therefore, either $S = 0$, in which case we are at a global minimum (which we assumed was not the case) or S is orthogonal to Q. It follows that $-\lambda_i \mathbf{u}_i' Q = (\mathbf{u}_i' S')Q = \mathbf{u}_i'(S'Q) = \mathbf{u}_i' 0 = \mathbf{0}$ since $\mathbf{u}_i$ is an eigenvector of S and S is symmetric.

Part 3: To optimize the inner product (9), introduce Lagrange multipliers $\xi_i > 0$ for the norm constraints. Since $-S$ is symmetric, we can re-express the inner objective as

$$\underset{\substack{\mathbf{u}_1,\ldots,\mathbf{u}_k \\ \mathbf{u}_i'\mathbf{u}_j=0, i\neq j, \ \mathbf{u}_i\neq 0}}{\operatorname{argmin}} \sum_i \mathbf{u}_i'(-S)'\mathbf{u}_i - \sum_i \xi_i \mathbf{u}_i'\mathbf{u}_i.$$

Considering the gradients yields $\frac{\partial}{\partial \mathbf{u}_i} = -S\mathbf{u}_i - 2\xi_i \mathbf{u}_i = 0$, which implies $(-S)\mathbf{u}_i = 2\xi_i \mathbf{u}_i$; that is $\mathbf{u}_i$ is an eigenvector of $-S$ corresponding to eigenvalue $\lambda_i = 2\xi_i > 0$. $\qquad\square$

Corollary 1. *Let $Q \in \mathbb{R}^{p\times d}$. If (i) Q is a local minimum of $f(QQ')$ with $\operatorname{rank}(Q) < d$ or (ii) Q is a critical point of $f(QQ')$ with $\operatorname{rank}(Q) = p$, then QQ' is a solution of (7).*

Proof. First assume condition (i) holds and argue by contradiction. Assume QQ' is not a global optimum of (7), and let $\mathbf{u}_1 \in \mathbb{R}^p$ be as defined as in Proposition 1. Then, $f(QQ' + \beta \mathbf{u}_1 \mathbf{u}_1') < f(QQ')$ for a sufficiently small $\beta > 0$. Furthermore, since $\operatorname{rank}(Q) < d$, there exists an orthogonal matrix $V \in \mathbb{R}^{d\times d}$ such that QV has a zero column. Let $\hat{Q}_\alpha$ be the matrix obtained from QV by replacing this zero column by $\alpha \mathbf{u}_1$, $\alpha = \sqrt{\beta}$. Then $\lim_{\alpha \to 0} \hat{Q}_\alpha V' = QVV' = Q$. Moreover, since $\mathbf{u}_1$ is orthogonal to the columns of Q, it is also orthogonal to the columns of QV, so $\hat{Q}_\alpha V'(\hat{Q}_\alpha V')' = QV(QV)' + \alpha^2 \mathbf{u}_1 \mathbf{u}_1' = QQ' + \beta \mathbf{u}_1 \mathbf{u}_1'$. Therefore, $f(\hat{Q}_\alpha \hat{Q}_\alpha') = f(QQ' + \beta \mathbf{u}_1 \mathbf{u}_1') < f(QQ')$ for $Q_\alpha \in \mathbb{R}^{p\times d}$, hence Q is not a local optimum of f.

Next assume (ii). Since Q is a critical point of $f(QQ')$, $\nabla f(QQ')Q = 0$. Since Q has rank p, the null-space of $\nabla f(QQ')$ is of dimension p, yielding that $\nabla f(QQ') = 0$. Since $QQ' \succeq 0$ and f is convex, $C = QQ'$ is an optimum of (7). $\qquad\square$

To efficiently solve (7), we therefore propose the Iterative Local Algorithm (ILA) shown in Algorithm 1. ILA iteratively adds multiple columns to an initially

---

Algorithm 1 Iterative local algorithm (ILA)

1: **Input:** $L : \mathcal{C} \to \mathbb{R}, \beta > 0$
2: **Output:** Q, such that $QQ' = \min_{C:C \succeq 0} L(C) + \beta \operatorname{tr}(C)$
3: $Q \leftarrow 0, k \leftarrow 1, \epsilon \leftarrow 10^{-6}$ ▷ Note $L(QQ') + \operatorname{tr}(QQ')$ is evaluable without forming QQ'
4: **while** not converged **do**
5:    $\{\mathbf{u}_1, ..., \mathbf{u}_j\} \leftarrow$ up-to-k-top-positive-eigenvectors$(-\nabla L(QQ') - \beta I)$
6:    $\{\lambda_1, ..., \lambda_j\} \leftarrow$ up-to-k-top-positive-eigenvalues$(-\nabla L(QQ') - \beta I)$
7:    **if** $k = 0$ or $\lambda_1 \leq \epsilon$ **then** break   ▷ converged
8:    $k \leftarrow j$
9:    $U \leftarrow \sum_i \mathbf{u}_i \mathbf{u}_i'$
10:   $(a, b) \leftarrow \operatorname*{argmin}_{a \geq 0, b \geq 0} L(aQQ' + bU) + \beta a \operatorname{tr}(QQ') + \beta bk$ ▷ Line search
11:   $Q_{init} \leftarrow [\sqrt{a}Q, \sqrt{b}\mathbf{u}_1, ..., \sqrt{b}\mathbf{u}_k]$ ▷ Start local optimization from Q_{init}
12:   $Q \leftarrow$ locally_optimize$(Q_{init}, L(QQ') + \beta \operatorname{tr}(QQ'))$
13:   $k \leftarrow 2k$
14: **return** $C = QQ'$

---

empty Q and performs a local optimization over $Q \in \mathbb{R}^{p \times d}$ until convergence. The main advantage of this approach over simply setting $d = p$ is that good initial points are generated, and if $d^* \ll p$, then incrementally growing d optimizes over much smaller Q variables. Furthermore, one hopes that when the number of columns d of Q_{init} is at least d^*, ILA finds the global optimum. In particular, if the local optimizer in line 12 of ILA always returns a local optimum whose rank is smaller than d if $d > d^*$ (we call this a *nice local optimizer*), then the optimality of a rank-deficient local minimum implies that ILA finds the global optimum when $d > d^*$. While in theory we cannot guarantee such a behavior of the local algorithm, it always happened in our experiments, similarly to what was reported in earlier work (Journée et al. 2010; Haeffele et al. 2014).

The main novelty of ILA over previous approaches is in the initialization and expansion of columns in Q, which reduces the number of iterations from d^* to $O(\log d^*)$ for nice local optimizers. In particular, motivated by Proposition 1, to generate the candidate columns, ILA uses eigenvectors corresponding to the top k positive eigenvalues of $-\nabla L(C) - \beta I$ capped at 2^{i-1} columns on the ith iteration. Such an exponential search quickly covers the space of possible d, even when d^* is large, while still initially optimizing over smaller Q matrices. This approach can be significantly faster than the typical single column increment (Journée et al. 2010; Zhang et al. 2012), whose complexity typically grows linearly with d^*.[4]

Compared to earlier work, there are also small differences in the optimization: Zhang et al. (2012) do not constrain C to be positive semi-definite. Journée et al. (2010) assume an equality constraint on the trace of C; their Lagrange variable (i.e., regularization parameter) can therefore be negative. Finally, ILA more effi-

---

[4] One can create problems where adding single columns improves performance, but we observe in our experiments that the proposed approach is more effective in practice.

ciently exploits the local algorithm. The convergence analysis of Zhang et al. (2012) does not include local training. In practice, we find that solely using boosting (with the top eigenvector as the weak learner) without local optimization, results in much slower convergence.

Corollary 1 implies ILA solves (7) when the local optimizer avoids saddle points.

Corollary 2. *Suppose the local optimizer always finds a local optimum, where d is the number of columns in Q. Then ILA stops with a solution to (7) in line 12 with $\text{rank}(Q) < d$ or $d = p$. If, in addition, the local optimizer is* nice, *this happens for $d > d^*$.*

Due to the exponential search in ILA, the algorithm stops in essentially at most $\log(p)$ iterations when the local optimizer avoids saddle points, and in about $\log(d^*)$ iterations for *nice* local optimizers. However, ILA can potentially be slower if there are not enough eigenvectors to add in a given iteration; i.e., $j < k$ in line 5.

Similarly to (Journée et al. 2010; Zhang et al. 2012; Haeffele et al. 2014) we have found that the local optimizer always returns local minima in practice. However, all of these search-based algorithms risk strange behavior if the local optimizer returns saddle points. Note that even in this case, if d reaches p in any iteration, ILA finds an optimum by Corollary 1. However, there is no guarantee that the rank of Q is not reduced in the local optimization step. If this happens and Q is a local optimum, QQ' is optimal by Corollary 1 and the algorithm halts. Unfortunately, this is not the only possibility: in every iteration of ILA we obtain Q_{init} by increasing the rank of the previous Q, but the ranks might be subsequently reduced during the local optimization step. This creates the potential for a loop where $\text{rank}(Q)$ never reaches p.

Such potential effects of saddle points have not been considered in previous papers. However, we close this section by showing that ILA is still consistent under mild technical conditions on L, even if the local optimizer can get trapped in saddle points.

Proposition 2. *Suppose that f is ν-smooth; that is, $\|\nabla f(C+S) - \nabla f(C)\|_{\text{tr}} \leq \nu \rho(S)$ for all $C, S \in \mathbb{R}^{p \times p}$, $C, S \succeq 0$ and some $\nu \geq 0$, where $\rho(S)$ denotes the spectral norm of S. Assume furthermore, for simplicity, that $L(C) \geq 0$ for all $C \succeq 0$. If the local optimizer in line 12 always returns a Q such that $\nabla f(QQ')Q = 0$, then QQ' in ILA converges to the globally optimal solution of (7).*

Proof. Let Q_m and U_m denote the matrix Q and U in ILA when line 10 is executed the mth time, and let $Q_{init,m}$ denote Q_{init} obtained from Q_m. Note that $Q_{init,m} = \sqrt{a} Q_m + \sqrt{b} U_m$ and Q_{m+1} is obtained from $Q_{init,m}$ via local optimization in line 12. Furthermore, let $C_m = Q_m Q'_m$ and $C_{init,m} = Q_{init,m} Q'_{init,m} = a_m C_m + b_m U_m U'_m$.

If C_m is not a global optimum of (7), then $f(C_{init,m}) < f(C_m)$ by Proposition 1. Furthermore, we assume that the local optimizer in line 12 cannot increase the function value f of $C_{init,m}$, hence $f(C_{m+1}) \leq f(C_{init,m})$, and consequently $f(C_{m+1}) < f(C_m)$. Note that since $L(C_m) \geq 0$, we have $\|Q_m\|_F = \text{tr}(C_m) \leq f(C_0)$, thus the entries of C_m are uniformly bounded for all m. Therefore, $(C_m)_m$ has a convergent subsequence, and denote its limit point by $\widehat{C}$. We will show that $\widehat{C}$ is an optimal solution of (7) by verifying the KKT conditions (10) with $S = \nabla f(\widehat{C})$. First notice that $\widehat{C}$ is positive semi-definite, $\nabla f(\widehat{C})\widehat{C} = 0$ by continuity since $\nabla f(C_m)C_m = \nabla f(QQ')QQ' = 0$. Thus, we only need to verify that $\nabla f(\widehat{C})$ is positive semi-definite.

To show the latter, we first apply Lemma 1 (provided in the appendix) to obtain a lower bound ILA's progress:

$$f(C_{m+1}) \leq f(C_{init,m+1}) = f(aC_m + bU_mU'_m) \leq f(C_m + \hat{b}U_mU'_m)$$
$$\leq f(C_m) + \text{tr}((\hat{b}U_mU'_m)'\nabla f(C_m)) + \frac{\nu}{2}\rho(\hat{b}U_mU'_m)^2$$
$$= f(C_m) + \text{tr}(\hat{b}U'_m\nabla f(C_m)U_m) + \frac{\nu\hat{b}^2}{2} \qquad (11)$$

for any $\hat{b} \geq 0$, where the last equality holds since $U_mU'_m$ has k_m eigenvalues equal 1, and $p - k_m$ equal 0, where k_m denotes the number of columns of U_m. Now consider

$$\hat{b} = -\frac{\text{tr}(U'_m\nabla f(C_m)U_m)}{\nu} = \frac{\text{tr}(U'_m\Lambda_mU_m)}{\nu} = \frac{1}{\nu}\sum_{i=1}^{k_m}\lambda_{m,i},$$

where $\lambda_1 \geq \cdots \geq \lambda_{k_m} > 0$ are the eigenvalues of $-\nabla f(C_m)$, and Λ_m is the diagonal matrix of the eigenvalues padded with $p - m_k$ zeros. Then $\text{tr}(\hat{b}U'_m\nabla f(C_m)U_m) = -\nu\hat{b}^2$, hence (11) yields

$$f(C_m) - f(C_{m+1}) \geq \frac{\nu}{2}\hat{b}^2 = \frac{1}{\nu}\left(\sum_{i=1}^{k_m}\lambda_{m,i}\right)^2 \geq \frac{\lambda_{m,1}^2}{2\nu}.$$

By our assumptions, $f(C_0) \geq 0$, and so using the monotonicity of $f(C_m)$, we have

$$f(C_0) \geq \lim_{m\to\infty} f(C_0) - f(C_{m+1}) = \lim_{m\to\infty}\sum_{i=0}^{m} f(C_i) - f(C_{i+1}) \geq \frac{1}{2\nu}\sum_{m=0}^{\infty}\lambda_{m,1}^2.$$

Therefore, $\lim_{m\to\infty}\lambda_{m,1} = 0$. Thus, by continuity, $-\nabla f(\widehat{C})$ has no positive eigenvalues, implying that $\nabla f(\widehat{C})$ is positive semi-definite, concluding the proof. $\qquad\square$

5 Empirical Computational Complexity

To compare the exponential versus linear rank expansion strategies for ILA we first consider a standard metric learning problem. In this experiment to control

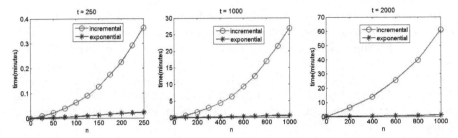

Fig. 2. Comparing the run time in minutes (y-axis) of linear versus exponential strategies in ILA as data dimension (x-axis) is increased. Left shows $t = 250$, middle shows $t = 1000$, and right shows $t = 2000$.

the rank of the solution, we generated synthetic data $X \in \mathbb{R}^{n \times t}$ from a standard normal distribution, systematically increasing the data dimension from $n = 1$ to $n = 1000$ and increasing the sample sizes from $t = 250$ to $t = 2000$. The training objective was set to

$$\min_{C \succeq 0} \|X'X - X'CX\|_F^2 + \beta \operatorname{tr}(C) \tag{12}$$

with a regularization parameter $\beta = 0.5$.

Figure 2 compares the run times of the linear versus exponential expansion strategies, both of which optimize over Q of increasing width rather than $C = QQ'$. Both methods used the same local optimizer but differed in how many new columns were generated for Q in ILA Line 8. For the smaller sample size $t = 250$, the exponential search already demonstrates an advantage as data dimension is increased. However, for larger sample sizes, the advantage of the exponential approach becomes even more pronounced. In this case, when n is increased from 0 to 1000 the run time of the linear expansion strategy goes from being about the same as of the exponential strategy to much slower. The trend indicates that the exponential search becomes more useful as the data dimension and number of samples increases.

6   Case Study: Multi-label Classification

Next, we evaluated ILA on a challenging problem setting—multi-label classification—with real data. In this setting one can view the labels themselves as objects to be co-embedded with data instances; given such an embedding, the multi-label classification of an input instance $\mathbf{x}$ can be determined by comparing the distance of its embedding to the embedded locations of each label. In particular, given a feature representation $\phi(\mathbf{x}) \in \mathbb{R}^n$ for data instances $\mathbf{x} \in \mathcal{X}$, we introduce a simple indicator feature map $\psi(\mathbf{y}) \in \mathbb{R}^m$ over $\mathbf{y} \in \mathcal{Y}$, which specifies a vector of all zeros with a single 1 in the entry corresponding to label $\mathbf{y}$. From a co-embedding perspective, the training problem then becomes to map the feature representations of both the input instances $\mathbf{x} \in \mathcal{X}$ and target labels $\mathbf{y} \in \mathcal{Y}$ into a common Euclidean space.

Based on this observation, we can then cast multi-label learning as an equivalent *metric learning* problem where one learns the inverse covariance C. Following the development in Section 3 (but here not using the threshold for $\mathbf{y}$ since it is not needed), the co-embedding parameters U, V and $\mathbf{u}_0$ can first be combined into a joint matrix $Q = [U, V, \mathbf{u}_0] \in \mathbb{R}^{p \times d}$, where $p = n + m + 1$. Then, as in (4), the co-embedding problem of optimizing U, V and $\mathbf{u}_0$ can be equivalently expressed as a metric learning problem of optimizing the inverse covariance $C = QQ' \in \mathbb{R}^{p \times p}$.

Training Objective. To develop a novel metric learning based approach to multi-label classification, we adopt a standard training loss that encourages small distances between an instance's embedding and the embeddings of its associated labels while encouraging large distances to embeddings of disassociated labels. In particular, we investigate the convex large margin loss suggested by Mirzazadeh et al. (2014) which was reported to yield good performance for multi-label classification (in a bilinear co-embedding model but not a metric learning model):

$$\min_{C \succeq 0} \beta \operatorname{tr}(C) + \sum_{x \in \mathcal{X}} \left[\operatorname*{sftmx}_{y \in \mathcal{Y}(\mathbf{x})} \tilde{h}(d_C(\mathbf{x}, \mathbf{y}) - t_C(\mathbf{x})) + \operatorname*{sftmx}_{\bar{\mathbf{y}} \in \bar{\mathcal{Y}}(\mathbf{x})} \tilde{h}(t_C(\mathbf{x}) - d_C(\mathbf{x}, \bar{\mathbf{y}})) \right] \quad (13)$$

where $\operatorname{sftmx}_{y \in \mathcal{Y}}(z_y) = \ln \sum_{y \in \mathcal{Y}} \exp(z_y)$, $t_C(x) = [\phi(\mathbf{x}), \mathbf{0}, -1] \, C \, [\phi(\mathbf{x}), \mathbf{0}, -1]'$, $d_C(\mathbf{x}, \mathbf{y}) = [\phi(\mathbf{x}), -\psi(\mathbf{y}), 0] \, C \, [\phi(\mathbf{x}), -\psi(\mathbf{y}), 0]'$ and $\tilde{h}(z) = (2 + z)_+^2 / 4$ if $0 \le z \le 2$; $(1 + z)_+$ otherwise. Here we are using $\mathcal{Y}(\mathbf{x}) \subset \mathcal{Y}$ to denote the subset of labels associated with $\mathbf{x}$, and $\bar{\mathcal{Y}}(\mathbf{x}) \subset \mathcal{Y}$ to denote the subset of labels disassociated with $\mathbf{x}$.

Note that in (13) we also use Frobenius norm regularization on the co-embedding parameters U, V and $\mathbf{u}_0$, which was shown in Section 3 to yield trace regularization of C: $\|U\|_F^2 + \|V\|_F^2 + \|\mathbf{u}_0\|_2^2 = \operatorname{tr}(U'U) + \operatorname{tr}(V'V) + \mathbf{u}_0' \mathbf{u}_0 = \operatorname{tr}(C)$.

Results. We investigate the behavior of ILA on five widely used multi-label classification data sets, summarized in Table 1. To establish the suitability of metric learning for multi-label classification, we evaluated test performance using three commonly used criteria for multi-label classification: Hamming score (Table 2), micro averaged F1 measure (Table 3) and macro averaged F1 measure (Table 4). Here β was chosen by cross-validation over

Table 1. Data properties for multi-label experiments. 1000 used for training and the rest for testing (2/3-1/3 split for Emotion).

Data set	examples	features	labels
Emotion	593	72	6
Scene	2407	294	6
Yeast	2417	103	14
Mediamill	3000	120	30
Corel5K	4609	499	30

$\{1, 0.5, 0.1, 0.05, 0.01, 0.005\}$. We compared the performance of the proposed approach against six standard competitors: BR(SMO), an independent SVM classifiers for each label (Platt 1998); BR(LOG), an independent logistic regression (LOG) classifiers for each label (Hastie et al. 2009); CLR(SMO) and CLR(LOG), the calibrated pairwise label ranking method of Fürnkranz et al. (2008) with SVM and LOG, respectively; and CC(SMO) and CC(LOG), a chain of SVM classifiers and a chain of logistic regression classifiers for multi-label classification by Read et al. (2011). The results in Tables 2–4 are averaged over 10 splits and demonstrate comparable performance to the best competitors consistently in all three criteria for all data sets.

Table 2. Comparison of ILA with competitors in terms of Hamming score.

	BR(SMO)	BR(LOG)	CLR(SMO)	CLR(LOG)	CC(SMO)	CC(LOG)	ILA
Emotion	80.9 ±1.0	77.1 ±1.2	79.9 ±0.7	76.0 ±1.4	79.0 ±0.9	75.2 ±1.1	80.2 ±0.8
Scene	88.7 ±0.4	81.9 ±0.6	89.7 ±0.3	85.7 ±0.4	88.9 ±0.4	80.9 ±0.4	88.0 ±0.5
Yeast	79.8 ±0.2	77.0 ±0.2	77.2 ±0.2	75.3 ±0.3	78.9 ±0.5	76.0 ±0.2	78.9 ±0.3
Mediamill	90.3 ±0.1	87.4 ±0.2	87.8 ±0.1	87.7 ±0.1	89.9 ±0.1	86.3 ±0.3	90.4 ±0.5
Corel5K	89.8 ±0.1	88.5 ±0.2	88.8 ±0.1	88.0 ±0.1	89.6 ±0.1	83.1 ±0.4	87.8 ±0.4

Table 3. Comparison of ILA with competitors in terms of Micro F1.

	BR(SMO)	BR(LOG)	CLR(SMO)	CLR(LOG)	CC(SMO)	CC(LOG)	ILA
Emotion	66.3 ±2.3	63.2 ±1.8	70.1 ± 1.2	64.5 ± 2.1	65.9 ± 1.8	60.3 ± 1.9	65.9 ± 1.3
Scene	66.8 ±1.0	49.5 ±1.5	72.2 ± 0.7	61.8 ± 1.3	68.8 ± 1.1	50.1 ± 1.1	65.9 ± 0.8
Yeast	63.2 ±0.3	62.0 ±0.4	65.0 ± 0.3	61.9 ± 0.4	63.7 ± 0.8	60.0 ± 0.4	62.4 ± 0.5
Mediamill	55.4 ±0.5	55.1 ±0.6	59.7 ± 0.4	58.7 ± 0.4	50.7 ± 0.9	53.1 ± 0.7	58.0 ± 0.7
Corel5K	21.9 ±0.7	17.4 ±0.5	27.6 ± 0.4	26.3 ± 0.5	21.9 ± 0.5	16.7 ± 0.6	21.9 ± 0.6

Table 4. Comparison of ILA with competitors in terms of Macro F1.

	BR(SMO)	BR(LOG)	CLR(SMO)	CLR(LOG)	CC(SMO)	CC(LOG)	ILA
Emotion	62.3 ±3.1	62.0 ±1.9	69.0 ±1.0	63.8 ±2.0	64.3 ±1.8	59.3 ±2.0	64.4 ±1.4
Scene	67.6 ±0.9	50.6 ±1.6	73.3 ±0.6	63.3 ±1.3	69.8 ±1.0	50.9 ±1.0	66.8 ±0.9
Yeast	32.9 ±0.7	41.9 ±0.8	40.3 ±0.6	42.6 ±0.7	35.1 ±0.4	40.4 ±0.4	37.8 ±0.8
Mediamill	10.0 ±0.4	29.9 ±0.7	21.4 ±0.7	31.7 ±0.8	8.9 ±1.0	29.5 ±0.8	16.2 ±0.9
Corel5K	17.8 ±0.4	11.6 ±0.4	21.4 ±0.5	22.0 ±0.5	17.6 ±0.5	14.4 ±0.6	17.8 ±0.6

Next, to also investigate the properties of the local optima achieved we ran local optimization from 1000 random initializations of Q at successive values for d, using $\beta = 1$. The values of the local optima we observed are plotted in Figure 3 as a function of d.[5] As expected, the local optimizer always achieves the globally optimal value when $d \geq d^*$. Interestingly, for $d < d^*$ we see that the initially wide diversity of local optimum values contracts quickly to a singleton, with values approaching the global minimum before reaching $d = d^*$. Although not displayed in the graphs, other useful properties can be observed. First, for $d \geq d^*$, the global optimum is achieved by local optimization under random initialization, but not with initialization to any of the critical points of smaller

---

[5] Note that Q is not unique since $C = QQ'$ is invariant to transform QR for orthonormal R.

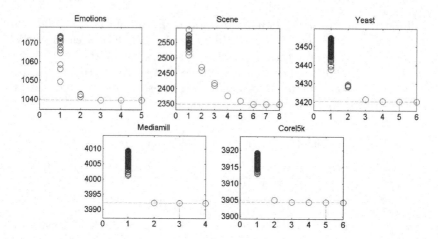

Fig. 3. Objective values achieved by local optimization given 1000 initializations of $Q \in \mathbb{R}^{p \times d}$. For small d a diversity of local minima are observed, but the set of local optima contracts rapidly as d increases, reaching a singleton at the global optimum by $d = d^*$.

d observed in Figure 3, which traps the optimization in a saddle point. Overall, empirically and theoretically, we find that ILA quickly finds global solutions for the multi-label objective, while typically producing good solutions before $d = d^*$.

7   Case Study: Tagging via Tensor Completion

Finally, we investigated Task 2 of the 2009 ECML/PKDD Discovery Challenge: a multi-relational problem involving users, items and tags, where users have tagged subsets of the items and the goal is to predict which tags the users will assign to other items. Here the training data is given in a tensor T, where $T(x, y, z) = 1$ indicates that x has tagged z with y, $T(x, y, z) = -1$ indicates that y is not a tag of z according to x, and $T(x, y, z) = 0$ denotes an unknown entry. The goal is to predict the unknown values, subject to a constraint that at most five tags can be active for any user-item pair. The "core at level 10" subsample reduces the data to 109, 192, 229 unique users, items, and tags respectively (Jäschke et al. 2008). The winner of this challenge (Rendle and Schmidt-Thieme 2009) used a multi-linear co-embedding model that assumed the completed tensor has a low rank structure.

Training Objective. To show that this multi-relational prediction problem can be tackled from the novel perspective of metric learning, we first express the problem in terms of a multi-way co-embedding where users, tags and items are mapped to a joint embedding space: $x \mapsto \sigma$, $y \mapsto \tau$ and $z \mapsto \rho$ where σ, τ, $\rho \in \mathbb{R}^d$. The training problem can then be expressed in terms of proximities between embeddings. In particular, following Rendle and Schmidt-Thieme (2009), we summarize the three-way interaction between a user, item and tag

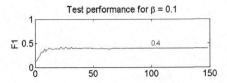

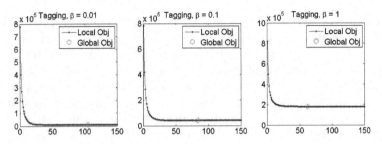

Fig. 4. F1 measure achieved by ILA on test data with an increasing number of columns (optimal rank is 84 in this case).

Fig. 5. Training objectives for $\beta \in \{0.01, 0.1, 1\}$ as a function of the rank of C, where the optimal ranks are 105, 84 and 62 respectively.

by the squared distance between the user and tag embeddings, and between the item and tag embeddings: $d(x, y, z) := d(x, y) + d(z, y) = \|\sigma - \tau\|^2 + \|\rho - \tau\|^2$. Given this definition, tags can be predicted from a given user-item pair (x, z) via

$$\hat{T}(x, y, z) = \begin{cases} 1 & \text{if } d(x, y, z) \text{ among smallest five } d(x, \cdot, z) \\ -1 & \text{otherwise} \end{cases}.$$

The training problem can be expressed as metric learning by exploiting a construction reminiscent of Section 3: the embedding vectors can conceptually be stacked in matrix factor $Q = [\sigma, \ \tau, \ \rho]'$, which defines the inverse covariance $C = QQ'$. To learn C, we use the same loss proposed by Rendle and Schmidt-Thieme (2009), regularized by the Frobenius norm over σ, τ and ρ (which again corresponds to trace regularization of C), yielding the convex training problem

$$\min_{C \succeq 0} \quad \beta \operatorname{tr}(C) + \sum_{x, z} \sum_{y \in tag(x, z)} \sum_{\bar{y} \notin tag(x, z)} L(d_C(x, z, \bar{y}) - d_C(x, z, y)). \quad (14)$$

Results. To establish the suitability of metric learning for multi-relational prediction, we first evaluated the test performance achieved on the down-sampled Discovery Challenge data. Figure 4 shows that ILA efficiently approaches the state of the art F1 performance of 0.42 reported by Mirzazadeh et al. (2014). Furthermore, we also investigated the behavior of local minima at different d by comparing the training objective values achieved by local optimization compared to the global minimum, here using $\beta \in \{0.01, 0.1, 1\}$. Figure 5 shows that although the optimal rank can be larger in this scenario, the properties of the local solutions become even more apparent: interestingly, the local minima approach the training global minimum at ranks much smaller than the optimum. These results further support the effectiveness of metric learning and the potential for ILA to solve these problems much more efficiently than standard semi-definite programming approaches.

8   Conclusion

We have demonstrated a unification of co-embedding and metric learning that enables a new perspective on several machine learning problems while expanding the range of applicability for metric learning methods. Additionally, by using recent insights from semi-definite programming theory, we developed a fast local optimization algorithm that is able to preserve global optimality while significantly improving the speed of existing methods. Both the framework and the efficient algorithm were investigated in different contexts, including metric learning, multi-label classification and multi-relational prediction—demonstrating their generality. The unified perspective and general algorithm show that a surprisingly large class of problems can be tackled from a simple perspective, while exhibiting a local-global property that can be usefully exploited to achieve faster training methods.

References

Akata, Z., Perronnin, F., Harchaoui, Z., Schmid, C.: Label-embedding for attribute-based classification. In: IEEE Conference on Computer Vision and Pattern Recognition (2013)

Bach, F., Mairal, J., Ponce, J.: Convex sparse matrix factorizations. CoRR (2008)

Bordes, A., Glorot, X., Weston, J., Bengio, Y.: Joint learning of words and meaning representations for open-text semantic parsing. In: Proceedings AISTATS (2012)

Bordes, A., Weston, J., Usunier, N.: Open question answering with weakly supervised embedding models. In: European Conference on Machine Learning (2014)

Burer, S., Monteiro, R.D.C.: A nonlinear programming algorithm for solving semidefinite programs via low-rank factorization. Math. Program. $95(2)$, 329–357 (2003)

Chechik, G., Shalit, U., Sharma, V., Bengio, S.: An online algorithm for large scale image similarity learning. In: Neural Information Processing Systems (2009)

Cheng, L.: Riemannian similarity learning. In: Internat. Conference on Machine Learning (2013)

Chopra, S., Hadsell, R., LeCun, Y.: Learning a similarity metric discriminatively, with application to face verification. In: IEEE Conf. on Computer Vision and Pattern Recogn. (2005)

Davis, J., Kulis, B., Jain, P., Sra, S., Dhillon, I.S.: Information-theoretic metric learning. In: International Conference on Machine Learning (2007)

Duan, L., Xu, D., Tsang, I.: Learning with augmented features for heterogeneous domain adaptation. In: International Conference on Machine Learning (2012)

Frome, A., Corrado, G.S., Shlens, J., Bengio, S., Dean, J., Mikolov, T., et al.: Devise: A deep visual-semantic embedding model. In: Neural Information Processing Systems (2013)

Fürnkranz, J., Hüllermeier, E., Loza Mencía, E., Brinker, K.: Multilabel classification via calibrated label ranking. Machine Learning **73**(2), 133–153 (2008)

Garreau, D., Lajugie, R., Arlot, S., Bach, F.: Metric learning for temporal sequence alignment. In: Neural Information Processing Systems (2014)

Globerson, A., Chechik, G., Pereira, F., Tishby, N.: Euclidean embedding of co-occurrence data. Journal of Machine Learning Research **8**, 2265–2295 (2007)

Globerson, A., Roweis, S.T.: Metric learning by collapsing classes. In: NIPS (2005)

Haeffele, B., Vidal, R., Young, E.: Structured low-rank matrix factorization: Optimality, algorithm, and applications to image processing. In: Proceedings ICML (2014)

Hastie, T., Tibshirani, R., Friedman, J.: The Elements of Statistical Learning: Data Mining, Inference and Prediction, 2nd edn. Springer (2009)

Huang, Z., Wang, R., Shan, S., Chen, X.: Learning euclidean-to-riemannian metric for point-to-set classification. In: IEEE Conference on Computer Vision and Pattern Recogn. (2014)

Jain, P., Kulis, B., Davis, J.V., Dhillon, I.S.: Metric and kernel learning using a linear transformation. Journal of Machine Learning Research **13**, 519–547 (2012)

Jäschke, R., Marinho, L.B., Hotho, A., Schmidt-Thieme, L., Stumme, G.: Tag recommendations in social bookmarking systems. AI Communications **21**(4), 231–247 (2008)

Journée, M., Bach, F.R., Absil, P.-A., Sepulchre, R.: Low-rank optimization on the cone of positive semidefinite matrices. SIAM Journal on Optimization **20**(5), 2327–2351 (2010)

Kulis, B.: Metric learning: A survey. Foundat. and Trends in Mach. Learn. **5**(4), 287–364 (2013)

Kulis, B., Saenko, K., Darrell, T.: What you saw is not what you get: Domain adaptation using asymmetric kernel transforms. In: Proceedings CVPR (2011)

Larochelle, H., Erhan, D., Bengio, Y.: Zero-data learning of new tasks. In: AAAI (2008)

Mirzazadeh, F., Guo, Y., Schuurmans, D.: Convex co-embedding. In: AAAI (2014)

Ngiam, J., Khosla, A., Kim, M., Nam, J., Lee, H., Ng, A.Y.: Multimodal deep learning. In: International Conference on Machine Learning (2011)

Platt, J.C.: Sequential minimal optimization: A fast algorithm for training support vector machines. Technical report, Advances in Kernel Methods (1998)

Read, J., Pfahringer, B., Holmes, G., Frank, E.: Classifier chains for multi-label classification. Machine Learning **85**(3), 333–359 (2011)

Rendle, S., Schmidt-Thieme, L.: Factor models for tag recommendation in bibsonomy. In: ECML/PKDD Discovery Challenge (2009)

Socher, R., Chen, D., Manning, C.D., Ng, A.: Reasoning with neural tensor networks for knowledge base completion. In: Advances in Neural Information Processing Systems (2013a)

Socher, R., Ganjoo, M., Manning, C.D., Ng, A.: Zero-shot learning through cross-modal transfer. In: Advances in Neural Information Processing Systems, pp. 935–943 (2013b)

Srivastava, N., Salakhutdinov, R.: Multimodal learning with deep boltzmann machines. In: Advances in Neural Information Processing Systems (2012)

Weinberger, K., Saul, L.: Distance metric learning for large margin nearest neighbor classification. Journal of Machine Learning Research **10**, 207–244 (2009)

Weinberger, K., Saul, L.K.: Fast solvers and efficient implementations for distance metric learning. In: International Conference on Machine Learning (2008)

Weston, J., Bengio, S., Usunier, N.: Large scale image annotation: learning to rank with joint word-image embeddings. Machine Learning **81**(1), 21–35 (2010)

Xie, P., Xing, E.: Multi-modal distance metric learning. In: Proceedings IJCAI (2013)

Xing, E., Ng, A., Jordan, M., Russell, S.: Distance metric learning with application to clustering with side-information. In: Neural Information Processing Systems (2002)

Yamanishi, Y.: Supervised bipartite graph inference. In: Proceedings NIPS (2008)

Zhai, X., Peng, Y., Xiao, J.: Heterogeneous metric learning with joint graph regularization for cross-media retrieval. In: AAAI Conference on Artificial Intelligence (2013)

Zhang, H., Huang, T.S., Nasrabadi, N.M., Zhang, Y.: Heterogeneous multi-metric learning for multi-sensor fusion. In: International Conference on Information Fusion (2011)

Zhang, X., Yu, Y., Schuurmans, D.: Accelerated training for matrix-norm regularization: A boosting approach. In: Neural Information Processing Systems (2012)

A   An Auxiliary Lemma

Lemma 1.   *Suppose f is ν-smooth. Then for any positive semi-definite $C, S \in \mathbb{R}^{p \times p}$,*

$$f(C + S) \leq f(C) + \operatorname{tr}(S'\nabla f(C)) + \frac{\nu}{2}\rho(S)^2 . \tag{15}$$

Proof.   Define $h(\eta) = f(C + \eta S)$ for $\eta \in [0, 1]$. Note that $h(0) = f(C)$, $h(1) = f(C + S)$, and $h'(\eta) = \operatorname{tr}(S'\nabla f(C + \eta S))$ for any $\eta \in (0, 1)$. Then

$$f(C + S) - f(C) - \operatorname{tr}(S'\nabla f(C))$$

$$= h(1) - h(0) - \operatorname{tr}(S'\nabla f(C)) = \int_0^1 h'(\eta)d\eta - \operatorname{tr}(S'\nabla f(C))$$

$$= \int_0^1 \operatorname{tr}(S'\nabla f(C+\eta S))d\eta - \operatorname{tr}(S'\nabla f(C)) = \int_0^1 \operatorname{tr}\left(S'(\nabla f(C+\eta S) - \nabla f(C))\right)d\eta$$

$$\leq \int_0^1 \rho(S)\|\nabla f(C+\eta S) - \nabla f(C)\|_{\operatorname{tr}} d\eta \leq \int_0^1 \nu\rho(S)\rho(\eta S)\eta = \int_0^1 \nu\eta\rho(S)^2 d\eta = \frac{\nu}{2}\rho(S)^2$$

where the first inequality holds by the Cauchy-Schwarz inequality, and the second by the Lipschitz condition on ∇f. Reordering the inequality establishes the lemma.   □

Large Scale Learning and Big Data

Adaptive Stochastic Primal-Dual Coordinate Descent for Separable Saddle Point Problems

Zhanxing Zhu$^{(\boxtimes)}$ and Amos J. Storkey

Institute of Adaptive Neural Computation, School of Informatics,
The University of Edinburgh, Edinburgh EH8 9AB, UK
{zhanxing.zhu,a.storkey}@ed.ac.uk

Abstract. We consider a generic convex-concave saddle point problem with a *separable* structure, a form that covers a wide-ranged machine learning applications. Under this problem structure, we follow the framework of primal-dual updates for saddle point problems, and incorporate stochastic block coordinate descent with *adaptive* stepsizes into this framework. We theoretically show that our proposal of adaptive stepsizes potentially achieves a sharper linear convergence rate compared with the existing methods. Additionally, since we can select "mini-batch" of block coordinates to update, our method is also amenable to *parallel* processing for large-scale data. We apply the proposed method to regularized empirical risk minimization and show that it performs comparably or, more often, better than state-of-the-art methods on both synthetic and real-world data sets.

Keywords: Large-scale optimization · Parallel optimization · Stochastic coordinate descent · Convex-concave saddle point problems

1 Introduction

The generic convex-concave saddle point problem is written as

$$\min_{\mathbf{x}\in\mathbb{R}^d} \max_{\mathbf{y}\in\mathbb{R}^q} \left\{ L(\mathbf{x},\mathbf{y}) = g(\mathbf{x}) + \langle \mathbf{x}, \mathbf{Ky}\rangle - \phi^*(\mathbf{y}) \right\}, \tag{1}$$

where $g(\mathbf{x})$ is a proper convex function, $\phi^*(\cdot)$ is the convex conjugate of a convex function $\phi(\cdot)$, and matrix $\mathbf{K} \in \mathbb{R}^{d\times q}$. Many machine learning tasks reduce to solving a problem of this form [3,6]. As a result, this saddle problem has been widely studied [1,2,4,5,14,16].

One important subclass of the general convex concave saddle point problem is where $g(\mathbf{x})$ or $\phi^*(\mathbf{y})$ exhibits an additive separable structure. We say $\phi^*(\mathbf{y})$ is *separable* when $\phi^*(\mathbf{y}) = \frac{1}{n}\sum_{i=1}^{n} \phi_i^*(\mathbf{y}_i)$, with $\mathbf{y}_i \in \mathbb{R}^{q_i}$, and $\sum_{i=1}^{n} q_i = q$. Separability for $g(\cdot)$ is defined likewise. To keep the consistent notation for the machine learning applications discussed later, we introduce matrix $\mathbf{A}$ and let

© Springer International Publishing Switzerland 2015
A. Appice et al. (Eds.): ECML PKDD 2015, Part I, LNAI 9284, pp. 645–658, 2015.
DOI: 10.1007/978-3-319-23528-8_40

$\mathbf{K} = \frac{1}{n}\mathbf{A}$. Then we partition matrix $\mathbf{A}$ into n column blocks $\mathbf{A}_i \in \mathbb{R}^{d \times q_i}$, $i = 1, \ldots, n$, and $\mathbf{Ky} = \frac{1}{n}\sum_{i=1}^{n}\mathbf{A}_i\mathbf{y}_i$, resulting in a problem of the form

$$\min_{\mathbf{x} \in \mathbb{R}^d} \max_{\mathbf{y} \in \mathbb{R}^q} \left\{ L(\mathbf{x}, \mathbf{y}) = g(\mathbf{x}) + \frac{1}{n}\sum_{i=1}^{n} (\langle \mathbf{x}, \mathbf{A}_i\mathbf{y}_i \rangle - \phi_i^*(\mathbf{y}_i)) \right\} \tag{2}$$

for $\phi^*(\cdot)$ separable. We call any problem of the form (1) where $g(\cdot)$ or $\phi^*(\cdot)$ has separable structure, a Separable Convex Concave Saddle Point (*Sep-CCSP*) problem. Eq. (2) gives the explicit form for when $\phi^*(\cdot)$ is separable.

In this work, we further assume that each $\phi_i^*(\mathbf{y}_i)$ is γ-strongly convex, and $g(\mathbf{x})$ is λ-strongly convex, i.e.,

$$\phi_i^*(\mathbf{y}_i') \geq \phi_i^*(\mathbf{y}_i) + \nabla\phi^*(\mathbf{y}_i)^T (\mathbf{y}_i' - \mathbf{y}_i) + \frac{\gamma}{2}\|\mathbf{y}_i' - \mathbf{y}_i\|_2^2, \quad \forall \mathbf{y}_i, \mathbf{y}_i' \in \mathbb{R}^{q_i}$$

$$g(\mathbf{x}') \geq g(\mathbf{x}) + \nabla g(\mathbf{x})^T (\mathbf{x}' - \mathbf{x}) + \frac{\lambda}{2}\|\mathbf{x}_i' - \mathbf{x}_i\|_2^2, \quad \forall \mathbf{x}, \mathbf{x}' \in \mathbb{R}^d,$$

where we use ∇ to denote both the gradient for smooth function and subgradient for non-smooth function. When the strong convexity cannot be satisfied, a small strongly convex perturbation can be added to make the problem satisfy the assumption [15].

One important instantiation of the Sep-CCSP problem in machine learning is the regularized empirical risk minimization (ERM, [3]) of linear predictors,

$$\min_{\mathbf{x} \in \mathbb{R}^d} \left\{ J(\mathbf{x}) = \frac{1}{n}\sum_{i=1}^{n} \phi_i(\mathbf{a}_i^T\mathbf{x}) + g(\mathbf{x}) \right\}, \tag{3}$$

where $\mathbf{a}_1, \ldots, \mathbf{a}_n \in \mathbb{R}^d$ are the feature vectors of n data samples, $\phi_i(\cdot)$ corresponds the convex loss function w.r.t. the linear predictor $\mathbf{a}_i^T\mathbf{x}$, and $g(\mathbf{x})$ is a convex regularization term. Many practical classification and regression models fall into this regularized ERM formulation, such as linear support vector machine (SVM), regularized logistic regression and ridge regression, see [3] for more details.

Reformulating the above regularized ERM by employing conjugate dual of the function $\phi_i(\cdot)$, i.e.

$$\phi_i^*(\mathbf{a}_i^T\mathbf{x}) = \max_{y_i \in \mathbb{R}} \langle \mathbf{x}, y_i\mathbf{a}_i \rangle - \phi_i^*(y_i), \tag{4}$$

leads directly to the following Sep-CCSP problem

$$\min_{\mathbf{x} \in \mathbb{R}^d} \max_{\mathbf{y} \in \mathbb{R}^n} g(\mathbf{x}) + \frac{1}{n}\sum_{i=1}^{n} (\langle \mathbf{x}, y_i\mathbf{a}_i \rangle - \phi_i^*(y_i)). \tag{5}$$

Comparing with the general form, we note that the matrix $\mathbf{A}_i$ in (2) is now a vector $\mathbf{a}_i$. For solving the general saddle point problem (1), many primal-dual algorithms can be applied, such as [1,2,4,5,16]. In addition, the saddle point problem we consider can also be formulated as a composite function minimization problem and then solved by Alternating Direction Method of Multipliers (ADMM) methods [9].

To handle the Sep-CCSP problem particularly for regularized ERM problem (5), Zhang and Xiao [15] proposed a stochastic primal-dual coordinate descent (SPDC) method. SPDC applies stochastic coordinate descent method [8,10,11] into the primal-dual framework, where in each iteration a random subset of dual coordinates are updated. This method inherits the efficiency of stochastic coordinate descent for solving large-scale problems. However, they use a conservative constant stepsize during the primal-dual updates, which leads to an unsatisfying convergence rate especially for unnormalized data.

In this work, we propose an *adaptive* stochastic primal-dual coordinate descent (*AdaSPDC*) method for solving the Sep-CCSP problem (2), which is a non-trivial extension of SPDC. By carefully exploiting the structure of individual subproblem, we propose an adaptive stepsize rule for both primal and dual updates according to the chosen subset of coordinate blocks in each iteration. Both theoretically and empirically, we show that AdaSPDC could yield a significantly better convergence performance than SPDC and other state-of-the-art methods.

The remaining structure of the paper is as follows. Section 2 summarizes the general primal-dual framework our method and SPDC are based on. Then we elaborate our method AdaSPDC in Section 3, where both the theoretical result and its comparison with SPDC are provided. In Section 4, we apply our method into regularized ERM tasks, and experiment with both synthetic and real-world datasets, and we show the superiority of AdaSPDC over other competitive methods empirically. Finally, Section 5 concludes the work.

2   Primal-dual Framework for Convex-Concave Saddle Point Problems

Chambolle and Pock [1] proposed a first-order primal-dual method for the CCSP problem (1). We refer this algorithm as PDCP. The update of PDCP in the $(t+1)$th iteration is as follows:

$$\mathbf{y}^{t+1} = \operatorname{argmin}_{\mathbf{y}} \phi^*(\mathbf{y}) - \langle \overline{\mathbf{x}}^t, \mathbf{K}\mathbf{y} \rangle + \frac{1}{2\sigma} \|\mathbf{y} - \mathbf{y}^t\|_2^2 \tag{6}$$

$$\mathbf{x}^{t+1} = \operatorname{argmin}_{\mathbf{x}} g(\mathbf{x}) + \langle \mathbf{x}, \mathbf{K}\mathbf{y}^{t+1} \rangle + \frac{1}{2\tau} \|\mathbf{x} - \mathbf{x}^t\|_2^2 \tag{7}$$

$$\overline{\mathbf{x}}^{t+1} = \mathbf{x}^{t+1} + \theta(\mathbf{x}^{t+1} - \mathbf{x}^t). \tag{8}$$

When the parameter configuration satisfies $\tau\sigma \leq 1/\|\mathbf{K}\|^2$ and $\theta = 1$, PDCP could achieve $O(1/T)$ convergence rate for general convex function $\phi^*(\cdot)$ and $g(\cdot)$, where T is total number of iterations. When $\phi^*(\cdot)$ and $g(\cdot)$ are both strongly convex, a linear convergence rate can be achieved by using a more scheduled stepsize. PDCP is a batch method and non-stochastic, i.e., it has to update all the dual coordinates in each iteration for Sep-CCSP problem, which will be computationally intensive for large-scale (high-dimensional) problems.

SPDC [15] can be viewed as a stochastic variant of the batch method PDCP for handling Sep-CCSP problem. However, SPDC uses a conservative constant stepsize for primal and dual updates. Both PDCP and SPDC do not consider the structure of matrix $\mathbf{K}$ and only apply constant stepsize for all coordinates of primal and dual variables. This might limit their convergence performance in reality.

Based on this observation, we exploit the structure of matrix $\mathbf{K}$ (i.e., $\frac{1}{n}\mathbf{A}$) and propose an adaptive stepsize rule for efficiently solving Sep-CCSP problem. A better linear convergence rate could be yielded when $\phi_i^*(\cdot)$ and $g(\cdot)$ are strongly convex. Our algorithm will be elaborated in the following section.

3   Adaptive Stochastic Primal-Dual Coordinate Descent

As a non-trivial extension of SPDC [15], our method AdaSPDC solves the Sep-CCSP problem (2) by using an adaptive parameter configuration. Concretely, we optimize $L(\mathbf{x}, \mathbf{y})$ by alternatively updating the dual and primal variables in a principled way. Thanks to the separable structure of $\phi(\mathbf{y})$, in each iteration we can randomly select m blocks of dual variables whose indices are denoted as S_t, and then we only update these selected blocks in the following way,

$$\mathbf{y}_i^{t+1} = \text{argmin}_{\mathbf{y}_i}\left[\phi_i(\mathbf{y}_i) - \left\langle\overline{\mathbf{x}}^t, \mathbf{A}_i\mathbf{y}_i\right\rangle + \frac{1}{2\sigma_i}\|\mathbf{y}_i - \mathbf{y}_i^t\|_2^2\right], \text{ if } i \in S_t. \quad (9)$$

For those coordinates in blocks not selected, $i \notin S_t$, we just keep $\mathbf{y}_i^{t+1} = \mathbf{y}_i^t$. By exploiting the structure of individual $\mathbf{A}_i$, we configure the stepsize parameter of the proximal term σ_i *adaptively*

$$\sigma_i = \frac{1}{2R_i}\sqrt{\frac{n\lambda}{m\gamma}}, \quad (10)$$

where $R_i = \|\mathbf{A}_i\|_2 = \sqrt{\mu_{\max}\left(\mathbf{A}_i^T\mathbf{A}_i\right)}$, with $\|\cdot\|_2$ is the spectral norm of a matrix and $\mu_{\max}(\cdot)$ to denote the maximum singular value of a matrix.

Our step size is different from the one used in SPDC [15], where R is a constant $R = \max\{\|\mathbf{a}_i\|_2 : i = 1, \ldots, n\}$ (since SPDC only considers ERM problem, the matrix $\mathbf{A}_i$ is a feature vector $\mathbf{a}_i$).

Remark. Intuitively, R_i in AdaSPDC can be understood as the coupling strength between the i-th dual variable block and primal variable, measured by the spectral norm of matrix $\mathbf{A}_i$. Smaller coupling strength allows us to use larger stepsize for the current dual variable block without caring too much about its influence on primal variable, and vice versa. Compared with SPDC, our proposal of an adaptive coupling strength for the chosen coordinate block directly results in larger step size, and thus helps to improve convergence speed.

In the stochastic dual update, we also use an intermediate variable $\bar{\mathbf{x}}^t$ as in PDCP algorithm, and we will describe its update later.

Since we assume $g(\mathbf{x})$ is not separable, we update the primal variable as a whole,

$$\mathbf{x}^{t+1} = \text{argmin}_{\mathbf{x}} \left[g(\mathbf{x}) + \left\langle \mathbf{x}, \mathbf{r}^t + \frac{1}{m} \sum_{j \in S_t} \mathbf{A}_j(\mathbf{y}_j^{t+1} - \mathbf{y}_j^t) \right\rangle + \frac{1}{2\tau^t} \|\mathbf{x} - \mathbf{x}^t\|_2^2 \right].$$

(11)

The proximal parameter τ^t is also configured *adaptively*,

$$\tau^t = \frac{1}{2R_{\max}^t} \sqrt{\frac{m\gamma}{n\lambda}},$$

(12)

where $R_{\max}^t = \max\{R_i | i \in S_t\}$, compared with constant R used in SPDC. To account for the incremental change after the latest dual update, an auxiliary variable $\mathbf{r}^t = \frac{1}{n} \sum_{i=1}^n \mathbf{A}_i \mathbf{y}_i^t$ is used and updated as follows

$$\mathbf{r}^{t+1} = \mathbf{r}^t + \frac{1}{n} \sum_{j \in S_t} \mathbf{A}_j \left(\mathbf{y}_j^{t+1} - \mathbf{y}_j^t\right).$$

(13)

Finally, we update the intermediate variable $\bar{\mathbf{x}}$, which implements an extrapolation step over the current $\mathbf{x}^{t+1}$ and can help to provide faster convergence rate [1,7].

$$\bar{\mathbf{x}}^{t+1} = \mathbf{x}^{t+1} + \theta^t(\mathbf{x}^{t+1} - \mathbf{x}^t),$$

(14)

where θ^t is configured adaptively as

$$\theta^t = 1 - \frac{1}{n/m + R_{\max}^t \sqrt{(n/m)/(\lambda\gamma)}},$$

(15)

which is contrary to the constant θ used in SPDC.

The whole procedure for solving Sep-CCSP problem (2) using AdaSPDC is summarized in Algorithm 1. There are several notable characteristics of our algorithms.

- Compared with SPDC, our method uses adaptive step size to obtain faster convergence (will be shown in Theorem 1), while the whole algorithm does not bring any other extra computational complexity. As demonstrated in the experiment Section 4, in many cases, AdaSPDC provides significantly better performance than SPDC.
- Since, in each iteration, a number of block coordinates can be chosen and updated independently (with independent evaluation of individual step size), this directly enables parallel processing, and hence use on modern computing clusters. The ability to select an arbitrary number of blocks can help to make use of all the computation structure available as effectively as possible.

Algorithm 1. AdaSPDC for Separable Convex-Concave Saddle Point Problems

---
1: **Input:** number of blocks picked in each iteration m and number of iterations T.
2: **Initialize:** $\mathbf{x}^0$, $\mathbf{y}^0$, $\bar{\mathbf{x}}^0 = \mathbf{x}^0$, $\mathbf{r}^0 = \frac{1}{n}\sum_{i=1}^n \mathbf{A}_i \mathbf{y}_i^0$
3: **for** $t = 0, 1, \ldots, T-1$ **do**
4:     Randomly pick m dual coordinate blocks from $\{1, \ldots, n\}$ as indices set S_t, with
        the probability of each block being selected equal to m/n.
5:     According to the selected subset S_t, compute the adaptive parameter configura-
        tion of σ_i, τ^t and θ^t using Eq. (10), (12) and (15), respectively.
6:     **for** each selected block in parallel **do**
7:         Update the dual variable block using Eq.(9).
8:     **end for**
9:     Update primal variable using Eq.(11).
10:    Extrapolate primal variable block using Eq.(14).
11:    Update the auxiliary variable $\mathbf{r}$ using Eq.(13).
12: **end for**

3.1   Convergence Analysis

We characterise the convergence performance of our method in the following
theorem.

Theorem 1. *Assume that each $\phi_i^*(\cdot)$ is γ-strongly convex, and $g(\cdot)$ is λ-strongly
convex, and given the parameter configuration in Eq. (10), (12) and (15), then
after T iterations in Algorithm 1, the algorithm achieves the following convergence performance*

$$\left(\frac{1}{2\tau^T} + \lambda\right)\mathbb{E}\left[\|\mathbf{x}^T - \mathbf{x}^\star\|_2^2\right] + \mathbb{E}\left[\|\mathbf{y}^T - \mathbf{y}^\star\|_\nu^2\right]$$

$$\leq \left(\prod_{t=1}^T \theta^t\right)\left(\left(\frac{1}{2\tau^T} + \lambda\right)\|\mathbf{x}^0 - \mathbf{x}^\star\|_2^2 + \|\mathbf{y}^0 - \mathbf{y}^\star\|_{\nu'}^2\right), \tag{16}$$

*where $(\mathbf{x}^\star, \mathbf{y}^\star)$ is the optimal saddle point, $\nu_i = \frac{1/(4\sigma_i) + \gamma}{m}$, $\nu_i' = \frac{1/(2\sigma_i) + \gamma}{m}$, and
$\|\mathbf{y}^T - \mathbf{y}^\star\|_\nu^2 = \sum_{i=1}^n \nu_i \|\mathbf{y}_i^T - \mathbf{y}_i^\star\|_2^2$.*

Since the proof of the above is technical, we provide it in the full version of this
paper [17].

In our proof, given the proposed parameter θ^t, the critical point for obtaining
a sharper linear convergence rate than SPDC is that we configure τ^t and σ_i as
Eq. (12) and (10) to guarantee the positive definiteness of the following matrix
in the t-th iteration,

$$\mathbf{P} = \begin{bmatrix} \frac{m}{2\tau^t}\mathbf{I} & -\mathbf{A}_{S_t} \\ -\mathbf{A}_{S_t}^T & \frac{1}{2\mathrm{diag}(\boldsymbol{\sigma}_{S_t})} \end{bmatrix}, \tag{17}$$

where $\mathbf{A}_{S_t} = [\ldots, \mathbf{A}_i, \ldots] \in \mathbb{R}^{d \times mq_i}$ and $\mathrm{diag}(\boldsymbol{\sigma}_{S_t}) = \mathrm{diag}(\ldots, \sigma_i \mathbf{I}_{q_i}, \ldots)$ for $i \in
S_t$. However, we found that the parameter configuration to guarantee the positive
definiteness of $\mathbf{P}$ is not unique, and there exist other valid parameter configurations besides the proposed one in this work. We leave the further investigation
on other potential parameter configurations as future work.

3.2 More Comparison with SDPC

Compared with SPDC [15], AdaSPDC follows the similar primal-dual framework. The crucial difference between them is that AdaSPDC proposes a larger stepsize for both dual and primal updates, see Eq. (10) and (12) compared with SPDC's parameter configuration given in Eq.(10) in [15], where SPDC applies a large constant $R = \max\{\|\mathbf{a}_i\|_2 : i = 1, \ldots, n\}$ while AdaSPDC uses a more adaptive value of R_i and $R_{\max}^t$ for t-th iteration to account for the different coupling strength between the selected dual coordinate block and primal variable. This difference directly means that AdaSPDC can potentially obtain a significantly sharper linear convergence rate than SPDC, since the decay factor θ^t of AdaSPDC is smaller than θ in SPDC (Eq.(10) in [15]) , see Theorem 1 for AdaSPDC compared with SPDC (Theorem 1 in [15]). The empirical performance of the two algorithms will be demonstrated in the experimental Section 4.

To mitigate the problem that SPDC uses a large R, the authors of SPDC proposes to non-uniformly sample the the dual coordinate to update in each iteration according to the norm of the each $\mathbf{a}_i$. However, as we show later in the empirical experiments, this non-uniform sampling does not work very well for some datasets. By configuring the adaptive stepsize explicitly, our method AdaSPDC provides a better solution for unnormalized data compared with SPDC, see Section 4 for more empirical evidence.

Another difference is that SPDC only considers the regularized ERM task, i.e., only handling the case that each $\mathbf{A}_i$ is a feature vector $\mathbf{a}_i$, while AdaSPDC extends that $\mathbf{A}_i$ can be a matrix so that AdaSPDC can cover a wider range of applications than SPDC, i.e. in each iteration, a number of *block* coordinates could be selected while for SPDC only a number of coordinates are allowed.

4 Empirical Results

In this section, we appy AdaSPDC to several regularized empirical risk minimization problems. The experiments are conducted to compare our method AdaSPDC with other competitive stochastic optimization methods, including SDCA [13], SAG [12], SPDC with uniform sampling and non-uniform sampling [15]. In order to provide a fair comparison with these methods, in each iteration only one dual coordinate (or data instance) is chosen, i.e., we run all the methods sequentially. To obtain results that are independent of the practical implementation of the algorithm, we measure the algorithm performance in term of objective suboptimality w.r.t. the effective passes to the entire data set.

Each experiment is run 10 times and the average results are reported to show statistical consistency. We present all the experimental results we have done for each application.

4.1 Ridge Regression

We firstly apply our method AdaSPDC into a simple ridge regression problem with synthetic data. The data is generated in the same way as Zhang and Xiao [15];

$n = 1000$ i.i.d. training points $\{\mathbf{a}_i, b_i\}_{i=1}^n$ are generated in the following manner,

$$b = \mathbf{a}^T \mathbf{x}^\circ + \epsilon, \quad \mathbf{a} \sim \mathcal{N}(\mathbf{0}, \boldsymbol{\Sigma}), \quad \epsilon \sim \mathcal{N}(0, 1),$$

where $\mathbf{a} \in \mathbb{R}^d$ and $d = 1000$, and the elements of the vector $\mathbf{x}^\circ$ are all ones. The covariance matrix $\boldsymbol{\Sigma}$ is set to be diagonal with $\Sigma_{jj} = j^{-2}$, for $j = 1, \ldots, d$. Then the ridge regression tries to solve the following optimization problem,

$$\min_{\mathbf{x} \in \mathbb{R}^d} \left\{ J(\mathbf{x}) = \frac{1}{n} \sum_{i=1}^n \frac{1}{2} (\mathbf{a}_i^T \mathbf{x} - b_i)^2 + \frac{\lambda}{2} \|\mathbf{x}\|_2^2 \right\}. \tag{18}$$

The optimal solution of the above ridge regression can be found as

$$\mathbf{x}^\star = \left(\mathbf{A}\mathbf{A}^T + n\lambda \mathbf{I}_d \right)^{-1} \mathbf{A}b.$$

By employing the conjugate dual of quadratic loss (crossref, Eq. (4)), we can reformulate the ridge regression as the following Sep-CCSP problem,

$$\min_{\mathbf{x} \in \mathbb{R}^d} \max_{\mathbf{y} \in \mathbb{R}^n} \frac{\lambda}{2} \|\mathbf{x}\|_2^2 + \frac{1}{n} \sum_{i=1}^n \left(\langle \mathbf{x}, y_i \mathbf{a}_i \rangle - \left(\frac{1}{2} y_i^2 + b_i y_i \right) \right). \tag{19}$$

It is easy to figure out that $g(\mathbf{x}) = \lambda/2 \|\mathbf{x}\|_2^2$ is λ-strongly convex, and $\phi_i^*(y_i) = \frac{1}{2} y_i^2 + b_i y_i$ is 1-strongly convex.

Thus, for ridge regression, the dual update in Eq. (9) and primal update in Eq. (11) of AdaSPDC have closed form solutions as below,

$$y_i^{t+1} = \frac{1}{1 + 1/\sigma_i} \left(\langle \overline{\mathbf{x}}^t, \mathbf{a}_i \rangle + b_i + \frac{1}{\sigma_i} y_i \right), \text{ if } i \in S_t$$

$$\mathbf{x}^{t+1} = \frac{1}{\lambda + 1/\tau^t} \left(\frac{1}{\tau^t} \mathbf{x}^t - \left(\mathbf{r}^t + \frac{1}{m} \sum_{j \in S_t} \mathbf{a}_j (y_j^{t+1} - y_j^t) \right) \right)$$

The algorithm performance is evaluated in term of objective suboptimality (measured by $J(\mathbf{x}^t) - J(\mathbf{x}^\star)$) w.r.t. number of effective passes to the entire datasets. Varying values of regularization parameter λ are experimented to demonstrate algorithm performance with different degree of ill-conditioning, $\lambda = \{10^{-3}, 10^{-4}, 10^{-5}, 10^{-6}\}$.

Fig. 1 shows algorithm performance with different degrees of regularization. It is easy to observe that AdaSPDC converges substantially faster than other compared methods, particularly for ill-conditioned problems. Compared with SPDC and its variant with non-uniform sampling, the usage of adaptive stepsize in AdaSPDC significantly improves convergence speed. For instance, in the case with $\lambda = 10^{-6}$, AdaSPDC achieves 100 times better suboptimality than both SPDC and its variant SPDC with non-uniform sampling after 300 passes.

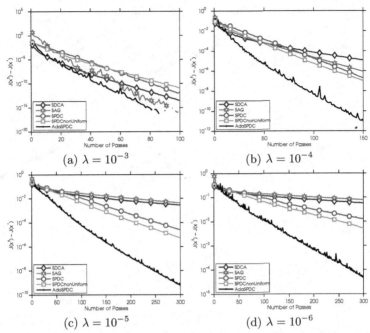

Fig. 1. Ridge regression with synthetic data: comparison of convergence performance w.r.t. the number of passes. Problem size: $d = 1000, n = 1000$. We evaluate the convergence performance using objective suboptimality, $J(\mathbf{x}^t) - J(\mathbf{x}^\star)$.

4.2 Binary Classification on Real-world Datasets

We now compare the performance of our method AdaSPDC with other competitive methods on several real-world data sets. Our experiments focus on the freely-available benchmark data sets for binary classification, whose detailed information are listed in Table 1. The *w8a*, *covertype* and *url* data are obtained from the LIBSVM collection[1]. The *quantum* and *protein* data sets are obtained from KDD Cup 2004[2]. For all the datasets, each sample takes the form $(\mathbf{a}_i, b_i)$

Table 1. Benchmark datasets used in our experiments for binary classification.

Datasets	Number of samples	Number of features	Sparsity
w8a	49,749	300	3.9%
covertype	20,242	47,236	0.16%
url	2,396,130	3,231,961	0.0018%
quantum	50,000	78	43.44%
protein	145,751	74	99.21%

---

[1] http://www.csie.ntu.edu.tw/~cjlin/libsvmtools/datasets/binary.html
[2] http://osmot.cs.cornell.edu/kddcup/datasets.html

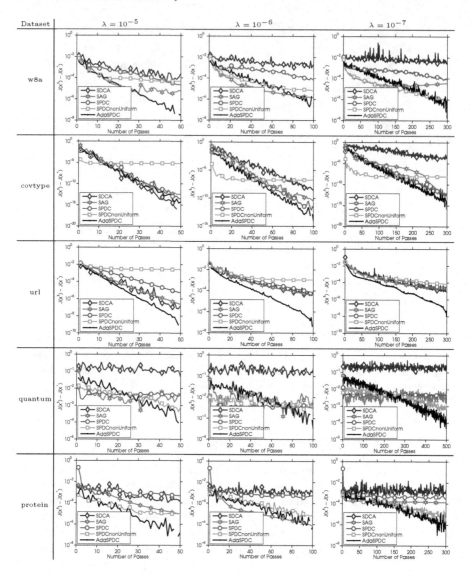

Fig. 2. Comparison of algorithm performance with smooth Hinge loss.

with $\mathbf{a}_i$ is the feature vector and b_i is the binary label -1 or 1. We add a bias term to the feature vector for all the datasets. We aim to minimize the regularized empirical risk with following form

$$J(\mathbf{x}) = \frac{1}{n} \sum_{i=1}^{n} \phi_i(\mathbf{a}_i^T \mathbf{x}) + \frac{\lambda}{2} \|\mathbf{x}\|_2^2 \tag{20}$$

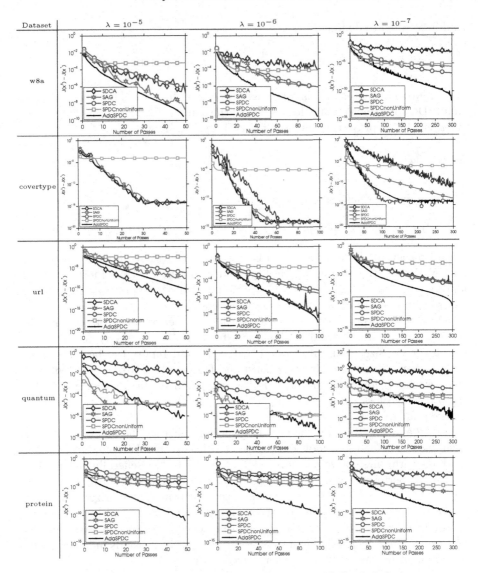

Fig. 3. Comparison of algorithm performance with Logistic loss.

To provide a more comprehensive comparison between these methods, we experiment with two different loss function $\phi_i(\cdot)$, smooth Hinge loss [13] and logistic loss, described in the following.

Smooth Hinge loss (with smoothing parameter $\gamma = 1$.)

$$\phi_i(z) = \begin{cases} 0 & \text{if } b_i z \geq 1, \\ 1 - \frac{\gamma}{2} - b_i z & \text{if } b_i z \leq 1 - \gamma \\ \frac{1}{2\gamma}(1 - b_i z)^2 & \text{otherwise.} \end{cases}$$

And its conjugate dual is

$$\phi_i^*(y_i) = b_i y_i + \frac{1}{2} y_i^2, \text{ with } b_i y_i \in [-1, 0].$$

We can observe that $\phi_i^*(y_i)$ is γ-strongly convex with $\gamma = 1$. The dual update of AdaSPDC for smooth Hinge loss is nearly the same with ridge regression except the necessity of projection into the interval $b_i y_i \in [-1, 0]$.

Logistic loss

$$\phi_i(z) = \log\left(1 + \exp(-b_i z)\right),$$

whose conjugate dual has the form

$$\phi_i^*(y_i) = -b_i y_i \log(-b_i y_i) + (1 + b_i y_i) \log(1 + b_i y_i) \text{ with } b_i y_i \in [-1, 0].$$

It is also easy to obtain that $\phi_i^*(y_i)$ is γ-strongly convex with $\gamma = 4$. Note that for logistic loss, the dual update in Eq. (9) does not have a closed form solution, and we can start from some initial solution and further apply several steps of Newton's update to obtain a more accurate solution.

During the experiments, we observe that the performance of SAG is very sensitive to the stepsize choice. To obtain best results of SAG, we try different choices of stepsize in the interval $[1/16L, 1/L]$ and report the best result for each dataset, where L is Lipschitz constant of $\phi_i(\mathbf{a}_i^T \mathbf{x})$, $1/16L$ is the theoretical stepsize choice for SAG and $1/L$ is the suggested empirical choice [12]. For smooth Hinge loss, $L = \max_i\{\|\mathbf{a}_i\|_2, i = 1, \ldots, n\}$, and for logistic loss, $L = \frac{1}{4}\max_i\{\|\mathbf{a}_i\|_2, i = 1, \ldots, n\}$.

Fig. 2 and Fig. 3 depict the algorithm performance on the different methods with smooth Hinge loss and logistics loss, respectively. We compare all these methods with different values of $\lambda = \{10^{-5}, 10^{-6}, 10^{-7}\}$. Generally, our method AdaSPDC performs consistently better or at least comparably with other methods, and performs especially well for the tasks with small regularized parameter λ. For some datasets, such as covertype and quantum, SPDC with non-uniform sampling decreases the objective faster than other methods in early epochs, however, cannot achieve comparable results with other methods in later epochs, which might be caused by its conservative stepsize.

5   Conclusion and Future Work

In this work, we propose Adaptive Stochastic Primal-Dual Coordinate Descent (AdaSPDC) for separable saddle point problems. As a non-trivial extension of a recent work SPDC [15], AdaSPDC uses an adaptive step size choices for both primal and dual updates in each iteration. The design of the step size for our method AdaSPDC explicitly and adaptively models the coupling strength between chosen block coordinates and primal variable through the spectral norm of each $\mathbf{A}_i$. We theoretically characterise that AdaSPDC holds a sharper linear convergence rate than SDPC. Additionally, we demonstrate the superiority of the proposed AdaSPDC method on ERM problems through extensive experiments on both synthetic and real-world data sets.

An immediate further research direction is to investigate other valid parameter configurations for the extrapolation parameter θ, and the primal and dual step sizes τ and σ both theoretically and empirically. In addition, discovering the potential theoretical connections with other stochastic optimization methods will also be enlightening.

Acknowledgments. Z. Zhu is supported by China Scholarship Council/University of Edinburgh Joint Scholarship. The authors would like to thank Jinli Hu for insightful discussion on the proof of Theorem 1.

References

1. Chambolle, A., Pock, T.: A first-order primal-dual algorithm for convex problems with applications to imaging. Journal of Mathematical Imaging and Vision **40**(1), 120–145 (2011)
2. Esser, E., Zhang, X., Chan, T.: A general framework for a class of first order primal-dual algorithms for convex optimization in imaging science. SIAM Journal on Imaging Sciences **3**(4), 1015–1046 (2010)
3. Hastie, T., Tibshirani, R., Friedman, J.: The elements of statistical learning, vol. 2. Springer (2009)
4. He, B., Yuan, X.: Convergence analysis of primal-dual algorithms for a saddle-point problem: from contraction perspective. SIAM Journal on Imaging Sciences **5**(1), 119–149 (2012)
5. He, Y., Monteiro, R.D.: An accelerated hpe-type algorithm for a class of composite convex-concave saddle-point problems. Optimization-online preprint (2014)
6. Jacob, L., Obozinski, G., Vert, J.P.: Group lasso with overlap and graph lasso. In: Proceedings of the 26th Annual International Conference on Machine Learning, pp. 433–440. ACM (2009)
7. Nesterov, Y.: Introductory lectures on convex optimization: A basic course, vol. 87. Springer (2004)
8. Nesterov, Y.: Efficiency of coordinate descent methods on huge-scale optimization problems. SIAM Journal on Optimization **22**(2), 341–362 (2012)
9. Ouyang, Y., Chen, Y., Lan, G., Pasiliao Jr., E.: An accelerated linearized alternating direction method of multipliers. SIAM Journal on Imaging Sciences **8**(1), 644–681 (2015)
10. Richtárik, P., Takáč, M.: Iteration complexity of randomized block-coordinate descent methods for minimizing a composite function. Mathematical Programming **144**(1–2), 1–38 (2014)
11. Richtárik, P., Takáč, M.: Parallel coordinate descent methods for big data optimization. Mathematical Programming, 1–52 (2012)
12. Schmidt, M., Roux, N.L., Bach, F.: Minimizing finite sums with the stochastic average gradient. arXiv preprint arXiv:1309.2388 (2013)
13. Shalev-Shwartz, S., Zhang, T.: Stochastic dual coordinate ascent methods for regularized loss. The Journal of Machine Learning Research **14**(1), 567–599 (2013)

14. Tseng, P.: On accelerated proximal gradient methods for convex-concave optimization. submitted to SIAM Journal on Optimization (2008)
15. Zhang, Y., Xiao, L.: Stochastic primal-dual coordinate method for regularized empirical risk minimization. In: International Conference of Machine Learning (2015)
16. Zhu, M., Chan, T.: An efficient primal-dual hybrid gradient algorithm for total variation image restoration. UCLA CAM Report, pp. 08–34 (2008)
17. Zhu, Z., Storkey, A.J.: Adaptive stochastic primal-dual coordinate descent for separable saddle point problems. arXiv preprint arXiv:1506.04093 (2015)

Hash Function Learning via Codewords

Yinjie Huang[1]([✉]), Michael Georgiopoulos[1], and Georgios C. Anagnostopoulos[2]

[1] Department of Electrical Engineering and Computer Science,
University of Central Florida, 4000 Central Florida Blvd, Orlando, FL 32816, USA
`yhuang@eecs.ucf.edu, michaelg@ucf.edu`
[2] Department of Electrical and Computer Engineering,
Florida Institute of Technology, 150 W University Blvd, Melbourne, FL 32901, USA
`georgio@fit.edu`

Abstract. In this paper we introduce a novel hash learning framework that has two main distinguishing features, when compared to past approaches. First, it utilizes codewords in the Hamming space as ancillary means to accomplish its hash learning task. These codewords, which are inferred from the data, attempt to capture similarity aspects of the data's hash codes. Secondly and more importantly, the same framework is capable of addressing supervised, unsupervised and, even, semi-supervised hash learning tasks in a natural manner. A series of comparative experiments focused on content-based image retrieval highlights its performance advantages.

Keywords: Hash function learning · Codeword · Support vector machine

1 Introduction

With the explosive growth of web data including documents, images and videos, content-based image retrieval (CBIR) has attracted plenty of attention over the past years [1]. Given a query sample, a typical CBIR scheme retrieves samples stored in a database that are most similar to the query sample. The similarity is gauged in terms of a pre-specified distance metric and the retrieved samples are the nearest neighbors of the query point w.r.t. this metric. However, exhaustively comparing the query sample with every other sample in the database may be computationally expensive in many current practical settings. Additionally, most CBIR approaches may be hindered by the sheer size of each sample; for example, visual descriptors of an image or a video may number in the thousands. Furthermore, storage of these high-dimensional data also presents a challenge.

Considerable effort has been invested in designing hash functions transforming the original data into compact binary codes to reap the benefits of a potentially fast similarity search; note that hash functions are typically designed to preserve certain similarity qualities between the data. For example, approximate nearest neighbors (ANN) search [2] using compact binary codes in Hamming space was shown to achieve sub-liner searching time. Storage of the binary code is, obviously, also much more efficient.

© Springer International Publishing Switzerland 2015
A. Appice et al. (Eds.): ECML PKDD 2015, Part I, LNAI 9284, pp. 659–674, 2015.
DOI: 10.1007/978-3-319-23528-8_41

Existing hashing methods can be divided into two categories: *data-independent* and *data-dependent*. The former category does not use a data-driven approach to choose the hash function. For example, Locality Sensitive Hashing (LSH) [3] randomly projects and thresholds data into the Hamming space for generating binary codes, where closely located (in terms of Euclidean distances in the data's native space) samples are likely to have similar binary codes. Furthermore, in [4], the authors proposed a method for ANN search using a learned Mahalanobis metric combined with LSH.

On the other hand, *data-dependent methods* can, in turn, be grouped into supervised, unsupervised and semi-supervised learning paradigms. The bulk of work in data-dependent hashing methods has been performed so far following the supervised learning paradigm. Recent work includes the Semantic Hashing [5], which designs the hash function using a Restricted Boltzmann Machine (RBM). Binary Reconstructive Embedding (BRE) in [6] tries to minimize a cost function measuring the difference between the original metric distances and the reconstructed distances in the Hamming space. Minimal Loss Hashing (MLH) [7] learns the hash function from pair-wise side information and the problem is formulated based on a bound inspired by the theory of structural Support Vector Machines [8]. In [9], a scenario is addressed, where a small portion of sample pairs are manually labeled as similar or dissimilar and proposes the Label-regularized Max-margin Partition algorithm. Moreover, Self-Taught Hashing [10] first identifies binary codes for given documents via unsupervised learning; next, classifiers are trained to predict codes for query documents. Additionally, Fisher Linear Discriminant Analysis (LDA) is employed in [11] to embed the original data to a lower dimensional space and hash codes are obtained subsequently via thresholding. Also, Boosting based Hashing is used in [12] and [13], in which a set of weak hash functions are learned according to the boosting framework. In [14], the hash functions are learned from triplets of side information; their method is designed to preserve the relative relationship reflected by the triplets and is optimized using column generation. Finally, Kernel Supervised Hashing (KSH) [15] introduces a kernel-based hashing method, which seems to exhibit remarkable experimental results.

As for unsupervised learning, several approaches have been proposed: Spectral Hashing (SPH) [16] designs the hash function by using spectral graph analysis with the assumption of a uniform data distribution. [17] proposed Anchor Graph Hashing (AGH). AGH uses a small-size anchor graph to approximate low-rank adjacency matrices that leads to computational savings. Also, in [18], the authors introduce Iterative Quantization, which tries to learn an orthogonal rotation matrix so that the quantization error of mapping the data to the vertices of the binary hypercube is minimized.

To the best of our knowledge, the only approach to date following a semi-supervised learning paradigm is Semi-Supervised Hashing (SSH) [19] [20]. The SSH framework minimizes an empirical error using labeled data, but to avoid over-fitting, its model also includes an information theoretic regularizer that utilizes both labeled and unlabeled data.

In this paper we propose *Supervised Hash Learning (*SHL) (* stands for all three learning paradigms), a novel hash function learning approach, which sets itself apart from past approaches in two major ways. First, it uses a set of Hamming space codewords that are learned during training in order to capture the intrinsic similarities between the data's hash codes, so that same-class data are grouped together. Unlabeled data also contribute to the adjustment of codewords leveraging from the inter-sample dissimilarities of their generated hash codes as measured by the Hamming metric. Due to these codeword-specific characteristics, a major advantage offered by *SHL is that it can naturally engage supervised, unsupervised and, even, semi-supervised hash learning tasks using a single formulation. Obviously, the latter ability readily allows *SHL to perform transductive hash learning.

In Sec. 2, we provide *SHL's formulation, which is mainly motivated by an attempt to minimize the within-group Hamming distances in the code space between a group's codeword and the hash codes of data. With regards to the hash functions, *SHL adopts a kernel-based approach. The aforementioned formulation eventually leads to a minimization problem over the codewords as well as over the Reproducing Kernel Hilbert Space (RKHS) vectors defining the hash functions. A quite noteworthy aspect of the resulting problem is that the minimization over the latter parameters leads to a set of Support Vector Machine (SVM) problems, according to which each SVM generates a single bit of a sample's hash code. In lieu of choosing a fixed, arbitrary kernel function, we use a simple Multiple Kernel Learning (MKL) approach (*e.g.* see [21]) to infer a good kernel from the data. We need to note here that Self-Taught Hashing (STH) [10] also employs SVMs to generate hash codes. However, STH differs significantly from *SHL; its unsupervised and supervised learning stages are completely decoupled, while *SHL uses a single cost function that simultaneously accommodates both of these learning paradigms. Unlike STH, SVMs arise naturally from the problem formulation in *SHL.

Next, in Sec. 3, an efficient Majorization-Minimization (MM) algorithm is showcased that can be used to optimize *SHL's framework via a Block Coordinate Descent (BCD) approach. The first block optimization amounts to training a set of SVMs, which can be efficiently accomplished by using, for example, LIBSVM [22]. The second block optimization step addresses the MKL parameters, while the third one adjusts the codewords. Both of these steps are computationally fast due to the existence of closed-form solutions.

Finally, in Sec. 5 we demonstrate the capabilities of *SHL on a series of comparative experiments. The section emphasizes on supervised hash learning problems in the context of CBIR, since the majority of hash learning approaches address this paradigm. We also included some preliminary transductive hash learning results for *SHL as a proof of concept. Remarkably, when compared to other hashing methods on supervised learning hash tasks, *SHL exhibits the best retrieval accuracy for all the datasets we considered. Some clues to *SHL's superior performance are provided in Sec. 4.

2   Formulation

In what follows, [·] denotes the Iverson bracket, *i.e.*, [predicate] = 1, if the predicate is true, and [predicate] = 0, if otherwise. Additionally, vectors and matrices are denoted in boldface. All vectors are considered column vectors and $\cdot^T$ denotes transposition. Also, for any positive integer K, we define $\mathbb{N}_K \triangleq \{1, \ldots, K\}$.

Central to hash function learning is the design of functions transforming data to compact binary codes in a Hamming space to fulfill a given machine learning task. Consider the Hamming space $\mathbb{H}^B \triangleq \{-1, 1\}^B$, which implies B-bit hash codes. *SHL addresses multi-class classification tasks with an arbitrary set $\mathcal{X}$ as sample space. It does so by learning a hash function $\mathbf{h} : \mathcal{X} \to \mathbb{H}^B$ and a set of G labeled codewords $\boldsymbol{\mu}_g$, $g \in \mathbb{N}_G$ (each codeword representing a class), so that the hash code of a labeled sample is mapped close to the codeword corresponding to the sample's class label; proximity is measured via the Hamming distance. Unlabeled samples are also able to contribute to learning both the hash function and the codewords as it will demonstrated in the sequel. Finally, a test sample is classified according to the label of the codeword closest to the sample's hash code.

In *SHL, the hash code for a sample $x \in \mathcal{X}$ is eventually computed as $\mathbf{h}(x) \triangleq \operatorname{sgn} \mathbf{f}(x) \in \mathbb{H}^B$, where the signum function is applied component-wise. Furthermore, $\mathbf{f}(x) \triangleq [f_1(x) \ldots f_B(x)]^T$, where $f_b(x) \triangleq \langle w_b, \phi(x) \rangle_{\mathcal{H}_b} + \beta_b$ with $w_b \in \Omega_{w_b} \triangleq \{w_b \in \mathcal{H}_b : \|w_b\|_{\mathcal{H}_b} \leq R_b, R_b > 0\}$ and $\beta_b \in \mathbb{R}$ for all $b \in \mathbb{N}_B$. In the previous definition, $\mathcal{H}_b$ is a RKHS with inner product $\langle \cdot, \cdot \rangle_{\mathcal{H}_b}$, induced norm $\|w_b\|_{\mathcal{H}_b} \triangleq \sqrt{\langle w_b, w_b \rangle_{\mathcal{H}_b}}$ for all $w_b \in \mathcal{H}_b$, associated feature mapping $\phi_b : \mathcal{X} \to \mathcal{H}_b$ and reproducing kernel $k_b : \mathcal{X} \times \mathcal{X} \to \mathbb{R}$, such that $k_b(x, x') = \langle \phi_b(x), \phi_b(x') \rangle_{\mathcal{H}_b}$ for all $x, x' \in \mathcal{X}$. Instead of a priori selecting the kernel functions k_b, MKL [21] is employed to infer the feature mapping for each bit from the available data. In specific, it is assumed that each RKHS $\mathcal{H}_b$ is formed as the direct sum of M common, pre-specified RKHSs $\mathcal{H}_m$, i.e., $\mathcal{H}_b = \bigoplus_m \sqrt{\theta_{b,m}} \mathcal{H}_m$, where $\boldsymbol{\theta}_b \triangleq [\theta_{b,1} \ldots \theta_{b,M}]^T \in \Omega_\theta \triangleq \{\boldsymbol{\theta} \in \mathbb{R}^M : \boldsymbol{\theta} \succeq \mathbf{0}, \|\boldsymbol{\theta}\|_p \leq 1, p \geq 1\}$, $\succeq$ denotes the component-wise $\geq$ relation, $\|\cdot\|_p$ is the usual l_p norm in $\mathbb{R}^M$ and m ranges over $\mathbb{N}_M$. Note that, if each preselected RKHS $\mathcal{H}_m$ has associated kernel function k_m, then it holds that $k_b(x, x') = \sum_m \theta_{b,m} k_m(x, x')$ for all $x, x' \in \mathcal{X}$.

Now, assume a training set of size N consisting of labeled and unlabeled samples and let $\mathcal{N}_L$ and $\mathcal{N}_U$ be the index sets for these two subsets respectively. Let also l_n for $n \in \mathcal{N}_L$ be the class label of the nth labeled sample. By adjusting its parameters, which are collectively denoted as $\boldsymbol{\omega}$, *SHL attempts to reduce the distortion measure

$$E(\boldsymbol{\omega}) \triangleq \sum_{n \in \mathcal{N}_L} d\left(\mathbf{h}(x_n), \boldsymbol{\mu}_{l_n}\right) + \sum_{n \in \mathcal{N}_U} \min_g d\left(\mathbf{h}(x_n), \boldsymbol{\mu}_g\right) \tag{1}$$

where d is the Hamming distance defined as $d(\mathbf{h}, \mathbf{h}') \triangleq \sum_b [h_b \neq h_b']$. However, the distortion E is difficult to directly minimize. As it will be illustrated further below, an upper bound $\bar{E}$ of E will be optimized instead.

In particular, for a hash code produced by *SHL, it holds that $d(\mathbf{h}(x), \boldsymbol{\mu}) = \sum_b [\mu_b f_b(x) < 0]$. If one defines $\bar{d}(\mathbf{f}, \boldsymbol{\mu}) \triangleq \sum_b [1 - \mu_b f_b]_+$, where $[u]_+ \triangleq \max\{0, u\}$ is the hinge function, then $d(\operatorname{sgn} \mathbf{f}, \boldsymbol{\mu}) \leq \bar{d}(\mathbf{f}, \boldsymbol{\mu})$ holds for every

$\mathbf{f} \in \mathbb{R}^B$ and any $\boldsymbol{\mu} \in \mathbb{H}^B$. Based on this latter fact, it holds that

$$E(\boldsymbol{\omega}) \leq \bar{E}(\boldsymbol{\omega}) \triangleq \sum_g \sum_n \gamma_{g,n} \bar{d} \left(\mathbf{f}(x_n), \boldsymbol{\mu}_g \right) \tag{2}$$

where

$$\gamma_{g,n} \triangleq \begin{cases} [g = l_n] & n \in \mathcal{N}_L \\ [g = \arg\min_{g'} \bar{d} \left(\mathbf{f}(x_n), \boldsymbol{\mu}_{g'} \right)] & n \in \mathcal{N}_U \end{cases} \tag{3}$$

It turns out that $\bar{E}$, which constitutes the model's loss function, can be efficiently minimized by a three-step algorithm, which delineated in the next section.

3   Learning Algorithm

The next proposition allows us to minimize $\bar{E}$ as defined in Eq. (2) via a MM approach [23], [24].

Proposition 1. *For any *SHL parameter values $\boldsymbol{\omega}$ and $\boldsymbol{\omega}'$, it holds that*

$$\bar{E}(\boldsymbol{\omega}) \leq \bar{E}(\boldsymbol{\omega}|\boldsymbol{\omega}') \triangleq \sum_g \sum_n \gamma'_{g,n} \bar{d} \left(\mathbf{f}(x_n), \boldsymbol{\mu}_g \right) \tag{4}$$

where the primed quantities are evaluated on $\boldsymbol{\omega}'$ and

$$\gamma'_{g,n} \triangleq \begin{cases} [g = l_n] & n \in \mathcal{N}_L \\ [g = \arg\min_{g'} \bar{d} \left(\mathbf{f}'(x_n), \boldsymbol{\mu}'_{g'} \right)] & n \in \mathcal{N}_U \end{cases} \tag{5}$$

Additionally, it holds that $\bar{E}(\boldsymbol{\omega}|\boldsymbol{\omega}) = \bar{E}(\boldsymbol{\omega})$ for any $\boldsymbol{\omega}$. In summa, $\bar{E}(\cdot|\cdot)$ majorizes $\bar{E}(\cdot)$.

Its proof is relative straightforward and is based on the fact that for any value of $\gamma'_{g,n} \in \{0, 1\}$ other than $\gamma_{g,n}$ as defined in Eq. (3), the value of $\bar{E}(\boldsymbol{\omega}|\boldsymbol{\omega}')$ can never be less than $\bar{E}(\boldsymbol{\omega}|\boldsymbol{\omega}) = \bar{E}(\boldsymbol{\omega})$.

The last proposition gives rise to a MM approach, where $\boldsymbol{\omega}'$ are the current estimates of the model's parameter values and $\bar{E}(\boldsymbol{\omega}|\boldsymbol{\omega}')$ is minimized with respect to $\boldsymbol{\omega}$ to yield improved estimates $\boldsymbol{\omega}^*$, such that $\bar{E}(\boldsymbol{\omega}^*) \leq \bar{E}(\boldsymbol{\omega}')$. This minimization can be achieved via a BCD.

Proposition 2. *Minimizing $\bar{E}(\cdot|\boldsymbol{\omega}')$ with respect to the Hilbert space vectors, the offsets β_p and the MKL weights $\boldsymbol{\theta}_b$, while regarding the codeword parameters as constant, one obtains the following B independent, equivalent problems:*

$$\inf_{\substack{w_{b,m} \in \mathcal{H}_m, m \in \mathbb{N}_M \\ \beta_b \in \mathbb{R}, \boldsymbol{\theta}_b \in \Omega_\theta, \mu_{g,b} \in \mathbb{H}}} C \sum_g \sum_n \gamma'_{g,n} \left[1 - \mu_{g,b} f_b(x_n) \right]_+$$

$$+ \frac{1}{2} \sum_m \frac{\|w_{b,m}\|^2_{\mathcal{H}_m}}{\theta_{b,m}} \qquad b \in \mathbb{N}_B \tag{6}$$

where $f_b(x) = \sum_m \langle w_{b,m}, \phi_m(x) \rangle_{\mathcal{H}_m} + \beta_b$ and $C > 0$ is a regularization constant.

The proof of this proposition hinges on replacing the (independent) constraints of the Hilbert space vectors with equivalent regularization terms and, finally, performing the substitution $w_{b,m} \leftarrow \sqrt{\theta_{b,m}} w_{b,m}$ as typically done in such MKL formulations (*e.g.* see [21]). Note that Prob. (6) is jointly convex with respect to all variables under consideration and, under closer scrutiny, one may recognize it as a binary MKL SVM training problem, which will become more apparent shortly.

First block minimization: By considering $w_{b,m}$ and β_b for each b as a single block, instead of directly minimizing Prob. (6), one can instead maximize the following problem:

Proposition 3. *The dual form of Prob. (6) takes the form of*

$$\sup_{\boldsymbol{\alpha}_b \in \Omega_{a_b}} \quad \boldsymbol{\alpha}_b^T \mathbf{1}_{NG} - \frac{1}{2} \boldsymbol{\alpha}_b^T \mathbf{D}_b [(\mathbf{1}_G \mathbf{1}_G^T) \otimes \mathbf{K}_b] \mathbf{D}_b \boldsymbol{\alpha}_b \quad b \in \mathbb{N}_B \tag{7}$$

where $\mathbf{1}_K$ stands for the all ones vector of K elements ($K \in \mathbb{N}$), $\boldsymbol{\mu}_b \triangleq [\mu_{1,b} \dots \mu_{G,b}]^T$, $\mathbf{D}_b \triangleq \mathrm{diag}\,(\boldsymbol{\mu}_b \otimes \mathbf{1}_N)$, $\mathbf{K}_b \triangleq \sum_m \theta_{b,m} \mathbf{K}_m$, where $\mathbf{K}_m$ is the data's m^{th} kernel matrix, $\Omega_{a_b} \triangleq \{\boldsymbol{\alpha} \in \mathbb{R}^{NG} : \boldsymbol{\alpha}_b^T (\boldsymbol{\mu}_b \otimes \mathbf{1}_N) = 0, 0 \preceq \boldsymbol{\alpha}_b \preceq C\boldsymbol{\gamma}'\}$ and $\boldsymbol{\gamma}' \triangleq [\gamma'_{1,1}, \dots, \gamma'_{1,N}, \gamma'_{2,1}, \dots, \gamma'_{G,N}]^T$.

Proof. After eliminating the hinge function in Prob. (6) with the help of slack variables $\xi_{g,n}^b$, we obtain the following problem for the first block minimization:

$$\min_{\substack{w_{b,m}, \beta_b \\ \xi_{g,n}^b}} \quad C \sum_g \sum_n \gamma'_{g,n} \xi_{g,n}^b + \frac{1}{2} \sum_m \frac{\|w_{b,m}\|_{\mathcal{H}_m}^2}{\theta_{b,m}}$$

$$\text{s.t.} \quad \xi_{g,n}^b \geq 0$$

$$\xi_{g,n}^b \geq 1 - (\sum_m \langle w_{b,m}, \phi_m(x) \rangle_{\mathcal{H}_m} + \beta_b) \mu_{g,b} \tag{8}$$

Due to the Representer Theorem (*e.g.*, see [25]), we have that

$$w_{b,m} = \theta_{b,m} \sum_n \eta_{b,n} \phi_m(x_n) \tag{9}$$

where n is the training sample index. By defining $\boldsymbol{\xi}_b \in \mathbb{R}^{RG}$ to be the vector containing all $\xi_{g,n}^b$'s, $\boldsymbol{\eta}_b \triangleq [\eta_{b,1}, \eta_{b,2}, \dots, \eta_{b,N}]^T \in \mathbb{R}^N$ and $\boldsymbol{\mu}_b \triangleq [\mu_{1,b}, \mu_{2,b}, \dots, \mu_{G,b}]^T \in \mathbb{R}^G$, the vectorized version of Prob. (8) in light of Eq. (9) becomes

$$\min_{\boldsymbol{\eta}_b, \boldsymbol{\xi}_b, \beta_b} \quad C\boldsymbol{\gamma}' \boldsymbol{\xi}_b + \frac{1}{2} \boldsymbol{\eta}_b^T \mathbf{K}_b \boldsymbol{\eta}_b$$

$$\text{s.t.} \quad \boldsymbol{\xi}_b \succeq \mathbf{0}$$

$$\boldsymbol{\xi}_b \succeq \mathbf{1}_{NG} - (\boldsymbol{\mu}_b \otimes \mathbf{K}_b) \boldsymbol{\eta}_b - (\boldsymbol{\mu}_b \otimes \mathbf{1}_N) \beta_b \tag{10}$$

where γ' and $\mathbf{K}_b$ are defined in Prop. 3. From the previous problem's Lagrangian $\mathcal{L}$, one obtains

$$\frac{\partial \mathcal{L}}{\partial \xi_b} = 0 \Rightarrow \begin{cases} \lambda_b = C\gamma' - \alpha_b \\ 0 \preceq \alpha_b \preceq C\gamma' \end{cases} \tag{11}$$

$$\frac{\partial \mathcal{L}}{\partial \beta_b} = 0 \Rightarrow \alpha_b^T(\mu_b \otimes \mathbf{1}_N) = 0 \tag{12}$$

$$\frac{\partial \mathcal{L}}{\partial \eta_b} = 0 \overset{\exists \mathbf{K}_b^{-1}}{\Rightarrow} \eta_b = \mathbf{K}_b^{-1}(\mu_b \otimes \mathbf{K}_b)^T \alpha_b \tag{13}$$

where α_b and λ_b are the dual variables for the two constraints in Prob. (10). Utilizing Eq. (11), Eq. (12) and Eq. (13), the quadratic term of the dual problem becomes

$$\begin{aligned}
(\mu_b \otimes \mathbf{K}_b)\mathbf{K}_b^{-1}(\mu_b^T \otimes \mathbf{K}_b) &= \\
&= (\mu_b \otimes \mathbf{K}_b)(1 \otimes \mathbf{K}_b^{-1})(\mu_b^T \otimes \mathbf{K}_b) \\
&= (\mu_b \otimes \mathbf{I}_{N \times N})(\mu_b^T \otimes \mathbf{K}_b) \\
&= (\mu_b \mu_b^T) \otimes \mathbf{K}_b
\end{aligned} \tag{14}$$

Eq. (14) can be further manipulated as

$$\begin{aligned}
(\mu_b \mu_b^T) \otimes \mathbf{K}_b &= \\
&= [(\operatorname{diag}(\mu_b)\mathbf{1}_G)(\operatorname{diag}(\mu_b)\mathbf{1}_G)^T] \otimes \mathbf{K}_b \\
&= [\operatorname{diag}(\mu_b)(\mathbf{1}_G \mathbf{1}_G^T)\operatorname{diag}(\mu_b)] \otimes [\mathbf{I}_N \mathbf{K}_b \mathbf{I}_N] \\
&= [\operatorname{diag}(\mu_b) \otimes \mathbf{I}_N][(\mathbf{1}_G \mathbf{1}_G^T) \otimes \mathbf{K}_b][\operatorname{diag}(\mu_b) \otimes \mathbf{I}_N] \\
&= [\operatorname{diag}(\mu_b \otimes \mathbf{1}_N)][(\mathbf{1}_G \mathbf{1}_G^T) \otimes \mathbf{K}_b][\operatorname{diag}(\mu_b \otimes \mathbf{1}_N)] \\
&= \mathbf{D}_b[(\mathbf{1}_G \mathbf{1}_G^T) \otimes \mathbf{K}_b]\mathbf{D}_b
\end{aligned} \tag{15}$$

The first equality stems from the identity $\operatorname{diag}(v)\mathbf{1} = v$ for any vector v, while the third one stems form the mixed-product property of the Kronecker product. Also, the identity $\operatorname{diag}(v \otimes \mathbf{1}) = \operatorname{diag}(v) \otimes \mathbf{I}$ yields the fourth equality. Note that $\mathbf{D}_b$ is defined as in Prop. 3. Taking into account Eq. (14) and Eq. (15), we reach the dual form stated in Prop. 3. $\qquad \square$

Given that $\gamma'_{g,n} \in \{0, 1\}$, one can easily now recognize that Prob. (7) is an SVM training problem, which can be conveniently solved using software packages such as LIBSVM. After solving it, obviously one can compute the quantities $\langle w_{b,m}, \phi_m(x) \rangle_{\mathcal{H}_m}$, β_b and $\|w_{b,m}\|^2_{\mathcal{H}_m}$, which are required in the next step.

Second block minimization: Having optimized over the SVM parameters, one can now optimize the cost function of Prob. (6) with respect to the MKL parameters θ_b as a single block using the closed-form solution mentioned in Prop.

Algorithm 1 Optimization of Prob. (6)

Input: Bit Length B, Training Samples X containing labeled or unlabled data.
Output: $\boldsymbol{\omega}$.
1. Initialize $\boldsymbol{\omega}$.
2. **While Not Converged**
3.   **For each bit**
4. $\gamma'_{g,n} \leftarrow Eq.$ (5).
5.   Step 1: $w_{b,m} \leftarrow Eq.$ (7).
6. $\beta_b \leftarrow Eq.$ (7).
7.   Step 2: Compute $\|w_{b,m}\|^2_{\mathcal{H}_m}$.
8. $\theta_{b,m} \leftarrow Eq.$ (16).
9.   Step 3: $\mu_{g,b} \leftarrow Eq.$ (17).
10.   **End For**
11. **End While**
12. Output $\boldsymbol{\omega}$.

---

2 of [21] for $p > 1$ and which is given next.

$$\theta_{b,m} = \frac{\|w_{b,m}\|^{\frac{2}{p+1}}_{\mathcal{H}_m}}{\left(\sum_{m'} \|w_{b,m'}\|^{\frac{2p}{p+1}}_{\mathcal{H}_{m'}}\right)^{\frac{1}{p}}}, \quad m \in \mathbb{N}_M, b \in \mathbb{N}_B. \tag{16}$$

Third block minimization: Finally, one can now optimize the cost function of Prob. (6) with respect to the codewords by mere substitution as shown below.

$$\inf_{\mu_{g,b} \in \mathbb{H}} \sum_n \gamma_{g,n} \left[1 - \mu_{g,b} f_b(x_n)\right]_+ \quad g \in \mathbb{N}_G, b \in \mathbb{N}_B \tag{17}$$

On balance, as summarized in Algorithm 1, for each bit, the combined MM/BCD algorithm consists of one SVM optimization step, and two fast steps to optimize the MKL coefficients and codewords respectively. Once all model parameters $\boldsymbol{\omega}$ have been computed in this fashion, their values become the current estimate (*i.e.*, $\boldsymbol{\omega}' \leftarrow \boldsymbol{\omega}$), the $\gamma_{g,n}$'s are accordingly updated and the algorithm continues to iterate until convergence is established[1]. Based on LIBSVM, which provides $\mathcal{O}(N^3)$ complexity [26], our algorithm offers the complexity $\mathcal{O}(BN^3)$ per iteration , where B is the code length and N is the number of instances.

4 Insights to Generalization Performance

The superior performance of *SHL over other state-of-the-art hash function learning approaches featured in the next section can be explained to some extent

---

[1] A MATLAB® implementation of our framework is available at
https://github.com/yinjiehuang/StarSHL

by noticing that *SHL training attempts to minimize the normalized (by B) expected Hamming distance of a labeled sample to the correct codeword, which is demonstarted next. We constrain ourselves to the case, where the training set consists only of labeled samples (*i.e.*, $N = \mathcal{N}_L$, $\mathcal{N}_U = 0$) and, for reasons of convenience, to a single-kernel learning scenario, where each code bit is associated to its own feature space $\mathcal{H}_b$ with corresponding kernel function k_b. Also, due to space limitations, we provide the next result without proof.

Lemma 1. *Let $\mathcal{X}$ be an arbitrary set, $\mathcal{F} \triangleq \{\mathbf{f} : x \mapsto \mathbf{f}(x) \in \mathbb{R}^B, \ x \in \mathcal{X}\}$, $\Psi : \mathbb{R}^B \to \mathbb{R}$ be L-Lipschitz continuous w.r.t $\|\cdot\|_1$, then*

$$\hat{\Re}_N \left(\Psi \circ \mathcal{F}\right) \leq L\hat{\Re}_N \left(\|\mathcal{F}\|_1\right) \tag{18}$$

where $\circ$ stands for function composition, $\hat{\Re}_N(\mathcal{G}) \triangleq \frac{1}{N} \mathbb{E}_\sigma \left\{\sup_{g \in \mathcal{G}} \sum_n \sigma_n g(x_n, l_n)\right\}$ is the empirical Rademacher complexity of a set $\mathcal{G}$ of functions, $\{x_n, l_n\}$ are i.i.d. samples and σ_n are i.i.d random variables taking values with $Pr\{\sigma_n = \pm 1\} = \frac{1}{2}$.

To show the main theoretical result of our paper with the help of the previous lemma, we will consider the sets of functions

$$\bar{\mathcal{F}} \triangleq \{\mathbf{f} : x \mapsto [f_1(x), ..., f_B(x)]^T, f_b \in \mathcal{F}_b, b \in \mathbb{N}_B\} \tag{19}$$

$$\mathcal{F}_b \triangleq \{f_b : x \mapsto \langle w_b, \phi_b(x)\rangle_{\mathcal{H}_b} + \beta_b, \ \beta_b \in \mathbb{R} \text{ s.t. } |\beta_b| \leq M_b,$$

$$w_b \in \mathcal{H}_b \text{ s.t. } \|w_b\|_{\mathcal{H}_b} \leq R_b, \ b \in \mathbb{N}_B\} \tag{20}$$

Theorem 1. *Assume reproducing kernels of $\{\mathcal{H}_b\}_{b=1}^B$ s.t. $k_b(x, x') \leq r^2$, $\forall x, x' \in \mathcal{X}$. Then for a fixed value of $\rho > 0$, for any $\mathbf{f} \in \bar{\mathcal{F}}$, any $\{\boldsymbol{\mu}_l\}_{l=1}^G$, $\boldsymbol{\mu}_l \in \mathbb{H}^B$ and any $\delta > 0$, with probability $1 - \delta$, it holds that:*

$$er\left(\mathbf{f}, \boldsymbol{\mu}_l\right) \leq \hat{er}\left(\mathbf{f}, \boldsymbol{\mu}_l\right) + \frac{2r}{\rho B \sqrt{N}} \sum_b R_b + \sqrt{\frac{\log\left(\frac{1}{\delta}\right)}{2N}} \tag{21}$$

where $er\left(\mathbf{f}, \boldsymbol{\mu}_l\right) \triangleq \frac{1}{B}\mathbb{E}\{d\left(\mathrm{sgn}\left(\mathbf{f}(x), \boldsymbol{\mu}_l\right)\right)\}$, $l \in \mathbb{N}_G$ is the true label of $x \in \mathcal{X}$, $\hat{er}\left(\mathbf{f}, \boldsymbol{\mu}_l\right) \triangleq \frac{1}{NB} \sum_{n,b} Q_\rho\left(f_b(x_n)\mu_{l_n,b}\right)$, where $Q_\rho(u) \triangleq \min\left\{1, \max\left\{0, 1 - \frac{u}{\rho}\right\}\right\}$.

Proof. Notice that

$$\frac{1}{B}d\left(\mathrm{sgn}\left(\mathbf{f}(x), \boldsymbol{\mu}_l\right)\right) = \frac{1}{B}\sum_b [f_b(x)\mu_{l,b} < 0] \leq \frac{1}{B}\sum_b Q_\rho\left(f_b(x)\mu_{l,b}\right)$$

$$\Rightarrow \mathbb{E}\left\{\frac{1}{B}d\left(\mathrm{sgn}\left(\mathbf{f}(x), \boldsymbol{\mu}_l\right)\right)\right\} \leq \mathbb{E}\left\{\frac{1}{B}\sum_b Q_\rho\left(f_b(x)\mu_{l,b}\right)\right\} \tag{22}$$

Consider the set of functions

$$\Psi \triangleq \{\psi : (x,l) \mapsto \frac{1}{B} \sum_b Q_\rho \left(f_b(x) \mu_{l,b} \right), \mathbf{f} \in \bar{\mathcal{F}}, \mu_{l,b} \in \{\pm 1\}, l \in \mathbb{N}_G, b \in \mathbb{N}_B \}$$

Then from Theorem 3.1 of [27] and Eq. (22), $\forall \psi \in \Psi$, $\exists \delta > 0$, with probability at least $1 - \delta$, we have:

$$er\left(\mathbf{f}, \boldsymbol{\mu}_l\right) \leq \hat{er}\left(\mathbf{f}, \boldsymbol{\mu}_l\right) + 2\Re_N(\Psi) + \sqrt{\frac{\log\left(\frac{1}{\delta}\right)}{2N}} \tag{23}$$

where $\Re_N(\Psi)$ is the Rademacher complexity of Ψ. From Lemma 1, the following inequality between empirical Rademacher complexities is obtained

$$\hat{\Re}_N(\Psi) \leq \frac{1}{B\rho} \hat{\Re}_N \left(\left\| \bar{\mathcal{F}}_\mu \right\|_1 \right) \tag{24}$$

where $\bar{\mathcal{F}}_\mu \triangleq \{(x,l) \mapsto [f_1(x)\mu_{l,1}, ..., f_B(x)\mu_{l,B}]^T, f \in \bar{\mathcal{F}} \text{ and } \mu_{l,b} \in \{\pm 1\}\}$. The right side of Eq. (24) can be upper-bounded as follows

$$\hat{\Re}_N \left(\left\| \bar{\mathcal{F}}_\mu \right\|_1 \right) = \frac{1}{N} \mathbb{E}_\sigma \left\{ \sup_{\mathbf{f} \in \bar{\mathcal{F}}, \{\boldsymbol{\mu}_{l_n}\} \in \mathbb{H}^B} \sum_n \sigma_n \sum_b |\mu_{l_n,b} f_b(x_n)| \right\}$$

$$= \frac{1}{N} \mathbb{E}_\sigma \left\{ \sup_{\mathbf{f} \in \bar{\mathcal{F}}} \sum_n \sigma_n \sum_b |f_b(x_n)| \right\}$$

$$= \frac{1}{N} \mathbb{E}_\sigma \left\{ \sup_{w_b \in \mathcal{H}_b, \|w_b\|_{\mathcal{H}_b} \leq R_b, |\beta_b| \leq M_b} \sum_n \sigma_n \sum_b | \langle w_b, \phi_b(x) \rangle_{\mathcal{H}_b} + \beta_b | \right\}$$

$$= \frac{1}{N} \mathbb{E}_\sigma \left\{ \sup_{w_b \in \mathcal{H}_b, \|w_b\|_{\mathcal{H}_b} \leq R_b, |\beta_b| \leq M_b} \sum_n \sigma_n \sum_b | \langle w_b, \text{sgn}(\beta_b)\phi_b(x) \rangle_{\mathcal{H}_b} + |\beta_b| | \right\}$$

$$= \frac{1}{N} \mathbb{E}_\sigma \left\{ \sup_{|\beta_b| \leq M_b} \sum_b [R_b \sqrt{\boldsymbol{\sigma}^T K_b \boldsymbol{\sigma}} + |\beta_b| \sum_n \sigma_n] \right\}$$

$$= \frac{1}{N} \mathbb{E}_\sigma \left\{ \sum_b R_b \sqrt{\boldsymbol{\sigma}^T K_b \boldsymbol{\sigma}} \right\} \overset{\text{Jensen's Ineq.}}{\leq} \frac{1}{N} \sum_b R_b \sqrt{\mathbb{E}_\sigma \{ \boldsymbol{\sigma}^T K_b \boldsymbol{\sigma} \}}$$

$$= \frac{1}{N} \sum_b R_b \sqrt{\text{trace}\{K_b\}} \leq \frac{r}{\sqrt{N}} \sum_b R_b \tag{25}$$

From Eq. (24) and Eq. (25) we obtain $\hat{\Re}_N(\Psi) \leq \frac{r}{\rho B \sqrt{N}} \sum_b R_b$. Since $\Re_N(\Psi) \triangleq \mathbb{E}_s \left\{ \hat{\Re}_N(\Psi) \right\}$, where $\mathbb{E}_s$ is the expectation over the samples, we have

$$\Re_N(\Psi) \leq \frac{r}{\rho B \sqrt{N}} \sum_b R_b \tag{26}$$

The final result is obtained by combining Eq. (23) and Eq. (26). $\qquad\square$

It can be observed that, minimizing the loss function of Prob. (6), in essence, also reduces the bound of Eq. (21). This tends to cluster same-class hash codes around the correct codeword. Since samples are classified according to the label of the codeword that is closest to the sample's hash code, this process may lead to good recognition rates, especially when the number of samples N is high, in which case the bound becomes tighter.

5 Experiments

5.1 Supervised Hash Learning Results

In this section, we compare *SHL to other state-of-the-art hashing algorithms: Kernel Supervised Learning (KSH) [15], Binary Reconstructive Embedding (BRE) [6], single-layer Anchor Graph Hashing (1-AGH) and its two-layer version (2-AGH) [17], Spectral Hashing (SPH) [16] and Locality-Sensitive Hashing (LSH) [3].

Five datasets were considered: *Pendigits* and *USPS* from the *UCI Repository*, as well as *Mnist*, *PASCAL07* and *CIFAR-10*. For *Pendigits* (10, 992 samples, 256 features, 10 classes), we randomly chose 3, 000 samples for training and the rest for testing; for *USPS* (9, 298 samples, 256 features, 10 classes), 3000 were used for training and the remaining for testing; for *Mnist* (70, 000 samples, 784 features, 10 classes), 10, 000 for training and 60, 000 for testing; for *CIFAR-10* (60, 000 samples, 1, 024 features, 10 classes), 10, 000 for training and the rest for testing; finally, for *PASCAL07* (6878 samples, 1, 024 features after down-sampling the images, 10 classes), 3, 000 for training and the rest for testing.

For all the algorithms used, average performances over 5 runs are reported in terms of the following two criteria: (i) retrieval precision of s-closest hash codes of training samples; we used $s = \{10, 15, \ldots, 50\}$. (ii) Precision-Recall (PR) curve, where retrieval precision and recall are computed for hash codes within a Hamming radius of $r \in \mathbb{N}_B$.

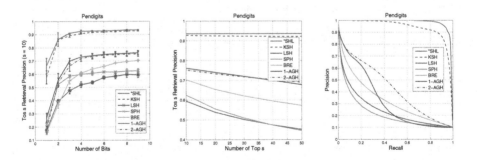

Fig. 1. The top s retrieval results and Precision-Recall curve on *Pendigits* dataset over *SHL and 6 other hashing algorithms. (view in color)

The following *SHL settings were used: SVM's parameter C was set to 1000; for MKL, 11 kernels were considered: 1 normalized linear kernel, 1 normalized polynomial kernel and 9 Gaussian kernels. For the polynomial kernel, the bias was set to 1.0 and its degree was chosen as 2. For the bandwidth σ of the Gaussian kernels the following values were used: $[2^{-7}, 2^{-5}, 2^{-3}, 2^{-1}, 1, 2^1, 2^3, 2^5, 2^7]$.

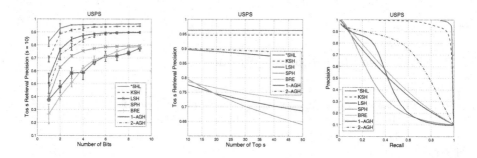

Fig. 2. The top s retrieval results and Precision-Recall curve on *USPS* dataset over *SHL and 6 other hashing algorithms. (view in color)

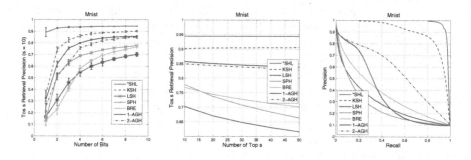

Fig. 3. The top s retrieval results and Precision-Recall curve on *Mnist* dataset over *SHL and 6 other hashing algorithms. (view in color)

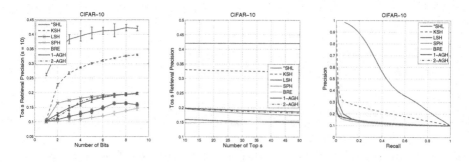

Fig. 4. The top s retrieval results and Precision-Recall curve on *CIFAR-10* dataset over *SHL and 6 other hashing algorithms. (view in color)

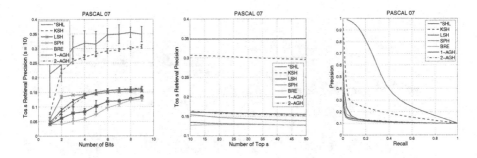

Fig. 5. The top s retrieval results and Precision-Recall curve on *PASCAL07* dataset over *SHL and 6 other hashing algorithms. (view in color)

Regarding the MKL constraint set, a value of $p = 2$ was chosen. For the remaining approaches, namely KSH, SPH, AGH, BRE, parameter values were used according to recommendations found in their respective references. All obtained results are reported in Fig. 1 through Fig. 5.

We clearly observe that *SHL performs best among all the algorithms considered. For all the datasets, *SHL achieves the highest top-10 retrieval precision. Especially for the non-digit datasets (*CIFAR-10, PASCAL07*), *SHL achieves significantly better results. As for the PR-curve, *SHL also yields the largest areas under the curve. Although noteworthy results were reported in [15] for KSH, in our experiments *SHL outperformed it across all datasets. Moreover, we observe that supervised hash learning algorithms, except BRE, perform better than unsupervised variants. BRE may need a longer bit length to achieve better performance as implied by Fig. 1 and Fig. 3. Additionally, it is worth pointing out that *SHL performed remarkably well for short big lengths across all datasets.

It must be noted that AGH also yielded good results, compared with other unsupervised hashing algorithms, perhaps due to the anchor points it utilizes as side information to generate hash codes. With the exception of *SHL and KSH, the remaining approaches exhibit poor performance for the non-digit datasets we considered.

When varying the top-s number between 10 and 50, once again with the exception of *SHL and KSH, the performance of the remaining approaches deteriorated in terms of top-s retrieval precision. KSH performs slightly worse, when s increases, while *SHL's performance remains robust for *CIFAR-10* and *PSACAL07*. It is worth mentioning that the two-layer AGH exhibits better robustness than its single-layer version for datasets involving images of digits. Finally, Fig. 6 shows some qualitative results for the *CIFAR-10* dataset. In conclusion, in our experimentation, *SHL exhibited superior performance for every code length we considered.

Fig. 6. Qualitative results on CIFAR-10. Query image is "Car". The remaining 15 images for each row were retrieved using 45-bit binary codes generated by different hashing algorithms .

5.2 Transductive Hash Learning Results

As a proof of concept, in this section, we report a performance comparison of our framework, when used in an inductive versus a transductive [28] mode. Note that, to the best of our knowledge, no other hash learning approaches to date accommodate transductive hash learning in a natural manner like *SHL. For illustration purposes, we used the *Vowel* and *Letter* datasets. We randomly chose 330 training and 220 test samples for the *Vowel* and 300 training and 200

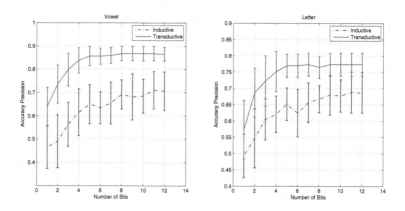

Fig. 7. Accuracy results between Inductive and Transductive Learning.

test samples for the *Letter*. Each scenario was run 20 times and the code length (B) varied from 4 to 15 bits. The results are shown in Fig. 7 and reveal the potential merits of the transductive *SHL learning mode across a range of code lengths.

6 Conclusions

In this paper we considered a novel hash learning framework with two main advantages. First, its Majorization-Minimization (MM)/Block Coordinate Descent (BCD) training algorithm is efficient and simple to implement. Secondly, this framework is able to address supervised, unsupervised and, even, semi-supervised learning tasks in a unified fashion. In order to show the merits of the method, we performed a series of experiments involving 5 benchmark datasets. In these experiments, a comparison between *Supervised Hash Learning (*SHL) to 6 other state-of-the-art hashing methods shows *SHL to be highly competitive.

Acknowledgments. Y. Huang was supported by a Trustee Fellowship provided by the Graduate College of the University of Central Florida. Additionally, M. Georgiopoulos acknowledges partial support from NSF grants No. 0806931, No. 0963146, No. 1200566, No. 1161228, and No. 1356233. Finally, G. C. Anagnostopoulos acknowledges partial support from NSF grant No. 1263011. Any opinions, findings, and conclusions or recommendations expressed in this material are those of the authors and do not necessarily reflect the views of the NSF.

References

1. Datta, R., Joshi, D., Li, J., Wang, J.Z.: Image retrieval: Ideas, influences, and trends of the new age. ACM Computing Surveys 40(2), 5:1–5:60 (2008)
2. Torralba, A., Fergus, R., Weiss, Y.: Small codes and large image databases for recognition. In: Proceedings of Computer Vision and Pattern Recognition, pp. 1–8 (2008)
3. Gionis, A., Indyk, P., Motwani, R.: Similarity search in high dimensions via hashing. In: Proceedings of the International Conference on Very Large Data Bases, pp. 518–529 (1999)
4. Kulis, B., Jain, P., Grauman, K.: Fast similarity search for learned metrics. IEEE Transactions on Pattern Analysis and Machine Intelligence 31(12), 2143–2157 (2009)
5. Salakhutdinov, R., Hinton, G.: Semantic hashing. International Journal of Approximate Reasoning 50(7), 969–978 (2009)
6. Kulis, B., Darrell, T.: Learning to hash with binary reconstructive embeddings. In: Proceedings of Advanced Neural Information Processing Systems, pp. 1042–1050 (2009)
7. Norouzi, M., Fleet, D.J.: Minimal loss hashing for compact binary codes. In: Proceedings of the International Conference on Machine Learning, pp. 353–360 (2011)
8. Yu, C.N.J., Joachims, T.: Learning structural svms with latent variables. In: Proceedings of the International Conference on Machine Learning, pp. 1169–1176 (2009)

9. Mu, Y., Shen, J., Yan, S.: Weakly-supervised hashing in kernel space. In: Proceedings of Computer Vision and Pattern Recognition, pp. 3344–3351 (2010)
10. Zhang, D., Wang, J., Cai, D., Lu, J.: Self-taught hashing for fast similarity search. In: Proceedings of the International Conference on Research and Development in Information Retrieval, pp. 18–25 (2010)
11. Strecha, C., Bronstein, A., Bronstein, M., Fua, P.: Ldahash: Improved matching with smaller descriptors. IEEE Transactions on Pattern Analysis and Machine Intelligence 34(1), 66–78 (2012)
12. Shakhnarovich, G., Viola, P., Darrell, T.: Fast pose estimation with parameter-sensitive hashing. In: Proceedings of the International Conference on Computer Vision, pp. 750–757 (2003)
13. Baluja, S., Covell, M.: Learning to hash: Forgiving hash functions and applications. Data Mining and Knowledge Discovery 17(3), 402–430 (2008)
14. Li, X., Lin, G., Shen, C., van den Hengel, A., Dick, A.R.: Learning hash functions using column generation. In: Proceedings of the International Conference on Machine Learning, pp. 142–150 (2013)
15. Liu, W., Wang, J., Ji, R., Jiang, Y.G., Chang, S.F.: Supervised hashing with kernels. In: Proceedings of Computer Vision and Pattern Recognition, pp. 2074–2081 (2012)
16. Weiss, Y., Torralba, A., Fergus, R.: Spectral hashing. In: Proceedings of Advanced Neural Information Processing Systems, pp. 1753–1760 (2008)
17. Liu, W., Wang, J., Kumar, S., Chang, S.F.: Hashing with graphs. In: Proceedings of the International Conference on Machine Learning, pp. 1–8 (2011)
18. Gong, Y., Lazebnik, S.: Iterative quantization: A procrustean approach to learning binary codes. In: Proceedings of Computer Vision and Pattern Recognition, pp. 817–824 (2011)
19. Wang, J., Kumar, S., Chang, S.F.: Sequential projection learning for hashing with compact codes. In: Proceedings of the International Conference on Machine Learning, pp. 1127–1134 (2010)
20. Wang, J., Kumar, S., Chang, S.F.: Semi-supervised hashing for large-scale search. IEEE Transactions on Pattern Analysis and Machine Intelligence 34(12), 2393–2406 (2012)
21. Kloft, M., Brefeld, U., Sonnenburg, S., Zien, A.: lp-norm multiple kernel learning. Journal of Machine Learning Research 12, 953–997 (2011)
22. Chang, C.C., Lin, C.J.: LIBSVM: A library for support vector machines. ACM Transactions on Intelligent Systems and Technology 2, 27:1–27:27 (2011). Software available at http://www.csie.ntu.edu.tw/~cjlin/libsvm
23. Hunter, D.R., Lange, K.: A tutorial on mm algorithms. The American Statistician 58(1), 30–37 (2004)
24. Hunter, D.R., Lange, K.: Quantile regression via an mm algorithm. Journal of Computational and Graphical Statistics 9(1), 60–77 (2000)
25. Schölkopf, B., Herbrich, R., Smola, A.J.: A generalized representer theorem. In: Helmbold, D.P., Williamson, B. (eds.) COLT/EuroCOLT 2001. LNCS (LNAI), vol. 2111, pp. 416–426. Springer, Heidelberg (2001)
26. List, N., Simon, H.U.: Svm-optimization and steepest-descent line search. In: Proceedings of the Conference on Computational Learning Theory (2009)
27. Mohri, M., Rostamizadeh, A., Talwalkar, A.: Foundations of Machine Learning. The MIT Press (2012)
28. Vapnik, V.N.: Statistical Learning Theory. Wiley-Interscience (1998)

HierCost: Improving Large Scale Hierarchical Classification with Cost Sensitive Learning

Anveshi Charuvaka[✉] and Huzefa Rangwala

George Mason University, Fairfax, USA
acharuva@gmu.edu, rangwala@cs.gmu.edu

Abstract. Hierarchical Classification (HC) is an important problem with a wide range of application in domains such as music genre classification, protein function classification and document classification. Although several innovative classification methods have been proposed to address HC, most of them are not scalable to web-scale problems. While simple methods such as top-down "pachinko" style classification and flat classification scale well, they either have poor classification performance or do not effectively use the hierarchical information. Current methods that incorporate hierarchical information in a principled manner are often computationally expensive and unable to scale to large datasets. In the current work, we adopt a cost-sensitive classification approach to the hierarchical classification problem by defining misclassification cost based on the hierarchy. This approach effectively decouples the models for various classes, allowing us to efficiently train effective models for large hierarchies in a distributed fashion.

1 Introduction

Categorizing entities according to a hierarchy of general to specific classes is a common practice in many disciplines. It can be seen as an important aspect of various fields such as bioinformatics, music genre classification, image classification and more importantly document classification [18]. Often the data is curated manually, but with exploding sizes of databases, it is becoming increasingly important to develop automated methods for hierarchical classification of entities.

Several classification methods have been developed over the past several years to address the problem of Hierarchical Classification (HC). One straightforward approach is to simply use multi-class or binary classifiers to model the relevant classes and disregard the hierarchical information. This methodology has been called *flat* classification scheme in HC literature [18]. While flat classification can be competitive, an important research directions is to improve the classification performance by incorporating the hierarchical structure of the classes in the learning algorithm. Another simple data decomposition approach trains local classifiers for each of the classes defined according to the hierarchy, such that the trained model can be used in a top-down fashion to take the most relevant path in testing. This top-down approach trains each classifier on a smaller dataset and

© Springer International Publishing Switzerland 2015
A. Appice et al. (Eds.): ECML PKDD 2015, Part I, LNAI 9284, pp. 675–690, 2015.
DOI: 10.1007/978-3-319-23528-8_42

is quite efficient in comparison to flat classification, which generally train one-vs-rest classifiers on the entire dataset. However, a severe drawback of this approach is that if a prediction error is committed at a higher level, then the classifier selects a wrong prediction path, making it impossible to recover from the errors at lower levels. Due to this error propagation, sometimes, sever degradation in performance has been noted for the top-down classifier in comparison to flat classifier [9]. A review of HC in several application domains can be found in a recent survey by Silla Jr. et al. [18].

In recent years, researchers have shown more interest in large scale classification where the number of categories, number of instances, as well as the number of features are large. This has been highlighted by large scale hierarchical text classification competitions such as LSHTC[1] [15] and BioASQ [2], which pose several interesting challenges. Firstly, since these problems deal with several thousands of classes, scalability of the methods is a crucial requirement. Secondly, in spite of having large number of total training examples, many categories have few positive training samples. For example, 76% of the class-labels in the Yahoo! Directory have 5 or fewer positive instances [11], and 72% in the Open Directory Project have fewer than 4 positive instances [9]. This data sparsity brings about two issues: (i) due to the lack of sufficient examples, the learned models tend to be less robust, and (ii) due to the large skew in positive and negative class distributions, the performance of smaller classes tends to deteriorate severely as the mis-predictions tend to favor larger classes.

In this paper, we try to address two main issues of large scale hierarchical classification, class imbalance and training efficiency, by extending the flat classification approach using cost sensitive training examples. Although regularization methods which constrain the learned models to be close to neighboring classes according to the hierarchy have been effective, they induce large scale optimization problems which require specialized solutions [3]. Instead, by re-defining the problem from regularization based approach to a cost sensitive classification approach (based on similar assumptions) tends to decouple the training of the models which can be trained in parallel fashion. We study various methods to incorporate cost-sensitive information into hierarchical classification and empirically evaluate their performance on several datasets. Finally, since any instance based cost sensitive method can be used as a base classifier, the HC problem can benefit from advancements in cost-sensitive classification.

2   Definitions and Notations

In this section, we discuss the notations commonly used in this paper. $\mathcal{N}$ denotes the set of all the nodes in the hierarchy, and $\mathcal{T} \subset \mathcal{N}$ denotes the set of terminal nodes to which examples are assigned. $\mathbf{w}_n$ denotes the model learnt for class $n \in \mathcal{N}$. $(\mathbf{x}_i, l_i)$ denotes the ith example where $\mathbf{x}_i \in \mathbb{R}^D$ and $l_i \in \mathcal{T}$. The number of examples is denote by N. We use y_i^n to denote the binary label used in the

---

[1] http://lshtc.iit.demokritos.gr
[2] http://bioasq.org

learning algorithm for $\mathbf{w}_n$. For the training example $(\mathbf{x}_i, l_i)$ we set $y_i^n = 1$ iff $l_i = n$ and $y_i^n = -1$ otherwise. $\gamma(a, b)$ denotes the graph distance between classes $a, b \in \mathcal{N}$ in the hierarchy, which is defined as the number of edges in the undirected path between nodes a and b. We use c_i^n to denote the cost of example i in training of the model for class n. To simplify the notation, in some places, we drop the super-script explicitly indicating the class, and use y_i , c_i and $\mathbf{w}$ in place of y_i^n, c_i^n and $\mathbf{w}_n$ where the class is implicitly understood to be n. L is used to denote a generic loss function. In the current work, logistic loss function is used, which is defined as $L(y, f(\mathbf{x})) = \log(1 + \exp(-yf(\mathbf{x})))$.

3   Motivation and Related Work

In this section, we discuss the motivation for the approach taken in this paper and examine various related methods proposed in the literature for addressing the hierarchical classification problem.

A few large margin methods have been proposed as cost sensitive extensions to the multi-class classification problem. Dekel et al. [5] proposed a large margin method where the margin is defined with respect to the tree distance. Although their method shows improvement on tree-error, the performance degrades with respect to misclassification error. The methods proposed by Cai et al. [2] and more recently by Chen et al. [4], also make an argument in favor of modifying the misclassification error by making it dependent on the hierarchy. Both these methods can be seen as special cases of a more general large margin structured output prediction method proposed by Tsochantaridis et al. [20]. Although all these methods try to incorporate cost sensitive losses based on the hierarchy, they formulate a global optimization problem where the models for all the classes are learned jointly and are not scalable to large scale classification problems.

Several methods try to incorporate the bias that categories which are semantically related according to the hierarchy should also be similar with respect to the learned models. McCallum et al. [13] show that for Naive Bayes classifier, smoothing the parameter estimates of the data-sparse children nodes with the parameter estimates of parent nodes, using a technique known as shrinkage, produces more robust models. Other models in this class of methods typically incorporate this assumption using parent child regularization or hierarchy based priors [9,13,17]. In one of the prototypical models in this class of works, which extends Support Vector Machines (SVM) and Logistic Regression (LR) [9], the objective function takes the form given in (1),

$$\min_{\mathbf{w}_1, \dots, \mathbf{w}_{|\mathcal{N}|}} \sum_{n \in \mathcal{N}} \frac{1}{2} \left\| \mathbf{w}_n - \mathbf{w}_{\pi(n)} \right\|_2^2 + C \sum_{n \in \mathcal{T}} \sum_{i=1}^{N} L\left(y_i^n, \mathbf{w}_n^T \mathbf{x}_i\right) \tag{1}$$

where, $\pi(n)$ represents the parent of the class n according to the provided hierarchy. The loss function L has been modeled as logistic loss or hinge loss. Note that the loss is defined only on the terminal nodes $\mathcal{T}$, and the non-terminal node $\mathcal{N} - \mathcal{T}$, are introduced only as a means to facilitate regularization. Since the weights associated with different classes are coupled in the optimization problem, Gopal et al. [9] used a distributed implementation of block coordinate descent where each block of variables corresponds to $\mathbf{w}_n$ for a particular class n. The model weights are learned similarly to standard LR or SVM for the leaf nodes $n \in \mathcal{T}$, with the exception that the weights are shrunk towards parents instead of towards the zero vector by the regularizer. For an internal non-leaf node, the weights updates are averages of the other nodes which are connected to it according to the hierarchy, i.e., the parents and children in the hierarchy.

The kind of regularization discussed above can be compared to the formulations proposed in transfer and multi-task learning (MTL) literature [7], where externally provided task relationships can be utilized to constrain the jointly learned model weights to be similar to each other. In the case of HC, the task relationship are explicitly provided as hierarchical relationships. However, one significant difference between the application of this regularization between HC and MTL is that the sets of examples in MTL for different tasks are, in general, disjoint. Whereas, in the case of HC, the examples which are classified as positive for one class are negative for all other classes except those which belong to the ancestors of that class. Therefore, even though these models impose similarity between siblings indirectly through the parent, when their respective models are trained, the negative and positive examples are flipped. Hence, the opposing forces for examples and regularization are acting simultaneously during the learning of these models. However, due to the regularization strength being imposed by the hierarchy, the net effect is that the importance of misclassifying the examples coming for nearby classes is down-weighted. This insight can be directly incorporated into the learning algorithm by defining the loss of nearby negative examples for a class, where "near" is defined with respect to the hierarchy, to be less severe than the examples which are farther. This yields a simple cost sensitive classification method where the misclassification cost is directly proportional to the distance between the nodes of the classes, which is the key contribution of our work. With respect to prediction, there are only two classes for each trained model, but the misclassification costs of negative examples are dependent on the nodes from which they originate.

In this framework for HC, we essentially decouple the learning for each node of the hierarchy and train the model for each one independently. Thus, rendering scalability to this method. Instead of jointly formulating the learning of model parameters for all the classes, we turn the argument around from that of regularizing the model weights towards those of the neighboring models, to the rescaling the loss of example depending on the class relationships. A similar argument was made in the case of multi-task transfer learning by Saha et al. [16], where, in place of joint regularization of model weights, as is typically done in multi-task learning [8], they augment the target tasks with examples from source tasks. However, the losses for the two sets of examples are scaled differently.

4   Methods

As shown in some previous works [1,9], the performance of flat classification has been found to be very competitive, especially for large scale problems. Although, the top-down classification method is efficient in training, it fares poorly with respect to classification performance due to error propagation. Hence, in this work, we extend the flat classification methodology to deal with HC. We use the one-vs-all approach for training, where for each class n, to which examples are assigned, we learn a classification model with weight $\mathbf{w}_n$. Note, that it is unnecessary to train the models for non-terminal classes, as they only serve as virtual labels in the hierarchy. Once the models for each terminal class are trained, we perform prediction for input example $\mathbf{x}$ as per (2)

$$\hat{y} = \mathbf{argmax}_n \ \mathbf{w}_n^T \mathbf{x} \tag{2}$$

The essential idea is to formulate the learning algorithm such that the mispredictions on negative examples coming from nearby classes are treated as being less severe. We encode this assumption through cost sensitive classification. Standard regularized binary classification models, such as SVMs and Logistic Regression, minimize an objective function consisting of loss and regularization terms as shown in (3).

$$\min_{\mathbf{w}} \underbrace{\sum_{i=1}^{N} L\left(y_i, f\left(\mathbf{x}_i \mid \mathbf{w}\right)\right)}_{loss} + \rho \underbrace{R\left(\mathbf{w}\right)}_{regularizer} \tag{3}$$

where $\mathbf{w}$ denotes the learned model weights. The loss function L, which is generally a convex approximation of zero-one loss, measures how well the model fits the training examples. Here, each example is considered to be equally important. As per the arguments made previously, we modify the importance of correctly classifying examples according to the hierarchy using example based misclassification costs. For models such as logistic regression, incorporating example based costs into the learning algorithm is simply a matter of scaling the loss by a constant positive value. Assuming that the classifier is being learned for class n, we can write the cost sensitive objective function as shown in (4).

$$\min_{\mathbf{w}} \sum_{i=1}^{N} c_i L\left(y_i, \mathbf{w}^T \mathbf{x}_i\right) + \rho R\left(\mathbf{w}\right) \tag{4}$$

Here, c_i denotes the cost associated with misclassification of the ith example. Although, this scaling works for the smooth loss function of Logistic Regression, it is not as straightforward in the case of non-smooth loss functions such as hinge loss [12]. Therefore, using the formulation given in (4), for each of the models, we can formulate the objective function for class n as a cost sensitive logistic regression problem where the cost of the example $\mathbf{x}_i$ for the binary classifier of class n depends on how far the actual label $l_i \in \mathcal{T}$ is from n, according to the hierarchy. Additionally, to deal with the issue of rare categories, we can also increase the cost of the positive examples for data-sparse classes thus mitigating the effects of highly skewed datasets. Since our primary motivation is to argue in favor of using hierarchical cost sensitive learning instead of more expensive regularization models, we only concentrate on logistic loss, which is easier to handle, from optimization perspective, than non-smooth losses such as SVM's hinge loss. The central issue, now, is that of defining the appropriate costs for the positive and negative examples based on the distance of the examples from the true class according to the hierarchy and the number of examples available for training the classifiers. In the following section we discuss the selection of costs for negative and positive examples.

4.1   Cost Calculations

Hierarchical Cost. Hierarchical costs impose the requirement that the misclassification of negative examples that are farther away from the training class according to the hierarchy should be penalized more severely. Encoding this assumption, we define the following instantiations of hierarchical costs. We assume the class under consideration is denoted by n.

Tree Distance (TrD): In (5), we define the cost of negative examples as the undirected graphical distance, $\gamma(n, l_i)$, between the class n and l_i, the class label of example x_i. We call this cost *Tree Distance (TrD)*. We define $\gamma_i \equiv \gamma(n, l_i)$ and $\gamma_{max} = \max_{j \in \mathcal{T}} \gamma_j$. Since dissimilarity increases with increasing γ_i, the cost is a monotonically increasing function of γ_i.

$$c_i = \begin{cases} \gamma_{max} & l_i = n \\ \gamma_i & l_i \neq n \end{cases} \tag{5}$$

Number of Common Ancestors (NCA): In some applications, where the depth (distance of a node from the root node) of all terminal labels is not uniform, a better definition of similarity might be the number of common ancestor between two nodes. This is encoded in *NCA* costs, represented in (6). In the definition, α_i is used to denote the number of common ancestors between the pair of nodes l_i and n. Unlike γ_i which is a monotonically increasing function of dissimilarity, α_i is a monotonically increasing function of similarity. $\alpha_{max} = \max_{j \in \mathcal{T}} \alpha_j$.

$$c_i = \begin{cases} \alpha_{max} + 1 & l_i = n \\ \alpha_{max} - \alpha_j + 1 & l_i \neq n \end{cases} \tag{6}$$

Exponentiated Tree Distance (ExTrD): Finally, in some cases, especially for deep hierarchies, the tree distances can be large, and therefore, in order to shrink the values of cost into a smaller range, we define *ExTrD* in (7), where $k > 1$, can be tuned according to the hierarchy. Through tuning we found that on our dataset, the range of values $1.1 \leq k \leq 1.25$ of works well.

$$c_i = \begin{cases} k^{\gamma_{\max}} & l_i \neq n \\ k^{\gamma_i} & l_i \neq n \end{cases} \tag{7}$$

In all these cases, we set the cost of the positive class to the maximum cost of any example.

Imbalance Cost. In certain cases, especially for large scale hierarchical text classification, some classes are extremely small with respect to the number of positive examples available for training. In these cases, the learned decision boundary might favor the larger classes. Therefore, to deal with this imbalance in the class distributions, we increase the cost of misclassifying rare classes. This has the effect of mitigating the influence of skew in the data distributions of abundant and rare classes. We call the cost function incorporating this as *Imbalance Cost (IMB)*, which is given in (8). We noticed that using cost such as inverse of class size diminishes the performance. Therefore, we use a squashing function inspired by logistic function $f(x) = L / [1 + \exp -k(x - x_0)]$, which would not severely disadvantage very large classes.

$$c_i = 1 + L / \left[1 + \exp \left(|N_i - N_0|_+ \right) \right] \tag{8}$$

where $|a|_+ = \max(a, 0)$ and N_i is the number of examples belonging to class denoted by l_i. The value of c_i lies in the range $(1, L/2 + 1)$. We can use a tunable parameter N_0, which can be intuitively interpreted as the number of examples required to build a "good" model, above which increasing the cost does not have a significant effect or might adversely affect the classification performance. In our experiments, we used $N_0 = 10$ and $L = 20$.

In order to combine the Hierarchical Costs with the Imbalance Costs, we simply multiply the contributions of both the costs. We also experimented with several other hierarchical cost variants, which are not discussed here due to space constraints.

4.2 Optimization

Since we are dealing with large scale classification problems, we need an efficient optimization method which relies only on the first order information to solve the learning problem given in (9).

$$\min_{\mathbf{w}} \left[f(\mathbf{w}) = \sum_{i=1}^{N} c_i \log \left(1 + \exp \left(-y_i \mathbf{w}^T \mathbf{x}_i \right) \right) + \rho \|\mathbf{w}\|_2^2 \right] \tag{9}$$

Since the cost values c_i are predefined positive scalars, we can adapt any method used to solve the standard regularized Logistic Regression (LR). We use accelerated gradient descent due to its efficiency and simplicity. The ordinary gradient descent method has a convergence rate of $O\left(1/k\right)$, where k is the number of iterations. Accelerated gradient method improves the convergence rate to $O\left(1/k^2\right)$ by additionally using the gradient information from the previous iteration [14]. The complete algorithm to solve the cost-sensitive binary logistic regression is provided in Line 1. We describe the notations and expressions used in the algorithm below.

N is the number of examples; $X \in \mathbb{R}^{N \times D}$ denotes the data matrix; $\mathbf{y} \in \{\pm 1\}^N$ is the binary label vector for all examples; $\rho \in \mathbb{R}_+$ is the regularization parameter; $\mathbf{c} = (c_1, c_2, \ldots, c_N) \in \mathbb{R}_+^N$ denotes the cost vector, where c_i is the cost for example i ; $\mathbf{w} \in \mathbb{R}^D$ denotes the weight vector learned by the classifier; $f\left(\mathbf{w}\right)$ denotes the objective function value, given in (10)

$$f\left(\mathbf{w}\right) = \mathbf{c}^T \left(\log\left[1 + \exp\left(X\mathbf{w}\right)\right]\right) + \rho\left\|\mathbf{w}\right\|_2^2 \tag{10}$$

$\nabla \mathbf{f}$ is the gradient of f w.r.t. $\mathbf{w}$, which is defined in (11), where $(\mathbf{y} \circ \mathbf{c})$ denotes the vector obtained from the element-wise product of $\mathbf{y}$ and $\mathbf{c}$. Similarly $\exp(\cdot)$ and division in (11) are element-wise operators.

$$\nabla \mathbf{f}\left(\mathbf{w}\right) = 2\mathbf{w} + X^T \left(\frac{-\mathbf{y} \circ \mathbf{c}}{1 + \exp\left\{(X\mathbf{w}) \circ \mathbf{y}\right\}}\right) \tag{11}$$

$\hat{f}_\lambda\left(\mathbf{u}, \mathbf{w}\right)$, described in (12), is the quadratic approximation of $f\left(\mathbf{u}\right)$ at $\mathbf{w}$ using approximation constant or step size λ. The appropriate step size in each iteration is found using line search.

$$\hat{f}_\lambda\left(\mathbf{u}, \mathbf{w}\right) = f\left(\mathbf{w}\right) + \left(\mathbf{u} - \mathbf{w}\right)^T \nabla \mathbf{f}\left(\mathbf{w}\right) + 1/2\lambda\left\|\mathbf{u} - \mathbf{w}\right\|_2^2 \tag{12}$$

4.3   Dealing with Hierarchical Multi-label Classification

HC problems are trivially multi-label problems because every example belonging to a class also inherits the labels of the ancestor classes. But in the current context, we call a problem as hierarchical multi-label problem if an example can be assigned multiple labels such that neither is an ancestor nor descendant of the other.

In the case of single label classification, we perform prediction as per (2), which selects only a single label per example. A trivial extension to multi-label classification can be done by choosing a threshold of 0 such that we assign label n to example $\mathbf{x}$ if $\mathbf{w}_n^T\mathbf{x} > 0$ as in the case of binary classification. However, a better strategy is to optimize the threshold t_n for each class using a validation set, such that the label n is assigned to the test example if $\mathbf{w}_n^T\mathbf{x} > t_n$. This strategy is called SCut method [21]. Other strategies such as learning a thresholding function $t\left(\mathbf{w}_1^T\mathbf{x}, \mathbf{w}_2^T\mathbf{x}, \ldots, \mathbf{w}_M^T\mathbf{x}\right)$ using the margin scores [9] might improve the results, but they are somewhat more expensive to tune for large scale problems. The SCut method can tune the threshold independently of all other classes. In cases where we do not have sufficient examples to tune the threshold, i.e. the class has a single training example, we set the threshold to $t_n = 0$.

Algorithm 1. Accelerated Gradient Method for Cost Sensitive LR

Data: $X, \mathbf{y}, \mathbf{c}, \rho, \beta \in (0,1), max\_iter$
Result: w
Let $\lambda_0 := 1; \mathbf{w}_{-1} = \mathbf{w}_0 = \mathbf{0}$;
for $k = 1 \ldots max\_iter$ **do**
    $\theta^k = \frac{k-1}{k+2}$
    $\lambda = \lambda_{k-1}$
    while *TRUE* **do**
        $\mathbf{w} = \mathbf{u}_k - \lambda \nabla f(\mathbf{u}_{k-1})$
        if $f(\mathbf{w}) \leq \hat{f}_\lambda(\mathbf{u}, \mathbf{w})$ **then**
            $\lambda_k = \lambda$
            $\mathbf{w}_k = \mathbf{w}$
            break
        else
            $\lambda = \beta\lambda$
        end
    end
    if *converged* **then**
        break
    end
end
return $\mathbf{w}_k$

The second issue that we must deal with is the definition of cost based on hierarchical distances and class sizes. With respect to the training of a class n, an example $\mathbf{x}_i$ might be associated with multiple labels $l_1, l_2 \ldots, l_K$. In this case the tree distance γ_i is not uniquely defined. Hence, we must aggregate the values of $\gamma(n, l_1), \ldots, \gamma(n, l_K)$. One strategy is to use an average of the values, but we found that the taking the minimum works a little better. Similarly we can use a minimum of of the number of common ancestors to all target labels for *NCA* costs.

Finally, since an example is associated with multiple class labels, the class size N_i of the examples is also not uniquely defined, in this case as well, we use the the effective size as the minimum size out of all the labels associated with $\mathbf{x}_i$ for our *IMB* cost. It also makes intuitive sense, because we are trying to upweight the rare classes, and the rarest class should be given precedence in terms of the cost definition.

5 Experimental Evaluations

5.1 Datasets

The details of the datasets used for our experimental evaluations are provided in Table 1. **CLEF** [6] is a dataset comprising of medical images annotated with hierarchically organized Image Retrieval in Medical Applications (IRMA) codes. The task is to predict the IRMA codes from image features. Images are described with 80 features extracted using a technique called local distribution of edges.

IPC is a collection of patent documents classified according to the International Patent Classification (IPC) System[3]. **DMOZ-small, DMOZ-2010** and **DMOZ-2012** are hierarchical text classification datasets released as part of PASCAL Large Scale Hierarchical Text Classification Challenge (LSHTC)[4] [15]. For LSHTC datasets except DMOZ-small, labels of the test datasets are not available, but certain classification metrics can be obtained through their online evaluation system. **RCV1-v2** [10] is a multi-label text classification dataset extracted from Reuters corpus of manually categorized newswire stories. **RCV1** is multi-label and non-mandatory leaf node predication [18] dataset, while the rest of the datasets are single label datasets with examples assigned only to leaf nodes. All the hierarchies used in the experiments are tree-based. For all the text datasets, raw term frequencies were converted to term weights using Cornell ltc term weighting [10].

Table 1. Dataset Statistics.

	CLEF	DMOZ SMALL	IPC	RCV1	DMOZ 2010	DMOZ 2012
Num. Training Examples	10000	4463	46324	23149	128710	383408
Num. Test Examples	1006	1858	28926	781265	34880	103435
Num. Features	80	51033	345479	48728	381580	348548
Num. Nodes	97	2388	553	117	17222	13963
Num. Terminal Nodes	63	1139	451	101	12294	11947
Max. Depth of Leaf Nodes	4	6	4	6	6	6
Avg. Labels per Example	1	1	1	3.18	1	1

5.2 Evaluation Metrics

We evaluate the prediction using the following standard performance measures used in HC literature. The set based measures Micro-F_1 and Macro-F_1 are shown below.

$$\text{Micro-}F_1 = (2PR)/(P+R) \qquad (13)$$

$$\text{Macro-}F_1 = \frac{1}{|\mathcal{T}|}\sum_{t\in\mathcal{T}} 2P_t R_t/(P_t + R_t) \qquad (14)$$

$\mathcal{T}$ denotes is the set of class labels, P_t and R_t are the precision and recall values for class $t \in \mathcal{T}$. P and R are the overall precision and recall values for the all the classes taken together. Micro-F_1 gives equal weight to all the examples therefore it favors the classes with more number of examples. In the case of single label classification, Micro-F_1 is equivalent to accuracy. Macro-F_1 gives equal weight

---

[3] http://www.wipo.int/classifications/ipc/en/
[4] http://lshtc.iit.demokritos.gr/

to all the classes irrespective of their size. Hence, the performance on the smaller categories is also taken into consideration.

Set based measures do not consider the distance of misclassification with respect to the true label of the example, but in general, it is reasonable to assume in most cases that misclassifications that are closer to the actual class are less severe than misclassifications that are farther from the true class with respect to the hierarchy. Hierarchical measures, therefore, take the distances between the actual and predicted class into consideration. The hierarchical measures, described in eqs. (15) to (17), are Hierarchical Precision (hP), Hierarchical Recall (hR), and their harmonic mean, Hierarchical F_1 (hF_1) respectively. These are hierarchical extensions of standard precision and recall scores. Tree-induced Error (TE) [19], given in (18), measures the average hierarchical distance between the actual and predicted labels.

$$hP = \sum_{i=1}^{N} \left| \mathcal{A}\left(l_i\right) \cap \mathcal{A}\left(\hat{l}_i\right) \right| / \sum_{i=1}^{N} \left| \mathcal{A}\left(\hat{l}_i\right) \right| \tag{15}$$

$$hR = \sum_{i=1}^{N} \left| \mathcal{A}\left(l_i\right) \cap \mathcal{A}\left(\hat{l}_i\right) \right| / \sum_{i=1}^{N} \left| \mathcal{A}\left(l_i\right) \right| \tag{16}$$

$$hF_1 = 2 \cdot hP \cdot hR / \left(hP + hR\right) \tag{17}$$

$$TE = \frac{1}{N} \sum_{i=1}^{N} \gamma\left(l_i, \hat{l}_i\right) \tag{18}$$

where, $\hat{l}_i$ and l_i are the predicted label and true labels of example i, respectively. $\gamma\left(a, b\right)$ the graph distance between a and b according to the hierarchy. $\mathcal{A}\left(l\right)$ denotes the set that includes the node l and all its ancestors except the root node. For TE lower values are better, whereas for all other measures higher values are better.

For multi-label classification, where each l_i is a set of micro-labels, we redefine graph distance and ancestors as: $\gamma_{ml}(l_i, \hat{l}_i) = \left|\hat{l}_i\right|^{-1} \sum_{a \in \hat{l}_i} \min_{b \in l_i} \gamma\left(a, b\right)$ and $\mathcal{A}_{ml}\left(l\right) = \cup_{a \in l} \mathcal{A}\left(a\right)$.

5.3   Experimental Details

For all the experiments, the regularization parameter is tuned using a validation set. The model is trained for a range of values 10^k with appropriate values for k selected depending on the dataset. Using the best parameter selected on validation set, we retrained the models on the entire training set and measured the performance on a held out test set. The source code implementing the methods discussed in this paper is available on our website [5]. The experiments were performed on computers with Dell C8220 processors with dual Intel Xeon E5-2670 8 core CPUs and 64 GB memory.

---

[5] http://cs.gmu.edu/~mlbio/HierCost/

5.4   Methods for Comparison

In our experimental evaluations, we compare our cost sensitive hierarchical classification methods with the following hierarchical and flat classification methods proposed in the literature.

Logistic Regression (LR). One-vs-rest binary logistic regression is used in the conventional flat classification setting. For single label classification, we assign test examples to the class which achieves best classification score.

Hierarchical Regularization for LR (HRLR). This method proposed by Gopal et al. [9], extends flat classification using recursive regularization based on hierarchical relationships. Since we used exactly the same setup as the authors, in terms of training and test datasets, we are reporting their classification scores directly from [9].

Top-Down Logistic Regression (TD). This denotes Top-down logistic regression model, where we train a one-vs-rest multi-class classifier at each internal node. At testing time, the predictions are made starting from the root node. At each internal node, the highest scoring child node is selected until we reach a leaf node.

5.5   Results

In this section, we present experimental comparisons of various cost sensitive learning strategies with other baseline methods. We provide separate comparisons of different cost based improvements on smaller datasets, and finally compare our best method with the competing methods. In the tables, statistically significant results for Micro-F_1 and Macro-F_1 [22] are marked with either † or ‡ which correspond to p-values < 0.05 and < 0.001 respectively.

In Table 2, we compare LR with various *hierarchical costs* defined in Section 4.1. The results show a uniform improvement in all the metrics reported. There was a statistically significant improvement in Micro-F_1, especially for DMOZ Small, IPC and RCV1 datasets. Macro-F_1 scores were also improved, but due to the presence of only a small number of categories in CLEF and RCV1 datasets, statistical significance could not be established, except for ExTrD.

In Table 3 we compare the effect of introducting *imbalance costs*, discussed in Section 4.1, on standard LR and hierarchical costs. In IMB+LR only the imbalance cost is used, in others, we use the product of costs derived from IMB strategy and the corresponding hierarchical costs. We also measured the significance of the improvement over the corresponding results from Table 2. Only for DMOZ Small, which has a large number of classes with few examples, imbalance costs further improve the results significantly for all the methods. On CLEF, IPC and RCV1, where majority of the classes have sufficient number of examples for training, the results did not improve significantly in most cases. Overall, the IMB+ExTrD method provides more robust improvements.

The final comparison of our best method (IMB+ExTrD, which we call *Hier-Cost* in the following) against various baseline methods is presented in Table 4.

Table 2. Performance comparison of hierarchical costs.

		Micro-F_1 ($\uparrow$)	Macro-F_1 ($\uparrow$)	hF_1 ($\uparrow$)	TE ($\downarrow$)
CLEF	LR	79.82	53.45	85.24	0.994
	TrD	80.02	55.51	**85.39**	0.984
	NCA	80.02	57.48	85.34	0.986
	ExTrD	**80.22**	**57.55**†	85.34	**0.982**
DMOZ SMALL	LR	46.39	30.20	67.00	3.569
	TrD	**47.52**‡	**31.37**‡	**68.26**	**3.449**
	NCA	47.36‡	31.20‡	68.12	3.460
	ExTrD	47.36‡	31.19‡	68.20	3.456
IPC	LR	55.04	48.99	72.82	1.974
	TrD	55.24‡	50.20‡	73.21	1.954
	NCA	**55.33**‡	**50.29**‡	**73.28**	**1.949**
	ExTrD	55.31‡	**50.29**‡	73.26	1.951
RCV1	LR	78.43	60.37	80.16	0.534
	TrD	79.46‡	60.61	82.83	0.451
	NCA	**79.74**‡	60.76	**83.11**	**0.442**
	ExTrD	79.33‡	**61.74**†	82.91	0.466

The evaluations on Dmoz 2010 and Dmoz 2012 datasets are blind and the predictions have to be uploaded to LSHTC website in order to obtain the scores. For Dmoz 2012, Tree Errors are not available and for Dmoz 2010, the hF_1 are not available. For HRLR, we do not have access to the predictions, hence, we could only report the values for Micro-F_1 and Macro-F_1 scores from [9]. Statistical significance tests compare the results of *HierCost* with LR. These tests could not be performed on LSHTC datasets due to non-availability of the true labels on test sets. As seen in Table 4, *HierCost* improves upon the baseline LR results as well as the results reported in [9], in most cases, especially the Macro-F_1 scores. The results of *HierCost* are better on most measures. TD performs worst on average on set-based measures. In fact, only on Dmoz 2012 dataset, TD is competitive, on the rest, the results are much worse than the *flat* LR classifier and its hierarchical extensions. On hierarchical measure, however, TD outperformed *flat* classifiers on some datasets.

In Table 5, we report the run-times comparisons of TD, LR and *HierCost*. We trained the models in parallel for different classes and computed the sum of run-times for each training instance. In theory, the run-times of LR and *Hier-Cost* should be equivalent, because they solve similar optimization problems. However, minor variations in the run-times were observed because of the variations in optimal regularization penalties, which influences the convergence of the optimization algorithm. The runtimes of flat methods were significantly worse than TD, which is efficient in terms of training, but at considerable loss in classification performance. Although, we do not measure the training times of HRLR, based on the experience from a similar problem [3], the recursive model take between 3-10 iterations for convergence. In each iteration, the models for all the terminal labels need to be trained hence each iteration is about as expensive as

Table 3. Peformance comparison with imbalance cost included.

		Micro-F_1 ($\uparrow$)	Macro-F_1 ($\uparrow$)	hF_1 ($\uparrow$)	TE ($\downarrow$)
CLEF	IMB + LR	79.52	53.11	85.19	1.002
	IMB + TrD	79.92	52.84	85.59	0.978
	IMB + NCA	79.62	51.89	85.34	0.994
	IMB + ExTrD	**80.32**	**58.45**	**85.69**	**0.966**
DMOZ SMALL	IMB + LR	48.55‡	32.72‡	68.62	3.406
	IMB + TrD	49.03‡	33.21‡	69.41	3.334
	IMB + NCA	48.87‡	33.27‡	69.37	3.335
	IMB + ExTrD	**49.03‡**	**33.34‡**	**69.54**	**3.322**
IPC	IMB + LR	55.04	49.00	72.82	1.974
	IMB + TrD	55.60‡	**50.45†**	73.56	1.933
	IMB + NCA	55.33	50.29	73.28	1.949
	IMB + ExTrD	**55.67‡**	50.42	**73.58**	**1.931**
RCV1	IMB + LR	78.59‡	60.77	81.27	0.511
	IMB + TrD	**79.63‡**	61.04	**83.13**	**0.435**
	IMB + NCA	79.61	61.04	82.65	0.458
	IMB + ExTrD	79.22	**61.33**	82.89	0.469

Table 4. Performance comparison of HierCost with other baseline methods.

		Micro-F_1 ($\uparrow$)	Macro-F_1 ($\uparrow$)	hF_1 ($\uparrow$)	TE ($\downarrow$)
	TD	73.06	34.47	79.32	1.366
CLEF	LR	79.82	53.45	85.24	0.994
	HRLR	80.12	55.83	-	-
	HierCost	**80.32**	**58.45†**	**85.69**	**0.966**
	TD	40.90	24.15	**69.99**	**3.147**
DMOZ SMALL	LR	46.39	30.20	67.00	3.569
	HRLR	45.11	28.48	-	-
	HierCost	**49.03‡**	**33.34‡**	69.54	3.322
	TD	50.22	43.87	69.33	2.210
IPC	LR	55.04	48.99	72.82	1.974
	HRLR	55.37	49.60	-	-
	HierCost	**55.67‡**	**50.42†**	**73.58**	**1.931**
	TD	77.85	57.80	**88.78**	0.524
RCV1	LR	78.43	60.37	80.16	0.534
	HRLR	**81.23**	55.81	-	-
	HierCost	79.22‡	**61.33**	82.89	**0.469**
	TD	38.86	26.29	-	3.867
DMOZ 2010	LR	45.17	30.98	-	3.400
	HRLR	45.84	**32.42**	-	-
	HierCost	**45.87**	32.41	-	**3.321**
	TD	51.65	**30.48**	73.33	-
DMOZ 2012	LR	51.72	27.19	72.53	-
	HRLR	53.18	20.04	-	-
	HierCost	**53.36**	28.47	**73.79**	-

Table 5. Total training runtimes (in mins).

	TD-LR	LR	HierCost
CLEF	<1	<1	<1
DMOZ SMALL	4	41	40
IPC	27	643	453
RCV1	20	29	48
DMOZ 2010	196	15191	20174
DMOZ 2012	384	46044	50253

a single run of LR. In addition, the distributed recursive models require communication between the training machines which incurs an additional overhead.

6   Conclusions

In this paper, we have argued that the methods that extend flat classification using hierarchical regularization, can be viewed in a complementary way as weighting the losses on the negative examples depending on dissimilarity between the positive and negative classes. The approach proposed in this paper, incorporates this insight directly into the loss function by scaling the loss function according to the dissimilarity between the classes with respect to the hierarchy, thus obviating the need for recursive regularization and iterative model training. At the same time, this approach also makes parallelization trivial. Our method also mitigates the adverse effects of imbalance in the training data by upweighting the loss for examples from smaller classes, thus, significantly improving their classification results. Our experimental results show that the proposed method is able to efficiently incorporate hierarchical information by transforming the hierarchical classification problem into an example based cost sensitive classification problem. In future work, we would like to evaluate the benefits of cost sensitive classification using large margin classifiers such as support vector machines.

Acknowledgement. This work was supported by NSF IIS-1447489 and NSF career award to Huzefa Rangwala (IIS-1252318). The experiments were run on ARGO (http://orc.gmu.edu).

References

1. Babbar, R., Partalas, I., Gaussier, E., Amini, M.R.: On flat versus hierarchical classification in large-scale taxonomies. In: Advances in Neural Information Processing Systems, pp. 1824–1832 (2013)
2. Cai, L., Hofmann, T.: Hierarchical document categorization with support vector machines. In: ACM International Conf. on Information & Knowledge Management, pp. 78–87 (2004)

3. Charuvaka, A., Rangwala, H.: Approximate block coordinate descent for large scale hierarchical classification. In: ACM SIGAPP Symposium on Applied Computing (2015)
4. Chen, J., Warren, D.: Cost-sensitive learning for large-scale hierarchical classification. In: ACM International Conf. on Information & Knowledge Management, pp. 1351–1360 (2013)
5. Dekel, O., Keshet, J., Singer, Y.: Large margin hierarchical classification. In: International Conf. on Machine Learning, p. 27 (2004)
6. Dimitrovski, I., Kocev, D., Loskovska, S., Džeroski, S.: Hierarchical annotation of medical images. Pattern Recognition 44(10), 2436–2449 (2011)
7. Evgeniou, T., Micchelli, C., Pontil, M.: Learning multiple tasks with kernel methods. Journal of Machine Learning Research 6(1), 615–637 (2005)
8. Evgeniou, T., Pontil, M.: Regularized multi-task learning. In: ACM SIGKDD International Conf. on Knowledge Discovery and Data Mining, pp. 109–117 (2004)
9. Gopal, S., Yang, Y.: Recursive regularization for large-scale classification with hierarchical and graphical dependencies. In: ACM SIGKDD International Conf. on Knowledge Discovery and Data Mining, pp. 257–265 (2013)
10. Lewis, D.D., Yang, Y., Rose, T.G., Li, F.: Rcv1: A new benchmark collection for text categorization research. The Journal of Machine Learning Research 5, 361–397 (2004)
11. Liu, T.Y., Yang, Y., Wan, H., Zeng, H.J., Chen, Z., Ma, W.Y.: Support vector machines classification with a very large-scale taxonomy. ACM SIGKDD Explorations Newsletter 7(1), 36–43 (2005)
12. Masnadi-Shirazi, H., Vasconcelos, N.: Risk minimization, probability elicitation, and cost-sensitive svms. In: International Conf. on Machine Learning, pp. 759–766 (2010)
13. McCallum, A., Rosenfeld, R., Mitchell, T.M., Ng, A.Y.: Improving text classification by shrinkage in a hierarchy of classes. In: International Conf. on Machine Learning, vol. 98, pp. 359–367 (1998)
14. Nesterov, Y.: Introductory lectures on convex optimization, vol. 87. Springer Science & Business Media (2004)
15. Partalas, I., Kosmopoulos, A., Baskiotis, N., Artieres, T., Paliouras, G., Gaussier, E., Androutsopoulos, I., Amini, M.R., Galinari, P.: Lshtc: A benchmark for large-scale text classification. arXiv preprint arXiv:1503.08581 (2015)
16. Saha, B., Gupta, S., Phung, D., Venkatesh, S.: Multiple task transfer learning with small sample sizes. Knowledge and Information Systems, 1–28 (2015)
17. Shahbaba, B., Neal, R.M., et al.: Improving classification when a class hierarchy is available using a hierarchy-based prior. Bayesian Analysis 2(1), 221–237 (2007)
18. Silla Jr., C.N., Freitas, A.A.: A survey of hierarchical classification across different application domains. Data Mining and Knowledge Discovery 22(1–2), 31–72 (2011)
19. Sokolova, M., Lapalme, G.: A systematic analysis of performance measures for classification tasks. Information Processing & Management 45(4), 427–437 (2009)
20. Tsochantaridis, I., Joachims, T., Hofmann, T., Altun, Y.: Large margin methods for structured and interdependent output variables. Journal of Machine Learning Research 6, 1453–1484 (2005)
21. Yang, Y.: A study of thresholding strategies for text categorization. In: ACM SIGIR Conf. on Research and Development in Information Retrieval, pp. 137–145 (2001)
22. Yang, Y., Liu, X.: A re-examination of text categorization methods. In: ACM SIGIR Conf. on Research and Development in Information Retrieval, pp. 42–49 (1999)

Large Scale Optimization with Proximal Stochastic Newton-Type Gradient Descent

Ziqiang Shi[(⊠)] and Rujie Liu

Fujitsu Research and Development Center,Beijing, China
{shiziqiang,rjliu}@cn.fujitsu.com

Abstract. In this work, we generalized and unified two recent completely different works of Jascha [10] and Lee [2] respectively into one by proposing the **proximal stochastic Newton-type gradient** (PROX-TONE) method for optimizing the sums of two convex functions: one is the average of a huge number of smooth convex functions, and the other is a nonsmooth convex function. Our PROXTONE incorporates second order information to obtain stronger convergence results, that it achieves a linear convergence rate not only in the value of the *objective* function, but also for the *solution*. The proofs are simple and intuitive, and the results and technique can be served as a initiate for the research on the proximal stochastic methods that employ second order information. The methods and principles proposed in this paper can be used to do logistic regression, training of deep neural network and so on. Our numerical experiments shows that the PROXTONE achieves better computation performance than existing methods.

1   Introduction and Problem Statement

In this work, we consider the problems of the following form:

$$\min_{x \in \mathbb{R}^p} f(x) := \frac{1}{n} \sum_{i=1}^n g_i(x) + h(x), \tag{1}$$

where g_i is a smooth convex loss function associated with a sample in a training set, and h is a non-smooth convex penalty function or a regularizer. Let $g(x) = \frac{1}{n} \sum_{i=0}^n g_i(x)$. We assume the optimal value $f^\star$ is attained at some optimal solution $x^\star$, not necessarily unique. Problems of this form often arise in machine learning, such as the least-squares regression, the Lasso, the elastic net, the logistic regression, and deep neural network.

For optimizing (1), the standard and popular *proximal full gradient method* (ProxFG) uses iterations of the form

$$x^{k+1} = \arg\min_{x \in \mathbb{R}^p} \left\{ \nabla g(x_k)^T x + \frac{1}{2\alpha_k} \|x - x_{k-1}\|^2 + h(x) \right\}, \tag{2}$$

© Springer International Publishing Switzerland 2015
A. Appice et al. (Eds.): ECML PKDD 2015, Part I, LNAI 9284, pp. 691–704, 2015.
DOI: 10.1007/978-3-319-23528-8_43

where α_k is the step size at the k-th iteration. Under standard assumptions the sub-optimality achieved on iteration k of the ProxFG method with a constant step size is given by

$$f(x^k) - f(x^*) = O(\frac{1}{k}).$$

When f is strongly-convex, the error satisfies [11]

$$f(x^k) - f(x^*) = O((\frac{L - \mu_g}{L + \mu_h})^k),$$

where L is the Lipschitz constant of $f(x)$, μ_g, and μ_h are the convexity parameters of $g(x)$ and $h(x)$ respectively. These notations mentioned here will be detailed in Section 1.1. This result in a linear convergence rate, which is also known as a geometric or exponential rate because the error is cut by a fixed fraction on each iteration.

Unfortunately, the ProxFG methods can be unappealing when n is large or huge because its iteration cost scales linearly in n. When the number of components n is very large, then each iteration of (2) will be very expensive since it requires computing the gradients for all the n component functions g_i, and also their average.

To overcome this problem, researchers proposed the *proximal stochastic gradient descent* methods (ProxSGD), whose main appealing is that they have an iteration cost which is independent of n, making them suited for modern problems where n may be very large. The basic ProxSGD method for optimizing (1), uses iterations of the form

$$x_k = \text{prox}_{\alpha_k h}\big(x_{k-1} - \alpha_k \nabla g_{i_k}(x_{k-1})\big), \qquad (3)$$

where at each iteration an index i_k is sampled uniformly from the set $\{1, ..., n\}$. The randomly chosen gradient $\nabla g_{i_k}(x_{k-1})$ yields an unbiased estimate of the true gradient $\nabla g(x_{k-1})$ and one can show under standard assumptions that, for a suitably chosen decreasing step-size sequence $\{\alpha_k\}$, the ProxSGD iterations have an expected sub-optimality for convex objectives of [1]

$$\mathbb{E}[f(x^k)] - f(x^*) = O(\frac{1}{\sqrt{k}})$$

and an expected sub-optimality for strongly-convex objectives of

$$\mathbb{E}[f(x^k)] - f(x^*) = O(\frac{1}{k}).$$

In these rates, the expectations are taken with respect to the selection of the i_k variables.

Besides these first order method, there is another group of methods, called *proximal Newton-type* methods, which converge much faster, but need more memory and computation to obtain the second order information about the objective function. These methods are always limited to small-to-medium scale problems that require a high degree of precision. For optimizing (1), proximal Newton-type methods [2] that incorporate second order information use iterations of the form $x^{k+1} \leftarrow x^k + \Delta x^k$, here Δx^k is obtained by

$$\Delta x^k = \arg\min_{d \in \mathbb{R}^p} \nabla g(x^k)^T d + \frac{1}{2} d^T H_k d + h(x^k + d), \tag{4}$$

where H_k denotes an approximation to $\nabla^2 g(x_k)$. According to the strategies for choosing H_k, we obtain different method, such as *proximal Newton method* (ProxN) when we choose H_k to be $\nabla^2 g(x^k)$; *proximal quasi-Newton method* (ProxQN) when we build an approximation to $\nabla^2 g(x_k)$ using changes measured in ∇g according to a quasi-Newton strategy [2]. Indeed if we compared (4) with (2), it can be seen ProxN is the ProxFG with scaled proximal mappings.

Based on the related background introduced above, now we can describe our approaches and findings. The primary contribution of this work is the proposal and analysis of a new algorithm that we call the proximal stochastic Newton-type gradient (PROXTONE, pronounced /prok stone/) method, a stochastic variant of the ProxN method. The PROXTONE method has the low iteration cost as that of ProxSGD methods, but achieves the convergence rates like the ProxFG method stated above. The PROXTONE iterations take the form $x^{k+1} \leftarrow x^k + t_k \Delta x^k$, where Δx^k is obtained by

$$\Delta x^k \leftarrow \arg\min_d d^T (\nabla_k + H_k x^k) + \frac{1}{2} d^T H_k d + h(x^k + d), \tag{5}$$

here $\nabla_k = \frac{1}{n} \sum_{i=1}^n \nabla_k^i$, $H_k = \frac{1}{n} \sum_{i=1}^n H_k^i$, and at each iteration a random index j and corresponding H_{k+1}^j is selected, then we set

$$\nabla_{k+1}^i = \begin{cases} \nabla g_i(x^{k+1}) - H_{k+1}^i x^{k+1} & \text{if } i = j, \\ \nabla_{k+1}^i & \text{otherwise.} \end{cases}$$

and $H_{k+1}^i \leftarrow H_k^i \ (i \neq j)$.

That is, like the ProxFG and ProxN methods, the steps incorporates a gradient with respect to each function; but, like the ProxSGD method, each iteration only computes the gradient with respect to a single example and the cost of the iterations is independent of n. Despite the low cost of the PROXTONE iterations, we show in this paper that the PROXTONE iterations have a linear convergence rate for strongly-convex objectives, like the ProxFG method. That is, by having access to j and by keeping a memory of the approximation for the Hessian matrix computed for the objective funtion, this iteration achieves a faster convergence rate than is possible for standard ProxSGD methods.

Besides PROXTONE, there are a large variety of approaches available to accelerate the convergence of ProxSGD methods, and a full review of this immense literature would be outside the scope of this work. Several recent work considered various special cases of (1), and developed algorithms that enjoy the linear convergence rate, such as ProxSDCA [8], MISO [3], SAG [7], ProxSVRG [11], SFO [10], and ProxN [2]. All these methods converge with an exponential rate in the value of the objective function, except that the ProxN achieves superlinear rates of convergence for the *solution*, however it is a batch mode method. Shalev-Shwartz and Zhang's ProxSDCA [8, 9] considered the case where the component functions have the form $g_i(x) = \phi_i(a_i^T x)$ and the Fenchel conjugate functions of ϕ_i and h can be computed efficiently. Schimidt et al.'s SAG [7] and Jascha et al.'s SFO [10] considered the case where $h(x) \equiv 0$.

Different from above related methods, our PROXTONE is a extension of the SFO and ProxN to a proximal stochastic Newton-type method for solving the more *general* nonsmooth (compared to ProxSDCA, SAG and SFO) class of problems defined in (1). PROXTONE makes connections between two completely different approaches. It achieves a linear convergence rate not only in the value of the objective function, but also for the *solution*. We now outline the rest of the study. Section 2 presents the main algorithm and gives an equivalent form in order for the ease of analysis. Section 3 states the assumptions underlying our analysis and gives the main results; we first give a linear convergence rate in *function value* (weak convergence) that applies for any problem, and then give a strong linear convergence rate for the *solution*, however with some additional conditions. We report some experimental results in Section 4 and provide concluding remarks in Section 5.

1.1   Notations and Assumptions

In this paper, we assume the function $h(x)$ is lower semi-continuous and convex, and its effective domain, $\mathrm{dom}(h) := \{x \in \mathbb{R}^p \mid h(x) < +\infty\}$, is closed. Each $g_i(x)$, for $i = 1, \ldots, n$, is differentiable on an open set that contains $\mathrm{dom}(h)$, and their gradients are Lipschitz continuous, that is, there exist $L_i > 0$ such that for all $x, y \in \mathrm{dom}(h)$,

$$\|\nabla g_i(x) - \nabla g_i(y)\| \leq L_i \|x - y\|. \tag{6}$$

Then from the Lemma 1.2.3 and its proof in Nesterov's book [5], for $i = 1, \ldots, n$, we have

$$|g_i(x) - g_i(y) - \nabla g_i(y)^T(x - y)| \leq \frac{L_i}{2}\|x - y\|^2. \tag{7}$$

A function $f(x)$ is called μ-strongly convex, if there exist $\mu \geq 0$ such that for all $x \in \mathrm{dom}(f)$ and $y \in \mathbb{R}^p$,

$$f(y) \geq f(x) + \xi^T(y - x) + \frac{\mu}{2}\|y - x\|^2, \quad \forall \xi \in \partial f(x). \tag{8}$$

The *convexity parameter* of a function is the largest μ such that the above condition holds. If $\mu = 0$, it is identical to the definition of a convex function. The strong convexity of $f(x)$ in (1) may come from either $g(x)$ or $h(x)$ or both. More precisely, let $g(x)$ and $h(x)$ have convexity parameters μ_g and μ_h respectively, then $\mu \geq \mu_g + \mu_h$. From Lemma B.5 in [3] and (8), we have

$$f(y) \geq f(x^*) + \frac{\mu}{2}\|y - x^*\|^2. \tag{9}$$

2 The PROXTONE Method

In this section we present the Proximal Stochastic Newton-type Gradient Descent (PROXTONE) algorithm for solving problems of the form (1). There are two key steps in the algorithm: (step 2) the regularized quadratic model (5) is solved to give a search direction; (step 4) the component function $g_j(x)$ is sampled randomly and the regularized quadratic model (5) is updated using this selected function. Once these key steps have been performed, the current point x_k is updated to give a new point x_{k+1}, and the process is repeated.

We summarize the PROXTONE method of (5) in Algorithm 1, while a thorough description of each of the key steps in the algorithm will follow in the rest of this section. It can be easily checked that if $n = 1$, then it becomes the determined proximal Newton-type methods proposed by Lee and Sun et al. [2] for minimizing composite functions:

$$\min_{x \in \mathbb{R}^p} f(x) := g(x) + h(x) \tag{10}$$

by (4). Thus PROXTONE is indeed a generalization of ProxN [2].

Algorithm 1. PROXTONE: A generic PROXimal sTOchastic NEwton-type gradient descent method

---

Input: start point $x^0 \in \text{dom } f$; for $i \in \{1, 2, .., n\}$, let $H^i_{-1} = H^i_0$ be a positive definite approximation to the Hessian of $g_i(x)$ at x^0, $\nabla^i_{-1} = \nabla^i_0 = \nabla g_i(x^0) - H^i_0 x^0$; and $\nabla_0 = \frac{1}{n} \sum_{i=1}^n \nabla^i_0$, $H_0 = \frac{1}{n} \sum_{i=0}^n H^i_0$.

1: **repeat**

2: Solve the subproblem for a search direction:$\triangle x^k \leftarrow \arg\min_d d^T(\nabla_k + H_k x^k) + \frac{1}{2} d^T H_k d + h(x^k + d)$.

3: Update: $x^{k+1} = x^k + \triangle x^k$.

4: Sample j from $\{1, 2, .., n\}$, use the $\nabla g_j(x^{k+1})$ and H^j_{k+1}, which is a positive definite approximation to the Hessian of $g_j(x)$ at x^{k+1}, to update the ∇^i_{k+1} ($i \in \{1, 2, .., n\}$): $\nabla^j_{k+1} \leftarrow \nabla g_j(x^{k+1}) - H^j_{k+1} x^{k+1}$, while leaving all other ∇^i_{k+1} and H^i_{k+1} unchanged: $\nabla^i_{k+1} \leftarrow \nabla^i_k$ and $H^i_{k+1} \leftarrow H^i_k$ ($i \neq j$) ; and finally obtain ∇_{k+1} and H_{k+1} by $\nabla_{k+1} \leftarrow \frac{1}{n} \sum_{i=1}^n \nabla^i_{k+1}$, $H_{k+1} \leftarrow \frac{1}{n} \sum_{i=1}^n H^i_{k+1}$.

5: **until** stopping conditions are satisfied.

Output: x^k.

It is also a generalization of recent work by Jascha [10], whose SFO is the special case of our PROXTONE with $h(x) \equiv 0$. Our algorithm in Jascha's style is summarized in Algorithm 2 which is equivalent to the original PROXTONE. To see the equivalence, keep in mind that $G^k(x)$ in Algorithm 2 is a quadratic function, we only need to check the following equations:

$$\nabla^2 G^k(x) = \frac{1}{n} \sum_{i=1}^n H^i_k \quad \text{and} \quad \nabla G^k(x) = \frac{1}{n} \sum_{i=1}^n \nabla g_i(x) + \frac{1}{n} \sum_{i=1}^n (x - x^k)^T H^i_k,$$

and

$$\nabla_k + H_k x^k = \frac{1}{n} \sum_{i=1}^{n} [\nabla g_i(x^{\theta_{i,k}}) + (x^k - x^{\theta_{i,k-1}})^T H^i_{\theta_{i,k}}]. \tag{11}$$

In following analysis, we will not distinguish these two forms of PROXTONE from each other.

Algorithm 2. PROXTONE in a form that is easy to analyze

Input: start point $x^0 \in \text{dom } f$; for $i \in \{1, 2, .., n\}$, let $g_i^0(x) = g_i(x^0) + (x - x^0)^T \nabla g_i(x^0) + \frac{1}{2}(x - x^0)^T H_0^i(x - x^0)$, where the notation H_0^i ($i \in \{1, 2, .., n\}$) are totally the same as they in Algorithm 1; and $G^0(x) = \frac{1}{n} \sum_{i=1}^{n} g_i^0(x)$.
1: **repeat**
2: Solve the subproblem for new approximation of the solution:

$$x^{k+1} \leftarrow \arg\min_x \left[G^k(x) + h(x) \right]. \tag{12}$$

3: Sample j from $\{1, 2, .., n\}$, and update the quadratic models or surrogate functions:

$$g_j^{k+1}(x) = g_j(x^{k+1}) + (x - x^{k+1})^T \nabla g_j(x^{k+1}) + \frac{1}{2}(x - x^{k+1})^T H_{k+1}^i(x - x^{k+1}), \tag{13}$$

while leaving all other $g_i^{k+1}(x)$ unchanged: $g_i^{k+1}(x) \leftarrow g_i^k(x)$ ($i \neq j$); and $G^{k+1}(x) = \frac{1}{n} \sum_{i=1}^{n} g_i^{k+1}(x)$.
4: **until** stopping conditions are satisfied.
Output: x^k.

To better understand this method, we make the following illustration and observations.

2.1 The Regularized Quadratic Model in Algorithm 2

For fixed $x \in \mathbb{R}^p$, we define a regularized piecewise quadratic approximation of $f(x)$ as follows:

$$G^k(x) + h(x) = \frac{1}{n} \sum_{i=1}^{n} g_i^k(x) + h(x)$$

where $g_i^k(x)$ is the quadratic model for $g_i(x)$

$$g_i^k(x)$$
$$= g_i(x^{\theta_{i,k}}) + (x - x^{\theta_{i,k}})^T \nabla g_i(x^{\theta_{i,k}}) + \frac{1}{2}(x - x^{\theta_{i,k}})^T H_{\theta_{i,k}}^i(x - x^{\theta_{i,k}}), \tag{14}$$

here $\theta_{i,k}$ is a random variable which have the following conditional probability distribution in each iteration:

$$\mathbb{P}(\theta_{i,k} = k|j) = \frac{1}{n} \quad \text{and} \quad \mathbb{P}(\theta_{i,k} = \theta_{i,k-1}|j) = 1 - \frac{1}{n}, \tag{15}$$

and $H^i_{\theta_{i,k}}$ is any positive definite matrix, which possibly depends on $x^{\theta_{i,k}}$. Then at each iteration the search direction is found by solving the subproblem (12).

One of the crucial ideas of this algorithm is that the component function to be used for updating the search direction at each iteration is chosen randomly. This allows the function to be selected very quickly. After the component function $g_j(x)$ selected and updated by (13), while leaving all other $g_i^{k+1}(x)$ unchanged.

2.2 The Hessian Approximation

Arguably, the most important feature of this method is that the regularized quadratic model (12) incorporates second order information in the form of a positive definite matrix H^i_k. This is key because, at each iteration, the user has complete freedom over the choice of H^i_k. A few suggestions for the choice of H^i_k include: the simplest option is $H^i_k = I$ that no second order information is employed; $H^i_k = \nabla^2 g_i(x_k)$ provides the most accurate second order information, but it is (potentially) more computationally expensive to work with.

3   Convergence Analysis

In this section we provide convergence theory for the PROXTONE algorithm. Under the standard assumptions, we now state our convergence result.

Theorem 1. *Suppose $\nabla g_i(x)$ is Lipschitz continuous with constant $L_i > 0$ for $i = 1, ..., n$, and $L_i I \preceq mI \preceq H^i_k \preceq MI$ for all $i = 1, ..., n$, $k \geq 1$. $h(x)$ is strongly convex with $\mu_h \geq 0$. Let $L_{max} = \{L_1, ..., L_n\}$, then the PROXTONE iterations satisfy for $k \geq 1$:*

$$\mathbb{E}[f(x^k)] - f^* \leq \frac{M + L_{max}}{2}[\frac{1}{n}\frac{M + L_{max}}{2\mu_h + m} + (1 - \frac{1}{n})]^k \|x^* - x^0\|^2. \qquad (16)$$

The ideas of the proof is closed related to that of MISO by Mairal [3] and for completeness we give a simple version in the appendix.

We have the following remarks regarding the above result:

- In order to satisfy $\mathbb{E}[f(x^k)] - f^* \leq \epsilon$, the number of iterations k needs to satisfy

$$k \geq (\log \rho)^{-1} \log[\frac{2\epsilon}{(M + L_{max})\|x^* - x^0\|^2}],$$

where $\rho = \frac{1}{n}\frac{M+L_{max}}{2\mu_h+m} + (1 - \frac{1}{n})$.
- Inequality (16) gives us a reliable stopping criterion for the PROXTONE method.

At this moment, we see that the expected quality of the output of PROXTONE is good. However, in practice we are not going to run this method many times on the same problem. What is the probability that our single run can give us also a good result. Since $f(x^k) - f^* \geq 0$, Markov's inequality and Theorem 1 imply that for any $\epsilon > 0$,

$$\text{Prob}\Big(f(x^k) - f^* \geq \epsilon\Big) \leq \frac{\mathbb{E}[f(x^k) - f^*]}{\epsilon} \leq \frac{(M + L_{max})\rho^k \|x^* - x^0\|^2}{2\epsilon}.$$

Thus we have the following high-probability bound.

Corollary 1. *Suppose the assumptions in Theorem 1 hold. Then for any $\epsilon > 0$ and $\delta \in (0, 1)$, we have*

$$\text{Prob}\big(f(x^k) - f(x^*) \leq \epsilon\big) \geq 1 - \delta,$$

provided that the number of iterations k satisfies

$$k \geq \log\left(\frac{(M + L_{max})\|x^* - x^0\|^2}{2\delta\epsilon}\right) \Big/ \log\left(\frac{1}{\rho}\right).$$

Based on Theorem 1 and its proof, we give a deeper and stronger result that the PROXTONE achieves a linear convergence rate for the solution.

Theorem 2. *Suppose $\nabla g_i(x)$ and $\nabla^2 g_i$ are Lipschitz continuous with constant $L_i > 0$ and $K_i > 0$ respectively for $i = 1, ..., n$, $h(x)$ is strongly convex with $\mu_h \geq 0$. Let $L_{max} = \{L_1, ..., L_n\}$ and $K_{max} = (1/n)\sum_{i=1}^{n} L_i$. If $H_{\theta_{i,k}}^i = \nabla^2 g_i(x^{\theta_{i,k}})$ and $L_i I \preceq mI \preceq H_k^i \preceq MI$, then PROXTONE converges exponentially to x^* in expectation:*

$$\mathbb{E}[\|x^{k+1} - x^*\|]$$
$$\leq \left(\frac{K_{max} + 2L_{max}}{m} \frac{M + L_{max}}{2\mu_h + m} + \frac{2L_{max}}{m}\right)\left[\frac{1}{n}\frac{M + L_{max}}{2\mu_h + m} + (1 - \frac{1}{n})\right]^{k-1}\|x^* - x^0\|^2.$$

In order to satisfy $\mathbb{E}[\|x^{k+1} - x^*\|] \leq \epsilon$, the number of iterations k needs to satisfy

$$k \geq (\log \rho)^{-1} \log\Big[\frac{\epsilon}{C\|x^* - x^0\|^2}\Big],$$

where ρ is as before and $C = \frac{K_{max} + 2L_{max}}{m} \frac{M + L_{max}}{2\mu_h + m} + \frac{2L_{max}}{m}$.

Due to the Markov's inequality, Theorem 2 implies the following result.

Corollary 2. *Suppose the assumptions in Theorem 2 hold. Then for any $\epsilon > 0$ and $\delta \in (0, 1)$, we have*

$$\text{Prob}\big(\|x^{k+1} - x^*\| \geq \epsilon\big) \geq 1 - \delta,$$

provided that the number of iterations k satisfies

$$k \geq \log\left(\frac{((K_{max} + 2L_{max})(M + L_{max}) + 2L_{max}(2\mu_h + m))\|x^* - x^0\|^2}{m(2\mu_h + m)\delta\epsilon}\right) \Big/ \log\left(\frac{1}{\rho}\right).$$

4 Numerical Experiments

The technique proposed in this paper has wide applications, it can be used to do least-squares regression, the Lasso, the elastic net, and the logistic regression. Furthermore the principle of PROXTONE can also be applies to do nonconvex optimization problems, such as training of deep convolutional network and so on.

In this section we present the results of some numerical experiments to illustrate the properties of the PROXTONE method. We focus on the sparse regularized logistic regression problem for binary classification: given a set of training examples $(a_1, b_1), \ldots, (a_n, b_n)$ where $a_i \in \mathbb{R}^p$ and $b_i \in \{+1, -1\}$, we find the optimal predictor $x \in \mathbb{R}^p$ by solving

$$\min_{x \in \mathbb{R}^p} \quad \frac{1}{n} \sum_{i=1}^{n} \log\big(1 + \exp(-b_i a_i^T x)\big) + \lambda_1 \|x\|_2^2 + \lambda_2 \|x\|_1,$$

where λ_1 and λ_2 are two regularization parameters. We set

$$g_i(x) = \log(1 + \exp(-b_i a_i^T x) + \lambda_1 \|x\|_2^2, \qquad h(x) = \lambda_2 \|x\|_1, \tag{17}$$

and

$$\lambda_1 = 1E - 4, \qquad \lambda_2 = 1E - 4.$$

In this situation, the subproblem (12) become a lasso problem, which can be effectively and accurately solved by the proximal algorithms [6].

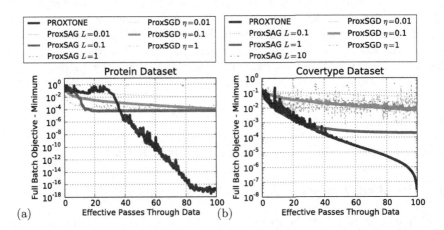

Fig. 1. A comparison of PROXTONE to competing optimization techniques for two datasets. The bold lines indicate the best performing hyperparameter for each optimizer.

We used some publicly available data sets. The *protein* data set was obtained from the KDD Cup 2004[1]; the covertype data sets were obtained from the LIBSVM Data[2].

[1] http://osmot.cs.cornell.edu/kddcup
[2] http://www.csie.ntu.edu.tw/~cjlin/libsvmtools/datasets

The performance of PROXTONE is compared with some related algorithms:

- ProxSGD: We used a constant step size that gave the best performance among all powers of 10.
- ProxSAG: This is a proximal version of the SAG method, with the trailing number providing the Lipschitz constant.

The results of the different methods are plotted for the first 100 effective passes through the data in Figure 1. The PROXTONE iterations seem to achieve the best of all.

5   Conclusions

This paper introduces a proximal stochastic method called PROXTONE for minimizing regularized finite sums. For nonsmooth and strongly convex problems, we show that PROXTONE not only enjoys the same linear rates as those of MISO, SAG, ProxSVRG and ProxSDCA, but also showed that the *solution* of this method converges in exponential rate too. There are some directions that the current study can be extended. In this paper, we have focused on the theory and the convex experiments of PROXTONE; it would be meaningful to also make clear the implementation details and do the numerical evaluation to nonconvex problems [10]. Second, combine with randomized block coordinate method [4] for minimizing regularized convex functions with a huge number of varialbes/coordinates. Moreover, due to the trends and needs of big data, we are designing distributed/parallel PROXTONE for real life applications. In a broader context, we believe that the current paper could serve as a basis for examining the method on the proximal stochastic methods that employ second order information.

Appendix

In this Appendix, we give the proofs of the two propositions.

A   Proof of Theorem 1

Since in each iteration of the PROXTONE, we have (14) and (15), that yields

$$\mathbb{E}[\|x^* - x^{\theta_{i,k}}\|^2] = \frac{1}{n}\mathbb{E}[\|x^* - x^k\|^2] + (1 - \frac{1}{n})\mathbb{E}[\|x^* - x^{\theta_{i,k-1}}\|^2]. \tag{18}$$

Since $0 \preceq H^i_{\theta_{i,k}} \preceq MI$ and $\nabla^2 g^k_i(x) = H^i_{\theta_{i,k}}$, by Theorem 2.1.6 of [5] and the assumption, $\nabla g^k_i(x)$ and $\nabla g_i(x)$ are Lipschitz continuous with constant M and L_i respectively, and further $\nabla g^k_i(x) - \nabla g_i(x)$ is Lipschitz continuous with constant $M + L_i$ for $i = 1, \ldots, n$. This together with (7) yields

$$|[g^k_i(x) - g_i(x)] - [g^k_i(y) - g_i(y)] - \nabla[g^k_i(y) - g_i(y)]^T(x - y)| \leq \frac{M + L_i}{2}\|x - y\|^2.$$

Applying the above inequality with $y = x^{\theta_{i,k}}$, and using the fact that $\nabla[g_i^k(x^{\theta_{i,k}})] = \nabla[g_i(x^{\theta_{i,k}})]$ and $g_i^k(x^{\theta_{i,k}}) = g_i(x^{\theta_{i,k}})$, we have

$$|g_i^k(x) - g_i(x)| \leq \frac{M + L_i}{2}\|x - x^{\theta_{i,k}}\|^2.$$

Summing over $i = 1, \ldots, n$ yields

$$[G^k(x) + h(x)] - [g(x) + h(x)] \leq \frac{1}{n}\sum_{i=1}^n \frac{M + L_i}{2}\|x - x^{\theta_{i,k}}\|^2. \qquad (19)$$

Then by the Lipschitz continuity of $\nabla g_i(x)$ and the assumption $L_i I \preceq mI \preceq H_k^i$, we have

$g_i(x)$
$$\leq g_i(x^{\theta_{i,k}}) + \nabla g_i(x^{\theta_{i,k}})^T(x - x^{\theta_{i,k}})| + \frac{L_i}{2}\|x - x^{\theta_{i,k}}\|^2$$
$$\leq g_i(x^{\theta_{i,k}}) + (x - x^{\theta_{i,k}})^T \nabla g_i(x^{\theta_{i,k}}) + \frac{1}{2}(x - x^{\theta_{i,k}})^T H_{\theta_{i,k}}^i(x - x^{\theta_{i,k}}) = g_i^k(x),$$

and thus, by summing over i yields $g(x) \leq G^k(x)$, and further by the optimality of x^{k+1}, we have

$$f(x^{k+1}) \leq G^k(x^{k+1}) + h(x^{k+1}) \leq G^k(x) + h(x)$$
$$\leq f(x) + \frac{1}{n}\sum_{i=1}^n \frac{M + L_i}{2}\|x - x^{\theta_{i,k}}\|^2 \qquad (20)$$

Since $mI \preceq H_{\theta_{i,k}}$ and $\nabla^2 g_i^k(x) = H_{\theta_{i,k}}$, by Theorem 2.1.11 of [5], $g_i^k(x)$ is m-strongly convex. Since $G^k(x)$ is the average of $g_i^k(x)$, thus $G^k(x) + h(x)$ is $(m + \mu_h)$-strongly convex, we have

$$f(x^{k+1}) + \frac{m + \mu_h}{2}\|x - x^{k+1}\|^2 \leq G^k(x^{k+1}) + h(x^{k+1}) + \frac{m + \mu_h}{2}\|x - x^{k+1}\|^2$$
$$\leq G^k(x) + h(x)$$
$$= f(x) + [G^k(x) + h(x) - f(x)]$$
$$\leq f(x) + \frac{1}{n}\sum_{i=1}^n \frac{M + L_i}{2}\|x - x^{\theta_{i,k}}\|^2.$$

By taking the expectation of both sides and let $x = x^*$ yields

$$\mathbb{E}[f(x^{k+1})] - f^* \leq \mathbb{E}[\frac{1}{n}\sum_{i=1}^n \frac{M + L_i}{2}\|x^* - x^{\theta_{i,k}}\|^2] - \mathbb{E}[\frac{m + \mu_h}{2}\|x^* - x^{k+1}\|^2].$$

We have

$$\frac{\mu_h}{2}\|x^{k+1} - x^*\|^2 \leq \mathbb{E}[f(x^{k+1})] - f^*$$
$$\leq \mathbb{E}[\frac{1}{n}\sum_{i=1}^n \frac{M + L_{max}}{2}\|x - x^{\theta_{i,k}}\|^2] - \mathbb{E}[\frac{m + \mu_h}{2}\|x - x^{k+1}\|^2].$$

thus

$$\|x^{k+1} - x^*\|^2 \le \frac{M + L_{max}}{2\mu_h + m} \mathbb{E}[\frac{1}{n}\sum_{i=1}^{n}\|x^* - x^{\theta_{i,k}}\|^2]. \tag{21}$$

then we have

$$\mathbb{E}[\frac{1}{n}\sum_{i=1}^{n}\|x^* - x^{\theta_{i,k}}\|^2] = \frac{1}{n}\|x^k - x^*\|^2 + (1 - \frac{1}{n})\mathbb{E}[\frac{1}{n}\sum_{i=1}^{n}\|x^* - x^{\theta_{i,k-1}}\|^2]$$

$$\le \frac{1}{n}\|x^k - x^*\|^2 + (1 - \frac{1}{n})\mathbb{E}[\frac{1}{n}\sum_{i=1}^{n}\|x^* - x^{\theta_{i,k-1}}\|^2]$$

$$\le [\frac{1}{n}\frac{M + L_{max}}{2\mu_h + m} + (1 - \frac{1}{n})]\mathbb{E}[\frac{1}{n}\sum_{i=1}^{n}\|x^* - x^{\theta_{i,k-1}}\|^2]$$

$$\le [\frac{1}{n}\frac{M + L_{max}}{2\mu_h + m} + (1 - \frac{1}{n})]^k \mathbb{E}[\frac{1}{n}\sum_{i=1}^{n}\|x^* - x^{\theta_{i,0}}\|^2]$$

$$\le [\frac{1}{n}\frac{M + L_{max}}{2\mu_h + m} + (1 - \frac{1}{n})]^k \|x^* - x^0\|^2.$$

Thus we have $\mathbb{E}[f(x^{k+1})] - f^* \le \frac{M + L_{max}}{2}[\frac{1}{n}\frac{M + L_{max}}{2\mu_h + m} + (1 - \frac{1}{n})]^k \|x^* - x^0\|^2.$

B   Proof of Theorem 2

We first examine the relations between the search directions of ProxN and PROXTONE.

By (4), (5) and Fermat's rule, Δx_{ProxN}^k and Δx^k are also the solutions to

$$\Delta x_{ProxN}^k = \arg\min_{d \in \mathbb{R}^p} \nabla g(x^k)^T d + (\Delta x_{ProxN}^k)^T H_k d + h(x^k + d),$$

$$\Delta x^k = \arg\min_{d \in \mathbb{R}^p} (\nabla_k + H_k x^k)^T d + (\Delta x^k)^T H_k d + h(x^k + d).$$

Hence Δx^k and Δx_{ProxN}^k satisfy

$$\nabla g(x^k)^T \Delta x^k + (\Delta x_{ProxN}^k)^T H_k \Delta x^k + h(x^k + \Delta x^k)$$
$$\ge \nabla g(x^k)^T \Delta x_{ProxN}^k + (\Delta x_{ProxN}^k)^T H_k \Delta x_{ProxN}^k + h(x^k + \Delta x_{ProxN}^k)$$

and

$$(\nabla_k + H_k x^k)^T \Delta x_{ProxN}^k + (\Delta x^k)^T H_k \Delta x_{ProxN}^k + h(x^k + \Delta x_{ProxN}^k)$$
$$\ge (\nabla_k + H_k x^k)^T \Delta x^k + (\Delta x^k)^T H_k \Delta x^k + h(x^k + \Delta x^k).$$

We sum these two inequalities and rearrange to obtain

$$(\Delta x^k)^T H_k \Delta x^k - 2(\Delta x_{ProxN}^k)^T H_k \Delta x^k + (\Delta x_{ProxN}^k)^T H_k \Delta x_{ProxN}^k$$
$$\le (\nabla_k + H_k x^k - \nabla g(x^k))^T (\Delta x_{ProxN}^k - \Delta x^k).$$

The assumptions $mI \preceq H_{\theta_{i,k}}$ yields that $mI \preceq H_k$, together with (11) we have

$$m\|\Delta x^k - \Delta x^k_{ProxN}\|^2 \tag{22}$$

$$\leq \|\frac{1}{n}\sum_{i=1}^{n}(\nabla g_i(x^{\theta_{i,k}}) - \nabla g_i(x^k) - (x^{\theta_{i,k}} - x^k)^T H^i_{\theta_{i,k}})\|\|(\Delta x^k - \Delta x^k_{ProxN})\|.$$

Since $\nabla^2 g_i$ is Lipschitz continuous with constant $K_i > 0$, by Lemma 1.2.4 of [5] we have

$$\|\nabla g_i(x^{\theta_{i,k}}) - \nabla g_i(x^k) - (x^{\theta_{i,k}} - x^k)^T H^i_{\theta_{i,k}}\| \leq \frac{K_i}{2}\|x^{\theta_{i,k}} - x^k\|^2. \tag{23}$$

Then from (22) and (23), we have

$$\|\Delta x^k - \Delta x^k_{ProxN}\| \leq \frac{K_{max}}{2mn}\sum_{i=1}^{n}\|x^{\theta_{i,k-1}} - x^k\|^2. \tag{24}$$

Since the ProxN method converges q-quadratically (cf. Theorem 3.3 of [2]),

$$\|x^{k+1} - x^\star\|$$
$$\leq \|x^k + \Delta x^k_{ProxN} - x^\star\| + \|\Delta x^k - \Delta x^k_{ProxN}\|$$
$$\leq \frac{K_{max}}{m}\|x^k - x^\star\|^2 + \|\Delta x^k - \Delta x^k_{ProxN}\|. \tag{25}$$

Thus from (24) and (25), we have almost surely that

$$\|x^{k+1} - x^\star\|$$
$$\leq \frac{K_{max}}{m}\|x^k - x^\star\|^2 + \frac{L_{max}}{2mn}\sum_{i=1}^{n}\|x^{\theta_{i,k-1}} - x^k\|^2$$
$$\leq \frac{K_{max}}{m}\|x^k - x^\star\|^2 + \frac{L_{max}}{mn}\sum_{i=1}^{n}2\|x^{\theta_{i,k-1}} - x^\star\|^2 + \frac{L_{max}}{mn}\sum_{i=1}^{n}2\|x^\star - x^k\|^2.$$

Then by (21), we have

$$\|x^{k+1} - x^\star\| \leq (\frac{K_{max} + 2L_{max}}{m}\frac{M + L_{max}}{2\mu_h + m} + \frac{2L_{max}}{m})\mathbb{E}[\frac{1}{n}\sum_{i=1}^{n}\|x^{\theta_{i,k}} - x^\star\|^2]$$

which yieds

$$\|x^{k+1} - x^\star\| \leq (\frac{K_{max} + 2L_{max}}{m}\frac{M + L_{max}}{2\mu_h + m} + \frac{2L_{max}}{m})[\frac{1}{n}\frac{M + L_{max}}{2\mu_h + m}$$
$$+ (1 - \frac{1}{n})]^k\|x^\star - x^0\|^2.$$

References

1. Bertsekas, D.P.: Incremental gradient, subgradient, and proximal methods for convex optimization: a survey. Optimization for Machine Learning 2010, 1–38 (2011)
2. Lee, J., Sun, Y., Saunders, M.: Proximal newton-type methods for convex optimization. In: Advances in Neural Information Processing Systems, pp. 836–844 (2012)
3. Mairal, J.: Optimization with first-order surrogate functions. arXiv preprint arXiv:1305.3120 (2013)
4. Nesterov, Y.: Efficiency of coordinate descent methods on huge-scale optimization problems. SIAM Journal on Optimization **22**(2), 341–362 (2012)
5. Nesterov, Y.: Introductory Lectures on Convex Optimization: A Basic Course. Kluwer, Boston (2004)
6. Parikh, N., Boyd, S.: Proximal algorithms. Foundations and Trends in Optimization **1**(3), 123–231 (2013)
7. Schmidt, M., Roux, N.L., Bach, F.: Minimizing finite sums with the stochastic average gradient. arXiv preprint arXiv:1309.2388 (2013)
8. Shalev-Shwartz, S., Zhang, T.: Proximal stochastic dual coordinate ascent. arXiv preprint arXiv:1211.2717 (2012)
9. Shalev-Shwartz, S., Zhang, T.: Stochastic dual coordinate ascent methods for regularized loss. The Journal of Machine Learning Research **14**(1), 567–599 (2013)
10. Sohl-Dickstein, J., Poole, B., Ganguli, S.: Fast large-scale optimization by unifying stochastic gradient and quasi-newton methods. In: Proceedings of the 31st International Conference on Machine Learning (ICML 2014), pp. 604–612 (2014)
11. Xiao, L., Zhang, T.: A proximal stochastic gradient method with progressive variance reduction. arXiv preprint arXiv:1403.4699 (2014)

Author Index

Printed in the United States
By Bookmasters